Organic Reactions

Organic Reactions

VOLUME 25

JOHN WILEY & SONS, INC.
NEW YORK · LONDON · SYDNEY · TORONTO

Published by John Wiley & Sons, Inc.

Library of Congress Catalogue Card Number: 42-20265

ISBN 0-471-01741-8
Printed in the United States of America.

10 9 8 7 6 5 4 3 2 1

PREFACE TO THE SERIES

In the course of nearly every program of research in organic chemistry the investigator finds it necessary to use several of the better-known synthetic reactions. To discover the optimum conditions for the application of even the most familiar one to a compound not previously subjected to the reaction often requires an extensive search of the literature; even then a series of experiments may be necessary. When the results of the investigation are published, the synthesis, which may have required months of work, is usually described without comment. The background of knowledge and experience gained in the literature search and experimentation is thus lost to those who subsequently have occasion to apply the general method. The student of preparative organic chemistry faces similar difficulties. The textbooks and laboratory manuals furnish numerous examples of the application of various syntheses, but only rarely do they convey an accurate conception of the scope and usefulness of the processes.

For many years American organic chemists have discussed these problems. The plan of compiling critical discussions of the more important reactions thus was evolved. The volumes of *Organic Reactions* are collections of chapters each devoted to a single reaction, or a definite phase of a reaction, of wide applicability. The authors have had experience with the processes surveyed. The subjects are presented from the preparative viewpoint, and particular attention is given to limitations, interfering influences, effects of structure, and the selection of experimental techniques. Each chapter includes several detailed procedures illustrating the significant modifications of the method. Most of these procedures have been found satisfactory by the author or one of the editors, but unlike those in *Organic Syntheses* they have not been subjected to careful testing in two or more laboratories.

Each chapter contains tables that include all the examples of the reaction under consideration that the author has been able to find. It is inevitable, however, that in the search of the literature some examples will be missed, especially when the reaction is used as one step in an extended synthesis. Nevertheless, the investigator will be able to use the tables and their accompanying bibliographies in place of most or all of the literature search so often required.

Because of the systematic arrangement of the material in the chapters and the entries in the tables, users of the books will be able to find information desired by reference to the table of contents of the appropriate chapter. In the interest of economy the entries in the indices have been kept to a minimum, and, in particular, the compounds listed in the tables are not repeated in the indices.

The success of this publication, which will appear periodically, depends upon the cooperation of organic chemists and their willingness to devote time and effort to the preparation of the chapters. They have manifested their interest already by the almost unanimous acceptance of invitations to contribute to the work. The editors will welcome their continued interest and their suggestions for improvements in *Organic Reactions.*

Chemists who are considering the preparation of a manuscript for submission to Organic Reactions are urged to write either secretary before they begin work.

SPECIAL PREFACE TO VOLUME 25

It is to be noted that the binding of this volume is special; indeed, it is silver to celebrate the 25th issue. I call to the attention of all readers the pages that follow this preface—"Some Recollections of Thirty-eight Years and Twenty-five Volumes." I am very grateful to A. H. Blatt for the preparation of this history of *Organic Reactions* in that I volunteered his services. Such is the way this publication functions, and it tells readers why they know the publication itself, but not the editors.

In the 38 years since this publication was first discussed, the presentation of chemical results has undergone an immense change, and the number and variety of review publications have been ever increasing. Yet, due to the sound planning of the original Board, *Organic Reactions* has remained almost constant in its style and in its worldwide acceptance. In its history, this publication has reviewed 130 distinct synthetic organic reactions, and these articles have been submitted by 184 authors. As of December 31, 1976, a total of 216,615 volumes of *Organic Reactions* have been purchased.

In the preface, the procedures followed are clearly described, but it should be called to the attention of the reader that almost all the effort required to produce a volume has been volunteered. During the same period many other related publications have been initiated and discontinued. The continuation of *Organic Reactions* can be directly traced to the standards found in the initial volume.

It is evident that a volunteer organization does not last 38 years without the generosity of many people. In the present case credit should go to the authors who donated their time and effort and to the Editorial Board for their generous contributions. There are very few ways in which organic chemists can repay, in part, what they took from the field during the course of their studies, but contributions to *Organic Reactions* do offer one such pathway. For this publication to continue under its present format it is essential that the oncoming generations continue to make such volunteer contributions.

During the period this volume was in page proof, Professor Louis F. Fieser, a member of the initial Board of Editors, died. His contributions to organic chemistry are known by all our readers, but those of us who had the privilege of working with him appreciate the special chemical insight we gained from him. He will be missed but his contributions to this publication and to me, personally, will be ever remembered.

WILLIAM G. DAUBEN
EDITOR-IN-CHIEF

August 1977
Department of Chemistry
University of California
Berkeley, California

SOME RECOLLECTIONS OF THIRTY-EIGHT YEARS AND TWENTY-FIVE VOLUMES

THE BEGINNING

The reasons for starting *Organic Reactions* and the ways in which it was planned to accomplish its goals were stated so well in the preface to the series that the reader is referred to that preface (pp. v–vi of this volume) for them.

The decision to undertake the preparation and presentation of "critical disucssions of the more important (synthetic) reactions" was made at a meeting of the editors of *Organic Syntheses* and representatives of John Wiley & Sons during the Eighth National Organic Chemistry Symposium at St. Louis in December 1939. At that meeting the organizational setup was agreed upon, the operating procedures were roughed out, and the topics and authors were selected for Volume 1. These actions were formalized by the incorporation of *Organic Reactions* in Illinois on August 1, 1942, for educational and research purposes, with Roger Adams, Harold R. Snyder, Werner E. Bachmann, John R. Johnson, and Louis F. Fieser as directors, and by the appearance later that year of Volume 1. Roger Adams was elected president and served as President and Editor-in-Chief until he was succeeded in both positions by Arthur C. Cope in 1960 with the publication of Volume 11. He remained an active member of the Editorial Board until his death in 1971. Professor Cope in turn was succeeded in 1969 with the publication of Volume 17 by the present Editor-in-Chief and President, William G. Dauben.

The close relationship of *Organic Reactions* to *Organic Syntheses*, Roger Adams, and John Wiley & Sons is obvious; the great value of that relationship is equally obvious to all who have been connected with the series as editors and authors.

THE OPERATING PROCEDURES

The editors chose the topics to be covered and the authors who would be asked to write the chapters. These decisions, together with

EDITORS-IN-CHIEF

ROGER ADAMS

ARTHUR C. COPE

WILLIAM G. DAUBEN

discussions of work in progress, were the bases of an annual working dinner for the editors.

Invitations to write chapters were extended by the Editor-in-Chief, and those who knew Roger Adams will not be surprised that few invitations were declined. Correspondence with authors and publishers between the annual meetings was handled by one of the editors who also served as secretary.

There was no formal rule on an editor's length of service; the informal understanding with the original board of five was that they would serve through the publication of five volumes.

Each draft manuscript when received was duplicated (no photocopy then!) and copies were sent to the editors. Every editor read each draft manuscript for accuracy and completeness of the chemistry involved and for clarity and effectiveness of presentation. Comments on these

ORIGINAL EDITORIAL BOARD

LOUIS F. FIESER

WERNER E. BACHMAN

JOHN R. JOHNSON

HAROLD R. SNYDER

matters were sent in writing by each editor to the Responsible Editor—an editor who had either volunteered or had been "volunteered" (army style) to follow the manuscript, with the aid of the Secretary and the Editor-in-Chief, through to publication or rejection.

The Responsible Editor summarized the comments tactfully and sent the summary to the author to enable him to take advantage of the comments in revising the manuscript.

The revised manuscript was read by the Responsible Editor and, if it met with his approval and that of the Editor-in-Chief, was copy edited and sent to the publisher. If the revised manuscript was not considered satisfactory it was referred to the Editorial Board, which would recommend rejection, acceptance, additional revision, or assistance by a co-author or an editor. The number of rejections was small but the amount of work done by authors and editors was often considerable.

The publisher also copy edited each accepted manuscript and sent it along for composition. Galley proofs were read by all editors, and their marked galleys sent to the Secretary who combined the corrections on a set of master galleys that served as the basis for page proofs. All page proofs were read by all editors. Finally, one editor prepared the index and checked the front material, which had been made up by the publisher.

The authors, of course, also read galley and page proofs.

* * *

The editorial process just described means that *Organic Reactions*, in publishers' jargon, is "tightly edited." Tight editing is the heart of quality in multiauthor books, but it makes heavy demands on authors and editors—really the same group because each author is an associate editor of the volume in which his chapter appears. To reduce these demands on individuals and still obtain maximum benefits of tight editing, a mixture of constraints and assistance has been developed.

The size of the editorial board has been almost doubled, and there are two secretaries who have taken the responsibility for correspondence and part of the responsibility for handling proofs.

Only actual errors ("the yield was 178%") are corrected without consulting the author. It was agreed at the outset that the editors' function is to help the author say what he wants to say in such a way that it would be difficult for a reader to misunderstand. To achieve this, changes in wording or arrangement are *suggested* for the author's consideration together with the reasons for the suggestions. Acceptance by the author depends on the cogency of the editors' letters. This has avoided asking for changes that represent an editor's preference for or dislike of a word or phrase—a practice that, justifiably, irritates authors. The author–editor relationship is not that of adversaries but one of people working toward a common goal. It is probably for this reason that so many authors over the years have written to thank the editors for their assistance.

Organic Reactions helps authors in making the literature search, pays for typing manuscripts, and furnishes each author with reprints of his article. Small recompense, but organic chemistry is the spur.

A final comment: the greatest difficulty we have had with manuscripts is the authors' tendency to write for the chemist who knows most about the reaction being reviewed rather than for the chemist who wants to learn about the reaction. To give a fictitious illustration: a chapter on the Wittig reaction should be written, not for Professor

ADVISORY BOARD

HOMER ADKINS

A. H. BLATT

VIRGIL BOEKELHEIDE

T. L. CAIRNS

DONALD J. CRAM

DAVID Y. CURTIN

ADVISORY BOARD

JOHN FRIED

HERBERT O. HOUSE

WILLY LEIMGRUBER

FRANK C. McGREW

BLAINE C. McKUSICK

CARL NIEMANN

Wittig but for chemists who want to familiarize themselves with the Wittig reaction. When authors keep the readers in mind, the results are better for both authors and readers.

THE PRODUCT

In a verbless sentence: 25 volumes, more than 216,600 copies as of the end of 1976, about 140 synthetic reactions (including the updates) discussed by almost 185 authors. The largest number of chapters, 12, was in the first volume; the smallest number, 1, was in volume 16; the average number of chapters per volume is slightly more than 5.

The majority of chapters originated in the United States. However, chapters have come from nine other countries and from five of the seven continents. It is not surprising that the great majority of chapters came from universities. However, the support from the chemical industry has been both consistent and significant. Volume 1 contained two chapters by chemists in industry, Volume 21 was written entirely by industrial research chemists, and almost a third of the chapters in the series have come from the same source. With the publication of Volume 5, the first editor from industry joined the Editorial Board and since that time there has always been at least one editor from industry. *Organic Reactions* is truly a joint effort of organic chemists, worldwide, industrial, and academic.

The scope—a term familiar to all users of the series—extends from such "classical" reactions as Cannizzaro, Hofmann, and Perkin to such modern ones as the Ritter reaction, the Wittig reaction, and the Ramburg-Bäcklund rearrangement. It includes older reactions that have acquired renewed interest, such as hydrocyanation and the synthesis of cyclopropanes using the zinc-copper couple.

The list of reactions, which can be found as an index in the most recently published volumes, is an overview of synthetic organic chemistry, and the series has become the most widely used one in the graduate training of organic chemists.

The problem of keeping *Organic Reactions* up to date was recognized early and has been the subject of frequent discussions that have not been limited to the editors. The present plan, adopted until a better one is found, began with Volume 22. It is to present a brief report on important new developments, together with a critical, but not complete, survey of the literature since the earlier report. The nature and extent of the coverage are specified in each update.

One change in *Organic Reactions* over the years deserves separate

EDITORIAL BOARD

comment. The average chapter length in Volume 1 was 33 pages. The average chapter length for the series approximates 80 pages. The change reflects the increase in the amount of organic chemistry published beginning after World War II, with the declassification and publication of research done during the war. This increase, however, was small compared with the increase that resulted from the introduction and spread of instrumentation. The black boxes increased the effective lives of organic chemists as did the discovery of logarithms earlier for astronomers. By the 1950s the publication rate was more than doubling each decade and, as the number and capabilities of the instruments grew, they not only produced more data but data that was previously unavailable and that permitted a better understanding of both structure and mechanism. The changes are illustrated in part by a comparison of two pages reproduced from Volume 1 and Volume 23 (see pages xx and xxi). The changes also resulted in *Organic Reactions* becoming not only necessary but essential.

BY-PRODUCTS

It was a fad some years ago to display signs in executive suites that read "This is a non-profit business but it wasn't planned that way". *Organic Reactions* has been just the reverse. It has been profitable, but the profits do not go to any individual. It was clear by the late 1940s that the series was needed and royalties had accumulated, so that the pleasant problem of how best to use them for education and research needed to be faced. The first action was to invest the royalties, so that the dividends and interest would smooth as much as possible the irregularities of income. This problem arose because the royalties were much higher in the year a volume was published than in other years and the series was not amenable to annual publication. The investments were successful, thanks to the acumen of Roger Adams and Arthur Cope. Income was stabilized sufficiently to permit a planned program, which began in traditional fashion with contributions to fellowship funds memorializing chemists who had been closely associated with *Organic Reactions* and its older sister *Organic Syntheses*. Along with these actions, but in a happier vein, was a series of annual dinners for the authors who had written chapters for the series. These dinners, held during American Chemical Society meetings, served to say thank you to the guests and enabled them to get to know each other as persons as well as chemists.

The annual dinners were so well attended and so successful that they destroyed their own usefulness as the number of authors grew larger

EDITORIAL BOARD

ANDREW S. KENDE

JAMES A. MARSHALL

JERROLD MEINWALD

BARRY M. TROST

BORIS WEINSTEIN

and larger. Finally in the late 1960s these dinners were reluctantly replaced by dinners for chemists who were writing chapters for forthcoming volumes. The smaller dinners, which are by now a tradition, have two nonchemical features of interest. First, they anticipated the new feminism by including women as authors and as guests. Second, they have made it moot whether prospective authors write more slowly than their predecessors did.

I must digress for one other one-time action. In 1951 during the 75th anniversary of the American Chemical Society, *Organic Reactions* had a small dinner to renew pre-World War II associations and to talk shop. Informally the dinner is remembered not only for the excellence of food, wines, and conversation, but also for the fact that the guests came from six continents and the percentage of Nobel Laureates, past and future, among them was very high.

Larger aid to chemistry and eduction followed a suggestion by John D. Roberts to sell up to a complete set of *Organic Reactions* at half the list price to graduate students, postdoctoral fellows, and research associates in chemistry, that is, to those who had not accepted what could be expected to be a permanent job. To avoid complications with foreign exchange, the offer was limited to colleges and universities in the United States and Canada. Later it was extended to include undergraduate majors and majors in related fields such as biochemistry. The difference between the amount a student paid and the actual cost of the volumes was made up by *Organic Reactions.* About 22,500 copies have been sold between December 1956, when the offer began, and December 31, 1976. Not a single student has failed to pay for his or her purchase.

Finally a personal touch. At one (deliberately dateless) meeting of the corporation, the directors, in a mood more expansive and expensive than prudent, changed the offer to $50.00 for a set of volumes. The result was a flood of orders that broke all records and threatened to bankrupt the corporation. Quick action returned the offer to its original form and the sale at half price has continued ever since.

The other by-product, a joint action of *Organic Syntheses* and *Organic Reactions,* is the funding of the Roger Adams Award—a major biennial award in honor of the founder of both organizations.

Two more items should be added to end these recollections. The first is that the appearance of a chapter in *Organic Reactions* serves to initiate work on the reaction discussed—an observation that has been verified by checking the number of references to the reaction for a few years before and after the appearance of the chapter. The second is that of all the by-products, the most rewarding to authors, editors, and

A page from Volume 1 of *Organic Reactions*

THE CLEMMENSEN REDUCTION

Formula	Compound	Method	Yield*	Reference†
$C_{14}H_{22}O_2$	1,1′-Ethynylenebiscyclohexanol	I	—	260
	R.P.‡ Δ^{11}-Dodecahydrophenanthrene			
$C_{14}H_{12}O_3$	β-1-Naphthoylpropionic acid	III	86	465
		III	—	370
		I	70	121
		I	64	78
$C_{14}H_{12}O_3$	β-2-Naphthoylpropionic acid	III	91	465
		I	70	121
		I	81	78
$C_{14}H_{12}O_3$	β-1- and 2-Naphthoylpropionic acids	IV	78	371
$C_{14}H_{12}O_3$	Benzyl 2,4-dihydroxyphenyl ketone	I	70	55
		I	—	53
$C_{14}H_{12}O_3$	2,4-Dihydroxy-3-methylbenzophenone	I	—	376
$C_{14}H_{16}O_3$	β-1-Tetroylpropionic acid	I	—	44
$C_{14}H_{16}O_3$	β-2-Tetroylpropionic acid	I	—	35
$C_{14}H_{16}O_3$	5-*p*-Anisyl-2-methylcyclohexa-1,3-dione	I	85	211
$C_{14}H_{18}O_3$	6,7-Dimethoxy-2,3-dimethyl-1-tetralone	I	11	122
$C_{14}H_{18}O_3$	α,α-Diethyl-β-benzoylpropionic acid	I	60	306
$C_{14}H_{18}O_3$	β-Ethyl-β-methyl-γ-benzoylbutyric acid	I	—	230
$C_{14}H_{18}O_3$	Ethyl α-methyl-β-*p*-toluylpropionate	I	—	36
$C_{14}H_{18}O_3$	Ethyl β-methyl-β-*p*-toluylpropionate	I	—	36
$C_{14}H_{18}O_3$	β-4-*t*-Butylbenzoylpropionic acid	III	84	465
$C_{14}H_{18}O_3$	6-Hydroxy-2,2,5,7,8-pentamethylchromanon	II	66	314
$C_{14}H_{20}O_3$	2,4-Dihydroxyphenyl *n*-heptyl ketone	I	—	53
$C_{14}H_{20}O_3$	Ketolactone from dihydroisoalantolactone	I	—	109
		I	68	113
$C_{14}H_{26}O_3$	13-Ketomyristic acid	I	—	160
$C_{14}H_8O_4$	Alizarin	I	73	3
	R.P.‡ Hexahydroanthracene			
$C_{14}H_{10}O_4$	3-Benzoyl-2,6-dihydroxybenzaldehyde	I	—	376
$C_{14}H_{12}O_4$	Benzyl 2,4,6-trihydroxyphenyl ketone	I	65	56
$C_{14}H_{12}O_4$	β-2-Hydroxy-3-naphthoylpropionic acid	III	—	313
$C_{14}H_{14}O_4$	6-Butyryl-5-hydroxy-4-methylcoumarin	II	—	340
$C_{14}H_{14}O_4$	8-Butyryl-7-hydroxy-4-methylcoumarin	I	—	424
$C_{14}H_{14}O_4$	6-Acetyl-8-ethyl-5-hydroxy-4-methylcoumarin	I	—	427
$C_{14}H_{18}O_4$	δ-4-Ethoxybenzoylvaleric acid	I	—	183
$C_{14}H_{18}O_4$	β-2-Ethyl-4-methoxy-5-methylbenzoylpropionic acid	I	96	337
$C_{14}H_{18}O_4$	β-5-Ethyl-4-methoxy-2-methylbenzoylpropionic acid	I	80	337
$C_{14}H_{18}O_4$	1,5-Di-*n*-butyryl-2,4-dihydroxybenzene	I	—	54
$C_{14}H_{18}O_5$	α,β-Dimethyl-β-3,4-dimethoxybenzoylpropionic acid	I	—	122
$C_{14}H_{18}O_6$	Diethyl-bicyclo(2:2:2)octadionedicarboxylate	IV	42	349
$C_{14}H_{19}O_2Cl$	5-Chloro-2-hydroxyphenyl *n*-heptyl ketone	I	—	165
$C_{14}H_{19}O_3Cl$	5-Chloro-2,4-dihydroxyphenyl *n*-heptyl ketone	II	—	167

* Q, yield reported as quantitative; G, yield reported as good; P, yield reported as poor. A dash indicates that the yield is not reported.

† Reference numbers refer to the bibliography on pp. 201–209.

‡ Reduction product.

A page from Volume 23 of *Organic Reactions*

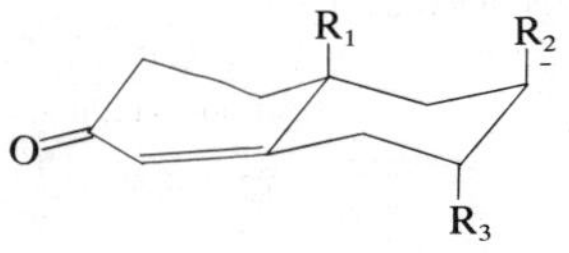

42

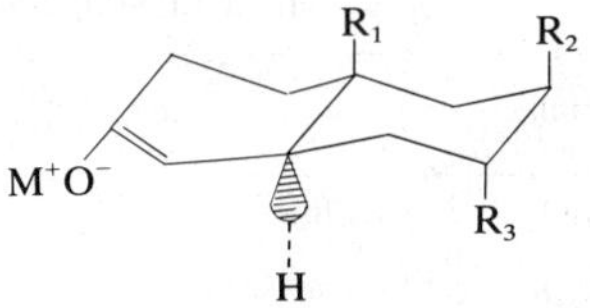

43

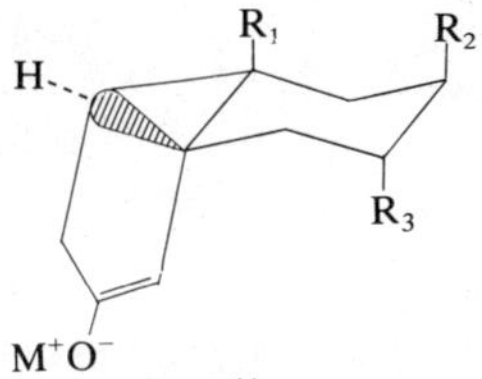

44

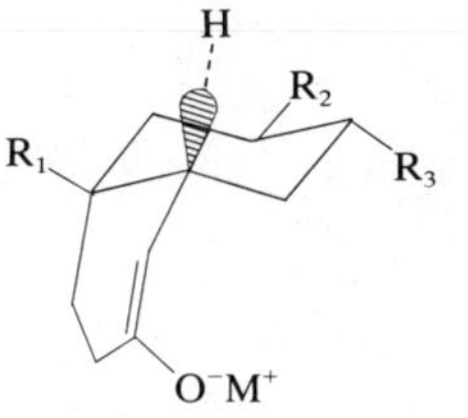

45

predicted from simple analysis of nonbonded interactions in the two stereo-electronically allowed reduction products; and, second, in systems in which a significant amount of strain must be introduced in order for protonation to occur axially, transition states resembling **45** in which the new C—H bond forms quasi-equatorially to the enolate ring may become important.

In connection with the first point, it may be noted that reductions of many 1(9)-octalin-**2**-ones yield *trans* products with a high degree of stereoselectivity.[78] For example, it was pointed out that 1(9)-octalin-2-one **(25)**

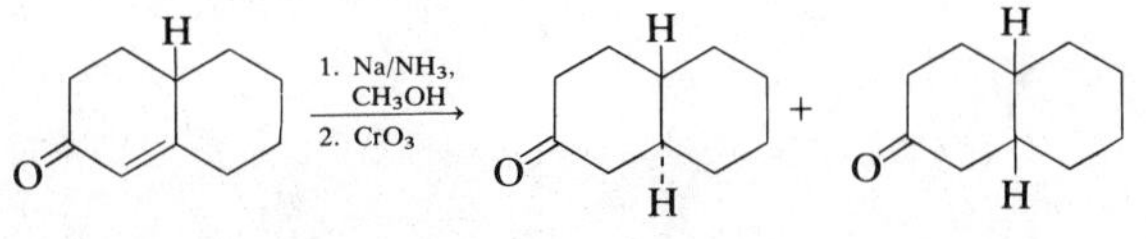

1. Na/NH_3, CH_3OH
2. CrO_3

25 **46** (99%) **47** (1%)

yielded a 99/1 mixture of the *trans*- and *cis*-decalones **46** and **47** on reduction with sodium in liquid ammonia, whereas analysis of nonbonded interactions in the corresponding 1(2)-enolates **48** and **49** indicated that the

ORGANIC REACTIONS DINNER: Hotel Vanderbilt, New York City, September 6, 1951

Left to right: F. C. McGrew (DuPont); C. Niemann (California Institute of Technology); H. R. Snyder (University of Illinois); Roger Adams (University of Illinois and (Editor-in-Chief)); A. C. Cope (Massachusetts Institute of Technology); A. H. Blatt (Queens College).

secretaries are the associations that have resulted from working together.

* * *

I am indebted to several friends and associates who read these pages and made suggestions. I am particularly indebted to Harold R. Snyder, a member of the original Board of Editors, both for his suggestions and for correcting my too often vagrant memories of the early days. The errors of fact and emphasis that remain are mine.

A. H. BLATT

CONTENTS

Organic Reactions

CHAPTER 1

THE RAMBERG-BÄCKLUND REARRANGEMENT

Leo A. Paquette*

The Ohio State University, Columbus, Ohio

CONTENTS

* The invaluable assistance of Donna Canode and Carol Rose in the meticulous typing of this manuscript is warmly appreciated.

INTRODUCTION

The halogen atoms of α-halo sulfones, in contrast to halogen atoms α to other electron-withdrawing functionalities,[1] show marked resistance to substitution by external nucleophiles.[2–7] Seemingly, polar, steric, and field effects combine to repel nucleophilic species.[1,6–8] However, the same α-sulfonyl halogen atoms are capable of facile intramolecular 1,3 elimination, leading to replacement of the sulfonyl group by a carbon–carbon double bond with loss of halide and sulfite ions. This extrusion process, frequently referred to as the α-halo sulfone[9,10] or Ramberg-Bäcklund rearrangement[11,12] after its discover-

[1] F. G. Bordwell and W. T. Brannen, Jr., *J. Amer. Chem. Soc.*, **86,** 4645 (1964).
[2] F. Raschig and W. Prahl, *Ann. Chem.*, **448,** 307 (1926).
[3] T. Thompson and T. S. Stevens, *J. Chem. Soc.*, **1932,** 69.
[4] W. M. Ziegler and R. Connor, *J. Amer. Chem. Soc.*, **62,** 2596 (1940).
[5] T. B. Johnson and I. B. Douglass, *J. Amer. Chem. Soc.*, **63,** 1571 (1941).
[6] F. G. Bordwell and G. D. Cooper, *J. Amer. Chem. Soc.*, **73,** 5184, 5187 (1951).
[7] F. G. Bordwell and B. B. Jarvis, *J. Org. Chem.*, **33,** 1182 (1968).
[8] C. Y. Meyers, *Tetrahedron Lett.*, **1962,** 1125.
[9] L. A. Paquette, *Mechanisms of Molecular Migrations*, Vol. I, B. S. Thyagarajan, Ed., Interscience, New York, 1968, pp. 121–156.
[10] L. A. Paquette, *Accounts Chem. Res.*, **1,** 209 (1968).
[11] F. G. Bordwell, *Organosulfur Chemistry*, M. J. Janssen, Ed., Interscience, New York, 1967, Chap. 16.
[12] F. G. Bordwell, *Accounts Chem. Res.*, **3,** 281 (1970).

ers,[13] has found broad utility in olefin synthesis. The reaction is generally applicable and easy to use.

Its earliest application dealt with the preparation of alkenes, the *cis* isomers of which predominated. Whereas such molecules may be more readily available by other methods, none of the alternative procedures offers the added option of specifically replacing the olefinic hydrogen atoms with deuterium by merely conducting the rearrangement in deuterated solvents. Moreover, because the α-halo sulfones undergo this transformation in alkaline solution, further rearrangement of the initially formed alkenes is precluded. Therefore the SO_2 group in the starting sulfone is invariably replaced cleanly and unequivocally by the π bond.

H
SO_2CH_2Cl
NaOD
D_2O
$=CD_2$

Br
SO_2CH_3
C
+
CH_2
SO_2CH_3
Br
$NaOCH_3$
DMSO
$=CH_2$

The rearrangement is generally applicable to molecules containing the minimal structural requirements of a sulfonyl group, an α-halogen atom, and at least one α'-hydrogen atom, even in systems leading to small-ring cycloalkenes. Di- and tri-halo sulfones behave analogously. Conformational constraints, adverse hybridization characteristics, and excessive strain are known to deter the reaction. Inasmuch as these features arise only in special circumstances, they do not necessarily detract from its usefulness.

This chapter summarizes the more important advances in our understanding of α-halo sulfone rearrangements, particular consideration being given to the nature of the intermediates, scope of the available synthetic alternatives, optimization of experimental conditions, and effect of structural features on reactivity. Closely related transformations are also discussed.

MECHANISM

Kinetic data for the reactions of α-halo sulfones in base conform to a second-order rate expression for release of halide ion that is first-order in sulfone and first-order in hydroxide.[6,14,15] These findings

[13] L. Ramberg and B. Bäcklund, *Ark. Kemi Mineral. Geol.*, **13A,** No. 27 (1940); *C.A.*, **34,** 4725 (1940).

[14] F. G. Bordwell and J. M. Williams, *J. Amer. Chem. Soc.*, **90,** 435 (1968).

[15] L. A. Paquette and L. S. Wittenbrook, *J. Amer. Chem. Soc.*, **90,** 6783 (1968).

suggested a pre-equilibrium involving the α-halo sulfone and its carbanions, but the direct involvement of such intermediates in that reaction leading to the observed olefins required further proof.[16,17] The possibility of α,α elimination with formation of α-sulfonyl carbenes was readily dismissed.[10] Additionally, by treating α-bromobenzyl benzyl sulfone with sodium methoxide in methanol-OD and stopping the rearrangement after one half-life, deuterium exchange at the α and α' positions was shown to be complete; thereby H-D exchange at a later stage was ruled out.[14] The unusually large leaving-group effect observed with these same molecules ($C_6H_5CH_2SO_2CHXC_6H_5$), a Br:Cl rate ratio of 620 at 0°, can best be rationalized by a carbanion mechanism.[14] The positive ρ value for rearrangement of the series $ArCHXSO_2CH_3$, indicative of extensive C—X bond breaking in the transition state, supports this thinking.[18] The experimental evidence has long been considered compatible with the mechanistic scheme in Eq. 1.

$$\text{H—C(SO}_2\text{)C—X} + \text{B}^- \underset{k_{-1}}{\overset{k_1}{\rightleftharpoons}} \text{BH} + {}^-\text{C(SO}_2\text{)C—X} \xrightarrow[\text{Slow}]{k_2} \text{X}^- + \text{C—C (episulfone, SO}_2\text{)} \xrightarrow[\text{Fast}]{k_3} \text{C=C} + \text{SO}_2 \qquad \text{(Eq. 1)}$$

Although the alkaline conditions required for rearrangement are sufficiently strenuous to preclude isolation of intermediate episulfones, the availability of both symmetrical and unsymmetrical episulfones by alternative syntheses[19] has made it possible to establish that the decomposition of these three-membered sulfones under customary rearrangement conditions is markedly stereospecific (except when very strong bases such as *t*-butoxide are used).[20–22] Consequently, the stereochemistry of the alkene has long been considered to be determined at the ring-closure stage.

[16] J. Hine, R. Wiesboeck, and O. B. Ramsay, *J. Amer. Chem. Soc.*, **83,** 1222 (1961).
[17] R. Breslow, *Tetrahedron Lett.*, **1964,** 399.
[18] F. G. Bordwell and M. D. Wolfinger, *J. Org. Chem.*, **39,** 2521 (1974).
[19] N. H. Fischer, *Synthesis*, **1970,** 393.
[20] N. P. Neureiter, *J. Amer. Chem. Soc.*, **88,** 558 (1966).
[21] F. G. Bordwell, J. M. Williams, Jr., E. B. Hoyt, Jr., and B. B. Jarvis, *J. Amer. Chem. Soc.*, **90,** 429 (1968).
[22] N. Tokura, T. Nagai, and S. Matsumura, *J. Org. Chem.*, **31,** 349 (1966).

Three further refinements of the above mechanism have been accorded much attention more recently: stereochemistry of deprotonation, configurational aspects and timing of halide displacement, and stereochemical factors involved in episulfone formation. The ready conversion of 1-chloro-9-thiabicyclo[3.3.1]nonane 9,9-dioxide (**1**) with its W-plan arrangement of α-Cl and α'-H to $\Delta^{1,5}$-bicyclo[3.3.0]octene with aqueous potassium hydroxide at 100° requires inversion of configuration at both reacting centers (Eq. 2).[22–24] Although this system is

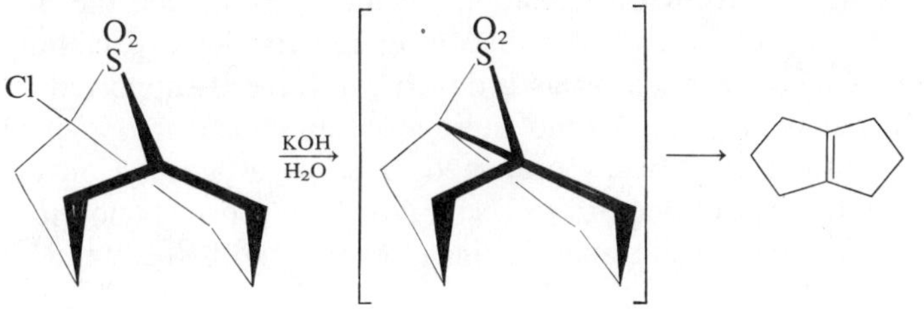

(Eq. 2)

highly constrained, evidence suggests that acyclic α-halo sulfones follow a comparable course (Eq. 3).[25,26] For example, deprotonation of

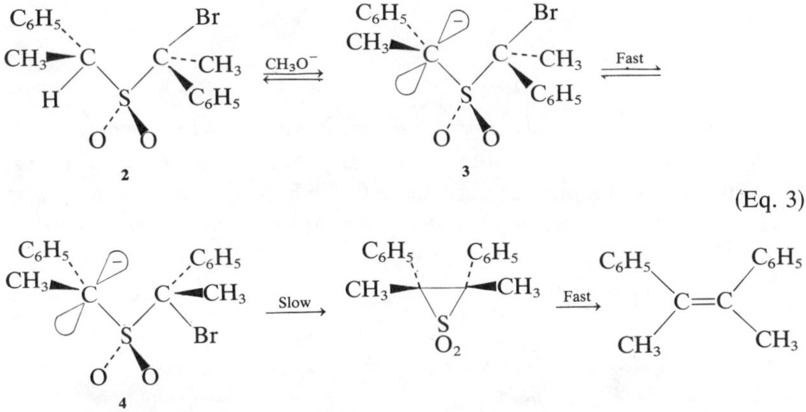

(Eq. 3)

erythro-α-bromo sulfone (**2**) probably occurs selectively from the conformation in which the proton is flanked by the two sulfonyl oxygens to give an "effectively planar" carbanion capable of maintaining asymmetry. Since rapid rotation about the S—C_α bond is possible, anions **3** and **4** are rapidly interconverted and closure to the episulfone occurs

[23] L. A. Paquette and R. W. Houser, *J. Amer. Chem. Soc.*, **91,** 3870 (1969).
[24] E. J. Corey and E. Block, *J. Org. Chem.*, **34,** 1233 (1969).
[25] F. G. Bordwell, E. Doomes, and P. W. R. Corfield, *J. Amer. Chem. Soc.*, **92,** 2581 (1970).
[26] F. G. Bordwell and E. Doomes, *J. Org. Chem.*, **39,** 2526 (1974).

with inversion at the C–Br center. A comparable high degree of stereoselectivity with inversion at both reaction sites has been observed in the reaction of *meso*-α-bromobenzylsulfone with triphenylphosphine.[27]

A concerted pathway for the 1,3-elimination reaction is more difficult to rule out experimentally but no longer appears possible. Such a mechanism requires strict coplanar alignment of the H—C—SO_2—C—Br atoms in the transition state and would be expected to exhibit a negative entropy of activation. In the cases studied the $\Delta S^{\ddagger}$ terms have been positive and thus do not agree with this expectation. That the one-stage mechanism has little driving force is supported by the kinetic behavior of α-bromo sulfone **5** in which the W-plan alignment of atoms is sterically preferred or easily attained. Conversion of the α-bromo sulfone **5** into acenaphthylene proceeds no more rapidly than for open-chain analogs, in agreement with the stepwise mechanism.[28]

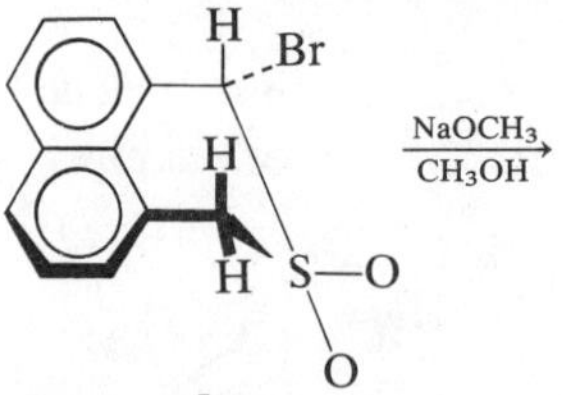

H Br H H S—O O $\xrightarrow[CH_3OH]{NaOCH_3}$

5

Steric requirements do exist, however, as revealed by the conformational requirement for ring contraction in 2-chloro-2,7-dihydro-3,4,5,6-dibenzothiepin 1,1-dioxides (**6**). Where R = R′ = H and R = CH_3, R′ = H, conversion into the phenanthrenes proceeds quantitatively.[29] In contrast, methyl groups at the R′ positions effectively

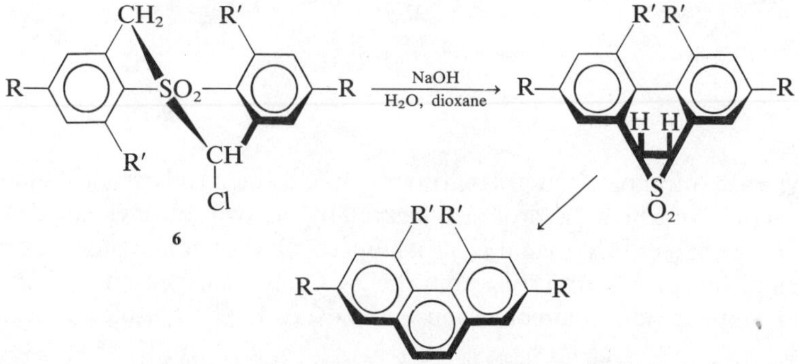

CH_2 R′ R— SO_2 —R R′ CH Cl $\xrightarrow[H_2O,\ dioxane]{NaOH}$ R′ R′ R— —R H H S O_2

6

R′ R′ R— —R

[27] F. G. Bordwell, B. B. Jarvis, and P. W. R. Corfield, *J. Amer. Chem. Soc.*, **90**, 5298 (1968).
[28] F. G. Bordwell and E. Doomes, *J. Org. Chem.*, **39**, 2531 (1974).
[29] L. A. Paquette, *J. Amer. Chem. Soc.*, **86**, 4085 (1964).

prohibit coplanarity of the benzene rings with the result that episulfone formation cannot be realized and intramolecular S_N2 displacement is not possible.

The most remarkable mechanistic aspect of the Ramberg-Bäcklund rearrangement is undoubtedly the stereochemistry of the olefinic products formed from acyclic α-halo sulfones. The products are consistently rich in the *cis* isomer. Since stereochemistry develops at the rate-determining cyclization step (Eqs. 1 and 3), features peculiar to the episulfone-forming transition state are most important. The causes for preferential *cis* positioning of the pair of R groups in the three-membered ring have been variously attributed to London forces,[20] preferential formation of one of the two possible diastereomeric carbanions in a higher equilibrium concentration,[10,15,26] and steric attraction.[30]

SCOPE AND LIMITATIONS

The Halogenation Step

When a sulfide possessing at least one hydrogen at an α-carbon atom is treated with one equivalent of chlorine,[31–33] sulfuryl chloride,[29,34–37] or N-chlorosuccinimide (NCS)[38–44] in an inert solvent, α-chloro sulfides are produced. Of these, N-chlorosuccinimide is most effective and has been used almost exclusively in recent years. The process is ionic, involving S-chlorosulfonium salts as initial intermediates. Conversion into the α-chloro sulfide may occur via an Elcb-type mechanism (path a) when X^- is the weakly basic chloride ion, or an E2 related process (path b) when the more basic succinimidyl anion is involved (Eq. 4).[42] In actuality, it is entirely possible that a continuum exists between these two extremes. These mechanisms imply that directive effects should gain importance, and they do.

30 R. Hoffmann, C. C. Levin, and R. A. Moss, *J. Amer. Chem. Soc.*, **95**, 629 (1973).

31 H. Böhme, H. Fischer. and R. Frank, *Ann. Chem.*, **563**, 54 (1949).

32 H. Böhme and H. J. Gran, *Ann. Chem.*, **581**, 133 (1953).

33 H. Richtzenhaim and B. Alfredsson, *Chem. Ber.*, **86**, 142 (1953).

34 W. E. Truce, G. H. Birum, and E. T. McBee, *J. Amer. Chem. Soc.*, **74**, 3594 (1952).

35 F. G. Bordwell and B. M. Pitt, *J. Amer. Chem. Soc.*, **77**, 572 (1955).

36 F. G. Bordwell, G. D. Cooper, and H. Morita, *J. Amer. Chem. Soc.*, **79**, 376 (1957).

37 L. A. Paquette, L. S. Wittenbrook, and K. Schreiber, *J. Org. Chem.*, **33**, 1080 (1968).

38 D. L. Tuleen and T. B. Stephens, *Chem. Ind.* (London), **1966**, 1555.

39 D. L. Tuleen and V. C. Markum, *J. Org. Chem.*, **32**, 204 (1967).

40 D. L. Tuleen and D. N. Buchanan, *J. Org. Chem.*, **32**, 495 (1967).

41 D. L. Tuleen, *J. Org. Chem.*, **32**, 4006 (1967).

42 D. L. Tuleen and T. B. Stephens, *J. Org. Chem.*, **34**, 31 (1969).

43 D. L. Tuleen and R. H. Bennett, *J. Heterocycl. Chem.*, **6**, 115 (1969).

44 E. Vilsmaier and W. Sprügel, *Ann. Chem.*, **747**, 151 (1971).

$$R'\!-\!\overset{\overset{Cl}{|}}{\underset{+}{S}}\!-\!CH_2R \;\; X^- \;\xrightleftharpoons{a}\; HX + R'\!-\!\overset{\overset{Cl}{|}}{\underset{+}{S}}\!-\!\underset{-}{C}HR \longleftrightarrow R'\!-\!\overset{\overset{Cl}{|}}{S}\!=\!CHR \longrightarrow R'\!-\!S\!-\!\overset{\overset{Cl}{|}}{C}HR$$

$$R'\!-\!\overset{\overset{Cl}{|}}{\underset{+}{S}}\!-\!CH_2R \;\; X^- \;\xrightarrow{b}\; HX + R'\!-\!\underset{+}{S}\!=\!CHR \;\; Cl^- \longleftrightarrow R'\!-\!S\underset{+}{C}HR \;\; Cl^- \longrightarrow R'\!-\!S\!-\!\overset{\overset{Cl}{|}}{C}HR \qquad \text{(Eq. 4)}$$

For example, chlorination of sulfides **7**[6,45] and **8**[46] leads exclusively to α-chlorobenzyl sulfides. Chlorinated isomers **9** and **10** are formed in the ratio 1.6:1 when benzyl *p*-chlorobenzyl sulfide is treated with an equimolar quantity of N-chlorosuccinimide in carbon tetrachloride solution.[39]

$$\underset{\mathbf{7}}{C_6H_5CH_2SCH_3} \xrightarrow{SO_2Cl_2} C_6H_5\underset{\underset{Cl}{|}}{C}HSCH_3$$

$$\mathbf{8} \xrightarrow{Cl_2} \text{1-chloroisothiochroman}$$

$$C_6H_5CH_2SCH_2C_6H_4Cl\text{-}p \xrightarrow{NCS} \underset{\mathbf{9}}{C_6H_5CH_2S\underset{\underset{Cl}{|}}{C}HC_6H_4Cl\text{-}p} + \underset{\mathbf{10}}{C_6H_5\underset{\underset{Cl}{|}}{C}HSCH_2C_6H_4Cl\text{-}p}$$

Polychlorination of dimethyl sulfide results in all of the hydrogen atoms of one methyl group being replaced by chlorine before the second group is attacked.[34] Chlorine therefore exerts a powerful directive influence that can be advantageous in the preparation of α,α-dichloro (*e.g.*, **11**)[47] and α,α,α-trichloro sulfides (*e.g.*, **12**).[48] Only two exceptions to this general behavior have been reported. Sulfuryl chloride chlorination of dibenzothiepin affords a mixture of dichloro sulfides **13** and **14** in the ratio 3.6:1.[49] It has been proposed that the strong inductive effect of the α-chloro substituent in the intermediate

[45] H. Böhme and H. J. Gran, *Ann. Chem.*, **577,** 68 (1952).

[46] H. Böhme, L. Tils, and B. Unterhalt, *Chem. Ber.*, **97,** 179 (1964).

[47] L. A. Paquette, L. S. Wittenbrook, and V. V. Kane, *J. Amer. Chem. Soc.*, **89,** 4487 (1967).

[48] L. A. Paquette and L. S. Wittenbrook, *J. Amer. Chem. Soc.*, **90,** 6790 (1968).

[49] L. A. Paquette, *J. Amer. Chem. Soc.*, **86,** 4089 (1964).

$$C_6H_5CH_2SCH_2Cl \xrightarrow{SO_2Cl_2} C_6H_5CH_2SCHCl_2$$

11

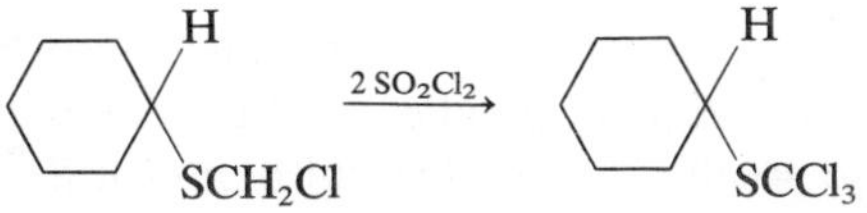

$$\text{cyclohexyl}(H)SCH_2Cl \xrightarrow{2\ SO_2Cl_2} \text{cyclohexyl}(H)SCCl_3$$

12

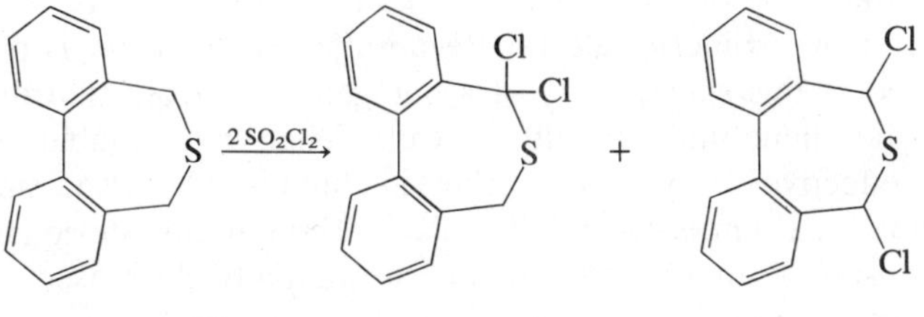

$\xrightarrow{2\ SO_2Cl_2}$ **13** + **14**

α-chloro sulfide is offset by steric factors in the transition state for α,α-dichlorination leading to substitution in the opposite direction. The second case concerns p-nitrobenzyl chloromethyl sulfide (**15**), which experiences halogen substitution at both α-carbon atoms.[37] No other substituent on the phenyl ring was able to effect this crossover.

$$p\text{-}O_2NC_6H_4CH_2SCH_2Cl$$

15

The stereochemistry of α chlorination has been examined in propellanes such as sulfides **16** and **17**.[50] The prevailing directive effects are

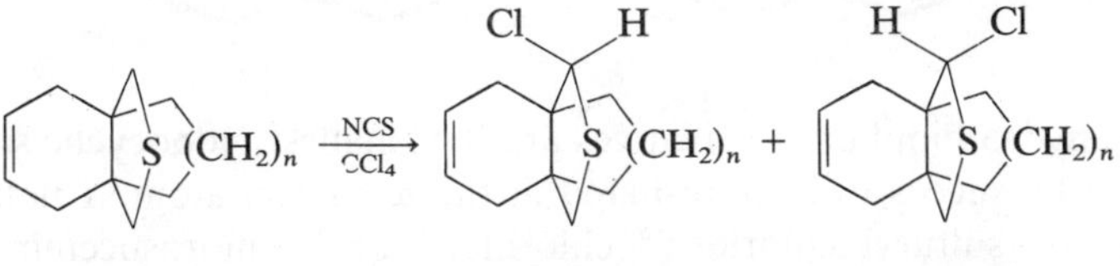

16 $\xrightarrow[CCl_4]{NCS}$ (Cl syn) + (Cl anti)

$n=2$: (44%) (56%)
$n=3$: (72%) (28%)

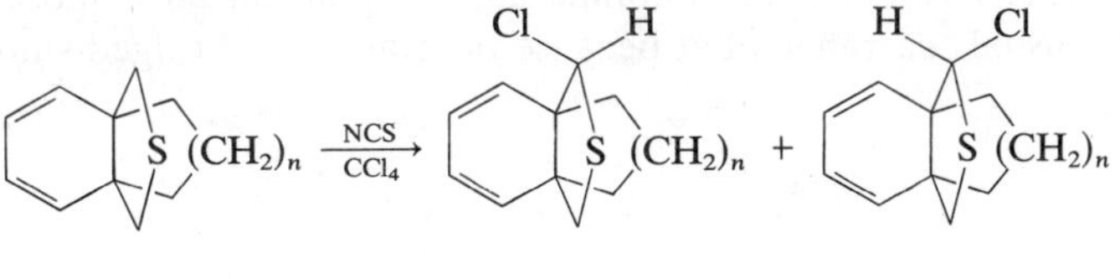

17 $\xrightarrow[CCl_4]{NCS}$ (Cl syn) + (Cl anti)

$n=2$: (35%) (65%)
$n=3$: (65%) (35%)

[50] L. A. Paquette, R. E. Wingard, Jr., J. C. Philips, G. L. Thompson, L. K. Read, and J. Clardy, *J. Amer. Chem. Soc.*, **93,** 4508 (1971).

such that the 11-thia[4.4.3]propellanes ($n = 2$) experience preferential entry of the chlorine atom from what may be considered the more hindered side. This directive effect is reversed in the 12-thia[5.4.3]-propellane series ($n = 3$). Although steric factors do gain importance, electronic effects also appear significant in examples that have sites of unsaturation close to the developing sulfonium ion.[50] Thus it can be expected that first-formed chlorosulfonium salt **18** can experience added stabilization of both a steric and an electronic nature that is not available to the isomeric salt **19**. Because stereochemistry is probably determined at this stage, entry of a halogen atom from that direction *anti* to the diene unit should prevail. When one double bond is present, effective π orbital overlap is diminished. Also, since the polymethylene chain is extended to five carbon atoms, steric effects in the chlorosulfonium salt **19** will be substantially decreased. As expected, both factors lead to more *syn* chlorination.

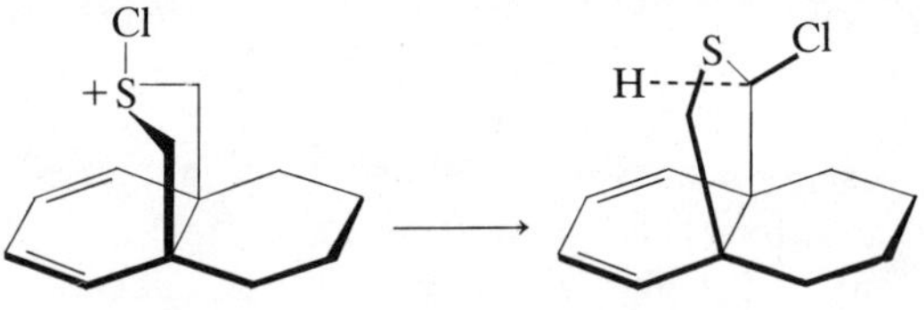

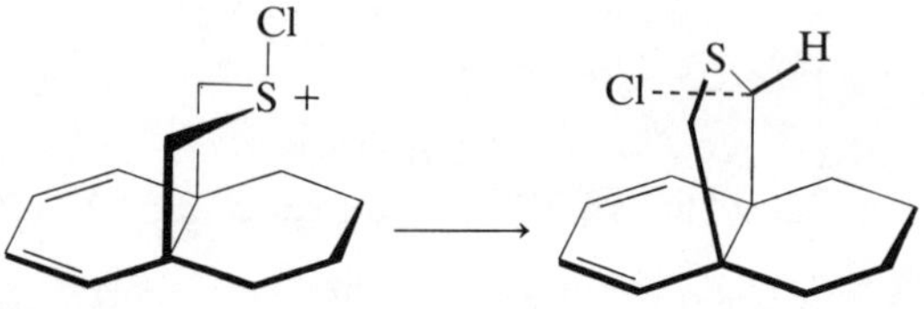

18

19

Tetrahydrothiophene derivatives are the smallest monocyclic sulfides that can be successfully chlorinated at the α-carbon atom. Thietane is cleaved by sulfuryl chloride,[35] chlorine,[51] or N-chlorosuccinimide,[38] forming 3-chloropropanesulfenyl chloride or N-(3-chloropropylthio)-succinimide. Evidently, nucleophilic attack upon ion **20** is more rapid than the usual rearrangement because of strain. The bridged sulfide **21**

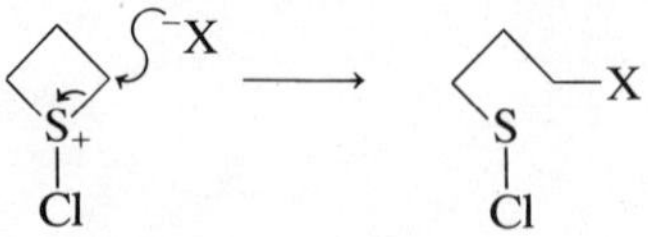

20

[51] J. M. Stewart and C. H. Burnside, *J. Amer. Chem. Soc.*, **75**, 243 (1953).

behaves anomalously,[23] giving the β-chloro sulfide **22** rather than the customary α-chlorination product. 8-Thiabicyclo[3.2.1]octane shows comparable reactivity.[23] Because of Bredt's rule, loss of a bridgehead proton is made difficult and fragmentation results. The medium-ring sulfenyl chloride produced can proceed to product by intramolecular electrophilic attack upon the newly generated double bond; the episulfonium ion intermediate requires attack by chloride ion with inversion of configuration.

$$\mathbf{21} \xrightarrow{SO_2Cl_2} [\text{chlorosulfonium } Cl^-] \xrightarrow{-HCl} [\text{sulfenyl chloride}] \longrightarrow [\text{episulfonium ion} + Cl^-] \longrightarrow \mathbf{22}$$

Because of the exceptional ease with which they undergo hydrolysis,[34] α-chloro sulfides cannot be satisfactorily prepared in moist solvents. Anhydrous conditions are mandatory for maximum yields. Such solvents as carbon tetrachloride and benzene have proven particularly useful. Protic solvents are to be avoided. Conversion of phenyl propargyl sulfide into the α-methoxy-substituted product with t-butyl hypochlorite in methanol illustrates the consequences of using a protic solvent.[52]

$$C_6H_5SCH_2C{\equiv}CH \xrightarrow[CH_3OH]{t\text{-}C_4H_9OCl} C_6H_5SCH(OCH_3)C{\equiv}CH$$

The hydrolysis reaction is applicable, however, to the conversion of halides into aldehydes and ketones.[41,53]

$$C_6H_{11}CH_2Br \xrightarrow[2.\ NCS]{1.\ C_6H_5S^-} C_6H_{11}CH(Cl)SC_6H_5 \xrightarrow{H_2O} C_6H_{11}CHO$$

[52] L. Skattebøl, B. Boulette, and S. Solomon, *J. Org. Chem.*, **32**, 3111, 3726 (1967).

[53] R. A. Snow, Ohio State University, unpublished observations.

Other Approaches to α-Halo Sulfides

An alternative, useful route to α-halo sulfides is condensation of aldehydes with mercaptans in the presence of one of the hydrogen halides (Refs. 15,31,45,47,54–58). When a source of formaldehyde is used, one-carbon atom homologation results. Higher aldehydes lead to α-halo sulfides having more substitution, whereas ketones are converted under these conditions into the corresponding mercaptals. Many of the simpler mercaptans are presently commercially available; others can be made from the parent halide and thiourea.

$$n\text{-}C_3H_7SH + (CH_2O)_x \xrightarrow{HBr} n\text{-}C_3H_7SCH_2Br$$

$$C_6H_5CH_2SH \xrightarrow[HCl]{C_2H_5CHO} C_6H_5CH_2SCHClC_2H_5$$

This method is a useful adjunct to the sulfide chlorination procedure for two principal reasons. First, the mercaptan–aldehyde condensation scheme results in homologation by whatever number of carbon atoms is desired, whereas direct halogenation does not. These additional carbon atoms may be linear or branched. Secondly, by judicious selection of starting materials, isomeric α-chloro sulfides can be prepared without the danger of intercontamination and the need for physical separation. Syntheses of isomers **23** and **24** are exemplary.[58]

$$C_6H_5CH_2SH + p\text{-}CH_3C_6H_4CHO \xrightarrow[CH_2Cl_2]{HCl} \underset{\mathbf{23}}{C_6H_5CH_2SCHClC_6H_4CH_3\text{-}p}$$

$$p\text{-}CH_3C_6H_4CH_2SH + C_6H_5CHO \xrightarrow[CH_2Cl_2]{HCl} \underset{\mathbf{24}}{C_6H_5CHClSCH_2C_6H_4CH_3\text{-}p}$$

Oxidation of α-Halo Sulfides to α-Halo Sulfones

Because they are susceptible to hydrolysis, α-chloro sulfides cannot be satisfactorily oxidized in aqueous solutions. Under anhydrous conditions, however, α-chloro sulfones can be isolated in high yields. Chromic anhydride in glacial acetic acid was one of the earliest reagents used;[34] another was dried solutions of 40% peracetic acid in

[54] H. Böhme, *Chem. Ber.*, **69**, 1610 (1936).

[55] L. A. Walter, L. H. Goodson, and R. J. Fosbinder, *J. Amer. Chem. Soc.*, **67**, 655, 657 (1945).

[56] L. A. Paquette, *J. Amer. Chem. Soc.*, **86**, 4383 (1964).

[57] N. P. Neureiter, *J. Org. Chem.*, **30**, 1313 (1965).

[58] L. A. Paquette and L. S. Wittenbrook, *J. Amer. Chem. Soc.*, **89**, 4483 (1967).

methylene chloride.[57] More convenient oxidants are *m*-chloroperbenzoic acid in chloroform,[29,49,56] and ethereal monoperphthalic acid.[33,50,59] The greater ease of removal of unreacted monoperphthalic acid by bicarbonate extraction after oxidation is a distinct advantage.

Because sulfur in the sulfide state is oxidized to the sulfone with exceptional rapidity, α-chloro sulfones having π bonds located elsewhere in the molecule can be prepared without affecting such functionality. The behavior of the unsaturated sulfide **25** is an example.[60]

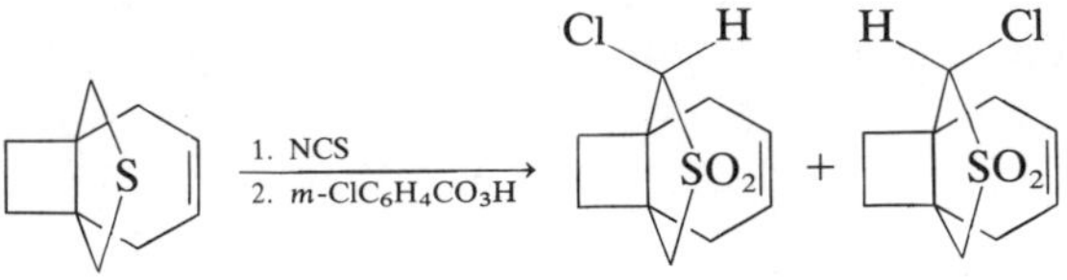

1. NCS
2. $m\text{-ClC}_6H_4CO_3H$

25

Alternative Routes to α-Halo Sulfones

Many other routes to α-halo sulfones are available. Since the starting materials in these routes are the parent sulfones, arenesulfinate salts, or the like, the method of choice may sometimes be predicated simply on the availability of starting materials. However, there are certain restrictions on the application of these alternative routes as summarized below.

Halogenation of α-Sulfonyl Carbanions

As might be expected from the nucleophilicity of α-sulfonyl carbanions, such intermediates are quite reactive toward elemental bromine and iodine (Refs. 4, 14, 24, 57, 61, 62) as well as other electrophiles such as N-chlorosuccinimide,[63] cyanogen bromide,[24] and trichloromethanesulfonyl chloride.[24] Dicyclopropyl sulfone, for example, is converted into α-chloro sulfones **26** and **27** when treated with slightly

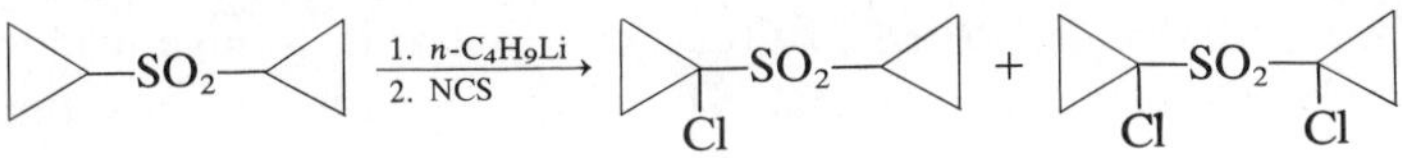

1. $n\text{-C}_4H_9Li$
2. NCS

26 **27**

[59] L. A. Paquette, R. H. Meisinger, and R. E. Wingard, Jr., *J. Amer. Chem. Soc.*, **95,** 2230 (1973).

[60] L. A. Paquette and R. W. Houser, *J. Amer. Chem. Soc.*, **93,** 4522 (1971).

[61] J. Ficini and G. Stork, *Bull. Soc. Chim. Fr.*, **1964,** 723.

[62] L. A. Paquette and U. Jacobsson, Ohio State University, unpublished observations.

[63] L. A. Paquette and R. W. Houser, *J. Org. Chem.*, **36,** 1015 (1971).

more than 1 equiv of *n*-butyllithium in tetrahydrofuran at room temperature followed by N-chlorosuccinimide at 0°.[60] The action of *t*-butyllithium and cyanogen bromide on bicyclic sulfone **28** leads to

SO_2 → 1. *t*-C_4H_9Li 2. BrCN → SO_2 Br

28

incorporation of bridgehead bromine.[24] Thiabarbaralane 1,1-dioxide (**29**) is readily iodinated at both α positions when its dianion is exposed to excess iodine.[62]

O_2 S → 1. *n*-C_4H_9Li 2. I_2 → I O_2 S I

29

Keto sulfone **30** is capable of selective bromination at its most acidic carbon atom.[61] With dialkyl sulfones, mixtures of α-halo isomers result if the starting materials are unsymmetrical.[57] In all cases it is possible to arrest reaction at the halogenation stage without the interference of α-halo sulfone rearrangement by controlling both the amount of base and the temperature.

O $SO_2C_4H_9$-*n* → 1. NaH, C_6H_6 2. Br_2, 0° → O Br $SO_2C_4H_9$-*n*

30

A somewhat less direct method was used to prepare α-chloro sulfones **1** and **31**.[23] In these examples the derived α-sulfonyl carbanions were treated with excess sulfuryl chloride at 0° to give bridgehead sulfonyl chlorides such as **32**. These compounds in turn undergo

SO_2 → 1. *n*-C_4H_9Li 2. SO_2Cl_2 → SO_2 SO_2Cl → Heat → SO_2 Cl

32 **31**

smooth thermal decomposition at 160° (0.05 mm) with evolution of sulfur dioxide and formation of the desired α-chloro sulfone.

In the only existing example of four-membered ring halogenation, 3,3-dimethylthietane 1,1-dioxide has been converted in 87% yield into its α-bromo derivative by sequential reaction with n-butyllithium in tetrahydrofuran at −78° and the indicated cyclic bromomalonate.[63a]

CH_3 / CH_3 / SO_2 —(1. n-C_4H_9Li; 2. Br, CH_3, O, O, CH_3, CH_3, O, O)→ Br, CH_3, CH_3, SO_2

Halogenative Decarboxylation

The capacity of α-carboxyalkyl sulfones for halogenative decarboxylation was recognized before the turn of the century.[64] If the sulfone carboxylic acids are of the type $ArSO_2CH_2CO_2H$, α,α-dihalomethyl sulfones result; more substituted derivatives such as $ArSO_2$-$CHRCO_2H$, on the other hand, give monohalo products. The corresponding sulfides used as starting materials are readily prepared by reaction of mercaptides with chloroacetic acids or of halides with mercaptoacetic acid salts. This procedure appears general and has been applied in the preparation of α,α'-dihalosulfones as well.[65–69]

$$C_6H_5CH_2SO_2CH_2CO_2H \xrightarrow[\text{aq HOAc, 50°}]{Br_2} C_6H_5CH_2SO_2CHBr_2$$

$$\triangleright\!-SO_2CH(C_6H_5)CO_2H \xrightarrow[\text{CCl}_4\text{, Heat}]{\text{N-iodosuccinimide}} \triangleright\!-SO_2CHIC_6H_5$$

Both elemental halogens and N-halosuccinimides can serve as the electrophilic reagent. The N-halosuccinimides no doubt merely provide low concentrations of X_2. Since the reaction does not proceed unless one or two hydrogen atoms are present on the carbon atom bearing

[63a] J. P. Marino, *Chem. Commun.*, **1973,** 861.

[64] C. M. Suter, *The Organic Chemistry of Sulfur,* Wiley, New York, 1940, pp. 678–680.

[65] F. Scholnic, Ph.D. Dissertation, University of Pennsylvania, 1955.

[66] L. A. Carpino and L. V. McAdams, III, *J. Amer. Chem. Soc.*, **87,** 5804 (1965).

[67] L. A. Carpino, L. V. McAdams, R. H. Rynbrandt, and J. W. Spiewak, *J. Amer. Chem. Soc.*, **93,** 476 (1971).

[68] F. G. Bordwell, E. B. Hoyt, Jr., B. B. Jarvis, and J. M. Williams, Jr., *J. Org. Chem.*, **33,** 2030 (1968).

[69] F. G. Bordwell, M. D. Wolfinger, and J. B. O'Dwyer, *J. Org. Chem.*, **39,** 2516 (1974).

the RSO_2 and CO_2H groups, a mechanism involving rate-limiting enolization followed by rapid halogenation and subsequent rapid decarboxylation is quite likely.[69]

$$C_6H_5CH(CO_2H)SO_2CH(CO_2H)C_6H_5 \xrightarrow{Br_2} C_6H_5CHBrSO_2CHBrC_6H_5$$

This method can also serve as a route to bromomethyl alkyl or aryl sulfones, since the dibromomethyl sulfones can be readily reduced with aqueous sulfite ion.[68–70]

$$m\text{-}O_2NC_6H_4CH_2SO_2CHBr_2 + SO_3^{2-} \xrightarrow{\text{Aq } C_2H_5OH} m\text{-}O_2NC_6H_4CH_2SO_2CH_2Br$$

Alkylation of Sulfinate Salts

Reaction of sodium arenesulfinates with chloroform or bromoform in the presence of aqueous base affords dichloromethyl and dibromomethyl sulfones in good yield.[71] Alkylsulfinates generally behave poorly under these conditions. Sodium *t*-butylsulfinate is an apparent exception since it provides the dichloromethyl sulfone in 55% yield. However, with bromoform only a small yield of dibromo product can be isolated.[71]

$$p\text{-}CH_3C_6H_4SO_2^-Na^+ \xrightarrow[\text{KOH, }H_2O]{CHCl_3} p\text{-}CH_3C_6H_4SO_2CHCl_2$$

Because no reaction occurs in the absence of strong base, the C–S bond-forming step is viewed as the result of attack by electrophilic dihalocarbenes (which are generated *in situ*) upon thiosulfinate anions.

When dichloroacetic acid replaces the haloforms, monochloromethyl sulfones are formed directly as the result of facile decarboxylation of the intermediate carboxylic acid.[72]

$$p\text{-}CH_3CONHC_6H_4SO_2^-Na^+ \xrightarrow{Cl_2CHCO_2H} p\text{-}CH_3CONHC_6H_4SO_2CH_2Cl$$

Cycloaddition of Halosulfenes and Diazo Compounds

When highly reactive halosulfenes are generated *in situ* at low temperature (usually from the corresponding sulfonyl chloride and triethylamine) in the presence of diazomethane, cycloaddition with loss of nitrogen takes place to give 2-halothiirane 1,1-

[70] C. J. Kelley, Ph.D. Dissertation, Indiana University, 1970.

[71] W. Middelbos, G. Strating, and B. Zwanenburg, *Tetrahedron Lett.*, **1971,** 351.

[72] B. C. Jain, B. H. Iyer, and P. C. Guba, *J. Indian Chem. Soc.*, **24,** 220 (1947).

dioxides.[47,67,73,74] Such episulfones are formed as intermediates in the Ramberg-Bäcklund rearrangement of dihalo sulfones and are extremely labile to base.

$$ClCH_2SO_2Cl \xrightarrow{(C_2H_5)_3N} [ClCH{=}SO_2] \xrightarrow[(C_2H_5)_2O,\ -5°]{CH_2N_2} \text{(episulfone: ring } SO_2\text{, H, Cl)}$$

$$CH_3CH(Br)SO_2Cl \xrightarrow{(C_2H_5)_3N} [CH_3C(Br){=}SO_2] \xrightarrow{CH_2N_2} \text{(episulfone: ring } SO_2\text{, Br, } CH_3\text{)}$$

Miscellaneous Methods

Benzylic sulfones undergo moderately efficient α bromination when heated with N-bromosuccinimide (NBS) in refluxing carbon tetrachloride containing benzoyl peroxide.[26]

$$C_6H_5CH(CH_3)SO_2CH_2CH_3 \xrightarrow[CCl_4,\ (C_6H_5CO)_2O_2]{NBS} C_6H_5C(CH_3)(Br)SO_2CH_2CH_3$$

Reaction of trichloromethyl sulfones with trialkyl phosphites has been reported to give only 1,1-dichloroalkyl sulfones.[75] A reinvestigation of this reaction has shown, however, that the primary products are both alkylated and phosphorylated sulfones.[70] Because phosphorylated sulfones undergo further reaction with the phosphite, a complex mixture frequently results. At this stage of development, therefore, the process has marginal synthetic utility.

$$R'SO_2CCl_3 + P(OR)_3 \rightarrow R'SO_2CCl_2R + (RO)_2P(O)Cl$$

In an innovative approach, methyl α-bromovinyl sulfone, readily prepared by bromination-dehydrobromination of methyl vinyl sulfone, has been used in cycloaddition reactions.[76] Although mixtures of epimers can arise in certain adducts, subsequent Ramberg-Bäcklund rearrangement with introduction of an exocyclic methylene group negates the stereochemical issue. An allene synthon is thereby provided.

[73] L. A. Carpino and R. H. Rynbrandt, *J. Amer. Chem. Soc.*, **88,** 5682 (1966).

[74] L. A. Paquette and L. S. Wittenbrook, *Org. Synth.*, **49,** 18 (1969).

[75] K. Szabo, U.S. Patent 3,106,585 [*C.A.*, **60,** 2841b (1964)]; U.S. Patent 3,294,845 [*C.A.*, **67,** 11210c (1967)].

[76] J. C. Philips and M. Oku, *J. Amer. Chem. Soc.*, **94,** 1012 (1972).

$CH_2{=}CH{-}CH{=}CH_2 + CH_2{=}C(Br)SO_2CH_3 \longrightarrow$ 1-bromo-1-(methylsulfonyl)cyclohex-3-ene

Rearrangements of α-Halo Sulfones

The original studies of Ramberg and Bäcklund were carried out in 2 *N* sodium hydroxide solution at 100°. Six monohalogenated sulfones underwent loss of hydrogen halide and extrusion of sulfur dioxide under these conditions. The yields varied from 51% with chloromethyl ethyl sulfone to 90% with α-bromoethyl *n*-propyl sulfone.[13] Since this work preceded the now commonplace use of gas chromatographic techniques, olefin stereochemistry had to be established by chemical methods. The *cis* isomers of 2-butene and 2-pentene predominanted as shown by the physical properties of the respective dibromides, *e.g.*, *dl*-2,3-dibromobutane was obtained upon bromination of the isolated 2-butene.[13] More refined studies have verified these conclusions and have shown that the stereochemistry of the particular reaction leading to 2-butene is remarkably insensitive to changes in the nature of the base and solvent, with the exception of potassium *t*-butoxide.[20,77] With this base a profound change in stereochemistry occurs and *trans*-2-butene predominates, the likely result (see independent experiments)

trans-$CH_3CD{=}CDCH_3$ (Major) $\xleftarrow[t\text{-}C_4H_9OD]{t\text{-}C_4H_9OK}$ *cis*-2-butene episulfone (CH_3, H / CH_3, H; S, O_2) $\xrightarrow[\text{or Heat}]{\text{NaOH, }H_2O\text{ or NaOD, }D_2O}$ *cis*-$CH_3CH{=}CHCH_3$

of epimerization of kinetically favored *cis*-2-butene episulfone to its *trans* isomer before loss of sulfur dioxide.[20] Since the equilibrium for 2-butene is approximately 30:70 in favor of the *trans* form, there is an unusual preference for formation of the thermodynamically less stable isomer under most conditions.

[77] N. P. Neureiter and F. G. Bordwell, *J. Amer. Chem. Soc.*, **85**, 1209 (1963).

$$ClCH_2SO_2C_2H_5 \rightarrow CH_2{=}CHCH_3 \quad (51\%)$$

$$CH_3CHClSO_2C_2H_5 \rightarrow CH_3CH{=}CHCH_3 \quad (75\%)$$

$$CH_3CHBrSO_2C_2H_5 \rightarrow CH_3CH{=}CHCH_3 \quad (85\%)$$

$$CH_3CHBrSO_2C_3H_7\text{-}n \rightarrow CH_3CH{=}CHC_2H_5 \quad (90\%)$$

$$C_2H_5CHBrSO_2C_2H_5 \rightarrow CH_3CH{=}CHC_2H_5 \quad (80\%)$$

$$CH_3CHBrSO_2C_3H_7\text{-}i \rightarrow CH_3CH{=}C(CH_3)_2 \quad (82\%)$$

In solutions 2 *N* in hydroxide ion, a number of dialkyl α-halo sulfones in which both alkyl groups and the halogen atoms were varied have yielded *cis* olefins predominantly (Table I). Because the rates of many of the α-chloro sulfone reactions are quite comparable, it would appear that release of chloride ion under pseudo first-order conditions is not subject to dramatic change over a relatively wide range of alkyl substitution. Intriguingly, the *cis* isomer is consistently formed in significantly greater relative yield when the leaving group is attached to the more bulky of the two alkyl groups within a given isomeric pair. Such data point to differing steric requirements of the alkyl groups attached to the nucleophilic carbon atom and the carbon atom which is the seat of displacement.

α-Halo sulfones of higher molecular weight are not readily soluble in aqueous base and changes in the medium (and base) are required. When complications are not present, use of dioxane as a co-solvent is recommended. Alternatively, sodium methoxide in methanol has many advantages. Both media allow kinetic analysis of the rearrangement.

TABLE I. Rearrangement of Dialkyl α-Halo Sulfones in 2 *N* Hydroxide Solution[15,20]

Sulfone	Product	Yield, %	*cis*, %	*trans*, %
α-Chloroethyl ethyl	2-Butene	76	78.8	21.2
α-Bromoethyl ethyl	2-Butene	85	79.5	20.5
α-Iodoethyl ethyl	2-Butene	87	78.2	21.8
α-Chloropropyl ethyl	2-Pentene	57	71.3	28.7
α-Bromopropyl ethyl	2-Pentene	71	70	30
α-Chloroethyl propyl	2-Pentene	74	65.8	34.3
α-Bromoethyl propyl	2-Pentene	85	60	40
α-Chlorobutyl ethyl	2-Hexene	63	69.4	30.6
α-Chloroethyl butyl	2-Hexene	76	66.2	33.8
α-Chloropropyl propyl	3-Hexene	—	57	43
α-Iodopropyl propyl	3-Hexene	57	55.5	44.5
α-Chloroisobutyl ethyl	4-Methyl-2-pentene	41	59.4	40.6
α-Chloroethyl isobutyl	4-Methyl-2-pentene	70	51.0	49.0
α-Iodobutyl butyl	4-Octene	50	52.5	47.5

When the episulfone intermediates possess unusually acidic α-sulfonyl protons, isomerization to the *trans* isomer will generally occur even with the weaker base systems. Benzyl α-chlorobenzyl sulfone, for example, is converted exclusively into *trans*-stilbene with sodium hydroxide in aqueous dioxane.[6] Not unexpectedly, therefore, *cis*-diphenylepisulfone (**33**) experiences epimerization and deuterium exchange before loss of sulfur dioxide.[21,22]

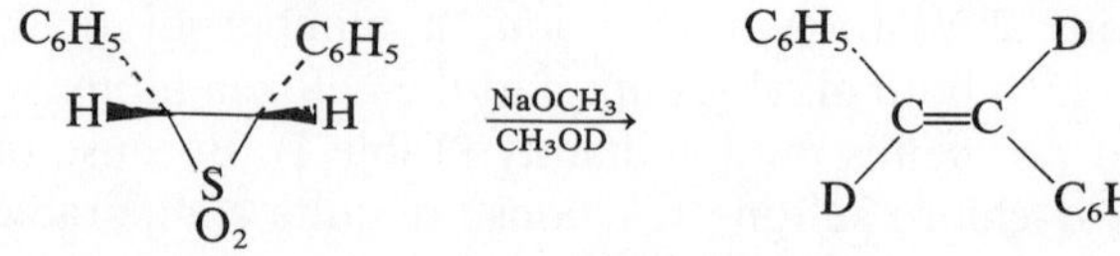

C_6H_5 C_6H_5 H H S O_2 **33** $\xrightarrow[CH_3OD]{NaOCH_3}$ C_6H_5(D)C=C(D)C_6H_5

Only more recently has attention been given to cyclic α-halo sulfones, perhaps because of early discouraging reports from two laboratories. A poor yield of cyclopentene was reported from the reaction of α-bromothiacyclohexane 1,1-dioxide **34** with base.[78] Additionally, the attempted conversion of α-bromo sulfone **35** into a

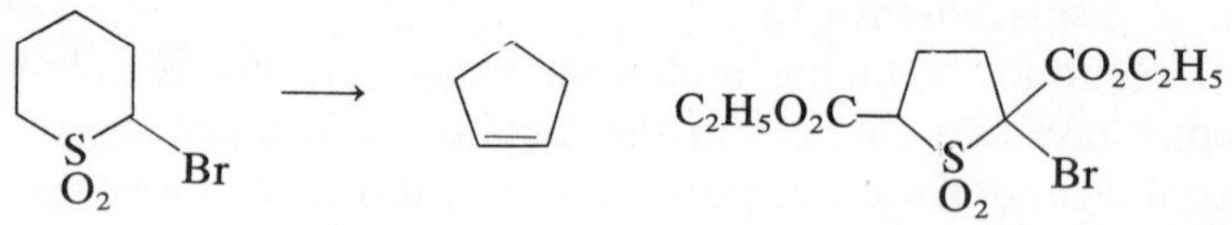

34 → cyclopentene; **35** ($C_2H_5O_2C$, $CO_2C_2H_5$, Br, S, O_2)

cyclobutene derivative failed.[79] However, the α-bromo sulfone **34** has now been converted into cyclopentene in 82% yield,[80] and the reactions of α-halo sulfones (**1**) and (**5**) further illustrate the formation of five-membered ring olefins. Cyclobutene derivatives can also be readily synthesized as shown by the conversion of the α-chloro sulfone (**36**) into the cyclobutene (**37**), although modifications in the nature of the

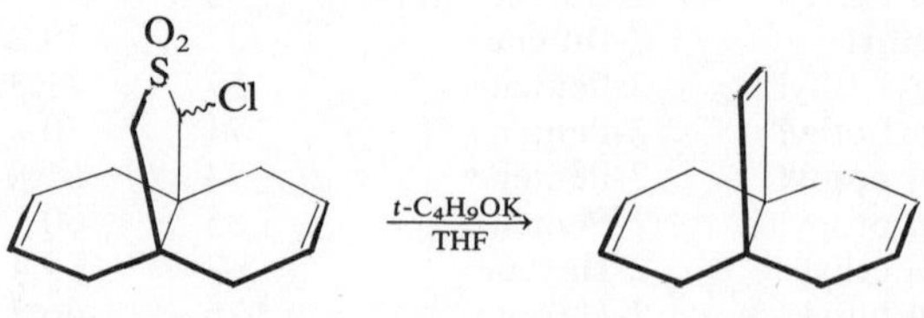

36 $\xrightarrow[THF]{t\text{-}C_4H_9OK}$ **37**

[78] I. Mischon, Ph.D. Dissertation, Technische Hochschule zu Karlsruhe, Germany, 1955.

[79] T. Bacchetti and A. G. Arnaboldi, *Atti Accad. Naz. Lincei Cl. Sci. Fis., Mat., Natur., Rend.*, **15,** 75 (1953) [*C.A.*, **49,** 2301b (1955)].

[80] J. W. Williams, Ph.D. Dissertation, Northwestern University, 1966.

base and the medium are necessary.[60,81–88] For example, α-chloro sulfone **36** is recovered in high yield after prolonged heating with 2 *N* sodium hydroxide in dioxane. In contrast, the cyclobutene **37** forms rapidly even at 0° when potassium *t*-butoxide in dry tetrahydrofuran is used. Lack of rearrangement in aqueous medium may be due to solvation. Rotational freedom in aliphatic sulfones permits conformations in which the highly solvated α-sulfonyl carbanion sphere can easily adopt a geometry favorable to the S_N2 displacement of halide ion; such a situation is more difficult to achieve in more conformationally fixed cyclic systems. But reduction in the solvation properties of the solvent probably results in a proportional decrease in nonbonded interactions near the carbanion and lower transition-state energy requirements for the 1,3 elimination. The combination of potassium *t*-butoxide and anhydrous tetrahydrofuran has proved most useful in these situations.

The α-halo sulfone rearrangement can be extended to compounds carrying electronegative substituents adjacent to the halogen atom.[61,65,89] As expected from earlier considerations, the only olefins isolated in reactions where geometric isomers are possible are those having the *trans* configuration. If a cyclic ketone part structure is involved as in the α-bromo sulfone **38**, initial reaction occurs by attack

O Br $SO_2CH_2C_3H_7$-*n* $\xrightarrow{OH^-}$ ^-O_2C H Br $SO_2CH_2C_3H_7$-*n* ⟶

38

$^-O_2C(CH_2)_4CH{=}CHC_3H_7$-*n*

of hydroxide ion at the carbonyl carbon resulting in ring cleavage and formation of an α-bromosulfonyl carboxylic acid. This intermediate subsequently is transformed into the structurally derived olefin. Again, the double bond cleanly replaces the sulfonyl group. Evidence for this mechanistic pathway was obtained by performing the ring opening

[81] L. A. Paquette and J. C. Philips, *Tetrahedron Lett.*, **1967,** 4645.
[82] L. A. Paquette and J. C. Philips, *Chem. Commun.*, **1969,** 680.
[83] L. A. Paquette and J. C. Philips, *J. Amer. Chem. Soc.*, **91,** 3973 (1969).
[84] L. A. Paquette, J. C. Philips, and R. E. Wingard, Jr., *J. Amer. Chem. Soc.*, **93,** 4516 (1971).
[85] L. A. Paquette and R. W. Houser, *J. Amer. Chem. Soc.*, **93,** 4522 (1971).
[86] L. A. Paquette and G. L. Thompson, *J. Amer. Chem. Soc.*, **94,** 7118 (1972).
[87] L. A. Paquette, R. E. Wingard, Jr., and J. M. Photis, *J. Amer. Chem. Soc.*, **96,** 5801 (1974).
[88] L. A. Paquette, R. K. Russell, and R. L. Burson, *J. Amer. Chem. Soc.*, **97,** 6124 (1975).
[89] I. Shahak and E. D. Bergmann, *Isr. J. Chem.*, **8,** 589 (1970).

under milder conditions. The α-bromosulfonyl carboxylic acid so produced underwent ready conversion into the unsaturated acid when treated with warm alkali.[61]

$$C_6H_5CH_2SO_2CHBrCH_2CO_2H \xrightarrow[H_2O]{NaOH} C_6H_5CH{=}CHCH_2CO_2H$$

$$C_6H_5CH_2SO_2CHBrSO_2CH_2C_6H_5 \xrightarrow[C_2H_5OH]{NaOC_2H_5} C_6H_5CH_2SO_2CH{=}CHC_6H_5$$

Adverse hybridization characteristics deter rearrangement. α-Chlorodicyclopropyl sulfone **26** illustrates this point nicely. The sulfone is recovered intact after prolonged exposure to refluxing solutions of aqueous 1.2 *N* potassium hydroxide and methanolic sodium methoxide.[63] In the presence of potassium *t*-butoxide in tetrahydrofuran at room temperature, dehydrochlorination and subsequent Michael addition are kinetically preferred. No bicyclopropylidene is formed because the cyclopropyl halogen atom is unreactive toward intramolecular displacement (I strain).

26 $\xrightarrow[THF]{t\text{-}C_4H_9OK}$ [cyclopropenyl–SO_2–cyclopropyl] ⟶ 2-(t-C_4H_9O),2-H-cyclopropyl–SO_2–cyclopropyl

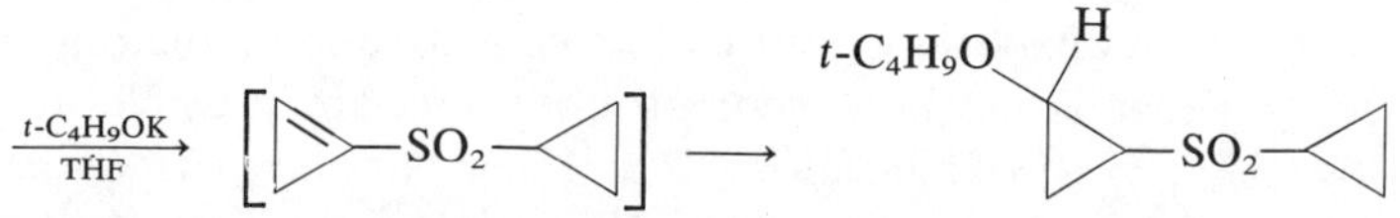

Attempts to extend the Ramberg-Bäcklund reaction to the synthesis of unsaturated propellanes having highly strained central bonds gave a marked decrease in yield of cyclic olefin.[85, 90, 91] Excessive ring strain such as is incorporated in the α-chloro sulfone **39** and the bis-(α-chloro sulfone) **40** acts as a deterrent to C—C bond formation and episulfone ring closure. A principal side reaction in these instances is α-*t*-butoxy sulfone production when potassium *t*-butoxide is the base. The transient generation of zwitterionic intermediates has been implicated in such reactions.[85]

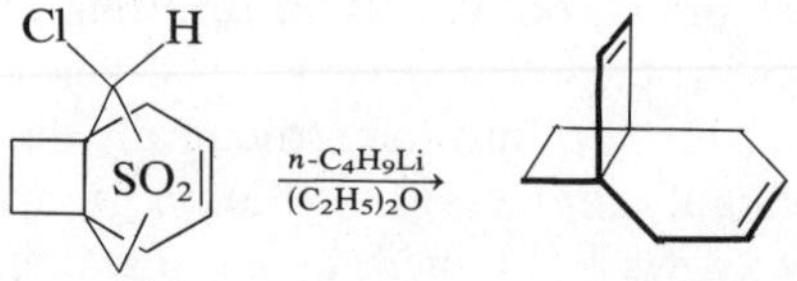

39 $\xrightarrow[(C_2H_5)_2O]{n\text{-}C_4H_9Li}$ (1.4%)

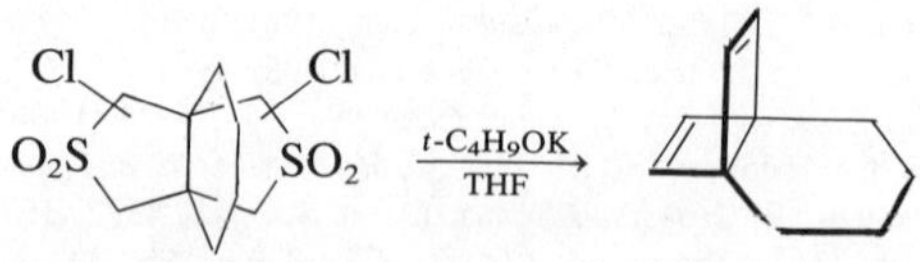

40 $\xrightarrow[THF]{t\text{-}C_4H_9OK}$ (1%)

[90] K. Weinges and K. Klessing, *Chem. Ber.*, **107**, 1915 (1974).
[91] K. Weinges and K. Klessing, *Chem. Ber.*, **107**, 1925 (1974).

Preparation of Deuterated Olefins

Hydrogen atoms α to the sulfonyl group exchange with the alkaline medium much more rapidly than the olefinic product is formed. Consequently, in a deuterated environment, essentially complete H/D exchange is established before C—C bond formation and elimination of sulfur dioxide. Therefore it is possible to prepare conveniently olefins which are highly deuterated exclusively at the double bond.[29,57,81,87] This unusually specific process would seem to hold much synthetic potential.

Cl
SO_2
NaOD
D_2O–Dioxane
D
D

H Cl
SO_2
NaOD
D_2O–Dioxane
D Cl
SO_2
D_2
t-C_4H_9OK
THF
D
D

The Bishomoconjugative Rearrangement

Although dipolar intermediates may routinely intervene in α-halo sulfone rearrangements, such zwitterions gain importance when strain factors become excessive and a good ionizing solvent is involved.[85] A proximate diene unit can serve as an intramolecular trap of such species and trigger an alternative reaction pathway which has been termed the bishomoconjugative rearrangement.[59,92] The behavior of the α-chloro sulfone **41** illustrates the process. When exposed briefly to potassium t-butoxide in dimethyl sulfoxide, this mixture of epimers is rapidly isomerized to sulfone **42**. The combined experimental evidence provides no information on whether base-initiated elimination of hydrogen chloride from the α-chloro sulfone **41** proceeds from the zwitterion stage to the episulfone; notwithstanding, subsequent intramolecular cycloaddition provides a cis^2-bishomobenzene which experiences rapid valence isomerization to give the final product. This

[92] L. A. Paquette, R. E. Wingard, Jr., and R. H. Meisinger, *J. Amer. Chem. Soc.*, **93,** 1047 (1971).

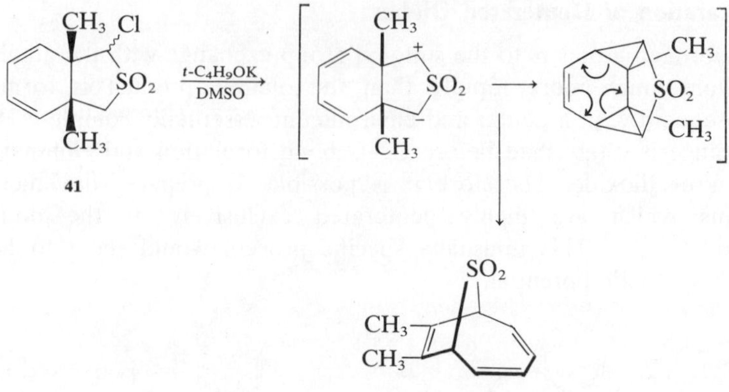

41

42

variation of the normal rearrangement permits access to 9-thiabicyclo[4.2.1]nona-2,4,7-triene 9,9-dioxides whose preparation cannot be realized readily by other methods.

Rearrangement of Sulfones Induced by Potassium Hydroxide/Carbon Tetrachloride

Sulfones possessing α-hydrogen atoms are readily α-chlorinated by carbon tetrachloride in the presence of potassium hydroxide and t-butyl alcohol.[93,94] The reactivity of a given sulfone depends on the concentration and nucleophilicity of the α-sulfonyl carbanion formed, whereas the ultimate course of the reaction depends largely on the structure of the substrate. The tendency is for multiple α chlorinations to occur faster than monochlorination because a covalently bound chlorine atom has a positive effect on carbanion stability. Consequently the concentration of the chlorinated carbanion is increased markedly and its nucleophilicity is not appreciably decreased. However, in sulfones possessing both α- and α'-hydrogen atoms, 1,3 elimination frequently occurs faster than introduction of a second chlorine atom and Ramberg-Bäcklund products result.

When this situation prevails, Meyers' method[93,94] is a valuable adjunct to the methods described above, principally because the sulfone starting materials are sometimes readily available. Moreover, the

$$C_6H_5CH_2SO_2CH(CH_3)C_6H_5 \xrightarrow[t\text{-}C_4H_9OH]{KOH,\ CCl_4} C_6H_5CH{=}C(CH_3)C_6H_5$$

(52%)

[93] C. Y. Meyers, A. M. Malte, and W. S. Matthews, *J. Amer. Chem. Soc.*, **91,** 7510 (1969).
[94] C. Y. Meyers, A. M. Malte, and W. S. Matthews, *Quart. Rep. Sulfur Chem.*, **5,** 229 (1970).

$$(C_6H_5)_2CHSO_2C_3H_7\text{-}i \xrightarrow[t\text{-}C_4H_9OH]{KOH, CCl_4} (C_6H_5)_2C{=}C(CH_3)_2 \quad (100\%)$$

process is a "one-flask" procedure. A drawback to the reaction is that dichlorocarbene is sometimes generated concomitantly. The carbene can react with the desired olefinic product, forming dichlorocyclopropanes. However, this does not always happen. Di-*s*-alkyl, benzylic and benzhydryl sulfones are particularly well-suited to this one-step olefin synthesis as illustrated in the accompanying reactions.[93–99] Allylic and cyclic sulfones having primary α-carbon atoms also provide simple alkenes, but di-primary-alkyl sulfones are transformed into *cis*-dialkylethylene sulfonic acid salts because they are first converted into dichloro sulfones.[100–102]

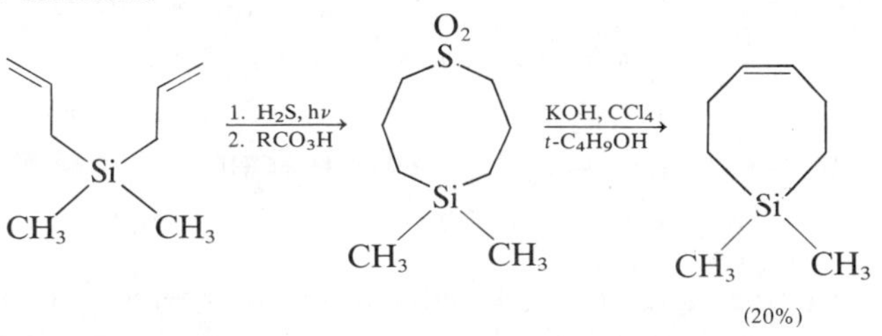

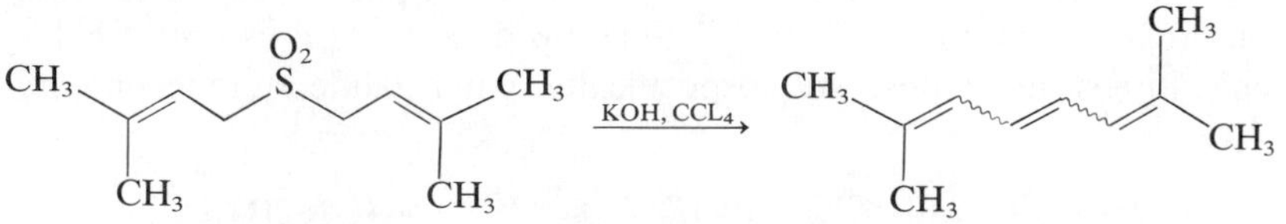

1. H_2S, $h\nu$; 2. RCO_3H; KOH, CCl_4, t-C_4H_9OH (20%)

KOH, CCL_4

Rearrangements of α,α- and α,α′-Dihalo Sulfones

Reactions of α,α- and α,α'-dihalo sulfones with base customarily give three types of products: acetylenes, vinyl chlorides, and α,β-unsaturated sulfonic acids. The nature of the product(s) formed depends upon the structure of the sulfone and the reaction conditions (Refs. 13, 47, 58, 65, 67, 68, 103, 104). The rearrangements proceed

95 C. Y. Meyers, W. S. Matthews, G. J. McCollum, and J. C. Branca, *Tetrahedron Lett.*, **1974,** 1105.

96 J. Kattenberg, E. R. de Waard, and H. O. Huisman, *Tetrahedron Lett.*, **1973,** 1481.

97 J. Kattenberg, E. R. de Waard, and H. O. Huisman, *Tetrahedron*, **30,** 3177 (1974).

98 G. Büchi and R. M. Freidinger, *J. Amer. Chem. Soc.*, **96,** 3332 (1974).

99 K. E. Koenig, R. A. Felix, and W. P. Weber, *J. Org. Chem.*, **39,** 1539 (1974).

100 C. Y. Meyers and L. L. Ho, *Tetrahedron Lett.*, **1972,** 4319.

101 C. Y. Meyers and I. Sataty, *Tetrahedron Lett.*, **1972,** 4323.

102 C. Y. Meyers, L. L. Ho, G. J. McCollum, and J. Branca, *Tetrahedron Lett.*, **1973,** 1843.

103 L. A. Paquette and L. S. Wittenbrook, *Chem. Commun.*, **1966,** 471.

104 F. G. Bordwell, J. M. Williams, Jr., and B. B. Jarvis, *J. Org. Chem.*, **33,** 2026 (1968).

again by 1,3 elimination to generate haloepisulfone intermediates (*e.g.*, **43**, Eq. 5) which may suffer either customary expulsion of sulfur dioxide to give vinyl halides or base-induced dehydrohalogenation. As with episulfones, it is not yet possible to isolate the resulting thiirene dioxides **44** from these rearrangements. However, weaker bases such

$$CH_3CHClSO_2CHClCH_3 \xrightarrow[H_2O]{KOH} CH_3CH{=}C(CH_3)SO_3^-K^+$$

$$C_6H_5CH_2SO_2CHBr_2 \xrightarrow[CH_3OH]{NaOCH_3} C_6H_5C{\equiv}CH + C_6H_5CH{=}CHBr + C_6H_5CH{=}CHSO_3^-Na^+$$

$$C_6H_5(CH_2)_3SO_2CHCl_2 \xrightarrow[\text{Aq dioxane}]{NaOH} C_6H_5CH_2CH_2C{\equiv}CH + C_6H_5CH_2CH_2CH{=}CHCl + (C_6H_5)(Na^{+-}O_3S)C{=}CH_2$$

as triethylamine and triethylenediamine do permit isolation of diphenylthiirene dioxides.[66, 67, 105] The ratio $k_{elim}/k_{dehydrohal}$ determines the relative amount of vinyl halide produced. Also, since 2-halo-episulfones decompose stereospecifically, vinyl halide stereochemistry

$$RCH_2SO_2CCl_2R' \xrightarrow{Base} RCH^-SO_2CCl_2R' \longrightarrow \underset{\mathbf{43}}{RCH{-}C(Cl)(R')\ (\text{episulfone},\ SO_2)}$$

$$\mathbf{43} \xrightarrow{-SO_2} RCH{=}CHCl \qquad \text{(Eq. 5)}$$

$$\mathbf{43} \xrightarrow[-HCl]{Base} \underset{\mathbf{44}}{RC{=}CR'\ (\text{thiirene},\ SO_2)}$$

$$\mathbf{44} \xrightarrow{-OH^-} RCH{=}C(R')SO_3^- \qquad \mathbf{44} \xrightarrow{-SO_2} RC{\equiv}CR'$$

[105] J. C. Philips, J. V. Swisher, D. Haidukewych, and O. Morales, *Chem. Commun.*, **1971,** 22.

is determined during the intramolecular cyclization. The yields of vinyl chlorides can vary widely but predictably. For example, whereas benzyl α,α-dichlorobenzyl sulfone gives a mixture of isomeric 1-chloro-1,2-diphenylethylenes in 4% yield, 2,2-dichloro-2,7-dihydro-3,4-5,6-dibenzothiepin 1,1-dioxide affords a 47.6% yield of 9-chlorophenanthrene under identical conditions. These observations agree with the expectation that intermediate **46** should have more driving force than chloroepisulfone **45** for sulfur dioxide expulsion because of the incipient gain in phenanthrene resonance energy. In short, the thermal and base-initiated reactions of the halo episulfones are always competitive.

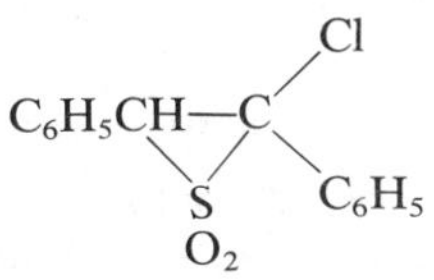

Cl, C_6H_5CH—C, S, O_2, C_6H_5 — **45**

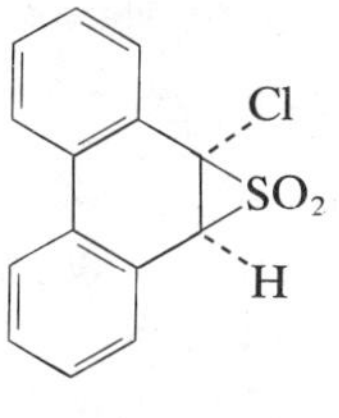

Cl, SO_2, H — **46**

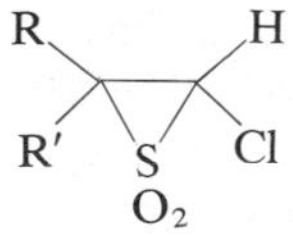

R, H, R′, S, O_2, Cl — **47**

When haloepisulfone dehydrohalogenation is prevented for structural reasons, as in the chloro episulfone **47**, such intermediates are not convertible into thiirene dioxides but two-way reactivity persists. As shown by the behavior of the dichloro sulfones **48** and **49**, cheletropic

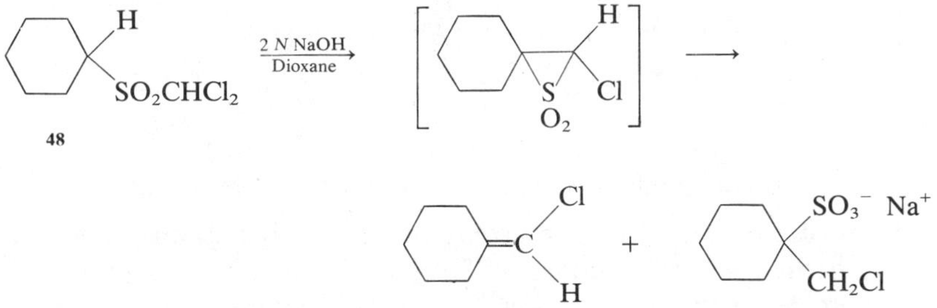

48 (H, SO_2CHCl_2) $\xrightarrow[\text{Dioxane}]{2\,N\ \text{NaOH}}$ [H, S, O_2, Cl] ⟶

C(Cl)(H) (60%) + $SO_3^-\ Na^+$, CH_2Cl (20%)

loss of sulfur dioxide with formation of vinyl chlorides continues as an important pathway. Competing with this transformation is a process characteristic of this subclass of episulfone intermediates, *viz.*, attack of hydroxide ion at the sulfur atom with C—S bond cleavage to give the more stable carbanion. Since in these examples the chlorine substituent has the greatest stabilizing influence, chloromethyl sulfonates result.

$$(CH_3)_2CHSO_2CHCl_2 \xrightarrow[\text{Dioxane}]{2\,N\text{ NaOH}} \left[\begin{matrix} CH_3 & & H \\ & \text{C—C} & \\ CH_3 & SO_2 & Cl \end{matrix} \right] \longrightarrow$$

49

$$(CH_3)_2C{=}CHCl \ (44\%) \quad + \quad (CH_3)_2C(SO_3^-\ Na^+)CH_2Cl \ (37\%)$$

In alkaline solution, thiirene dioxides suffer analogous fragmentations. Their thermal decomposition is the mechanism by which acetylenes are produced, while base attack with cleavage of a C—S bond leads to vinyl sulfonic acids.

A complication sometimes encountered with α,α-dibromo sulfones[65] is contamination of the acetylene by the corresponding alkene. Results like those from the α,α-dibromo sulfones (**50**) and (**51**) have been explained by reduction of the dibromosulfones to their monobromo derivatives by sulfite ion (formed during the rearrangement) and subsequent olefin production in the conventional manner. Since α,α-dichloro sulfones are reduced very slowly if at all under the reaction conditions, their use is recommended over the dibromides.

$$\underset{\mathbf{50}}{C_6H_5CH_2SO_2CHBr_2} \xrightarrow[\text{Dioxane}]{2\,N\text{NaOH}} \underset{(28\%)}{C_6H_5C{\equiv}CH} + \underset{(26\%)}{C_6H_5CH{=}CH_2}$$

$$\underset{\mathbf{51}}{\alpha\text{-}C_{10}H_7CH_2SO_2CHBr_2} \xrightarrow[\text{Dioxane}]{2\,N\text{NaOH}} \underset{(29\%)}{\alpha\text{-}C_{10}H_7C{\equiv}CH} + \underset{(36\%)}{\alpha\text{-}C_{10}H_7CH{=}CH_2}$$

Rearrangements of α,α,α-Trichloromethyl Sulfones

Marked susceptibility to rearrangement has also been noted for α,α,α-trichloromethyl sulfones.[106] Very small amounts of neutral products are generated, except when the dichloro episulfone intermediate is *gem*-dialkyl substituted. Otherwise, sulfonic acid formation predominates. The structural type of sulfonate is strictly dependent upon possible chlorothiirene dioxide intervention. Using α,α,α-trichloromethyl sulfone (**52**) and related molecules where dehydrochlorination of the dichloro episulfone intermediate is not structurally feasible, dichloromethyl sulfonates result. When access can be gained to a

[106] L. A. Paquette and L. S. Wittenbrook, *J. Amer. Chem. Soc.*, **90,** 6790 (1968).

$$\underset{\mathbf{52}}{(CH_3)_2CHSO_2CCl_3} \xrightarrow[\text{THF}]{2\,N\,\text{NaOH}} \underset{(73\%)}{(CH_3)_2C(SO_3^-Na^+)(CHCl_2)} + \underset{(10\%)}{(CH_3)_2C{=}CCl_2} + \underset{(\text{Trace})}{(CH_3)_2C{=}CHCl}$$

transient unsaturated three-membered ring as with the α,α,α-trichloro sulfone (**53**), unsaturated sulfonates are isolated.

$$\underset{\mathbf{53}}{C_6H_5CH_2CH_2SO_2CCl_3} \xrightarrow[\text{THF}]{2\,N\,\text{NaOH}} \underset{(37\%)}{C_6H_5CH_2C(SO_3H){=}CHCl}$$

$$+ \underset{(22\%)}{C_6H_5CH_2C(SO_3H){=}CCl_2} + \underset{(\text{Trace})}{C_6H_5CH_2CH{=}CHCl}$$

$$+ \underset{(\text{Trace})}{C_6H_5CH_2CH{=}CCl_2}$$

RELATED TRANSFORMATIONS

The α-halo sulfone rearrangement has proved unsatisfactory when applied to paracyclophane systems having $—CH_2SCH_2—$ bridges because of severe conformational restraints that impede the intramolecular S_N2 displacement. In actuality, the rigid geometry of such molecules requires ring contraction and sulfur extrusion to produce the new carbon-carbon bond in a frontside manner. The Stevens rearrangement, but not the Sommelet, can meet such requirements and, when this reaction is used with a subsequent Hofmann elimination, sulfide linkages are readily transformed into carbon-carbon double bonds. The potential of this sequence has been exploited in the

S Me Me S $\xrightarrow[\text{2. NaH, THF}]{\text{1. }(CH_3O)_2\overset{+}{C}H\,BF_4^-}$ MeS Me Me SMe $\xrightarrow[\text{2. }t\text{ }C_4H_9OK\text{, THF}]{\text{1. }(CH_3O)_2\overset{+}{C}HBF_4^-}$

Me Me $\rightleftharpoons$ Me Me

synthesis of polycyclic systems.[107–116] Some representative examples are shown in the accompanying reactions. Dimethoxycarbonium fluoroborate seems to give the cleanest products and to be most convenient to use for S-methylation.[111]

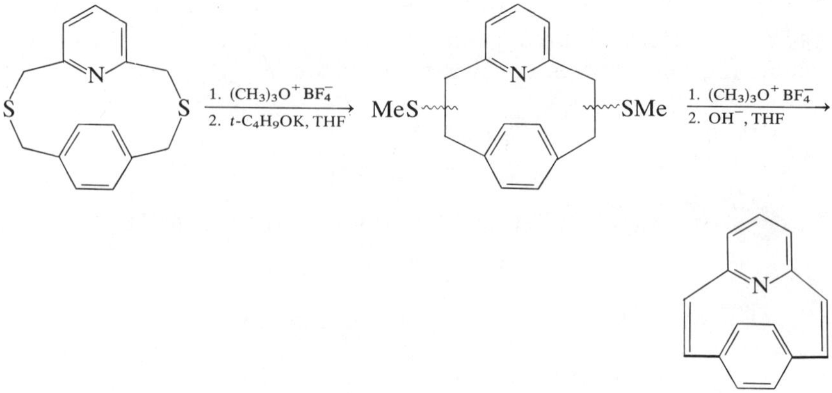

$$\text{[dithia-pyridinophane]} \xrightarrow[\text{2. } t\text{-}C_4H_9OK\text{, THF}]{\text{1. }(CH_3)_3O^+BF_4^-} \text{MeS}\sim\sim\text{SMe} \xrightarrow[\text{2. }OH^-\text{, THF}]{\text{1. }(CH_3)_3O^+BF_4^-}$$

Benzyl phenethyl sulfone (**54**) is converted into 1,3-diphenylpropene and stilbene when treated with ethylmagnesium bromide in dry xylene at 105° for 19 hours.[117] The behavior of 2,5-diphenyltetrahydrothiophene dioxide under comparable conditions is somewhat more controlled, 1,2-diphenylcyclobutene being formed in 47% crude yield.[118]

$$\underset{\mathbf{54}}{C_6H_5CH_2SO_2CH_2CH_2C_6H_5} \xrightarrow[\text{Xylene, heat}]{C_2H_5MgBr} \underset{\substack{cis\ (25.6\%)\\ trans\ (27.3\%)}}{C_6H_5CH{=}CHCH_2C_6H_5} + \underset{\substack{cis\ (6.3\%)\\ trans\ (23.1\%)}}{C_6H_5CH{=}CHC_6H_5}$$

A versatile scheme based on the reductive ring contraction of 2,5-dialkyltetrahydrothiophene dioxide anions with lithium aluminum hydride in refluxing dioxane has recently been reported.[119] A wide variety

[107] R. H. Mitchell and V. Boekelheide, *Tetrahedron Lett.*, **1970,** 1197.
[108] V. Boekelheide and P. H. Anderson, *Tetrahedron Lett.*, **1970,** 1207.
[109] R. H. Mitchell and V. Boekelheide, *J. Amer. Chem. Soc.*, **92,** 3510 (1970).
[110] V. Boekelheide and R. A. Hollins, *J. Amer. Chem. Soc.*, **92,** 3512 (1970).
[111] R. H. Mitchell and V. Boekelheide, *J. Amer. Chem. Soc.*, **96,** 1547 (1974).
[112] V. Boekelheide, P. H. Anderson, and T. A. Hylton, *J. Amer. Chem. Soc.*, **96,** 1558 (1974).
[113] V. Boekelheide, K. Galuszko, and K. S. Seto, *J. Amer. Chem. Soc.*, **96,** 1578 (1974).
[114] J. Lawson, R. Du Vernet, and V. Boekelheide, *J. Amer. Chem. Soc.*, **95,** 957 (1973).
[115] J. R. Davy and J. A. Reiss, *Chem. Commun.*, **1973,** 806.
[116] P. J. Jessup and J. A. Reiss, *Tetrahedron Lett.*, **1975,** 1453.
[117] R. M. Dodson, P. P. Schlangen, and E. L. Mutsch, *Chem. Commun.*, **1965,** 352.
[118] R. M. Dodson and A. G. Zielske, *Chem. Commun.*, **1965.** 353.
[119] J. M. Photis and L. A. Paquette, *J. Amer. Chem. Soc.*, **96,** 4715 (1974).

C_2H_5MgBr, $(C_2H_5)_2O$, C_6H_6 (47%)

of 1,2-dialkylcyclobutenes can thus be prepared.[119–122] Since the starting sulfones are conveniently synthesized from the corresponding succinic anhydride, the overall conversion can be thought of as an

1. 2 n-C_4H_9Li 2. Excess CH_3I; 1. n-C_4H_9Li 2. $LiAlH_4$, dioxane, heat (72%)

anhydride → 1,2-dialkylcyclobutene transformation. As α substitution on the sulfolane ring is decreased, there is a parallel decrease in the efficiency of the ring contraction. Reduction of sulfone to sulfide is now not as sterically inhibited and can compete effectively.

$C_6H_5CH_2$–(sulfolane)–$CH_2C_6H_5$ → 1,2-bis($CH_2C_6H_5$)cyclobutene

Several steps → Several steps

[120] L. A. Paquette, J. S. Ward, R. A. Boggs, and W. B. Farnham, *J. Amer. Chem. Soc.*, **97**, 1101 (1975).

[121] L. A. Paquette, J. M. Photis, and G. D. Ewing, *J. Amer. Chem. Soc.*, **97**, 3538 (1975).

[122] J. M. Photis, Ohio State University, unpublished observations.

Sulfonyl 1,3-dicarbanions are subject to oxidation by copper(II) salts to form alkenes.[123] Again, no mechanistic details are yet available, but two-electron oxidation to afford a thiirane 1,1-dioxide intermediate appears likely.

$$[C_2H_5O_2CCH(CH_3)]_2SO_2 \xrightarrow[\substack{2.\ n\text{-}C_4H_9Li \\ 3.\ CuCl_2}]{1.\ (i\text{-}C_3H_7)_2N^-Li^+} \underset{(35\%)}{C_2H_5O_2CC(CH_3){=}C(CH_3)CO_2C_2H_5}$$

SYNTHETIC UTILITY

The need for olefins in many synthetic organic chemical problems is obvious. In this connection, the Ramberg-Bäcklund rearrangement of α-chloro sulfones can be advantageous in five broad chemical schemes. They involve the coupling of two residues through a sulfide linkage (usually indirectly) with subsequent introduction of the double bond, homologization of olefins, conversion of mercaptans into homologous terminal alkenes, synthesis of olefins deuterated exclusively at the vinyl positions, and preparation of strained cycloalkenes.

In the first scheme mixed sulfides (such as **55**) are widely available and typify the bonding of two differing groups through a sulfide linkage. As long as the requirement of one α and one α' hydrogen atom is met, desulfurization with π bond formation can occur. For purposes of degradative analysis, the removal of sulfur dioxide from an alkyl sulfone with retention of the carbon skeleton may be desirable. This also can be conveniently done.[57]

$$\underset{\mathbf{55}}{RCH_2SCH_2R'} \xrightarrow[\substack{2.\ RCO_3H \\ 3.\ Base}]{1.\ NCS} RCH{=}CHR'$$

$$n\text{-}C_3H_7SO_2C_3H_7\text{-}n \xrightarrow[3.\ OH^-,\ H_2O]{\substack{1.\ n\text{-}C_4H_9Li \\ 2.\ I_2}} \underset{(47\%)}{C_2H_5CH{=}CHC_2H_5}$$

A reaction sequence is also available for converting mercaptans possessing at least one α-hydrogen atom into the higher homologous terminal alkene.[56] For example, n-hexylmercaptan upon treatment with formaldehyde and hydrogen chloride affords in good yield the chloromethyl sulfide (**56**) which can be oxidized and rearranged to 1-heptene (54% overall). The ease with which mercaptans condense with aldehydes of higher molecular weight than formaldehyde to give α-chloro sulfides (*vide supra*) should enable similar conversion of mercaptans into internal olefins of any desired length.

[123] J. S. Grossert, J. Buter, E. W. H. Asveld, and R. M. Kellogg, *Tetrahedron Lett.*, **1974,** 2805.

This sequence is then combined with the free-radical addition of hydrogen sulfide[124] or thiolacetic acid to an olefin[125] for olefin homologization.[57] The synthesis of 3-methyl-1-pentene from 2-methyl-1-butene illustrates the technique.

$$CH_3(CH_2)_4CH_2SH \xrightarrow[\text{HCl}]{(CH_2O)_x} CH_3(CH_2)_4CH_2SCH_2Cl \xrightarrow[\text{2. OH}^-,\,H_2O]{\text{1. RCO}_3H} CH_3(CH_2)_4CH{=}CH_2$$

56

$$C_2H_5C(CH_3){=}CH_2 \xrightarrow[h\nu]{H_2S} sec\text{-}C_4H_9CH_2SH \xrightarrow[\text{HCl}]{(CH_2O)_x}$$

$$sec\text{-}C_4H_9CH_2SCH_2Cl \xrightarrow[\text{2. OH}^-,\,H_2O]{\text{1. }CH_3CO_3H} sec\text{-}C_4H_9CH{=}CH_2$$

Other uses of α-halo sulfone rearrangements have been noted earlier.

The synthetic potential of the α,α-dihalosulfone rearrangement has been little explored. However, a number of useful transformations are available. For example, by combining four synthetic manipulations, a mercaptan should be converted into a terminal acetylene of one additional carbon atom (Eq. 6). Additionally, when these reactions are

$$RCH_2SH \xrightarrow[\text{2. NCS; 3. RCO}_3H]{\text{1. }(CH_2O)_x,\ HCl} RCH_2SO_2CHCl_2 \xrightarrow[t\text{-}C_4H_9OH]{t\text{-}C_4H_9OK} RC{\equiv}CH \quad \text{(Eq. 6)}$$

$$RCH{=}CH_2 \xrightarrow[\text{steps}]{\text{Five}} RCH_2C{\equiv}CH \quad \text{(Eq. 7)}$$

combined with free-radical addition of hydrogen sulfide or thiolacetic acid to olefins, an α-olefin should be converted into its homologous terminal acetylene (Eq. 7). The procedure is especially attractive because of the commercial availability of many α olefins.

Lastly, further modification is possible that will provide symmetrically substituted acetylenes containing double the number of carbon atoms as the primary or secondary halide precursors (Eq. 8).

$$2\ RCH_2Br \xrightarrow[\text{3. RCO}_3H;\ \text{4. }t\text{-}C_4H_9OK,\ t\text{-}C_4H_9OH]{\text{1. }Na_2S;\ \text{2. 2 NCS}} RC{\equiv}CR \quad \text{(Eq. 8)}$$

EXPERIMENTAL PROCEDURES

The following examples of the α-halo sulfone rearrangement have been carefully chosen to illustrate useful and general experimental

[124] For example, see H. L. Goering, D. I. Relyea, and D. W. Larsen, *J. Amer. Chem. Soc.*, **78,** 348 (1956).

[125] For example, see F. G. Bordwell and W. A. Hewett, *J. Amer. Chem. Soc.*, **79,** 3493 (1957).

procedures. In particular, they serve as a guide to those few experimental factors that require care in execution. Preparations of chloro sulfones used as starting materials in the Ramlung-Bäcklund rearrangement are given first.

***n*-Butyl α-Chloroethyl Sulfone (Hydrogen-Chloride-Promoted Aldehyde–Thiol Condensation).**[15] Into a mixture of 9.7 g (0.22 mol) of paraldehyde and 18.0 g (0.20 mol) of *n*-butyl mercaptan cooled to −15° was introduced a gentle stream of anhydrous hydrogen chloride gas. During the hydrogen chloride uptake (2–3 hours), the temperature was not allowed to increase above −5°. Excess calcium chloride was added and the mixture was tumbled at −10° on a rotary evaporator with gradually decreasing pressure until water and hydrogen chloride removal appeared complete. Methylene chloride was added and the solvent partially removed under vacuum to remove final traces of hydrogen chloride. The solution was filtered and the filtrate added dropwise to a stirred ethereal solution of monoperphthalic acid cooled to 0°. After the addition was complete, the reaction mixture was allowed to stir overnight at ambient temperature. Filtration of the insoluble phthalic acid gave the crude α-chloro sulfone in ethereal solution. To remove dissolved phthalic acid, the solution was washed with saturated sodium bicarbonate solution and then with water. The ethereal layer was separated, dried, and evaporated. The residue was fractionally distilled to give 22.7 g (62%) of *n*-butyl α-chloroethyl sulfone as a colorless liquid, bp 74–75° (0.5 mm); n_D^{26} 1.4669; infrared (neat) μ: 7.57 and 8.80; proton magnetic resonance (in chloroform-*d*) δ: 4.86 (quartet, $J = 7.1$ Hz, 1*H*, —$SO_2C\underline{H}Cl$—), 3.2 (triplet, $J = 7.5$ Hz, 2*H*, —$SO_2C\underline{H}_2$—), and 0.8–2.1 (complex pattern, 10H).

11-Chloro-12-thia[4.4.3]propella-3,8-diene 12,12-Dioxide (Chlorination with N-Chlorosuccinimide).[50] A mechanically stirred slurry of 400 g (1.67 mol) of freshly recrystallized (ethanol) sodium sulfide nonahydrate in 1.2 l of anhydrous hexamethylphosphoramide (distilled from calcium hydride and stored over molecular sieves) was heated to 97° and *ca.* 20 mm and the aqueous distillate (max bp 83°) was collected and discarded. The slurry was cooled to ambient temperature, 225 g (0.644 mol) of *cis*-9,10-bis(methanesulfonyloxymethyl)-$\Delta^{2,6}$-hexalin was added in one portion, and the dark-green reaction mixture was stirred at 120° for 18 hours. The brownish contents were cooled, treated with 1.5 l of water, and extracted with three 1-l portions of ether. The combined organic extracts were washed with water (three 1-l portions) and saturated brine (250 ml) before drying and evaporation. The resulting tan semisolid was kept under vacuum at

0.1 mm until complete crystallization occurred. The solid was dissolved in pentane and passed through a column of neutral, activity I alumina. Removal of the pentane afforded 121 g (98.8%) of 12-thia[4.4.3]-propella-3,8-diene as a waxy white solid; proton magnetic resonance (in chloroform-*d*) δ: 5.45 (multiplet, 4*H*, olefinic), 2.76 (singlet, 4*H*, —$C\underline{H}_2S$—), and 2.08 (multiplet, 8*H*, allyl).

A mixture of 9.8 g (0.0733 mol) of N-chlorosuccinimide and 14.1 g (0.0733 mol) of this sulfide in 150 ml of carbon tetrachloride was refluxed under nitrogen for 1.5 hours. The reaction mixture was cooled and the succinimide removed by filtration. The carbon tetrachloride was evaporated and the residual oil was dissolved in 50 ml of ether and cooled to 0°. To this magnetically stirred solution was added dropwise a standardized ethereal solution containing 0.147 mol of monoperphthalic acid. The reaction mixture was stirred at room temperature for 6 hours and then processed as above to give 15.0 g (79%) of colorless solid, mp 111–113° (from ether at −20°); proton magnetic resonance (in chloroform-*d*) δ: 5.59 (multiplet, 4H, olefinic), 5.05 (singlet, 1H, >CHCl), 3.42, 3.21, 3.11, and 2.89 (AB quartet, $J_{AB} = 13$ Hz, 2*H*, —$C\underline{H}_2SO_2$—), and 1.59–2.82 (multiplet, 8*H*, allyl).

2-Chloro-2,7-dihydro-3,4-5,6-dibenzothiepin 1,1-Dioxide (Chlorination with Sulfuryl Chloride).[29] To a stirred solution of 9.5 g (0.045 mol) of 2,7-dihydro-3,4-5,6-dibenzothiepin in 40 ml of carbon tetrachloride was added in small portions over a 10-minute period a solution of 7.5 g (0.055 mol) of sulfuryl chloride in 20 ml of carbon tetrachloride. When the addition was completed, the solution was warmed to 60° and kept at that temperature for approximately 1 hour. After cooling, the solvent was evaporated under reduced pressure and the residual yellow solid was dissolved in 150 ml of chloroform. To this stirred solution was added portionwise with external cooling 15.5 g (0.09 mol) of *m*-chloroperbenzoic acid. As the reaction proceeded, the resulting *m*-chlorobenzoic acid crystallized from the solution. The reaction mixture, after standing overnight, was filtered, shaken well with aqueous bicarbonate solution, dried, and evaporated to give 7.8 g (62.8%) of crude product, mp 196–198°. Recrystallization of this material from benzene–hexane afforded pure α-chloro sulfone as white prisms, mp 204.5–205.5°.

1-Bromo-7-thiabicyclo[2.2.1]heptane 7,7-Dioxide (Halogenation of an α-Sulfonyl Carbanion).[24] A solution of 4.00 g (27.4 mmol) of 7-thiabicyclo[2.2.1]heptane 7,7-dioxide in 200 ml of tetrahydrofuran (distilled from lithium aluminum hydride) was cooled in a nitrogen atmosphere to −78° and treated all at once with 20 ml of 1.97 *M* *t*-butyllithium (39.4 mmol). After stirring at −70° for 30 minutes, the yellow solution was added in 20-ml portions during 30 minutes to a vigorously stirred solution of 7 g of cyanogen bromide (66 mmol) in 150 ml of anhydrous ether at −112° in a nitrogen atmosphere. The reaction mixture was stored overnight at −50°, warmed to 0°, and washed several times with 10% aqueous potassium hydroxide and saturated aqueous sodium hydrogen sulfite. The organic phase was dried over anhydrous magnesium sulfate, concentrated to a minimal volume, and treated with excess *n*-pentane to precipitate 5.35 g of crude product. Recrystallization of the crude solid from methanol gave 4.87 g (76%) of colorless α-bromo sulfone, mp 149–151°; infrared μ 7.58, 7.71, 8.70, and 8.82; proton magnetic resonance (in trifluoroacetic acid) δ: 3.25 (singlet, 1*H*, $>C\underline{H}SO_2$—) and 2.40 (multiplet, 8*H*).

cis- and _trans_-2-Hexene (Removal of Volatile Alkene during Its Formation).[15] A 1.85-g (10-mmol) sample of *n*-butyl α-chloroethyl sulfone was placed in a 100-ml reaction flask fitted with a magnetic stirrer, nitrogen gas inlet tube, and 3-in. Vigreux column atop of which was a Bantamware distillation head. To the sulfone was added in one portion 60 ml of 2 *N* sodium hydroxide solution (120 mmol). The reaction mixture was slowly warmed to a gentle reflux (vigorous stirring), and the volatile product was carried over by a slow (one to three bubbles per second) nitrogen stream into a pear-shaped flask (surrounded by a solid carbon dioxide–acetone bath) attached to the distillation head. After 2.5 hours the contents of the receiver were removed, accurately weighed (0.48 g), and checked by gas–liquid chromatography (a 22 ft × 0.25 in. aluminum column packed with 15% β,β′-oxydipropionitrile on Chromosorb P) at ambient temperature. Again at 4.5 (0.15 g) and 6.0 hours (<0.01 g) the volatile product was removed, weighed, and checked by gas–liquid chromatography. Reactions were generally allowed to proceed 0.5–1 hour beyond the time when no additional product was observed. The total weight of olefin in this case was 0.62 g or 74%. Only the isomeric *cis*- and *trans*-2-hexenes were produced. The olefins were readily identified by comparison of retention times with authentic samples and by peak enhancement. Infrared and nmr spectra of isolated samples confirmed the structural assignments.

[4.4.2]Propella-3,8,11-triene (Potassium _t_-Butoxide as Base).[84] To an ice-cold, magnetically stirred solution of 4.64 g (0.0179 mol) of 11-chloro-12-thia[4.4.3]propella-3,8-diene 12,12-dioxide in 50 ml of anhydrous tetrahydrofuran was added 6.0 g (0.0537 mol) of powdered potassium *t*-butoxide in one portion. The reaction mixture was stirred magnetically for 5 hours at 0° under a nitrogen atmosphere. A mixture of 40 ml of water and 30 ml of ether was added to the reaction mixture. The organic layer was separated, washed with water (2× 20 ml), and dried. Concentration of the filtrate and short-path distillation of the residue afforded 1.51 g (53%) of [4.4.2]propella-3,8,11-triene as a colorless oil: bp 75° (5 mm); infrared (neat) cm^{-1}: 3003, 2875, 2793, and 1645; proton magnetic resonance (in chloroform-*d*) δ: 5.75 (multiplet, 6*H*, olefinic) and 2.02 (multiplet, 8*H*, allyic).

9,10-Dideuteriophenanthrene (Specific Deuteration of Olefinic Products α to Original Sulfonyl Group).[29] A mixture of 5.5 g (0.02 mol) of 2-chloro-2,7-dihydro-3,4-5,6-dibenzothiepin 1,1-dioxide, 28 ml of approximately 25% sodium deuteroxide in deuterium oxide, and 20 ml of dioxane was stirred at room temperature for 0.5 hour and refluxed with stirring for 4.5 hours. The mixture was cooled and diluted with 350 ml of ice water. The precipitated product was filtered and dried to give 1.65 g (91.7%) of pale-yellow solid, mp 92.5–94.5°. Two recrystallizations of this material from ethanol gave shiny white flakes of pure 9,10-dideuteriophenanthrene, mp 98.5–99.5°, containing 19.97 atom of excess D (theory, 20 atom % excess D).

1,1-Dimethyl-1-sila-4-cycloheptene (Direct Olefination of Sulfones).[99] In a dry one-necked, 100-ml round-bottomed flask equipped with a reflux condenser and a magnetic stirring bar were placed 0.22 g (1.07 mmol) of 1,1-dimethyl-1-sila-5-thiacyclooctane 5,5-dioxide, 3.0 g (0.065 mol) of powdered potassium hydroxide, 15 ml of dry *t*-butyl alcohol, and 35 ml of dry carbon tetrachloride. The mixture was heated for 12 hours at 50° while it was stirred. The reaction mixture was transferred to a separatory funnel and 100 ml of ether was added. The organic layer was extracted with three equal volumes of water, dried, filtered, and evaporated. The residue was bulb-to-bulb distilled to give 50% of 1,1-dimethyl-1-sila-4-cycloheptene, final purification of which was effected on an 18 ft× 0.25 in. Apiezon L column at 130°; infrared cm^{-1}: 1645; proton magnetic resonance (in chlorodifluoromethane) δ: 5.8 (multiplet, 2H), 2.3 (multiplet (4H), 0.67 (multiplet, 4H), and 0.06 (singlet, 6H).

5,10-Dihydro-5,10-epithiobenzocyclooctene 11,11-Dioxide (Reaction in Dimethyl Sulfoxide Solution; Bishomoconjugative Rearrangement).[59] To a magnetically stirred solution of 11-chloro-12-thia[4.4.3]propella-2,4,7,9-tetraene in 30 ml of dry dimethyl sulfoxide under nitrogen was added 112 mg (1.00 mmol) of potassium *t*-butoxide, and the deep-green solution was stirred for 30 minutes.

With cooling, 50 ml of water was added and the solution was acidified with concentrated hydrochloric acid. The aqueous solution was extracted with chloroform and the extract washed with water and brine and dried. Evaporation of the solvent and recrystallization from methanol–chloroform gave 137 mg (68.4%) of 5,10-dihydro-5,10-epithiobenzocyclooctene 11,11-dioxide as a white crystalline solid, mp 269–271°; ultraviolet (95% ethanol), nm max (ϵ): 267 (2280) shoulder, 273 (2470), and 278 (2280) shoulder; proton magnetic resonance (in dimethyl sulfoxide-*d*) δ: 7.43 (singlet, 4H aryl C$\underline{H}$), 5.93 (broad singlet, 4*H*, vinylic C$\underline{H}$), and 4.90–5.07 (multiplet, 2*H*, bridgehead *H*).

9-Chlorophenanthrene and Phenanthrene-9-sulfonic Acid (α,α-Dihalosulfone Rearrangement and Isolation of Sulfonic Acid).[49] A mixture of 3.1 g (0.01 mol) of 2,2-dichloro-2,7-dihydro-3,4-5,6-dibenzothiepin 1,1-dioxide, 25 ml of dioxane, and 15 ml of 2 *N* sodium hydroxide solution was heated at reflux with stirring for 16 hours. Upon cooling, 60 ml of water was added and the aqueous solution was extracted with dichloromethane (3 × 30 ml), acidified with concentrated hydrochloric acid, and treated with 1.5 g (0.015 mol) of *p*-toluidine. The mixture was warmed with scratching until the resulting oil had crystallized. After it had cooled, the solid was filtered and dried to afford 1.7 g (47.3%) of the *p*-toluidine salt of phenanthrene-9-sulfonic acid, mp 210–216°. Recrystallization of this salt from water yielded pale-yellow needles, mp 231–232°, spectral properties identical with those described in the literature.

The combined organic layers were dried, filtered, and evaporated to give 1.0 g (47.6%) of 9-chlorophenanthrene, mp 49–50°. Recrystallization from aqueous ethanol afforded pure white needles, mp 52–53°, identical with an authentic specimen.

2,3-Diphenylthiirene 1,1-Dioxide (Isolation of Strained Three-Membered Ring Intermediate).[67] To 63.6 g of crude α,α'-dibromodibenzyl sulfone (mp 130–145°) dissolved in 400 ml of dichloromethane was added 50.5 g of triethylamine and the solution was refluxed gently for 3 hours. The solution was washed with 3 *N* hydrochloric acid (2 × 200 ml) followed by 100 ml of water. Evaporation of the dichloromethane by distillation from a water bath at 30° with the aid of a water aspirator followed by washing the residue with 50 ml of cold ethanol (5°) gave 34.1 g (90%) of crude sulfone, mp 116–126° (dec). Recrystallization from benzene gave 26.5 g (70%) of 2,3-diphenylthiirene 1,1-dioxide as a snow-white solid, mp 116–126° (varies with rate of heating); infrared (carbon tetrachloride) μ: 7.85 and 8.59; ultraviolet (95% ethanol), nm max (log ϵ): 222.5 (4.26), 296

(4.34), 307 (4.41), 322 (4.27); proton magnetic resonance (in chloroform-*d*) δ: 7.55 (multiplet, aryl C$\underline{H}$).

TABULAR SURVEY

Compounds submitted to the conditions of the Ramberg-Bäcklund reaction are collected in Tables II–VI. An attempt has been made to include all α-halo sulfone rearrangements reported through mid-1975; however, since there is no systematic method of searching the literature for this reaction, the completeness of the tabulation cannot be guaranteed.

Within each table the substrates are listed in order of increasing complexity. A dash (—) in the yield column means that the yield was not reported. Where more than one product is formed in a reaction, the products are arranged roughly in order of complexity.

The abbreviations used in the tables are: DME, 1,2-dimethoxyethane; DMSO, dimethyl sulfoxide; THF, tetrahydrofuran.

TABLE II. Base-Promoted Rearrangements of Preformed α-Halo Sulfones

No. of C Atoms	Substrate	Base	Solvent	Temperature, °C	Product(s) and Yield(s) (%)	Refs.
			A. Acyclic Compounds			
C_3	$ClCH_2SO_2C_2H_5$	2 *N* NaOH	H_2O	100	$CH_2{=}CHCH_3$ (51)	13
C_4	$CH_3CHClSO_2C_2H_5$	2 *M* KOH	H_2O	100	$CH_3CH{=}CHCH_3$ (76) (78.1 *cis*; 21.9 *trans*)[a]	20
		2 *M* NaOH	H_2O	100	" (75; 78.8 *cis*; 21.2 *trans*)[a]	13, 15, 20
		2 *M* KOH	H_2O	60	" (—; 81.5 *cis*; 18.5 *trans*)[a]	20
		2 *N* KOH	H_2O	100	" (76; 78.1 *cis*; 21.9 *trans*)[a]	20
		1 *M* $Ba(OH)_2$	H_2O	100	" (66; 78.6 *cis*; 21.4 *trans*)[a]	20
		2 *M* LiOH	H_2O	100	" (—; 76.1 *cis*; 23.9 *trans*)[a]	20
		1 *M* n-C_4H_9ONa	n-C_4H_9OH	117	" (74; 74.6 *cis*; 25.4 *trans*)[a]	20
		0.75 *M* C_6H_5Li	C_6H_6	80	" (—; 73.3 *cis*; 26.7 *trans*)[a]	20
		0.75 *M* C_6H_5Li	C_6H_6	—	" (—; 78.1 *cis*; 21.9 *trans*)[a]	20
		1.0 *M* t-C_4H_9OK	t-C_4H_9OH	~93	" (82; 22.6 *cis*; 77.4 *trans*)[a]	20
		1.3 *M* t-C_4H_9OK	t-C_4H_9OH	82	" (—; 18.6 *cis*; 81.4 *trans*)[a]	20
		1.0 *M* t-C_4H_9OK	Toluene	110	" (—; 24.5 *cis*; 75.5 *trans*)[a]	20
		C_6H_5Li	Xylene	120	" (80; 73.3 *cis*; 26.7 *trans*)[a]	15
	$CH_3CHBrSO_2C_2H_5$	2 *N* NaOH	H_2O	100	$CH_3CH{=}CHCH_3$ (85) (mostly *cis*)	13
		2 *N* KOH	H_2O	85	" (85) (79.5 *cis*; 20.5 *trans*)[a]	20
	$CH_3CHISO_2C_2H_5$	2 *N* KOH	H_2O	75	$CH_3CH{=}CHCH_3$ (87) (78.2 *cis*; 21.8 *trans*)[a]	20
	$C_2H_5SO_2CHBrCO_2C_2H_5$	4 *N* $NaOC_2H_5$	C_2H_5OH	80	$CH_3CH{=}CHCO_2H$ (78)	89
	$C_2H_5O_2CCH_2SO_2CHBrCO_2C_2H_5$	4 *N* $NaOC_2H_5$	C_2H_5OH	80	$C_2H_5O_2CCH{=}CHCO_2C_2H_5$ (92)	89
	$CH_3O_2CCHBrSO_2CH_2CO_2CH_3$	NaOH	H_2O-dioxane	Reflux	E-$HO_2CCH{=}CHCO_2H$ (50)	79

C_5	$CH_3CHClSO_2C_3H_7$-*n*	2 *N* KOH	H_2O	100	$CH_3CH_2CH{=}CHCH_3$ (74) (65.8 *cis*; 34.3 *trans*)[a]	20
	$CH_3CHBrSO_2C_3H_7$-*n*	2 *N* NaOH	H_2O	100	$CH_3CH{=}CHCH_2CH_3$ (90) (mostly *cis*)	13
		2 *N* KOH	H_2O	85	$CH_3CH_2CH{=}CHCH_3$ (85) (60 *cis*; 40 *trans*)[a]	20
	$C_2H_5CHClSO_2C_2H_5$	2 *N* KOH	H_2O	100	$CH_3CH_2CH{=}CHCH_3$ (57) (71.3 *cis*; 28.7 *trans*)[a]	20
	$C_2H_5CHBrSO_2C_2H_5$	2 *N* NaOH	H_2O	100	$CH_3CH{=}CHC_2H_5$ (80) (mostly *cis*)	13
		2 *N* KOH	H_2O	85	$C_2H_5CH{=}CHCH_3$ (71) (70 *cis*; 30 *trans*)[a]	20
	$CH_3CHBrSO_2C_3H_7$-*i*	2 *N* NaOH	H_2O	100	$CH_3CH{=}C(CH_3)_2$ (82)	13
C_6	$C_2H_5CHISO_2C_3H_7$-*n*	2 *N* KOH	H_2O	75	$C_2H_5CH{=}CHC_2H_5$ (57) (55.5 *cis*; 44.5 *trans*)[a]	20
	n-$C_3H_7CHClSO_2C_2H_5$	NaOH	H_2O	100	*n*-$C_3H_7CH{=}CHCH_3$ (63) (69.4 *cis*; 30.6 *trans*)[a]	15
		C_6H_5Li	Xylene	120	" (30) (68.0 *cis*; 32.0 *trans*)[a]	15
	$CH_3CHClSO_2C_4H_9$-*n*	NaOH	H_2O	75	$CH_3CH{=}CHC_3H_7$-*n* (76) (66.2 *cis*; 33.8 *trans*)[a]	15
		C_6H_5Li	Xylene	100	$CH_3CH{=}CHC_3H_7$-*n* (30) (63.8 *cis*; 36.2 *trans*)[a]	15
	i-$C_3H_7CHClSO_2C_2H_5$	NaOH	H_2O	100	*i*-$C_3H_7CH{=}CHCH_3$ (41) (59.4 *cis*; 40.6 *trans*)[a]	15
		C_6H_5Li	Xylene	120	" (59) (59.0 *cis*; 41.0 *trans*)[a]	15
	$CH_3CHClSO_2C_4H_9$-*i*	NaOH	H_2O	100	*i*-$C_3H_7CH{=}CHCH_3$ (70) (51.0 *cis*: 49.0 *trans*)[a]	15
		C_6H_5Li	Xylene	120	" (56) (51.7 *cis*; 48.3 *trans*)[a]	15

Note: References 126–131 are on p. 71.

[a] The percentages for the individual isomers are normalized to 100%.

TABLE II. BASE-PROMOTED REARRANGEMENTS OF PREFORMED α-HALO SULFONES (*Continued*)

No. of C Atoms	Substrate	Base	Solvent	Tempera-ture, °C	Product(s) and Yield(s) (%)	Refs.
			A. Acyclic Compounds			
C_6 (*contd.*)	n-$C_4H_9SO_2CHBrCO_2C_2H_5$	4 *N* $NaOC_2H_5$	C_2H_5OH	80	n-$C_3H_7CH{=}CHCO_2H$ (68)	89
C_7	n-$C_6H_{13}SO_2CH_2Cl$	NaOH	H_2O	100	n-$C_5H_{11}CH{=}CH_2$ (77.6)	56
C_8	n-$C_3H_7CHISO_2C_4H_9$-n	2 *N* KOH	H_2O	75	n-$C_3H_7CH{=}CHC_3H_7$-n (50) (52.5 *cis*; 47.5 *trans*)[a]	20
C_9	$C_6H_5CH_2SO_2CHBrCH_3$	2 *N* NaOH	H_2O	100	$C_6H_5CH{=}CHCH_3$ (50)	65
	$C_6H_5CH_2SO_2CHBrCO_2CH_3$	4 *N* $NaOC_2H_5$	C_2H_5OH	80	$C_6H_5CH{=}CHCO_2H$ (83)	89
	$C_6H_5CH_2SO_2CHBrCO_2C_2H_5$	4 *N* $NaOC_2H_5$	C_2H_5OH	80	$C_6H_5CH{=}CHCO_2H$ (79)	89
	n-$C_3H_7CH_2SO_2CHBr(CH_2)_3CO_2H$	2 *N* KOH	H_2O	100	n-$C_3H_7CH{=}CH(CH_2)_3CO_2H$ (65)	61
C_{10}	$(CH_3)(H)(H)C{-}SO_2{-}C(CH_3)(C_6H_5)(Br)$	$NaOCH_3$	CH_3OD	65	*E*-$C_6H_5C(CH_3){=}CDCH_3$ (43), *Z*-$C_6H_5C(CH_3){=}CDCH_3$ (57)	26
	$(C_6H_5)(CH_3)(H)C{-}SO_2{-}C(CH_3)(Br)(H)$	$NaOCH_3$	CH_3OD	65	*E*-$C_6H_5C(CH_3){=}CDCH_3$ (23), *Z*-$C_6H_5C(CH_3){=}CDCH_3$ (77)	26
	$(CH_3)(C_6H_5)(H)C{-}SO_2{-}C(CH_3)(H)(Br)$	$NaOCH_3$	CH_3OH	65	*E*-$C_6H_5C(CH_3){=}CHCH_3$ (70), *Z*-$C_6H_5C(CH_3){=}CHCH_3$ (30)	26

	$C_6H_5CH_2SO_2CBr(CH_3)CO_2H$	2 *N* NaOH	H_2O	100	$C_6H_5CH{=}C(CH_3)CO_2H$ (56)	65
	$C_6H_5CH_2SO_2C(CH_3)BrCO_2H$	4 *N* $NaOC_2H_5$	C_2H_5OH	80	$C_6H_5CH{=}C(CH_3)CO_2H$ (88)	89
	$C_6H_5CH_2SO_2CHBrCH_2CO_2H$	2 *N* NaOH	H_2O	100	$C_6H_5CH{=}CHCH_2CO_2H$ (66)	65
	$C_6H_5CH_2SO_2CHBrCOCH_3$	4 *N* $NaOC_2H_5$	C_2H_5OH	80	$C_6H_5CH{=}CHCOCH_3$ (68)	89
C_{13}	$C_6H_5CH_2SO_2C(C_2H_5)BrCO_2C_2H_5$	4 *N* $NaOC_2H_5$	CH_2OH	80	$C_6H_5CH{=}C(C_2H_5)CO_2H$ (84)	89
C_{14}	$C_6H_5CHClSO_2CH_2C_6H_5$	NaOH	H_2O–dioxane	87	*E*-$C_6H_5CH{=}CHC_6H_5$ (100)	20
	$C_6H_5CHBrSO_2CH_2C_6H_5$	$NaOCH_3$	CH_3OD	25	*E*-$C_6H_5CD{=}CDC_6H_5$ (89)	14
C_{15}	$(C_6H_5)_2CHSO_2CHBrCO_2C_2H_5$	4 *N* $NaOC_2H_5$	C_2H_5OH	80	$(C_6H_5)_2C{=}CHCO_2H$ (90)	89
	$C_6H_5CH_2SO_2CHBrSO_2CH_2C_6H_5$	4 *N* $NaOC_2H_5$	C_2H_5OH	80	$C_6H_5CH_2SO_2CH{=}CHC_6H_5$ (98)	89
	$C_6H_5CH_2SO_2CHBrCOC_6H_5$	4 *N* $NaOC_2H_5$	C_2H_5OH	80	$C_6H_5CH{=}CHCOC_6H_5$ (73)	89
C_{16}	$C_6H_5(CH_3)(Br)C{-}SO_2{-}C(CH_3)(C_6H_5)H$	$NaOCH_3$	CH_3OH	25	*E*-$C_6H_5C(CH_3){=}C(CH_3)C_6H_5$ (100)	26
	$C_6H_5(CH_3)(Br)C{-}SO_2{-}C(CH_3)(H)C_6H_5$	$NaOCH_3$	CH_3OH	25	*Z*-$C_6H_5C(CH_3){=}C(CH_3)C_6H_5$ (93), *E*-$C_6H_5C(CH_3){=}C(CH_3)C_6H_5$ (7)	26
	$C_6H_5COCH_2SO_2CHBrCOC_6H_5$	4 *N* $NaOC_2H_5$	C_2H_5OH	80	$C_6H_5COCH{=}CHCOC_6H_5$ (89)	89
C_{18}	$(C_6H_5CH_2SO_2CHBrCH_2)_2$	2 *N* NaOH	H_2O	100	$C_6H_5CH{=}CHCH_2CH_2CH{=}CHC_6H_5$ (64)	65
	$CH_3O_2C(CH_2)_8SO_2CHBr(CH_2)_7CO_2CH_3$	NaOH	H_2O-dioxane	Reflux	*E*-$HO_2C(CH_2)_7CH{=}CH(CH_2)_7CO_2H$ (Low)	79

Note: References 126–131 are on p. 71.

[a] The percentages for the individual isomers are normalized to 100%.

TABLE II. Base-Promoted Rearrangements of Preformed α-Halo Sulfones (*Continued*)

No. of C Atoms	Substrate	Base	Solvent	Temperature, °C	Product(s) and Yield(s) (%)	Refs.
			B. Cyclic Compounds			
C_6	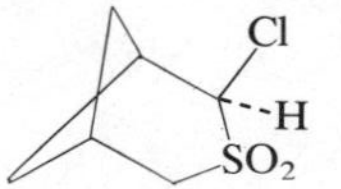 Cl, H, SO_2	t-$C_5H_{11}ONa$	$(C_6H_5)_2O$	[b]	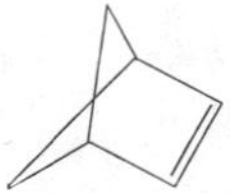 (68)	126
C_7	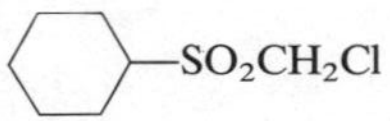 SO_2CH_2Cl	NaOH	H_2O	100	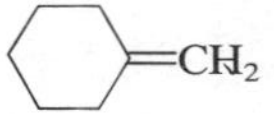 $=CH_2$ (80)	56
	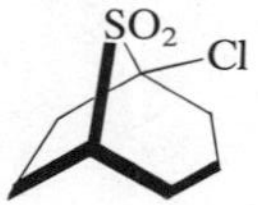 SO_2, Cl	KOH	H_2O	100	No reaction	23
		t-C_4H_9OK	t-C_4H_9OH	50–100	Polymer	23
C_8	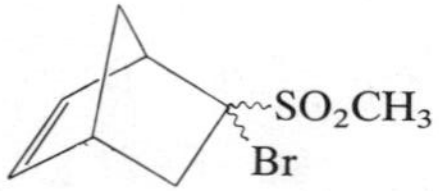 SO_2CH_3, Br	$NaOCH_3$	DMSO	25	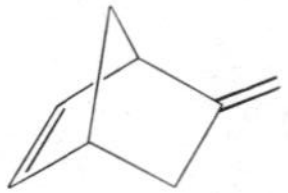 (>90)	76
		2 *N* NaOH	H_2O	100	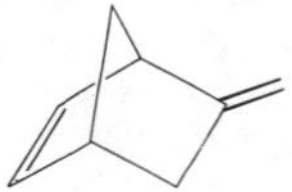 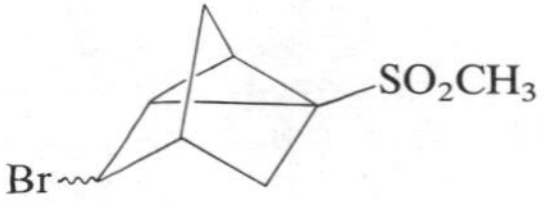 (-),	
					Br, SO_2CH_3 (-)	76

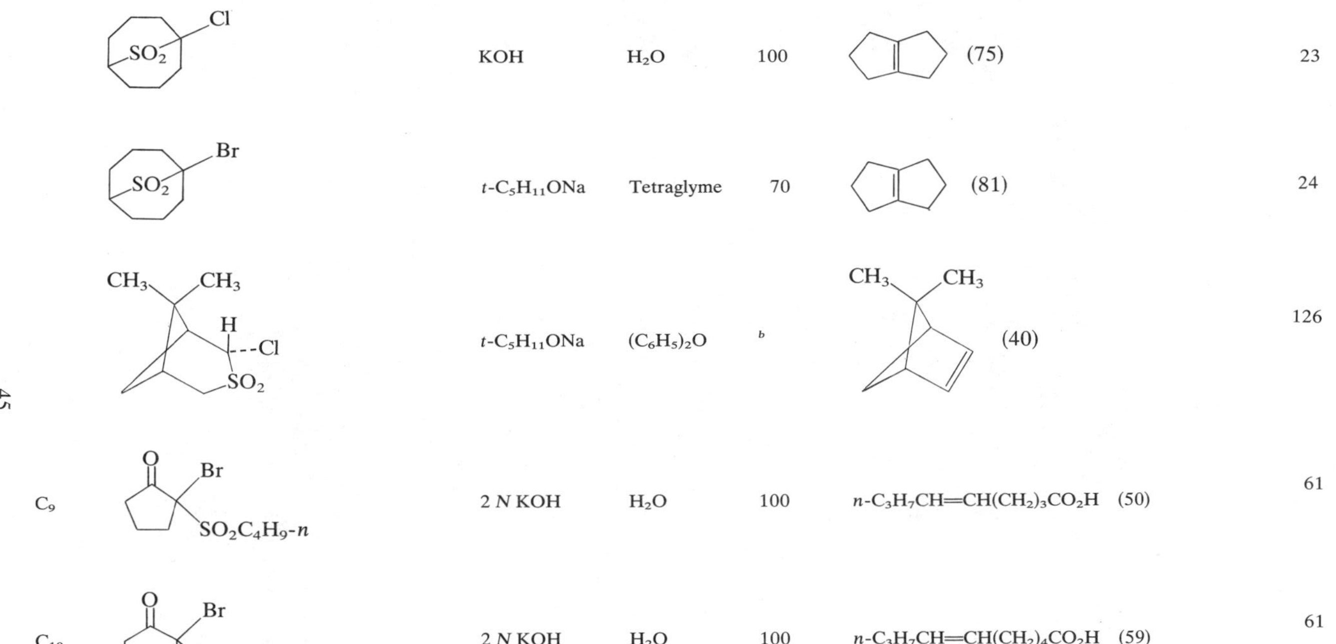

		KOH	H_2O	100	(75)	23
		t-$C_5H_{11}ONa$	Tetraglyme	70	(81)	24
		t-$C_5H_{11}ONa$	$(C_6H_5)_2O$	[b]	(40)	126
C_9		2 *N* KOH	H_2O	100	n-$C_3H_7CH{=}CH(CH_2)_3CO_2H$ (50)	61
C_{10}		2 *N* KOH	H_2O	100	n-$C_3H_7CH{=}CH(CH_2)_4CO_2H$ (59)	61

Note: References 126–131 are on p. 71.

[b] The product was isolated by direct distillation from the reaction mixture under reduced pressure.

TABLE II. Base-Promoted Rearrangements of Preformed α-Halo Sulfones (*Continued*)

No. of C Atoms	Substrate	Base	Solvent	Temperature, °C	Product(s) and Yield(s) (%)	Refs.
			B. Cyclic Compounds			
C_{10} (*contd.*)	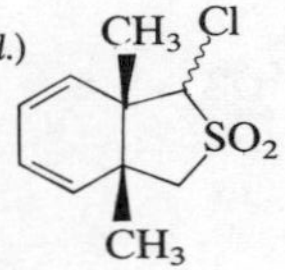 [structure: CH_3, Cl, SO_2, CH_3]	t-C_4H_9OK	DMSO	25–35	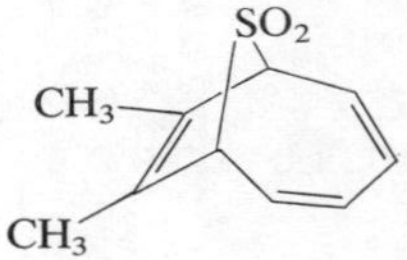 [structure: SO_2, CH_3, CH_3] (27)	59
	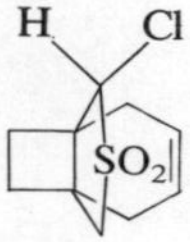 [structure: H, Cl, SO_2]	t-C_4H_9OK	THF	0	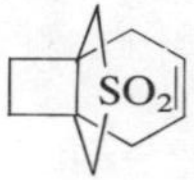 [structure: SO_2] (60)	60
					[structure: t-C_4H_9O, H, SO_2] (11)	
		n-C_4H_9Li	$(C_2H_5)_2O$	−78	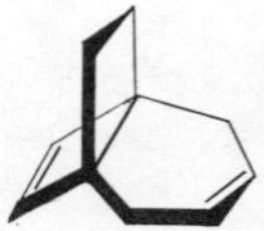 [structure] (2.6)	60

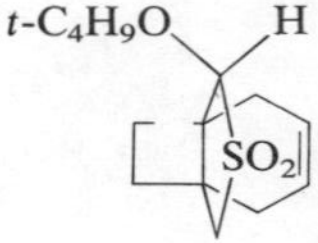

Reactant	Base	Solvent	Temp. (°C)	Product(s) and Yield(s) (%)	Refs.
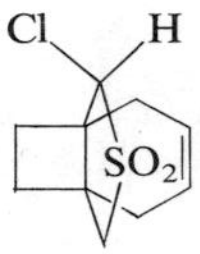 Cl, H, SO_2	t-C_4H_9OK	THF	0	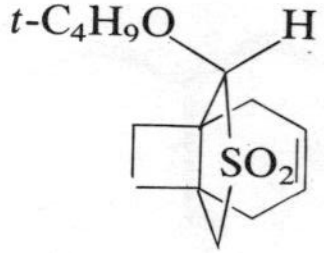 t-C_4H_9O, H, SO_2 (40)	60
	n-C_4H_9Li	$(C_2H_5)_2O$	−78	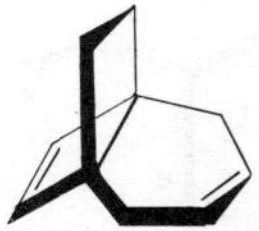 (1.4)	60
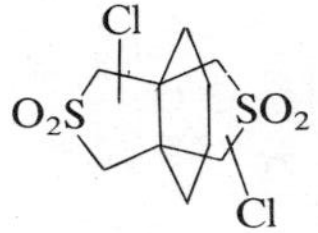 Cl, O_2S, SO_2, Cl	t-C_4H_9OK	THF	−15	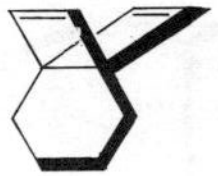 (1)	90
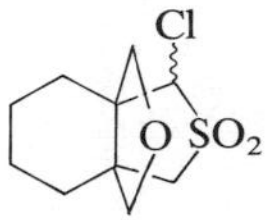 Cl, O, SO_2	t-C_4H_9OK	THF	Reflux	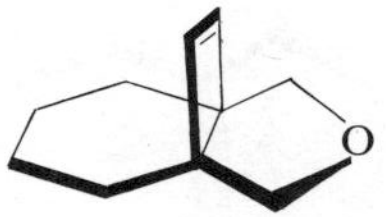 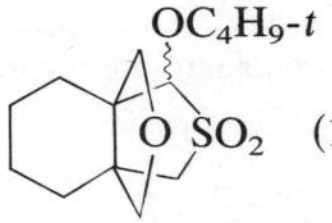 O (23-26), OC_4H_9-t, O, SO_2 (12.9-14.5)	91

Note: References 126–131 are on p. 71.

TABLE II. BASE-PROMOTED REARRANGEMENTS OF PREFORMED α-HALO SULFONES (*Continued*)

No. of C Atoms	Substrate	Base	Solvent	Temperature, °C	Product(s) and Yield(s) (%)	Refs.
			B. Cyclic Compounds			
C_{10} (*contd.*)	Cl, SO_2, O	t-C_4H_9OK	THF	65	O (39)	88
	Cl, SO_2, O	t-C_4H_9OK	THF	65	O (27)	88
	Cl, SO_2, SO_2	t-C_4H_9OK	THF	0	SO_2 (46.3)	91
	Cl, SO_2, SO_2	t-C_4H_9OK	THF	25	SO_2 (50)	88

C_{11}	t-C_4H_9OK	DME	25	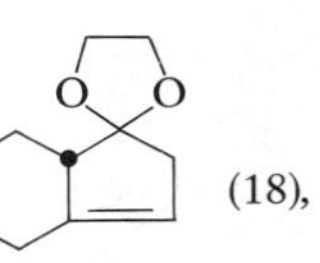 (18),	97
				(35)	
	t-C_4H_9OK	DMSO	25	(62)	97
	t-C_4H_9OK	DME	25	(48),	97
				(25)	

Note: References 126–131 are on p. 71.

TABLE II. BASE-PROMOTED REARRANGEMENTS OF PREFORMED α-HALO SULFONES (*Continued*)

No. of C Atoms	Substrate	Base	Solvent	Temperature, °C	Product(s) and Yield(s) (%)	Refs.
			B. Cyclic Compounds			
C_{11} (*contd.*)	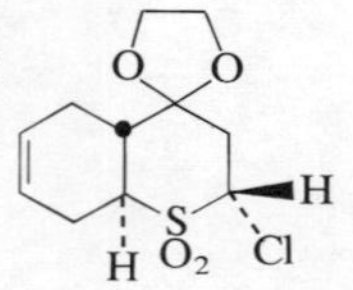	$t\text{-}C_4H_9OK$	DME	25	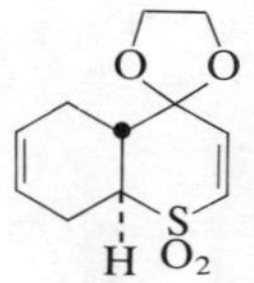 (55)	97
	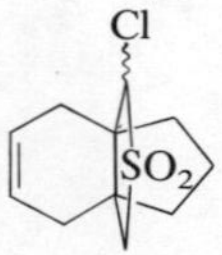	$t\text{-}C_4H_9OK$	THF	65	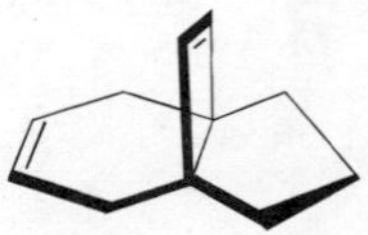 (41)	127
C_{12}	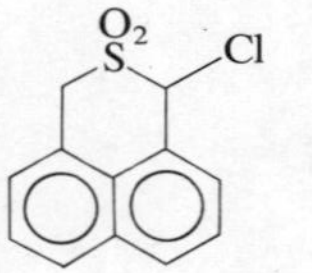	$NaOCH_3$	CH_3OD	25	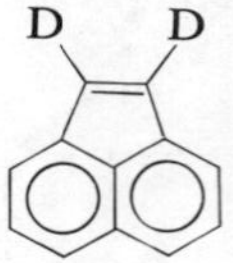 (High)	28

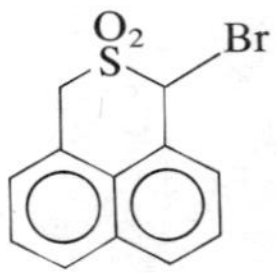	$NaOCH_3$	CH_3OH	25	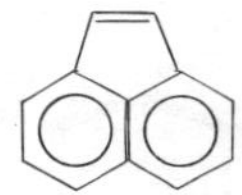 (High)	28
	$NaOCH_3$	CH_3OD	25	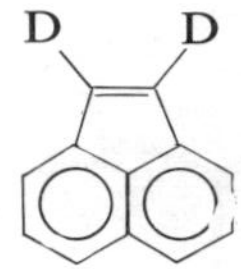 (High)	28
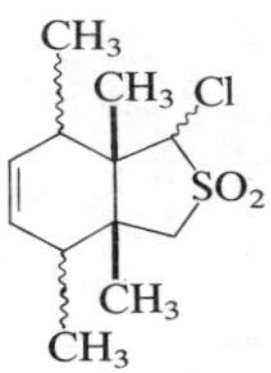	$t\text{-}C_4H_9OK$	THF	65	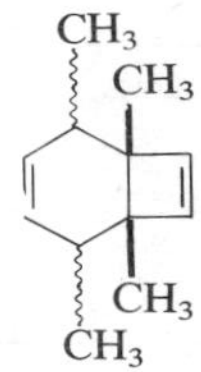 (49.6)	122
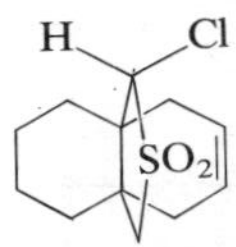	$t\text{-}C_4H_9OK$	THF	25	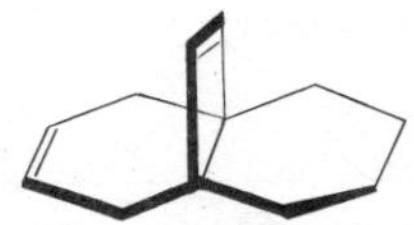 (46.5)	84

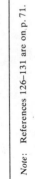

Note: References 126–131 are on p. 71.

TABLE II. BASE-PROMOTED REARRANGEMENTS OF PREFORMED α-HALO SULFONES (*Continued*)

No. of C Atoms	Substrate	Base	Solvent	Temperature, °C	Product(s) and Yield(s) (%)	Refs.
			B. Cyclic Compounds			
C_{12} (*contd.*)	H, Cl; SO_2	t-C_4H_9OK	THF	25	(55.6)	84
	D, Cl; SO_2; D, D	t-C_4H_9OK	THF	65	D, D (81)	87
	Cl; D_2; SO_2; D_2	t-C_4H_9OK	THF	65	D_2, D_2 (37.1)	122

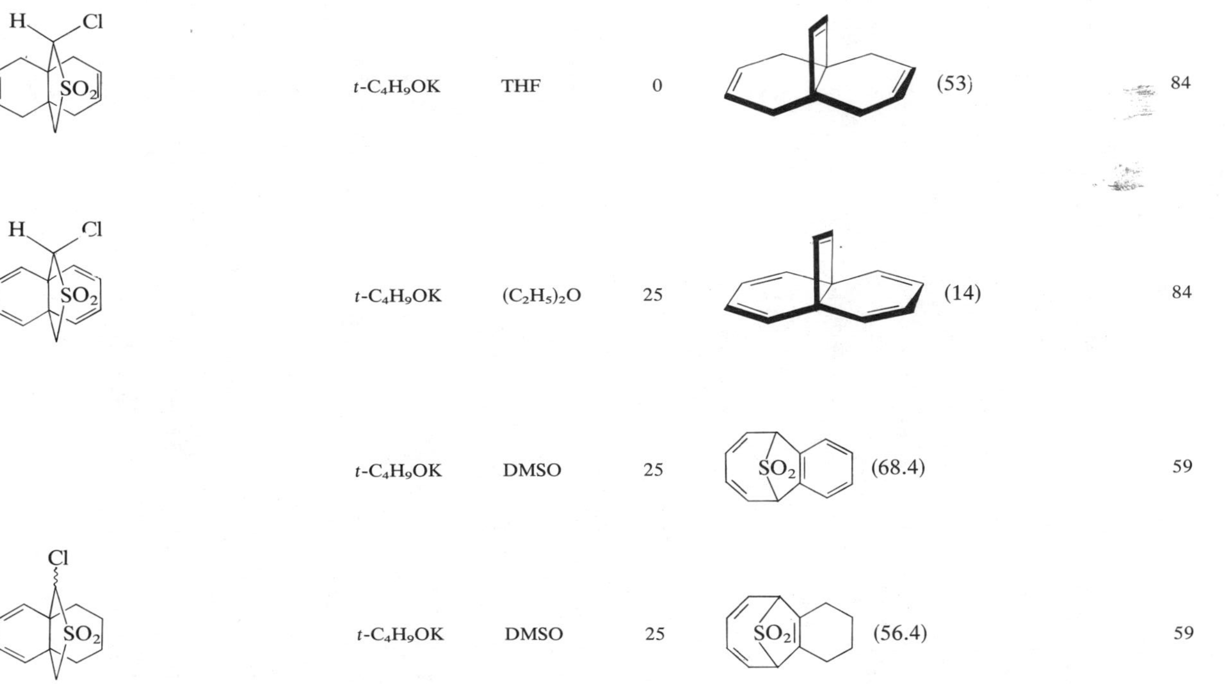

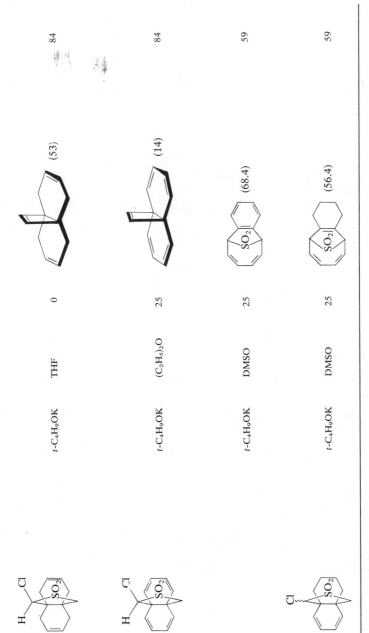

	$t\text{-}C_4H_9OK$	THF	0	(53)	84
	$t\text{-}C_4H_9OK$	$(C_2H_5)_2O$	25	(14)	84
	$t\text{-}C_4H_9OK$	DMSO	25	(68.4)	59
	$t\text{-}C_4H_9OK$	DMSO	25	(56.4)	59

Note: References 126–131 are on p. 71.

TABLE II. Base-Promoted Rearrangements of Preformed α-Halo Sulfones (*Continued*)

No. of C Atoms	Substrate	Base	Solvent	Temperature, °C	Product(s) and Yield(s) (%)	Refs.
			B. Cyclic Compounds			
C_{13}	[structure: Cl, H, S, O_2, H, O]	t-C_4H_9OK	DME	25	[structure: O] (18), [structure: S, H, O_2, O] (8)	97
		t-C_4H_9OK	DMSO	25	[structure: O] (42)	97

Reactant	Base	Solvent	Temp.	Product(s) (Yield %)	Ref.
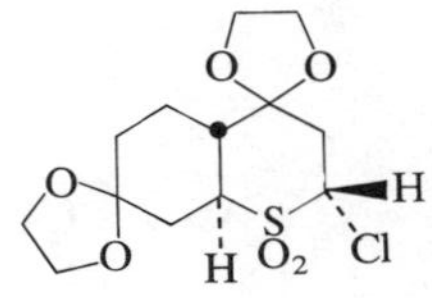 (ketal sulfone, H, Cl)	t-C_4H_9OK	DMSO	25	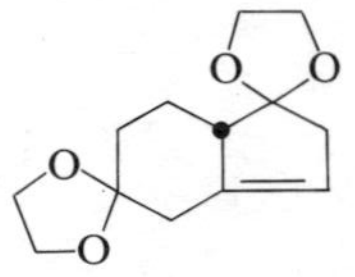 (20), 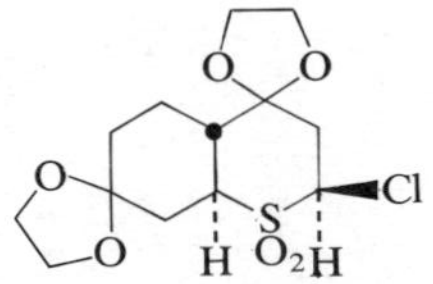 (55)	97
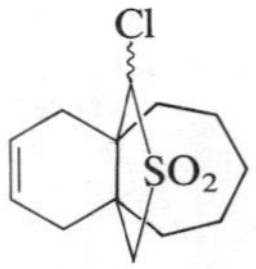 (Cl, SO_2 bicyclic)	t-C_4H_9OK	THF	25	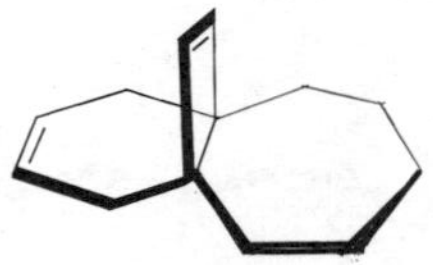 (66)	84
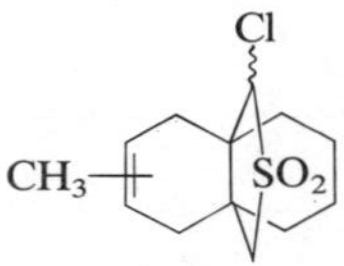 CH_3 (Cl, SO_2 bicyclic)	t-C_4H_9OK	THF	65	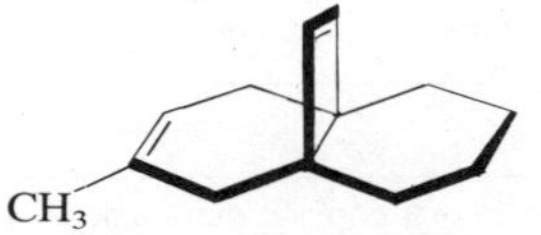 CH_3 (46.5)	122

Note: References 126–131 are on p. 71.

TABLE II. BASE-PROMOTED REARRANGEMENTS OF PREFORMED α-HALO SULFONES (*Continued*)

No. of C Atoms	Substrate	Base	Solvent	Temperature, °C	Product(s) and Yield(s) (%)	Refs.
			B. Cyclic Compounds			
C_{13} (*contd.*)	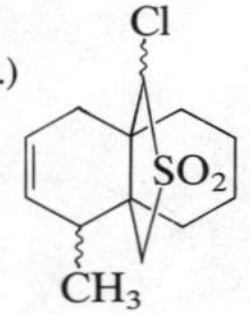 Cl, SO_2, CH_3	t-C_4H_9OK	THF	65	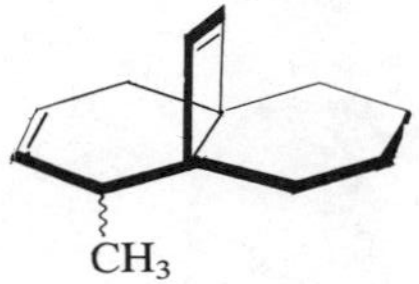 CH_3 (47.6)	122
	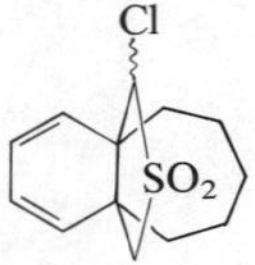 Cl, SO_2	t-C_4H_9OK	DMSO	25	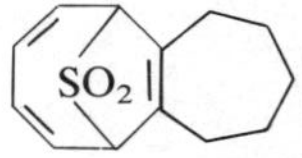 SO_2 (55.8)	59
C_{14}	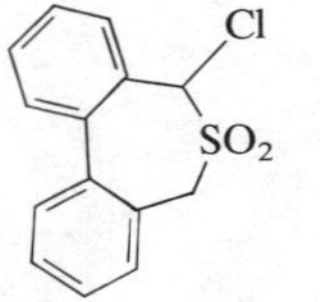 Cl, SO_2	NaOH	H_2O–dioxane	87	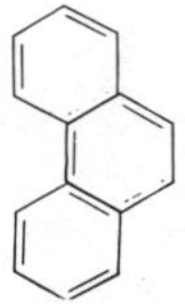 (95.8)	29
		NaOD	D_2O–dioxane	87	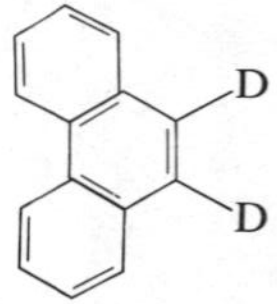 D, D (91.7)	29

	Reactant	Base	Solvent	Temp.	Product (yield, %)	Ref.
	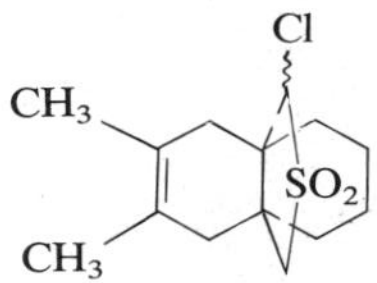 CH_3, CH_3, Cl, SO_2	t-C_4H_9OK	THF	65	 CH_3, CH_3 (39.4)	122
C_{16}	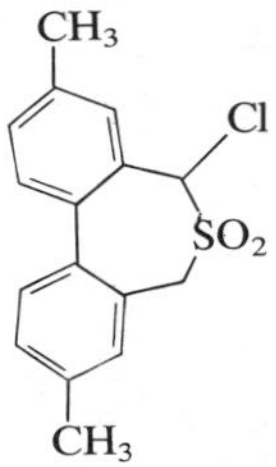 CH_3, Cl, SO_2, CH_3	NaOH	H_2O–dioxane	87	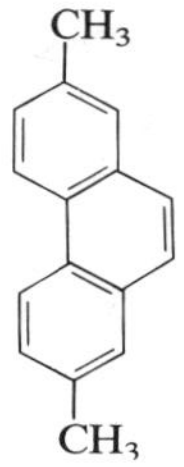 CH_3, CH_3 (92)	29
		NaOD	D_2O–dioxane	87	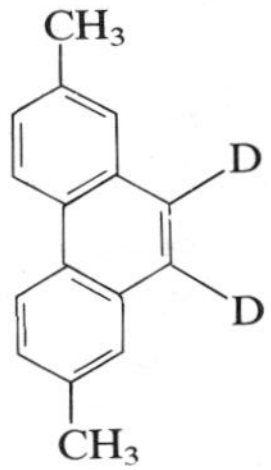 CH_3, D, D, CH_3 (90.7)	29

Note: References 126–131 are on p.71.

TABLE II. BASE-PROMOTED REARRANGEMENTS OF PREFORMED α-HALO SULFONES (*Continued*)

No. of C Atoms	Substrate	Base	Solvent	Tempera- ture, °C	Product(s) and Yield(s) (%)	Refs.
			B. Cyclic Compounds			
C_{16} (*contd.*)	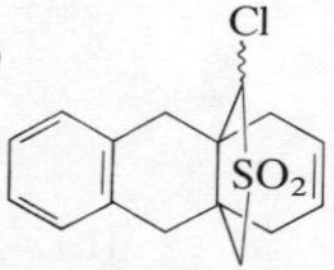 Cl, SO_2	t-C_4H_9OK	THF	65	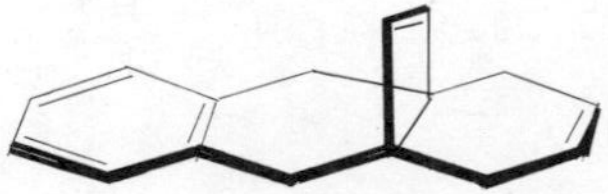 $(57)^c$	128
	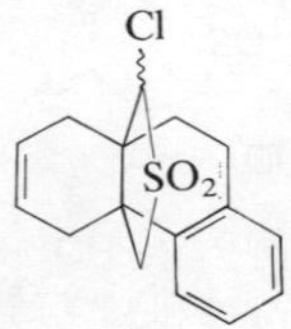 Cl, SO_2	t-C_4H_9OK	THF	65	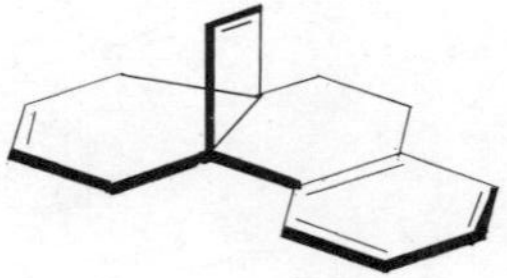 $(51.7)^c$	129

Note: References 126–131 are on p. 71.
[c] The yield is based on the starting sulfide.

TABLE III. Direct Conversion of Sulfones into Olefins with Potassium Hydroxide–Carbon Tetrachloride-*t*-Butyl Alcohol

No. of C Atoms	Substrate	Product(s) and Yield(s) (%)	Refs.
C_4	$(C_2H_5)_2SO_2$	*E*-$CH_3CH{=}C(CH_3)SO_3^-K^+$ (70)	100
C_5	$CH_3CHClSO_2C_3H_7$-*n*	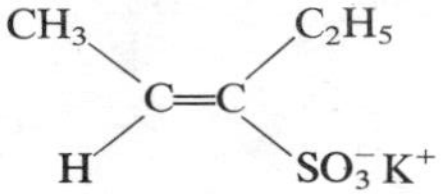 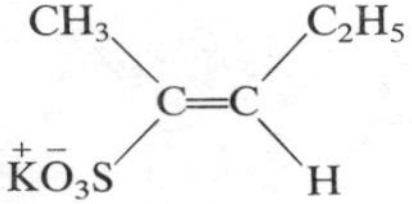 $CH_3(H)C{=}C(C_2H_5)SO_3^-K^+$ (45), $CH_3(\overset{+}{K}\overset{-}{O_3}S)C{=}C(C_2H_5)H$ (55)	102
	$C_2H_5SO_2CHClC_2H_5$	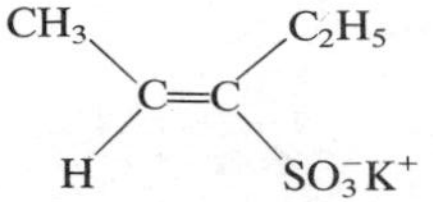 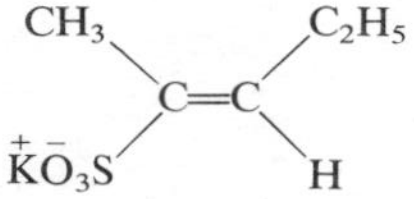 $CH_3(H)C{=}C(C_2H_5)SO_3^-K^+$ (45), $CH_3(\overset{+}{K}\overset{-}{O_3}S)C{=}C(C_2H_5)H$ (55)	102
C_6	$(n\text{-}C_3H_7)_2SO_2$	*E*-$C_2H_5CH{=}C(C_2H_5)SO_3^-K^+$ (>70)	100
C_8	$(n\text{-}C_4H_9)_2SO_2$	*n*-$C_3H_7CH{=}C(n\text{-}C_3H_7)SO_3^-K^+$ (75)	102
	$(sec\text{-}C_4H_9)_2SO_2$	$C_2H_5(CH_3)C{=}C(CH_3)C_2H_5$, 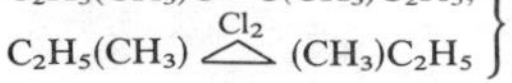 $C_2H_5(CH_3)$ $\overset{Cl_2}{\triangle}$ $(CH_3)C_2H_5$ (90)	93
	$(n\text{-}C_4H_9)_2SO_2$	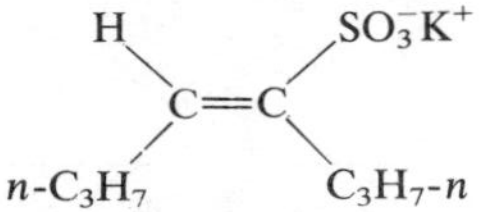 $H(n\text{-}C_3H_7)C{=}C(SO_3^-K^+)C_3H_7\text{-}n$ (>70)	100
	 SO_2	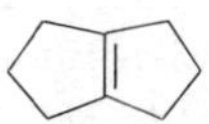 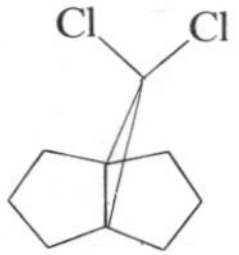 (45), Cl Cl (50)	93

Note: References 126–131 are on p.71.

TABLE III. Direct Conversion of Sulfones into Olefins with Potassium Hydroxide–Carbon Tetrachloride-*t*-Butyl Alcohol (*Continued*)

No. of C Atoms	Substrate	Product(s) and Yield(s) (%)	Refs.
C_8 (*contd.*)	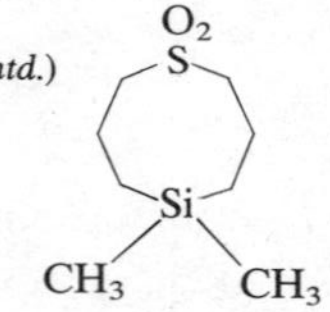 O_2S / Si ring with CH_3, CH_3	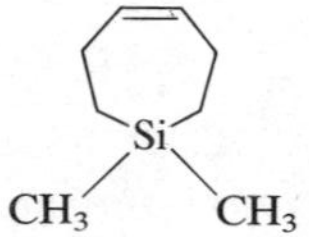 Si ring with CH_3, CH_3 (50)	99
	n-$C_3H_7CHClSO_2C_4H_9$-n	n-$C_3H_7CH{=}C(SO_3^-K^+)C_3H_7$-$n$ (>90)	102
C_{10}	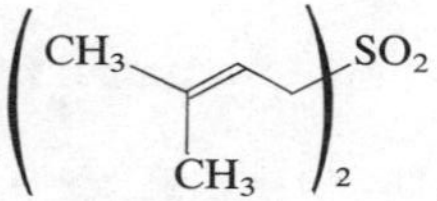 $((CH_3)_2C{=}CHCH_2)_2SO_2$	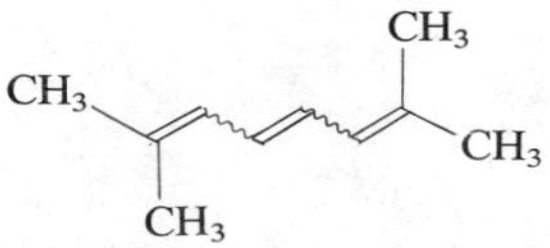 $(CH_3)_2C{=}CH{-}CH{=}CH{-}CH{=}C(CH_3)_2$ (3:1 isomer mixture) (78)	98
	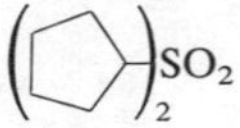 $(C_5H_9)_2SO_2$	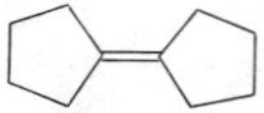 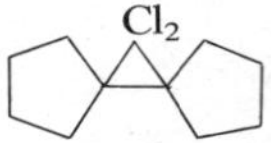 $C_5H_8{=}C_5H_8$ (30), Cl_2 dispiro compound (45)	93

	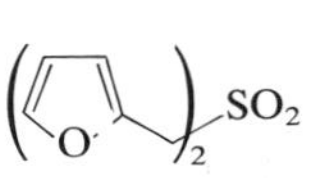 SO_2 (furyl)$_2$	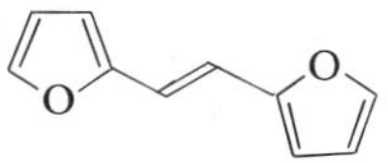 (18)	98
C_{12}	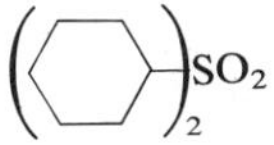 SO_2 (cyclohexyl)$_2$	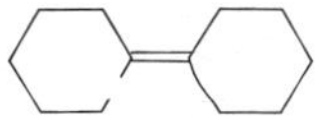 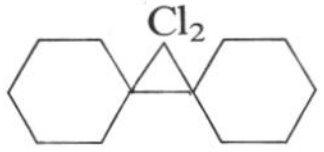 (32), Cl_2 (60)	93
	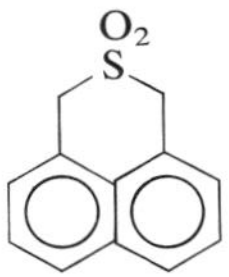 O_2 S	 (45)	93
	$(2\text{-PyridylCH}_2)_2SO_2$	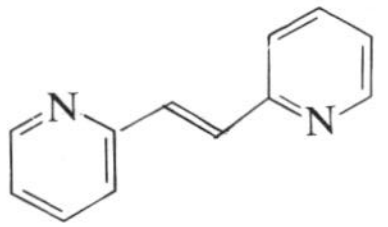 N N (Excellent)	130
C_{16}	$(n\text{-}C_8H_{17})_2SO_2$	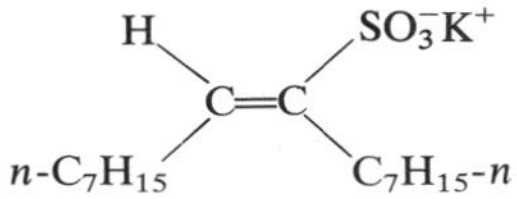 H, $SO_3^-K^+$, C=C, $n\text{-}C_7H_{15}$, $C_7H_{15}\text{-}n$ (>70)	100

Note: References 126–131 are on p. 71.

TABLE III. Direct Conversion of Sulfones into Olefins with Potassium Hydroxide–Carbon Tetrachloride-*t*-Butyl Alcohol (*Continued*)

Atoms	Substrate	Product(s) and Yield(s) (%)	Refs.
C_{14} (*contd.*)	$(C_6H_5CH_2)_2SO_2$	*E*-$C_6H_5CH{=}CHC_6H_5$ (100)	93
C_{15}	$(C_6H_5)_2CHSO_2C_2H_5$	$(C_6H_5)_2C{=}CHCH_3$ (>96)	95
C_{16}	$[C_6H_5CH(CH_3)]_2SO_2$	$C_6H_5(CH_3)C{=}C(CH_3)C_6H_5$ *dl* (100; 28 *cis*; 72 *trans*) *meso* (100; 90 *cis*; 10 *trans*)	93
C_{18}	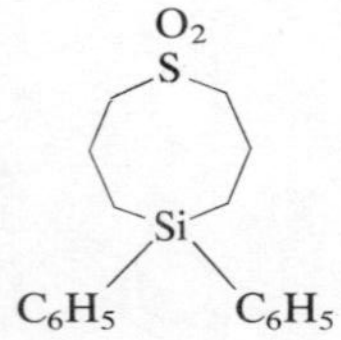 O_2 S, Si, C_6H_5 C_6H_5	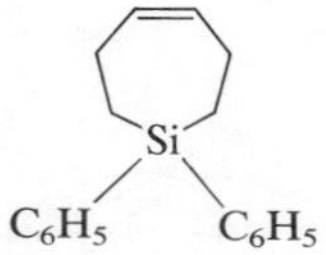 Si, C_6H_5 C_6H_5 (40)	99
C_{20}	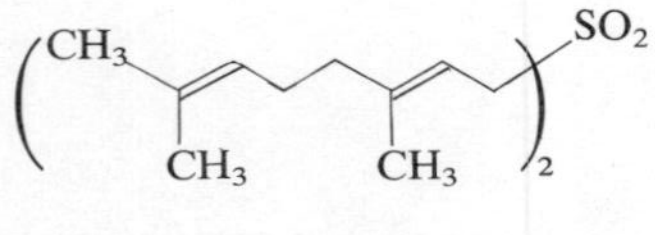 (CH_3, CH_3, CH_3)$_2SO_2$	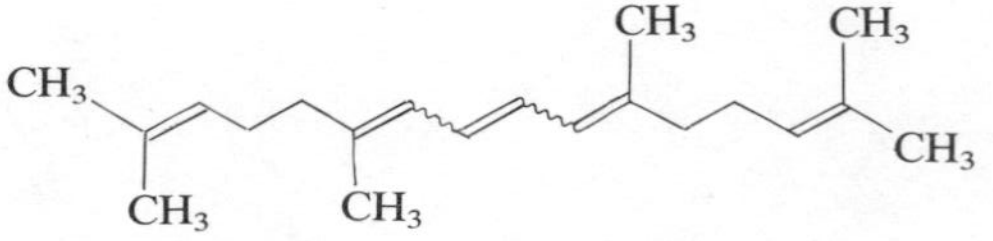 CH_3, CH_3, CH_3, CH_3, CH_3, CH_3 (69)	98
C_{26}	$[(C_6H_5)_2CH]_2SO_2$	$(C_6H_5)_2C{=}C(C_6H_5)_2$ (>96)	95
C_{30}	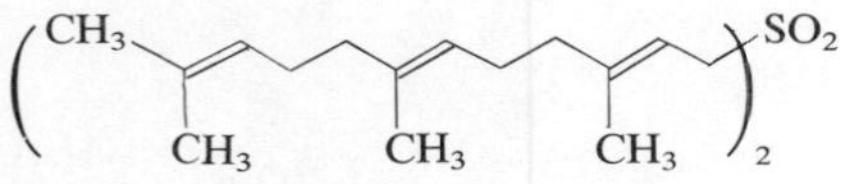 (CH_3, CH_3, CH_3, CH_3)$_2SO_2$	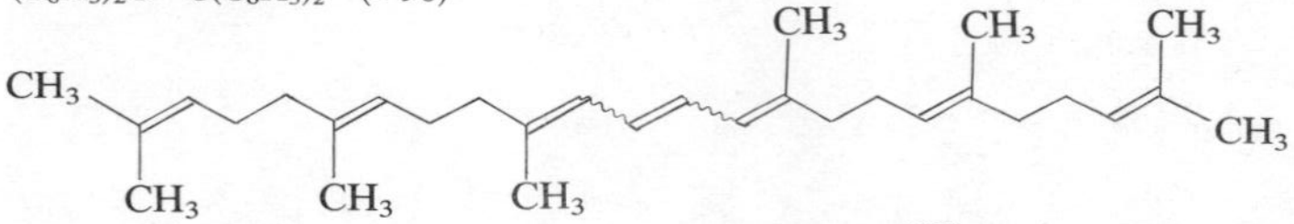 CH_3, CH_3, CH_3, CH_3, CH_3, CH_3, CH_3, CH_3 (89)	98

Note: References 126–131 are on p. 71.

TABLE IV. Base-Promoted Rearrangements of α,α- and α,α'-Dihalo Sulfones

No. of C Atoms	Substrate	Base	Solvent	Temperature, °C	Product(s) and Yield(s) (%)	Refs.
C_4	$(CH_3CHCl)_2SO_2$	2 *N* NaOH	H_2O	100	$CH_3CH{=}C(CH_3)SO_3^-Na^+$ (—)	13
	$i\text{-}C_3H_7SO_2CHCl_2$	NaOH	H_2O	100	$(CH_3)_2C{=}CHCl$ (44), $(CH_3)_2C(CH_2Cl)SO_3^-Na^+$ (37)	47
C_7	$n\text{-}C_6H_{13}SO_2CHCl_2$	NaOH	H_2O	100	$n\text{-}C_5H_{11}C{\equiv}CH$ (39), $n\text{-}C_5H_{11}CH{=}CHCl$ (30), $(n\text{-}C_5H_{11})(Na^{+-}O_3S)C{=}CH_2$ (20)	47
		$t\text{-}C_4H_9OK$	$t\text{-}C_4H_9OH$	84	$n\text{-}C_5H_{11}C{\equiv}CH$ (46)	47
	$c\text{-}C_6H_{11}SO_2CHCl_2$	NaOH	H_2O–dioxane	87	$c\text{-}C_6H_{10}{=}CHCl$ (60), $1\text{-}(CH_2Cl)\text{-}c\text{-}C_6H_{10}\text{-}1\text{-}SO_3^-\ Na^+$ (20)	47
		$t\text{-}C_4H_9OK$	$t\text{-}C_4H_9OH$	84	$c\text{-}C_6H_{10}{=}CHCl$ (60)	47
C_8	$C_6H_5CH_2SO_2CHCl_2$	NaOH	H_2O–dioxane	87	$C_6H_5C{\equiv}CH$ (56), $E\text{-}C_6H_5CH{=}CHSO_3^-K^+$ (31)	47
		KOH	H_2O–dioxane	87	$C_6H_5C{\equiv}CH$ (61), $E\text{-}C_6H_5CH{=}CHSO_3^-K^+$ (28)	47
		$t\text{-}C_4H_9OK$	$t\text{-}C_4H_9OH$	84	$C_6H_5C{\equiv}CH$ (76)	47
	$C_6H_5CH_2SO_2CHBr_2$	NaOH	H_2O–dioxane	0	$C_6H_5CH{=}CH_2$ (10), $C_6H_5C{\equiv}CH$ (20), $C_6H_5CH{=}CHSO_3H$ (40)	68
			H_2O–dioxane	85	$C_6H_5CH{=}CH_2$ (15), $C_6H_5C{\equiv}CH$ (35), $C_6H_5CH{=}CHSO_3H$ (35)	68
		$NaOCH_3$	CH_3OH	0	$C_6H_5CH{=}CH_2$ (0), $C_6H_5C{\equiv}CH$ (Trace), $C_6H_5CH{=}CHSO_3H$ (75), $C_6H_5CH{=}CHBr$ (8)	68

Note: References 126–131 are on p. 71.

TABLE IV. Base-Promoted Rearrangements of α,α- and α,α'-Dihalo Sulfones (*Continued*)

No. of C Atoms	Substrate	Base	Solvent	Temperature, °C	Product(s) and Yield(s) (%)	Refs.
C_8 (*contd.*)	$C_6H_5CH_2SO_2CHBr_2$	$NaOCH_3$	CH_3OH	65	$C_6H_5CH{=}CH_2$ (8), $C_6H_5C{\equiv}CH$ (15), $C_6H_5CH{=}CHSO_3H$ (50), $C_6H_5CH{=}CHBr$ (10)	68
		t-C_4H_9OK	t-C_4H_9OH	25	$C_6H_5CH{=}CH_2$ (Trace), $C_6H_5C{\equiv}CH$ (45), $C_6H_5CH{=}CHSO_3H$ (40), $C_6H_5CH{=}CHBr$ (0)	68
		t-C_4H_9OK	t-C_4H_9OH	83	$C_6H_5CH{=}CH_2$ (Trace), $C_6H_5C{\equiv}CH$ (70), $C_6H_5CH{=}CHSO_3H$ (15), $C_6H_5CH{=}CHBr$ (0)	68
		NaOH	H_2O	100	$C_6H_5CH{=}CH_2$ (26), $C_6H_5C{\equiv}CH$ (28), $C_6H_5CH{=}CHSO_2H$ (—), $C_6H_5CH{=}CHBr$ (—)	68
C_9	$C_6H_5CH_2CH_2SO_2CHCl_2$	NaOH	H_2O–dioxane	87	$C_6H_5CH{=}C{=}CH_2$ (3), $C_6H_5CH_2CH{=}CHCl$ (17), $(C_6H_5CH_2)(\overset{+}{Na}\,\overset{-}{O_3}S)C{=}CH_2$ (17)	47
C_{10}	$C_6H_5(CH_2)_3SO_2CHCl_2$	NaOH	H_2O–dioxane	87	$C_6H_5(CH_2)_2C{\equiv}CH$ (50), $C_6H_5(CH_2)_2CH{\equiv}CHCl$ (14), $(C_6H_5(CH_2)_2)(\overset{+}{Na}\,\overset{-}{O_3}S)C{=}CH_2$ (19)	47
		t-C_4H_9OK	t-C_4H_9OH	84	$C_6H_5(CH_2)_2C{\equiv}CH$ (2.5), $C_6H_5(CH_2)_2CH{=}CH_2$ (9) $C_6H_5(CH_2)_2CH{=}CHCl$ (0.6)	47
	o-$C_6H_4(CH_2SO_2CHBr_2)_2$	2 *N* NaOH	H_2O	100	o-$C_6H_4(C{\equiv}CH)_2$ (10.3)	65
C_{12}	α-$C_{10}H_7CH_2SO_2CHBr_2$	2 *N* NaOH	H_2O	100	α-$C_{10}H_7C{\equiv}CH$ (30), α-$C_{10}H_7CH{=}CH_2$ (36)	65

C_{14}	$C_6H_5CH_2SO_2C(Cl)_2C_6H_5$	NaOH	H_2O–dioxane	87	$C_6H_5CH{=}C(C_6H_5)SO_3H$ (70.6), $C_6H_5C{\equiv}CC_6H_5$ (17), $C_6H_5CH{=}C(Cl)C_6H_5$ (6)	49
		$(C_2H_5)_3N$	DMSO	25	2,3-$(C_6H_5)_2$-thiirene 1,1-dioxide (O_2S ring, C_6H_5 on each ring carbon) (>90)	105
	p-$ClC_6H_4CH_2SO_2C(Cl)_2C_6H_4Cl$-$p$	$(C_2H_5)_3N$	DMSO	25	2,3-bis(p-ClC_6H_4)thiirene 1,1-dioxide (O_2S ring, p-ClC_6H_4 and C_6H_4Cl-p on the ring carbons) (>90)	105
	$(C_6H_5CHBr)_2SO_2$	NaOH	H_2O–dioxane	0	$C_6H_5C{\equiv}CC_6H_5$ (13), $C_6H_5CH{=}C(C_6H_5)SO_3H$ (–)	104
			H_2O–dioxane	85	,, (20), ,, (75)	104
		$NaOCH_3$	CH_3OH	0	,, (17), ,, (75)	104
			CH_3OH	65	,, (28), ,, (65)	104
		t-C_4H_9OK	t-C_4H_9OH	25	,, (26), ,, (65)	104
			t-C_4H_9OH	83	,, (28), ,, (65)	104

Note: References 126–131 are on p. 71.

TABLE IV. BASE-PROMOTED REARRANGEMENTS OF α,α- AND α,α'-DIHALO SULFONES (*Continued*)

No. of C Atoms	Substrate	Base	Solvent	Temperature, °C	Product(s) and Yield(s) (%)	Refs.
C_{14} (*contd.*)		$(C_2H_5)_3N$	CH_2Cl_2	Reflux	2,3-Diphenylthiirene 1,1-dioxide (C_6H_5, C_6H_5, SO_2 ring) (90)	67
	$(C_6H_5)_2CHSO_2CHBr_2$	NaOH	H_2O–dioxane	0	$(C_6H_5)_2C{=}CH_2$ (8), $(C_6H_5)_2C{=}CHBr$ (27), $(C_6H_5)_2C{=}CHSO_3H$ (60)	68
			H_2O–dioxane	85	$(C_6H_5)_2C{=}CH_2$ (30), $(C_6H_5)_2C{=}CHBr$ (45), $(C_6H_5)_2C{=}CHSO_3H$ (20)	68
		$NaOCH_3$	CH_3OH	0	$(C_6H_5)_2C{=}CH_2$ (Trace), $(C_6H_5)_2C{=}CHBr$ (22), $(C_6H_5)_2C{=}CHSO_3H$ (73)	68
			CH_3OH	65	$(C_6H_5)_2C{=}CH_2$ (3), $(C_6H_5)_2C{=}CHBr$ (65), $(C_6H_5)_2C{=}CHSO_3H$ (30)	68
		t-C_4H_9OK	t-C_4H_9OH	25	$(C_6H_5)_2C{=}CH_2$ (10), $(C_6H_5)_2C{=}CHBr$ (20), $(C_6H_5)_2C{=}CHSO_3H$ (65)	68
			t-C_4H_9OH	83	$(C_6H_5)_2C{=}CH_2$ (15), $(C_6H_5)_2C{=}CHBr$ (25), $(C_6H_5)_2C{=}CHSO_3H$ (58)	68
	$(C_6H_5)_2CHSO_2CHCl_2$	NaOH	H_2O–dioxane	0	$(C_6H_5)_2C{=}CH_2$ (0), $(C_6H_5)_2C{=}CHCl$ (43), $(C_6H_5)_2C{=}CHSO_3H$ (55)	68
			H_2O–dioxane	85	$(C_6H_5)_2C{=}CH_2$ (Trace), $(C_6H_5)_2C{=}CHCl$ (77), $(C_6H_5)_2C{=}CHSO_3H$ (20)	68
		$NaOCH_3$	CH_3OH	0	$(C_6H_5)_2C{=}CH_2$ (0), $(C_6H_5)_2C{=}CHCl$ (48), $(C_6H_5)_2C{=}CHSO_3H$ (50)	68

			CH_3OH	65	$(C_6H_5)_2C{=}CH_2$ (Trace), $(C_6H_5)_2C{=}CHCl$ (72), $(C_6H_5)_2C{=}CHSO_3H$ (25)	68
		t-C_4H_9OK	t-C_4H_9OH	25	$(C_6H_5)_2C{=}CH_2$ (0), $(C_6H_5)_2C{=}CHCl$ (35), $(C_6H_5)_2C{=}CHSO_3H$ (60)	68
			t-C_4H_9OH	83	$(C_6H_5)_2C{=}CH_2$ (4), $(C_6H_5)_2C{=}CHCl$ (35), $(C_6H_5)_2C{=}CHSO_3H$ (60)	68
	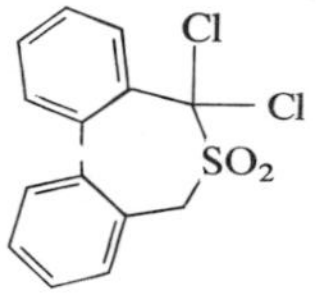 Cl, Cl, SO_2	NaOH	H_2O–dioxane	87	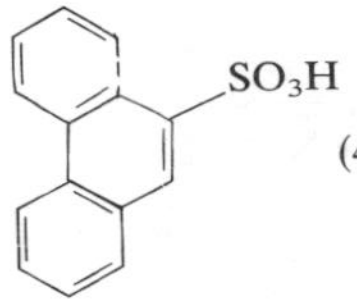 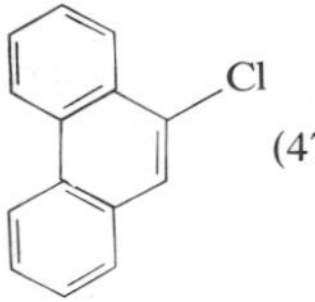 SO_3H (47.3), Cl (47.6)	49
C_{15}	$C_6H_5CH_2SO_2CCl_2C_6H_4CH_3$-$p$	NaOH	H_2O–dioxane	87	p-$CH_3C_6H_4CH{=}C(C_6H_5)SO_3H$, $C_6H_5CH{=}C(C_6H_4CH_3$-$p)SO_3H$, (combined 70), p-$CH_3C_6H_4C{\equiv}CC_6H_5$ (21), mixture of vinyl chlorides (5)	58

Note: References 126–131 are on p. 71.

TABLE IV. BASE-PROMOTED REARRANGEMENTS OF α,α- AND α,α'-DIHALO SULFONES (*Continued*)

No. of C Atoms	Substrate	Base	Solvent	Temperature, °C	Product(s) and Yield(s) (%)	Refs
C_{15} (*contd.*)	p-$CH_3C_6H_4CH_2SO_2C(Cl)_2C_6H_5$	NaOH	H_2O–dioxane	87	p-$CH_3C_6H_4CH{=}C(C_6H_5)SO_3H$, $C_6H_5CH{=}C(C_6H_4CH_3$-$p)SO_3H$ (combined 70), p-$CH_3C_6H_4C{\equiv}CC_6H_5$ (20), mixture of vinyl chlorides (5)	58
C_{16}	p-$CH_3C_6H_4CH_2SO_2C(Cl)_2C_6H_4CH_3$-$p$	$(C_2H_5)_3N$	DMSO	25	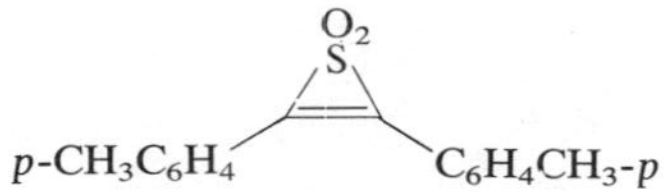 O_2 S; p-$CH_3C_6H_4$; $C_6H_4CH_3$-p (>90)	105
	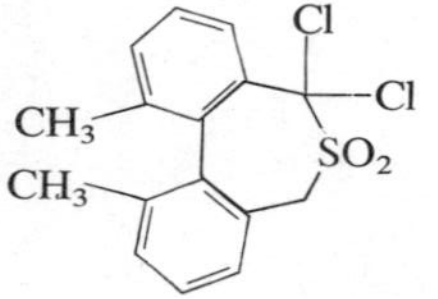 Cl; Cl; CH_3; CH_3; SO_2	NaOH	H_2O–dioxane	87	No reaction	49

References 126–131 are on p. 71.

TABLE V. Base-Promoted Rearrangements of α,α,α-Trichloromethyl Sulfones

No. of C Atoms	Substrate	Base	Solvent	Temperature, °C	Product(s) and Yield(s) (%)	Refs.
C_4	$(CH_3)_2CHSO_2CCl_3$	NaOH	H_2O–THF	65	$(CH_3)_2C{=}CCl_2$ (14), $(CH_3)_2C(SO_3H)CHCl_2$ (72)	48
			H_2O	100	$(CH_3)_2C{=}CCl_2$ (8), $(CH_3)_2C(SO_3H)CHCl_2$ (86)	48
C_7	1-H-1-SO_2CCl_3-cyclohexane (cyclohexyl, H, SO_2CCl_3)	NaOH	H_2O–THF	65	cyclohexylidene$=CHCl$ (1), cyclohexylidene$=CCl_2$ (51), 1-SO_3H-1-$CHCl_2$-cyclohexane (29)	48
C_8	$C_6H_5CH_2SO_2CCl_3$	NaOH	H_2O–THF	65	$C_6H_5C{\equiv}CH$ (2), $C_6H_5CH{=}CCl_2$ (<1), $C_6H_5C(SO_3H){=}CHCl$ (27) $C_6H_5COCH_2SO_3H$ (39)	48
C_9	$C_6H_5(CH_2)_2SO_2CCl_3$	NaOH	H_2O–THF	65	$C_6H_5CH_2CH{=}CHCl$ (Trace), $C_6H_5CH_2CH{=}CCl_2$ (Trace), $C_6H_5CH_2C(SO_3H){=}CHCl$ (37), $C_6H_5CH_2C(SO_3H){=}CCl_2$ (22)	48

Note: References 126–131 are on p. 71.

TABLE V. BASE-PROMOTED REARRANGEMENTS OF α,α,α-TRICHLOROMETHYL SULFONES (*Continued*)

No. of C Atoms	Substrate	Base	Solvent	Temperature, °C	Product(s) and Yield(s) (%)	
C_{10}	$C_6H_5(CH_2)_3SO_2CCl_3$	NaOH	H_2O–THF	65	$C_6H_5(CH_2)_2C{\equiv}CH$ (1), $C_6H_5(CH_2)_2CH{=}CHCl$ (1), $C_6H_5(CH_2)_2CH{=}CCl_2$ (1), $C_6H_5(CH_2)_2C(SO_3H){=}CHCl$ (66)	48

TABLE VI. TETRAHALO SULFONE REARRANGEMENTS

No. of C Atoms	Substrate	Halophile	Solvent	Temperature, °C	Product(s) and Yield(s) (%)	Refs.
C_4	$CH_3CBr_2SO_2CBr_2CH_3$	$(C_6H_5)_3P$	CH_2Cl_2	0	2,3-dimethylthiirene 1,1-dioxide: CH_3, CH_3, S, O_2 (50)	131
C_6	$C_2H_5CBr_2SO_2CBr_2C_2H_5$	$[(CH_3)_2N]_3P$	CH_2Cl_2	−70	2,3-diethylthiirene 1,1-dioxide: C_2H_5, C_2H_5, S, O_2 (89)	131

Note: References 126–131 are on p. 71.

REFERENCES TO TABLES II–VI

[126] R. G. Carlson and K. D. Mory, *Tetrahedron Lett.*, **1975,** 947.
[127] R. E. Wingard, Jr., R. K. Russell, and L. A. Paquette, *J. Amer. Chem. Soc.*, **96,** 7474 (1974).
[128] T. Kempe, Ohio State University, unpublished observations.
[129] T. G. Wallis, Ohio State University, unpublished observations.
[130] H. J. J.-B. Martel and M. Rasmussen, *Tetrahedron Lett.*, **1971,** 3843.
[131] L. A. Carpino and J. R. Williams, *J. Org. Chem.*, **39,** 2320 (1974).

CHAPTER 2

SYNTHETIC APPLICATIONS OF PHOSPHORYL-STABILIZED ANIONS

William S. Wadsworth, Jr.*

South Dakota State University, Brookings, South Dakota

CONTENTS

PAGE

* I wish to thank Ms. Carolyn Sidor of E. I. du Pont de Nemours and Company for much help with the literature search.

INTRODUCTION

The successful synthesis of a wide variety of unsaturated compounds by treatment of phosphoranes with carbonyl compounds (the Wittig reaction) has stimulated the search for other synthetic methods that

$$R_3\overset{+}{P}\text{—}\overset{-}{C}HR^1 + R^2R^3CO \rightarrow R^2R^3C\text{=}CHR^1 + R_3P(O)$$

employ organophosphorus reagents. One of the most fruitful results has been the discovery that phosphoryl-stabilized carbanions have wide applicability in the preparation of unsaturated compounds and offer significant advantages over conventional procedures.

$$R_2\overset{\overset{O}{\|}}{P}\overset{-}{C}HR^1 \longleftrightarrow R_2\overset{\overset{O^-}{|}}{P}\text{=}CHR^1 \xrightarrow{R^2R^3CO} R^2R^3C\text{=}CHR^1 + R_2P(O)O^-$$

This review discusses the chemistry of carbanions stabilized by delocalization of their negative charge by means of a P(O) group. Carbanions obtained by treating phosphine oxides, $R_2P(O)CH_2R^1$, phosphinates, $RP(OR)(O)CH_2R^1$, and phosphonates, $(RO)_2P(O)CH_2R^1$, with a base have been explored in depth. In particular, the phosphonates have found great popularity because of their availability and ease of application. The use of Wittig-type reagents in synthesis has been reviewed and is not mentioned except by way of comparison.[1]

Carbanions prepared from phosphinates have no known advantage over those prepared from phosphonates or phosphine oxides. Since the phosphinate carbanions are more difficult to prepare than either of the latter two, the phosphinate carbanions have claimed little attention and are not considered further. Those carbanions obtained from phosphine oxides will be considered only in the way of comparison.

Anions that have their negative charge located on an atom adjacent to a P(O) group gain stability owing to delocalization of the charge through the phosphoryl group. Although the exact nature of the delocalization is not well defined, the ability of phosphorus to become pentacovalent by use of its *d* orbitals is undoubtedly a contributing factor. It should be noted that the P(O) group is generally not so effective as the carbonyl group in stabilizing a negative charge. If other stabilizing groups are not present, formation of P(O)-stabilized carbanions generally requires more basic conditions than does the removal of a proton from a carbon atom alpha to a carbonyl group.

Horner and co-workers were the first to investigate the synthetic utility of P(O)-stabilized carbanions.[2] Under relatively vigorous basic conditions (potassium *t*-butoxide in refluxing toluene) they were able to obtain olefins by adding aldehydes to phosphine oxides having a hydrogen atom on the carbon atom adjacent to phosphorus. Under similar conditions the same workers obtained stilbene from the treatment of diethyl benzylphosphonate with benzaldehyde and ethyl 5-phenyl-2,4-pentadienoate from the proper starting phosphonate and benzaldehyde. Later ethyl cinnamate was obtained in low yield upon

$$(C_2H_5O)_2P(O)CH{=}CHCH_2CO_2C_2H_5 + C_6H_5CHO \xrightarrow[C_6H_5CH_3]{t\text{-}C_4H_9OK} C_6H_5CH{=}CHCH{=}CHCO_2C_2H_5 + (C_2H_5O)_2PO_2H$$

treatment of ethyl diethylphosphonoacetate with benzaldehyde in the presence of piperidine.[3] It was incorrectly assumed that the product

[1] U. Schöllkopf, *Angew. Chem.*, **71,** 260 (1959).

[2] L. Horner, H. Hoffman, H. G. Wippel, and G. Klahre, *Chem. Ber.*, **92,** 2499 (1959).

[3] S. Patai and A. Schwartz, *J. Org. Chem.*, **25,** 1232 (1960).

arose upon hydrolysis and subsequent decomposition of a benzylidene intermediate, $C_6H_5CH{=}C(CO_2C_2H_5)P(O)(OC_2H_5)_2$. The utility of phosphonate carbanions as synthetic intermediates was expanded by the discovery that, as olefin-forming reagents, phosphonates have certain advantages over both phosphoranes and phosphine oxides.[4] As a result, the method has found favor because of the availability of reagents, ease of workup, and convenient reaction conditions. In particular, the reactivity of phosphonate carbanions has been taken advantage of in numerous ways. As indicated by a number of reviews, these carbanions have become one of the most frequently employed organophosphorus reagents.[5–9]

The utility of phosphoryl-stabilized anions is apparent from the variety of materials that can be prepared by suitable structural modifications. Besides simple monoolefins and polyenes, a partial list would include allenes, unsaturated amides, aldehydes, esters, sulfides, sulfones, and nitrogen compounds such as imines, isocyanates, and ketenimines. In addition to their convenience for introducing unsaturation, these stabilized anions offer routes to cyclopropanes and heterocyclic systems. In this chapter, reactions that are of unusual interest are discussed separately; more routine reactions are included only in the tabular survey.

Olefin formation by means of phosphonate carbanions has been referred to as the Horner-Emmons or Wadsworth-Emmons modification of the Wittig reaction, whereas the use of phosphine oxide or phosphinate carbanions is generally referred to as the Horner modification. In order to prevent confusion and to avoid misdirecting credit, proper names are omitted and the synthesis of an olefin by phosphonate carbanions is termed phosphonate-olefin formation. Likewise, the term phosphine oxide-olefin formation is employed where appropriate.

MECHANISM

Although phosphonate anions and, to a much lesser extent, phosphine oxide anions have been used in a number of diverse syntheses, olefin formation has commanded by far the most attention. The mechanism of this reaction is discussed in detail. Mechanisms of other

[4] W. S. Wadsworth, Jr., and W. D. Emmons, *J. Amer. Chem. Soc.*, **83,** 1733 (1961).
[5] J. Boutagy and R. Thomas, *Chem. Rev.*, **74,** 87 (1974).
[6] G. Sturtz, *Colloq. Int. Cent. Nat. Rech. Sci.*, **1970,** 217 [*C.A.*, **74,** 22929q (1971)].
[7] A. V. Dombrovskii and V. A. Dombrovskii, *Russ. Chem. Rev.*, **35,** 733 (1966).
[8] H. O. Huisman, *Chem. Weekbl.*, **59,** 133 (1963).
[9] A. Hantz, *Stud. Cercet. Chim*, **1968,** 665 [*C.A.*, **70,** 46516w (1969)].

sequences that involve phosphoryl-stabilized anions are elaborated upon as presented.

The mechanism widely accepted for phosphonate-olefin formation is analogous to that of the Wittig reaction. The first step, the formation

$$(C_2H_5O)_2P(O)CHRY + B^- \rightleftharpoons (C_2H_5O)_2P(O)\bar{C}RY + BH$$

$$(C_2H_5O)_2P(O)\bar{C}RY + R^1CHO$$

k'_E, $-k'_E$; k'_T, $-k'_T$

$(C_2H_5O)_2P(O)$—C(R)(Y)—C(H)(R^1)—$\bar{O}$ (*erytho*) $\rightleftharpoons$ ($R = H, Y = CN$, $R^1 = C_6H_5$) $(C_2H_5O)_2P(O)$—C(R)(Y)—C(R^1)(H)—$\bar{O}$ (*threo*)

k_{E^1} ↓ ; k_{T^1} ↓

cyclic $(C_2H_5O)_2P(O^-)$—C(R)(Y)—C(H)(R^1)—O ; cyclic $(C_2H_5O)_2P(O^-)$—C(R)(Y)—C(R^1)(H)—O

↓ ; ↓

$R^1(H)C{=}C(Y)(R)$ + $(C_2H_5O)_2P(O)O^-$ (Z) ; $R^1(H)C{=}C(R)(Y)$ + $(C_2H_5O)_2P(O)O^-$ (E)

(Eq. 1)

of the anion stabilized by an electron-withdrawing group, Y, and the P(O) function, may or may not be an equilibrium process depending upon the base used. With sodium hydride, anion formation is irreversible. The second stage is a reversible aldol condensation that gives two possible diastereoisomeric oxyanions. There is no direct evidence that cyclic intermediates are actually involved, but the oxyanions are thought to decompose via a *syn* elimination in an irreversible fashion to give olefins. Thus the *erythro* form should lead stereospecifically to the Z olefin, while the *threo* form should lead to the E olefin (Eq. 1). The driving force for the elimination might be the formation of the new phosphorus to oxygen bond in the phosphate.

There is strong supporting evidence for the intermediacy of an oxyanion, particularly where β-hydroxyphosphonates (protonated oxyanions) have been isolated. For example, anions have been formed from alkylphosphonates in which the phosphonate does not contain an additional α-electron-withdrawing group.[10] A β-hydroxyphosphonate is obtained in high yield upon addition of benzophenone to the anion prepared from treatment of the alkylphosphonate with n-butyllithium, and subsequent hydrolysis. Neither the adduct nor its conjugate base,

$$CH_3CH_2P(O)(OCH_3)_2 \xrightarrow[\text{2. } (C_6H_5)_2CO \text{ 3. } H_2O]{\text{1. } n\text{-}C_4H_9Li} (C_6H_5)_2C(OH)CH(CH_3)P(O)(OCH_3)_2$$

however, undergoes cycloelimination to form an olefin. In contrast, isolated β-hydroxyphosphonic acid bisamides have been reported to form quantitatively in a similar manner and undergo elimination upon heating.[11] Also, when heated, the diastereomers of the β-hydroxy adducts yield olefins in a stereospecific manner. The *cis* form of 1-phenylpropene is obtained by refluxing a toluene solution of the major diastereomer of the appropriate β-hydroxyphosphonic acid bisamide.

$$C_6H_5CH(OH)CH(CH_3)P(O)[N(CH_3)_2]_2 \xrightarrow{\text{Heat}} C_6H_5CH{=}CHCH_3 + [(CH_3)_2N]_2PO_2H$$

The oxyanions obtained from α-lithio salts of alkylphosphonothioates and ketones but not aldehydes decompose to olefins under milder conditions.[10]

$$CH_3P(S)(OCH_3)_2 \xrightarrow[(C_6H_5)_2CO]{n\text{-}C_4H_9Li} (C_6H_5)_2C(O^-)CH_2P(S)(OCH_3)_2 \longrightarrow (C_6H_5)_2C{=}CH_2 + (CH_3O)_2P(S)O^-$$

No sodium or lithium salt of an oxyanion intermediate has been isolated where the alkyl portion carries an additional α-electron-withdrawing group, Y = CN, CO_2R, or COR. Under the normal basic conditions of phosphonate-olefin formation these oxyanions, even at −100°, spontaneously decompose into olefins. Consequently, upon hydrolysis of reaction mixtures, β-hydroxyphosphonates are not isolated. The rate enhancement of cycloelimination by an electron-

[10] E. J. Corey and G. T. Kwiatkowski, *J. Amer. Chem. Soc.*, **88**, 5654 (1966).

[11] E. J. Corey and G. T. Kwiatkowski, *J. Amer. Chem. Soc.*, **88**, 5652 (1966).

withdrawing group might be interpreted as indicating that the transition state for the elimination has considerable ionic character.[12] It has been reasoned that, since magnesium salts are more covalent than sodium or lithium salts, the intermediate oxyanions with magnesium as the cation should be less reactive and more stable than the comparable sodium or lithium salts.[13] Such is the case, for upon condensation of the organomagnesium derivative of a phosphonitrile with benzaldehyde followed by hydrolysis, a mixture of separable diastereomers is obtained. Whereas condensation of the phosphonate (R = H) with benzaldehyde under normal phosphonate-olefin formation conditions gives a mixture of *cis* and *trans* cinnamonitriles in the ratio 15/85,

$$(C_6H_5O)_2P(O)CH(R)CN \xrightarrow[\text{2. } C_6H_5CHO \text{ 3. } H_2O]{\text{1. } i\text{-}C_3H_7MgCl} (C_2H_5O)_2P(O)\text{—}C(R)(CN)\text{—}C(H)(C_6H_5)\text{—OH} + (C_2H_5O)_2P(O)\text{—}C(R)(CN)\text{—}C(C_6H_5)(H)\text{—OH}$$

thermal decomposition of the first adduct (R = H) affords a *cis* to *trans* ratio of 80/20 and the second (R = H) yields a mixture of the same compounds in the ratio 10/90. Thus cycloelimination from the β-hydroxyphosphonates is highly stereospecific and occurs primarily in a *syn* fashion.

Using a different approach for the preparation of β-hydroxyphosphonates, it has been found that reduction of an α-ketophosphonate prepared by condensation of an anion with an acyl halide gives a mixture of diastereomers which, because of instability, cannot be separated.[14] Conversion of the mixture into an anion with sodium ethoxide in ethanol gives the *trans* ester in yields of over 90%

[12] G. Sturtz, *Bull. Soc. Chim. Fr.*, **1964,** 2349.
[13] B. Deschamps, G. Lefebvre, and J. Seyden-Penne, *Tetrahedron,* **28,** 4290 (1972).
[14] G. Durrant and J. K. Sutherland, *J. Chem. Soc., Perkin I,* **1972,** 2582.

(Eq. 2). Treatment of the mixture of alcohols with acetic anhydride

$$(C_2H_5O)_2P(O)\bar{C}HCO_2C_2H_5 + n\text{-}C_5H_{11}COCl \longrightarrow n\text{-}C_5H_{11}CO\underset{\displaystyle (O)P(OC_2H_5)_2}{\underset{|}{C}}HCO_2C_2H_5 \xrightarrow[C_2H_5OH]{H_2,\ Pd}$$

$$n\text{-}C_5H_{11}CH(OH)CH(CO_2C_2H_5)P(O)(OC_2H_5)_2 \xrightarrow[C_2H_5OH]{NaOC_2H_5} \begin{matrix} n\text{-}C_5H_{11} & & H \\ & C{=}C & \\ H & & CO_2C_2H_5 \end{matrix} \quad \text{(Eq. 2)}$$

and pyridine yields the dehydration product, a vinyl phosphonate (Eq. 3). More recently the successful synthesis and purification of dias-

$$n\text{-}C_5H_{11}CH(OH)CH(CO_2C_2H_5)P(O)(OC_2H_5)_2 \xrightarrow[C_5H_5N]{(CH_3CO)_2O} n\text{-}C_5H_{11}CH{=}C(CO_2C_2H_5)P(O)(OC_2H_5)_2 \quad \text{(Eq. 3)}$$

tereomers of an α-carbethoxy-β-hydroxyphosphonate has been reported.[15] The latter were prepared in a manner similar to their nitrile counterparts: condensation of the magnesium salt of a phosphonoester with benzaldehyde followed by hydrolysis.

syn Elimination from a β-hydroxyphosphine oxide has recently been employed in a novel method of olefin inversion.[16] It is possible to convert *cis* into *trans* olefins by an overall four-step sequence. For example, pure *trans*-1-methylcyclooctene was prepared from the *cis* isomer (Eq. 4). The success of the method supports the argument in favor of a stereospecific *syn* elimination.

$$\begin{matrix} R & & R \\ & C{=}C & \\ H & & H \end{matrix} \longrightarrow \begin{matrix} R & & R \\ & \overset{}{\underset{O}{C{-}C}} & \\ H & & H \end{matrix} \xrightarrow[2.\ AcOH{-}H_2O_2]{1.\ (C_6H_5)_2PLi} \begin{matrix} H & & P(O)(C_6H_5)_2 \\ R & C{-}C & H \\ HO & & R \end{matrix} \xrightarrow{NaH} \begin{matrix} R & & H \\ & C{=}C & \\ H & & R \end{matrix} \quad \text{(Eq. 4)}$$

[15] T. Bottin-Strzalko, *Tetrahedron*, **29**, 4199 (1973).

[16] A. J. Bridges and G. H. Whitham, *Chem. Commun.*, **1974,** 142.

From early studies, phosphonate-olefin formation was believed to yield only *trans* isomers.[2,17] In the many subsequent reports of phosphonate-olefin formation in which *cis* isomer was detected, the stereochemistry of the reaction still favored the *trans* isomer. Indeed in many reactions the *trans* isomer is produced with almost complete exclusion of the *cis*. A number of workers have found that by modifying either the structure of the anion or carbonyl co-reactant the final isomer ratio can be made to vary. The treatment of citral with an anion having hydrogen or a methyl group on the carbon atom next to phosphorus gives only a *trans* ester.[18] Upon increasing the size of the phosphonate side chain ($R = C_2H_5$) appreciable quantities of both *cis* and *trans* isomers are obtained.

$$(CH_3)_2C{=}CH(CH_2)_2C(CH_3){=}CHCHO + (C_2H_5O)_2P(O)\bar{C}RCOR^1 \rightarrow C_9H_{15}CH{=}CRCOR^1$$

A number of reactions have been reported in which the *cis*/*trans* ratio is found to vary with the size of the substituents attached to the carbonyl carbon atom. The isomer ratio of unsaturated nitriles obtained by treating diethyl cyanomethylphosphonate carbanion with various alkyl aryl ketones has been studied quantitatively.[19] When the alkyl group is unbranched, the *trans* or E isomer predominates. Branching of the alkyl group or *ortho* substitution on the aryl group increases the proportion of *cis* or Z isomer formed. This latter isomer predominates with the *t*-butyl

$$(C_2H_5O)_2P(O)\bar{C}HCN + C_6H_5COR \longrightarrow \underset{cis\text{ or Z}}{\overset{H\quad\;\; R}{\underset{NC\quad\;\; C_6H_5}{C{=}C}}} + \underset{trans\text{ or E}}{\overset{H\quad\;\; C_6H_5}{\underset{NC\quad\;\; R}{C{=}C}}}$$

group. In the reaction of diethyl carboethoxylalkyl phosphonate carbanions with aliphatic aldehydes the product isomer ratios depend upon the size of both R and R^1.[20] Again, when R is hydrogen, only *trans*

$$(C_2H_5O)_2P(O)\bar{C}(R)CO_2C_2H_5 + R^1CHO \rightarrow R^1CH{=}C(R)CO_2C_2H_5$$

isomers are obtained from all aldehydes. On the other hand, condensation of an anion that has an α-methyl group with isobutyraldehyde produces an isomer mixture containing 65% of the Z isomer.

[17] L. Horner, W. Klink, and H. Hoffman, *Chem. Ber.*, **96,** 3133 (1963).
[18] K. Sasaki, *Bull. Chem. Soc. Jpn.*, **39,** 2703 (1966).
[19] G. Jones and R. F. Maisey, *Chem. Commun.*, **1968,** 543.
[20] T. H. Kinstle and B. Y. Mandanas, *Chem. Commun.*, **1968,** 1969.

The effect of structural changes on the isomer ratio may best be explained by assuming a difference in stability, due primarily to bulk effects, between the two possible diastereomeric oxyanion intermediates. If the stereoselectivity of the aldol condensation is such that both oxyanions form and the condensation is reversible, then the amount of Z isomer formed should be dependent upon the relative magnitude of the various rate constants. With aliphatic aldehydes it is assumed that the condensation is essentially irreversible when the chains are branched, whereas with benzaldehyde the condensation is highly reversible and the product ratio is thermodynamically controlled.

The reversibility of the condensation between anion and carbonyl reactant has been amply documented. In a novel approach, β-hydroxyphosphonate oxyanions have been prepared *in situ* by nucleophilic addition of hydroxide ion to esters of α-cyanovinylphosphonic acids.[21] The oxyanions underwent both elimination and retroaldolization. The latter process, which is the reverse of the first stage of phosphonate-olefin formation, was established by detection of benzophenone in the hydrolysis mixture.

$$(C_6H_5)_2C{=}C(CN)P(O)(OC_2H_5)_2 + OH^- \longrightarrow (C_6H_5)_2\overset{O^-}{\overset{|}{C}}CH(CN)P(O)(OC_2H_5)_2 \xrightarrow{H_2O}$$

$$(C_2H_5O)_2P(O)CH_2CN + \underset{(20\%)}{(C_6H_5)_2CO} + (C_6H_5)_2C{=}CHCN + (C_2H_5O)_2PO_2H$$

The degree of retroaldolization varies with the nature of the substituents in the potential ketone or aldehyde. For example, aliphatic aldehydes are produced in only trace amounts upon addition of hydroxide ion to α-cyanovinyl phosphonates. To provide further evidence for the reversibility of the first step, the hydroxide ion addition was carried out in the presence of benzaldehyde.[22] Mixtures of cinnamonitrile and diphenylacrylonitrile were obtained, again indicating the reversible formation of phosphonate carbanion. In a similar fashion both diastereomers of a β-hydroxyphosphonate, when treated with base under phosphonate-olefin formation conditions, revert partially to aldehydes and phosphonate carbanions as evidenced by the detection of aldehyde, R^1CHO ($R^1 = C_6H_5$), in the basic media.[13] Also, both diastereomers ($R = H$, $Y = CN$, $R^1 = C_6H_5$), when converted into their oxyanion in the presence of an excess of aldehyde ($R^2 = p\text{-}ClC_6H_4$), produce a mixture of olefins. The reversibility is not total, for the ratio of $R^1CH{=}CRY$ to $R^2CH{=}CRY$, starting with a single diastereomer,

[21] D. Danion and R. Carrie, *Tetrahedron Lett.*, **1968,** 4537.

[22] D. Danion and R. Carrie, *Tetrahedron,* **28,** 4223 (1972).

is greater than that found for a direct reaction between phosphonate carbanion and a mixture containing an equivalent ratio of aldehydes.

$$R^1CH(OH)CRYP(O)(OC_2H_5)_2 \xrightarrow{NaH} R^1CH(O^-)CRYP(O)(OC_2H_5)_2 \longrightarrow R^1CH{=}CRY$$

$$R^1CH(O^-)CRYP(O)(OC_2H_5)_2 \rightleftharpoons R^1CHO + (C_2H_5O)_2P(O)\bar{C}RY \xrightarrow{R^2CHO} R^2CH{=}CRY$$

It has been reported recently that the formation of oxyanions from phosphonoesters and benzaldehyde is less complete than the formation of oxyanions from phosphononitriles and benzaldehyde.[15] Interaction between the phenyl and nitrile or ester group may develop during oxyanion formation. Since the ester group is more bulky than the nitrile, it causes a rate depression. The bulk effect would be felt particularly in the *erythro* diastereomer as it yields the *cis* or Z olefin. It would not be so great in the less hindered *threo* form. Because both k_E1 and k_T1 are less for the phosphonoesters than for phosphononitriles, retroaldolization becomes more pronounced in the former. The stereoselectivity of olefin formation depends at least in part upon the k_E1/k_T1 ratio. Since the ratio would be expected to be smaller for the ester, the ester would produce more of the *trans* olefin than the nitrile. In the formation of cinnamates by phosphonate-olefin formation, the *trans* isomer is formed almost exclusively.

A further complication of the proposed mechanism is that certain diastereomeric oxyanions, $R = H$, $R^1 = C_6H_5$, $Y = CN$ (see Eq. 1), interconvert directly.[13] Epimerization of an oxyanion is possible only when α hydrogen is present. As a consequence the stereospecificity of cycloelimination from a single β-hydroxyphosphonate diastereomer under basic conditions is less when $R = H$ than when $R = CH_3$. Where epimerization does occur, the equilibrium between diastereomers is presumed to be rapid and the total stereochemical balance depends only upon the k_E1/k_T1 ratio.

Although a number of investigators have stated that the stereochemistry of olefinic products prepared by phosphonate-olefin formation does not vary with polarity of the medium, recent data show that it does.[19,22–24] It has been observed that when epimerization of the diastereomeric oxyanions is impossible, as when there is only one α hydrogen in the starting phosphonate, the stereochemistry

[23] L. D. Bergelson and M. M. Shemyzkin, *Pure Appl. Chem.*, **9,** 271 (1964).

[24] D. H. Wadsworth, O. E. Schupp, E. J. Deus, and J. A. Ford, Jr., *J. Org. Chem.*, **30,** 680 (1965).

of the reaction does vary with the solvent.[25,26] The variations with solvent were attributed to a change in the degree of reversibility of the aldol condensation. Different isomeric ratios are obtained when the oxyanions are either highly associated with a cation or solvated and reversibility is appreciable, as compared with polar, aprotic solvents where the oxyanions are not solvated and reversibility is small.

In recent work it has been found that nucleophilic substitutions at phosphorus may proceed with both retention and inversion, the latter being enhanced by added cations.[27,28] The cations may stabilize a pentacovalent intermediate, which structurally is not dissimilar to the cyclic intermediate that may take part in phosphonate-olefin formation (see Eq. 1). Factors that stabilize the cyclic intermediate, such as the degree of coordination with a cation are influenced by solvent polarity, and may determine the stereochemical ratio of products formed.

In this regard a number of findings are relevant. Acetone adds to the anion of N,N,N′,N′-tetramethylallylphosphonodiamide at both possible positions when the cation is magnesium but only at the γ position when lithium is used.[29] Magnesium may differ from lithium in its ability to stabilize a cyclic pentacovalent intermediate anion that may revert to the alcohol upon hydrolysis. Recently the effects of cation and temperature have been described in some detail.[30] At 20°

$$P(O)[N(CH_3)_2]_2 \xrightarrow[\text{2. } MgBr_2]{\text{1. } n\text{-}C_4H_9Li} \xrightarrow[\text{4. } H^+]{\text{3. } (CH_3)_2CO} P(O)[N(CH_3)_2]_2 \text{ (OH)} + \text{OH} \; P(O)[N(CH_3)_2]_2$$

the Z/E ratio of isomeric α-methylcinnamonitriles produced from these reactants is changed from 25/75 with lithium as the cation to 60/40 with potassium. It is obvious that more attention needs to be given to this area, and perhaps a study of the effect of added excess cations on isomer ratios would be a good place to start.

[25] J. Seyden-Penne and G. Lefebvre, *C.R. Acad. Sci., Ser. C*, **1969,** 48 [*C.A.*, **71,** 90 618a (1969)].

[26] B. Deschamps, G. Lefebvre, A. Redjal, and J. Seyden-Penne, *Tetrahedron*, **29,** 2437 (1973).

[27] W. S. Wadsworth, Jr., S. Larsen, and H. L. Horten, *J. Org. Chem.*, **38,** 256 (1973).

[28] W. S. Wadsworth, Jr., *J. Org. Chem.*, **38,** 2921 (1973).

[29] E. J. Corey and D. E. Cane, *J. Org. Chem.*, **34,** 3053 (1969).

[30] A. Redjal and J. Seyden-Penne, *Tetrahedron Lett.*, **1974,** 1733.

In the preparation of the (±)-pyrethric acids the ratio of isomers varies with the size of the alkoxyl group attached to phosphorus as well as with temperature (Eq. 5).[31] Again the bulkiness of the ester alkyl

$$(RO)_2P(O)\bar{C}(CH_3)CO_2CH_3 + \text{OHC-cyclopropane-}CO_2C_4H_9\text{-}t \longrightarrow$$

$$CH_3O_2C\text{-alkene-cyclopropane-}CO_2C_4H_9\text{-}t + CH_3O_2C\text{-alkene-cyclopropane-}CO_2C_4H_9\text{-}t \quad \text{(Eq. 5)}$$

groups and their effect on the equilibrium between oxyanions is believed to be the cause. When the R groups attached to phosphorus as RO are methyl, the mixture of olefinic products contains 34.7% of the Z isomer; when R is isopropyl the percentage drops to 19%. Since the alkoxyl groups on phosphorus are quite far removed from the reaction center, it is difficult to explain why the effect arises only from a difference in the position of equilibrium of the oxyanions. A change in relative stability of possible cyclic intermediates and perhaps a change in their rates of decomposition would appear to be more comprehensible.

SCOPE AND LIMITATIONS

Since P(O)-stabilized anions have been employed primarily to widen and extend the scope of the Wittig reaction, their use is presented in some detail. The effect of structural modification in the phosphoryl-containing substrate is discussed and a presentation of selected syntheses that result from the reactions of many diverse carbonyl co-reactants follows.

The Phosphoryl Reagent

Perhaps because of the emphasis placed on comparisons with analogous Wittig reagents, few studies have been made of different P(O)-stabilizing groups with respect to their reactivity in olefin formation. The relative reactivity of phosphinates and phosphonates has

[31] T. Sugiyama, A. Kobayashi, and K. Yamashita, *Agri. Biol. Chem.* **36,** 565 (1972).

been determined, however, in one case.[32] Treatment of ethyl bis-(diethylphosphonomethyl)phosphinate with an aldehyde in the presence of base yields only diethylvinylphosphonate. No explanation is apparent for the preferential cleavage of the carbon-phosphorus bond of a phosphinate over that of a phosphonate.

$$\mathrm{RCHO} + (\mathrm{C_2H_9O})_2\mathrm{P(O)}\bar{\mathrm{C}}\mathrm{HP(O)(OC_2H_5)CH_2P(O)(OC_2H_5)_2} \longrightarrow$$

$$\mathrm{RCH{=}CHP(O)(OC_2H_5)_2} + (\mathrm{C_2H_5O})_2\mathrm{P(O)CH_2P(O)(OC_2H_5)O^-}$$

Owing to this limited study, broad conclusions are difficult to reach with regard to the relative reactivity of phosphonate carbanions, phosphinate carbanions, and phosphine oxide carbanions. Yet, comparisons can be drawn between reagents that contain the identical $\mathrm{P(O)^-}$ activating group but differ in the nature of the alkyl group attached directly to the phosphorus atom. Electron-withdrawing groups in $(\mathrm{C_2H_5O})_2\mathrm{P(O)CH_2R}$, *e.g.*, $\mathrm{R = CN}$, $\mathrm{CO_2R^1}$, or $\mathrm{COR^1}$, increase the reactivity of the phosphonate carbanion as a reactant in olefin formation. When electron-withdrawing groups are absent ($\mathrm{R = H}$ or alkyl), cycloelimination from isolable betaine intermediates or their conjugate acids does not occur under a wide range of conditions. Both α-lithioalkylphosphonothioates and α-lithioalkylphosphonic acid bisamides are exceptions to this observation. Although adducts formed from them and ketones do produce olefins in good yields, the adducts decompose under more stringent conditions than those required for intermediate betaines containing an additional α-electron-withdrawing group.[10, 11, 33–35]

The isolable adducts prepared from alkylphosphonothioate esters and aldehydes do not afford olefins even under drastic conditions. Since adducts from ketones do form olefins, a bulk effect is assumed. Similarly, it is reported that the anion from diethyl cyanomethylphosphonate when treated with aldehydes or ketones gives higher yields of olefins than the anion prepared from triethyl phosphonoacetate.[36] The effect is attributed to the greater bulk of the latter. Bulkiness is assumed to hinder oxyanion formation, although once the adduct is

[32] W. F. Gilmore and J. W. Hubert, *J. Org. Chem.*, **38,** 1423 (1973).
[33] E. J. Corey and G. T. Kwiatkowski, *J. Amer. Chem. Soc.*, **90,** 6816 (1968).
[34] E. J. Corey and G. T. Kwiatkowski, *J. Amer. Chem. Soc.*, **88,** 5653 (1966).
[35] E. J. Corey and J. I. Shulman, *J. Org. Chem.*, **35,** 777 (1970).
[36] A. K. Bose and R. T. Dahill, Jr., *J. Org. Chem.*, **30,** 505 (1965).

formed bulky groups aid oxyanion decomposition. Certain other phosphonate carbanions are inert to olefin formation owing to stringent steric requirements. For example, bridged ketones give only low yields of the expected α,β-unsaturated ester when treated with diethyl carbomethoxymethylphosphonate carbanion.[37] Similarly, it was shown

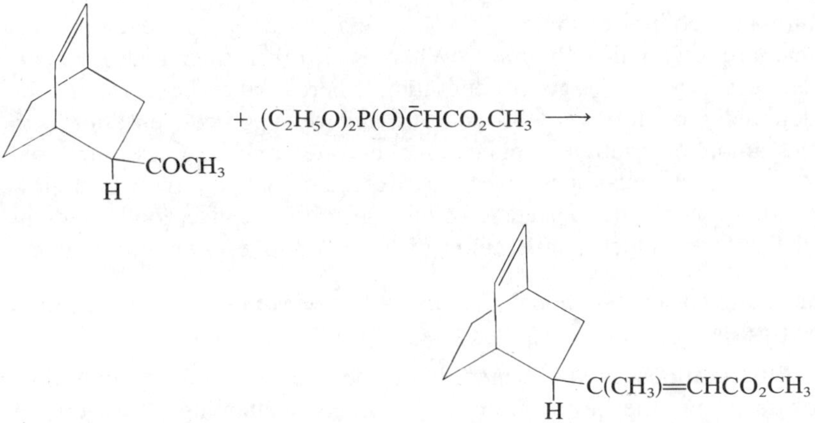

$$+ (C_2H_5O)_2P(O)\bar{C}HCO_2CH_3 \longrightarrow$$

COCH$_3$
H

C(CH$_3$)=CHCO$_2$CH$_3$
H

that 20-oxo-21-methyl steroids are unreactive toward the same carbanion although the 20-oxo-21-hydroxypregnanes react exothermally to give the corresponding cardenolids.[38] The latter reaction may result from a thermodynamically favorable ring formation that cannot occur in the absence of the 21-hydroxyl group.

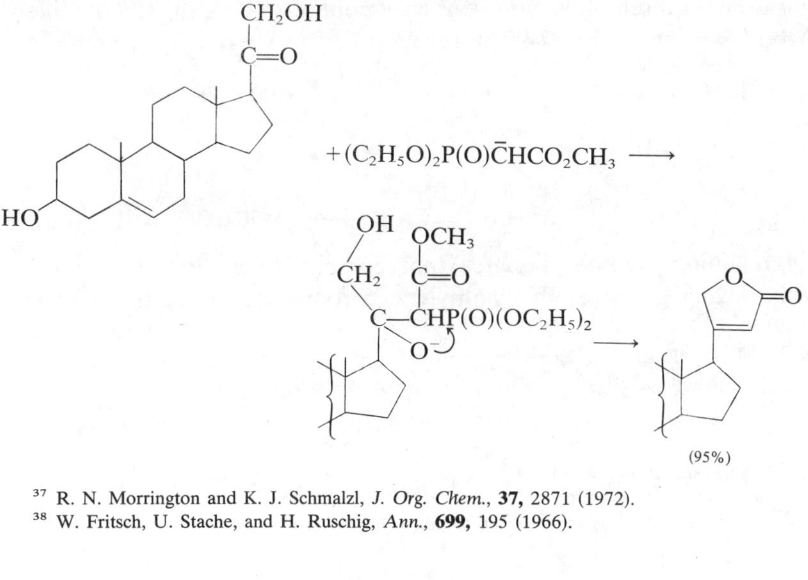

CH$_2$OH
C=O
HO

$$+ (C_2H_5O)_2P(O)\bar{C}HCO_2CH_3 \longrightarrow$$

OH OCH$_3$
CH$_2$ C=O
C—CHP(O)(OC$_2$H$_5$)$_2$
O$^-$

(95%)

[37] R. N. Morrington and K. J. Schmalzl, *J. Org. Chem.*, **37**, 2871 (1972).
[38] W. Fritsch, U. Stache, and H. Ruschig, *Ann.*, **699**, 195 (1966).

The effect of α-alkyl substituents on the reactivity of phosphonates has been determined.[39] As an example, both aromatic and aliphatic ketones give good yields of unsaturated amides upon reaction with phosphonacetamide anions (Eq. 6). Aliphatic aldehydes produce mix-

$$(C_2H_5O)_2P(O)\bar{C}RCONHR^1 + R^2R^3CO \rightarrow R^2R^3C{=}CRCONHR^1 \quad \text{(Eq. 6)}$$

tures of geometric isomers, while aromatic aldehydes yield products that are predominately *trans.* When R is other than hydrogen, aldehydes again undergo condensation whereas alicyclic ketones, undoubtedly for steric reasons, do not. Similarly, upon condensation with the anion of triethyl α-phosphonobutyrate, only the less hindered ketone, cyclohexanone, gives the expected olefin, although in low yield. As might be expected, ketones also do not give good yields of olefins when added to the anion of triethyl α-phosphonopropionate.

Introduction of Functional Groups by Modification of the Phosphoryl Substrate

Modifications in the structure of the phosphonate or phosphine oxide allow the preparation of products containing a variety of functional groups. Although α,β-unsaturated carbonyl compounds are commonly the end products of olefin formation, a practical synthesis of saturated ketones has been reported.[40] The route entails the preparation and subsequent hydrolysis of enamines, the latter prepared from the proper P(O)-activated carbanion and aldehyde, or ketone. The method may be used as a simple chain-lengthening reaction, especially since the required phosphonate is easily obtained by a slightly modified Arbuzov reaction (Eq. 7). The reaction has been extended by employ-

$$(R^1CH{=}\overset{+}{N}R_2)Cl^- + (C_2H_5O)_3P \longrightarrow (C_2H_5O)_2P(O)CH(NR_2)R^1$$

$$(C_2H_5O)_2P(O)CH(NR_2)R^1 \xrightarrow[\text{2. } R^2R^3CO]{\text{1. NaH}}$$

$$R_2NC(R^1){=}CR^2R^3 \xrightarrow{H_2O} R^1COCHR^2R^3 \quad \text{(Eq. 7)}$$

ing α-aminoalkylphosphonates that contain an additional electron-withdrawing group at the methylene carbon atom.[41] In this manner

$$(C_2H_5O)_2P(O)CH(NR_2)COR^1 \xrightarrow[\text{2. } R^2CHO]{\text{1. NaH}}$$

$$R^2CH{=}C(NR_2)COR^1 \xrightarrow{H_2O} R^2CH_2COCOR^1$$

[39] I. Shakak, J. Almog, and E. D. Bergmann, *Isr. J. Chem.*, **7**, 585 (1969).

[40] H. Gross and E. Hoft, *Angew. Chem., Int. Ed. Engl.*, **6**, 353 (1967).

[41] H. Gross and W. Buerger, *J. Prakt. Chem.*, **311**, 395 (1969).

α-ketonic esters and amides are obtained upon hydrolysis of the intermediate imine. β-Ketoesters are obtained from α-aminoalkylphosphonates by condensation with an α-ketoester.

$$(C_2H_5O)_2P(O)CH(NHC_6H_5)C_6H_4NO_2\text{-}p \xrightarrow[2.\ HCOCO_2R]{1.\ NaH}$$

$$p\text{-}O_2NC_6H_4C(NHC_6H_5){=}CHCO_2R \xrightarrow{H_2O} p\text{-}O_2NC_6H_4COCH_2CO_2R$$

Enamine phosphine oxides have been used to prepare α,β-unsaturated ketones and aldehydes. The technique entails the synthesis of α,β-unsaturated ketimines or aldimines by olefin formation followed by hydrolysis. Enamine phosphine oxides are prepared by adding a primary amine to a 1-alkynylphosphine oxide (Eq. 8).[42–44] Overall yields of

$$(C_6H_5)_2P(O)C{\equiv}CR + R^1NH_2 \longrightarrow (C_6H_5P)_2(O)CH{=}C(NHR^1)R \xrightarrow{NaH}$$

$$(C_6H_5)_2P(O)CH{\cdots}C(\overset{-}{N}R^1)R \xrightarrow{R^2R^3CO}$$

$$R^2R^3C{=}CHC(R){=}NR^1 \xrightarrow{H_3O^+} R^2R^3C{=}CHCOR \quad \text{(Eq. 8)}$$

ketones or aldehydes up to 70% are reported based on starting phosphine oxide. Enamine phosphonates are generated in a similar fashion by addition of a primary amine to a diethyl alkynyl-1-phosphonate. Upon olefin formation followed by hydrolysis, unsaturated ketones are formed. These syntheses are particularly attractive since R may be alkyl or aryl.

$$(C_2H_5O)_2P(O)C{\equiv}CR + R^1NH_2 \longrightarrow$$

$$(C_2H_5O)_2P(O)CH{=}C(NHR^1)R \xrightarrow[2.\ R^2R^3CO]{1.\ NaH}$$

$$R^2R^3C{=}CHC(R){=}NR^1 \xrightarrow{H_3O^+} R^2R^3C{=}CHCOR$$

An interesting synthetic sequence leads to formyl olefin formation from carbonyl compounds.[45] The method produces unsaturated aldehydes in

[42] N. A. Portnoy, C. J. Morrow, M. S. Chatta, J. C. Williams, and A. M. Aguiar, *Tetrahedron Lett.*, **1971,** 1397.

[43] N. A. Portnoy, C. J. Morrow, M. S. Chatta, J. C. Williams, and A. M. Aguiar, *Tetrahedron Lett.*, **1971,** 1401.

[44] N. A. Portnoy, K. S. Youg, and A. M. Agwiar, *Tetrahedron Lett.*, **1971,** 2559.

[45] W. Nagata and Y. Hayase, *Tetrahedron Lett.*, **1968,** 4359.

yields up to 86%. The required enamine phosphonate is synthesized by

$$(C_2H_5O)_2P(O)CR{=}CHNHC_6H_{11} \xrightarrow[\text{2. } R^1R^2CO]{\text{1. NaH}}$$

$$R^1R^2C{=}C(R)CH{=}NC_6H_{11} \xrightarrow{H_3O^+} R^1R^2C{=}C(R)CHO$$

treating diethyl formylmethylphosphonate with cyclohexylamine followed, if desired, by alkylation (Eq. 9). Formyl olefin formation by

$$(C_2H_5)_2P(O)CH_2CHO \xrightarrow{C_6H_{11}NH_2}$$

$$(C_2H_5O)_2P(O)CH_2CH{=}NC_6H_{11} \xrightarrow[RX]{NaH} (C_2H_5O)_2P(O)CR{=}CHNHC_6H_{11} \quad \text{(Eq. 9)}$$

this procedure is reported to have the advantages of simplicity, high efficiency, and selectivity. Thus, whereas diethyl carbethoxymethylphosphonate carbanion does not condense with sterically hindered steroidal 17-ketones, the enamine phosphonate carbanions do react to give exclusively the *trans* enals after hydrolysis.

$$\text{Oxazine-}CH_2P(O)(OC_2H_5)_2 \xrightarrow[R^1R^2CO]{NaH} \text{oxazine-}CH{=}CR^1R^2 \xrightarrow[BH_4^-]{CH_3I} \text{N-methyl tetrahydro-oxazine-}CH{=}CR^1R^2 \xrightarrow{H_3O^+} R^1R^2C{=}CHCHO$$

$$[\text{N-methyl oxazinium-}CH{=}CR^1R^2]^+ I^- \xrightarrow{R^3Li} \text{2-}R^3\text{-N-}CH_3\text{ oxazine-}CH{=}CR^1R^2 \xrightarrow{H_3O^+} R^1R^2C{=}CHCOR^3$$

$$[\text{N-methyl oxazinium-}CH{=}CR^1R^2]^+ I^- \xrightarrow{OH^-} CH_3NH\text{-}C(CH_3)_2CH_2CH(CH_3)O_2CCH{=}CR^1R^2 \xrightarrow{H_3O^+} R^1R^2C{=}CHCO_2H$$

Vinyl oxazines, prepared by phosphonate-olefin formation, serve, after quaternization, as precursors to unsaturated aldehydes, ketones, and acids.[46] The starting phosphonate is prepared from 2-chloromethyloxazine. Yields of unsaturated aldehydes and ketones are reported to vary from 65 to 80%.

Sulfides, Sulfones, and Ethers. Further modification of the phosphoryl-stabilized anion has allowed the synthesis of α,β-unsaturated sulfides and sulfones. Diethyl (*p*-bromophenylthio)methylphosphonate, prepared from a chloromethyl sulfide by means of an Arbuzov synthesis, is readily oxidized by permanganate to the sulfone. Both the sulfide and the sulfone react with aromatic aldehydes to give the corresponding *trans* alkenes.[39,47,48] Likewise, treatment of al-

$$(C_2H_5O)_2P(O)CH_2SC_6H_4Br\text{-}p \xrightarrow{KMnO_4} (C_2H_5O)_2P(O)CH_2SO_2C_6H_4Br\text{-}p$$

$$\downarrow \text{1. NaH} \;|\; \text{2. RCHO} \qquad\qquad \downarrow \text{1. NaH} \;|\; \text{2. RCHO}$$

$$RCH{=}CHSC_6H_4Br\text{-}p \qquad\qquad RCH{=}CHSO_2C_6H_4Br\text{-}p$$

dehydes or ketones with methylthiomethylphosphonate or methylsulfonylmethylphosphonate anions gives *trans* alkenes.[49] For reaction with aldehydes, a mixture of the sulfide, sodium hydride, and aldehyde is heated under reflux in benzene, while dimethylformamide and temperatures to 100° are needed to convert ketones into mercaptoalkenes. *trans*-Ethyl styryl sulfone, $C_2H_5SO_2CH{=}CHC_6H_5$, can be prepared stereospecifically in 84% yield from ethylsulfonylmethylphosphonate.[50] In a comparable Knoevenagel-type synthesis, condensation of aldehydes with alkylsulfonylacetic acids followed by decarboxylation affords low yields of products with uncertain geometry.

A number of vinyl sulfides have been prepared from diethyl methylthiomethylphosphonates, which when hydrolyzed in the presence of

$$(C_2H_5O)_2P(O)CH_2SCH_3 \xrightarrow[\text{2. RX}]{\text{1. } n\text{-}C_4H_9Li} (C_2H_5O)_2P(O)CH(R)SCH_3 \xrightarrow[\text{2. } R^1R^2CO]{\text{1. } n\text{-}C_4H_9Li}$$

$$R^1R^2\overset{O^-}{\underset{}{C}}-\overset{P(O)(OC_2H_5)_2}{\underset{SCH_3}{C}}-R \xrightarrow{50^\circ} R^1R^2C{=}C\begin{matrix} R \\ SCH_3 \end{matrix} \xrightarrow[H_2O]{HgCl_2} R^1R^2CHCOR$$

[46] G. R. Malone and A. I. Meyers, *J. Org. Chem.*, **39,** 623 (1974).
[47] I. Shahak and J. Almog, *Synthesis,* **1970,** 145.
[48] M. Green, *J. Chem. Soc.*, **1963,** 1324.
[49] I. Shahak and J. Almog, *Synthesis,* **1969,** 170.
[50] I. C. Popoff, J. L. Dever, and G. R. Leader, *J. Org. Chem.*, **34,** 1128 (1969).

mercuric chloride give ketones.[35] The method has utility, since the phosphonate precursor can be alkylated prior to olefin formation, thereby increasing the versatility of the synthesis. This synthesis has been extended by preparation of an allyl vinyl thioether, which in the presence of red mercuric oxide undergoes a thio-Claisen rearrangement to 2-cyclohexyl-3-pentenal.[51]

$$(C_2H_5O)_2P(O)CH_2SCH_2CH{=}CH_2 \xrightarrow[\text{2. cyclohexanone}]{\text{1. } n\text{-}C_4H_9Li} \text{cyclohexylidene}{=}CHSCH_2CH{=}CH_2 \xrightarrow{\text{HgO, Heat}} \text{1-}(CH_2CH{=}CH_2)\text{-cyclohexane-1-}CHO$$

As indicated, the alkylthiomethylphosphonate esters are acidic enough to form anions readily. Note that alkoxymethylphosphonates even in the presence of strong bases are not converted into carbanions. Oxygen, unlike bivalent sulfur, does not have the ability to stabilize an adjacent carbanion. Phosphonamides, in contrast to the analogous alkoxymethylphosphonates, can be converted into anions by strong bases. No explanation has been given for this difference. By means of the phosphonamide modification of phosphonate-olefin formation, vinyl ethers can be successfully prepared.[52]

$$ROCH_2P(O)[N(CH_3)_2]_2 \xrightarrow[\text{2. } R^1R^2CO]{\text{1. } n\text{-}C_4H_9Li} R^1R^2C{=}CHOR + [(CH_3)_2N]_2PO_2^-$$

Dienes. Olefin formation to produce dienes can be conducted by two methods: (1) by reaction of an unsaturated phosphonate carbanion with a saturated aldehyde or ketone or (2) by reaction of a saturated phosphonate carbanion with an unsaturated aldehyde or ketone. To produce conjugated dienes by the first procedure, an allylphosphonate is used. For allylphosphonates to be sufficiently acidic for anion formation under relatively mild conditions, they must contain an electron-withdrawing group at either the α or γ position. Thus the anion prepared from triethyl 4-phosphonocrotonate,

[51] E. J. Corey and J. I. Shulman, *J. Amer. Chem. Soc.*, **92,** 5522 (1970); see also S. J. Rhoads and N. R. Raulins, *Org. Reactions*, **22,** 1 (1975).

[52] G. Lavielle and D. Reisdorf, *C.R. Acad. Sci., Ser. C*, **272,** 100 (1971).

$(C_2H_5O)_2P(O)CH_2CH{=}CHCO_2C_2H_5$, gives a series of diene esters in 50–80% yield when treated with substituted benzaldehydes.[53] The method was reported to be superior to both the Reformatsky and Wittig reactions for the preparation of 5-phenyl-2,4-pentadienoates and produces better yields than the alternative method of treating carbomethoxymethylphosphonate carbanion with cinnamaldehydes. The preparation of 2,4-pentadienoates was extended to study the synthesis of several amides of vegetable origin.[54] Often the 1,3-diene may be modified by hydrolysis after its preparation.[55]

$$(C_2H_5O)_2P(O)CH_2CH{=}C(R)X \xrightarrow[2.\ R^1R^2CO]{1.\ NaH} R^1R^2C{=}CHCH{=}C(R)X \xrightarrow{H_3O^+} R^1R^2CHCH{=}CHCOR$$

$$X = OCH_3,\ SCH_3,\ N(CH_3)_2$$

An investigation has been made of the stereochemistry of olefin formation with respect to its effect on the geometry of the allylic double bond.[56] *cis*- and *trans*-Diethyl 3-carbomethoxy-2-methylprop-2-enyl-phosphonate were condensed with either *cis*-citral or benzaldehyde. The *trans* ester retains its stereochemical integrity in both

$$(C_2H_5O)_2P(O)CH_2C(CH_3){=}CHCO_2CH_3 + RCHO \xrightarrow{CH_3ONa} RCH{=}CHC(CH_3){=}CHCO_2CH_3$$

$R = C_6H_5$ or $(CH_3)_2C{=}CHCH_2CH_2C(CH_3){=}$ (cis-citral residue)

(Eq. 10)

cases to give only *trans* 2-esters (Eq. 10). On the other hand, owing to stereomutation during the reaction, the *cis* ester forms only 25% of the *cis* 2-ester; the remainder is *trans* (Eq. 11). Isomerization of the

$$(C_2H_5O)_2P(O)CH_2C(CH_3){=}CHCO_2CH_3\ (cis) + RCHO \xrightarrow{CH_3ONa}$$

$$RCH{=}CHC(CH_3){=}CHCO_2CH_3\ (cis) + RCH{=}CHC(CH_3){=}CHCO_2CH_3\ (trans)$$

(Eq. 11)

[53] L. M. Werbel, N. Headen, and E. F. Elslager, *J. Med. Chem.*, **10,** 366 (1967).
[54] R. S. Burden and L. Crombie, *J. Chem. Soc., C,* **1969,** 2477.
[55] G. Lavielle and G. Sturtz, *Bull. Soc. Chim. Fr.*, **1970,** 1369.
[56] G. Pattenden and B. C. L. Weedon, *J. Chem. Soc., C,* **1968,** 1984.

starting *cis* ester may have taken place after anion formation. The fact that the *trans* allylphosphonate apparently undergoes very little if any isomerization upon formation of its anion and subsequent olefination was taken advantage of in the synthesis of the dehydro analog of the C_{18}-*Cecropia* juvenile hormone.[57,58] The *trans*-2,*trans*-4,*trans*-6,*cis* 10-tetraene is obtained from the appropriate aldehyde in about 99%

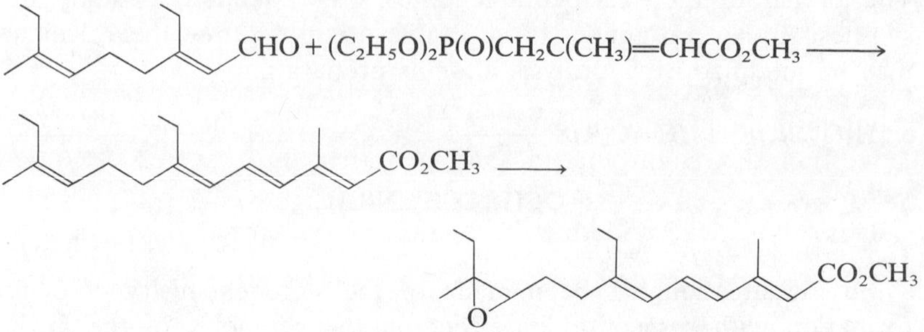

CHO + $(C_2H_5O)_2P(O)CH_2C(CH_3){=}CHCO_2CH_3$ ⟶

CO_2CH_3 ⟶

CO_2CH_3

O

purity under carefully chosen reaction conditions. The allylphosphonate carbanion is prepared prior to the addition of the aldehyde by treating the phosphonate with lithium diisopropylamide in a mixed hexamethylphosphoramide (HMPT)-tetrahydrofuran (THF) solvent.

The diene synthesis was also used in the preparation of the all *trans* (+)-trisporic acid β-methyl ester.[59] The condensation product was hydrolyzed to the β-keto ester.

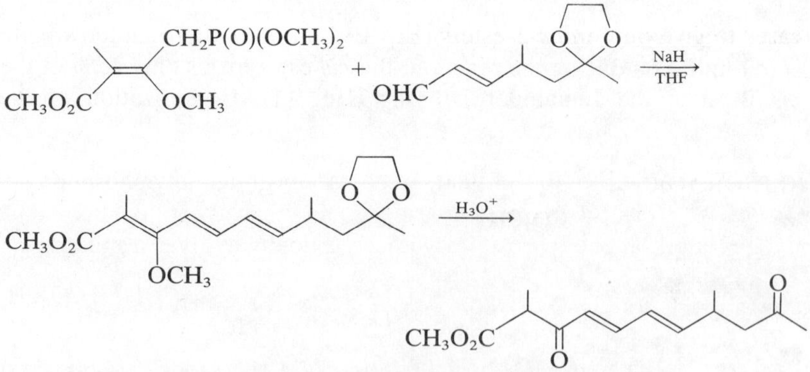

$CH_2P(O)(OCH_3)_2$ + O O

CH_3O_2C OCH_3 OHC $\xrightarrow[\text{THF}]{\text{NaH}}$

O O

CH_3O_2C $\xrightarrow{H_3O^+}$

OCH_3

O

CH_3O_2C

O

[57] E. J. Corey and B. W. Erickson, *J. Org. Chem.*, **39,** 821 (1974).

[58] E. J. Corey, J. A. Katzenellenbogen, S. A. Roman, and N. W. Gilman, *Tetrahedron Lett.*, **1971,** 1821.

[59] J. A. Edwards, V. Schwarz, J. Fajkos, M. L. Maddox, and J. H. Fried, *Chem. Commun.*, **1971,** 292.

Allylphosphonates that contain an electron-withdrawing group at the α position have also been successfully employed in diene synthesis.[60]

$$(CH_3)_2CHCH{=}C(CN)P(O)(OC_2H_5)_2 \xrightarrow{NaH}$$

$$(CH_3)_2C{=}CH\bar{C}(CN)P(O)(OC_2H_5)_2 \xrightarrow{p\text{-}O_2NC_6H_4CHO}$$

$$p\text{-}O_2NC_6H_4CH{=}C(CN)CH{=}C(CH_3)_2$$

The carbanion formed by treating an allylphosphonodiamide with n-butyllithium gives, upon addition of acetone, only the γ adduct.[29] By converting the anion into a Grignard reagent before addition of acetone the α adduct is made to predominate over the γ adduct in a 3/1 ratio. Only the α adduct decomposes to diene upon thermolysis.

$$[(CH_3)_2N]_2P(O)CH_2CH{=}CH_2 \xrightarrow[2.\ MgBr_2]{1.\ n\text{-}C_4H_9Li} \xrightarrow[4.\ H^+]{3.\ (CH_3)_2CO}$$

$$[(CH_3)_2N]_2P(O)CH{=}CHCH_2C(OH)(CH_3)_2 + [(CH_3)_2N]_2P(O)\underset{\displaystyle (CH_3)_2COH}{\underset{|}{C}}HCH{=}CH_2$$

More satisfactory results are obtained by reducing the steric requirements at the α position, thereby allowing almost exclusive formation of the α adduct, especially with unbranched ketones. If the two alkyl groups of the ketone are different, the diastereomers of the α adduct cannot be separated. In yet another example of addition of a carbonyl

$$\text{(cyclic } CH_3N\text{–}CH_2CH_2\text{–}NCH_3\text{)}P(O)CH_2CH{=}CH_2 \xrightarrow[2.\ RR^1CO]{1.\ n\text{-}C_4H_9Li} \text{(cyclic } CH_3N\text{–}CH_2CH_2\text{–}NCH_3\text{)}P(O)CH(\bar{O}CRR')CH{=}CH_2 \longrightarrow \underset{(75\%)}{RR^1C{=}CHCH{=}CH_2}$$

co-reactant to both the α and γ positions of an allylphosphonate carbanion, it is found that treatment of a carbanion containing a halogen atom at the γ position with benzaldehyde gives a mixture of

$$(CH_3O)_2P(O)CH_2CH{=}C(CH_3)Cl \xrightarrow[2.\ C_6H_5CHO]{1.\ NaH} C_6H_5CH{=}CHCH{=}C(CH_3)Cl +$$

$$\text{2-}CH_3\text{-5-}C_6H_5\text{-4-}[P(O)(OCH_3)_2]\text{-4,5-dihydrofuran} + C_6H_5\underset{\diagdown O \diagup}{CH{-}C}(CH_3)CH{=}CHP(O)(OCH_3)_2$$

[60] D. Danion and R. Carrie, *Tetrahedron Lett.*, **1971,** 3219.

products.[61] Both the diene and 2,3-dihydro-3-furylphosphonate arise from an α addition, whereas the 3,4-epoxy-1-butenylphosphonate is a consequence of γ addition.

Alkynes. Participation of an α-halophosphonate in phosphonate-olefin formation gives rise to alkynes.[4] Excess base is employed to dehydrohalogenate the intermediate vinyl halide. Intermediates need not be isolated to obtain yields of over 60% based on starting phosphonate. Diethyl α-bromo-p-nitrobenzylphosphonate, prepared

$$(C_2H_5O)_2P(O)CH_2CO_2C_2H_5 \xrightarrow[2.\ Br_2]{1.\ NaH}$$

$$(C_2H_5O)_2P(O)CHBrCO_2C_2H_5 \xrightarrow[2.\ C_2H_5CHO]{1.\ NaH}$$

$$C_6H_5CH{=}CBrCO_2C_2H_5 \xrightarrow{NaH} C_6H_5C{\equiv}CCO_2C_2H_5$$

by bromination of the parent phosphonate, gives an α-bromostilbene when treated with an aromatic aldehyde in the presence of one equivalent of base, but a diphenylacetylene in the presence of two equivalents of base.[62]

$$(C_2H_5O)_2P(O)CHBrC_6H_4NO_2\text{-}p + C_6H_5CHO \xrightarrow{C_2H_5ONa}$$

$$C_6H_5CH{=}CBrC_6H_4NO_2\text{-}p \xrightarrow{C_2H_5ONa} C_6H_5C{\equiv}CC_6H_4NO_2\text{-}p$$

Diphenylacetylene has been prepared from an acylated phosphine oxide.[63] Diphenylbenzylphosphine oxide in the presence of n-butyllithium can react with benzoyl chloride, and the product can be converted into its anion. The enolate form is in essence a phosphinoxy betaine that eliminates phosphinate to form diphenylacetylene.

$$(C_6H_5)_2P(O)CH_2C_6H_5 \xrightarrow[2.\ C_6H_5COCl]{1.\ n\text{-}C_4H_9Li} (C_6H_5)_2P(O)CH(C_6H_5)COC_6H_5 \xrightarrow{t\text{-}C_4H_9OK}$$

$$\begin{array}{l}(C_6H_5)_2P(O)CC_6H_5 \\ \qquad\quad \| \\ \quad \bar{O}{-}CC_6H_5\end{array} \longrightarrow C_6H_5C{\equiv}CC_6H_5 + (C_6H_5)_2PO_2^-$$

[61] G. Lavielle, *C.R. Acad. Sci., Ser. C*, **1970,** 86 [*C.A.*, **72,** 111184u (1970)].

[62] A. Yamaguchi and M. Okazaki, *Nippon Kagaku Kaishi*, **1972,** 2103 [*C.A.*, **78,** 29372w (1973)].

[63] S. T. D. Gough and S. Trippett, *J. Chem. Soc.*, **1962,** 2333.

Intramolecular Olefin Formation. A modification of phosphonate structure that permits intramolecular olefin formation has been reported.[64] The transformation of 20-oxo-21-acyloxysteroids to cardenolids is an example of a novel approach that may find wide

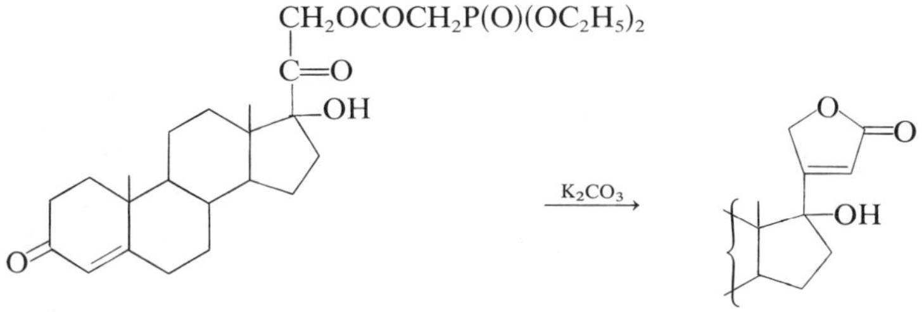

$CH_2OCOCH_2P(O)(OC_2H_5)_2$ $\xrightarrow{K_2CO_3}$

application for the preparation of unsaturated lactones. An intramolecular cyclization based on this procedure has been used to prepare alkyl pyrrolones and dihydropyridines.[65] The method has particular merit because of its flexibility with respect to location of substituents (Eq. 12). An interesting conversion of enol lactones into

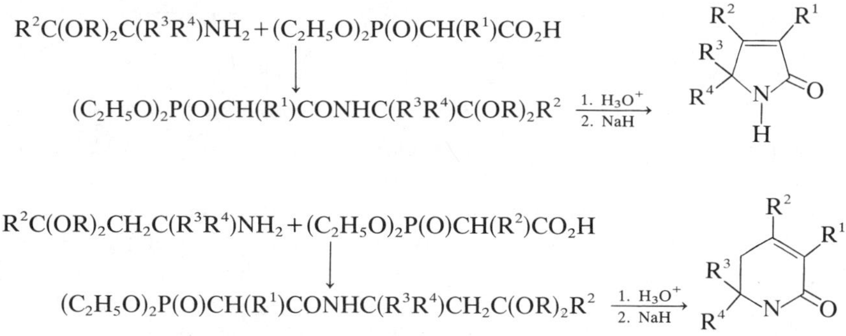

$$R^2C(OR)_2C(R^3R^4)NH_2 + (C_2H_5O)_2P(O)CH(R^1)CO_2H$$
$$\downarrow$$
$$(C_2H_5O)_2P(O)CH(R^1)CONHC(R^3R^4)C(OR)_2R^2 \xrightarrow[\text{2. NaH}]{\text{1. }H_3O^+}$$

$$R^2C(OR)_2CH_2C(R^3R^4)NH_2 + (C_2H_5O)_2P(O)CH(R^2)CO_2H$$
$$\downarrow$$
$$(C_2H_5O)_2P(O)CH(R^1)CONHC(R^3R^4)CH_2C(OR)_2R^2 \xrightarrow[\text{2. NaH}]{\text{1. }H_3O^+}$$

(Eq. 12)

α,β-unsaturated cyclic ketones has been discovered.[66,67] The carbanion is prepared from *n*-butyllithium before addition of the enol lactone

[64] H. G. Lehmann and R. Wiechert, *Angew. Chem., Int. Ed. Engl.*, **7,** 300 (1968).

[65] G. Stork and R. Matthews, *Chem. Commun.*, **1970,** 445.

[66] J. H. Fried, Ger. Pat., 1,812,124 [*C.A.*, **71,** 123725p (1969)].

[67] C. A. Henrick, E. Bohme, J. A. Edwards, and J. H. Fried, *J. Amer. Chem. Soc.*, **90,** 5926 (1968).

and, if desired, the intermediate phosphonate can be isolated. The reaction has been applied to the synthesis of steroids.

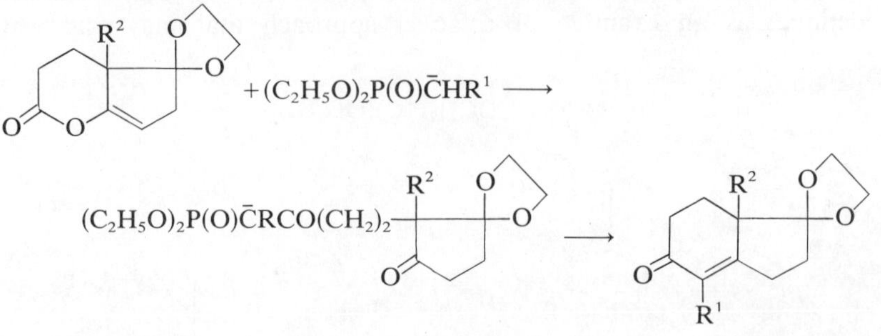

$$+ (C_2H_5O)_2P(O)\bar{C}HR^1 \longrightarrow$$

$$(C_2H_5O)_2P(O)\bar{C}RCO(CH_2)_2- \longrightarrow$$

Recently the same procedure has been used to prepare a key intermediate in the total synthesis of β-lactam antibiotics (Eq. 13).[68,69]

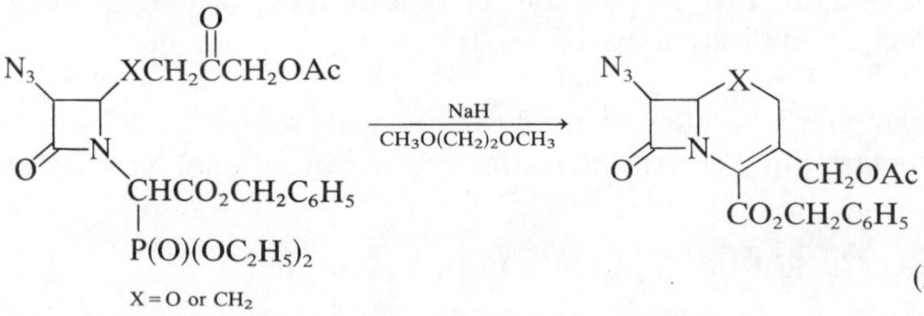

$$N_3 \quad XCH_2\overset{O}{\overset{\|}{C}}CH_2OAc \quad CHCO_2CH_2C_6H_5 \quad P(O)(OC_2H_5)_2 \xrightarrow[CH_3O(CH_2)_2OCH_3]{NaH} N_3 \quad CH_2OAc \quad CO_2CH_2C_6H_5$$

X = O or CH_2 (Eq. 13)

Bisphosphonates as Olefin-Forming Reagents. Addition of a second P(O)-activating group to a carbon atom greatly enhances the acidity of hydrogen atoms attached to the carbon. Carbanions obtained from compounds activated by two P(O) groups readily react with aldehydes or ketones to give a wide assortment of vinyl phosphorus compounds. The original report that tetraethylmethylenebisphosphonate carbanion forms substituted alkenephosphonates upon addition of a carbonyl compound has become the basis of a number of investigations.[4] Treatment of a methylenebisphosphonate carbanion with cinnamaldehyde, for example, gives 4-phenylbutadiene-1-phosphonate.[70] In an interesting extension of the synthesis, tetraethyl

$$[(C_2H_5O)_2P(O)]_2CH_2 + C_6H_5CH{=}CHCHO \xrightarrow{NaH} C_6H_5CH{=}CHCH{=}CHP(O)(OC_2H_5)_2$$

[68] R. N. Guthikonda, L. D. Cama, and B. G. Christensen, *J. Amer. Chem. Soc.*, **96,** 7584 (1974).
[69] L. D. Cama and B. G. Christensen, *J. Amer. Chem. Soc.*, **96,** 7582 (1974).
[70] H. G. Henning and D. Gloyma, *Z. Chem.*, **6,** 28 (1966).

dimethylaminomethylenebisphosphonate in the presence of base, forms an anion that, upon addition of a carbonyl co-reactant, produces 1-dimethylaminoalkenylphosphonates (Eq. 14). When hydrolyzed, the latter yield carboxylic acids.[71] The bisphosphonate is prepared by

$$2(C_2H_5O)_2P(O)H + (CH_3O)_2CHN(CH_3)_2 \xrightarrow{-2CH_3OH} [(C_2H_5O)_2P(O)]_2CHN(CH_3)_2$$

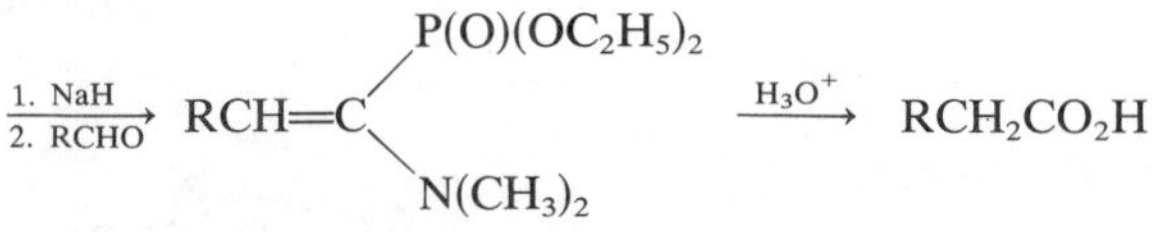

$$\xrightarrow[2.\ RCHO]{1.\ NaH} RCH{=}C\begin{matrix} P(O)(OC_2H_5)_2 \\ N(CH_3)_2 \end{matrix} \xrightarrow{H_3O^+} RCH_2CO_2H \qquad \text{(Eq. 14)}$$

heating diethyl phosphite with dimethylformamide acetal with concurrent removal of methanol. α-Ketophosphonates, which are known to hydrolyze readily to carboxylic acids, are assumed to be intermediates in the hydrolysis.

Methylene bisphosphonate carbanion has been employed to prepare phosphonic acids analogous to biologically important phosphate monoesters. The synthesis of the phosphonic acid analog of a pyridoxal phosphate is an illustration.[72] Under similar conditions blocked al-

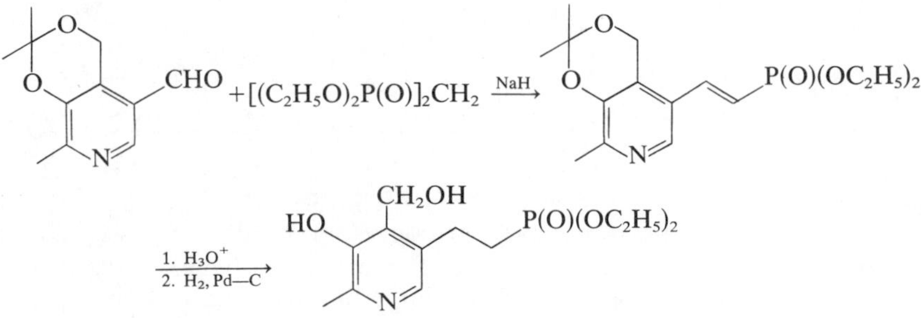

[Structure: isopropylidene-protected pyridine aldehyde (CHO) $+ [(C_2H_5O)_2P(O)]_2CH_2 \xrightarrow{NaH}$ isopropylidene-protected pyridine with $CH{=}CHP(O)(OC_2H_5)_2$]

$\xrightarrow[2.\ H_2,\ Pd\text{—}C]{1.\ H_3O^+}$ [Structure: pyridine bearing HO, CH_2OH and $CH_2CH_2P(O)(OC_2H_5)_2$]

dehyde sugars have been reported to yield *trans* unsaturated phosphonate sugars.[73] Interestingly, with ethylidene-D-erythose, the carbanion

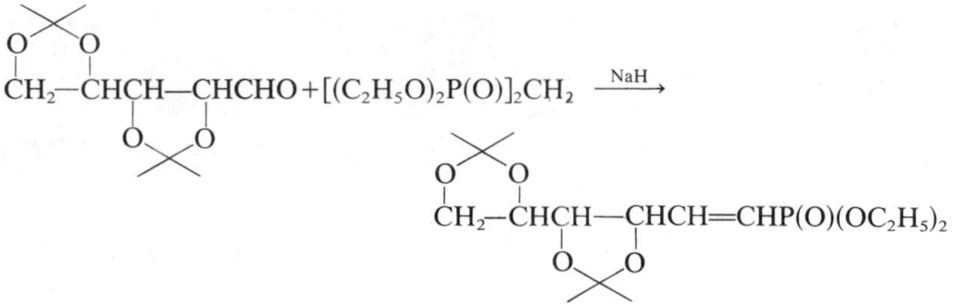

[Structure: $CH_2\text{—}CHCH\text{—}CHCHO$ with two isopropylidene (O–C–O) bridges] $+ [(C_2H_5O)_2P(O)]_2CH_2 \xrightarrow{NaH}$

[Structure: $CH_2\text{—}CHCH\text{—}CHCH{=}CHP(O)(OC_2H_5)_2$ with two isopropylidene (O–C–O) bridges]

[71] H. Gross and B. Costisella, *Angew. Chem., Int. Ed., Engl.*, **7,** 391 (1968).
[72] T. L. Hullar, *Tetrahedron Lett.*, **1967,** 4921.
[73] H. Paulsen, W. Bartsch, and J. Thiem, *Chem. Ber.*, **104,** 2545 (1971).

gives an open *trans* phosphonate whereas a comparable mixed Wittig-phosphonate reagent forms a cyclic unsaturated phosphonate.

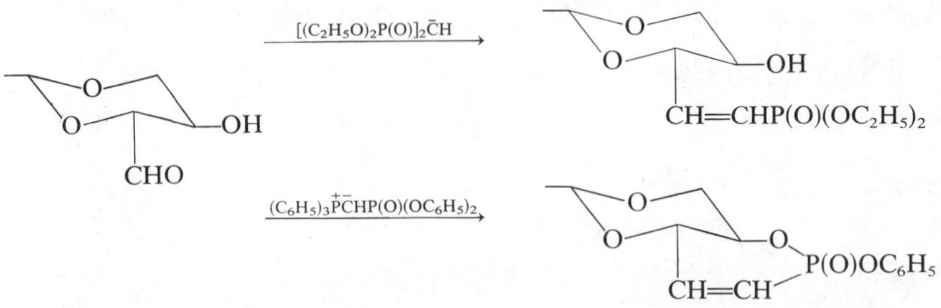

$[(C_2H_5O)_2P(O)]_2\bar{C}H$ → CH=CHP(O)$(OC_2H_5)_2$ (OH)

CHO, OH

$(C_6H_5)_3\overset{+}{P}\bar{C}HP(O)(OC_6H_5)_2$ → CH=CH–P(O)OC_6H_5 (cyclic)

The initial step in the synthesis of isosteric phosphonate analogs of nucleoside 3′-phosphates involves the condensation of tetraethylmethylenebisphosphonate with 1,2,5,6-di-O-isopropylidene-α-D-ribo-hexofuranos-3-ulose.[74] By a series of subsequent steps including catalytic reduction the product can be converted into the nucleotide analogs.

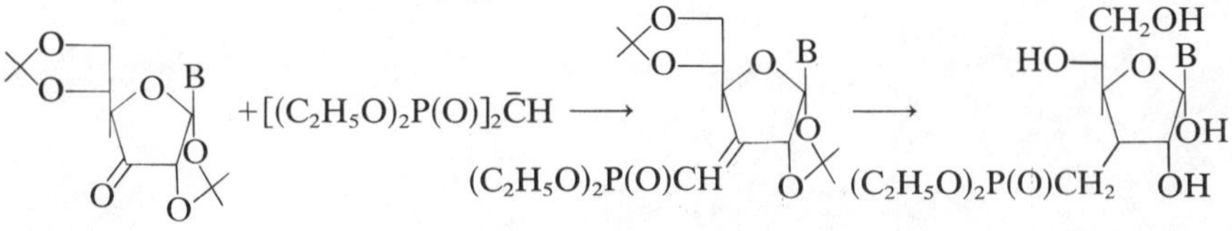

$+[(C_2H_5O)_2P(O)]_2\bar{C}H \longrightarrow$ $(C_2H_5O)_2P(O)CH$ $\longrightarrow$ $(C_2H_5O)_2P(O)CH_2$ (HO, CH_2OH, B, OH, OH)

B = adenine, guanine, cytosine, thymine, or uracil.

A number of examples of diolefin formation by means of bisphosphonates with the P(O) functions separated by more than one carbon atom have been reported.[75,76] Thus *trans,trans*-*p*-distyrylbenzene is prepared from *p*-xylylenebisphosphonate and benzaldehyde.[4] In a

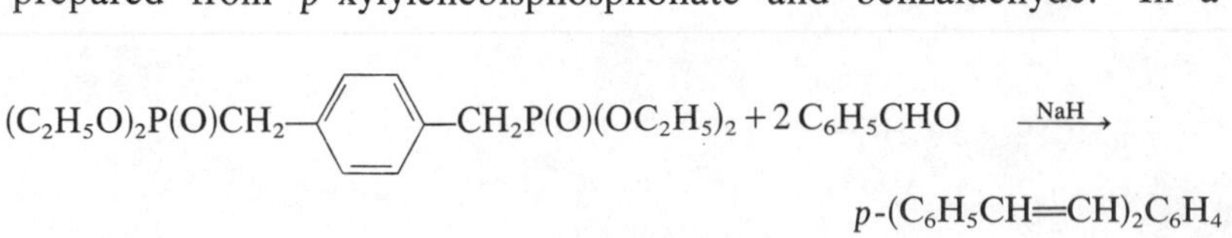

$$(C_2H_5O)_2P(O)CH_2\text{–}C_6H_4\text{–}CH_2P(O)(OC_2H_5)_2 + 2\,C_6H_5CHO \xrightarrow{NaH} p\text{-}(C_6H_5CH{=}CH)_2C_6H_4$$

[74] G. H. Jones, H. P. Albrecht, N. P. Damodaran, and J. G. Moffatt, *J. Amer. Chem. Soc.*, **92,** 5511 (1970).

[75] E. Muller, M. Sauerbier, D. Striechfuss, R. Thomas, W. Winter, G. Zoontsas, and J. Heiss, *Ann.*, **735,** 99 (1970).

[76] H. W. Whitlock, Jr., P. E. Sandviek, L. E. Overman, and P. B. Reichardt, *J. Org. Chem.*, **34,** 879 (1969).

similar manner, *o*-distyrylbenzene is formed, as is *trans,trans*-1,8-distyrylnaphthalene.[77]

$$(C_2H_5O)_2P(O)CH_2 \quad CH_2P(O)(OC_2H_5)_2 \text{ (1,8-naphthalene)} + 2\,C_6H_5CHO \xrightarrow[\text{DMF}]{CH_3ONa} \text{1,8-bis}(C_6H_5CH{=}CH)\text{naphthalene}$$

Phosphoramidate Anions. The phosphoramidate anions represent a class of P(O)-stabilized anions which, unlike those previously described, are not carbanions. Their chemistry is analogous to that of the iminophosphoranes (phosphinimines), $R_3\overset{+}{P}\overset{-}{N}R$.[78] Upon treatment with a variety of carbonyl reactants they yield unsaturated nitrogen compounds. This route is appealing because the phosphoramidates from which the anions are prepared by treatment with sodium hydride are readily available.[79] The mechanism here is probably similar to that discussed for phosphonate-olefin formation, *i.e.*, the production of an intermediate betaine. A single adduct, prepared by condensation of

$$RN{=}C{=}NR^1 \xleftarrow{R^1NCO} (C_2H_5O)_2P(O)\bar{N}R \xrightarrow{R^1CHO} R^1CH{=}NR$$

$$(C_2H_5O)_2P(O)\bar{N}R \xrightarrow{R^1R^2C{=}C{=}O} R^1R^2C{=}C{=}NR \qquad (C_2H_5O)_2P(O)\bar{N}R \xrightarrow{CO_2} RNCO \qquad (C_2H_5O)_2P(O)\bar{N}R \xrightarrow{CS_2} RNCS$$

chloral with a phosphoramidate, has been isolated.[80] In refluxing triethylamine the adduct does not decompose to phosphate and imine but yields chloroform and an N-phosphonoformamide (Eq. 15). The

$$(C_2H_5O)_2P(O)NHC_2H_5 + Cl_3CCHO \xrightarrow{H^+} (C_2H_5O)_2P(O)N(C_2H_5)CH(OH)CCl_3 \xrightarrow{Et_3N} CHCl_3 + (C_2H_5O)_2P(O)N(C_2H_5)CHO \quad \text{(Eq. 15)}$$

[77] J. Meinwald and J. W. Young, *J. Amer. Chem. Soc.*, **93,** 725 (1971).

[78] H. Staudinger and E. Hawser, *Helv. Chim. Acta*, **4,** 861 (1921).

[79] F. R. Atherton, H. T. Openshaw, and A. R. Todd, *J. Chem. Soc.*, **1945,** 660.

[80] K. N. Konorov, E. A. Guryler, and A. V. Chernova, *Izv. Akad. Nauk SSSR, Ser. Khim.*, **1968,** 587 [*C.A.*, **69,** 95864w (1968)].

phosphoramidate anion route to isocyanates has been utilized to prepare dialkylaminoisocyanates, which on formation dimerize to novel nitrogen ylids.[81]

$$(C_2H_5O)_2P(O)\bar{N}N(CH_3)_2 \xrightarrow{CO_2} 2(CH_3)_2NNCO \longrightarrow \begin{array}{c}(CH_3)_2\overset{+}{N}—N^- \\ | \quad\quad | \\ O{=}C\diagdown_{N}\diagup C{=}O \\ | \\ N(CH_3)_2\end{array}$$

Recent work has revealed that aminocarbodiimides also dimerize.[82]

$$(C_2H_5O)_2P(O)\bar{N}N(CH_3)_2 \xrightarrow{t\text{-}C_4H_9NCO}$$

$$t\text{-}C_4H_9N{=}C{=}NN(CH_3)_2 \longrightarrow \begin{array}{c}(CH_3)_2\overset{+}{N}—N^- \\ | \quad\quad | \\ t\text{-}C_4H_9N{=}C\diagdown_{N}\diagup C{=}NC_4H_9\text{-}t \\ | \\ N(CH_3)_2\end{array}$$

The Carbonyl Co-reactant

In the previous section the structural changes in the phosphoryl reagent that led to unusual results were considered. Changes in the carbonyl co-reactant that give products whose features deserve separate recognition are recorded in this section.

Phosphonate-olefin formation does not produce satisfactory yields with carbonyl compounds that are readily enolizable. Proton transfers may occur faster than carbonyl addition with subsequent isomerization, enolate condensation, or both. High yields of vinyl sulfides are obtained from the lithium salts of methylthiomethylphosphate esters and carbonyl compounds except when either acetophenone or cyclopentanone is employed.[35] With these two ketones there are obtained substantial amounts of base-catalyzed self-condensation products. A similar situation prevails, especially with cyclopentanone, when Wittig reagents are used.[83,84]

Dialdehydes and Diketones

As expected, diketones ($n = 0, 2, 4, 6, 7$) give unsaturated keto esters when treated with an equivalent of a phosphonate anion.[85]

[81] W. S. Wadsworth, Jr., and W. D. Emmons, *J. Org. Chem.*, **32,** 1279 (1967).

[82] W. S. Wadsworth, Jr., unpublished results.

[83] D. R. Coulson, *Tetrahedron Lett.*, **1964,** 3323.

[84] G. Wittig, W. Böll, and K. H. Krück, *Chem. Ber.* **95,** 2514 (1962).

[85] B. G. Kovalev, E. M. Al'tmark, and E. S. Lavrinenko, *Zh. Org. Khim.*, **1970,** 2187 [*C.A.*, **74,** 41827w (1971)].

Acetylacetone ($n = 1$) is more acidic than the phosphonate and is merely converted into its enolate anion by the carbanion. When $n = 5$ the diketone undergoes intramolecular aldol condensation.

$$CH_3CO(CH_2)_nCOCH_3 + (C_2H_5O)_2P(O)\bar{C}HCO_2C_2H_5 \rightarrow CH_3CO(CH_2)_nC(CH_3){=}CHCO_2C_2H_5$$

Selective olefin-bond formation has been observed with diketones.[86] Cholestane-3,7-dione yields exclusively the 3-carbethoxy-

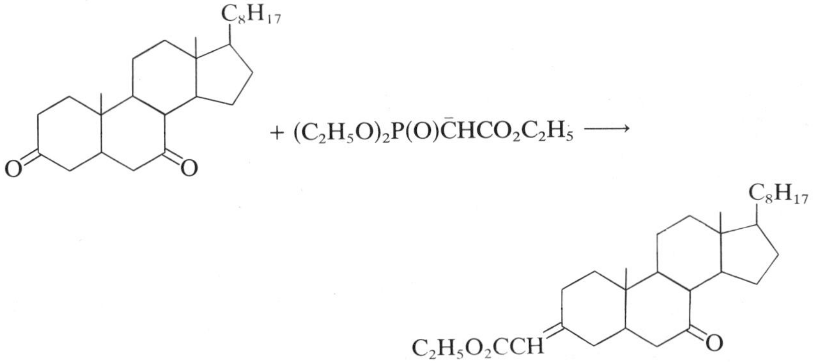

C_8H_{17}

$+ (C_2H_5O)_2P(O)\bar{C}HCO_2C_2H_5 \longrightarrow$

C_8H_{17}

$C_2H_5O_2CCH$

methylene analog when treated with triethyl phosphonoacetate and base. Presumably a steric effect prevents olefin formation at both sites. In contrast, a Wittig reagent attacks both carbonyl groups. More recently ketoaldehydes ($n = 3, 4$) have been condensed with a phosphonate carbanion to give monoolefins.[87] Only the aldehyde function undergoes olefin formation in the presence of an equivalent of the phosphonate, $R = COCH_3$ or $CH{=}CHCO_2C_2H_5$ (Eq. 16). When a compound con-

$$CH_3CO(CH_2)_nCHO + (C_2H_5O)_2P(O)\bar{C}HR \rightarrow CH_3CO(CH_2)_nCH{=}CHR \quad \text{(Eq. 16)}$$

taining both keto and acetal groups is employed, however, condensation occurs only at the keto group.

$$CH_3CO(CH_2)_nCH(OC_2H_5)_2 + (C_2H_5O)_2P(O)\bar{C}HR \rightarrow RCH{=}C(CH_3)(CH_2)_nCH(OC_2H_5)_2$$

[86] A. K. Bose and R. T. Dahill, Jr., *Tetrahedron Lett.*, **1963,** 959.

[87] B. G. Kovalev, and E. M. Al'tmark, *Zh. Org. Khim.*, **1972,** 1582 [*C.A.*, **77,** 163967z (1972)].

As expected, dialdehydes undergo diolefin formation in a normal manner.[75,88] An interesting cyclization takes place upon addition of

$$o\text{-}C_6H_4(CHO)_2 + 2\ 2,6\text{-}F_2C_6H_3\bar{C}HP(O)(OC_2H_5)_2 \longrightarrow o\text{-}C_6H_4(CH{=}CHC_6H_3F_2\text{-}2,6)_2$$

glutaric dialdehyde to a phosphonate carbanion.[89,90] Cyclohexanols, $R = CO_2C_2H_5$ or CN, are prepared in yields of over 40% (Eq. 17).

$$OHC(CH_2)_3CHO + (C_2H_5O)_2P(O)\bar{C}HR \longrightarrow$$

$$(C_2H_5O)_2P(O)CHRCH(O^-)(CH_2)_3CHO \longrightarrow$$

$$(C_2H_5O)_2P(O)\bar{C}RCH(OH)(CH_2)_3CHO \longrightarrow \text{2-R-cyclohex-2-en-1-ol} \qquad \text{(Eq. 17)}$$

Unsaturated Aldehydes and Ketones

As stated earlier, dienes can be synthesized from P(O)-stabilized carbanions and allylic phosphorus substrates. In an alternative route they can also be prepared by treatment of a saturated phosphonyl-stabilized anion with an unsaturated aldehyde or ketone. Thus, by warming a bisphosphonate with cinnamaldehyde in the presence of sodium hydride, a bisbutadienyl system is obtained (Eq. 18).[4] Likewise,

$$p\text{-}C_6H_4[CH_2P(O)(OC_2H_5)_2]_2 + 2\ C_6H_5CH{=}CHCHO \xrightarrow{\text{NaH}}$$

$$p\text{-}C_6H_4(CH{=}CHCH{=}CHC_6H_5)_2 \qquad \text{(Eq. 18)}$$

[88] G. V. Grinev, V. A. Dombrovskii, and L. A. Yanovskaya, *Izv. Akad. Nauk SSSR, Ser. Khim.*, **1972,** 635 [*C.A.*, **77,** 113982*e* (1972)].

[89] B. G. Kovalev and N. P. Dormidontova, *J. Gen. Chem. USSR*, **40,** 910 (1970).

[90] B. G. Kovalev and N. P. Dormidontova, *J. Gen. Chem. USSR*, **40,** 932 (1970).

a number of esters of 4-alkyl-2,4-pentadienoic acids have been prepared from 2-alkylacroleins, with minor amounts of a 5-ethoxy by-product accompanying the major product (Eq. 19).[91] Although ac-

$$CH_2{=}C(R)CHO + (C_2H_5O)_2P(O)CH_2CO_2C_2H_5 \xrightarrow[C_2H_5OH]{C_2H_5ONa}$$

$$\underset{(50\text{–}65\%)}{CH_2{=}C(R)CH{=}CHCO_2C_2H_5} + C_2H_5OCH_2CH(R)CH{=}CHCO_2C_2H_5 \quad \text{(Eq. 19)}$$

rolein itself furnishes a dienoic acid in low yield, the phosphonate method is apparently superior to the more common method of preparing vinylacrylic acids, namely, the amine-catalyzed decarboxylative condensation of an α,β-unsaturated aldehyde with malonic acid. By contrast, olefin formation produces only the 2,3-*trans* isomers. Michael addition of the phosphonate carbanion to the β-carbon atom of the 2-alkylacroleins or diene product may also compete with olefin formation and be a major factor in limiting the yield.

The Michael addition of phosphonate carbanions to α,β-unsaturated nitriles and esters is well known and, indeed, as early as 1952 it was extended to α,β-unsaturated ketones.[92,93] Thus triethyl phosphonoacetate, converted into its anion with sodium ethoxide in ethanol and allowed to react with benzalacetone, adds across the double bond in complete preference to olefin formation (Eq. 20). Under similar conditions monoalkylated phosphonoacetates also add in Michael fashion. It

$$(C_2H_5O)_2P(O)CH_2CO_2C_2H_5 + C_6H_5CH{=}CHCOCH_3 \xrightarrow[C_2H_5OH]{C_2H_5ONa}$$

$$(C_2H_5O)_2P(O)CH[CH(C_6H_5)CH_2COCH_3]CO_2C_2H_5 \quad \text{(Eq. 20)}$$

has been noted that chalcone (benzalacetophenone) gave only Michael adducts with phosphonate carbanions.[94,95] In another study it was found that both triethyl phosphonoacetate and diethyl benzylphosphonate react with chalcone either by the Michael or olefin formation routes depending upon operating conditions.[96] Olefin formation (R = C_6H_5 or $CO_2C_2H_5$) only is observed when sodium hydride and diglyme

[91] R. J. Sundberg, P. A. Bukowick, and F. O. Holcombe, *J. Org. Chem.*, **32,** 2938 (1967).

[92] A. N. Pudovik and N. M. Lebedeva, *J. Gen. Chem. USSR*, **22,** 2128 (1952).

[93] A. N. Pudovik and N. M. Lebedeva, *Dokl. Akad. Nauk. SSSR*, **90,** 799 (1953) [*C.A.*, **50,** 2429d (1956)].

[94] B. Fiszer and J. Michalski, *Rocz. Chem.*, **28,** 185 (1954) [*C.A.*, **49,** 9493e (1955)].

[95] B. Fiszer and J. Michalski, *Rocz. Chem.*, **34,** 1461 (1960) [*C.A.*, **55,** 15331g (1961)].

[96] E. D. Bergmann and A. Sulomonovici, *Tetrahedron*, **27,** 2675 (1971).

are employed as the base and condensing medium, respectively (Eq. 21). When sodium ethoxide in benzene is used, only Michael products

$$(C_2H_5O)_2P(O)CH_2R + C_6H_5CH{=}CHCOC_6H_5 \xrightarrow[\text{Diglyme}]{\text{NaH}}$$

$$C_6H_5CH{=}CHC(C_6H_5){=}CHR \quad \text{(Eq. 21)}$$

are observed.

$$(C_2H_5O)_2P(O)CH_2CO_2C_2H_5 + 2\ C_6H_5CH{=}CHCOC_6H_5 \xrightarrow[C_6H_6]{C_2H_5ONa}$$

$$(C_2H_5O)_2P(O)C[CH(C_6H_5)CH_2COC_6H_5]_2CO_2C_2H_5 \longrightarrow$$

[cyclohexene: C_6H_5, C_6H_5, $C_2H_5O_2C$, $(C_2H_5O)_2P(O)$, COC_6H_5, C_6H_5]

The presumed double Michael adduct intermediate from the phosphonate ester cyclizes spontaneously to the cyclohexene, whereas the di-adduct with benzylphosphonate is stable (Eq. 22). Both phos-

$$(C_2H_5O)_2P(O)CH_2C_6H_5 + 2\ C_6H_5CH{=}CHCOC_6H_5 \xrightarrow[C_6H_6]{C_2H_5ONa}$$

$$(C_2H_5O)_2P(O)C[CH(C_6H_5)CH_2COC_6H_5]_2C_6H_5 \quad \text{(Eq. 22)}$$

phonates give mono-Michael adducts with chalcone when sodium amide in either is employed, but when sodium in benzene is used, the anions give very low yields of the Michael adducts and again olefin formation predominates. It would appear that Michael adduct formation is favored when a proton source exists but not when anion formation is irreversible. When phosphonate anion formation is irreversible, addition by the anion across the double bond may be reversible. Since olefin formation is irreversible, the consequences of such a prior addition are not apparent. When a proton source is available, added salts might divert the reactions to olefin formation in preference to formation of a Michael adduct. This would be the result if a cyclic intermediate that can be stabilized by an added cation were involved in olefin formation. Indeed, the nature of the cation as well as the solvent may be important in determining the course of the reaction.

A diene has been prepared by adding triethyl phosphonoacetate anion to β-methyl lactol.[56,97] The lactol is obtained from the diethyl acetal of pyruvaldehyde. Since the *cis,trans* half ester is the only product,

[97] G. Pattenden, J. E. Way, and B. C. L. Weedon, *J. Chem. Soc., C.*, **1970**, 235.

indications are that in the last reaction there is complete retention of stereochemistry at the double bond derived from the lactol.

$$(C_2H_5O)_2CHCOCH_3 + (C_2H_5O)_2P(O)CH_2CO_2C_2H_5 \longrightarrow (C_2H_5O)_2CHC(CH_3){=}CHCO_2C_2H_5$$

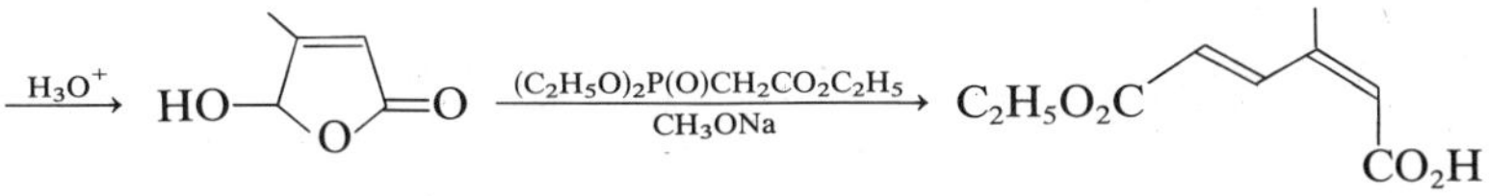

$\xrightarrow{H_3O^+}$ HO— (lactone) =O $\xrightarrow[CH_3ONa]{(C_2H_5O)_2P(O)CH_2CO_2C_2H_5}$ $C_2H_5O_2C$ … CO_2H

An unsaturated aldehyde has been used in phosphonate-olefin formation to modify steroids, as in the accompanying preparation of a bufodienolide.[98] Olefin formation with unsaturated aldehydes has also

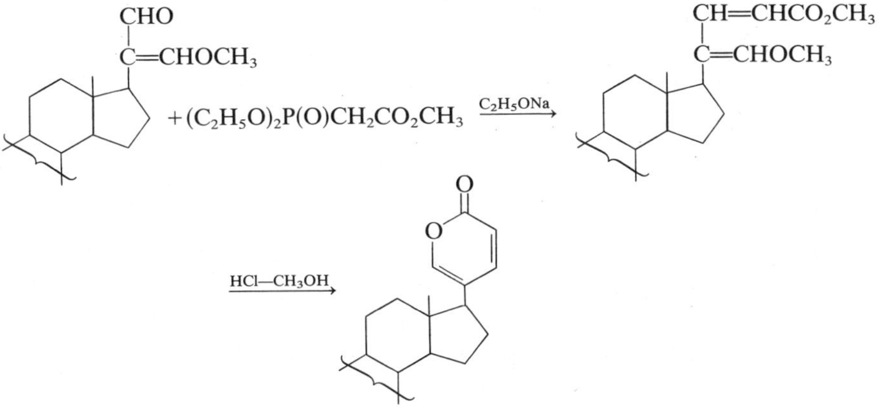

CHO, C=CHOCH$_3$ (steroid) $+ (C_2H_5O)_2P(O)CH_2CO_2CH_3 \xrightarrow{C_2H_5ONa}$ CH=CHCO$_2$CH$_3$, C=CHOCH$_3$ (steroid)

$\xrightarrow{HCl—CH_3OH}$ (bufodienolide)

found application in the syntheses of other natural products. A key step in the total synthesis of mycophenolic acid is typical (Eq. 23), as is

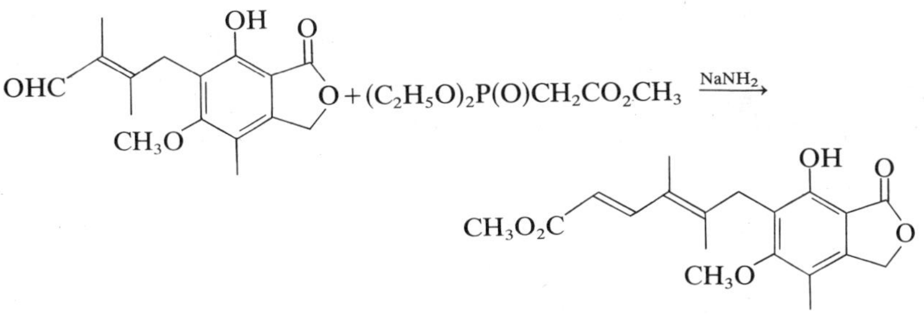

OHC … OH, O, CH$_3$O … O $+ (C_2H_5O)_2P(O)CH_2CO_2CH_3 \xrightarrow{NaNH_2}$ CH$_3$O$_2$C … OH, O, CH$_3$O … O

(Eq. 23)

[98] K. Radscheit, U. Stache, W. Haede, W. Fritsch, and H. Ruschig, *Tetrahedron Lett.*, **1969**, 3029.

the use of a phosphonate carbanion in the preparation of analogs of vitamin A.[99] The all *trans*-9,13-desmethyl vitamin A methyl ester is prepared from an unsaturated aldehyde and a γ-phosphonocrotonate.

$$+ (C_2H_5O)_2P(O)CH_2CH{=}CHCO_2CH_3 \xrightarrow{NaNH_2}$$

The all-*trans* starting aldehyde may also be prepared by an olefin bond route.

$$\xrightarrow[\substack{2.\ LiAlH_4 \\ 3.\ MnO_2}]{1.\ (C_2H_5O)_2P(O)\bar{C}HCO_2C_2H_5}$$

Esters

Certain esters have been found to act as substrates in phosphonate-olefin formation.[100] Diethyl oxalate reacts with triethyl phosphonoacetate carbanion to give both diethyl ethoxyfumarate and diethyl ethoxymaleate. It is felt that the fumarate might result from the reversible addition of ethanol to the originally formed maleate. The reaction can

$$(C_2H_5O)_2P(O)CH_2CO_2C_2H_5 + (CO_2C_2H_5)_2 \xrightarrow[(C_2H_5)_2O]{NaH} (C_2H_5O)_2P(O)CH(CO_2C_2H_5)\text{–}C(O^-)(OC_2H_5)(CO_2C_2H_5)$$

$$\longrightarrow C_2H_5OC(CO_2C_2H_5){=}CHCO_2C_2H_5 + C_2H_5OC(CO_2C_2H_5){=}CHCO_2C_2H_5$$

be extended to produce diethyl 2-fluoro-3-ethoxyfumarate and diethoxy fumarate. Neither cyanomethylphosphonate nor benzylphosphonate carbanions undergo similar olefin bond reactions. A few other

[99] P. J. Van Den Tempel and H. O. Huisman, *Tetrahedron*, **22**, 293 (1966).

[100] W. Grell and H. Machleidt, *Ann.*, **693**, 134 (1966).

$$(C_2H_5O)_2P(O)CHFCO_2C_2H_5 + (CO_2C_2H_5)_2 \xrightarrow[(C_2H_5)_2O]{NaH} (C_2H_5O)_2P(O){-}\underset{\underset{F}{|}}{\overset{\overset{CO_2C_2H_5}{|}}{\overset{^{-}OCOC_2H_5}{|}}{C}}CO_2C_2H_5 \longrightarrow$$

$$\begin{array}{c} C_2H_5OCCO_2C_2H_5 \\ \| \\ CFCO_2C_2H_5 \end{array}$$

$$(C_2H_5O)_2P(O)CH(OC_2H_5)CO_2C_2H_5 + (CO_2C_2H_5)_2 \xrightarrow[Dioxane]{NaH}$$

$$\begin{array}{c} C_2H_5OCCO_2C_2H_5 \\ \| \\ C_2H_5O_2CCOC_2H_5 \end{array}$$

esters may also act as substrates in phosphonate-olefin formation. Ethyl trifluoroacetate yields a ketal ester. The alcohol used in the final

$$(C_2H_5O)_2P(O)CH_2CO_2C_2H_5 + CF_3CO_2C_2H_5 \xrightarrow[Glyme]{NaH}$$

$$CF_3C(OC_2H_5){=}CHCO_2C_2H_5 \xrightarrow{C_2H_5OH} CF_3C(OC_2H_5)_2CH_2CO_2C_2H_5$$

step might come from a reversible condensation (Eq. 24). In contrast

$$(C_2H_5O)_2P(O)\bar{C}HCO_2C_2H_5 + CF_3CO_2C_2H_5 \xrightleftharpoons{-C_2H_5OH} (C_2H_5O)_2P(O)CH(CO_2C_2H_5)COCF_3 \quad \text{(Eq. 24)}$$

to the trifluoroacetate ester, trichloroacetate ester chlorinates the anion. The presence of the chloroanion is shown by treatment of the reaction mixture with isobutyraldehyde and isolation of the expected α-chlorolefin. The anion of triethyl phosphonoacetate with monochloroacetate leads merely to substitution.

$$(C_2H_5O)_2P(O)\bar{C}HCO_2C_2H_5 + Cl_3CCO_2C_2H_5 \xrightarrow{-C_2H_5OH}$$

$$\begin{array}{c} Cl_3CCO \\ | \\ ^{-}CCO_2C_2H_5 \\ | \\ (C_2H_5O)_2P(O) \end{array} \longrightarrow \begin{array}{c} Cl_2\bar{C}CO \\ | \\ ClCCO_2C_2H_5 \\ | \\ (C_2H_5O)_2P(O) \end{array} \xrightarrow{+C_2H_5OH}$$

$$Cl_2CHCO_2C_2H_5 + (C_2H_5O)_2P(O)\bar{C}(Cl)CO_2C_2H_5$$

Allenes

Allenes can be prepared by treating a phosphonate carbanion with a ketene.[4,101] The synthesis proceeds under milder conditions than the

$$(C_2H_5O)_2P(O)\bar{C}HCO_2C_2H_5 + (C_6H_5)_2C{=}CO \longrightarrow (C_6H_5)_2C{=}C{=}CHCO_2C_2H_5$$

comparable Wittig reaction in which phosphoranes are employed. Recently the preparation and optical resolution of some allenecarboxylic acids prepared by phosphonate-olefin formation have been reported.[102] Allenes can also be obtained by effecting an elimination from a betaine that possesses an allylic type of structure (Eq. 25).[103] In

$$(C_6H_5)_2P(O)C{\equiv}CR + R^1MgX \xrightarrow[(C_2H_5)_2O]{CuI} \left[(C_6H_5)_2P(O)\bar{C}{=}C\begin{matrix}R\\R^1\end{matrix}\right]MgX^+$$

$$\xrightarrow{C_6H_5CHO} (C_6H_5)_2P(-O-MgX-O-)CHC_6H_5 \text{ bridged via } C{=}C(R)(R^1) \xrightarrow{H_2O} (C_6H_5)_2P(-\bar{O}\cdots H-O-)CHC_6H_5 \text{ bridged via } C{=}C(R)(R^1)$$

$$\xrightarrow{NaH} C_6H_5CH{=}C{=}C\begin{matrix}R\\R^1\end{matrix} + (C_6H_5)_2P(O)\bar{O}Na^+ \qquad \text{(Eq. 25)}$$

this novel synthesis a vinyl phosphine oxide anion is treated with an aromatic aldehyde. The starting anion is generated by adding a Grignard reagent to a phosphine oxide alkyne.[104] Transformation of the allylic alcohol prepared from the intermediate magnesium salt to a sodium salt promoted decomposition to the allene. The sodium salt, in contrast to the copper or magnesium salt, is unstable.

Imines from Nitroso Substrates

The reaction of nitrosobenzenes with phosphonate anions has been described.[105] It is assumed that the mechanism for imine formation is

[101] S. D. Andrews, A. C. Day, and R. N. Inwood, *J. Chem. Soc., C*, **1969,** 2443.

[102] G. Kresze, W. Runge, and E. Ruch, *Ann.*, **756,** 112 (1972).

[103] M. Marszak, M. Simalty, and A. Seuleiman, *Tetrahedron Lett.*, **1974,** 1905.

[104] A. M. Aguiar and J. R. S. Irelan, *J. Org. Chem.*, **34,** 4030 (1969).

[105] H. Zimmer, P. J. Berez, and G. H. Heuer, *Tetrahedron Lett.*, **1968,** 171.

identical to that of olefin formation and a cyclic intermediate may take part. The preparation of amidines is illustrative.

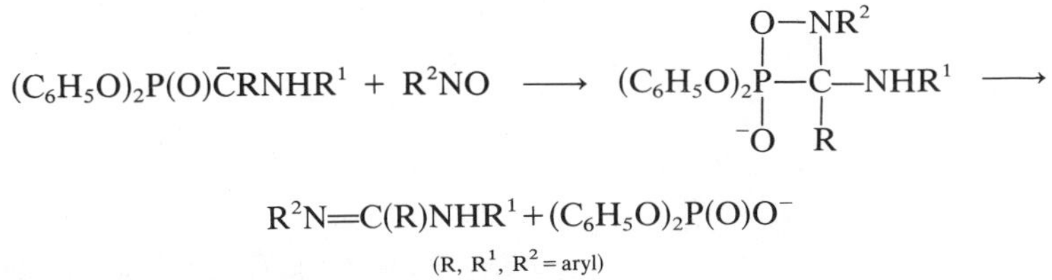

$$(C_6H_5O)_2P(O)\bar{C}RNHR^1 + R^2NO \longrightarrow (C_6H_5O)_2\overset{O-NR^2}{\underset{^-O}{P}}-\underset{R}{C}-NHR^1 \longrightarrow$$

$$R^2N{=}C(R)NHR^1 + (C_6H_5O)_2P(O)O^-$$

$(R, R^1, R^2 = \text{aryl})$

Condensation of nitroso compounds with phosphonate carbanions forms the basis of an interesting new pteridine synthesis.[106, 107] The first step, the formation of nitrogen–carbon unsaturation upon addition of the 4,6-diamino-5-nitrosopyrimidine to the anion, is followed by cyclization. Yields are high, and the method is valuable for it is unerring in its placement of substituents in the pyrazine ring.

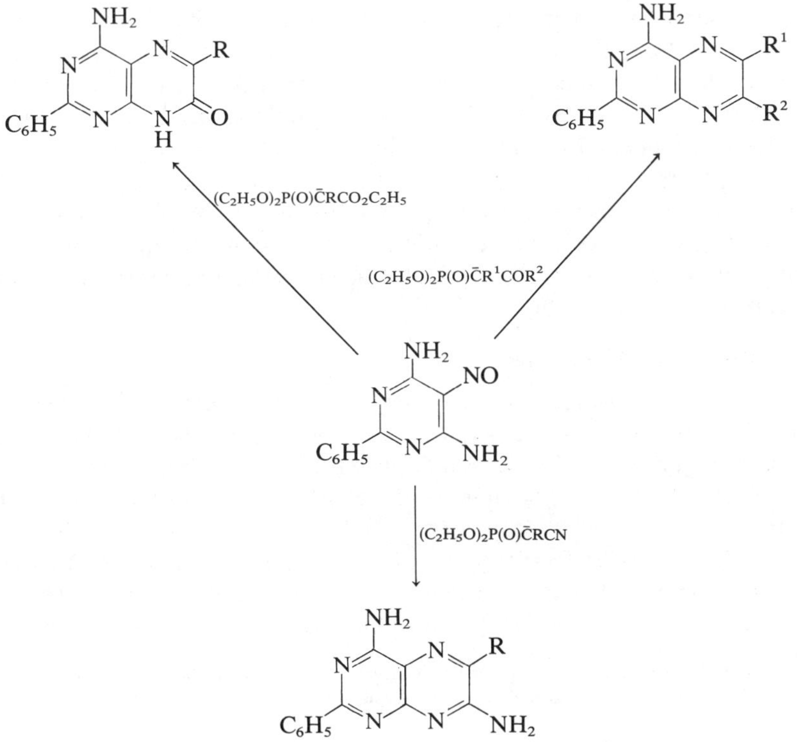

$(C_2H_5O)_2P(O)\bar{C}RCO_2C_2H_5$

$(C_2H_5O)_2P(O)\bar{C}R^1COR^2$

$(C_2H_5O)_2P(O)\bar{C}RCN$

[106] R. D. Youssetych and A. Kalmus, *Chem. Commun.*, **1970,** 1371.
[107] E. C. Taylor and B. E. Evans, *Chem. Commun.*, **1971,** 189.

Only the tertiary nitrosoalkanes, which unlike the primary and secondary are monomeric, react successfully with phosphonate carbanions.[108] Thus 2-methyl-2-nitrosopropane reacts with triethyl phosphonacetate or methyl (diethyl phosphono)-β-methylcrotonate in the presence of base to yield the expected aldimines.

$$(CH_3)_3CNO + (C_2H_5O)_2P(O)\bar{C}HCO_2C_2H_5 \rightarrow (CH_3)_3CN{=}CHCO_2C_2H_5$$

$$(CH_3)_3CNO + (C_2H_5O)_2P(O)\bar{C}HC(CH_3){=}CHCO_2CH_3 \rightarrow (CH_3)_3CN{=}CHC(CH_3){=}CHCO_2CH_3$$

Polymers

Attempts have been made to prepare conjugated polymers by means of the phosphonate-olefin formation reaction. Polymer formation has succeeded primarily by employing bisdialkylphosphonates and unsaturated dialdehydes as co-reactants.[109] It is claimed that water-soluble phosphorus-containing products can be prepared by carrying out the condensation in polar organic solvents such as dimethylformamide. The formation of polymers from the condensation of tetraethyl *p*-xylylenebisphosphonate with dialdehydes such as *p*-$C_6H_4(CHO)_2$, *p*-$C_6H_4(CH{=}CHCHO)_2$, 4,4′-biphenylylenedicarboxyaldehyde, and 2,4,6-octatrienedial has been described.[110] Tetraethyl 2-butylene-1,4-biphosphonate, tetraethyl 2,4-hexadiene-1,6-bisphosphonate, as well as ethylene-1,2-bis(diphenylphosphine oxide), have also been employed as substrates. Polymerization takes place under basic conditions in a variety of solvents. The usually chrome-yellow polymers have not been fully characterized.

Double-Bond Migrations

Many simple olefins formed by means of P(O)-stabilized anions lead to products that undergo subsequent reactions. Double-bond migration is observed under the basic conditions needed for phosphonate-olefin synthesis. It is, however, not so prevalent as in a phosphorane-olefin procedure.

Both 1-methyl-4-(carbethoxymethylene)phosphorane and N-methyl-4-carbethoxymethylenepiperidine isomerize under thermal conditions.[111] The tendency for the phosphine and the amine to rearrange is

[108] J. A. Maassen, A. J. W. Wajer, and J. de Boer, *Rec. Trav. Chim.*, **88**, 5 (1969).

[109] G. A. Lapitakii, S. M. Makin, and A. S. Chebotarev, Russ. Pat. 275, 408 [*C.A.*, **74**, 13593v (1971)].

[110] G. Drefahl, H. H. Hoerhold, and H. Wildner, Ger. (East) Pat. 51,436 (1966) [*C.A.*, **66**, 105346h (1967)].

[111] L. D. Quin, J. W. Russell, Jr., R. D. Prince, and H. E. Shook, Jr., *J. Org. Chem.*, **36**, 1495 (1971).

attributed in part to intramolecular catalysis at the basic centers. In

1-methyl-4-phosphorinanone + $(C_2H_5O)_2P(O)CH_2CO_2C_2H_5$ $\xrightarrow{\text{Base}}$ (exocyclic $HCCO_2C_2H_5$ product, P–CH_3) $\xrightarrow{\text{Heat}}$ (endocyclic $CH_2CO_2C_2H_5$ product, P–CH_3)

1-methyl-4-piperidone + $(C_2H_5O)_2P(O)CH_2CO_2C_2H_5$ $\xrightarrow{\text{Base}}$ (exocyclic $HCCO_2C_2H_5$ product, N–CH_3) $\xrightarrow{\text{Heat}}$ (endocyclic $CH_2CO_2C_2H_5$ product, N–CH_3)

contrast, cyclohexylideneacetate, prepared by phosphonate-olefin formation, does not thermally isomerize.[112] It has been reported that N-benzoylpiperidone, when treated with triethyl phosphonoacetate, yields both exo- and endo-cyclic unsaturation regardless of the base.[113] Here the exocyclic ester is the initial product. In a similar study of

N-(COC_6H_5)-3-ethyl-4-piperidone + $(C_2H_5O)_2P(O)\bar{C}HCO_2C_2H_5$ $\longrightarrow$ (exocyclic $C_2H_5O_2CCH$ product, N–COC_6H_5) + (endocyclic $CH_2CO_2C_2H_5$ product, N–COC_6H_5)

[112] W. S. Wadsworth, Jr. and W. D. Emmons, *Org. Syntheses*, **45**, 44 (1965).
[113] R. J. Sundberg and I. O. Holcombe, *J. Org. Chem.*, **34**, 3273 (1969).

double-bond migration, 2-(cyclohexen-1′-yl)cyclohexanone was found to produce an endocyclic nonconjugated ester when heated with triethyl phosphonoacetate carbanion, but the normal exocyclic product with diethyl cyanomethylphosphonate carbanion.[114] In contrast, the

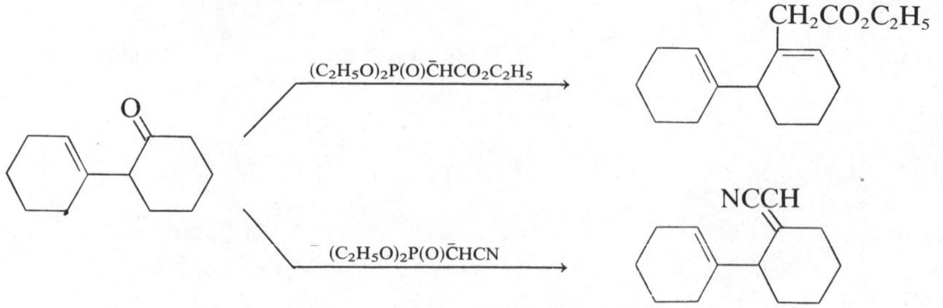

$(C_2H_5O)_2P(O)\bar{C}HCO_2C_2H_5$ → $CH_2CO_2C_2H_5$

$(C_2H_5O)_2P(O)\bar{C}HCN$ → NCCH

Reformatsky product was said to be the endocyclic conjugated ester. A β,γ-unsaturated ester was also reported as the major product obtained in one of the key steps in the total synthesis of anthramycin.

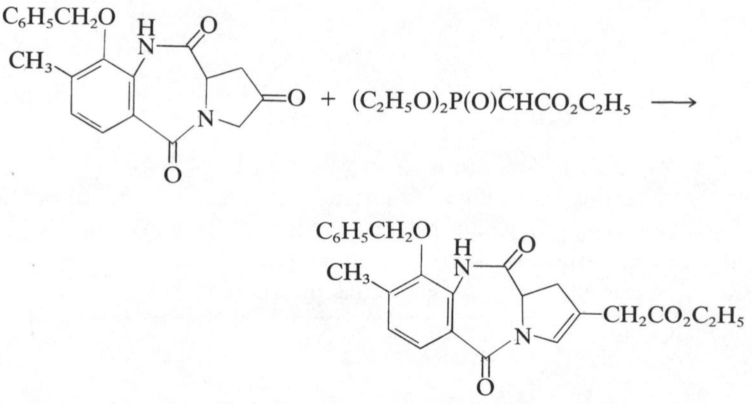

$C_6H_5CH_2O$, CH_3, H, N, O, N, =O + $(C_2H_5O)_2P(O)\bar{C}HCO_2C_2H_5$ ⟶

$C_6H_5CH_2O$, CH_3, H, N, O, N, $-CH_2CO_2C_2H_5$, O

In the synthesis of emetines it was noted that an intermediate ketone underwent phosphonate-olefin formation to give four products, the two epimers of both geometrical isomers of an exocyclic olefin (Eq. 26).[115] When sodium hydride is used as the base, only the *cis* and *trans* isomers of the epimerized product are obtained.[116] The different result with respect to products was originally ascribed to a difference in base strengths. Later the epimerization was correctly attributed to excess

[114] R. C. Gupta, S. C. Srivastava, P. K. Grover, and N. Anand, *Indian J. Chem.*, **1971,** 890.
[115] N. Whittaker, *J. Chem. Soc., C,* **1969,** 94.
[116] C. Szantay, L. Toke, and P. Kolonits, *J. Org. Chem.*, **31,** 1447 (1966).

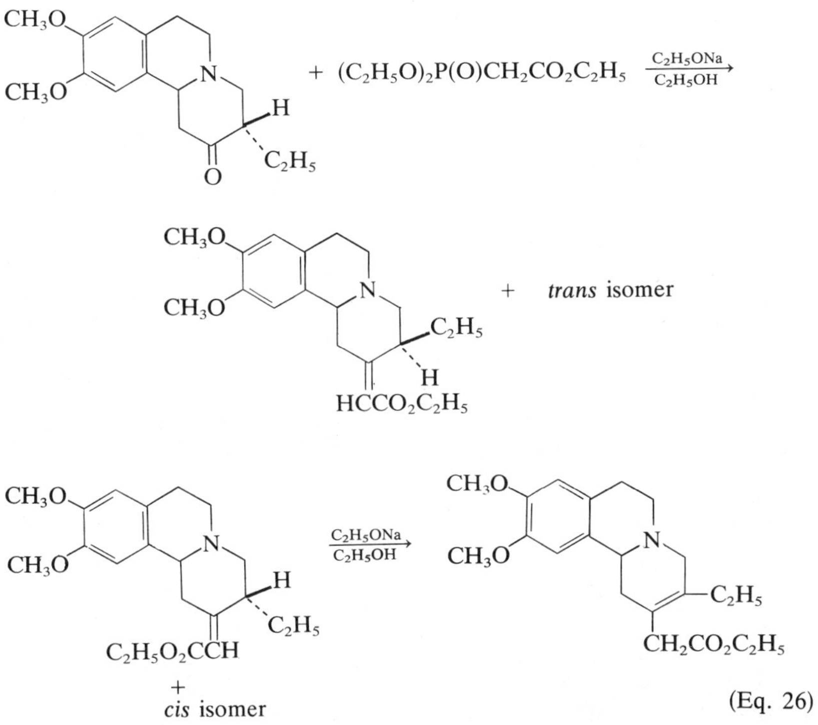

$+ (C_2H_5O)_2P(O)CH_2CO_2C_2H_5 \xrightarrow[C_2H_5OH]{C_2H_5ONa}$

$+$ *trans* isomer

$\xrightarrow[C_2H_5OH]{C_2H_5ONa}$

$+$
cis isomer

(Eq. 26)

base because only the geometrical isomers of the second unepimerized exocyclic olefin are formed when excess base is avoided regardless of the base used. In the presence of excess base, base-catalyzed isomerization of the starting material before olefin-bond formation is a distinct possibility. The exocyclic products in refluxing sodium ethoxide-ethanol isomerize to the endocyclic isomer.

Miscellaneous Examples of Olefin Formation

Phosphonate-olefin formation has been used in the preparation of 2-cyanomethyl-1-cyanomethylenecycloalkanes, which when treated with hydrogen bromide lead to functionally substituted azepines.[117]

The 3,4-dihydro-2H-pyranylethylenes undergo a Claisen rearrangement to yield cyclohexenes, a reaction that complements the usual Diels-Alder synthesis of these materials.[118,119] The starting materials,

[117] W. A. Nasutaricus and F. Johnson, *J. Org. Chem.*, **32,** 2367 (1967).

[118] G. Buchi and J. E. Powell, Jr., *J. Amer. Chem. Soc.*, **89,** 4560 (1967).

[119] G. Buchi and J. E. Powell, Jr., *J. Amer. Chem. Soc.*, **92,** 3126 (1970).

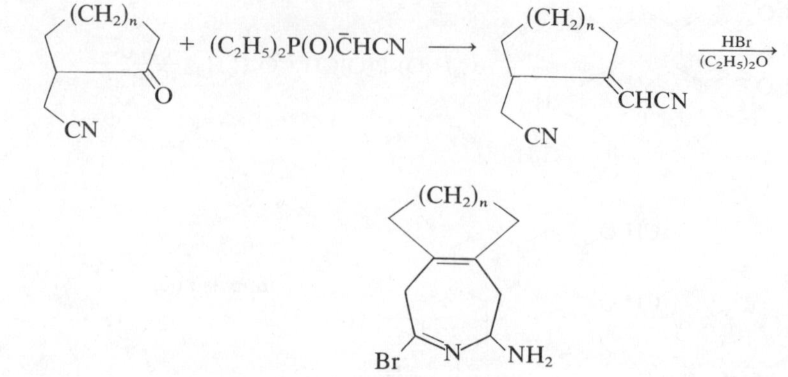

$$\text{2-(cyanomethyl)cycloalkanone } [(CH_2)_n] + (C_2H_5)_2P(O)\bar{C}HCN \longrightarrow \text{(CN-CH}_2\text{)cycloalkylidene-CHCN} \xrightarrow[(C_2H_5)_2O]{HBr} \text{2-bromo-7-amino azepine fused to } (CH_2)_n \text{ ring}$$

which are allyl vinyl ethers, are best prepared by treatment of ketene dimer with a phosphonate carbanion (Eq. 27). Claisen rearrangement

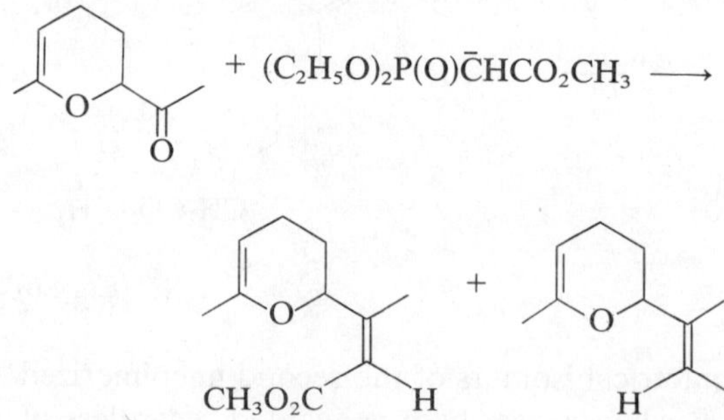

$$\text{2-acetyl-6-methyl-3,4-dihydro-2H-pyran} + (C_2H_5O)_2P(O)\bar{C}HCO_2CH_3 \longrightarrow$$

$$\text{pyranyl-C(CH}_3\text{)=C(H)CO}_2\text{CH}_3\ (CH_3O_2C, H) + \text{(H, } CO_2CH_3) \qquad \text{(Eq. 27)}$$

of the pure *trans* ester produces predominantly a *cis* cyclohexene, a result which would indicate that the stereochemical integrity of the ester is essentially preserved in the product (Eq. 28). The *cis* ester

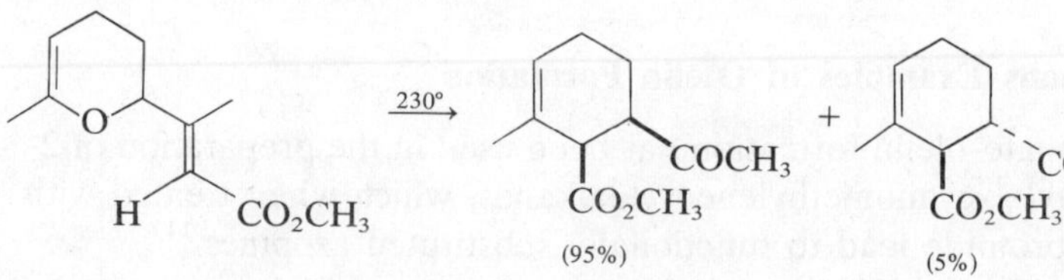

$$\text{pyranyl-C(CH}_3\text{)=CH(CO}_2\text{CH}_3) \xrightarrow{230^\circ} \text{cyclohexene(COCH}_3\text{, CO}_2\text{CH}_3) \ (95\%) + \text{cyclohexene(COCH}_3\text{, CO}_2\text{CH}_3) \ (5\%) \qquad \text{(Eq. 28)}$$

produces a more crowded transition state during rearrangement and thus is more stable than the *trans*. The method is of value since substituted cyclohexenes may be prepared in which the substituents have unambiguous positions, a result that is not always possible in the Diels-Alder synthesis.

A single case of asymmetric induction by means of phosphonate-olefin formation has been reported.[120] Upon condensation with 4-methylcyclohexanone, the carbanion from diethyl (–)-carbomenthoxymethylphosphonate produces an ester that hydrolyzes to an acid of 50% optical purity.

$$(C_2H_5O)_2P(O)CH_2CO_2\text{-menthyl}(-) + CH_3\text{-}C_6H_9{=}O \xrightarrow[2.\ H_3O^+]{1.\ \text{Base}} CH_3\text{-}C_6H_9{=}C(H)CO_2H$$

Phosphonate-Olefin Formation in Natural Product Syntheses

The high order of stereospecificity and the experimental simplicity of the phosphonate-olefin formation method have made it a popular route for the introduction of unsaturation concurrent with chain lengthening. Its development has occurred largely in natural product syntheses. Although a comprehensive review is not intended, an extension of the examples previously given is needed to acquaint the reader more fully with the scope of the reaction. The tabular survey is intended as a complete listing.

The formation of predominantly *trans* unsaturation has rendered phosphonate-olefin formation particularly attractive for the synthesis of isoprenoid compounds: terpenes, sesquiterpenes, diterpenes, and their oxygenated analogs. A recent series of papers, for example, describes the preparation of terpene alcohols in which olefin formation by means of the ethyl α-diethylphosphonopropionate carbanion is a key step.[121–127] Specifically, 2,6-dimethyl-3,6-epoxyoct-7-enol [(±)-lilac alcohol] is prepared in a number of steps in which a dioxolane obtained by phosphonate-olefin formation is a key intermediate.

[120] I. Tomoskozi and G. Janzso, *Chem. Ind.* (London), **1962,** 2085.
[121] O. P. Vig, A. Lal, G. Singh, and K. L. Matta, *Indian J. Chem.*, **1968,** 431.
[122] O. P. Vig and R. C. Anand, *Indian J. Chem.*, **1970,** 851.
[123] O. P. Vig, R. S. Bhatt, J. Kavr, and J. C. Kapur, *Indian J. Chem.*, **1973,** 37.
[124] O. P. Vig, J. P. Salota, M. P. Sharma, and S. D. Sharma, *Indian J. Chem.*, **1968,** 369.
[125] O. P. Vig, J. P. Salota, V. Baldev, and R. Bhagat, *Indian J. Chem.*, **1967,** 475.
[126] O. P. Vig, J. C. Kapur, J. Singh, and B. Vig, *Indian J. Chem.*, **1969,** 574.
[127] O. P. Vig, B. Vig, and R. C. Anand, *Indian J. Chem.*, **1969,** 1111.

$$OHCCH_2CH_2C(OCH_2CH_2O)CH_3 + (C_2H_5O)_2P(O)\bar{C}(CH_3)CO_2C_2H_5 \longrightarrow$$

$$C_2H_5O_2CC(CH_3){=}CHCH_2CH_2C(OCH_2CH_2O)CH_3 \xrightarrow{HCl}$$

$$C_2H_5O_2CC(CH_3){=}CHCH_2CH_2COCH_3 \xrightarrow{CH_2{=}CHMgBr}$$

$$C_2H_5O_2CC(CH_3){=}CHCH_2CH_2C(OH)(CH_3)CH{=}CH_2 \xrightarrow[C_6H_6]{NaI}$$

$$C_2H_5O_2CCH(CH_3)\text{-(tetrahydrofuran-2-yl)-}C(CH{=}CH_2)(CH_3) \xrightarrow{LiAlH_4}$$

$$HOCH_2CH(CH_3)\text{-(tetrahydrofuran-2-yl)-}C(CH{=}CH_2)(CH_3)$$

The first step in the total synthesis of the sesquiterpene, (±)-4-demethylaristolone, requires preparation of an α,β-unsaturated nitrile.[128, 129] The photolysis of several 1,6-dienes prepared by means

$$\text{2-}[CH{=}C(CH_3)_2]\text{-2-}CH_3\text{-cyclohexanone} + (C_2H_5O)_2P(O)\bar{C}HCN \longrightarrow$$

$$\text{1-}(=CHCN)\text{-2-}[CH{=}C(CH_3)_2]\text{-2-}CH_3\text{-cyclohexane}$$

of phosphonate-olefin formation led to head-to-head cyclization. For example, 5-ketohexanal has been converted by this procedure into the naturally occurring α-bourbonene.[130]

$$\text{(keto aldehyde, CHO)} + 2\ (C_2H_5O)_2P(O)\bar{C}HCOCH_3 \longrightarrow$$

$$\text{(diene: }{=}CHCOCH_3\text{, }CH{=}CHCOCH_3\text{)} \xrightarrow{h\nu} \text{(bicyclic: }COCH_3\text{, }COCH_3\text{)}$$

[128] E. Piers, W. DeWaal, and R. W. Britton, *Chem. Commun.*, **1968,** 188.

[129] E. Piers, W. DeWaal, and R. W. Britton, *Can. J. Chem.*, **47,** 4299 (1969).

[130] M. Brown, *J. Org. Chem.*, **33,** 162 (1968).

The phosphonate-olefin formation procedure has been used with notable success to prepare carotenoids and related compounds. The syntheses of *trans* retinoate and retinol with radioactive carbon at the ten position has been accomplished.[131] A phosphonate anion was used in an early stage of the synthesis.

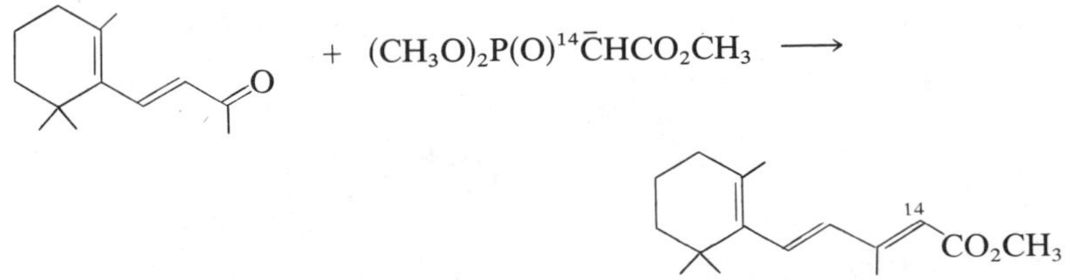

$$+ \ (CH_3O)_2P(O)^{14}\bar{C}HCO_2CH_3 \longrightarrow$$

$$^{14}\ CO_2CH_3$$

Phosphonate-olefin formation is especially suited to synthesis of ^{14}C-labelled isoprenoid alcohols. The label can be introduced near the end of the reaction sequence, and a high *trans/cis* ratio is obtained. An example is the condensation of all-*trans* farnesylacetone with labelled methyl diethyl phosphonoacetate-2-^{14}C.[132] The *trans* product can be separated from the *cis* and reduced to geranylgeraniol-2-^{14}C by means of lithium aluminum hydride. β-Carotene has been

$$RCH_2COCH_3 + (C_2H_5O)_2P(O)^{14}CH_2CO_2CH_3 \xrightarrow{NaH}$$

$$RCH_2C(CH_3){=}^{14}CHCO_2CH_3 \xrightarrow{LiAlH_4} RCH_2C(CH_3){=}^{14}CHCH_2OH$$

$$R = (CH_3)_2C{=}CH[CH_2CH_2C(CH_3){=}CHCH_2]_2$$

prepared in good yield by condensation of retinyl phosphonate with vitamin A aldehyde.[133]

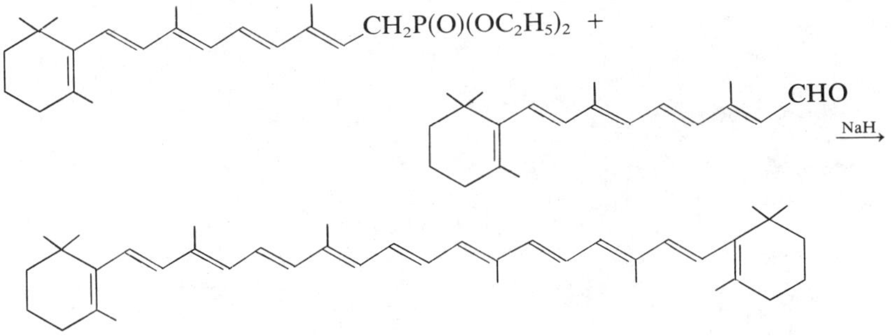

$$CH_2P(O)(OC_2H_5)_2 \ +$$

$$CHO \quad \xrightarrow{NaH}$$

[131] J. D. BuLock, S. A. Ovarrie, and D. A. Taylor, *J. Labelled Compd.*, **9**, 311 (1973).
[132] W. M. Walter, Jr., *J. Labelled Compd.*, **3**, 54 (1967).
[133] J. D. Surmatis and R. Thommen, *J. Org. Chem.*, **34**, 559 (1969).

Phosphonate-olefin formation has been employed extensively in the preparation of terpenoids that may possess insect juvenile hormone activity. An example is the final step in the synthesis of sirenin, a sex hormone.[134, 135] A similar route was used to prepare various

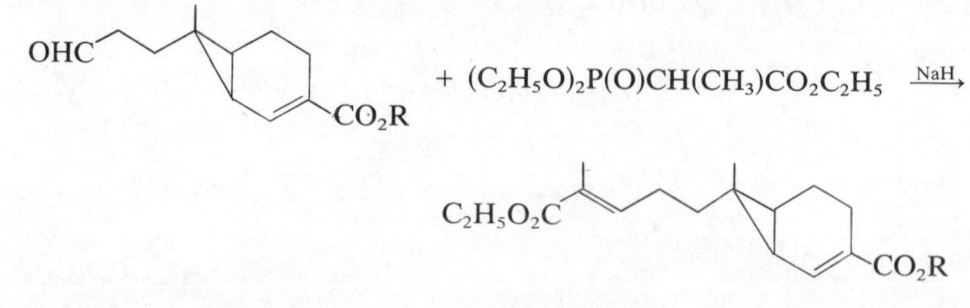

$$\text{OHC}\cdots\text{(bicyclic)}\cdots CO_2R + (C_2H_5O)_2P(O)CH(CH_3)CO_2C_2H_5 \xrightarrow{NaH} C_2H_5O_2C\cdots\text{(bicyclic)}\cdots CO_2R$$

juvenoids that have unusual structures and are remarkable for their exceptionally high specificity.[136] In their preparation, unsaturation is introduced by means of the olefin-formation reaction. A similar phosphonoacetate approach using 2-^{14}C-phosphonate was employed in the synthesis of a ^{14}C-labelled juvenile hormone needed for biological studies.[137] The juvenile hormone isolated from abdomens of adult male *Cecropia* moths, methyl 10,11-epoxy-7-ethyl-3,11-dimethyl-2,6-tridecandienoate, has been independently synthesized by a route in which phosphonate-olefin formation was used before final epoxidation.[138–142]

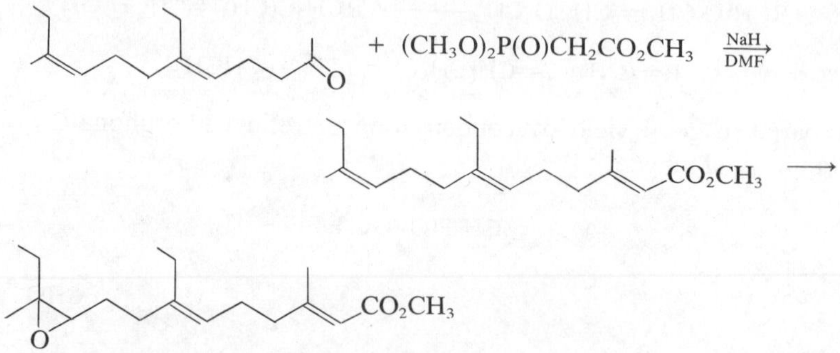

$$\text{(ketone)} + (CH_3O)_2P(O)CH_2CO_2CH_3 \xrightarrow[DMF]{NaH} \cdots CO_2CH_3 \longrightarrow \text{(epoxide)}\cdots CO_2CH_3$$

[134] K. Mori and M. Matsui, *Tetrahedron*, **26**, 2801 (1970).
[135] K. Mori and M. Matsui, *Tetrahedron Lett.*, **1969**, 4435.
[136] M. Romanuk and F. Sorm, *Coll. Czech. Chem. Commun.*, **38**, 2296 (1973).
[137] W. Hafferl, R. Zurflult, and L. Dunham, *J. Labelled Compd.*, **7**, 267 (1971).
[138] B. H. Braun, M. Jacobson, M. Schwarz, P. E. Sonnet, N. Wakabayashi, and R. M. Waters, *J. Econ. Entomol.*, **61**, 866 (1968) [*C.A.*, **69**, 58998x (1968)].
[139] G. W. K. Cavill, D. G. Laing, and P. J. Williams, *Aust. J. Chem.*, **22**, 2145 (1969).
[140] J. S. Cochrane and J. R. Hanson, *J. Chem. Soc., Perkin I*, **1972**, 361.
[141] K. H. Dahm, B. M. Trost, and H. Röller, *J. Amer. Chem. Soc.*, **89**, 5292 (1967).
[142] K. H. Dahm, H. Röller, and B. M. Trost, *Life Sci., Part 3*, **7**, 129 (1968).

The synthesis of the four geometrical isomers of (±)-pyrethric acid has been described.[143] The effect of substituents and geometry of the aldehyde on the final stereochemistry of the product is of interest. The *trans* aldehyde ester gives the *trans,cis* product ($R = CH_3$) with only 17% stereoselectivity, while the *cis* aldehyde ester gives the *cis,cis* compound with 45% selectivity (Eq. 29). The four separated *t*-butyl

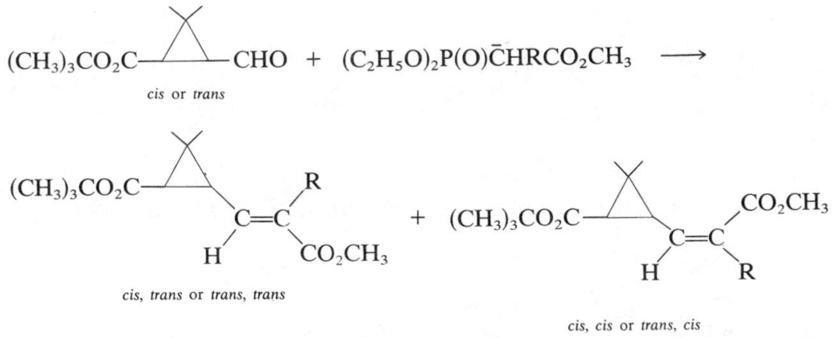

$(CH_3)_3CO_2C$—(cyclopropane)—CHO + $(C_2H_5O)_2P(O)\bar{C}HRCO_2CH_3$ ⟶

cis or *trans*

$(CH_3)_3CO_2C$—(cyclopropane)—CH=C(R)CO_2CH_3 + $(CH_3)_3CO_2C$—(cyclopropane)—CH=C(R)CO_2CH_3

cis, trans or *trans, trans*   *cis, cis* or *trans, cis*

(Eq. 29)

esters may be selectively hydrolyzed to the four geometrical isomers of (±)-pyrethric acid. The *trans* aldehyde ester condenses with methyl diethylphosphonacetate (R = H) to give the *trans,trans* product with more than 93% stereoselectivity. With diethyl 1-cyanoethylphosphonate, which is considered to allow rather loose steric control, the *trans,trans* product is formed with only 70% stereoselectivity. The

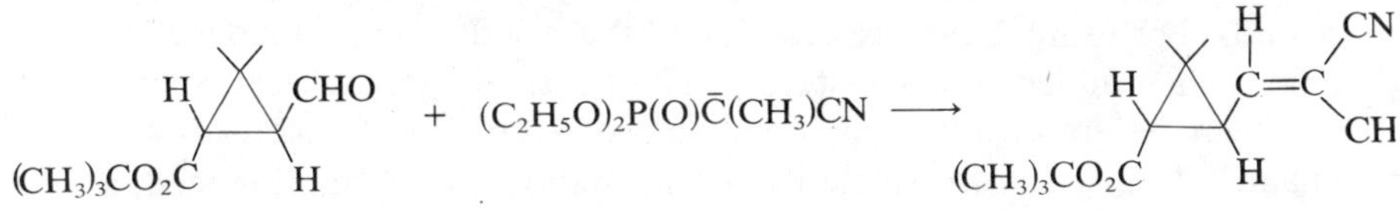

$(CH_3)_3CO_2C$(H)—(cyclopropane)—(H)CHO + $(C_2H_5O)_2P(O)\bar{C}(CH_3)CN$ ⟶ $(CH_3)_3CO_2C$(H)—(cyclopropane)—(H)C(H)=C(CN)CH_3

trans,cis (±)-pyrethric acid prepared from the *trans* aldehyde ester after slight structural modification has pronounced toxicity to horseflies. The yield of *trans,cis t*-butyl ester from the *trans* aldehyde ester starting material is increased from 17 to 61.5% by carrying out the phosphonate-olefin formation at low temperatures (−60° *vs.* room temperature).[31] The temperature effect is believed to reflect the lack of equilibration between the various betaines at low temperature.

Phosphonate-olefin formation has received significant application to prostaglandin chemistry. It has primarily been employed to complete

[143] K. Ueda and M. Matsui, *Agri. Biol. Chem.*, **34,** 1119 (1970).

the synthesis of the C-8 side chain.[144–150] The reaction has been used

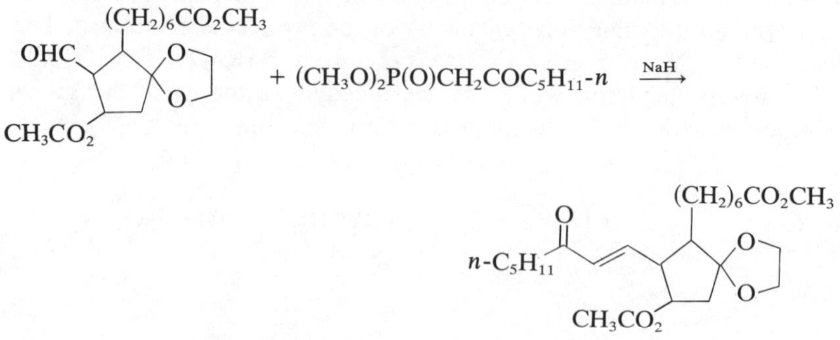

$(CH_2)_6CO_2CH_3$, OHC, CH_3CO_2 (dioxolane) + $(CH_3O)_2P(O)CH_2COC_5H_{11}$-$n$ $\xrightarrow{NaH}$ n-C_5H_{11}, O, $(CH_2)_6CO_2CH_3$, CH_3CO_2

in a synthesis of the C_{11} epimers of the natural E_1 and F_1 hormones.[151]

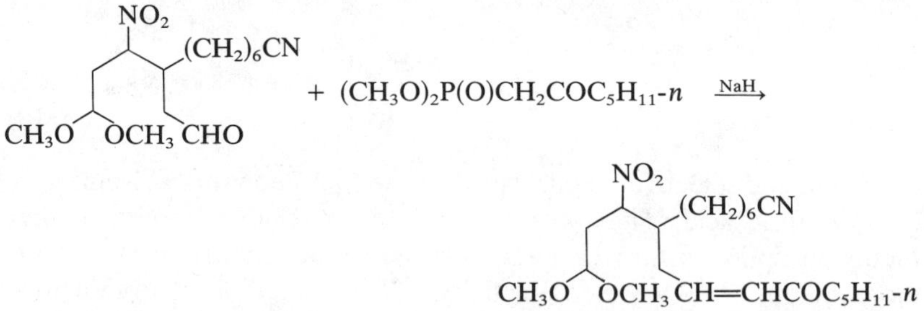

NO_2, $(CH_2)_6CN$, CH_3O OCH_3 CHO + $(CH_3O)_2P(O)CH_2COC_5H_{11}$-$n$ $\xrightarrow{NaH}$ NO_2, $(CH_2)_6CN$, CH_3O OCH_3 $CH{=}CHCOC_5H_{11}$-n

Treatment of steroid aldehydes and ketones with phosphonate anions has transformed them into potentially biologically active compounds. For example, it has been found that triethyl phosphonoacetate anion when added to 3-oxosteroids gives a single geometrical isomer that can be transformed in a number of steps into a cortical side chain.[86] The reason for single isomer formation is uncertain; indeed, a change in base from sodium hydride to potassium t-butoxide leads to

[144] D. Taub, R. D. Hoffsommer, C. H. Kuo, H. L. Slates, Z. S. Zelawski, and N. L. Wendler, *Chem. Commun.*, **1970,** 1258.

[145] H. L. Slates, Z. S. Zelawski, D. Taub, and N. L. Wendler, *Chem. Commun.*, **1972,** 304.

[146] B. J. Magerlein, Ger. Pat. 2,217,044 [*C.A.*, **78,** 42897p (1973)].

[147] E. J. Corey, S. Terashima, P. N. Ramwell, R. Jessup, N. M. Weinshenker, D. M. Floyd, and G. A. Crosby, *J. Org. Chem.*, **37,** 3043 (1972).

[148] M. Hayashi, H. Miyake, T. Tanouchi, S. Iguchi, Y. Iguchi, and F. Tanouchi, *J. Org. Chem.*, **38,** 1250 (1973).

[149] J. F. Bagli and T. Bogri, Ger. Pat. 2,231,244 [*C.A.*, **78,** 159057f (1973)].

[150] J. Himizu, S. Harigaya, A. Ishida, K. Yoshikawa, and M. Sato, Ger. Pat. 2,229,225 [*C.A.*, **78,** 111102q (1973)].

[151] E. J. Corey, I. Vlattas, N. H. Anderson, and K. Harding, *J. Amer. Chem. Soc.*, **90,** 3247 (1968).

formation of the opposite isomer as the major product. In a less

C_8H_{17}

O H + $(C_2H_5O)_2P(O)CH_2CO_2C_2H_5$ $\xrightarrow{NaH}$

$C_2H_5O_2CCH$ H

straightforward side-chain modification, cardenolides are obtained by condensing a C-20 ketone with diethyl cyanomethylphosphonate.[152–155] An iminolactone intermediate is postulated. The yield of cardenolide is substantially reduced as a result of formation of an α,β-unsaturated nitrile with the nitrile and acetoxy groups *trans*, a situation that prohibits ring closure. The cardenolides inhibit cell growth and

CH_2OCOCH_3
C=O

CH_3CO_2 + $(C_2H_5O)_2P(O)CH_2CN$ $\xrightarrow[2.\ H_3O^+]{1.\ NaH}$ CH_2OCOCH_3 C=CHCN

HN O + $\xrightarrow{CH_3OH+HCl}$ O O

(24–65%)

[152] G. R. Pettit, C. L. Herald, and J. P. Yardley, *J. Org. Chem.*, **35,** 1389 (1970).
[153] Farbwerke Hoechst, A. G., Fr. Pat. 1,491,081 (1967) [*C.A.*, **69,** 77606j (1968)].
[154] Farbwerke Hoechst, A. G., Neth. Pat. 6,607,315 (1966) [*C.A.*, **66,** 115884w (1967)].
[155] H. Kaneko and M. Okazaki, Jpn. Pat. 6,821,059 (1968) [*C.A.*, **70,** 58136k (1969)].

have an effect upon heart muscle. In summary, phosphonate-olefin formation is firmly established as a synthetic tool in natural product syntheses, a use which will undoubtedly become even more wide spread.

COMPARISON WITH OTHER METHODS

Of the numerous methods available for introducing unsaturation accompanied by chain elongation, olefin formation by means of phosphoryl-stabilized anions has a number of advantages. They are particularly noteworthy when a direct comparison is made with alternative procedures involving Wittig reagents (phosphoranes). The latter generally do not undergo smooth alkylation. In contrast, numerous examples of both alkylation and acylation of P(O)-stabilized carbanions have been reported (Eq. 30).[4,63,92,93,156–160] Thus a means is provided

$$(RO)_2P(O)\bar{C}HR^1 \xrightarrow{R^2X} (RO)_2P(O)CHR^1R^2 \xrightarrow[2.\ R^3R^4CO]{1.\ Base} R^3R^4C{=}CR^1R^2 + (RO)_2P(O)O^- \quad \text{(Eq. 30)}$$

for the elaboration of structure that greatly extends the utility of P(O)-stabilized carbanions in olefin formation. It is also apparent that phosphoryl-stabilized carbanions are more nucleophilic, the lower nucleophilicity of the phosphoranes arising from the delocalization of the negative charge by overlap of the filled orbital of negative carbon with an empty phosphorus $3d$ orbital. Delocalization is less in the P(O)-stabilized anions because of a smaller positive charge on phosphorus, a result of back donation from oxygen.

Direct comparisons between P(O)-stabilized carbanions and phosphoranes have been made.[17] A competitive reaction between diphenylbenzylphosphine oxide and triphenylbenzylphosphonium bromide was carried out by treating a mixture containing one equivalent of each with one equivalent of benzaldehyde. With one equivalent of base, triphenylphosphine oxide predominated in the product whereas with excess base diphenylphosphonic acid predominated. In the latter case

$$(C_6H_5)_3\overset{+}{P}CH_2C_6H_5\ Br^- + (C_6H_5)_2P(O)CH_2C_6H_5 \xrightarrow[C_6H_5CHO]{t\text{-}C_4H_9OK} C_6H_5CH{=}CHC_6H_5 + (C_6H_5)_3P(O) + (C_6H_5)_2P(O)OH$$

[156] A. N. Pudovik and N. M. Lebedeva, *J. Gen. Chem. USSR*, **25**, 1920 (1955).
[157] G. M. Kosolapoff, *J. Amer. Chem. Soc.*, **75**, 1500 (1953).
[158] B. R. Baker and G. J. Lowrens, *J. Med. Chem.*, **11**, 672 (1968).
[159] H. Hoffmann, *Angew. Chem.*, **71**, 379 (1959).
[160] S. Trippett and D. M. Walker, *Chem. Ind.* (London), **1961**, 990.

it was estimated that about 90% of the stilbene was produced by reaction of the phosphinoxy carbanion with aldehyde. The experiments provide evidence that, although the phosphonium salts are more acidic than the phosphine oxides, the carbanions from the latter are more nucleophilic and thus more reactive than Wittig reagents. Further evidence was provided by treatment of a phosphine oxide-phosphonium salt with benzaldehyde in the presence of excess base (Eq. 31).

$$[p\text{-}(C_6H_5)_2P(O)CH_2C_6H_4CH_2\overset{+}{P}(C_6H_5)_3]Br^- + (C_6H_5)_2CO \xrightarrow{t\text{-}C_4H_9OK}$$

$$[p\text{-}(C_6H_5)_2C{=}CHC_6H_4CH_2\overset{+}{P}(C_6H_5)_3]Br^- + (C_6H_5)_2PO_2H \qquad \text{(Eq. 31)}$$

Phosphonate carbanions with their negative charge stabilized by a carbonyl group react readily with ketones, whereas their phosphorane counterparts are much less reactive.[4,45,160] Treatment of a mixture of one equivalent of diethyl benzylphosphonate and one equivalent of benzyltriphenylphosphonium bromide with an equivalent of benzophenone and excess base led to olefin formation with 70% recovery of the unchanged phosphonium salt. This result provides good evidence

$$(C_2H_5O)_2P(O)CH_2C_6H_5 + [(C_6H_5)_3PCH_2C_6H_5]^+Br^- \xrightarrow[(C_6H_5)_2CO]{t\text{-}C_4H_9OK}$$

$$(C_6H_5)_2C{=}CHC_6H_5 + (C_6H_5)_3P(O) + (C_2H_5O)_2POH$$

that phosphonate carbanions are more nucleophilic than phosphoranes. In contrast, diphenyl triphenylphosphoranylidenemethylphosphonate, which is stable and capable of isolation, was found to give exclusively *trans* diphenyl vinylphosphonates when treated with aromatic aldehydes (Eq. 32).[161] Here the difference in nucleophilicity

$$p\text{-}ClC_6H_4CHO + (C_6H_5)_3\overset{+}{P}\overset{-}{C}HP(O)(OC_6H_5)_2 \rightarrow$$

$$p\text{-}ClC_6H_4CH{=}CHP(O)(OC_6H_5)_2 + (C_6H_5)_3P(O) \qquad \text{(Eq. 32)}$$

is immaterial and the results indicate that phosphine oxide elimination from a betaine is more facile than phosphate monoanion elimination.

Perhaps as a consequence of their relatively low reactivity in olefin formation, the Wittig reagents give more artifacts.[56] The Wittig reaction of a *trans* allylic phosphorane and *n*-hexanal (Eq. 33) yields not

$$n\text{-}C_5H_{11}CHO + (C_6H_5)_3P{=}CHC(CH_3){=}CHCO_2CH_3 \rightarrow$$

$$n\text{-}C_5H_{11}CH{=}CHC(CH_3){=}CHCO_2CH_3$$

$$+ n\text{-}C_5H_{11}CH{=}C(CO_2CH_3)C(CH_3){=}CH_2 \qquad \text{(Eq. 33)}$$

[161] G. H. Jones, E. K. Hamamurz, and J. G. Moffatt, *Tetrahedron Lett.*, **1968,** 5371.

only the four possible geometric isomers of the expected diene, but also both geometric isomers of the diene arising from γ condensation.[57] In contrast, the analogous *trans* phosphonate provided only the *trans*-2-*trans*-4, and *trans*-2-*cis*-4 isomers of the α condensation product in a 6/1 ratio. Also, double-bond migration especially from an exo- to an endo-cyclic position is more prevalent in products obtained from a Wittig reagent than in those obtained from phosphonate anions.[111]

The workup of reaction mixtures and subsequent isolation of products are relatively simple when phosphonate anions are employed. The desired unsaturated compounds can be readily separated from the highly water-soluble alkali metal dialkylphosphate salts. Indeed, when solvents of low polarity are used, the phosphate salt ordinarily precipitates and can be removed by filtration. In contrast, phosphine oxides, the by-products of the Wittig reaction, normally have solubilities similar to those of the olefins; hence elaborate procedures are often needed for their removal.

There is a distinct difference in the isomer content of olefins prepared by phosphonate-olefin formation as compared to those from Wittig reagents. The former usually give a much higher percentage of the *trans* isomer.[18,24,91] For example, the reaction of pivaldehyde with the phosphorane $(C_6H_5)_3P{=}CHC_7H_{15}$-*n* produces a product containing 98.5% of *cis*-1-*t*-butyl-1-nonene. In contrast the condensation of α-lithio-*n*-octylphosphonic acid bisamide with the same aldehyde yields the olefin in a 3/1 mixture of *trans* to *cis*.[34] Diphenyl benzylphosphine oxide with benzaldehyde in dimethylformamide gives a 3/97 ratio of *cis* and *trans* stilbenes, whereas the analogous phosphorane, $(C_6H_5)_3P{=}CHC_6H_5$, under identical conditions yields a 60/40 isomer ratio.[162] In a final example the preparation of 3,3,5-trimethyl-1-phenyl-1,4-hexadiene by the Wittig method gives a mixture of isomers in which the *cis* predominates in a ratio of 3/2, whereas phosphonate-olefin formation gives only the *trans* isomer.[163] The isomer ratio also

$(C_2H_5O)_2P(O)\bar{C}HC_6H_5$

CHO $\xrightarrow{(C_6H_5)_3P{=}CHC_6H_5}$ C_6H_5 + C_6H_5

[162] L. D. Bergelson, V. A. Vaver, L. I. Barsukov, and M. M. Shemyakin, *Izv. Akad. Nauk SSSR, Ser. Khim.*, **1966,** 506 [*C.A.*, **65,** 10615e (1966)].

[163] H. E. Zimmerman, P. Baechstrom, T. Johnson, and D. W. Kurtz, *J. Amer. Chem. Soc.*, **96,** 1459 (1974).

appears to vary to a greater degree with solvent polarity with Wittig reagents than with phosphonate anions.

In reactions from which diastereomeric β-hydroxyphosphonates or β-hydroxyphosphonamides can be isolated, separated, and decomposed to olefin stereospecifically, the advantages over Wittig reagents are obvious. Although attainment of predictable and complete control of stereochemistry by this method is within our grasp, the technique is in its infancy and will require more information than is now available.

Introduction of unsaturation with chain-lengthening can, of course, be accomplished by an older method, the Reformatsky reaction. The Reformatsky reaction is well known to produce, in many instances, low yields and numerous side products, especially those arising from double-bond migration.[114, 164, 165]

In a few cases, anions stabilized by groups other than the P(O) group have been reported to give carbonyl addition products that decompose to olefins. Sulfinamide dianions react with either aldehydes or ketones to give addition products that can be isolated.[166] Upon heating, the adducts decompose to olefins often in high yields. In a similar manner certain β-hydroxy sulfoxides can be converted into olefins, although

$$\mathrm{RCH_2SONHC_6H_4CH_3\text{-}}p \xrightarrow{2\,n\text{-}\mathrm{C_4H_9Li}} \mathrm{R\underset{\displaystyle Li}{\underset{|}{C}}HSO\underset{\displaystyle Li}{\underset{|}{N}}C_6H_4CH_3\text{-}}p \xrightarrow[2.\ \mathrm{H_2O}]{1.\ \mathrm{R^1R^2CO}}$$

$$\mathrm{R^1R^2\overset{\displaystyle OH}{\overset{|}{C}}\underset{\displaystyle R}{\underset{|}{C}}HSONHC_6H_4CH_3\text{-}}p \xrightarrow[\text{reflux}]{\mathrm{C_6H_6}} \mathrm{R^1R^2C{=}CHR + SO_2 +}\ p\text{-}\mathrm{CH_3C_6H_4NH_2}$$

high temperatures are required.[167] The procedure appears to offer no advantage over phosphonate-olefin formation.

$$\mathrm{CH_3SOCH_3} \xrightarrow{n\text{-}\mathrm{C_4H_9Li}} \mathrm{LiCH_2SOCH_3} \xrightarrow[\mathrm{H_2O}]{\mathrm{(C_6H_5)_2CO}} \mathrm{(C_6H_5)_2\overset{\displaystyle OH}{\overset{|}{C}}CH_2SOCH_3} \xrightarrow[0.3\ \mathrm{mm}]{160^\circ} \underset{(45\%)}{\mathrm{(C_6H_5)_2C{=}CH_2}}$$

[164] J. L. Rabinowitz and M. Zanger, *J. Labelled Compd.*, **8,** 657 (1972).

[165] K. Fujiwara, *Nippon Kagaku Zasshi*, **84,** 659 (1963) [*C. A.*, **60,** 401f (1964)]; see also M. W. Rathke, *Org. Reactions*, **22,** 423 (1975).

[166] E. J. Corey and T. Durst, *J. Amer. Chem. Soc.*, **88,** 5656 (1966).

[167] E. J. Corey and M. Chaykorsky, *J. Amer. Chem. Soc.*, **87,** 1345 (1965).

More recently the condensation of ethyl trimethylsilylacetate carbanion with carbonyl compounds has been reported to produce α,β-unsaturated carboxylic esters.[168] Yields of over 80% are obtainable

$$(CH_3)_3SiCH_2CO_2C_2H_5 \xrightarrow{(C_6H_{11})_2NLi}$$

$$[(CH_3)_3Si\bar{C}HCO_2C_2H_5]Li^+ \xrightarrow{R_1R_2CO} R_1R_2C{=}CHCO_2C_2H_5$$

even in the case of readily enolizable ketones, *e.g.*, cyclopentanone. With Wittig-type reactions, proton transfers often compete with carbonyl addition.

PREPARATION OF PHOSPHORYL-STABILIZED ANIONS

Besides ease of workup and relatively high yields, it is the availability of starting materials that often makes phosphonate-olefin formation the method of choice. Wittig reagents (phosphoranes) are commonly derived from phosphines which, in turn, often require Grignard reagents in their preparation.

$$PCl_3 + 3\,RMgX \longrightarrow R_3P + 3\,MgX_2$$

$$R_3P + R_2'CHX \longrightarrow (R_3PCHR_2')^+X^- \xrightarrow{-HX} R_3P{=}CR_2'$$

Triphenylphosphine is commercially available, as are a few trialkyl phosphines. Phosphine oxides also require the preparation of a phosphine, although the problem is further complicated in that the desired phosphine normally does not contain identical alkyl groups attached to phosphorus. Oxidation of the phosphine yields the oxide. Tertiary phosphine oxides may also be prepared directly by action of a Grignard reagent on phosphorus oxychloride. The synthesis has no appar-

$$3\,RMgX + POX_3 \rightarrow R_3P(O) + 3\,MgX_2$$

ent advantage over the first method and addition must be conducted in a progressive stepwise manner if the alkyl groups are not all the same. This procedure is often tedious. Perhaps the most efficient procedure for placing a desired alkyl group on phosphorus entails the isomerization of an ester of phosphinous acid. The reaction is an

$$R_2POR' + R''X \rightarrow R_2P(O)R'' + R'X$$

[168] K. Shimoji, H. Taguchi, K. Oshima, H. Yamamoto, and H. Nozaki, *J. Amer. Chem. Soc.*, **96**, 1620 (1974).

example of the general Michaelis-Arbuzov rearrangement that is more simply carried out with triesters of phosphorus acid (trialkyl phosphites).[169] The latter are much more readily available than the esters of phosphinous acid and are the genesis of the phosphonates.

Trialkyl phosphites are commercially available at reasonable cost. Their laboratory synthesis merely entails the addition of a phosphorus trihalide, usually the chloride, to an alcohol in the presence of an acid scavenger. The Michaelis–Arbuzov reaction produces phosphonates

$$3\,ROH + PCl_3 \rightarrow (RO)_3P + 3\,HCl$$

from phosphites by expansion of the valence shell of the phosphorus atom and is a prime example of this unique aspect of organophosphorus chemistry. The P(O) bond energy is sufficiently high that its attainment renders the reaction thermodynamically desirable and outweighs any detrimental factors. Triethyl or trimethyl phosphites are

$$(RO)_3P + R'X \rightarrow (RO)_2P(O)R' + RX$$

normally employed. A wide variety of alkyl halides, primarily those capable of undergoing bimolecular nucleophilic displacements, can be used successfully. The reactivity sequences are: acyl > primary alkyl > secondary alkyl; and iodide > bromide > chloride. A limitation of the Michaelis–Arbuzov reaction is the formation of isomeric vinylphosphates. The Michaelis-Arbuzov reaction has been adequately reviewed.[169,170]

The Perkow reaction is possible whenever α-halo-aldehydes, -ketones, or α-halo esters are used.[171] Here also valency expansion of phosphorus is involved, the initial attack of phosphorus being believed to be at oxygen rather than at carbon.

$$(RO)_3P + XCH_2COR^1 \rightarrow (RO)_2P(O)OC(R^1){=}CH_2$$

With α-trihalo-aldehydes and -ketones, and α-dihalo-aldehydes and -ketones, vinyl phosphates are the preferred products; with α-monohaloketones, phosphonate formation competes successfully and can be made to predominate. The ratio of phosphonate to vinylphosphate is dependent upon the halide used and the reaction temperature. Higher temperatures appear to favor phosphonate formation. α-Chloroketones give predominantly the vinyl analog, α-bromoketones give better yields of the phosphonate, and α-iodo-ketones produce phosphonates almost exclusively. Thus with carefully controlled conditions

[169] G. Kosolapoff, *Organophosphorus Compounds*, John Wiley, New York, 1950, Chap. 7.

[170] P. Crofts, *Quart. Rev.*, **12,** 341 (1958).

[171] F. W. Lichtenthaler, *Chem. Rev.*, **61,** 607 (1961).

those phosphonates normally required for the formation of P(O)-stabilized carbanions can be formed in high yield. Unfortunately, α-haloketones that have halogen on a secondary carbon react to a significantly greater extent by the Perkow reaction than do those with halogen on a primary carbon. By necessity, therefore, the synthesis of a 1-alkyl-2-ketophosphonate must be accomplished by the alkylation of a phosphonate carbanion and the phosphonate cannot be prepared directly by the Michaelis–Arbuzov reaction.

Trichloroacetates give both Perkow and Michaelis–Arbuzov products, whereas the α-monohalo-esters and -amides regardless of the halide appear to give predominantly the desired phosphonacetates. Since α-haloaldehydes do not yield phosphonates but react with trialkyl phosphites almost exclusively by the Perkow route, an indirect route must be employed to prepare formyl phosphonates. The acetal of α-bromoacetaldehyde can be employed in the normal Michaelis–Arbuzov manner followed by hydrolysis.[172]

$$(C_2H_5O)_3P + BrCH_2CH(OC_2H_5)_2 \xrightarrow{-C_2H_5Br}$$

$$(C_2H_5O)_2P(O)CH_2CH(OC_2H_5)_2 \xrightarrow[HCl]{H_2O}$$

$$(C_2H_5O)_2P(O)CH_2CHO + 2\,C_2H_5OH$$

Tetraalkyl phosphonosuccinates can be prepared by adding a trialkyl phosphite to a monoalkyl ester of maleic acid.[173,174] The phosphonates condense with aldehydes to give olefins in yields that are superior to those from comparable Stobbe syntheses.[175]

$$(CH_3O)_3P + \begin{matrix} HCCO_2CH_3 \\ \| \\ HCCO_2H \end{matrix} \longrightarrow (CH_3O)_2P(O)CH(CO_2CH_3)CH_2CO_2CH_3 \xrightarrow[RCHO]{NaH}$$

$$RCH{=}C(CO_2CH_3)CH_2CO_2CH_3$$

The Michaelis–Arbuzov preparation of phosphonates is supplemented by the Michaelis–Becker–Nylen reaction, nucleophilic substitution of halide ion by salts of dialkyl hydrogen phosphonates.

$$(RO)_2P(O)Na + R'X \rightarrow (RO)_2P(O)R' + NaX$$

Ordinarily the sodium or potassium salts of the diethyl esters are used. There are limitations. In some cases side products severely lower the

[172] A. I. Razumov and V. V. Moskva, *J. Gen. Chem. USSR*, **34,** 2589 (1964).
[173] R. Harvey, *Tetrahedron*, **22,** 2561 (1966).
[174] R. S. Ludinton, U.S. Pat. 3,400,102 (1968) [*C.A.*, **69,** 78112g (1968)].
[175] L. S. Melvin, Jr. and B. M. Trost, *J. Amer. Chem. Soc.*, **94,** 1790 (1972).

yield of the desired phosphonate. Where possible, depending upon the structure of the halide, elimination often competes with substitution. From α-haloketones a mixture of products is obtained in which a 1,2-epoxyalkylphosphonate is a major component.[170, 176–181]

$$(RO)_2P(O)Na + XCH_2COR_1 \xrightarrow{-NaX} (RO)_2P(O)\overset{R_1}{\overset{|}{C}}\underset{\diagdown\ \diagup \atop O}{}\!\!-CH_2 + (RO)_2P(O)OC(R_1){=}CH_2 + (RO)_2P(O)CH_2COR_1$$

In order to overcome epoxide and vinyl phosphate formation, a novel synthesis based on the Michaelis–Becker–Nylen reaction has been developed.[178,182] The acetal of an α-haloketone is converted into an enol ether by heat before treatment with the sodium salt of the phosphonate esters.

$$XCH_2C(OR^1)_2CH_2R \xrightarrow[-ROH]{Heat} XCH_2C(OR^1){=}CHR \xrightarrow{(C_2H_5O)_2P(O)Na}$$

$$(C_2H_5O)_2P(O)CH_2C(OR^1){=}CHR \xrightarrow[HCl]{H_2O} (C_2H_5O)_2P(O)CH_2COCH_2R$$

Ethynyl phosphine oxides and phosphonates can be prepared by yet another procedure. Di-*n*-butyl propynyl-1-phosphonate is obtained from the addition of an acetylenic Grignard reagent to dibutyl phosphorochloridate.[183, 184] A convenient preparation of dialkyl ethynylphos-

$$CH_3C{\equiv}CMgBr + (C_4H_9O)_2P(O)Cl \rightarrow (C_4H_9O)_2P(O)C{\equiv}CCH_3 + MgBrCl$$

phonates has been reported that employs the previously described Michaelis–Arbuzov reaction.[185] The acetylenic phosphonates and

$$(RO)_3P + ClC{\equiv}CSiMe_3 \longrightarrow (RO)_2P(O)C{\equiv}CSiMe_3 + RCl \xrightarrow{OH^-} (RO)_2P(O)C{\equiv}CH$$

[176] B. A. Arbuzov, V. S. Vinogradova, and N. A. Polezhaeva, *Dokl. Akad. Nauk SSSR*, **111,** 107 (1956) [*C.A.*, **51,** 8001g (1957)].

[177] B. A. Arbuzov, V. S. Vinogradova, and N. A. Polezhaeva, *Izv. Akad. Nauk SSSR, Otd. Khim. Nauk*, **41,** (1959) [*C.A.*, **53,** 15035e (1959)].

[178] H. Normant and G. Sturtz, *C.R. Acad. Sci., Ser. C*, **253,** 2366 (1961).

[179] R. F. Hudson, *Structure and Mechanism in Organophosphorus Chemistry*, Academic Press, London, 1965, Chap. 5.

[180] G. Sturtz, *Bull. Soc. Chim. Fr.*, **1964,** 2333.

[181] A. Meisters and J. M. Swan, *Aust. J. Chem.*, **18,** 159 (1956).

[182] H. Normant and G. Sturtz, *C.R. Acad. Sci., Ser. C*, **256,** 1800 (1963).

[183] B. G. Christensen, W. Leanza, T. R. Beattie, A. A. Patchett, B. H. Arison, R. E. Ormond, and F. A. Kuehl, *Science*, **166,** 123 (1969).

[184] N. N. Girota and N. L. Wendler, *Tetrahedron Lett.*, **1969,** 4647.

[185] D. W. Burt and P. Simpson, *J. Chem. Soc., C*, **1969,** 2273.

phosphine oxides are important because additions to the triple bond give rise to other classes of phosphonates and phosphine oxides that are capable of participating in olefin formation.

A final procedure for the preparation of organophosphorus compounds containing P(O) groups requires addition of dialkyl phosphites, $(RO)_2P(O)H$, or phosphonous acids, $R_2P(O)H$, to double bonds. Many substrates suitable for olefin synthesis have been prepared by this method.[186] In a similar fashion, under basic conditions diethyl phosphite adds successfully to azomethines.[187,188]

$$(C_2H_5O)_2P(O)H + RN{=}CHC_6H_5 \rightarrow (C_2H_5O)_2P(O)CH(C_6H_5)NHR$$

The utility of phosphonate-olefin formation and phosphine oxide-olefin formation is enhanced by the large number of procedures that have become available for the modification of the phosphoryl substrates. In this manner the range of olefins that can be prepared by these methods is greatly extended. Alkylation and acylation of anions have been mentioned.

A few additional procedures should be brought to the reader's attention. Preferential alkylation at the γ-carbon atom of a β-ketophosphonate takes place upon treatment of a 1,3-dianion with an alkyl halide.[189] Alkylation at the γ position appears to be highly specific. It has been reported that P(O)-stabilized anions undergo

$$(CH_3O)_2P(O)CH_2COCH_3 \xrightarrow[\text{THF}]{\text{NaH}} (CH_3O)_2P(O)\bar{C}HCOCH_3 \xrightarrow{n\text{-}C_4H_9Li}$$

$$(CH_3O)_2P(O)\bar{C}HCO\bar{C}H_2 \xrightarrow[2.\ H^+]{1.\ RX} (CH_3O)_2P(O)CH_2COCH_2R$$

$$(C_6H_5)_3P(O) + CH_3Li \longrightarrow (C_6H_5)_2P(O)CH_3 + C_6H_5Li \longrightarrow$$

$$(C_6H_5)_2P(O)\bar{C}H_2 \xrightarrow[2.\ H^+]{1.\ CO_2} (C_6H_5)_2P(O)CH_2CO_2H$$

carbonylation.[190,191] Carbonylation of anions may assume some importance in light of the recent report that phosphonate anions also undergo carbonylation to give phosphonoacetic acids. Salts of the

[186] A. N. Pudovik, *Usp. Khim.*, **23,** 547 (1954) [*C.A.*, **49,** 8788i (1955)].
[187] L. V. Hopkins, J. P. Vacik, and W. H. Shelver, *J. Pharm. Sci.*, **61,** 114 (1972).
[188] B. P. Lugovkin, *J. Gen. Chem. USSR*, **40,** 562 (1970).
[189] P. A. Grieco and C. S. Pogonowski, *J. Amer. Chem. Soc.*, **95,** 3071 (1973).
[190] D. Seyferth, D. E. Welch, and J. K. Heeren, *J. Amer. Chem. Soc.*, **86,** 1100 (1964).
[191] D. Seyferth and D. E. Welch, *J. Organometal. Chem.*, **2,** 1 (1964).

latter are active olefin-formation reagents and indicate the possibility of a direct synthesis of α,β-unsaturated acids with a dialkyl phosphite as the starting material.[192]

$$(RO)_2P(O)H \xrightarrow[2.\ CH_3I]{1.\ CH_3Li} (RO)_2P(O)CH_3 \xrightarrow[2.\ CO_2, H^+]{1.\ CH_3Li}$$

$$(RO)_2P(O)CH_2CO_2H \xrightarrow{2(i\text{-}C_3H_7)_2NLi} (RO)_2P(O)\bar{C}HCO_2^- \xrightarrow[2.\ H^+]{1.\ R_2CO} R_2C{=}CHCO_2H \quad (60\text{–}80\%)$$

Phosphonate carbanions condense with esters, a reaction that may in some cases be an alternative route to α-phosphonoketones. This method was used in a modification of the phosphonamide route to olefins.[11] The preparation of the starting methylphosphonic acid bisamide is included in the scheme. Similarly, properly substituted

$$CH_3Cl + PCl_3 \xrightarrow[H_2O]{AlCl_3} CH_3P(O)Cl_2 \xrightarrow{2(CH_3)_2NH} CH_3P(O)[N(CH_3)_2]_2 \xrightarrow{n\text{-}C_4H_9Li}$$

$$\bar{C}H_2P(O)[N(CH_3)_2]_2 \xrightarrow{C_6H_5CO_2CH_3} C_6H_5COCH_2P(O)[N(CH_3)_2]_2$$

phosphonate anions undergo Dieckmann cyclizations.[193]

$$(RO)_2P(O)CH(CO_2C_2H_5)(CH_2)_3CO_2C_2H_5 \xrightarrow{C_2H_5ONa}$$

$$\text{2-oxocyclopentane with } CO_2C_2H_5 \text{ and } P(O)(OR)_2 \text{ at C-1} \xrightarrow{H_2O} \text{2-oxocyclopentyl-}P(O)(OR)_2$$

Addition of an anion across an activated double bond is another method of modifying a phosphonate substrate.[93,156,194–196] Such Michael-type additions are typical of carbanions, so it is apparent that phosphonate carbanions are capable of undergoing normal carbanionic reactions.

$$(C_2H_5O)_2P(O)CH_2COCH_3 + CH_2{=}CHCO_2CH_3 \xrightarrow{C_2H_5ONa} (C_2H_5O)_2P(O)CH(COCH_3)CH_2CH_2CO_2CH_3$$

[192] G. A. Koppel and M. D. Kinnick, *Tetrahedron Lett.*, **1974,** 711.
[193] H. Henning and G. Petzold, *Z. Chem.*, **7,** 184 (1967).
[194] J. Michalski and S. Musierowicz, *Tetrahedron Lett.*, **1964,** 1187.
[195] A. N. Pudovik, G. E. Yastrebova and O. A. Pudovik, *J. Gen. Chem. USSR*, **40,** 462 (1970).
[196] M. Kirilov and L. van Huyen, *Tetrahedron Lett.*, **1972,** 4487.

The addition of amines to alkynyl-1-phosphonates and thiophosphonates has been described in a series of papers.[42–44,197,198] The adducts can be hydrolyzed to the useful β-ketophosphonates. If prim-

$$(C_2H_5O)_2P(S)C{\equiv}CR \xrightarrow{R_2NH} (C_2H_5O)_2P(S)CH{=}C(R)NR_2 \xrightarrow{H_3O^+} (C_2H_5O)_2P(S)CH_2COR$$

ary amines are employed, then the enamine phosphonates can be either hydrolyzed to ketones or, as stated earlier, converted into anions from which α,β-unsaturated ketimines can be prepared by olefin formation.

gem-Diphosphonates are best prepared by a double Michaelis–Arbuzov reaction. Treatment of methylene dibromide with two equivalents of a trialkyl phosphite, preferably triisopropyl phosphite, gives the best yields. The Michaelis–Becker–Nylen reaction is less satisfactory because disodium diethyl methylenebisphosphonate is the major product upon treatment of methylene diiodide with sodium diethyl phosphate. If the substitutions are carried out stepwise, then fair yields of the tetraalkyl ester can be obtained.[199] The preparation of *gem*-

$$(RO)_2P(O)Na + ClCH_2P(O)(OR)_2 \rightarrow CH_2[P(O)(OR)_2]_2 + NaCl$$

diphosphonates has been reviewed.[200] They are easily converted into their carbanions which can be alkylated, halogenated, and generally undergo all carbanionic-type reactions.[157] Thus structural modification of the diphosphonates allows a practical synthesis by olefin formation of a variety of α,β-unsaturated phosphonates.

Dialkyl phosphoroamidates are prepared by one of two methods, treatment of a dialkylphosphorochloridate with an amine or, more

$$(RO)_2P(O)Cl + 2\ RNH_2 \rightarrow (RO)_2P(O)NHR + R\overset{+}{N}H_3Cl^-$$

conveniently, by addition of dialkylphosphite to a primary amine and carbon tetrachloride.[79]

The base required for the conversion of a dialkyl phosphonate into its anion is dependent upon the acidity of the substrate. The acidity will increase with the inclusion of electron-withdrawing groups capable of aiding the P(O) function in delocalizing the negative charge. The carbonyl and cyano groups have been used extensively and such phosphonates are converted most conveniently into their anions by

[197] M. S. Chatta and A. M. Aguiar, *J. Org. Chem.*, **37,** 1845 (1972).

[198] M. S. Chatta and A. M. Aguiar, *Tetrahedron Lett.*, **1971,** 1419.

[199] G. Schwarzenbach and J. Zurc, *Monatsh. Chem.*, **81,** 202 (1950).

[200] J. D. Curry, D. A. Nicholson, and O. T. Quimby in *Topics in Phosphorus Chemistry*, E. J. Griffith and M. Grayson, Eds., Vol. 7, Interscience Publishers, New York, 1972, p. 39.

addition to a slurry of sodium hydride in an inert solvent such as dimethoxyethane. After hydrogen evolution, which normally takes place at room temperature, has ceased, the co-reactant is added; one then sees an almost immediate precipitation of sodium dialkylphosphate. With less acidic dialkylphosphonates other techniques and stronger bases must be employed. The anion of diethyl benzylphosphonate, for example, cannot be preformed. Although hydrogen is evolved upon warming a sodium hydride slurry that contains the phosphonate, subsequent addition of a carbonyl compound does not produce olefin. Under the higher temperatures needed for anion formation the benzylphosphonate is unstable as evidenced by the viscous intractable character of the final product. These undesirable consequences are overcome by preparing the benzylphosphonate anion in the presence of the carbonyl co-reactant. Under such circumstances excellent yields of phenyl-conjugated olefins are often obtained. Simple allylic phosphonates having no other electron-withdrawing groups are less acidic than benzylphosphonates.

Dialkyl phosphonates that contain no additional anion-stabilizing groups require bases stronger than the hydride ion for their conversion into anions. Thus *n*-butyllithium in tetrahydrofuran at −78° has been used to convert phosphonates such as $RCH_2P(O)(OCH_3)_2$, $RCH_2P(S)(OCH_3)_2$, $RCH_2P(O)(NR_2)_2$, $CH_3SCH(R)P(O)(OC_2H_5)_2$, and $CH_2{=}CHCH_2SCH(R)P(O)(OC_2H_5)_2$, where R is either hydrogen or an alkyl group, into their α-lithio derivatives.[10,11,33,35,51] With the exception of the first, the salts give olefins when treated with ketones. Other bases have been employed, *e.g.*, Grignard reagents, sodamide, potassium *t*-butoxide, phenyllithium, and alkali metals in inert solvents.[201] As pointed out earlier, the stability of the intermediate betaine prepared from the anion and carbonyl compound is highly dependent upon the nature of the metallic cation. Thus, although the lithium salt of diphenylbenzylphosphine oxide with benzaldehyde gives no olefin, the β-hydroxy intermediate can be isolated. Conversion of the intermediate into its potassium salt affords the olefin stilbene in good yield.[202] The potassium anion of benzylphosphine oxide, on the other hand, gives stilbene directly when treated with benzaldehyde.

$$(C_6H_5)_2P(O)CH_2C_6H_5 \xrightarrow{C_6H_5Li} (C_6H_5)_2P(O)\bar{C}HC_6H_5 \xrightarrow[2.\ H^+]{1.\ C_6H_5CHO}$$

$$(C_6H_5)_2P(O)CH(C_6H_5)CH(OH)C_6H_5 \xrightarrow{t\text{-}C_4H_9OK} C_6H_5CH{=}CHC_6H_5$$

[201] M. Kirklov and G. Petrov, *Monatsh. Chem.*, **1972,** 1651.

[202] J. J. Richards and C. V. Banks, *J. Org. Chem.*, **28,** 123 (1963).

For phosphonates of relatively high acidity, sodium ethoxide in ethanol often proves to be a suitable medium for olefin formation.[203] Sodium ethoxide in ethanol may have certain advantages in large-scale preparations. A distinct disadvantage is that an extraction step is required to remove the product.

It is apparent that weak bases such as amines cannot be employed. In such an environment Knoevenagel-type condensations with concurrent water formation appear to take precedence.

For the most part inert polar solvents are preferred where both the phosphonate carbanion and product are soluble and the sodium phosphonate is insoluble. Tetrahydrofuran, dimethoxyethane, hexamethylphosphoramide, and dimethylformamide are in this class. Benzene and cyclohexane have also been used.

REACTIONS THAT DO NOT GIVE OLEFINIC PRODUCTS

Cyclopropanes and Aziridines. There are a number of reactions of phosphoryl-stabilized anions that do not produce unsaturation and that deserve special attention. Perhaps the reaction that bears the closest resemblence to olefin formation is between phosphorus anions and epoxides to form cyclopropanes. The mechanism of the reaction is assumed to be similar to that of epoxides with phosphoranes.[4,204,205] The initial step is believed to be a displacement of the oxide bridge by the carbanion to give what may be a cyclic intermediate. The second step entails the cleavage of the P—C bond with subsequent inversion at carbon upon breakage of the C—O bond and ring closure. The reaction between optically active styrene oxide and phosphonate carbanions proceeds with inversion at the asymmetric center.[206] Since only

$$C_6H_5CH\overset{O}{—}CH_2 + (RO)_2P(O)\bar{C}HR^1 \longrightarrow \begin{matrix} {}^-O\diagdown\ \diagup(OR)_2 \\ P \\ O\diagup\ \diagdown CHR^1 \\ | \qquad\quad | \\ C_6H_5CH—CH_2 \end{matrix} \longrightarrow$$

$$C_6H_5\overset{CHR^1}{CH—CH_2} + (RO)_2P(O)O^-$$

trans-substituted cyclopropanes have been reported, a cyclic intermediate is highly suspect. However, a cyclopropane does not form

[203] K. Hejno and V. Jarolim, *Coll. Czech. Chem. Commun.*, **38**, 3511 (1973).

[204] L. Horner, Hoffman, W. Klink, H. Ertel, and V. G. Toscano, *Chem. Ber.*, **95**, 581 (1962), and references therein cited.

[205] D. B. Denney, J. J. Will, and M. J. Boskin, *J. Amer. Chem. Soc.*, **84**, 3944 (1962).

[206] I. Tomoskozi, *Tetrahedron*, **19**, 1969 (1963).

when an open-chain structure analogous to one expected as a logical intermediate is prepared. Thus attack by benzaldehyde at the β position of a phosphonate with a cyano group vicinal to phosphorus gives an intermediate betaine that does not decompose to a three-membered ring.[60] Instead a benzyl diethyl phosphate is isolated. This example

$$C_6H_5CH(CN)CH(CN)P(O)(OC_2H_5)_2 \xrightarrow[\text{NaH}]{C_6H_5CHO}$$

$$C_6H_5CH(O^-)C(CN)(C_6H_5)CH(CN)P(O)(OC_2H_5)_2 \xrightarrow{H_3O^+}$$

$$C_6H_5CH[OP(O)(OC_2H_5)_2]C(CN)(C_6H_5)CH_2CN$$

may not be representative because steric factors may prevent formation of a cyclic intermediate or its decomposition to a three-membered ring. Indeed, the yields of cyclopropanes prepared by the phosphonate-carbanion method appear to increase as the size of substituents in the epoxide become smaller.[182]

The greater nucleophilicity of the P(O)-stabilized carbanions over the phosphoranes becomes quite evident upon comparison of their reactivity toward epoxides. Thus triphenylcarbethoxymethylenephosphorane, $Ph_3P{=}CHCO_2C_2H_5$, on treatment with styrene oxide gives ethyl *trans*-2-phenylcyclopropanecarboxylate in 21% yield under forcing conditions (120°), whereas the same product is produced in 40% yield under much milder conditions with the corresponding phosphonate.[4,207] The reaction of tri-*n*-butyl carbethoxymethylenephosphorane with various epoxides has been investigated, and it has been found that the products are not cyclopropanes but predominantly α,β-unsaturated esters.[208,209] The formation of the latter involves either a hydride migration or, in the case of cyclohexene oxide, a ring contraction (Eq. 34). In contrast the analogous phosphonate carbanion produces the

$$CH_3(CH_2)_5\overset{O}{CH{-}CH_2} + (n\text{-}C_4H_9)_3P{=}CHCO_2C_2H_5 \longrightarrow CH_3(CH_2)_6CH{=}CHCO_2C_2H_5$$

$$\text{cyclohexene oxide} + (n\text{-}C_4H_9)_3P{=}CHCO_2C_2H_5 \longrightarrow \text{cyclopentyl-}CH{=}CHCO_2C_2H_5 \ (92\text{–}95\%) + \text{cyclopentylidene-}CHCH_2CO_2C_2H_5 \ (2\text{–}5\%) + \text{bicyclo[4.1.0]heptane-}CO_2C_2H_5 \ (3\%)$$

(Eq. 34)

[207] D. B. Denney and M. J. Boskin, *J. Amer. Chem. Soc.*, **81,** 6330 (1959).
[208] B. Rickborn and R. M. Gerkin, *J. Amer. Chem. Soc.*, **93,** 1693 (1971).
[209] R. M. Gerkin and B. Rickborn, *J. Amer. Chem. Soc.*, **89,** 5850 (1967).

cyclopropane, ethyl α-norcaranecarboxylate, in 58% yield when treated with cyclohexene oxide; there is no detectable rearrangement.[4]

$$(C_2H_5O)_2P(O)CH_2CO_2C_2H_5 + \text{cyclohexene oxide} \xrightarrow{NaH} \text{norcarane}{-}CO_2C_2H_5 + (C_2H_5O)_2PO_2^-$$

A number of cyclopropane syntheses are of interest, but one in particular deserves special mention. Treatment of an enamine phosphine oxide with an epoxide in the presence of base yields cyclopropyl ketimines. The products are subsequently hydrolyzed to the corresponding cyclopropyl ketones.[42–44]

$$(C_6H_5)_2P(O)CH{=}C(R^1)NHR \xrightarrow[\text{2. } R_2\text{-epoxide}]{\text{1. } n\text{-}C_4H_9Li} R^2{-}\triangle{-}C(R^1){=}NR \xrightarrow{H_3O^+} R^2{-}\triangle{-}COR^1 \ (40\text{–}70\%)$$

A variation of the cyclopropane synthesis that employs nitrones in place of epoxides has recently been reported.[210,211] The reaction, which in all probability also proceeds via a five-membered cyclic intermediate, represents a novel approach to the synthesis of aziridines.

$$\left[C_6H_5CH{=}\overset{+}{N}(O^-)CH_3 \rightleftharpoons C_6H_5CH_2{-}\overset{+}{N}(O^-){=}CH_2\right] + (C_2H_5O)_2P(O)CH_2CO_2C_2H_5 \xrightarrow{NaH}$$

$$\underset{(40\%)}{CH_2{-}CHCO_2C_2H_5 \ (N{-}CH_2C_6H_5)} + \underset{trans\ (20\%)}{C_6H_5CH{-}CHCO_2C_2H_5 \ (N{-}CH_3)}$$

Knoevenagel Condensations. Knoevenagel-type condensations between phosphonate carbanions and ketones and aldehydes have been

[210] E. Breuer and I. Ronen-Braunstein, *Chem. Commun.*, **1974,** 949.

[211] E. Breuer, S. Zbaida, J. Pesso, and S. Levi, *Tetrahedron Lett.*, **1975,** 3103.

investigated.[212–218] The condensations are conducted in the presence of only a trace of base. Benzene is commonly used as solvent and often water is continuously removed as formed. Numerous vinyl-

$$(C_2H_5O)_2P(O)CH_2COR + R^1R^2CO \xrightarrow{-H_2O} R^1R^2C{=}C(COR)P(O)(OC_2H_5)_2$$

phosphonates have been prepared from phosphonates having highly reactive methylene groups. Recently, similar condensations have been conducted in tetrahydrofuran with titanium tetrachloride and an organic base as catalysts.[219] Both triethyl-alkylidene and -arylidene phosphonoacetic acids and tetraalkyl-alkylidene and -arylidene methanebisphosphonic acids can be prepared by this method. The products are easily hydrogenated with sodium borohydride in ethanol or catalytically to the corresponding saturated compounds in nearly quantitative yields. The role of the titanium tetrachloride is to stabilize the intermediate betaine by forming a cyclic titanate ester. Such a structure prevents normal olefin bond formation by dephosphonation.

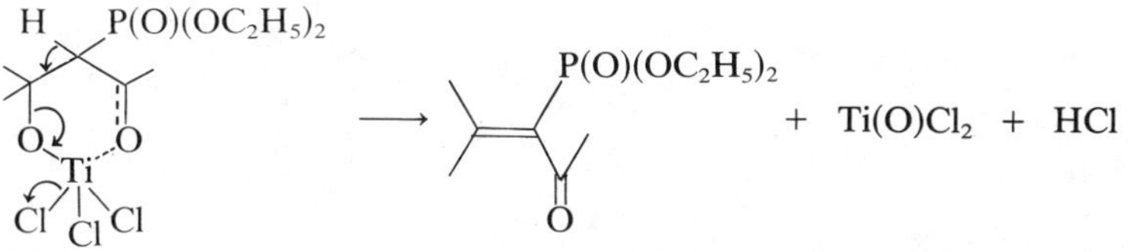

H P(O)(OC₂H₅)₂ ... O O Ti Cl Cl Cl $\longrightarrow$ P(O)(OC₂H₅)₂ ... O + Ti(O)Cl₂ + HCl

In a related condensation, epoxyphosphonates can be prepared by the reaction of the anion of diethyl chloromethylphosphonate with aldehydes and ketones.[220] In this analog of the Darzens synthesis,

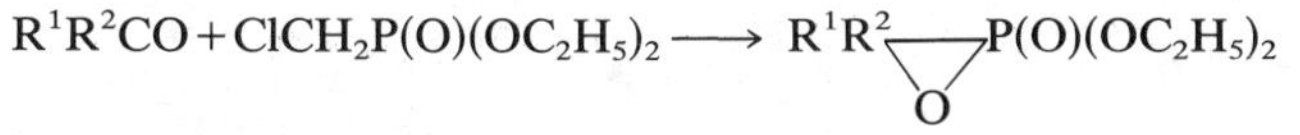

$$R^1R^2CO + ClCH_2P(O)(OC_2H_5)_2 \longrightarrow R^1R^2\underset{O}{C{-}C}HP(O)(OC_2H_5)_2$$

[212] A. N. Pudovik, G. E. Yastrebova, and V. I. Nikitina, *J. Gen. Chem. USSR*, **39,** 213 (1969).

[213] A. N. Pudovid, G. E. Yastrebova, and V. I. Nikitina, *J. Gen. Chem. USSR*, **38,** 300 (1968).

[214] A. N. Pudovik, G. E. Yastrebova, L. M. Leonteva, T. A. Zyablikova, and V. I. Nikitina, *J. Gen. Chem. USSR*, **39,** 1230 (1969).

[215] A. N. Pudovik, G. E. Yastrehova, and V. I. Nikitina, *J. Gen. Chem. USSR*, **37,** 519 (1967).

[216] A. N. Pudovik and N. M. Lebedeva, *Dokl. Akad. Nauk SSSR*, **90,** 799 (1953) [*C.A.*, **50,** 2429d (1956)].

[217] C. N. Robinson and J. F. Addison, *J. Org. Chem.*, **31,** 4325 (1966).

[218] D. Danion and R. Carrie, *Bull. Soc. Chim. Fr.*, **1972,** 1130.

[219] W. Lehnert, *Tetrahedron*, **30,** 301 (1974).

[220] V. F. Martynov and V. E. Timofeev, *J. Gen. Chem. USSR*, **32,** 3383 (1962).

elimination of chloride ion from the initially formed betaine adduct competes successfully with olefin formation.

A number of cyclizations are important. The Michael addition of a phosphonate carbanion to a β-trichloromethyl vinyl ester gives a new carbanion that subsequently loses chloride ion with ring closure.[100]

$$Cl_3CCH{=}CHCO_2R + (C_2H_5O)_2P(O)\bar{C}HCO_2C_2H_5 \longrightarrow (C_2H_5O)_2P(O)CH(CO_2C_2H_5)CH(CCl_3)\bar{C}HCO_2R \rightarrow$$

$$(C_2H_5O)_2P(O)CH(CO_2C_2H_5)\text{-cyclopropane}(Cl, Cl)(CO_2R)$$

Phosphonates with reactive methylene groups react with aryl azides or nitrile oxides under basic conditions with formation of heterocyclic phosphonates.[221] A novel ring contraction has been observed that

$$(C_2H_5O)_2P(O)\bar{C}HCN \xrightarrow{C_6H_5N_3} \text{5-}H_2N\text{-4-}P(O)(OC_2H_5)_2\text{-1-}C_6H_5\text{-1,2,3-triazole}$$

$$(C_2H_5O)_2P(O)\bar{C}HCN \xrightarrow{C_6H_5CNO} \text{5-}H_2N\text{-4-}P(O)(OC_2H_5)_2\text{-3-}C_6H_5\text{-isoxazole}$$

involves a phosphonate carbanion.[222] No mechanism was given.

$$\text{3,8-dichloro-dibenzo[c,f][1,2]diazepin-11-one} + (C_2H_5O)_2P(O)CH_2CH{=}CHCO_2C_2H_5 \xrightarrow{NaH} \text{dichlorobenzo[c]cinnoline}$$

EXPERIMENTAL CONDITIONS

The starting phosphonates are most easily obtained by the Michaelis–Arbuzov reaction. Normally the halides and phosphite are

[221] U. Heep, *Ann.*, **1973**, 578.

[222] R. J. Dubois and F. D. Popp, *Chem. Commun.*, **1968**, 675.

heated without a solvent until weight loss due to removal of the volatile alkyl halide ceases. Triethyl or trimethyl phosphite is the most convenient to employ because the corresponding bromides are very volatile. The product is purified by fractional distillation.

Except when the phosphonate intermediates are unstable, *e.g.*, the benzylphosphonate anion, the carbanions are formed before addition of the co-reactant. If the α hydrogens are sufficiently acidic, sodium hydride is the preferred base. The rate of evolution of hydrogen can be followed and the point at which conversion is complete determined. With α-trialkyl phosphonoacetates or dialkyl cyanomethylphosphonate, hydrogen evolution occurs at room temperature. With sodium hydride the production of a nonvolatile side product is avoided; therefore product workup is simplified. A 50% slurry of sodium hydride in mineral oil is commercially available. It is resistant to atmospheric moisture and is stable indefinitely.[223]

Although most aprotic solvents have been employed, ethers have found the most favor. The choice of solvent is predicated to a degree on the boiling point of the product; the solvent should be easily separated from the product by distillation. Glyme has been most widely used. Although certain protic solvents, *e.g.*, alcohols, have allowed favorable product yields, the by-product, a diester of sodium phosphate, owing to its solubility is not easily separated from the reaction mixture. Often it is desirable to separate the precipitated dialkylphosphate from the mixture prior to workup, especially if the product might be water sensitive.

Normally a solution of the co-reactant in the reaction solvent is added at room temperature. Precipitation usually occurs merely by stirring the reaction mixture and heating is unnecessary. Yields are usually higher if elevated temperatures can be avoided. Reaction mixtures are usually allowed to stir at room temperature for 24 hours to assure complete reaction.

The ease of workup is perhaps the most appealing aspect of the phosphonate method. The reaction mixture may be diluted with a large quantity of water, in which the by-product dialkyl phosphate is soluble; and, if a water-immiscible solvent is employed, the layers separated. If a reaction is carried out in glyme, the homogeneous solution obtained upon dilution with water is extracted with a hydrocarbon solvent. The product is obtained from the water-insoluble layer by evaporation after drying.

[223] Metal Hydrides Inc., Beverly, Massachussetts, supplies sodium hydride, 50–51% in mineral oil.

EXPERIMENTAL PROCEDURES

The example of the preparation of a dialkyl phosphonate has been chosen to illustrate a typical Michaelis–Arbuzov reaction. Since phosphonic acid bisamides may assume importance in the future, especially when the phosphonates themselves are not acidic enough to form anions, an example of their preparation is also included.

Procedures have been selected which make apparent the simplicity of synthesis. Although the base and solvent are often varied, procedures are uniform and straightforward.

Triethyl α-Phosphonopropionate.[224] Triethyl phosphite (17.5 g, 0.10 mol) was added dropwise to 18.1 g (0.10 mol) of ethyl α-bromopropionate at room temperature with stirring. After approximately one-fourth of the phosphite had been added, the mixture was warmed until an exothermic reaction started. The remaining triethyl phosphite was then added at a rate sufficient to maintain reflux. The ethyl bromide by-product was allowed to escape. The mixture was refluxed for 1 hour and distilled to give 14.6 g (59.4%) of product, bp 93–95° (0.85 mm). The coupling constant between phosphorus and the α-methine proton is 24 Hz ($CDCl_3$).

Methylphosphonic Acid Bis(dimethylamide).[33,225,226] Phosphorus trichloride (69.3 g, 0.48 mol), aluminum chloride (80 g, 0.60 mol), and methyl chloride (31 g, 0.61 mol) were mixed and the solution stirred with external cooling (to moderate the exothermic reaction) at room temperature for 4 hours. The mixture was dissolved in 900 ml of methylene chloride and 120 ml of water was added slowly with cooling. The solid was removed by filtration and the filtrate evaporated under reduced pressure. Distillation, bp 60 (11 mm), gave 33.5 g (53%) of methylphosphoric dichloride, mp 34°.

To dimethylamine (18.0 g, 0.40 mol) in 250 ml of ether was added, at 0° under nitrogen, 10.0 g (0.0753 mol) of methylphosphoric dichloride. The solution was stirred at 0° for 1 hour and at 25° for 3 hours. The dimethylamine hydrochloride precipitate was removed by filtration and the filtrate evaporated under vacuum. Distillation of the residue afforded methylphosphonic acid bis(dimethylamide) bp 74–75° (1.5 mm), 10.0 g (89%). The nmr spectrum ($CDCl_3$) showed a doublet at 1.61 ppm.

[224] A. E. Arbuzov and A. A. Durin, *J. Russ. Phys. Chem. Soc.*, **46,** 295 (1914) [*C.A.*, **8,** 2551 (1914)].

[225] A. M. Kinnear and E. A. Perren, *J. Chem. Soc.*, **1952,** 3437.

[226] G. M. Kosolapoff and L. B. Payne, *J. Org. Chem.*, **21,** 413 (1956).

Diethyl N-Dimethylaminophosphoramidate.[81] To a stirred solution of 138 g (1.0 mol) of diethyl phosphite, 154 g of carbon tetrachloride, and 250 ml of xylene was added dropwise at room temperature (external cooling was needed) 118 g (2.0 mol) of *unsym*-dimethylhydrazine. The mixture was stirred at room temperature for one-half hour and filtered. The filtrate was evaporated under vacuum and the residue distilled to give 120.5 g (61%) of product, bp 85° (0.25 mm).

Ethyl Cinnamylideneacetate.[227] Sodium hydride (50%) (2.4 g, 0.05 mol) was added to 100 ml of dry tetrahydrofuran. The slurry was cooled to 0°, and 12.5 g (0.05 mol) of triethyl phosphonocrotonate was added dropwise with stirring. After the addition the solution was stirred at 0° for 10 hours. To the brown solution, maintained below 0°, 5.3 g (0.05 mol of benzaldehyde was added dropwise and the solution stirred at room temperature for 7 hours. After evaporation of the solvent at reduced pressure, a gummy precipitate appeared. A large excess of water was added and the product was extracted with ether. The ether, after being dried over magnesium sulfate, was removed and the residue distilled, giving 4.1 g (41%) of pale-yellow liquid, bp 130–132° (1 mm), n_D^{20} 1.6120; infrared cm^{-1}: 1625 and 1600.

Methyl Deca-*trans*-2,*trans*-4-dienoate.[54] Methyl 4-diethylphosphonocrotonate (20.0 g, 0.085 mol) and *n*-hexanal (8.5 g, 0.085 mol) in 50 ml of redistilled dimethylformamide were vigorously stirred at room temperature and 5.4 g (0.1 mol) of sodium methoxide in dry methanol was added dropwise during 1 hour. The solution was diluted with an excess of water and then extracted several times with petroleum ether and the extracts washed with water. Evaporation and distillation gave 3.8 g (25%) of product, bp 71–73° (0.4 mm).

***trans*-Stilbene.**[4] To a slurry of 2.4 g of sodium hydride (0.05 mol of 50% dispersion) in 100 ml dry 1,2-dimethoxyethane was added 11.4 g (0.05 mol) of diethyl benzylphosphonate and 5.3 g (0.05 mol) of benzaldehyde. The mixture was heated slowly with stirring to 85°. At 70° there was a large evolution of gas and the appearance of a semisolid precipitate. The solution was refluxed for one-half hour, cooled, and diluted with a large excess of water. The solution was filtered and the precipitate recrystallized from alcohol to afford 5.6 g (62.9%) of product, mp 124–125°, spectroscopically identical with *trans*-stilbene.

Ethyl Cyclohexylideneacetate.[112] Full details for the preparation of ethyl cyclohexylideneacetate from the anion of triethyl phosphonoacetate and cyclohexanone in 70–80% yield are given in *Organic Syntheses.*[112]

[227] K. Sato, S. Mizuno, and M. Hirayama, *J. Org. Chem.*, **32,** 177 (1967).

Ethyl α-n-Butylacrylate.[4] To a slurry of 18.4 g of sodium hydride (0.38 mol of 50% dispersion) in 100 ml of dry 1,2-dimethoxyethane was added slowly at room temperature 86 g (0.384 mol) of triethyl phosphonoacetate. The solution was stirred until gas evolution had ceased. *n*-Butyl bromide (52.6 g, 0.38 mol) was added and the solution heated to 50° for 1 hour. After cooling to 10°, 18.4 g of sodium hydride (0.38 mol of 50% dispersion), was added all at once. The slurry was allowed to come slowly to room temperature with stirring, during which time rapid evolution of gas took place. After gas evolution had ceased, about 1 hour at room temperature, a mixture of 1.21 g (0.38 mol) of dry paraformaldehyde in 50 ml of dry 1,2-dimethoxyethane was added at 20° during 1 hour. The mixture, which contained a gummy precipitate, was stirred for 1 hour and diluted with 250 ml of water. The aqueous layer was extracted with two 100-ml portions of ether, the combined ether layers dried over magnesium sulfate, and the solvent evaporated under reduced pressure. When distilled, the residue (bp 68–69°, 100 mm), gave 38.4 g (60%) of ethyl *α*-*n*-butylacrylate; infrared cm^{-1}: 1640.

1-Carbethoxy-2-phenylcyclopropane.[4] Triethyl phosphonoacetate (22.4 g, 0.1 mol) was added dropwise with stirring to a slurry of 5.0 g of sodium hydride (0.104 mol of 50% dispersion) in 100 ml of dry 1,2-dimethoxyethane at 25°. After the addition the solution was stirred for 1 hour or until gas evolution had ceased. Styrene oxide, 12.0 g (0.1 mol), was added and the solution refluxed for 4 hours. After cooling, 250 ml of water was added and the aqueous solution extracted with two 100-ml portions of ether. The combined ether extracts were dried over magnesium sulfate, evaporated under reduced pressure, and the residue distilled to give 8.0 g of product, bp 100° (0.5 mm), in 42% yield. The distillate crystallized on standing (mp 37–38°); infrared cm^{-1}: 1020 (cyclopropane).

1,1-Diphenylethylene (Preparation of a β-Hydroxy Phosphonamide Adduct and Its Decomposition).[33] To a stirred solution of methylphosphonic acid bis(dimethylamide) (1.0 g, 6.67 mmol) in 15 ml of dry tetrahydrofuran, under nitrogen at −78° was added 4.35 ml (6.95 mmol) of a 1.6 *M* solution of *n*-butyllithium in hexane. To this solution with stirring at −78° was added 1.091 g (6 mmol) of benzophenone in one portion. The reaction mixture was stirred at −78° for one-half hour and then allowed to warm to room temperature, after which 10 ml of water was added and the mixture was extracted with ether. The extract was washed with water, dried over magnesium sulfate, and evaporated to yield a white solid. Recrystallization from pentane-ether afforded 1.89 g (95%) of the β-hydroxyphosphonamide adduct, mp 155–157°.

A solution of 1.55 g of the β-hydroxyphosphonamide adduct in 25 ml of benzene containing 3.2 g of silica gel was heated at reflux for 12 hours. Ether (25 ml) was added and the mixture filtered. Evaporation of the solvent and distillation of the residue afforded 0.785 g (93%) of 1,1-diphenylethylene spectroscopically identical with authentic material.

Dimethylaminoisocyanate.[81] To a slurry of 9.6 g (0.2 mol) of 50% sodium hydride in 200 ml of dry benzene was added 39.2 g (0.2 mol) of diethyl N-dimethylaminophosphoramidate. The mixture was stirred at room temperature for 12 hours during which time the stoichiometric quantity of hydrogen was evolved. Carbon dioxide was slowly passed through the mixture at 10°. When gas absorption had ceased, the mixture was allowed to come to room temperature. Upon long standing (6 days) a precipitate formed which was removed by filtration. The precipitate was extracted with 1 l of hot benzene from which the product was obtained by cooling and filtering. It could be purified further by sublimination at 130–140° (0.025 mm) to afford 5.5 g (32%) of colorless material, mp 178°. The hygroscopic product gave the correct elemental analysis for dimethylaminoisocyanate; however, titration and mass spectroscopy indicated that it was a dimer, mol wt 175.

TABULAR SURVEY

The tables are compiled on the basis of product type. Within each table the entries are arranged in order of increasing number of carbon atoms in the phosphoryl reagent. They are subdivided with the non-phosphonyl-containing reactant also listed in order of increasing carbon atom number. Where there is more than one reference and yields are reported, the highest yields are given along with the corresponding base and solvent. Data are taken from the first reference cited.

A number of abbreviations are used: glyme (dimethyl ether of ethylene glycol), THF (tetrahydrofuran), HMPT (hexamethylphosphoramide), DMSO (dimethyl sulfoxide), and ether (diethyl ether).

An attempt was made to report all pertinent reactions appearing in the literature through December 1975 and several additional ones noted in early 1976.

Because phosphoryl anions have found utility primarily as olefin-formation reagents, the tables are concerned primarily with this reaction. Table X is the exception since it includes a list of reactants and reagents for which, although anions are involved, phosphate elimination is not a condition for product formation.

TABLE I. Olefins from Alkyl (Arenyl) Phosphoryl Reagents

$$(R)_2P(O)CHR^1_2 + R^2R^3CO \xrightarrow{\text{Base}} R^2R^3C{=}CR^1_2 + (R)_2PO_2^-$$

No. of C Atoms	Reagent	Reactant	Base/Solvent	Product(s) and Yield(s) (%)	Refs.
C_3	$(CH_3O)_2P(S)CH_3$	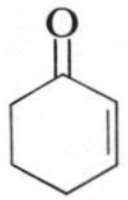 O	n-C_4H_9Li/THF	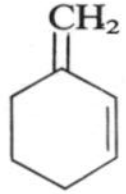 CH_2 (31)	10
		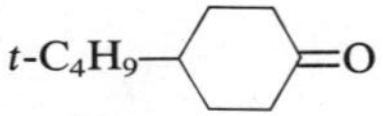 t-C_4H_9— =O	n-C_4H_9Li/THF	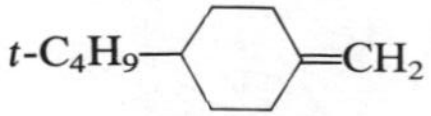 t-C_4H_9— $=CH_2$ (52)	10
		$(C_6H_5)_2CO$	n-C_4H_9Li/THF	$(C_6H_5)_2C{=}CH_2$ (81)	10
	$(CH_3O)_2P(O)CH_3$	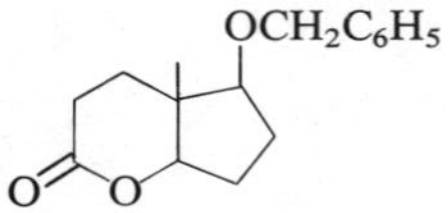 $OCH_2C_6H_5$ O O	n-C_4H_9Li/THF	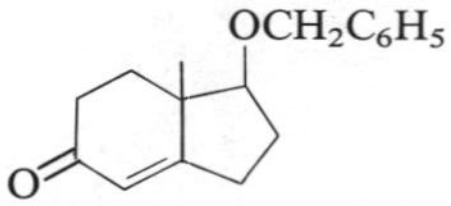 $OCH_2C_6H_5$ O (−)	66
		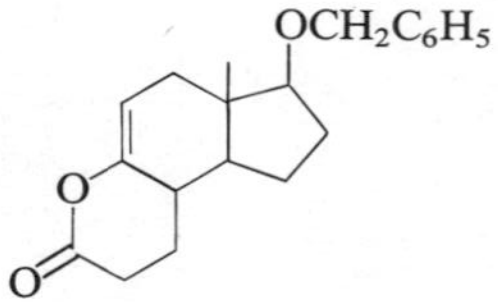 $OCH_2C_6H_5$ O O	n-C_4H_9Li/THF	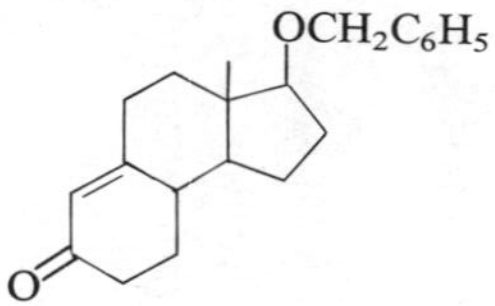 $OCH_2C_6H_5$ O (−)	66

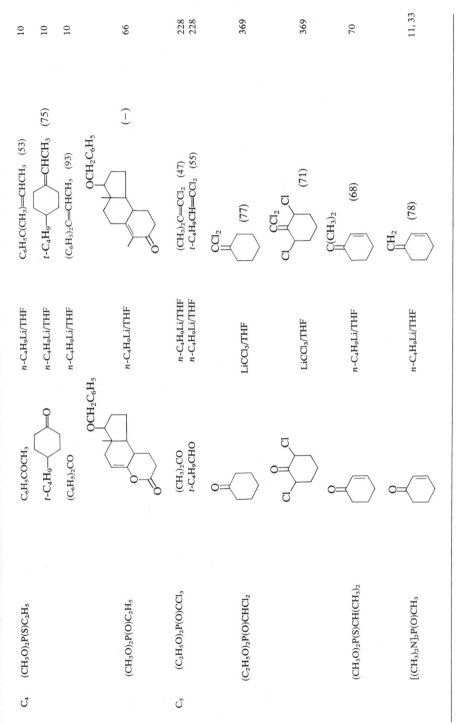

C_4	$(CH_3O)_2P(S)C_2H_5$	$C_6H_5COCH_3$	n-C_4H_9Li/THF	$C_6H_5C(CH_3){=}CHCH_3$ (53)	10
		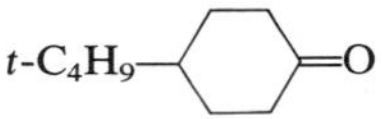	n-C_4H_9Li/THF	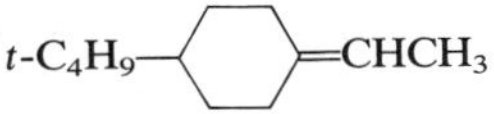 (75)	10
		$(C_6H_5)_2CO$	n-C_4H_9Li/THF	$(C_6H_5)_2C{=}CHCH_3$ (93)	10
	$(CH_3O)_2P(O)C_2H_5$	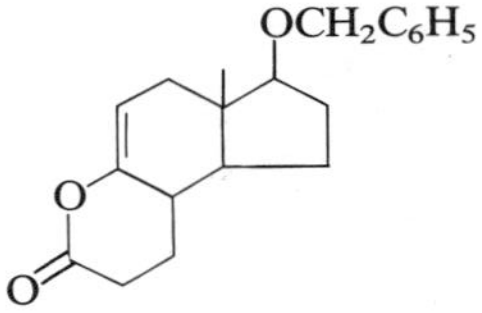	n-C_4H_9Li/THF	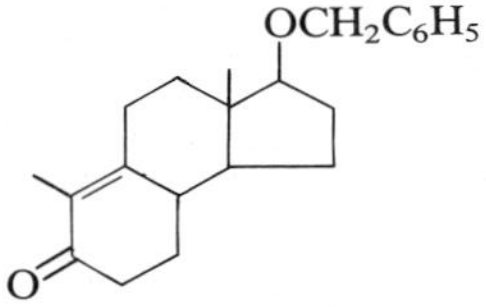 (−)	66
C_5	$(C_2H_5O)_2P(O)CCl_3$	$(CH_3)_2CO$ t-C_4H_9CHO	n-C_4H_9Li/THF n-C_4H_9Li/THF	$(CH_3)_2C{=}CCl_2$ (47) t-$C_4H_9CH{=}CCl_2$ (55)	228 228
	$(C_2H_5O)_2P(O)CHCl_2$	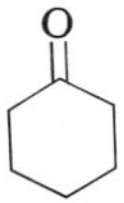	$LiCCl_3$/THF	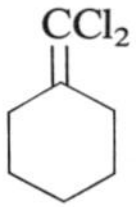 (77)	369
		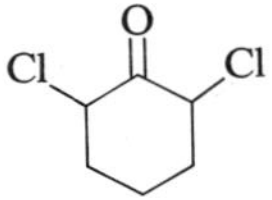	$LiCCl_3$/THF	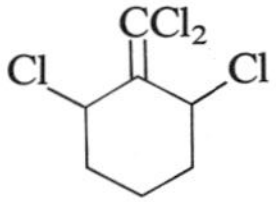 (71)	369
	$(CH_3O)_2P(S)CH(CH_3)_2$	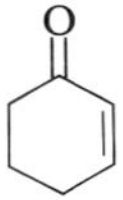	n-C_4H_9Li/THF	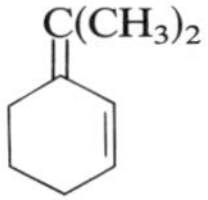 (68)	70
	$[(CH_3)_2N]_2P(O)CH_3$	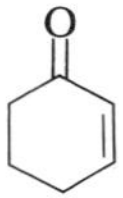	n-C_4H_9Li/THF	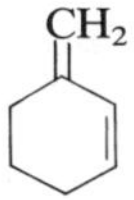 (78)	11, 33

Note: References 228–415 are on pp. 250–253.

TABLE I. Olefins from Alkyl (Arenyl) Phosphoryl Reagents

$$(R)_2P(O)CHR^1_2 + R^2R^3CO \xrightarrow{\text{Base}} R^2R^3C{=}CR^1_2 + (R)_2PO_2^- \quad (\textit{Continued})$$

No. of C Atoms	Reagent	Reactant	Base/Solvent	Product(s) and Yield(s) (%)	Refs.
C_5 (*Contd.*)		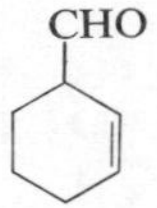 CHO	n-C_4H_9Li/THF	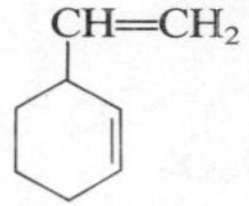 CH=CH_2 (67)	11, 33
		C_6H_5CHO	n-C_4H_9Li/THF	$C_6H_5CH{=}CH_2$ (53)	11, 33
	$(C_2H_5O)_2P(O)CHCl_2$	p-ClC_6H_4CHO	$LiCCl_3$/THF	p-$ClC_6H_4CH{=}CCl_2$ (67)	369
		p-FC_6H_4CHO	$LiCCl_3$/THF	p-$FC_6H_4CH{=}CCl_2$ (80)	369
		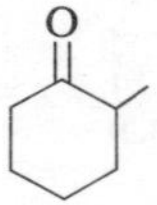 O	$LiCCl_3$/THF	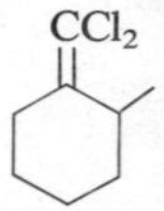 CCl_2 (78)	369
	$(C_2H_5O)_2P(O)CCl_3$	$C_6H_5COCH_3$	n-C_4H_9Li/THF	$C_6H_5(CH_3)C{=}CCl_2$ (69)	228
	$(C_2H_5O)_2P(O)CHCl_2$	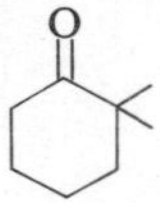 O	$LiCCl_3$/THF	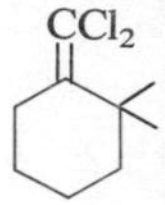 CCl_2 (84)	369
		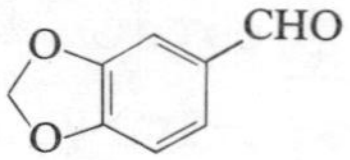 CHO	$LiCCl_3$/THF	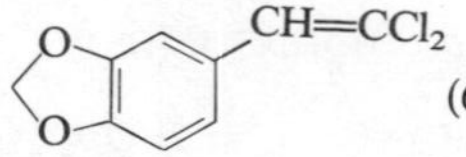 CH=CCl_2 (62)	369

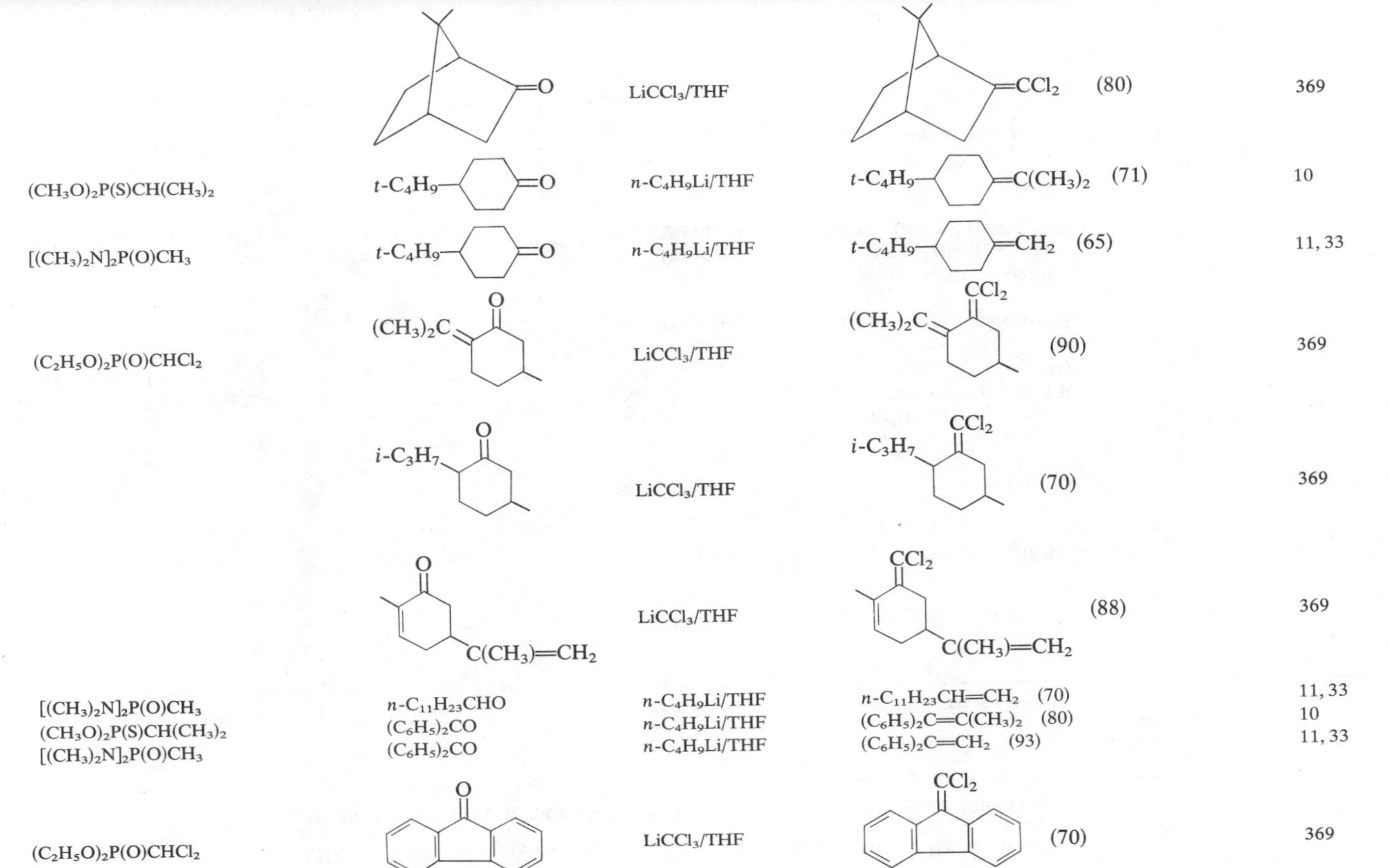

Reagent	Carbonyl Compound	Conditions	Product (Yield, %)	Refs.
		$LiCCl_3$/THF	(80)	369
$(CH_3O)_2P(S)CH(CH_3)_2$		n-C_4H_9Li/THF	(71)	10
$[(CH_3)_2N]_2P(O)CH_3$		n-C_4H_9Li/THF	(65)	11, 33
$(C_2H_5O)_2P(O)CHCl_2$		$LiCCl_3$/THF	(90)	369
		$LiCCl_3$/THF	(70)	369
		$LiCCl_3$/THF	(88)	369
$[(CH_3)_2N]_2P(O)CH_3$	n-$C_{11}H_{23}CHO$	n-C_4H_9Li/THF	n-$C_{11}H_{23}CH{=}CH_2$ (70)	11, 33
$(CH_3O)_2P(S)CH(CH_3)_2$	$(C_6H_5)_2CO$	n-C_4H_9Li/THF	$(C_6H_5)_2C{=}C(CH_3)_2$ (80)	10
$[(CH_3)_2N]_2P(O)CH_3$	$(C_6H_5)_2CO$	n-C_4H_9Li/THF	$(C_6H_5)_2C{=}CH_2$ (93)	11, 33
$(C_2H_5O)_2P(O)CHCl_2$		$LiCCl_3$/THF	(70)	369

Note: References 228–415 are on pp. 250–253.

TABLE I. OLEFINS FROM ALKYL (ARENYL) PHOSPHORYL REAGENTS

$(R)_2P(O)CHR^1_2 + R^2R^3CO \xrightarrow{\text{Base}} R^2R^3C{=}CR^1_2 + (R)_2PO_2^-$ (*Continued*)

No. of C Atoms	Reagent	Reactant	Base/Solvent	Product(s) and Yield(s) (%)	Refs.
C_6	$(CH_3O)_2P(S)C_4H_9$-*n*	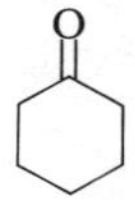 [cyclohexanone structure]	*n*-C_4H_9Li/THF	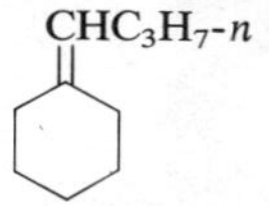 [cyclohexylidene structure]=CHC_3H_7-*n* (61)	10
		$(n\text{-}C_3H_7)_2CO$	*n*-C_4H_9Li/THF	$(CH_3CH_2CH_2)_2C{=}CHC_3H_7$-*n* (69)	10
	$[(CH_3)_2N]_2P(O)C_2H_5$	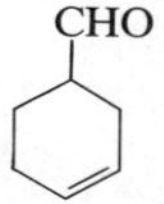 [cyclohex-3-enyl structure]–CHO	*n*-C_4H_9Li/THF	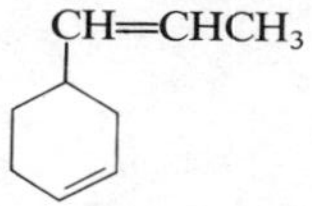 [cyclohex-3-enyl structure]–$CH{=}CHCH_3$ (79)	11, 33
		C_6H_5CHO	*n*-C_4H_9Li/THF	$C_6H_5CH{=}CHCH_3$ (90)	11, 33, 34
	$(CH_3O)_2P(O)CH_2CH{=}C(CH_3)Cl$	C_6H_5CHO	NaH/glyme	$C_6H_5CH{=}CHCH{=}C(CH_3)Cl$ (—)	61
	$[(CH_3)_2N]_2P(O)C_2H_5$	$C_6H_5COCH_3$	*n*-C_4H_9Li/THF	$C_6H_5C(CH_3){=}CHCH_3$ (90)	33
		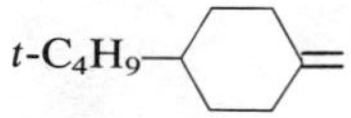 *t*-C_4H_9–[cyclohexanone structure]	*n*-C_4H_9Li/THF	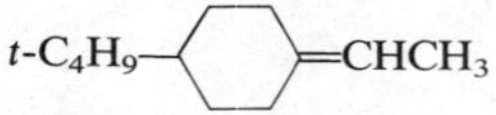 *t*-C_4H_9–[cyclohexylidene structure]=$CHCH_3$ (80)	11, 33
	$(CH_3O)_2P(S)C_4H_9$-*n*	$(C_6H_5)_2CO$	*n*-C_4H_9Li/THF	$(C_6H_5)_2C{=}CHC_3H_7$-*n* (88)	10
	$[(CH_3)_2N]_2P(O)C_2H_5$	$(C_6H_5)_2CO$	*n*-C_4H_9Li/THF	$(C_6H_5)_2C{=}CHCH_3$ (90)	11, 34
C_7	$[(CH_3)_2N]_2P(O)CH_2CH{=}CH_2$	$(CH_3)_2CO$	*n*-C_4H_9Li/DMF	$(CH_3)_2C{=}CHCH{=}CH_2$ (50)	29
	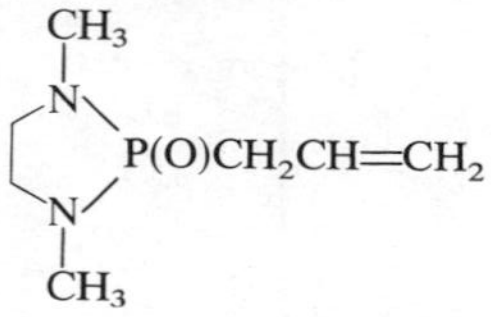 [1,3-dimethyl-1,3,2-diazaphospholidine structure, N-CH_3, N-CH_3]$P(O)CH_2CH{=}CH_2$	$(CH_3)_2CO$	*n*-C_4H_9Li/THF	$(CH_3)_2C{=}CHCH{=}CH_2$ (75)	29

Phosphonate	Carbonyl compound	Base/solvent	Product (yield %)	Refs.
	$CH_3COC_2H_5$	n-C_4H_9Li/THF	$C_2H_5C(CH_3){=}CHCH{=}CH_2$ (90)	29
	t-C_4H_9CHO	n-C_4H_9Li/THF	$(CH_3)_3CCH{=}CHCH{=}CH_2$ (50)	29
$(CH_3O)_2P(O)CH_2$-(2-thienyl)	3-thienyl-CHO	NaH/glyme	3-thienyl-CH=CH-2-thienyl (40)	229
	2-furyl-CHO	NaH/glyme	2-thienyl-CH=CH-2-furyl (65)	229
$(CH_3O)_2P(O)CH_2$-(3-thienyl)	3-thienyl-CHO	NaH/glyme	3-thienyl-CH=CH-3-thienyl (46)	229
$(CH_3O)_2P(O)CH_2$-(2-furyl)	2-furyl-CHO	NaH/glyme	2-furyl-CH=CH-2-furyl (52)	229
$(CH_3O)_2P(O)CH_2$-(3-thienyl)		NaH/glyme	CH_3-thienyl-CH=CH-thienyl (41)	229
$(CH_3O)_2P(O)CH_2$-(3-furyl)	5-CH_3-2-thienyl-CHO	NaH/glyme	5-CH_3-2-thienyl-CH=CH-thienyl (60)	229
	2-thienyl-$COCH_3$	NaH/glyme	2-thienyl-$C(CH_3){=}CH$-thienyl (25)	229
1,3-dimethyl-1,3,2-diazaphospholidine: $(CH_3N)_2P(O)CH_2CH{=}CH_2$	cyclohexanone	n-C_4H_9Li/THF	cyclohexylidene=$CHCH{=}CH_2$ (90)	29
	C_6H_5CHO	n-C_4H_9Li/THF	$C_6H_5CH{=}CHCH{=}CH_2$ (60)	29

Note: References 228–415 are on pp. 250–253.

TABLE I. Olefins from Alkyl (Arenyl) Phosphoryl Reagents

$(R)_2P(O)CHR^1_2 + R^2R^3CO \xrightarrow{\text{Base}} R^2R^3C{=}CR^1_2 + (R)_2PO_2^-$ (*Continued*)

No. of C Atoms	Reagent	Reactant	Base/Solvent	Product(s) and Yield(s) (%)	Refs.
C_7 (*Contd.*)	$(CH_3O)_2P(O)CH_2CH_2$-$CH{=}C(Cl)CH_3$	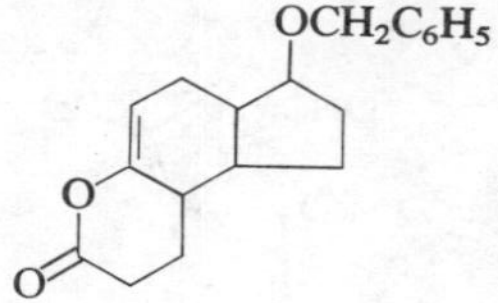 [structure: $OCH_2C_6H_5$]	n-C_4H_9Li/THF	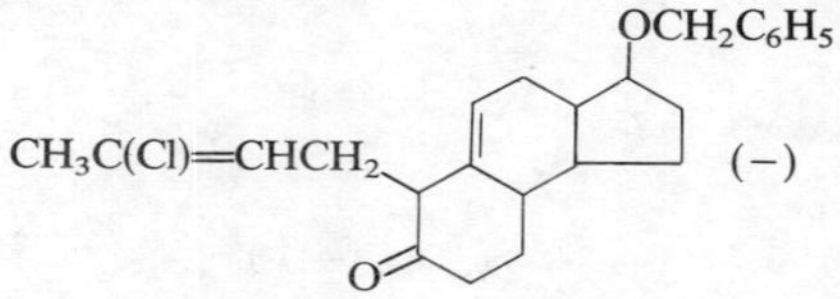 $CH_3C(Cl){=}CHCH_2$ [structure: $OCH_2C_6H_5$] (–)	66
C_8	$(CH_3O)_2P(O)CH_2$-$(CH_2)_2CH(OCH_3)_2$	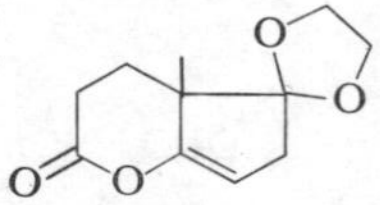 [structure]	n-C_4H_9Li/THF	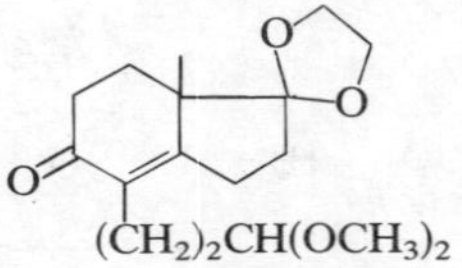 [structure: $(CH_2)_2CH(OCH_3)_2$] (–)	66
		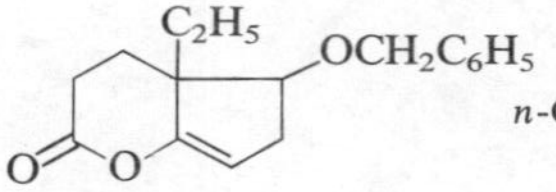 [structure: C_2H_5, $OCH_2C_6H_5$]	n-C_4H_9Li/THF	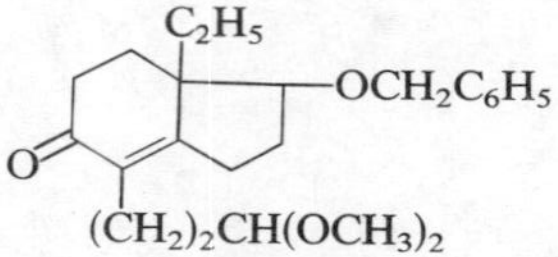 [structure: C_2H_5, $OCH_2C_6H_5$, $(CH_2)_2CH(OCH_3)_2$] (–)	66
C_9	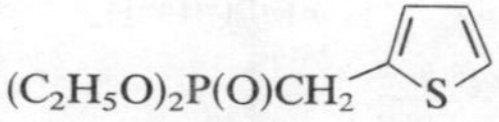 $(C_2H_5O)_2P(O)CH_2$-2-thienyl	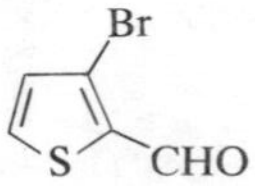 3-Br-thiophene-2-CHO	CH_3ONa/DMF	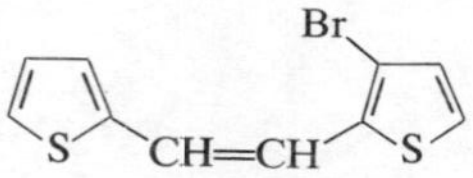 2-thienyl-$CH{=}CH$-(3-Br-2-thienyl) (–)	230
		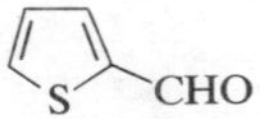 thiophene-2-CHO	NaH/glyme	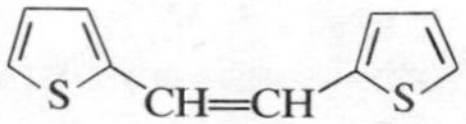 2-thienyl-$CH{=}CH$-2-thienyl (68)	229

	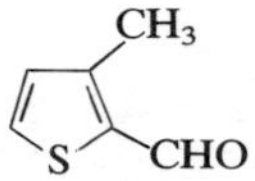 CH_3, CHO, S	NaH/glyme	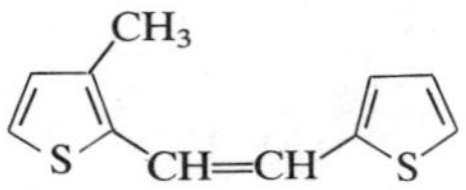 CH_3, S, CH=CH, S (40)	229
	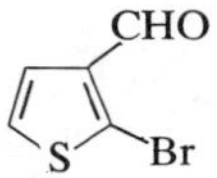 CHO, Br, S	CH_3ONa/DMF	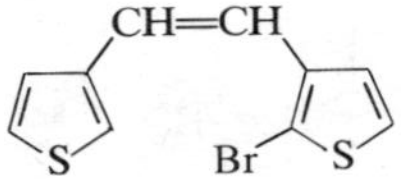 CH=CH, S, Br, S (–)	230
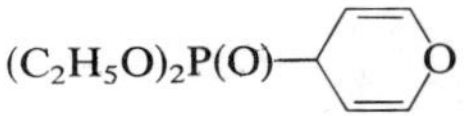 $(C_2H_5O)_2P(O)$–, O	C_6H_5CHO	t-C_4H_9OK/t-C_4H_9OH	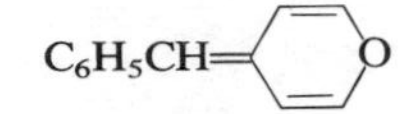 C_6H_5CH=, O (–)	231
 $(C_2H_5O)_2P(O)CH_2$, S	o-$HO_2CC_6H_4CHO$	CH_3ONa/DMF	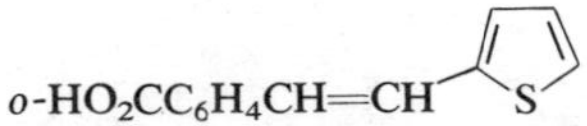 o-$HO_2CC_6H_4CH$=CH, S (85)	232
	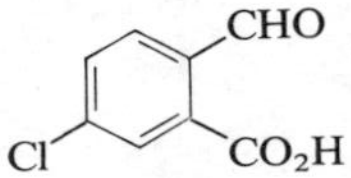 CHO, Cl, CO_2H	CH_3ONa/DMF	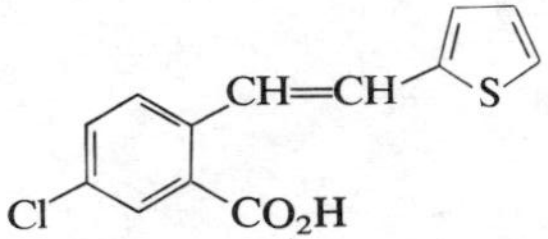 CH=CH, S, Cl, CO_2H (88)	232
	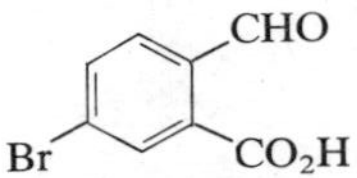 CHO, Br, CO_2H	CH_3ONa/DMF	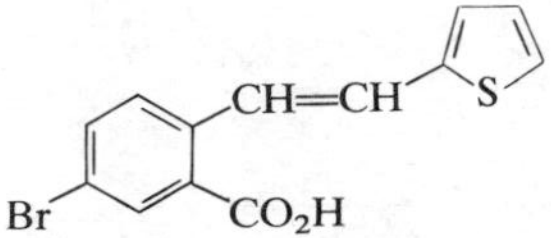 CH=CH, S, Br, CO_2H (72)	232
	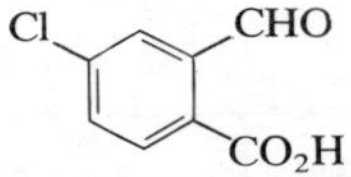 Cl, CHO, CO_2H	CH_3ONa/DMF	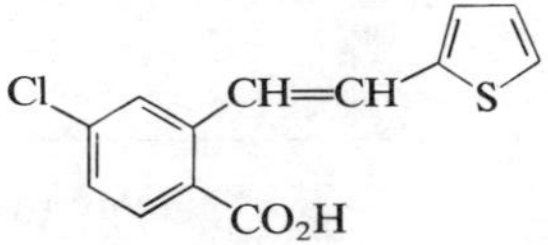 Cl, CH=CH, S, CO_2H (85)	232

Note: References 228–415 are on pp. 250–253.

TABLE I. Olefins from Alkyl (Arenyl) Phosphoryl Reagents

$$(R)_2P(O)CHR^1_2 + R^2R^3CO \xrightarrow{\text{Base}} R^2R^3C{=}CR^1_2 + (R)_2PO_2^- \quad (\textit{Continued})$$

No. of C Atoms	Reagent	Reactant	Base/Solvent	Product(s) and Yield(s) (%)	Refs.
C_9 (*Contd.*)	$(C_2H_5O)_2P(O)CH_2$–(2-thienyl) (*Contd.*)	4-CH_3O-2-CHO-$C_6H_3CO_2H$ (CHO, CO_2H, CH_3O)	CH_3ONa/DMF	CH_3O, CH=CH–(2-thienyl), CO_2H (82)	232
C_{10}	2,6-Cl-pyridin-3-yl $CH_2P(O)(OC_2H_5)_2$ (Cl, N, Cl)	tetrahydrofuran-2-CHO (O, CHO)	NaH/glyme	Cl, N, Cl, CH=CH–(tetrahydrofuran-2-yl) (47)	400
		N-CH_3-pyrrolidine-2-CHO (N, CH_3, CHO)	NaH/glyme	Cl, N, Cl, CH=CH–(N-CH_3-pyrrolidin-2-yl) (67)	400
		pyridine-3-CHO (N, CHO)	NaH/glyme	Cl, N, Cl, CH=CH–(3-pyridyl) (57)	400
	2,6-Br-pyridin-3-yl $CH_2P(O)(OC_2H_5)_2$ (Br, N, Br)	pyridine-3-CHO (N, CHO)	NaH/glyme	Br, N, Br, CH=CH–(3-pyridyl) (49)	400

$(C_2H_5O)_2P(O)CH_2(CH{=}CH)_2CH_3$	$2\text{-}C_5H_4N\text{-}CHO$	$NaNH_2$/THF	$2\text{-}C_5H_4N\text{-}CH{=}CH(CH{=}CH)_2CH_3$ (27)	233
	$4\text{-}OCH_3\text{-}2\text{-}C_5H_3N\text{-}CHO$	$NaNH_2$/THF	$4\text{-}OCH_3\text{-}2\text{-}C_5H_3N\text{-}CH{=}CH(CH{=}CH)_2CH_3$ (22)	233
$2,6\text{-}Cl_2\text{-}3\text{-}C_5H_2N\text{-}CH_2P(O)(OC_2H_5)_2$	C_6H_5CHO	NaH/glyme	$2,6\text{-}Cl_2\text{-}3\text{-}C_5H_2N\text{-}CH{=}CHC_6H_5$ (56)	400
$2,6\text{-}Cl_2\text{-}3\text{-}C_5H_2N\text{-}CH_2P(O)(OC_2H_5)_2$	C_6H_5CHO	NaH/glyme	$2,6\text{-}Cl_2\text{-}3\text{-}C_5H_2N\text{-}CH{=}CHC_6H_5$ (72)	400
$(C_2H_5O)_2P(O)CH_2CH(OC_2H_5)_2$	C_6H_5CHO	NaH/glyme	$C_6H_5CH{=}CHCH(OC_2H_5)_2$ (95)	234
$(C_2H_5O)_2P(O)CH_2\text{-}2\text{-}C_5H_4N$	C_6H_5CHO	NaOH, H_2O/CH_2Cl_2, $(n\text{-}C_4H_9)_4N^+I^-$	$C_6H_5CH{=}CH\text{-}2\text{-}C_5H_4N$ (71)	367
	$C_6H_5CH{=}CHCHO$	NaOH, H_2O/benzene, $(n\text{-}C_4H_9)_4N^+I^-$	$C_6H_5(CH{=}CH)_2\text{-}2\text{-}C_5H_4N$ (75)	382, 367

Note: References 228–415 are on pp. 250–253.

TABLE I. OLEFINS FROM ALKYL (ARENYL) PHOSPHORYL REAGENTS

$$(R)_2P(O)CHR^1_2 + R^2R^3CO \xrightarrow{\text{Base}} R^2R^3C{=}CR^1_2 + (R)_2PO_2^- \ (\textit{Continued})$$

No. of C Atoms	Reagent	Reactant	Base/Solvent	Product(s) and Yield(s) (%)	Refs.
C_{10} (*Contd.*)	2,6-dibromopyridin-3-yl–$CH_2P(O)(OC_2H_5)_2$ (Br, N, Br)	$C_6H_5CH{=}CHCHO$	NaH/glyme	2,6-dibromopyridin-3-yl–$(CH{=}CH)_2C_6H_5$ (Br, N, Br) (63)	400
	$(C_2H_5O)_2P(O)CH_2(CH{=}CH)_2CH_3$	1-$C(C_6H_5)_3$-pyrrolidine-2-CHO	$NaNH_2$/THF	1-$C(C_6H_5)_3$-pyrrolidin-2-yl–$CH{=}CH(CH{=}CH)_2CH_3$ (60)	235
C_{11}	$(C_2H_5O)_2P(O)CH_2C_6H_5$	C_2H_5CHO	NaH/cyclohexane	$C_6H_5CH{=}CHC_2H_5$ (30)	162
		2-furyl–CHO (O)	CH_3ONa/DMF	2-furyl–$CH{=}CHC_6H_5$ (O) (77)	236
		2-thienyl–CHO (S)	CH_3ONa/DMF	2-thienyl–$CH{=}CHC_6H_5$ (S) (84)	236
		2-pyridyl–CHO (N)	CH_3ONa/DMF	2-pyridyl–$CH{=}CHC_6H_5$ (N) (75)	236
		3-pyridyl–CHO (N)	CH_3ONa/DMF	3-pyridyl–$CH{=}CHC_6H_5$ (N) (58)	236

	C_6H_5CHO	t-C_4H_9OK/DMF	$C_6H_5CH{=}CHC_6H_5$ (80)	204, 4, 162, 236
$(C_2H_5O)_2P(O)CHBrC_6H_4NO_2$-$p$	C_6H_5CHO	C_2H_5ONa/C_2H_5OH	$C_6H_5CH{=}CBrC_6H_4NO_2$-p (—)	62
$(C_2H_5O)_2P(O)CH_2C_6F_5$	C_6F_5CHO	NaH/glyme	$C_6H_5CH{=}CHC_6F_5$ (63)	237, 396
$(C_2H_5O)_2P(O)CH_2C_6H_5$	$C_6H_5COCH_3$	CH_3ONa/DMF	$C_6H_5CH{=}C(CH_3)C_6H_5$ (—)	2
	p-$C_6H_4(CHO)_2$	t-C_4H_9OK/DMF	p-$C_6H_4(CH{=}CHC_6H_5)_2$ (78)	88, 204
	$(CH_3)_2C{=}CHC(CH_3)_2CHO$	NaH/glyme	$(CH_3)_2C{=}CHC(CH_3)_2CH{=}CHC_6H_5$ (—)	163
$(C_2H_5O)_2P(O)CH_2C_6H_3F_2$-2,6	o-$C_6H_4(CHO)_2$	CH_3ONa/DMF	o-$C_6H_4(CH{=}CHC_6H_3F_2\text{-}2,6)_2$ (14)	75
$(C_2H_5O)_2P(O)CH_2C_6H_5$	[structure: benzofuran, O] CHO	KOH/DMF	[structure: benzofuran, O] $CH{=}CHC_6H_5$ (−)	238
$(C_2H_5O)_2P(O)CH_2C_6H_4Cl$-$p$	[structure: benzofuran, O] CHO	KOH/DMF	[structure: benzofuran, O] $CH{=}CHC_6H_5Cl$-p (−)	238
$(C_2H_5O)_2P(O)CH_2C_6H_5$	3,5-$(CH_3O)_2C_6H_3CHO$	CH_3ONa/DMF	3,5-$(CH_3O)_2C_6H_3CH{=}CHC_6H_5$ (91.5)	239
	$C_6H_5CH{=}CHCHO$	t-C_4H_9OK/DMF	$C_6H_5(CH{=}CH)_2C_6H_5$ (80)	240, 204, 382
$(C_2H_5O)_2P(O)CH_2C_6H_4Br$-$p$	$C_6H_5CH{=}CHCHO$	NaOH, H_2O/C_6H_6 $(n\text{-}C_4H_9)_4N^+I^-$	p-$BrC_6H_4(CH{=}CH)_2C_6H_5$ (81)	382
$(C_2H_5O)_2P(O)CH_2C_6H_5$	p-$C_6H_4(CH{=}CHCHO)_2$	C_2H_5ONa/C_2H_5OH	p-$C_6H_4[(CH{=}CH)_2C_6H_5]_2$ (—)	88
	[structure: O-ring–benzene] CHO	NaH/glyme	[structure: O-ring–benzene] $CH{=}CHC_6H_5$ (−)	241
	$(C_6H_5)_2CO$	$NaNH_2/C_6H_6$	$(C_6H_5)_2C{=}CHC_6H_5$ (88)	17, 2, 204, 242, 243
	$C_6H_5CH{=}CHCOC_6H_5$	NaH/glyme	$C_6H_5CH{=}CHC(C_6H_5){=}CHC_6H_5$ (—)	96

Note: References 228–415 are on pp. 250–253.

TABLE I. Olefins from Alkyl (Arenyl) Phosphoryl Reagents

$$(R)_2P(O)CHR^1_2 + R^2R^3CO \xrightarrow{\text{Base}} R^2R^3C{=}CR^1_2 + (R)_2PO_2^-$$ (*Continued*)

No. of C Atoms	Reagent	Reactant	Base/Solvent	Product(s) and Yield(s) (%)	Refs.
C_{12}					
	$(C_2H_5O)_2P(O)CH_2$— (O, N ring)	C_2H_5CHO	NaH/glyme	(O, N ring) $CH{=}CHC_2H_5$ (75)	46
		$(CH_3)_2CO$	NaH/glyme	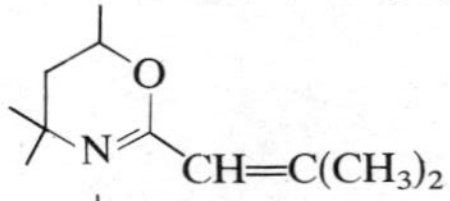 (O, N ring) $CH{=}C(CH_3)_2$ (73)	46
		O (ring)	NaH/glyme	(O, N ring) CH= (ring) (77)	46
		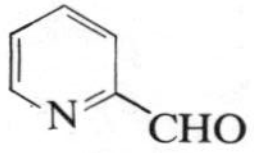 (N ring) CHO	NaH/glyme	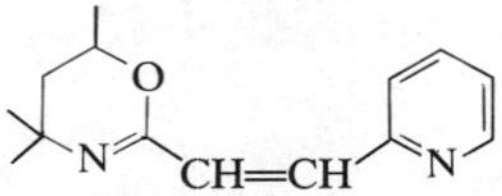 (O, N ring) CH=CH (N ring) (65)	46
	$(C_2H_5O)_2P(O)CH_2C_6H_4CN$-*p*	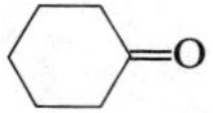 (ring) =O	NaH/glyme	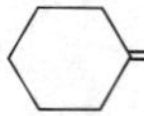 (ring) $=CHC_6H_4CN$-*p* (—)	413
	$(C_2H_5O)_2P(O)CH_2$— 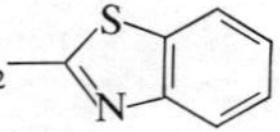 (S, N ring)	$CH_3(CH{=}CH)_2CHO$	$NaNH_2$/THF	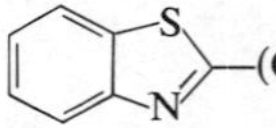 (S, N ring) —$(CH{=}CH)_3CH_3$ (48)	233

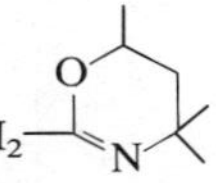

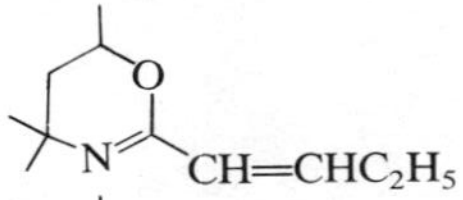

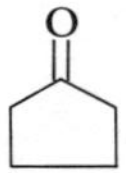

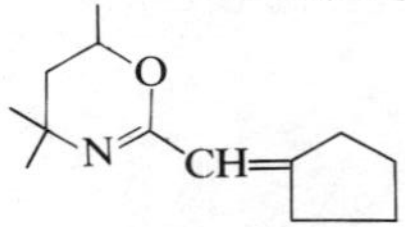

	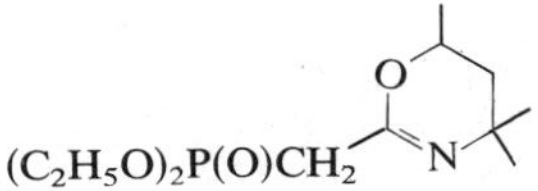 $(C_2H_5O)_2P(O)CH_2$ O N	C_6H_5CHO	NaH/glyme	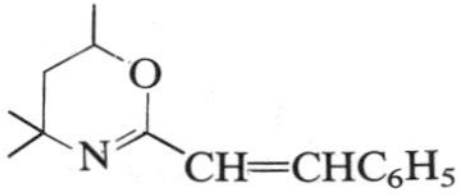 O N $CH{=}CHC_6H_5$ (80)	46
		n-$C_6H_{13}CHO$	NaH/glyme	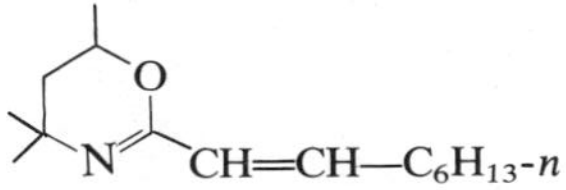 O N $CH{=}CH{-}C_6H_{13}$-n (72)	46
	$(C_2H_5O)_2P(O)CH(CH_3)C_6H_5$	C_6H_5CHO	NaH/DMF	$C_6H_5CH{=}C(CH_3)C_6H_5$ (61)	39
	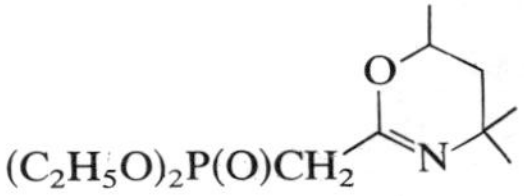 $(C_2H_5O)_2P(O)CH_2$ O N	$C_6H_5COCH_3$	NaH/glyme	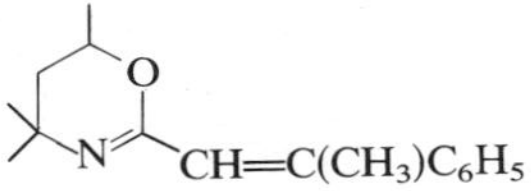 O N $CH{=}C(CH_3)C_6H_5$ (57)	46
	$(C_2H_5O)_2P(O)CH_2C_6H_4CN$-$p$	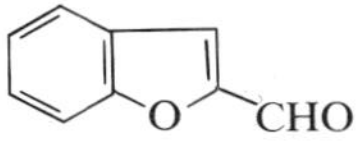 O CHO	KOH/DMF	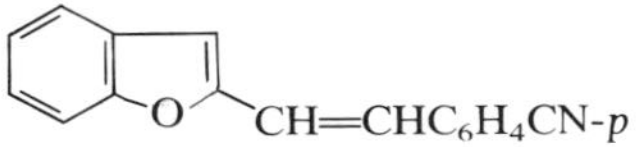 O $CH{=}CHC_6H_4CN$-p (−)	238
	$(C_2H_5O)_2P(O)CH_2C_6H_4CO_2H$-$p$	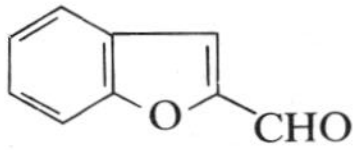 O CHO	KOH/DMF	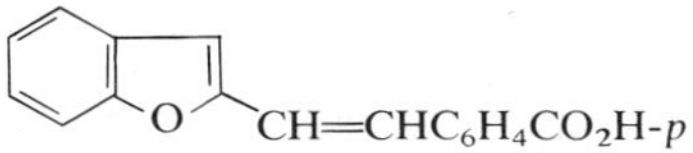 O $CH{=}CHC_6H_4CO_2H$-p (−)	238
C_{13}	 $(C_2H_5O)_2P(O)CH_2$ O N	$(C_6H_5)_2CO$	NaH/glyme	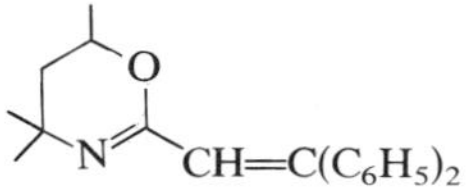 O N $CH{=}C(C_6H_5)_2$ (77)	46
	$(C_2H_5O)_2P(O)CH_2CH{=}CHC_6H_5$	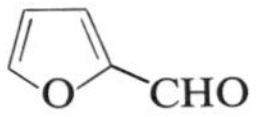 O CHO	NaOH, H_2O/C_6H_6, $(n$-$C_4H_9)_4N^+I^-$	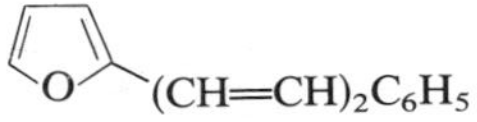 O $(CH{=}CH)_2C_6H_5$ (84)	382
		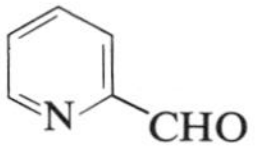 N CHO	NaOH, H_2O/C_6H_6, $(n$-$C_4H_9)_4N^+I^-$	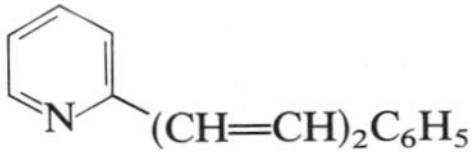 N $(CH{=}CH)_2C_6H_5$ (59)	382

Note: References 228–415 are on pp. 250–253.

TABLE I. Olefins from Alkyl (Arenyl) Phosphoryl Reagents

$$(RO)_2P(O)CH(R^1)CN + R^2R^3CO \xrightarrow{\text{Base}} R^2R^3C{=}CR^1_2 + (R)_2PO_2^-$$ (*Continued*)

No. of C Atoms	Reagent	Reactant	Base/Solvent	Product(s) and Yield(s) (%)	Refs.
C_{13} (*Contd.*)					
	$(C_2H_5O)_2P(O)CH_2$ N H	O N CH_3	NaH/DMF	N H CH= NCH_3 (−)	244
		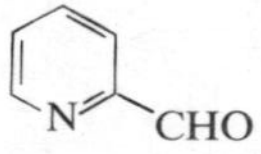 N CHO	NaH/DMF	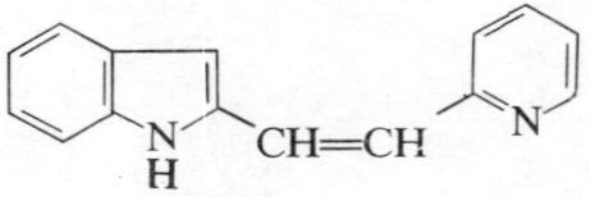 N H CH=CH N	244
	$(C_2H_5O)_2P(O)CH(CH_3)$ 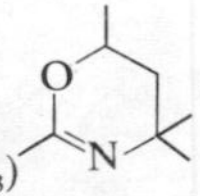 O N	C_6H_5CHO	NaH/glyme	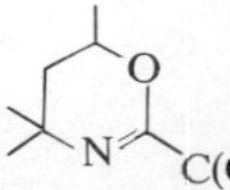 O N $C(CH_3){=}CHC_6H_5$ (75)	46
	$(C_2H_5O)_2P(O)CH_2CH{=}CHC_6H_5$	C_6H_5CHO	NaOH, H_2O/C_6H_6, $(n\text{-}C_4H_9)_4N^+I^-$	$C_6H_5(CH{=}CH)_2C_6H_5$ (70)	382
		$p\text{-}O_2NC_6H_4CHO$	NaOH, H_2O/C_6H_6, $(n\text{-}C_4H_9)_4N^+I^-$	$p\text{-}O_2NC_6H_4(CH{=}CH)_2C_6H_5$ (57)	382
		$C_6H_5CH{=}CHCHO$	NaOH, H_2O/C_6H_6, $(n\text{-}C_4H_9)_4N^+I^-$	$C_6H_5(CH{=}CH)_3C_6H_5$ (80)	382
	$(C_2H_5O)_2P(O)CH_2C_6H_4CO_2CH_3\text{-}p$	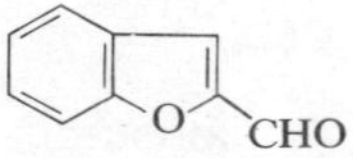 O CHO	KOH/DMF	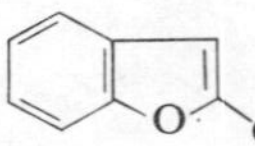 O $CH{=}CHC_6H_4CO_2CH_3\text{-}p$ (−)	238

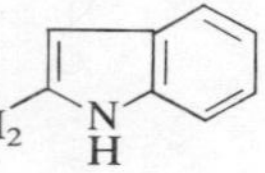

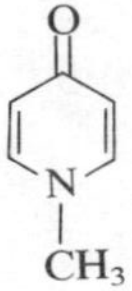

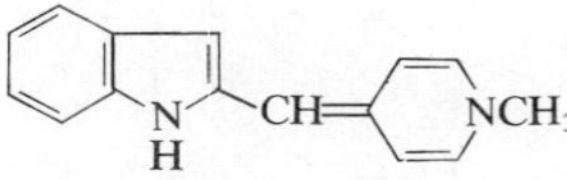

	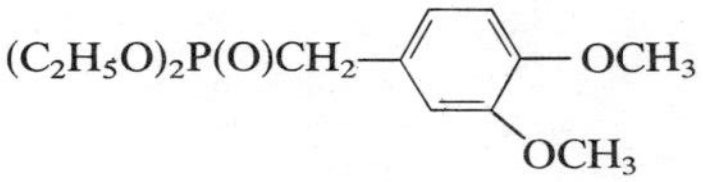 $(C_2H_5O)_2P(O)CH_2$–(ring)–OCH_3, OCH_3	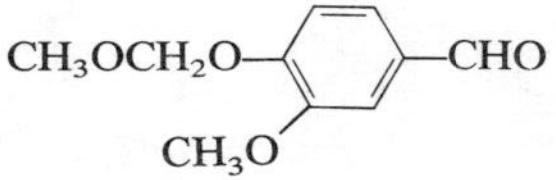 CH_3OCH_2O–(ring)–CHO, CH_3O	NaH/glyme	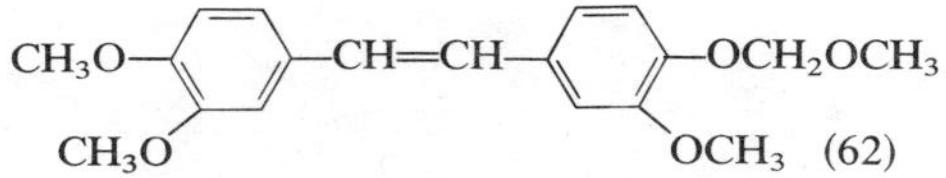 CH_3O–(ring)–CH=CH–(ring)–OCH_2OCH_3, CH_3O, OCH_3 (62)	245
	$(C_6H_5)_2P(O)CH_3$	$(C_6H_5)_2CO$	$NaNH_2/C_6H_6$	$(C_6H_5)_2C{=}CH_2$ (70)	243, 242
		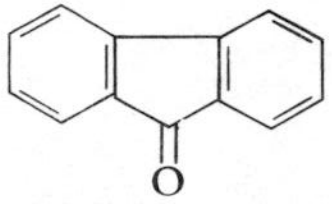	$NaNH_2/C_6H_6$	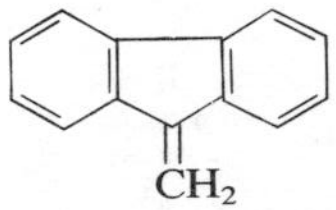 CH_2 (–)	243
C_{14}	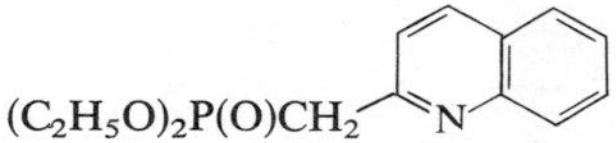 $(C_2H_5O)_2P(O)CH_2$–(N)	$CH_3(CH{=}CH)_2CHO$	$NaNH_2/THF$	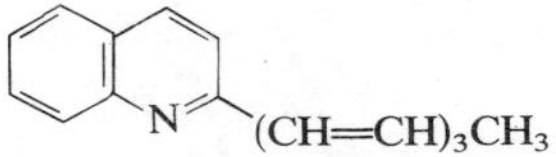 (N)–$(CH{=}CH)_3CH_3$ (60)	233
	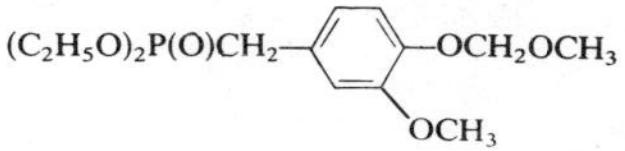 $(C_2H_5O)_2P(O)CH_2$–(ring)–OCH_2OCH_3, OCH_3	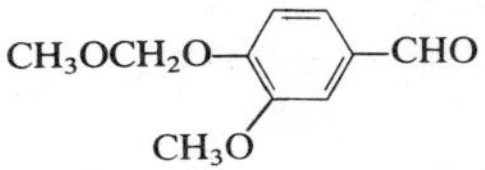 CH_3OCH_2O–(ring)–CHO, CH_3O	NaH/glyme	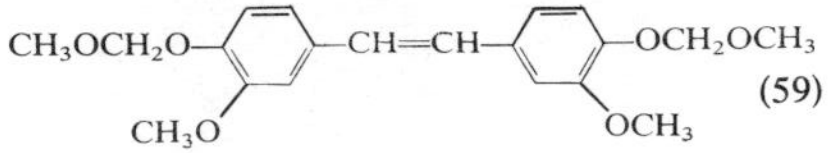 CH_3OCH_2O–(ring)–CH=CH–(ring)–OCH_2OCH_3, CH_3O, OCH_3 (59)	245
	$(C_6H_5)_2P(O)C_2H_5$	$(C_6H_5)_2CO$	$NaNH_2/C_6H_6$	$(C_6H_5)_2C{=}CHCH_3$ (—)	243, 202, 242
C_{15}	$(C_2H_5O)_2P(O)CH_2C_{10}H_7$-1	O_2	t-C_4H_9OK/DMF	1-$C_{10}H_7CH{=}CHC_{10}H_7$-1 (50)	246
	$(C_6H_5)_2P(O)CH_2C_2H_5$	C_6H_5CHO	NaH/DMF	$C_6H_5CH{=}CHC_2H_5$ (8)	162
	$(C_6H_5)_2P(O)CH_2CH{=}CH_2$	$(C_6H_5)_2CO$	t-C_4H_9OK/C_6H_6	$(C_6H_5)_2C{=}CHCH{=}CH_2$ (—)	242
C_{16}	$(C_6H_5)_2P(O)CH_2C_3H_7$-n	C_6H_5CHO	t-C_4H_9OK/C_6H_6	$C_6H_5CH{=}CHC_3H_7$-n (—)	242
		$(C_6H_5)_2CO$	t-C_4H_9OK/C_6H_6	$(C_6H_5)_2C{=}CHC_3H_7$-n (—)	242
	 $(C_6H_5)_2P(O)CH_2$–(cyclopropyl)	$(C_6H_5)_2CO$	t-C_4H_9OK/C_6H_6	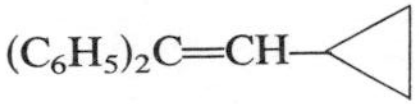 $(C_6H_5)_2C{=}CH$–(cyclopropyl) (55)	247
C_{17}	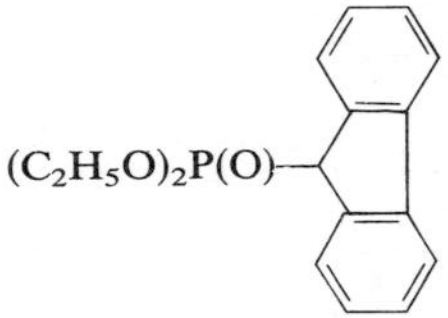 $(C_2H_5O)_2P(O)$–(fluorenyl)	CH_2O	NaH/glyme	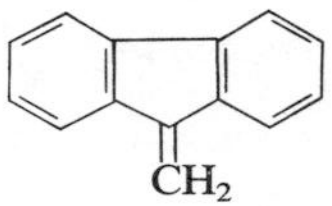 CH_2 (77)	4

Note: References 228–415 are on pp. 250–253.

TABLE I. OLEFINS FROM ALKYL (ARENYL) PHOSPHORYL REAGENTS

$(R)_2P(O)CHR^1_2 + R^2R^3CO \xrightarrow{\text{Base}} R^2R^3C{=}CR^1_2 + (R)_2PO_2^-$ (*Continued*)

No. of C Atoms	Reagent	Reactant	Base/Solvent	Product(s) and Yield(s) (%)	Refs.
C_{17} (*Contd.*)	$(C_2H_5O)_2P(O)CH_2C_6H_4C_6H_5$-*p*	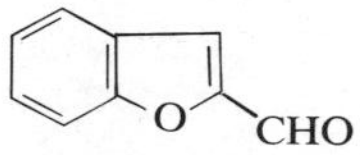 [structure] CHO	KOH/DMF	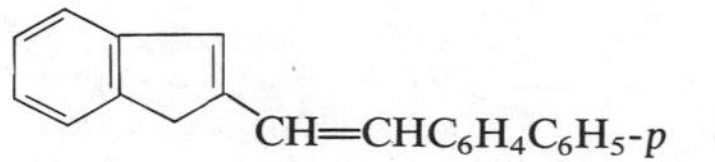 [structure] $CH{=}CHC_6H_4C_6H_5$-*p* (—)	238
C_{18}	$(C_6H_5)_2P(O)$—[structure] 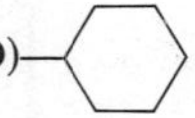	CH_3CHO	*n*-C_4H_9Li/ether	$(C_6H_5)_2P(O)$—[structure] OH (—) 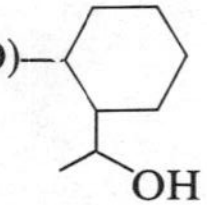	377
	$(C_2H_5O)_2P(O)$—[structure] NCH_3 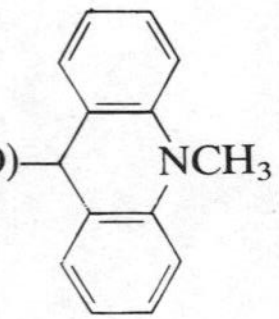	C_6H_5CHO	NaH/glyme	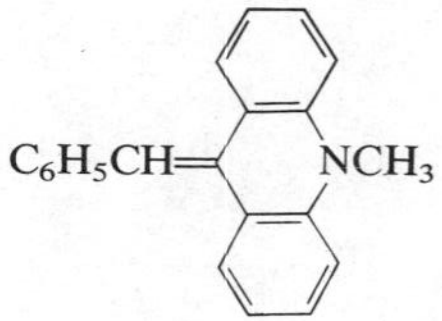 $C_6H_5CH{=}$[structure] NCH_3 (–)	248
		m-$O_2NC_6H_4CHO$	NaH/glyme	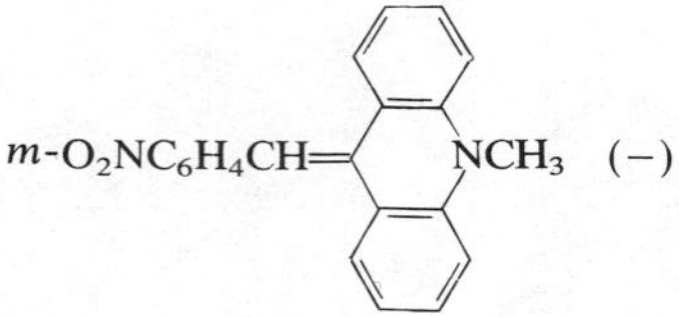 *m*-$O_2NC_6H_4CH{=}$[structure] NCH_3 (–)	248

		p-$CH_3OC_6H_4CHO$	NaH/glyme	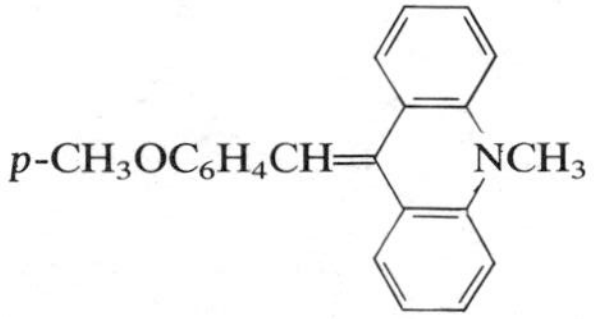 p-$CH_3OC_6H_4CH$=[10-methylacridan-9-ylidene, NCH_3] (–)	248
C_{19}	$(C_6H_5)_2P(O)CH_2C_6H_5$	O_2	t-C_4H_9OK/DMF	$C_6H_5CH{=}CHC_6H_5$ (88)	246
	$(C_6H_5)_2P(O)CH_2C_6H_3Cl_2$-3,4	O_2	t-C_4H_9OK/DMF	3,4-$Cl_2C_6H_3CH{=}CHC_6H_3Cl_2$-3,4 (80)	246
	$(C_6H_5)_2P(O)CH_2C_6H_5$	C_2H_5CHO	NaH/DMF	$C_6H_5CH{=}CHC_2H_5$ (23)	162
		$C_2H_5COCH_3$	t-C_4H_9OK/C_6H_6	$C_6H_5CH{=}C(CH_3)C_2H_5$ (54)	2
		i-$C_3H_7COCH_3$	t-C_4H_9OK/C_6H_6	i-$C_3H_7C(CH_3){=}CHC_6H_5$ (—)	2
		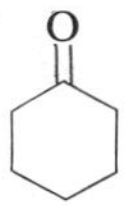 [cyclohexanone, C=O]	t-C_4H_9OK/C_6H_6	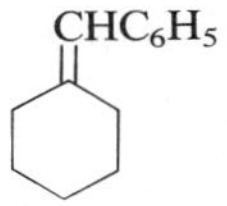 [cyclohexylidene]=CHC_6H_5 (47)	242
		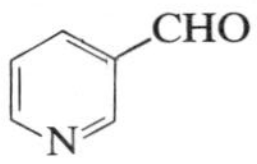 [pyridine-3-yl]-CHO	t-$C_4H_9OK/C_6H_5CH_3$	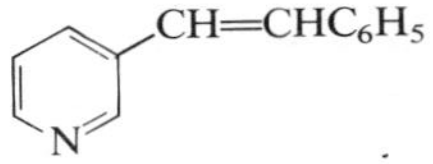 [pyridine-3-yl]-$CH{=}CHC_6H_5$ (58)	2
		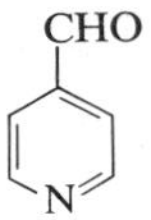 [pyridine-4-yl]-CHO	t-$C_4H_9OK/C_6H_5CH_3$	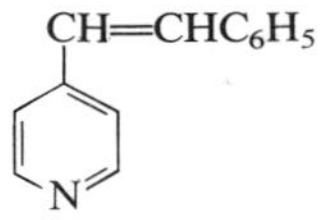 [pyridine-4-yl]-$CH{=}CHC_6H_5$ (47)	2
		C_6H_5CHO	t-C_4H_9OK/C_6H_6	$C_6H_5CH{=}CHC_6H_5$ (70)	162, 2, 204, 242
		$C_6H_5COCH_3$	t-C_4H_9OK/C_6H_6	$C_6H_5(CH_3)C{=}CHC_6H_5$ (60)	2, 242
		$C_6H_4(CHO)_2$-p	t-$C_4H_9OK/C_6H_5CH_3$	$C_6H_4(CH{=}CHC_6H_5)_2$-p (57)	2, 242
		$C_6H_5CH{=}CHCHO$	t-C_4H_9K/DMF	$C_6H_5CH{=}CHCH{=}CHC_6H_5$ (52)	2, 204, 242
		$(C_6H_5)_2CO$	t-C_4H_9OK/C_6H_6	$(C_6H_5)_2C{=}CHC_6H_5$ (70)	2, 204

Note: References 228–415 are on pp. 250–253.

TABLE I. Olefins from Alkyl (Arenyl) Phosphoryl Reagents

$$(R)_2P(O)CHR^1_2 + R^2R^3CO \xrightarrow{\text{Base}} R^2R^3C{=}CR^1_2 + (R)_2PO_2^-$$ *(Continued)*

No. of C Atoms	Reagent	Reactant	Base/Solvent	Product(s) and Yield(s) (%)	Refs.
C_{19} (*Contd.*)	$(C_6H_5)_2P(O)CH_2C_6H_5$ (*Contd.*)	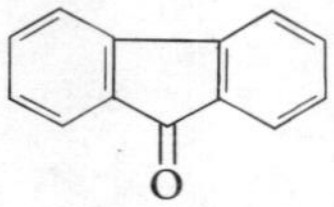 [structure] O	t-C_4H_9OK/C_6H_6	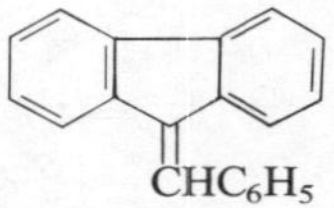 [structure] CHC_6H_5 (43)	2
		$C_6H_5CH{=}NC_6H_5$	t-C_4H_9OK/DMF	$C_6H_5CH{=}CHC_6H_5$ (73)	246
		$C_6H_5COCOC_6H_5$	t-$C_4H_9OK/C_6H_5CH_3$	$[C_6H_5CH{=}C(C_6H_5)]_2$ (50)	2, 242
		p-$OHCC_6H_4CH{=}NNHC_6H_5$	t-C_4H_9OK/C_6H_6	p-$C_6H_5CH{=}CHC_6H_4CH{=}NNHC_6H_5$ (76)	204
	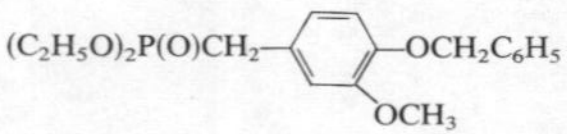 $(C_2H_5O)_2P(O)CH_2$–[structure]–$OCH_2C_6H_5$, OCH_3	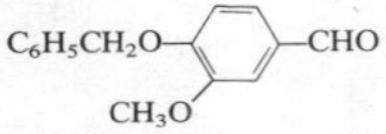 $C_6H_5CH_2O$–[structure]–CHO, CH_3O	NaH/glyme	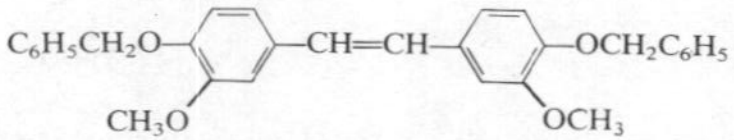 $C_6H_5CH_2O$–[structure]–CH=CH–[structure]–$OCH_2C_6H_5$, CH_3O, OCH_3 (58)	245
	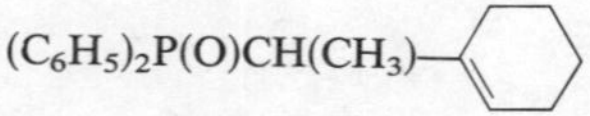 $(C_6H_5)_2P(O)CH(CH_3)$–[structure]	O_2	n-C_4H_9Li/ether	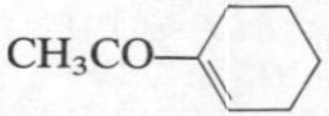 CH_3CO–[structure] (—)	377
		CH_3CHO	n-C_4H_9Li/ether	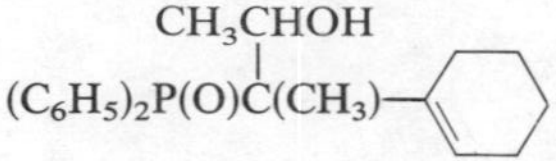 $(C_6H_5)_2P(O)C(CH_3)(CH_3CHOH)$–[structure] (—)	377
		C_6H_5CHO	NaH/DMF	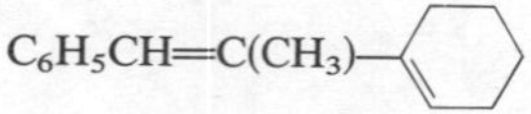 $C_6H_5CH{=}C(CH_3)$–[structure] (—)	377

		C_6H_5CHO	n-C_4H_9Li/ether	C_6H_5CHOH / $(C_6H_5)_2P(O)C(CH_3)$—(cyclohexenyl) (–)	377
	$(C_6H_5)_2P(O)CH(CH_3)C_6H_5$	C_6H_5CHO	t-C_4H_9OK/C_6H_6	$C_6H_5CH{=}C(CH_3)C_6H_5$ (—)	242
C_{22}	$(C_6H_5)_2P(O)CH(C_6H_5)C(CH_3){=}CH_2$	C_6H_5CHO	n-C_4H_9Li/ether	$(C_6H_5)_2P(O)C(C_6H_5){=}C(CH_3)$- $CH_2CH(C_6H_5)OH$ (—)	377
C_{23}	$(C_6H_5)_2P(O)CH_2C_{10}H_7$-1	O_2	t-C_4H_9OK/DMF	1-$C_{10}H_7CH{=}CHC_{10}H_7$-1 (90)	246
C_{24}	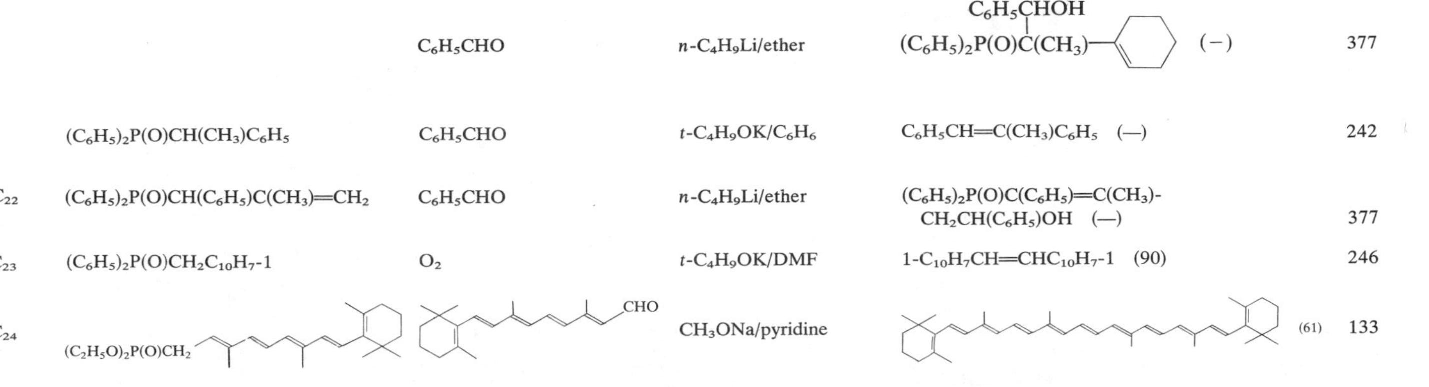 $(C_2H_5O)_2P(O)CH_2$	CHO	CH_3ONa/pyridine	(61)	133
C_{27}	$(C_6H_5)_2P(O)CH_2C(C_6H_5){=}CHC_6H_5$	O_2	t-C_4H_9OK/DMF	$C_6H_5CH{=}C(C_6H_5)$- $CH{=}CHC(C_6H_5){=}CHC_6H_5$ (16)	246
C_{28}	$(C_6H_5)_2P(O)(CH_2)_4P(O)(C_6H_5)_2$	$(C_6H_5)_2CO$	t-$C_4H_9OK/C_6H_5CH_3$	$(C_6H_5)_2C{=}CH(CH_2)_3P(O)(C_6H_5)_2$ (37)	249
C_{29}	$(C_6H_5)_2P(O)(CH_2)_6P(O)C_6H_5)_2$	$(C_6H_5)_2CO$	t-$C_4H_9OK/C_6H_5CH_3$	$(C_6H_5)_2C{=}CH(CH_2)_5P(O)(C_6H_5)_2$ (32)	249
C_{30}	$[p$-$(C_6H_5)_2P(O)$- $CH_2C_6H_4CH_2P(C_6H_5)_3]^+Br^-$	C_6H_5CHO	t-C_4H_9OK/DMF	$[p$-$C_6H_5CH{=}CHC_6H_4CH_2P(C_6H_5)_3]^+Br^-$ (—)	17
		$(C_6H_5)_2CO$	t-C_4H_9OK/DMF	$[p$-$(C_6H_5)_2C{=}CHC_6H_4CH_2P(C_6H_5)_3]^+Br^-$ (—)	17

Note: References 228–415 are on pp. 250–253.

TABLE II. Unsaturated Nitriles From Phosphoryl Reagents Containing a Cyano Group

$$(RO)_2P(O)CH(R^1)CN + R^2R^3CO \xrightarrow{\text{Base}} R^2R^3C{=}C(R^1)CN + (RO)_2PO_2^-$$

No. of C Atoms	Reagent	Reactant	Base/Solvent	Product(s) and Yield(s) (%)	Refs.
C_6	$(C_2H_5O)_2P(O)CH_2CN$	CH_2O	NaH/glyme	$CH_2{=}CHCN$ (—)	4
		CH_3CHO	$NaNH_2$/THF	$CH_3CH{=}CHCN$ (51)	367, 250
		$ClCH_2COCH_3$	$NaNH_2$/ether	$ClCH_2C(CH_3){=}CHCN$ (72)	165
		$(CH_3)_2CO$	$NaNH_2$/ether	$(CH_3)_2C{=}CHCN$ (61)	165, 367
		O CHO	$NaNH_2$/ether	O CH=CHCN (–)	165
		$(CH_3)_2C{=}CHCHO$	$NaNH_2$/ether	$(CH_3)_2C{=}CHCH{=}CHCN$ (—)	165
		$OHC(CH_2)_3CHO$	NaH/glyme	OH CN (40)	90
		$NCCH_2CH_2COCH_3$	NaH/glyme	$NCCH_2CH_2C(CH_3){=}CHCN$ (—)	251
		O	$NaNH_2$/ether	CHCN (82)	393, 165
		CHO N	$NaNH_2$/ether	CH=CHCN (—) N	165

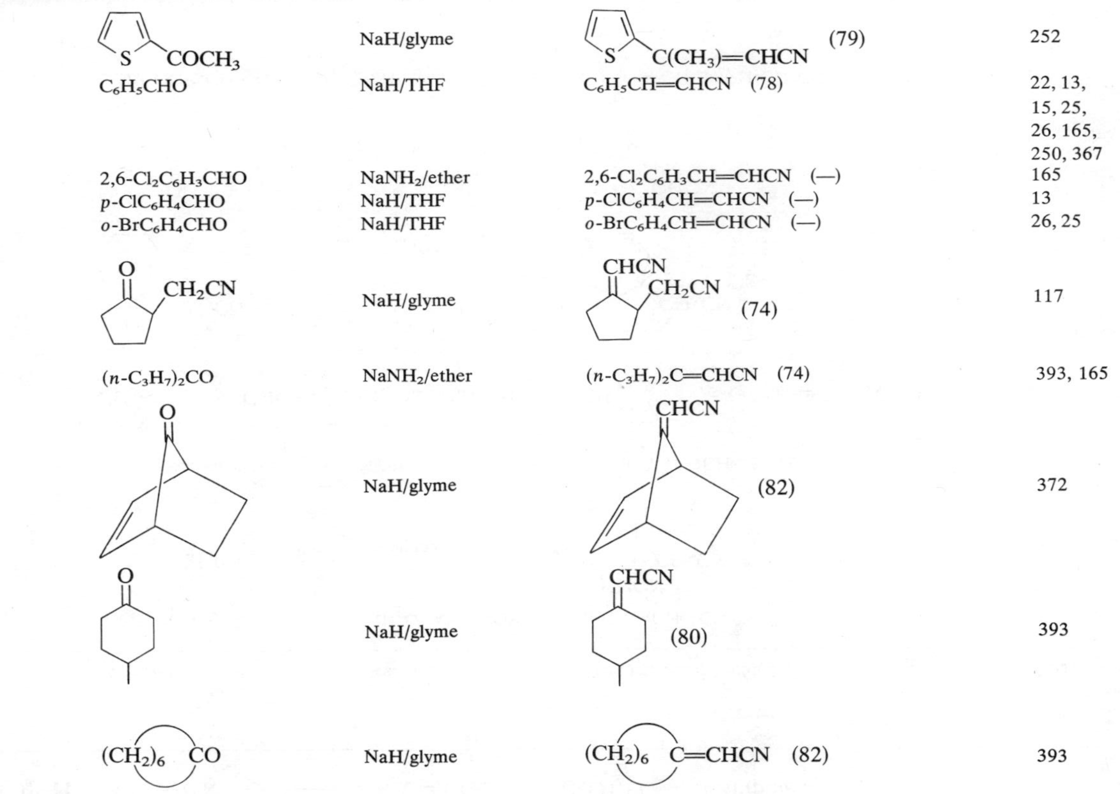

(2-thienyl) $COCH_3$	NaH/glyme	(2-thienyl) $C(CH_3){=}CHCN$ (79)	252
C_6H_5CHO	NaH/THF	$C_6H_5CH{=}CHCN$ (78)	22, 13, 15, 25, 26, 165, 250, 367
2,6-$Cl_2C_6H_3CHO$	$NaNH_2$/ether	2,6-$Cl_2C_6H_3CH{=}CHCN$ (—)	165
p-ClC_6H_4CHO	NaH/THF	p-$ClC_6H_4CH{=}CHCN$ (—)	13
o-BrC_6H_4CHO	NaH/THF	o-$BrC_6H_4CH{=}CHCN$ (—)	26, 25
O, CH_2CN (cyclopentanone)	NaH/glyme	CHCN, CH_2CN (74)	117
$(n\text{-}C_3H_7)_2CO$	$NaNH_2$/ether	$(n\text{-}C_3H_7)_2C{=}CHCN$ (74)	393, 165
O (bicyclic ketone)	NaH/glyme	CHCN (82)	372
O (cyclohexanone)	NaH/glyme	CHCN (80)	393
$(CH_2)_6$ CO	NaH/glyme	$(CH_2)_6$ $C{=}CHCN$ (82)	393

Note: References 228–415 are on pp. 250–253.

TABLE II. Unsaturated Nitriles From Phosphoryl Reagents Containing a Cyano Group

$$(RO)_2P(O)CH(R^1)CN + R^2R^3CO \xrightarrow{\text{Base}} R^2R^3C{=}C(R^1)CN + (RO)_2PO_2^- \quad (\textit{Continued})$$

No. of C Atoms	Reagent	Reactant	Base/Solvent	Product(s) and Yield(s) (%)	Refs.
C_6 (*Contd.*)	$(C_2H_5O)_2P(O)CH_2CN$ (*Contd.*)	p-$C_6H_4(CHO)_2$	C_2H_5ONa/C_2H_5OH	p-$C_6H_4(CH{=}CHCN)_2$ (—)	88
		O, CH_2CN	Na/glyme	CHCN, CH_2CN (61)	117
		$C_6H_5CH{=}CHCHO$	$NaNH_2$/ether	$C_6H_5CH{=}CHCH{=}CHCN$ (—)	165
		CH_3—, —$COCH_3$	CH_3ONa/C_2H_5OH	CH_3—, —$C(CH_3){=}CHCN$ (95)	253
		O, CH_2CN	NaH/glyme	CHCN, CH_2CN (55)	117
		O	NaH/glyme	CHCN (–)	240
		$C_6H_5COCH_2CH_2CN$	NaH/glyme	$C_6H_5C({=}CHCN)CH_2CH_2CN$ (53)	251

Reactant	Conditions	Product (Yield, %)	Refs.
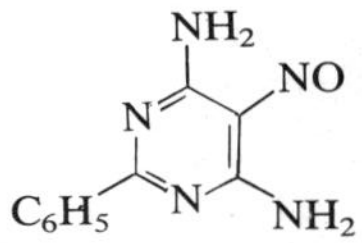 NH_2, NO, C_6H_5, N, N, NH_2	NaH/THF	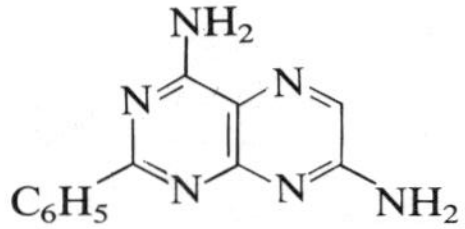 NH_2, N, N, N, N, C_6H_5, NH_2 (28)	107, 254
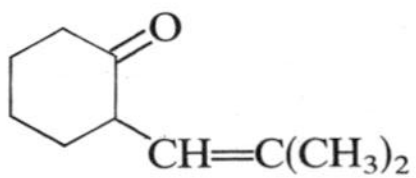 O, $CH{=}C(CH_3)_2$	NaH/DMSO	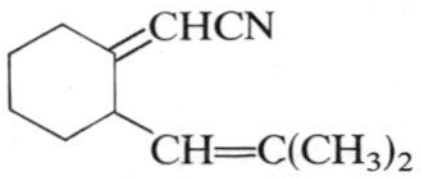 CHCN, $CH{=}C(CH_3)_2$ (95)	129
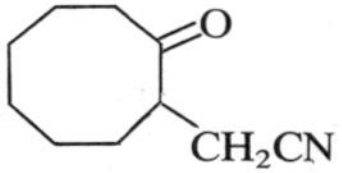 O, CH_2CN	NaH/glyme	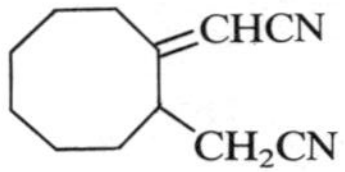 CHCN, CH_2CN (50)	117
$(CH_2)_9$ CO	NaH/glyme	$(CH_2)_9$ C=CHCN (78)	393
p-$C_6H_4(CH{=}CHCHO)_2$	C_2H_5ONa/C_2H_5OH	p-$C_6H_4[(CH{=}CH)_2CN]_2$ (—)	88
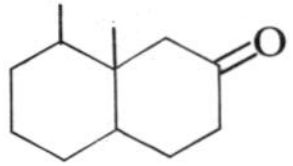 O	NaH/glyme	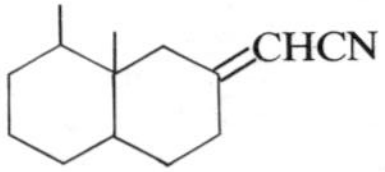 CHCN (79)	255
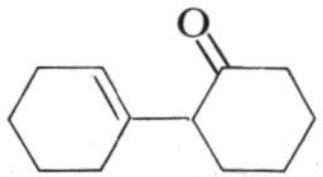 O	NaH/glyme	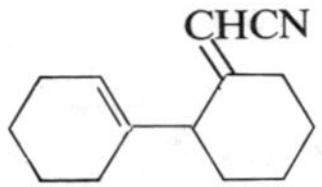 CHCN (41)	114
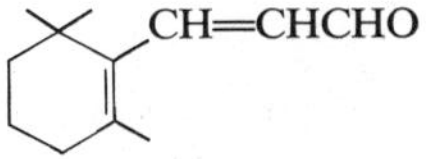 $CH{=}CHCHO$	$NaNH_2$/THF	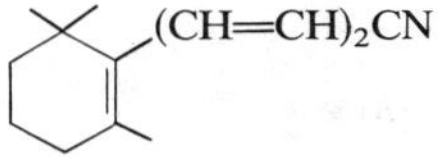 $(CH{=}CH)_2CN$ (−)	256

Note: References 228–415 are on pp. 250–253.

TABLE II. UNSATURATED NITRILES FROM PHOSPHORYL REAGENTS CONTAINING A CYANO GROUP

$(RO)_2P(O)CH(R^1)CN + R^2R^3CO \xrightarrow{\text{Base}} R^2R^3C{=}C(R^1)CN + (RO)_2PO_2^-$ (*Continued*)

No. of C Atoms	Reagent	Reactant	Base/Solvent	Product(s) and Yield(s) (%)	Refs.
C_6 (*Contd.*)	$(C_2H_5O)_2P(O)CH_2CN$ (*Contd.*)	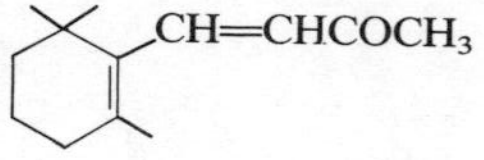 $CH{=}CHCOCH_3$	NaH/DMF	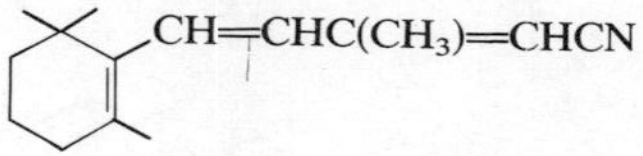 $CH{=}CHC(CH_3){=}CHCN$ (85)	257
		$(C_6H_5)_2CO$	$NaNH_2$/THF	$(C_6H_5)_2C{=}CHCN$ (5)	165, 22, 250
		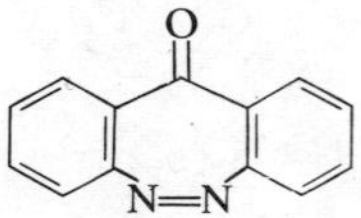 O, N=N	NaH/dioxane	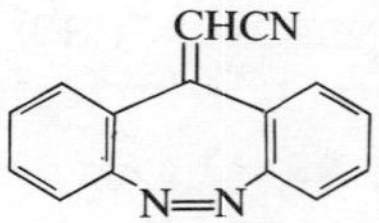 CHCN, N=N (100)	258
		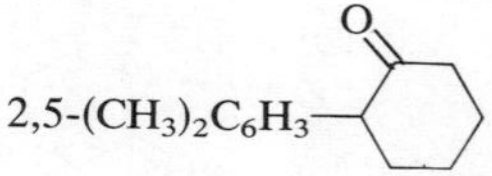 2,5-$(CH_3)_2C_6H_3$— (O)	NaH/glyme	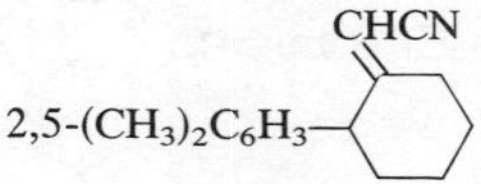 2,5-$(CH_3)_2C_6H_3$— (CHCN) (61)	259
		o-$C_6H_5CH_2SC_6H_4CHO$	NaH/glyme	o-$C_6H_5CH_2SC_6H_4CH{=}CHCN$ (—)	260
		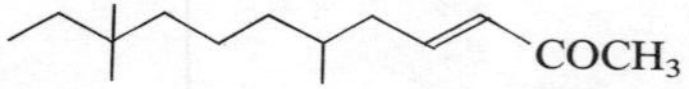 $COCH_3$	NaH/glyme	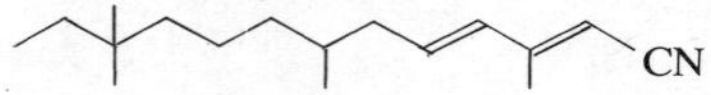 CN (−)	261

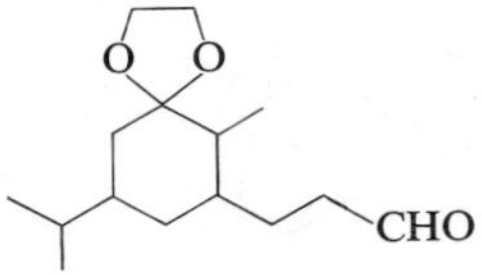

CHO	NaH/glyme	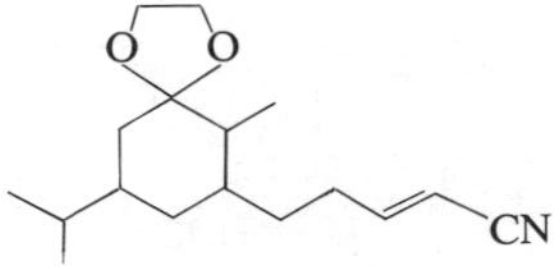 CN (86)	397
$C_6H_5CH(CN)CH_2COC_6H_5$	NaH/glyme	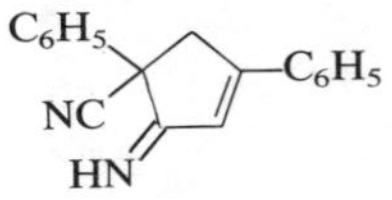 C_6H_5, NC, HN, C_6H_5 (−)	251
$NCCH_2CH(C_6H_5)COC_6H_5$	NaH/glyme	$NCCH_2C(C_6H_5){=}C(C_6H_5)CH_2CN$ (—) +	251
		$NCCH_2CH(C_6H_5)C(C_6H_5){=}CHCN$ (—)	251
O	NaH/glyme	CN (−)	262
$=CHCH{=}CHCOCH_3$	NaH/DMF	$=CHCH{=}CHC(CH_3){=}CHCN$ (85)	263
O, O	$NaNH_2$/THF	CHCN, NCCH (86)	86
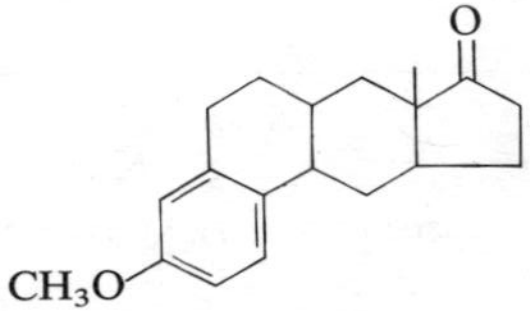 O, CH_3O	$NaNH_2$/THF	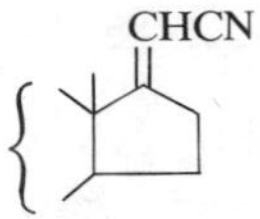 CHCN (84)	86

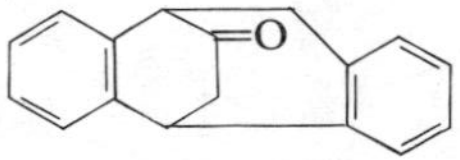

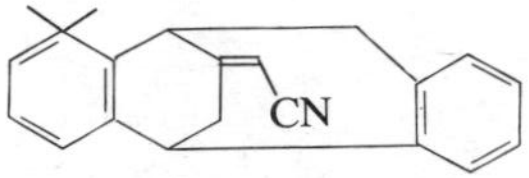

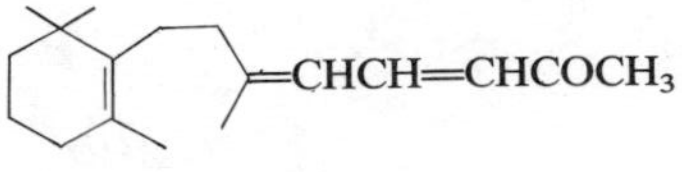

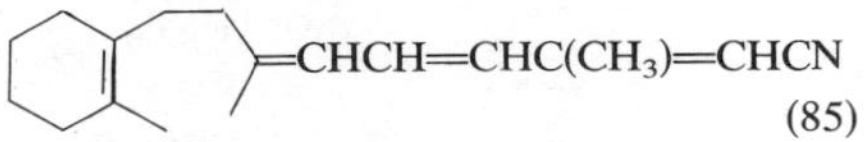

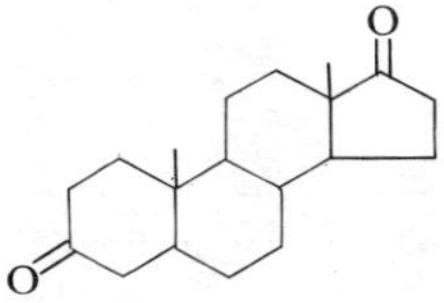

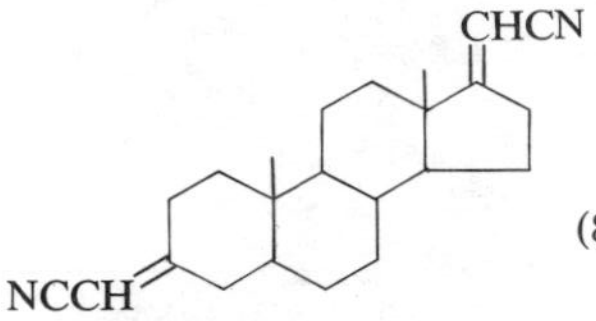

Note: References 228–415 are on pp. 250–253.

TABLE II. UNSATURATED NITRILES FROM PHOSPHORYL REAGENTS CONTAINING A CYANO GROUP

$$(RO)_2P(O)CH(R^1)CN + R^2R^3CO \xrightarrow{\text{Base}} R^2R^3C{=}C(R^1)CN + (RO)_2PO_2^-$$ (*Continued*)

No. of C Atoms	Reagent	Reactant	Base/Solvent	Product(s) and Yield(s) (%)	Refs.
C_6 (*Contd.*)	$(C_2H_5O)_2P(O)CH_2CN$ (*Contd.*)	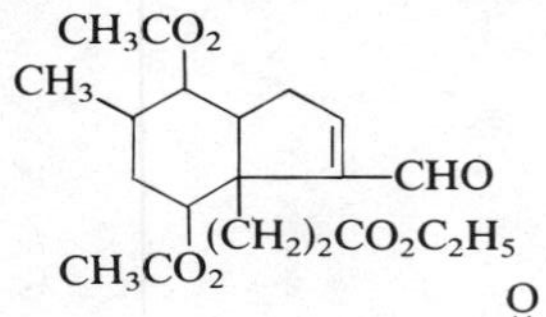 [structure: CH_3CO_2, CH_3, CHO, $(CH_2)_2CO_2C_2H_5$, CH_3CO_2]	NaH/glyme	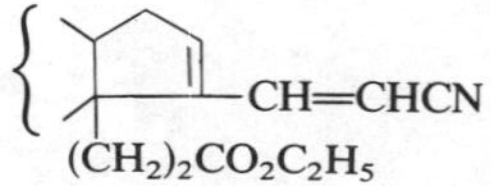 [structure] CH=CHCN, $(CH_2)_2CO_2C_2H_5$ (66)	264
		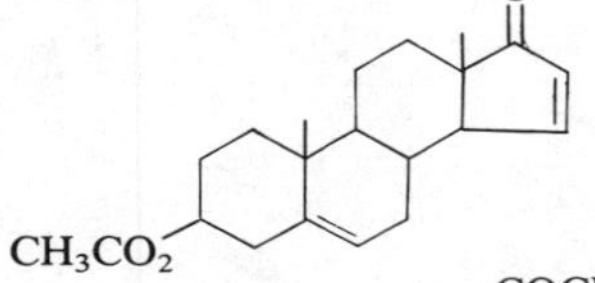 [structure: steroid, O, CH_3CO_2]	$NaNH_2$/THF	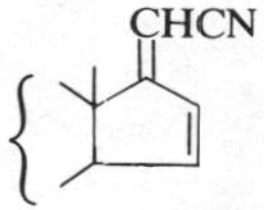 [structure] CHCN (67)	86
		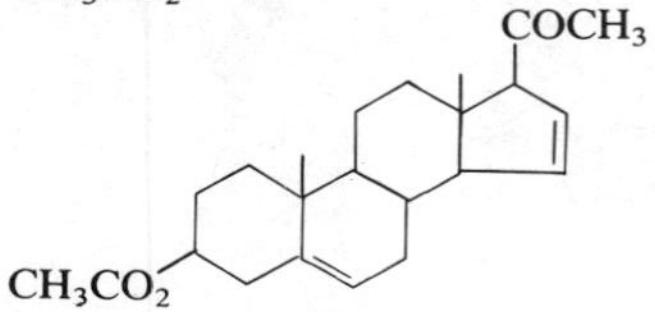 [structure: steroid, $COCH_3$, CH_3CO_2]	$NaNH_2$/THF	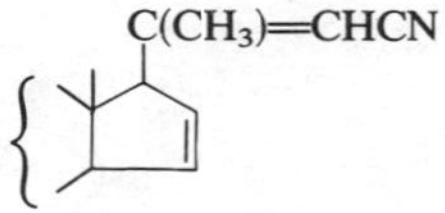 [structure] $C(CH_3){=}CHCN$ (82)	86
		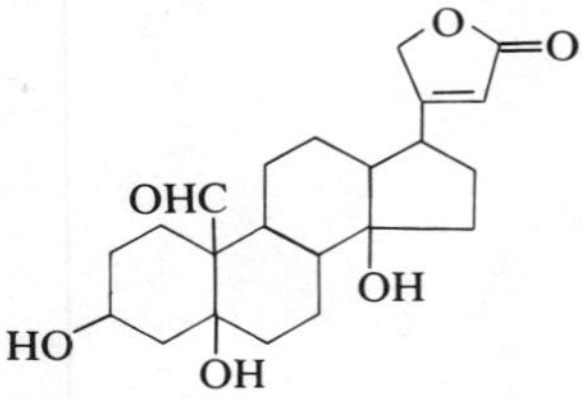 [structure: steroid lactone, OHC, OH, OH, HO, O]	$NaNH_2$/THF	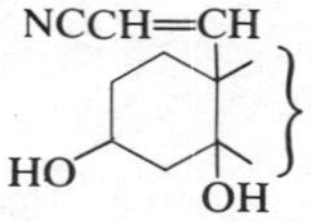 [structure] NCCH=CH, HO, OH (–)	265

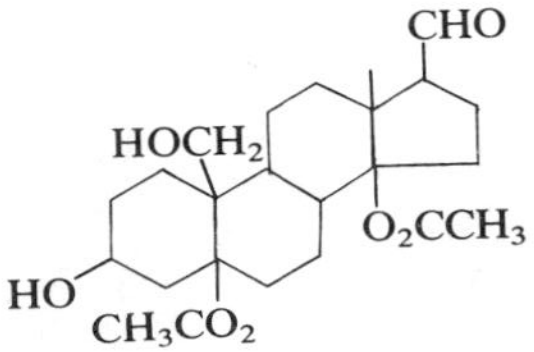

Reactant	Conditions	Product (Yield, %)		Refs.
HOCH$_2$, CHO, HO, CH$_3$CO$_2$, O$_2$CCH$_3$ (steroid)	t-C$_4$H$_9$OK/glyme	CH=CHCN, O$_2$CCH$_3$	(80)	266
CH$_3$O, CHO, CH$_3$CO$_2$ (steroid)	NaH/THF	CH$_3$O, CN	(–)	267
COCH$_2$OCOCH$_3$, CH$_3$CO$_2$ (steroid)	NaH/THF	CH$_2$OCOCH$_3$, C=CHCN	(72)	152
COCH$_2$OCOCH$_3$, CH$_3$CO$_2$ (Δ^5-steroid)	NaH/THF	CH$_2$OCOCH$_3$, C=CHCN	(18)	152

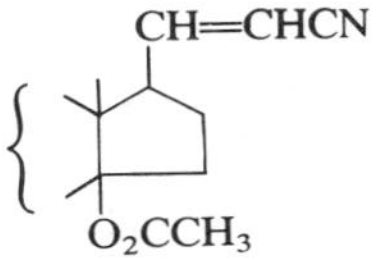

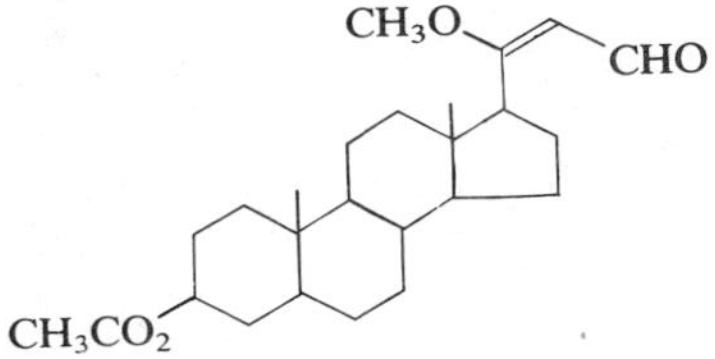

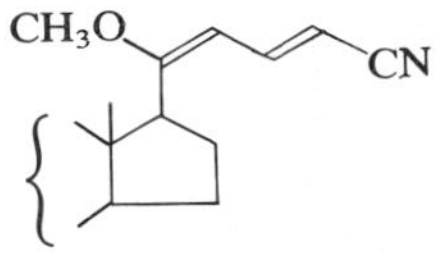

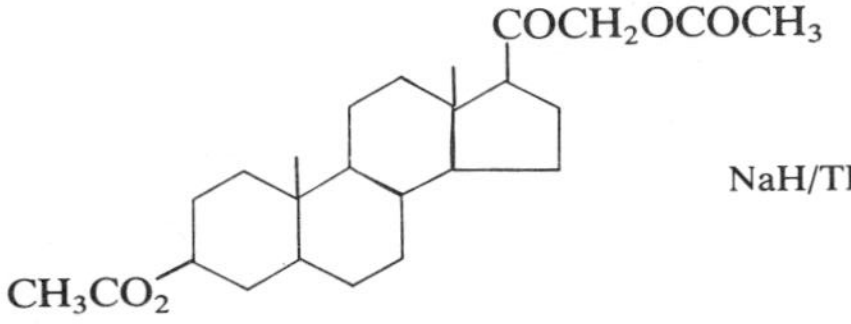

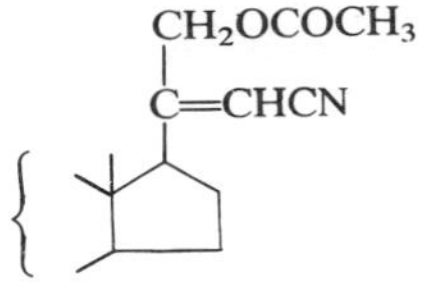

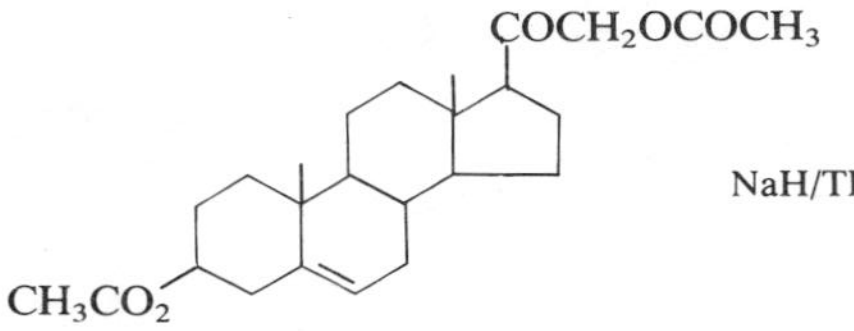

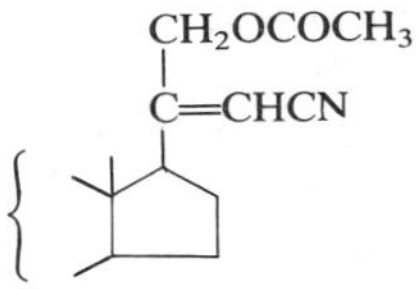

Note: References 228–415 are on pp. 250–253.

TABLE II. UNSATURATED NITRILES FROM PHOSPHORYL REAGENTS CONTAINING A CYANO GROUP

$$(RO)_2P(O)CH(R^1)CN + R^2R^3CO \xrightarrow{\text{Base}} R^2R^3C{=}C(R^1)CN + (RO)_2PO_2^-$$ (*Continued*)

No. of C Atoms	Reagent	Reactant	Base/Solvent	Product(s) and Yield(s) (%)	Refs.
C_6 (*Contd.*)	$(C_2H_5O)_2P(O)CH_2CN$ (*Contd.*)	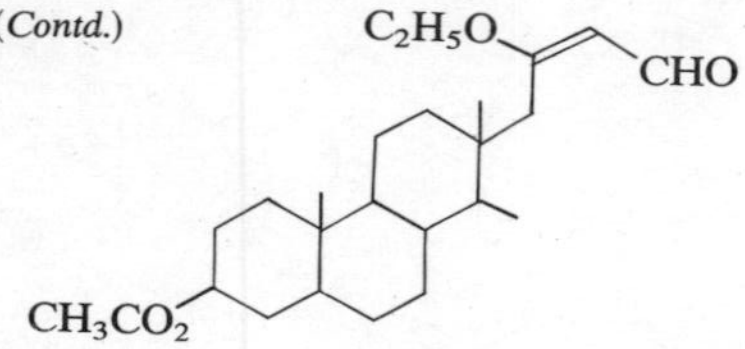 [structure: C_2H_5O, CHO, CH_3CO_2]	NaH/THF	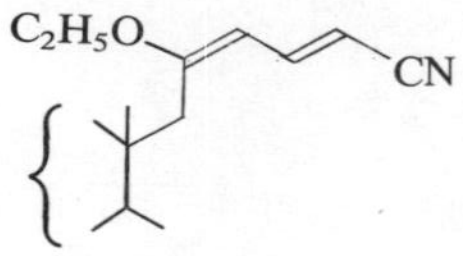 [structure: C_2H_5O, CN] (–)	267
		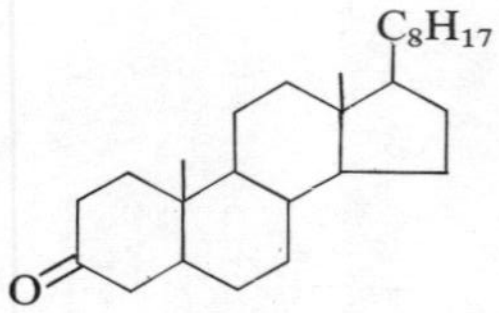 [structure: C_8H_{17}, O]	$NaNH_2$/THF	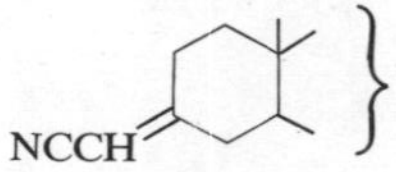 [structure: NCCH] (71)	36, 268
		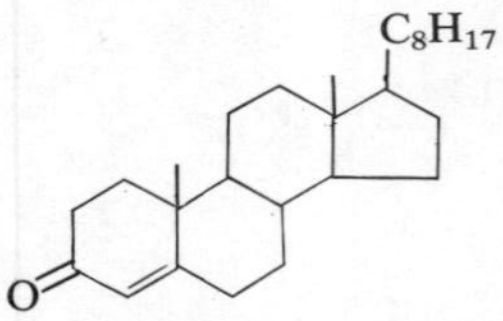 [structure: C_8H_{17}, O]	$NaNH_2$/THF	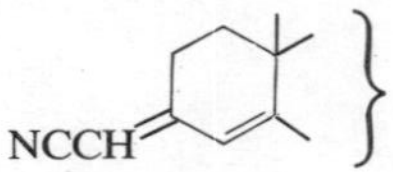 [structure: NCCH] (20)	36, 268
C_7	$(C_2H_5O)_2P(O)CH(CH_3)CN$	i-C_3H_7CHO	Na/THF	i-$C_3H_7CH{=}C(CH_3)CN$ (—)	30
		C_6H_5CHO	Na/THF	$C_6H_5CH{=}C(CH_3)CN$ (—)	13, 15, 25, 26, 30, 370

		p-ClC_6H_4CHO	NaH/THF	p-$ClC_6H_4CH{=}C(CH_3)CN$ (—)	13
		p-$CH_3OC_6H_4CHO$	Na/THF	p-$CH_3OC_6H_4CH{=}C(CH_3)CN$ (—)	30
		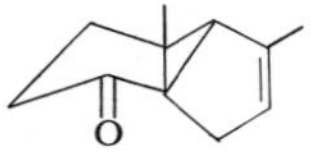	NaH/glyme	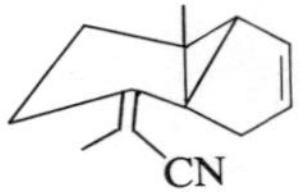 (94)	378
		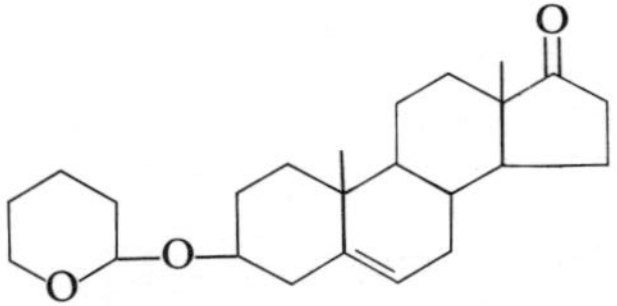	NaH/glyme	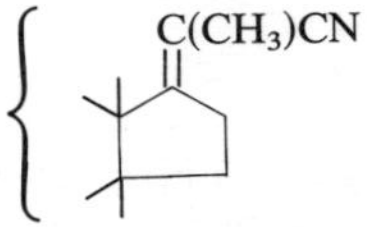 $C(CH_3)CN$ (72)	387
C_8	$(C_2H_5O)_2P(O)CH(CN)CH_2CN$	CH_3CHO	NaH/glyme	$CH_3CH{=}C(CN)CH_2CN$ (64)	384
		$(CH_3)_2CO$	NaH/glyme	$(CH_3)_2C{=}C(CN)CH_2CN$ (—)	384
		C_6H_5CHO	NaH/glyme	$C_6H_5CH{=}C(CN)CH_2CN$ (50)	384, 30
		p-ClC_6H_4CHO	NaH/glyme	p-$ClC_6H_4CH{=}C(CN)CH_2CN$ (50)	384
		$C_6H_5COCH_3$	NaH/glyme	$C_6H_5(CH_3)C{=}C(CN)CH_2CN$ (—)	384
		$(C_6H_5)_2CO$	NaH/glyme	$(C_6H_5)_2C{=}C(CN)CH_2CN$ (—)	384
C_{10}	$(C_2H_5O)_2P(O)CH(CN)CH{=}C(CH_3)_2$	p-$O_2NC_6H_4CHO$	NaH/glyme	p-$O_2NC_6H_4CH{=}C(CN)CH{=}C(CH_3)_2$ (—)	60
C_{12}	$(C_2H_5O)_2P(O)CH(C_6H_5)CN$	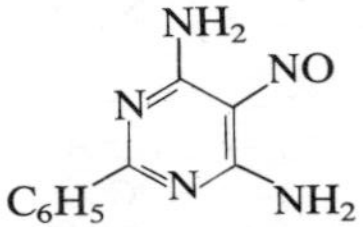	NaH/THF	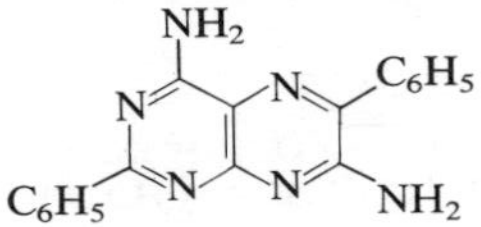 (27)	107, 254
C_{14}	$(C_2H_5O)_2P(O)CH(CN)CH(C_6H_5)CN$	CH_3CHO	NaH/glyme	$CH_3CH{=}C(CN)CH(C_6H_5)CN$ (78)	384
		$(CH_3)_2CO$	NaH/glyme	$(CH_3)_2C{=}C(CN)CH(C_6H_5)CN$ (85)	384
		C_6H_5CHO	NaH/glyme	$C_6H_5CH{=}C(CN)CH(C_6H_5)CN$ (35)	384
C_{15}	$(C_2H_5O)_2P(O)CH(CN)C(CH_3)(C_6H_5)CN$	CH_3CHO	NaH/glyme	$CH_3CH{=}C(CN)C(CH_3)(C_6H_5)CN$ (56)	384
		C_6H_5CHO	NaH/glyme	$C_6H_5CH{=}C(CN)C(CH_3)(C_6H_5)CN$ (57)	384

Note: References 228–415 are on pp. 250–253.

TABLE III. Unsaturated Esters from Phosphoryl Reagents Containing a Carboxylic Group

$$R_2P(O)CHR^1CO_2R^2 + R^3R^4CO \rightarrow R^3R^4C{=}CR^1CO_2R^2 + R_2PO_2^-$$

No. of C Atoms	Reagent	Reactant	Base/Solvent	Product(s) and Yield(s) (%)	Refs.
C_5	$(CH_3O)_2P(O)CH_2CO_2CH_3$	CF_3COCH_3	NaH/C_6H_6	$CF_3C(CH_3){=}CHCO_2CH_3$ (90)	376
		$C_2H_5COCH_3$	NaH/THF	$C_2H_5C(CH_3){=}CHCO_2CH_3$ (17)	141
		i-C_3H_7CHO	NaH/glyme	i-$C_3H_7CH{=}CHCO_2CH_3$ (65)	406, 269
		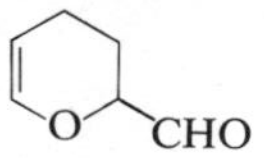 CHO	NaH/glyme	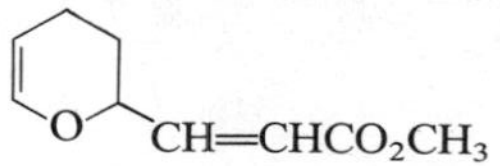 $CH{=}CHCO_2CH_3$ (−)	119
		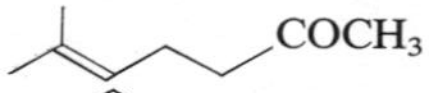 $COCH_3$	NaH/DMF	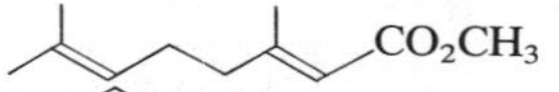 CO_2CH_3 (83)	138
		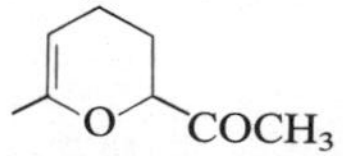 $COCH_3$	NaH/glyme	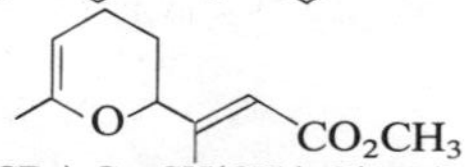 CO_2CH_3 (−)	118, 119
		$(CD_3)_2C{=}CH(CH_2)_2COCH_3$	NaH/THF	$(CD_3)_2C{=}CH(CH_2)_2C(CH_3){=}CH$-$CO_2CH_3$ (—)	270
		$(CH_3)_2C{=}CH(CH_2)_2COCF_3$	NaH/C_6H_6	$(CH_3)_2C{=}CH(CH_2)_2C(CF_3){=}CHCO_2CH_3$ (94)	376
		p-$CH_3C_6H_4COCH_3$	NaH/glyme	p-$CH_3C_6H_4C(CH_3){=}CHCO_2CH_3$ (64)	386
		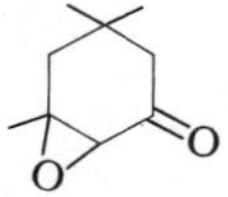 O	NaH/glyme	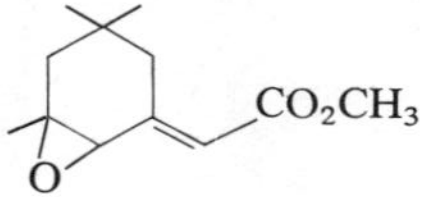 CO_2CH_3 (80)	391
		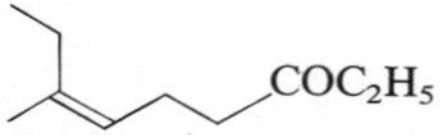 COC_2H_5	NaH/THF	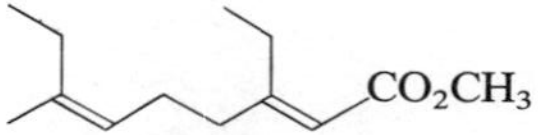 CO_2CH_3 (39)	141
		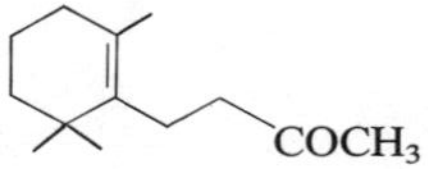 $COCH_3$	NaH/C_6H_6	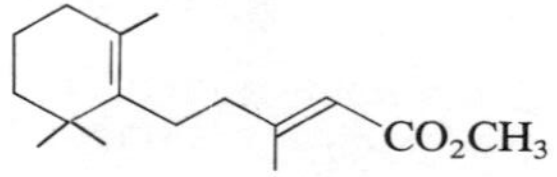 CO_2CH_3 (57)	271

Reagent	Carbonyl Compound	Base/Solvent	Product (Yield, %)	Refs.
	$(CH_3)_2C{=}CH(CH_2)_2$-$C(CH_3){=}CH(CH_2)_2COCF_3$	NaH/C_6H_6	$(CH_3)_2C{=}CH(CH_2)_2C(CH_3){=}CH(CH_2)_2$-$C(CF_3){=}CHCO_2CH_3$ (95)	376
	O	CH_3ONa/DMF	CO_2CH_3 (—)	409
$(CH_3O)_2P(O)^{14}CH_2CO_2CH_3$	$COCH_3$	NaH/THF	14 CO_2CH_3 (–)	131
	O $COCH_3$	NaH/THF	O 14 CO_2CH_3 (–)	137
$(CH_3O)_2P(O)CH_2CO_2CH_3$	$COCH_3$	NaH/THF	CO_2CH_3 (–)	272, 139, 273
	$COCH_3$	NaH/DMF	CO_2CH_3 (90)	138
	p-$C_6H_5OC_6H_4(CH_2)_2CHO$	NaH/glyme	p-$C_6H_5OC_6H_4(CH_2)_2CH{=}CHCO_2CH_3$ (—)	412
	$COCH_3$	NaH/glyme	CO_2CH_3 (–)	139
	$COCH_3$	NaH/glyme	CO_2CH_3 (91)	272
	14 $COCH_3$	NaH/THF	14 CO_2CH_3 (–)	131

Note: References 228–415 are on pp. 250–253.

TABLE III. UNSATURATED ESTERS FROM PHOSPHORYL REAGENTS CONTAINING A CARBOXYLIC GROUP

$R_2P(O)CHR^1CO_2R^2 + R^3R^4CO \rightarrow R^3R^4C{=}CR^1CO_2R^2 + R_2PO_2^-$ (*Continued*)

No. of C Atoms	Reagent	Reactant	Base/Solvent	Product(s) and Yield(s) (%)	Refs.
C_5 (*Contd.*)	$(CH_3O)_2P(O)CH_2CO_2CH_3$ (*Contd.*)	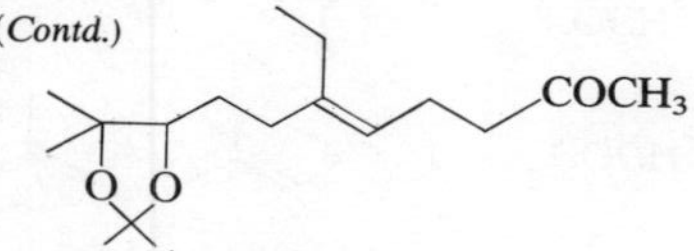 [structure] $COCH_3$	NaH/glyme	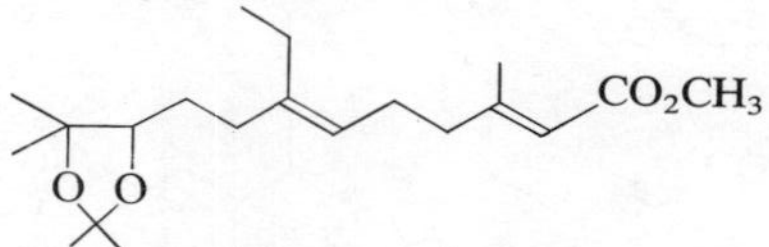 [structure] CO_2CH_3 (90)	139
		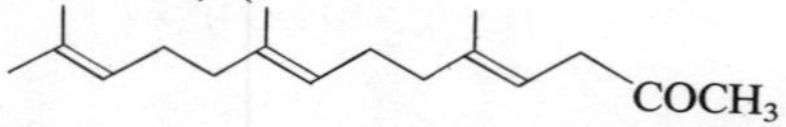 [structure] $COCH_3$	NaH/glyme	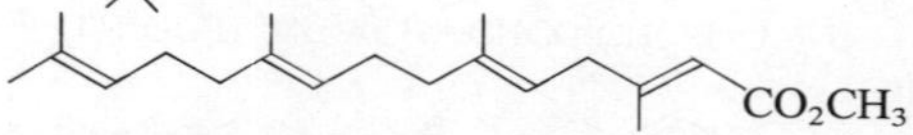 [structure] CO_2CH_3 (−)	274
C_6	$(CH_3O)_2P(O)CH_2CO_2C_2H_5$	$CH_2{=}C(CH_3)CH_2COCH_3$	$NaNH_2$/ether	$CH_2{=}C(CH_3)CH_2C(CH_3){=}CH$-$CO_2C_2H_5$ (—)	165
	$(CH_3O)_2P(O)CH(CH_3)CO_2CH_3$	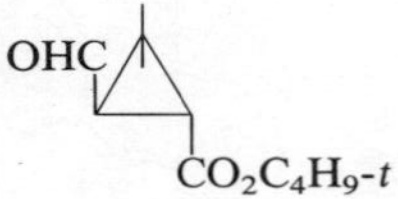 OHC [structure] $CO_2C_4H_9$-*t*	CH_3ONa/DMF	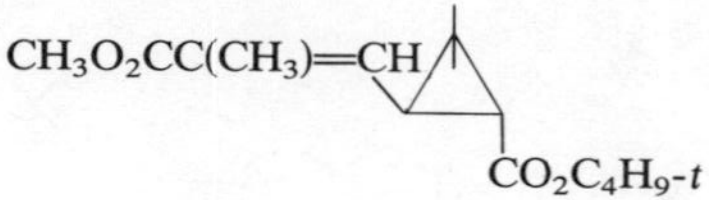 $CH_3O_2CC(CH_3){=}CH$ [structure] $CO_2C_4H_9$-*t* (42)	31
C_7	$(C_2H_5O)_2P(O)CH_2CO_2CH_3$	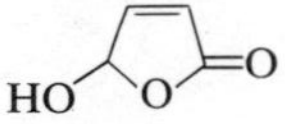 HO [structure] O	CH_3ONa/CH_3OH	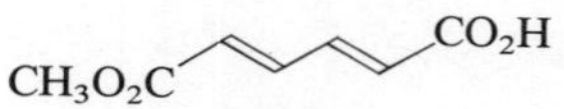 CH_3O_2C [structure] CO_2H (85)	56
		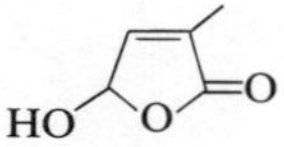 HO [structure] O	CH_3ONa/CH_3OH	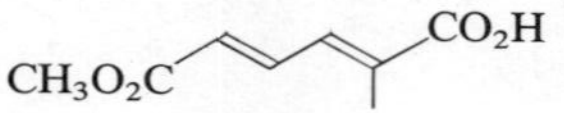 CH_3O_2C [structure] CO_2H (88)	56
		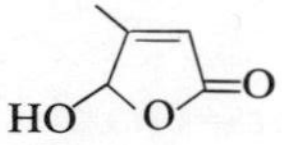 HO [structure] O	CH_3ONa/CH_3OH	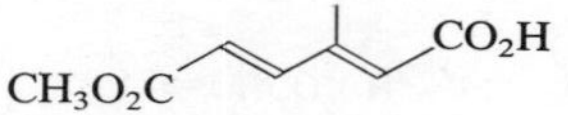 CH_3O_2C [structure] CO_2H (92)	56, 97

$CH_2{=}CH(CH_2)_2CHO$	CH_3ONa/CH_3OH	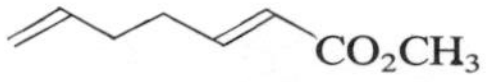 CO_2CH_3 (−)	275
$CH_3CH{=}CH(CH_2)_2CHO$	CH_3ONa/CH_3OH	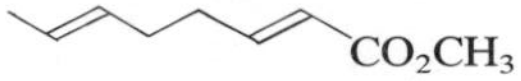 CO_2CH_3 (−)	275
$CH_3CO(CH_2)_2CH(OCH_3)_2$	NaH/glyme	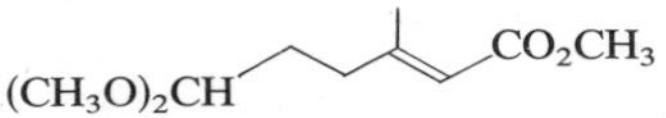 $(CH_3O)_2CH$ … CO_2CH_3 (−)	276
$(C_2H_5O)_2CHCOCH_3$	NaH/glyme	$(C_2H_5O)_2CHC(CH_3){=}CHCO_2CH_3$ (90)	56, 55
C_6H_5CHO	NaH/THF	$C_6H_5CH{=}CHCO_2CH_3$ (—)	15
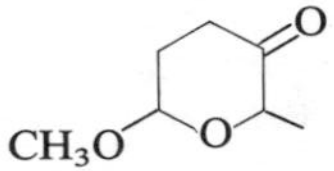 CH_3O, O, O	NaH/glyme	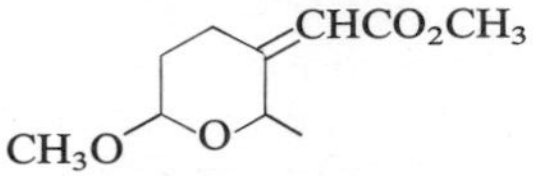 CH_3O, O, $CHCO_2CH_3$ (—)	404
t-$C_4H_9O(CH_2)_2COCH_3$	NaH/glyme	t-$C_4H_9O(CH_2)_2C(CH_3){=}CHCO_2CH_3$ (—)	415
p-$C_6H_4(CHO)_2$	C_2H_5ONa/C_2H_5OH	p-$C_6H_4(CH{=}CHCO_2CH_3)_2$ (—)	88
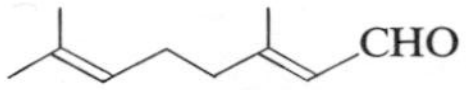 CHO	CH_3OH/DMF	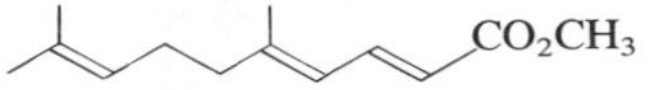 CO_2CH_3 (75)	55
$C_6H_5(CH_2)_3COCH_3$	NaH/glyme	$C_6H_5(CH_2)_3C(CH_3){=}CHCO_2CH_3$ (—)	411
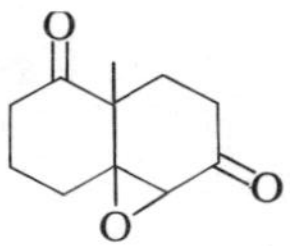 O, O, O	NaH/glyme	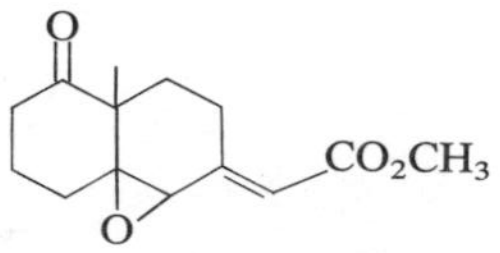 O, O, CO_2CH_3 (85)	391
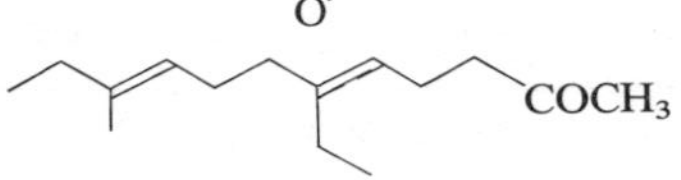 $COCH_3$	Na/C_6H_6	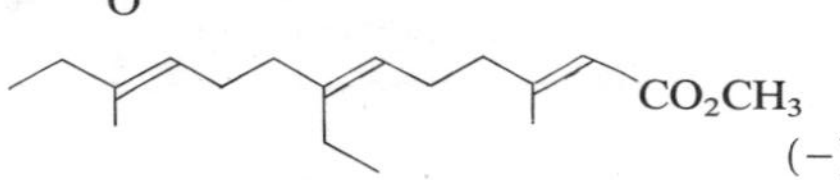 CO_2CH_3 (−)	140
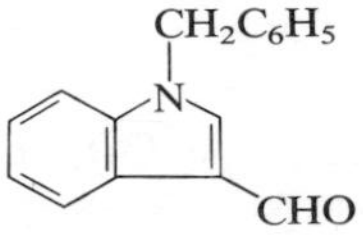 $CH_2C_6H_5$, N, CHO	NaH/glyme	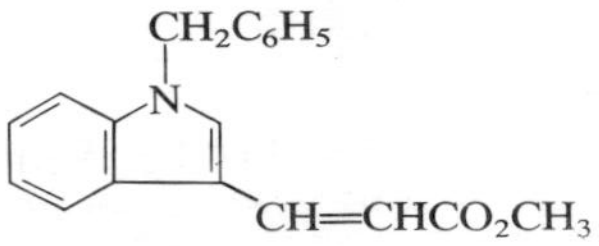 $CH_2C_6H_5$, N, $CH{=}CHCO_2CH_3$ (80)	277

Note: References 228–415 are on pp. 250–253.

TABLE III. Unsaturated Esters from Phosphoryl Reagents Containing a Carboxylic Group $R_2P(O)CHR^1CO_2R^2 + R^3R^4CO \rightarrow R^3R^4C{=}CR^1CO_2R^2 + R_2PO_2^-$ (*Continued*)

No. of C Atoms	Reagent	Reactant	Base/Solvent	Product(s) and Yield(s) (%)	Refs.
C_7 (*Contd.*)	$(C_2H_5O)_2P(O)CH_2CO_2CH_3$ (*Contd.*)	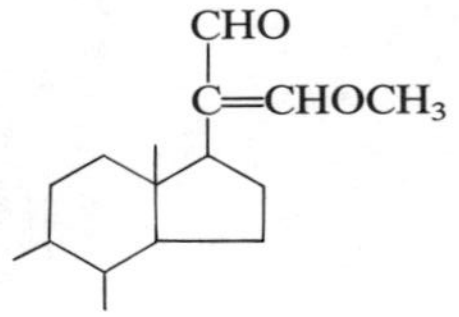 CHO, C=CHOCH₃ [structure]	NaH/glyme	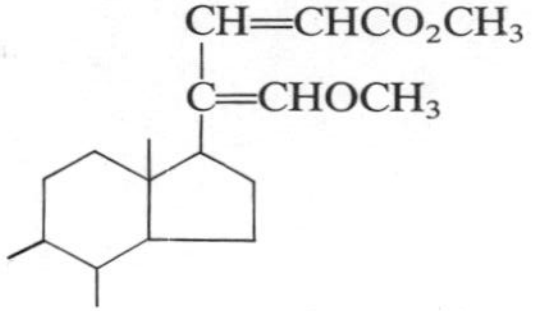 CH=CHCO₂CH₃, C=CHOCH₃ [structure] (–)	98
		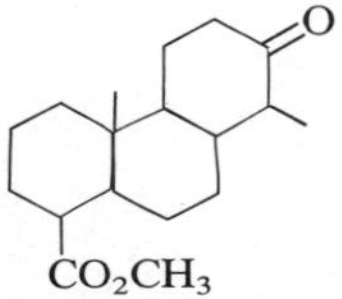 [structure] CO_2CH_3	NaH/THF	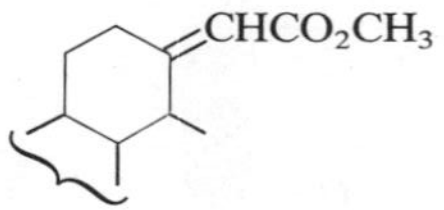 $CHCO_2CH_3$ [structure] (–)	278
		p-$C_6H_5C_6H_4(CH_2)_3COCH_3$	NaH/glyme	p-$C_6H_5C_6H_4(CH_2)_3C(CH_3){=}CHCO_2CH_3$ (—)	411
		p-$(n$-$C_6H_{13})C_6H_4$—$(CH_2)_3COCH_3$	NaH/glyme	p-$(n$-$C_6H_{13})C_6H_4(CH_2)_3C(CH_3){=}CHCO_2CH_3$ (—)	411
		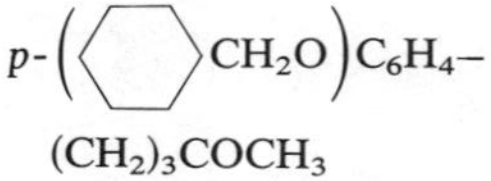 p-(cyclohexyl-$CH_2O)C_6H_4$–$(CH_2)_3COCH_3$	NaH/glyme	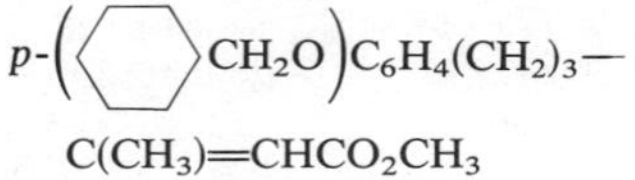 p-(cyclohexyl-$CH_2O)C_6H_4(CH_2)_3$—$C(CH_3){=}CHCO_2CH_3$ (—)	411
		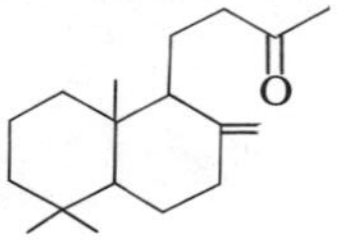 [structure]	NaH/DMSO	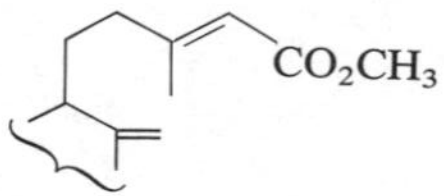 [structure] CO_2CH_3 (100)	388

	Reagent	Carbonyl compound	Conditions	Product (yield, %)	Refs.
	$(C_2H_5O)_2P(O)^{14}CH_2CO_2CH_3$	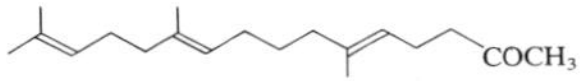 $COCH_3$	NaH/glyme	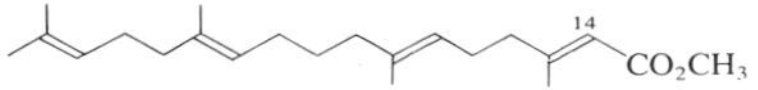 ^{14} CO_2CH_3 (74)	132
	$(C_2H_5O)_2P(O)CH_2CO_2CH_3$	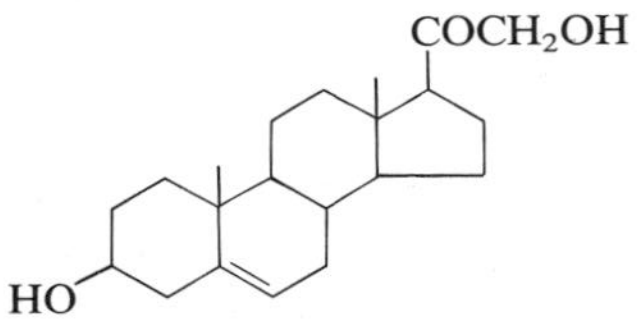 $COCH_2OH$, HO	NaH/glyme	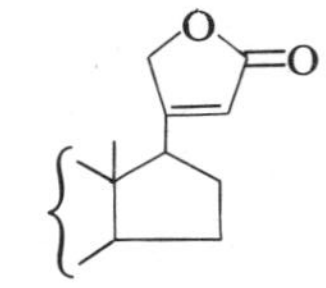 O, O (95)	38, 279
C_8	$(C_2H_5O)_2P(O)CH_2CO_2C_2H_5$	CH_3CHO	NaOH, H_2O/CH_2Cl_2, $(n\text{-}C_4H_9)N^+I^-$	$CH_3CH{=}CHCO_2C_2H_5$ (54)	367, 280
	$(CH_3O)_2P(O)CH(CO_2CH_3)$-$CH_2CO_2CH_3$	CH_3CHO	NaH/glyme	$CH_3CH{=}C(CO_2CH_3)CH_2CO_2CH_3$ (81)	390, 175
	$(C_2H_5O)_2P(O)CH_2CO_2C_2H_5$	CCl_3CHO	NaH/glyme	$CCl_3CH{=}CHCO_2C_2H_5$ (50)	4
		$CH_2{=}CHCHO$	NaH/glyme	$CH_2{=}CHCH{=}CHCO_2C_2H_5$ (14)	280
	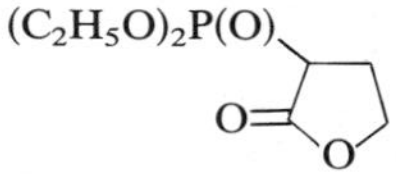 $(C_2H_5O)_2P(O)$, O=, O	CCl_3CHO	NaH/C_6H_6	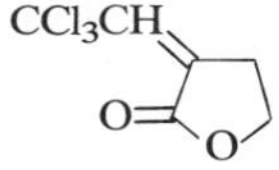 CCl_3CH, O=, O (100)	281
	$(C_2H_5O)_2P(O)CH_2CO_2C_2H_5$	$(CH_3)_2CO$	NaH/glyme	$(CH_3)_2C{=}CHCO_2C_2H_5$ (61)	4, 250 282, 283
		$ClCH_2COCH_3$	$NaNH_2$/ether	$ClCH_2C(CH_3){=}CHCO_2C_2H_5$ (58)	165
		$BrCH_2COCH_3$	NaH/ether	$BrCH_2C(CH_3){=}CHCO_2C_2H_5$ (20)	165
	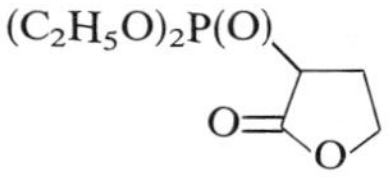 $(C_2H_5O)_2P(O)$, O=, O	$(CH_3)_2CO$	NaH/C_6H_6	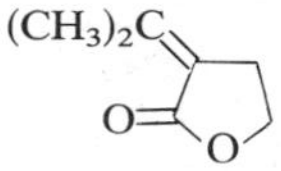 $(CH_3)_2C$, O=, O (89)	281
		C_2H_5CHO	NaH/C_6H_6	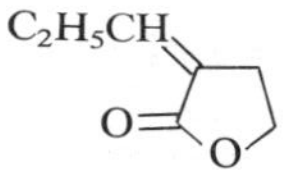 C_2H_5CH, O=, O (100)	281

Note: References 228–415 are on pp. 250–253.

TABLE III. Unsaturated Esters from Phosphoryl Reagents Containing a Carboxyl Group

$R_2P(O)CHR^1CO_2R^2 + R^3R^4CO \rightarrow R^3R^4C{=}CR^1CO_2R^2 + R_2PO_2^-$ (*Continued*)

No. of C Atoms	Reagent	Reactant	Base/Solvent	Product(s) and Yield(s) (%)	Refs.
C_8	$(C_2H_5O)_2P(O)CH_2CO_2C_2H_5$	$CH_2{=}CHCHO$	C_2H_5ONa/C_2H_5OH	$CH_2{=}CHCH{=}CHCO_2C_2H_5$ (4)	91
(*Contd.*)	$(C_2H_5O)_2P(O)CHBrCO_2C_2H_5$	$(CH_3)_2CO$	NaH/glyme	$(CH_3)_2C{=}C(Br)CO_2C_2H_5$ (65)	4
	$(C_2H_5O)_2P(O)CH(CO_2CH_3)NHCSH$	CH_3COCH_2Cl	NaH/glyme	S, N, CH_3, CO_2CH_3 (ring) (83)	371
	$(C_2H_5O)_2P(O)CH_2CO_2C_2H_5$	$CH_2{=}C(CH_3)CHO$	C_2H_5ONa/C_2H_5OH	$CH_2{=}C(CH_3)CH{=}CHCO_2C_2H_5$ (52)	91
		$CH_3COC_2H_5$	$NaNH_2$/ether	$CH_3C(C_2H_5)C{=}CHCO_2C_2H_5$ (58)	201, 280
		n-C_3H_7CHO	$NaNH_2$/ether	n-$C_3H_7CH{=}CHCO_2C_2H_5$ (67)	250, 284
		$CH_3CH{=}CHCHO$	$NaNH_2$/ether	$CH_3CH{=}CHCH{=}CHCO_2C_2H_5$ (56)	250, 280, 285
	$(CH_3O)_2P(O)CH(CO_2CH_3)$-$CH_2CO_2CH_3$	$CH_3CH{=}CHCHO$	NaH/glyme	$CH_3CH{=}CHCH{=}C(CO_2CH_3)$-$CH_2CO_2CH_3$ (73)	175, 390
	$(C_2H_5O)_2P(O)CH_2CO_2C_2H_5$	i-C_3H_7CHO	NaH/ether	i-$C_3H_7CH{=}CHCO_2C_2H_5$ (86)	4, 20, 280, 286, 287
		$(CH_3O)_2CHCHO$	$NaNH_2$/ether	$(CH_3O)_2CHCH{=}CHCO_2C_2H_5$ (63)	280
		$\triangleright$—CHO	NaH/glyme	$\triangleright$—$CH{=}CHCO_2C_2H_5$ (73)	285, 288
		$CH_3COCOCH_3$	NaH/glyme	$CH_3COC(CH_3){=}CHCO_2C_2H_5$ (—)	85

Reagent	Carbonyl Compound	Base/Solvent	Product (Yield %)	Refs.
	$(CH_3)_3CNO$	NaH/glyme	$(CH_3)_3CN{=}CHCO_2C_2H_5$ (80)	108
	$CF_3CO_2C_2H_5$	NaH/glyme	$CF_3C(OC_2H_5){=}CHCO_2C_2H_5$ (—)	100
$(C_2H_5O)_2P(O)CHClCO_2C_2H_5$	i-C_3H_7CHO	NaH/glyme	i-$C_3H_7CH{=}CClCO_2C_2H_5$ (70)	100
$(C_2H_5O)_2P(O)CHBrCO_2C_2H_5$	i-C_3H_7CHO	NaH/glyme	i-$C_3H_7CH{=}CBrCO_2C_2H_5$ (66)	4
$(C_2H_5O)_2P(O)CH(CH_3)CO_2CH_3$	i-C_3H_7CHO	NaH/glyme	i-$C_3H_7CH{=}C(CH_3)CO_2CH_3$ (83)	269
$(C_2H_5O)_2P(O)CH_2CO_2C_2H_5$	$(C_2H_5)_2CO$	$NaNH_2$/ether	$(C_2H_5)_2C{=}CHCO_2C_2H_5$ (64)	4, 289
	t-C_4H_9CHO	NaH/ether	t-$C_4H_9CH{=}CHCO_2C_2H_5$ (88)	287, 20, 290
	n-$C_3H_7COCH_3$	NaH/glyme	n-$C_3H_7(CH_3)C{=}CHCO_2C_2H_5$ (67)	286, 164, 291
	$CH_2{=}C(C_2H_5)CHO$	C_2H_5ONa/C_2H_5OH	$CH_2{=}C(C_2H_5)CH{=}CHCO_2C_2H_5$ (66)	91
	i-$C_3H_7COCH_3$	NaH/glyme	i-$C_3H_7C(CH_3){=}CHCO_2C_2H_5$ (81)	286, 285, 288, 292
	i-$C_3H_7CH_2CHO$	NaH/ether	i-$C_3H_7CH_2CH{=}CHCO_2C_2H_5$ (90)	287
	cyclopropyl—$COCH_3$	NaH/glyme	cyclopropyl—$C(CH_3){=}CHCO_2C_2H_5$ (83)	285, 288
$(CH_3O)_2P(O)CH(CO_2CH_3)$-$CH_2CO_2CH_3$	$(CH_3)_2C{=}CHCHO$	NaH/glyme	$(CH_3)_2C{=}CHCH{=}C(CO_2CH_3)$-$CH_2CO_2CH_3$ (60)	175, 390
$(C_2H_5O)_2P(O)CH_2CO_2C_2H_5$	methylcyclopropyl—CHO	NaH/glyme	methylcyclopropyl—$CH{=}CHCO_2C_2H_5$ (—)	285, 288
	2-furyl—CHO	NaH/glyme	2-furyl—$CH{=}CHCO_2C_2H_5$ (71)	284
	$OHC(CH_2)_3CHO$	NaH/glyme	ethyl 6-hydroxycyclohex-1-enecarboxylate (OH, $CO_2C_2H_5$) (40)	89

Note: References 228–415 are on pp. 250–253.

TABLE III. Unsaturated Esters from Phosphoryl Reagents Containing a Carboxylic Group $R_2P(O)CHR^1CO_2R^2 + R^3R^4CO \rightarrow R^3R^4C{=}CR^1CO_2R^2 + R_2PO_2^-$ (*Continued*)

No. of C Atoms	Reagent	Reactant	Base/Solvent	Product(s) and Yield(s) (%)	Refs.
C_8 (*Contd.*)	$(C_2H_5O)_2P(O)CH_2CO_2C_2H_5$ (*Contd.*)	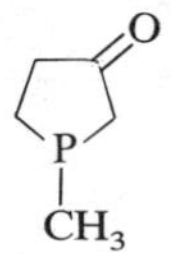 O; P; CH_3	NaH/glyme	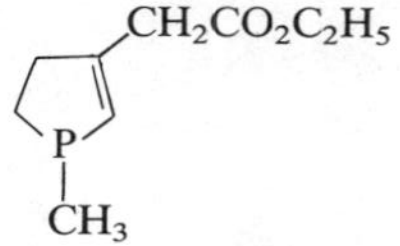 $CH_2CO_2C_2H_5$; P; CH_3 (63)	293
		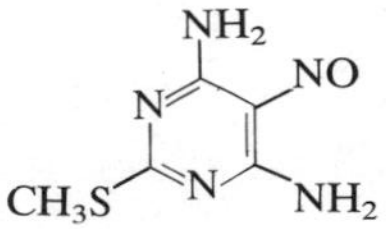 NH_2; NO; N; N; CH_3S; NH_2	NaH/glyme	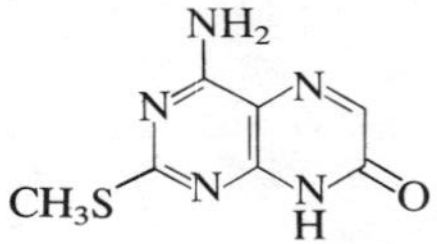 NH_2; N; N; N; N; H; O; CH_3S (82)	106
		$(CH_3O)_2P(O)OCH(CH_3)CHO$	NaH/glyme	$C_2H_5O_2CCH{=}CHCH(CH_3)OP(O)(OCH_3)_2$ (70)	385
		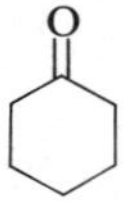 O	$NaNH_2$/ether	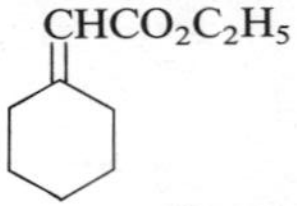 $CHCO_2C_2H_5$ (76)	250, 4, 289, 294
		$CH_3CO(CH_2)_2COCH_3$	NaH/glyme	$CH_3CO(CH_2)_2C(CH_3){=}CHCO_2C_2H_5$ (—)	85
		$CH_3(CH{=}CH)_2CHO$	NaH/glyme	$CH_3(CH{=}CH)_3CO_2C_2H_5$ (53)	280
		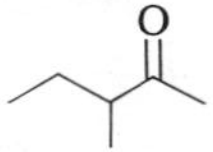 O	NaH/glyme	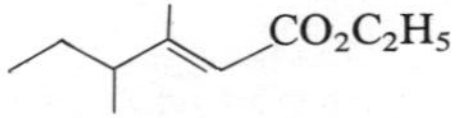 $CO_2C_2H_5$ (—)	398
		n-$C_5H_{11}CHO$	NaH/ether	n-$C_5H_{11}CH{=}CHCO_2C_2H_5$ (87)	287
		t-$C_4H_9CH_2CHO$	NaH/ether	t-$C_4H_9CH_2CH{=}CHCO_2C_2H_5$ (92)	287
		$C_2H_5O_2CCO_2C_2H_5$	NaH/ether	$C_2H_5O_2CC(OC_2H_5){=}CHCO_2C_2H_5$ (70)	100
		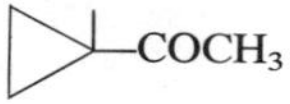 $COCH_3$	NaH/glyme	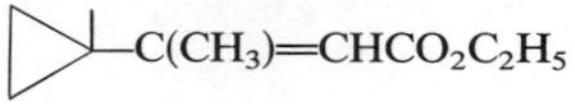 $C(CH_3){=}CHCO_2C_2H_5$ (60)	285, 288

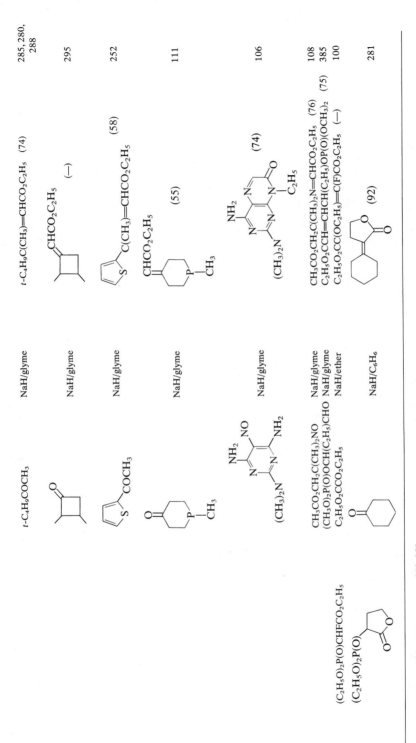

Phosphonate	Carbonyl compound	Base/solvent	Product (% yield)	Refs.
	$t\text{-}C_4H_9COCH_3$	NaH/glyme	$t\text{-}C_4H_9C(CH_3){=}CHCO_2C_2H_5$ (74)	285, 280, 288
	2,3-dimethylcyclobutanone	NaH/glyme	2,3-dimethylcyclobutylidene$=CHCO_2C_2H_5$ (—)	295
	2-thienyl$COCH_3$	NaH/glyme	2-thienyl$C(CH_3){=}CHCO_2C_2H_5$ (58)	252
	1-methyl-4-phosphorinanone ($P{-}CH_3$)	NaH/glyme	1-methyl-4-phosphorinanylidene$=CHCO_2C_2H_5$ (55)	111
	2-$(CH_3)_2N$-4,6-diamino-5-nitrosopyrimidine	NaH/glyme	2-$(CH_3)_2N$-4-NH_2-8-C_2H_5-pteridin-7-one (74)	106
	$CH_3CO_2CH_2C(CH_3)_2NO$	NaH/glyme	$CH_3CO_2CH_2C(CH_3)_2N{=}CHCO_2C_2H_5$ (76)	108
	$(CH_3O)_2P(O)OCH(C_2H_5)CHO$	NaH/glyme	$C_2H_5O_2CCH{=}CHCH(C_2H_5)OP(O)(OCH_3)_2$ (75)	385
$(C_2H_5O)_2P(O)CHFCO_2C_2H_5$	$C_2H_5O_2CCO_2C_2H_5$	NaH/ether	$C_2H_5O_2CC(OC_2H_5){=}C(F)CO_2C_2H_5$ (—)	100
$(C_2H_5O)_2P(O)$-(γ-butyrolactone-α-yl)	cyclohexanone	NaH/C_6H_6	α-cyclohexylidene-γ-butyrolactone (92)	281

Note: References 228–415 are on pp. 250–253.

TABLE III. Unsaturated Esters from Phosphoryl Reagents Containing a Carboxylic Group

$R_2P(O)CHR^1CO_2R^2 + R^3R^4CO \rightarrow R^3R^4C{=}CR^1CO_2R^2 + R_2PO_2^-$ (*Continued*)

No. of C Atoms	Reagent	Reactant	Base/Solvent	Product(s) and Yield(s) (%)	Refs.
C_8 (*Contd.*)	$(C_2H_5O)_2P(O)$-(2-oxotetrahydrofuran-3-yl)	C_6H_5NO	NaH/C_6H_6	C_6H_5NH-(2-oxo-2,5-dihydrofuran-3-yl) (94)	181
	$(C_2H_5O)_2P(O)CH_2CO_2C_2H_5$	2-methylcyclohexanone	NaH/glyme	2-methylcyclohexylidene=$CHCO_2C_2H_5$ (—)	296
		3-methylcyclohexanone	NaH/glyme	3-methylcyclohexylidene=$CHCO_2C_2H_5$ (70)	297
		$(n\text{-}C_3H_7)_2CO$	NaH/glyme	$(n\text{-}C_3H_7)_2C{=}CHCO_2C_2H_5$ (—)	297
		4-methylcyclohexanone	NaH/glyme	4-methylcyclohexylidene=$CHCO_2C_2H_5$ (—)	298
		C_6H_5CHO	NaH/glyme	$C_6H_5CH{=}CHCO_2C_2H_5$ (84)	4, 15, 201, 250, 284, 367
		$(CH_3O)_2P(O)OCH(C_3H_7\text{-}n)CHO$	NaH/glyme	$C_2H_5O_2CCH{=}CHCH(C_3H_7\text{-}n)OP(O)(OCH_3)_2$ (60)	385

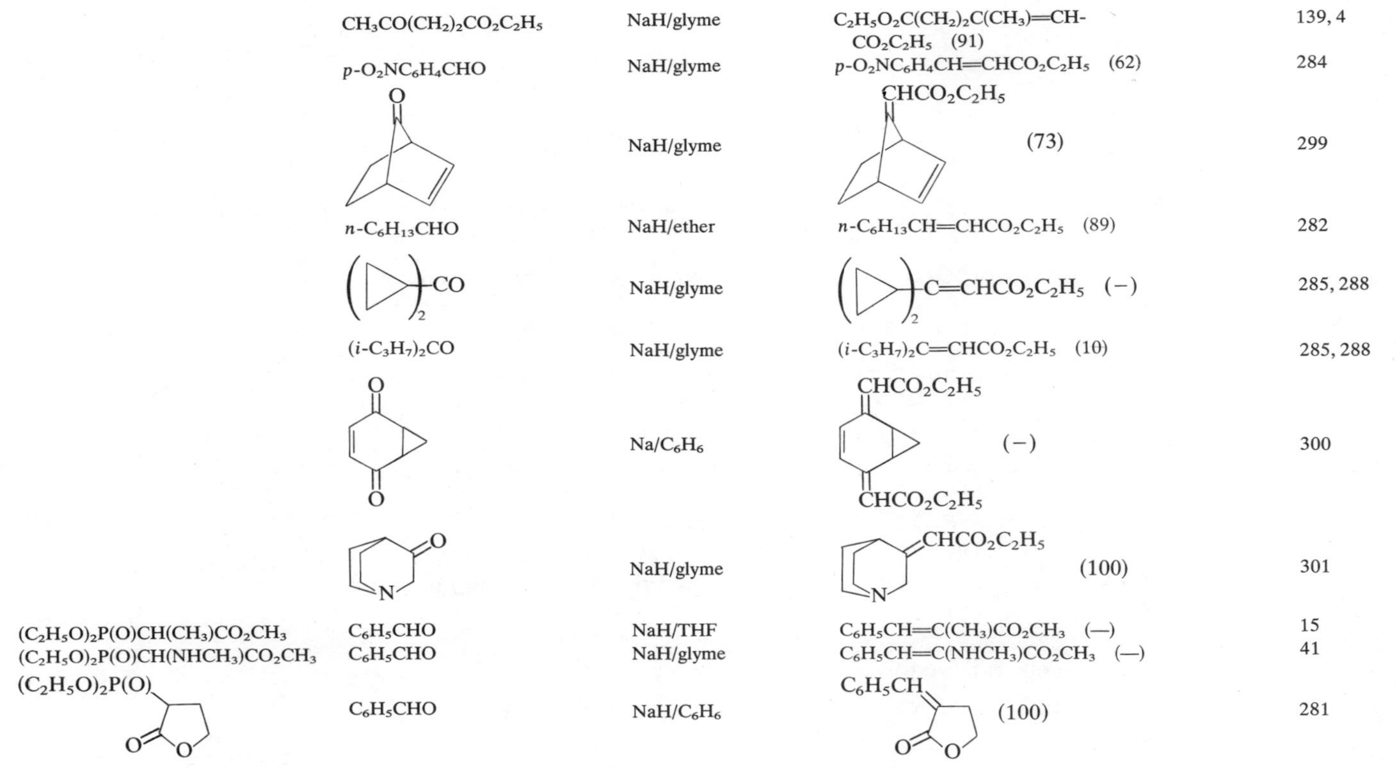

Phosphonate	Carbonyl compound	Base/Solvent	Product (Yield %)	Refs.
	$CH_3CO(CH_2)_2CO_2C_2H_5$	NaH/glyme	$C_2H_5O_2C(CH_2)_2C(CH_3){=}CH$-$CO_2C_2H_5$ (91)	139, 4
	p-$O_2NC_6H_4CHO$	NaH/glyme	p-$O_2NC_6H_4CH{=}CHCO_2C_2H_5$ (62)	284
	(structure)	NaH/glyme	(structure, $=CHCO_2C_2H_5$) (73)	299
	n-$C_6H_{13}CHO$	NaH/ether	n-$C_6H_{13}CH{=}CHCO_2C_2H_5$ (89)	282
	$(c\text{-}C_3H_5)_2CO$ (structure)	NaH/glyme	$(c\text{-}C_3H_5)_2C{=}CHCO_2C_2H_5$ (−)	285, 288
	$(i$-$C_3H_7)_2CO$	NaH/glyme	$(i$-$C_3H_7)_2C{=}CHCO_2C_2H_5$ (10)	285, 288
	(structure)	Na/C_6H_6	(structure, $CHCO_2C_2H_5$ / $CHCO_2C_2H_5$) (−)	300
	(structure)	NaH/glyme	(structure, $=CHCO_2C_2H_5$) (100)	301
$(C_2H_5O)_2P(O)CH(CH_3)CO_2CH_3$	C_6H_5CHO	NaH/THF	$C_6H_5CH{=}C(CH_3)CO_2CH_3$ (—)	15
$(C_2H_5O)_2P(O)CH(NHCH_3)CO_2CH_3$	C_6H_5CHO	NaH/glyme	$C_6H_5CH{=}C(NHCH_3)CO_2CH_3$ (—)	41
$(C_2H_5O)_2P(O)$ (structure)	C_6H_5CHO	NaH/C_6H_6	$C_6H_5CH{=}$ (structure) (100)	281

Note: References 228–415 are on pp. 250–253.

TABLE III. Unsaturated Esters from Phosphoryl Reagents Containing a Carboxylic Group
$R_2P(O)CHR^1CO_2R^2 + R^3R^4CO \rightarrow R^3R^4C{=}CR^1CO_2R^2 + R_2PO_2^-$ (*Continued*)

No. of C Atoms	Reagent	Reactant	Base/Solvent	Product(s) and Yield(s) (%)	Refs.
C_8 (*Contd.*)	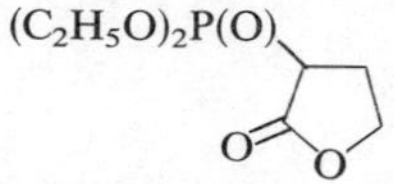 $(C_2H_5O)_2P(O)$ O O (*Contd.*)	p-$O_2NC_6H_5CHO$	NaH/C_6H_6	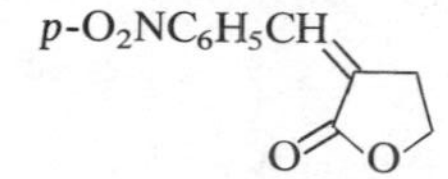 p-$O_2NC_6H_5CH$ O O (71)	281
	$(C_2H_5O)_2P(O)CH_2CO_2C_2H_5$	$C_6H_5COCH_3$	$NaNH_2$/ether	$C_6H_5C(CH_3){=}CHCO_2C_2H_5$ (77)	4, 201, 250, 284
		$CH_3(CH{=}CH)_3CHO$	$NaNH_2$/ether	$CH_3(CH{=}CH)_4CO_2C_2H_5$ (75)	280
		$CH_2{=}C(C_5H_{11}$-$n)CHO$	C_2H_5ONa/C_2H_5OH	$CH_2{=}C(C_5H_{11}$-$n)CH{=}CHCO_2C_2H_5$ (57)	91
		o-$CH_3C_6H_4CHO$	NaH/glyme	o-$CH_3C_6H_4CH{=}CHCO_2C_2H_5$ (—)	302
		p-$CH_3C_6H_4CHO$	NaH/glyme	p-$CH_3C_6H_4CH{=}CHCO_2C_2H_5$ (—)	302
		$(CH_3)_2C{=}CH(CH_2)_2COCH_3$	CH_3ONa/DMF	$(CH_3)_2C{=}CH(CH_2)_2C(CH_3){=}CH$-$CO_2C_2H_5$ (43)	303
		$(C_2H_5O)_2CHCH_2COCH_3$	NaH/glyme	$(C_2H_5O)_2CHCH_2C(CH_3){=}CHCO_2C_2H_5$ (31)	303
		i-$C_6H_{13}COCH_3$	NaH/glyme	i-$C_6H_{13}C(CH_3){=}CHCO_2C_2H_5$ (—)	304
		$CH_3CO(CH_2)_4COCH_3$	NaH/glyme	$CH_3CO(CH_2)_4C(CH_3){=}CHCO_2C_2H_5$ (—)	85
		$C_6H_{11}COCH_3$	NaH/glyme	$C_6H_{11}C(CH_3){=}CHCO_2C_2H_5$ (35)	305
		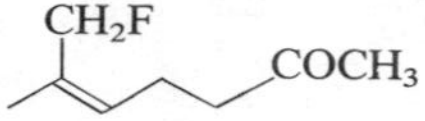 CH_2F $COCH_3$	NaH/ether	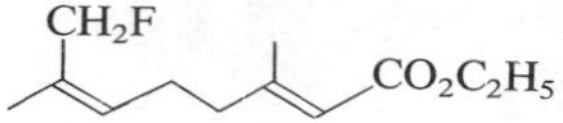 CH_2F $CO_2C_2H_5$ (72)	306
		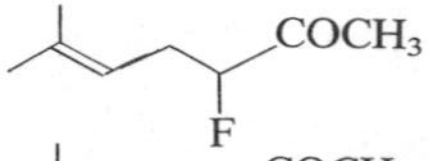 $COCH_3$ F	NaH/ether	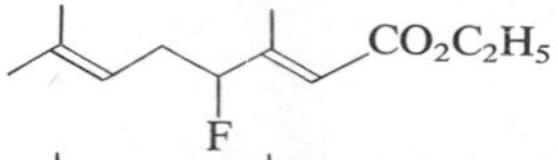 $CO_2C_2H_5$ F (–)	307
		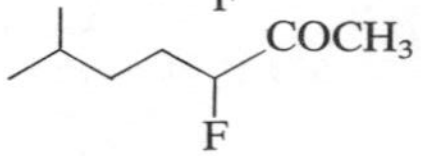 $COCH_3$ F	NaH/ether	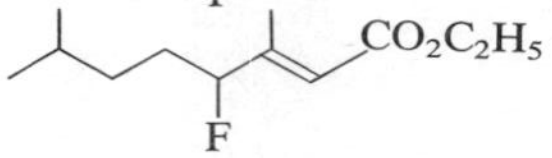 $CO_2C_2H_5$ F (–)	307

Reagent	Carbonyl Compound	Conditions	Product(s) (Yield, %)	Refs.
	$(CH_3)_2C{=}CHCH_2CH_2COCH_2F$	NaH/ether	$(CH_3)_2C{=}CHCH_2CH_2C(CH_2F){=}CHCO_2C_2H_5$ (–)	307
	3,5-Br$_2$-4-CH_3O-C_6H_2CHO	CH_3ONa/CH_3OH	3,5-Br$_2$-4-CH_3O-$C_6H_2CH{=}CHCO_2C_2H_5$ (–)	308
	2-$(C_2H_5)_2N$-4,6-$(NH_2)_2$-5-NO-pyrimidine	NaH/glyme	4-NH_2-2-$(C_2H_5)_2N$-8-C_2H_5-pteridin-7(8H)-one (71)	106
	p-$C_6H_4(CHO)_2$	C_2H_5ONa/C_2H_5OH	p-$C_6H_4(CH{=}CHCO_2C_2H_5)_2$ (—)	88
	1,4-dithiaspiro[4.5]decan-8-one	NaH/C_6H_6	1,4-dithiaspiro[4.5]decan-8-ylidene$=CHCO_2C_2H_5$ (63)	407
$(C_2H_5O)_2P(O)$-(2-oxotetrahydrofuran-3-yl)	o-$C_6H_4(CHO)_2$	NaH/C_6H_6	o-C_6H_4(CH=(2-oxotetrahydrofuran-3-ylidene))$_2$ (54)	281
	p-$C_6H_4(CHO)_2$	NaH/C_6H_6	p-C_6H_4(CH=(2-oxotetrahydrofuran-3-ylidene))$_2$ (71)	281

Note: References 228–415 are on pp. 250–253.

TABLE III. Unsaturated Esters from Phosphoryl Reagents Containing a Carboxylic Group

$R_2P(O)CHR^1CO_2R^2 + R^3R^4CO \rightarrow R^3R^4C{=}CR^1CO_2R^2 + R_2PO_2^-$ (*Continued*)

No. of C Atoms	Reagent	Reactant	Base/Solvent	Product(s) and Yield(s) (%)	Refs.
C_8 (*Contd.*)	$(C_2H_5O)_2P(O)CH_2CO_2C_2H_5$	$C_6H_5CH_2COCH_3$	NaH/glyme	$C_6H_5CH_2C(CH_3){=}CHCO_2C_2H_5$ (—)	309
		$C_6H_5CH_2COCH_3$	NaH/glyme	$C_6H_5CH{=}C(CH_3)CH_2CO_2C_2H_5$ (—)	309
		[tetrahydrofuran-2-yl]$(CH_2)_2COCH_3$	C_2H_5ONa/C_2H_5OH	[tetrahydrofuran-2-yl]$(CH_2)_2C(CH_3){=}CHCO_2C_2H_5$ (93)	203
		$(C_2H_5O)_2CH(CH_2)_2COCH_3$	NaH/glyme	$(C_2H_5O)_2CH(CH_2)_2C(CH_3){=}CHCO_2C_2H_5$ (—)	310
		p-$CH_3OC_6H_4COCH_3$	NaH/glyme	p-$CH_3OC_6H_4C(CH_3){=}CHCO_2C_2H_5$ (—)	311
		$C_6H_5CH{=}CHCHO$	$NaNH_2$/ether	$C_6H_5(CH{=}CH)_2CO_2C_2H_5$ (72)	250
		p-$(CH_3)_2NC_6H_4CHO$	NaH/glyme	p-$(CH_3)_2NC_6H_4CH{=}CHCO_2C_2H_5$ (81)	284
		n-$C_7H_{15}COCH_3$	NaH/glyme	n-$C_7H_{15}C(CH_3){=}CHCO_2C_2H_5$ (43)	280
		CH_3–[cyclohex-3-en-1,4-diyl]–$COCH_3$	NaH/glyme	CH_3–[cyclohex-3-en-1,4-diyl]–$C(CH_3){=}CHCO_2C_2H_5$ (75)	402
		$CH_3COCH_2CH_2CH(CHO)CH(CH_3)_2$	NaH/glyme	$C_2H_5O_2CCH{=}C(CH_3)CH_2CH_2CH(CH{=}CHCO_2C_2H_5)CH(CH_3)_2$ (50)	130
		$C_2H_5O_2C$–[3-oxocyclohexyl]	NaH/C_6H_6	$C_2H_5O_2C$–[cyclohexyl]$={}$$CHCO_2C_2H_5$ (3-position) (–)	312
	$(C_2H_5O)_2P(O)$–[γ-butyrolacton-α-yl]	$C_6H_5CH{=}CHCHO$	NaH/C_6H_6	$C_6H_5CH{=}CHCH{=}$[γ-butyrolacton-α-ylidene] (55)	281
	$(C_2H_5O)_2P(O)CH_2CO_2C_2H_5$	$C_6H_5C(CH_3)_2CHO$	NaH/glyme	$C_6H_5C(CH_3)_2CH{=}CHCO_2C_2H_5$ (—)	313

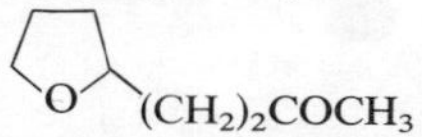

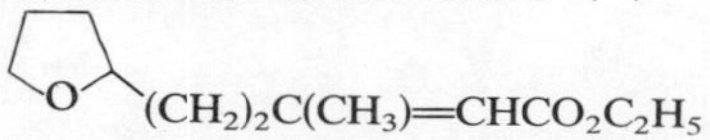

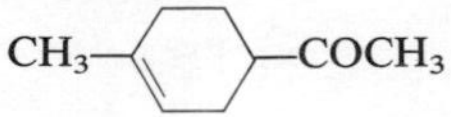

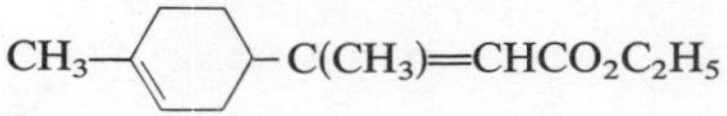

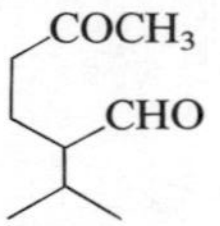

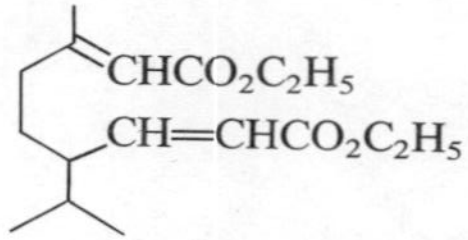

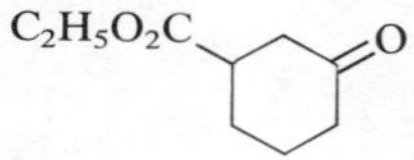

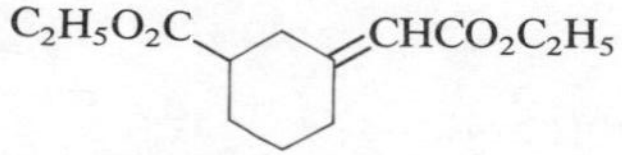

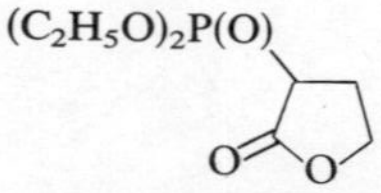

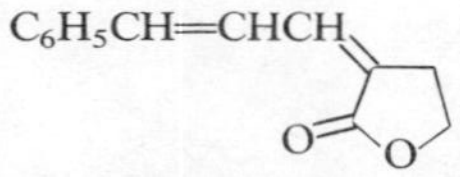

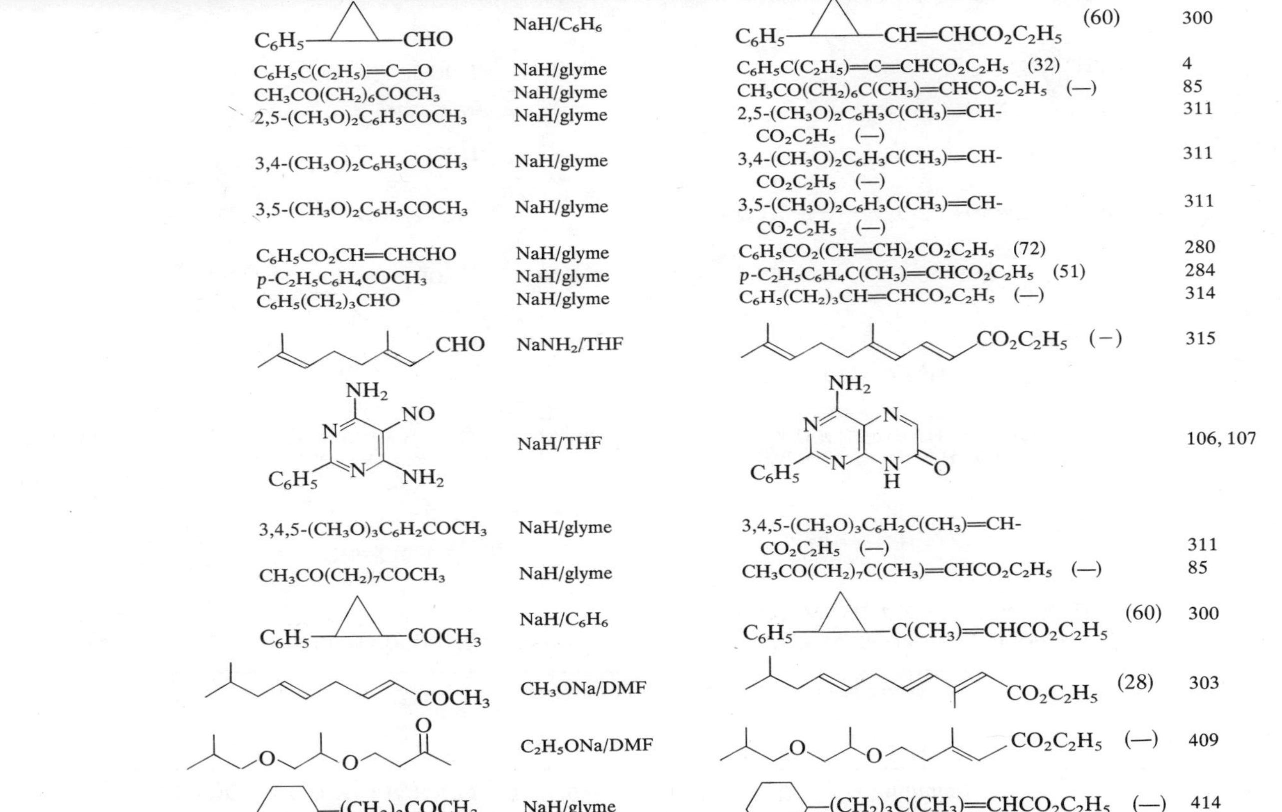

Reactant	Base/Solvent	Product (Yield, %)	Refs.
C_6H_5–(cyclopropyl)–CHO	NaH/C_6H_6	C_6H_5–(cyclopropyl)–$CH{=}CHCO_2C_2H_5$ (60)	300
$C_6H_5C(C_2H_5){=}C{=}O$	NaH/glyme	$C_6H_5C(C_2H_5){=}C{=}CHCO_2C_2H_5$ (32)	4
$CH_3CO(CH_2)_6COCH_3$	NaH/glyme	$CH_3CO(CH_2)_6C(CH_3){=}CHCO_2C_2H_5$ (—)	85
2,5-$(CH_3O)_2C_6H_3COCH_3$	NaH/glyme	2,5-$(CH_3O)_2C_6H_3C(CH_3){=}CH$-$CO_2C_2H_5$ (—)	311
3,4-$(CH_3O)_2C_6H_3COCH_3$	NaH/glyme	3,4-$(CH_3O)_2C_6H_3C(CH_3){=}CH$-$CO_2C_2H_5$ (—)	311
3,5-$(CH_3O)_2C_6H_3COCH_3$	NaH/glyme	3,5-$(CH_3O)_2C_6H_3C(CH_3){=}CH$-$CO_2C_2H_5$ (—)	311
$C_6H_5CO_2CH{=}CHCHO$	NaH/glyme	$C_6H_5CO_2(CH{=}CH)_2CO_2C_2H_5$ (72)	280
p-$C_2H_5C_6H_4COCH_3$	NaH/glyme	p-$C_2H_5C_6H_4C(CH_3){=}CHCO_2C_2H_5$ (51)	284
$C_6H_5(CH_2)_3CHO$	NaH/glyme	$C_6H_5(CH_2)_3CH{=}CHCO_2C_2H_5$ (—)	314
(structure)–CHO	$NaNH_2$/THF	(structure)–$CO_2C_2H_5$ (–)	315
(structure: NH_2, NO, N, N, C_6H_5, NH_2)	NaH/THF	(structure: NH_2, N, N, N, N, H, O, C_6H_5)	106, 107
3,4,5-$(CH_3O)_3C_6H_2COCH_3$	NaH/glyme	3,4,5-$(CH_3O)_3C_6H_2C(CH_3){=}CH$-$CO_2C_2H_5$ (—)	311
$CH_3CO(CH_2)_7COCH_3$	NaH/glyme	$CH_3CO(CH_2)_7C(CH_3){=}CHCO_2C_2H_5$ (—)	85
C_6H_5–(cyclopropyl)–$COCH_3$	NaH/C_6H_6	C_6H_5–(cyclopropyl)–$C(CH_3){=}CHCO_2C_2H_5$ (60)	300
(structure)–$COCH_3$	CH_3ONa/DMF	(structure)–$CO_2C_2H_5$ (28)	303
(structure: O, O, O)	C_2H_5ONa/DMF	(structure: O, O)–$CO_2C_2H_5$ (—)	409
(cyclohexyl)–$(CH_2)_3COCH_3$	NaH/glyme	(cyclohexyl)–$(CH_2)_3C(CH_3){=}CHCO_2C_2H_5$ (—)	414

Note: References 228–415 are on pp. 250–253.

TABLE III. Unsaturated Esters from Phosphoryl Reagents Containing a Carboxylic Group

$R_2P(O)CHR^1CO_2R^2 + R^3R^4CO \rightarrow R^3R^4C{=}CR^1CO_2R^2 + R_2PO_2^-$ (*Continued*)

No. of C Atoms	Reagent	Reactant	Base/Solvent	Product(s) and Yield(s) (%)	Refs.
C_8 (*Contd.*)	$(C_2H_5O)_2P(O)CH_2CO_2C_2H_5$ (*Contd.*)	*p*-$CH_3OC_6H_4COCO_2C_2H_5$	Na/C_6H_6	*p*-$CH_3OC_6H_4C(CO_2C_2H_5){=}CHCO_2C_2H_5$ (90)	316
		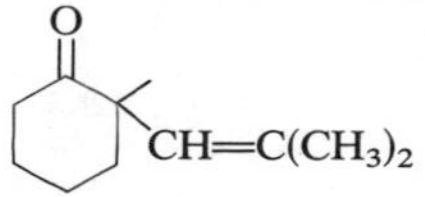 O, $CH{=}C(CH_3)_2$	NaH/glyme	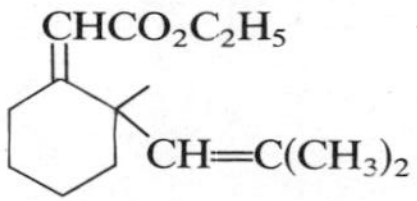 $CHCO_2C_2H_5$, $CH{=}C(CH_3)_2$ (−)	129
	$(C_2H_5O)_2P(O)CH(CH_3)CO_2CH_3$	 $(CH_3)_3COCO$, CHO	CH_3ONa/DMF	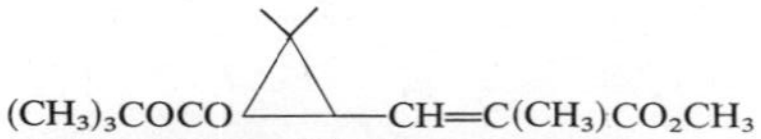 $(CH_3)_3COCO$, $CH{=}C(CH_3)CO_2CH_3$ (−)	143
	$(C_2H_5O)_2P(O)CH_2CO_2C_2H_5$	$C_6H_5CO_2(CH{=}CH)_2CHO$	NaH/glyme	$C_6H_5CO_2(CH{=}CH)_3CO_2C_2H_5$ (51)	280
		n-$C_{11}H_{23}CHO$	NaH/ether	*n*-$C_{11}H_{23}CH{=}CHCO_2C_2H_5$ (82)	287
		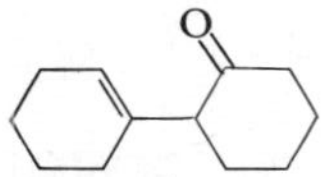 O	NaH/glyme	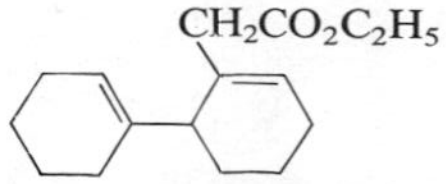 $CH_2CO_2C_2H_5$ (88)	114
		p-$C_6H_4(CH{=}CHCHO)_2$	C_2H_5ONa/C_2H_5OH	*p*-$C_6H_4[(CH{=}CH)_2CO_2C_2H_5]_2$ (—)	88
		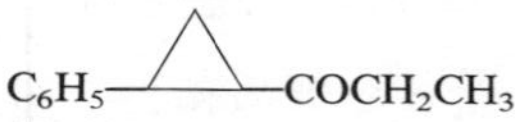 C_6H_5, $COCH_2CH_3$	Na/C_6H_6	 C_6H_5, $C(CH_2CH_3){=}CHCO_2C_2H_5$ (16)	300
		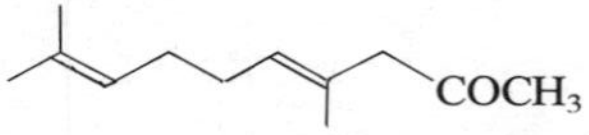 $COCH_3$	NaH/glyme	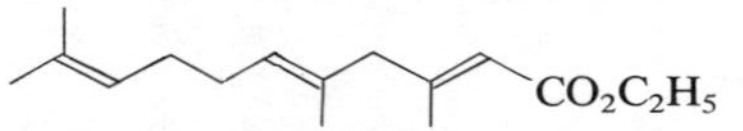 $CO_2C_2H_5$ (54)	280, 303
		$(CH_3)_2CHO(CH_2)_7COCH_3$	NaH/glyme	$(CH_3)_2CHO(CH_2)_7C(CH_3){=}CHCO_2C_2H_5$ (—)	317
		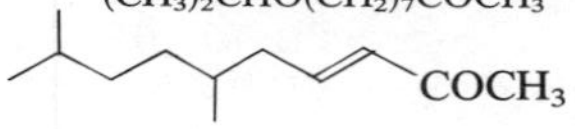 $COCH_3$	NaH/glyme	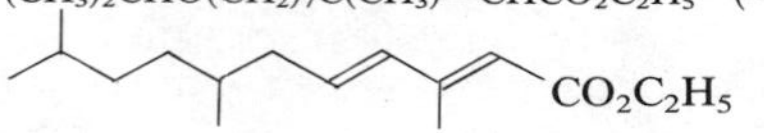 $CO_2C_2H_5$ (−)	318

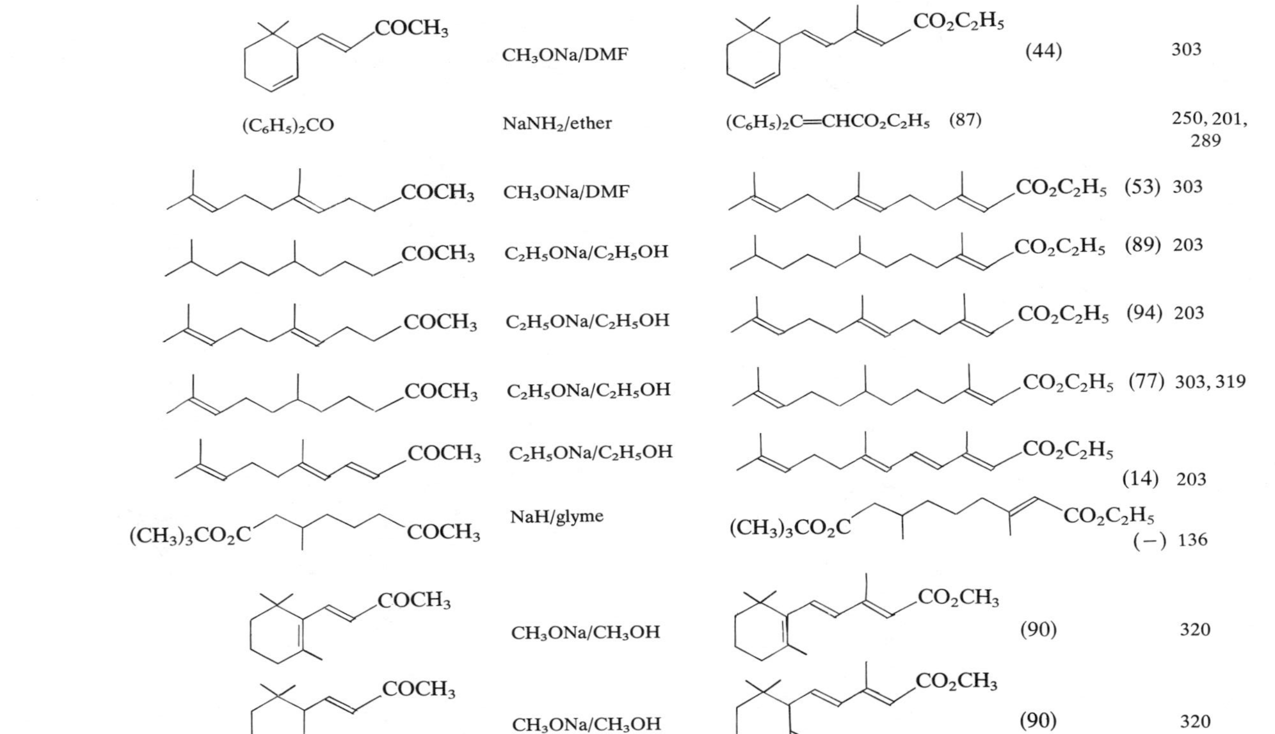

Note: References 228–415 are on pp. 250–253.

TABLE III. Unsaturated Esters from Phosphoryl Reagents Containing a Carboxylic Group $R_2P(O)CHR^1CO_2R^2 + R^3R^4CO \rightarrow R^3R^4C{=}CR^1CO_2R^2 + R_2PO_2^-$ (*Continued*)

No. of C Atoms	Reagent	Reactant	Base/Solvent	Product(s) and Yield(s) (%)	Refs.
C_8 (*Contd.*)	$(C_2H_5O)_2P(O)CH_2CO_2C_2H_5$ (*Contd.*)	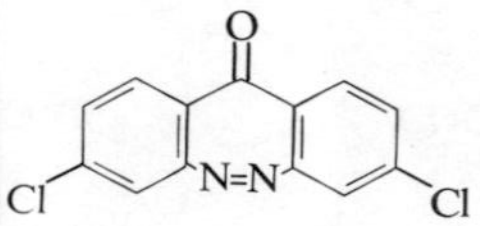 O; N=N; Cl; Cl	NaH/dioxane	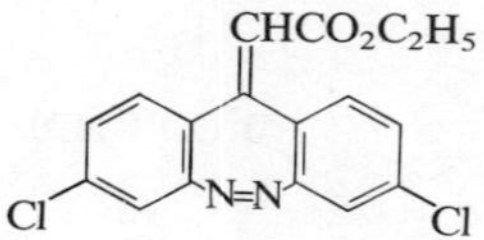 $CHCO_2C_2H_5$; N=N; Cl; Cl (29)	258
		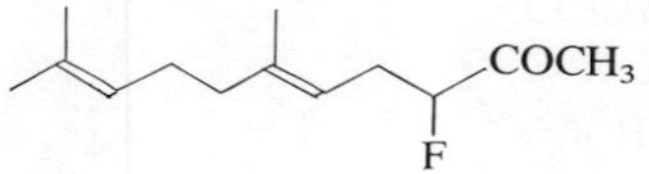 $COCH_3$; F	NaH/glyme	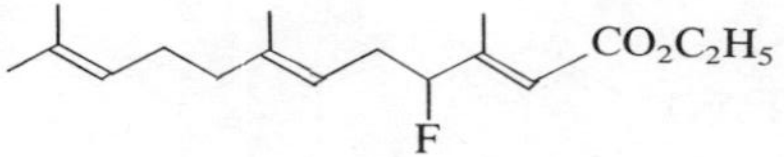 $CO_2C_2H_5$; F (–)	307
		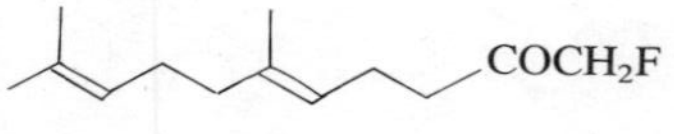 $COCH_2F$	NaH/glyme	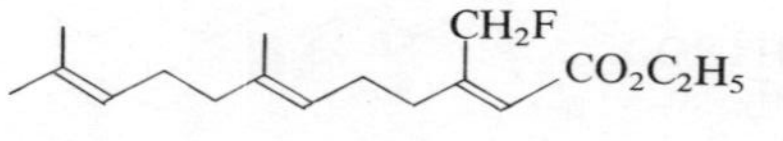 CH_2F; $CO_2C_2H_5$ (–)	307
	$(C_2H_5O)_2P(O)CH_2CONHC_2H_5$	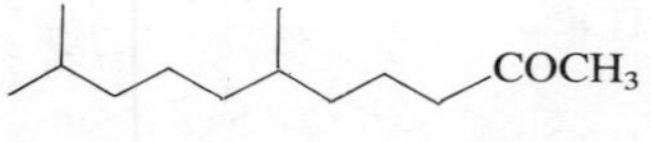 $COCH_3$	C_2H_5ONa/C_2H_5OH	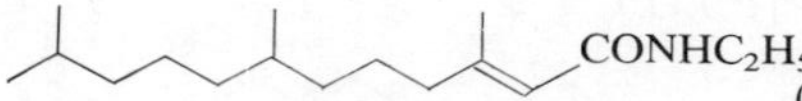 $CONHC_2H_5$ (62)	203
		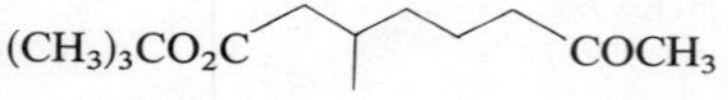 $(CH_3)_3CO_2C$; $COCH_3$	NaH/glyme	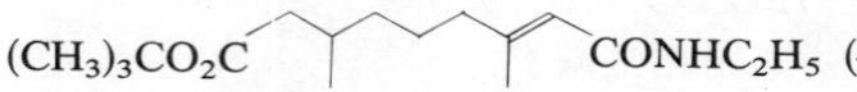 $(CH_3)_3CO_2C$; $CONHC_2H_5$ (—)	136
	$(C_2H_5O)_2P(O)CH_2CO_2C_2H_5$	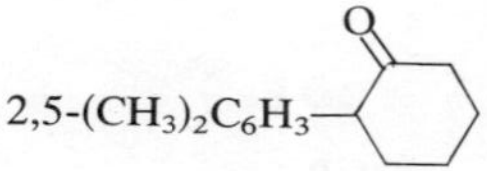 2,5-$(CH_3)_2C_6H_3$–; O	NaH/glyme	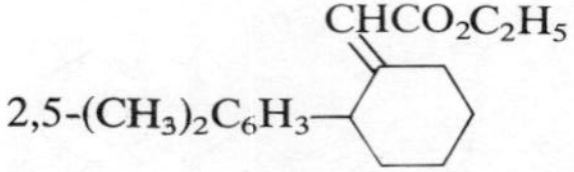 $CHCO_2C_2H_5$; 2,5-$(CH_3)_2C_6H_3$– (32)	259

Reagent	Carbonyl Compound	Base/Solvent	Product (Yield, %)	Refs.
	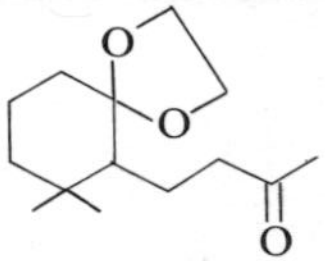 (ketal cyclohexane ketone)	NaH/DMSO	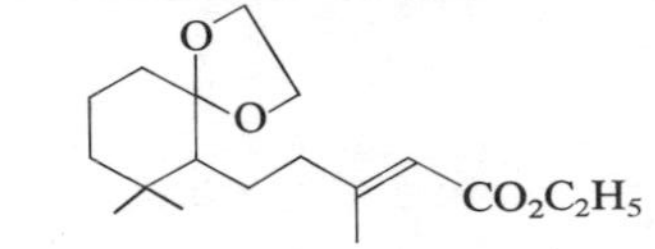 (ketal) $CO_2C_2H_5$ (−)	321
	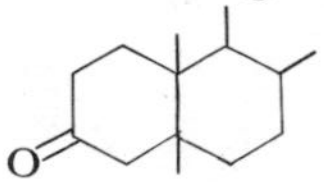 (decalone)	CH_3ONa/CH_3OH	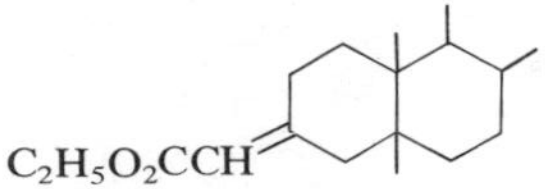 $C_2H_5O_2CCH$= (decalin) (−)	322
	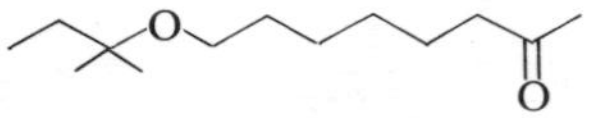 (t-amyl ether ketone)	NaH/glyme	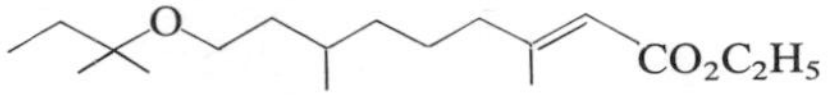 $CO_2C_2H_5$ (—)	410
	$C_6H_5CO_2(CH{=}CH)_3CHO$	NaH/glyme	$C_6H_5CO_2(CH{=}CH)_4CO_2C_2H_5$ (60)	280
	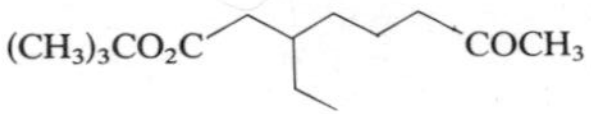 $(CH_3)_3CO_2C$ … $COCH_3$	NaH/glyme	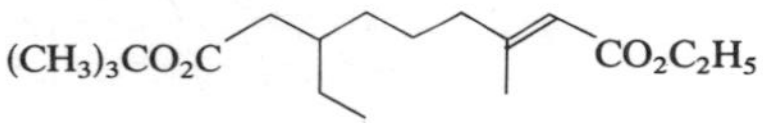 $(CH_3)_3CO_2C$ … $CO_2C_2H_5$ (−)	136
	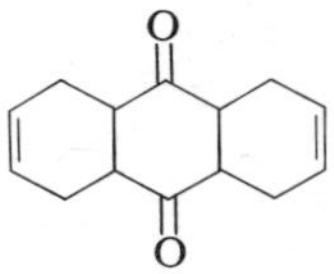 (hexahydroanthraquinone)	NaH/glyme	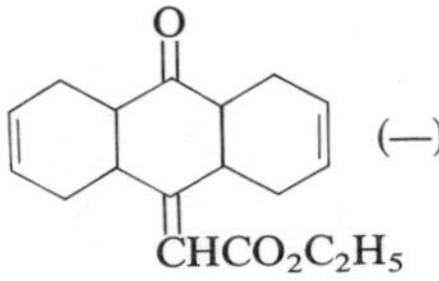 (—) $CHCO_2C_2H_5$	394
$(C_2H_5O)_2P(O)CH(CH_3)CO_2CH_3$	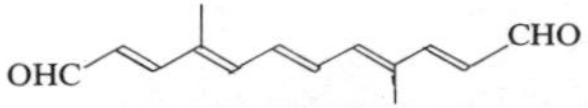 OHC … CHO	CH_3ONa/CH_3OH	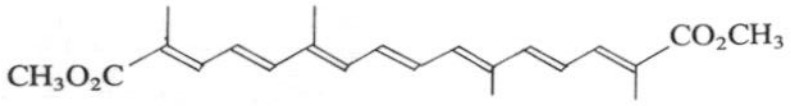 CH_3O_2C … CO_2CH_3 (−)	323
$(C_2H_5O)_2P(O)CH_2CO_2C_2H_5$	$C_6H_5CH_2CH_2COC_6H_5$	NaH/glyme	$C_6H_5CH_2CH_2C(C_6H_5){=}CHCO_2C_2H_5$ (—)	324
	$C_6H_5CH{=}CHCOC_6H_5$	NaH/glyme	$C_6H_5CH{=}CHC(C_6H_5){=}CHCO_2C_2H_5$ (—)	96
	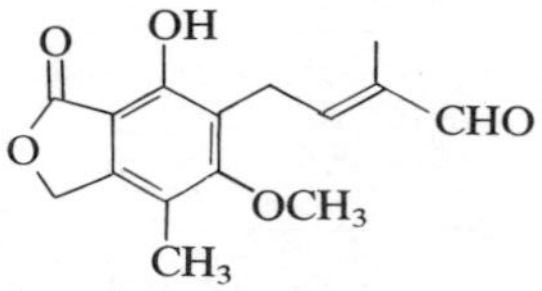 (O, OH, O, OCH_3, CH_3) CHO	NaH/glyme	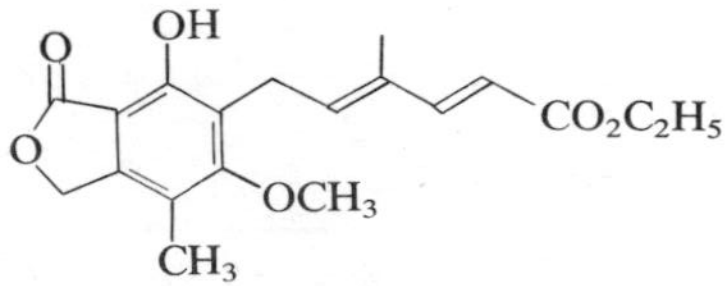 (O, OH, O, OCH_3, CH_3) $CO_2C_2H_5$ (−)	325
	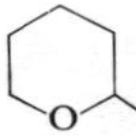 (tetrahydropyranyl) $O(CH_2)_3C(CH_3){=}CH(CH_2)_2COCH_3$	NaH/glyme	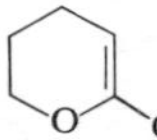 (dihydropyranyl) $O(CH_2)_3C(CH_3){=}CH(CH_2)_2C(CH_3){=}CHCO_2C_2H_5$ (61)	381

Note: References 228–415 are on pp. 250–253.

TABLE III. Unsaturated Esters from Phosphoryl Reagents Containing a Carboxylic Group $R_2P(O)CHR^1CO_2R^2 + R^3R^4CO \rightarrow R^3R^4C{=}CR^1CO_2R^2 + R_2PO_2^-$ (*Continued*)

No. of C Atoms	Reagent	Reactant	Base/Solvent	Product(s) and Yield(s) (%)	Refs.
C_8 (*Contd.*)	$(C_2H_5O)_2P(O)CH_2CO_2C_2H_5$ (*Contd.*)	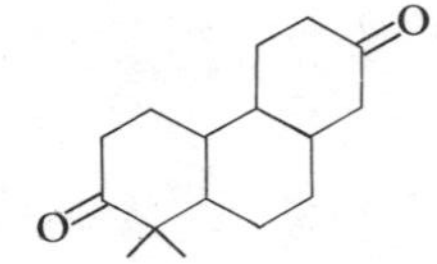 [structure]	NaH/DMF	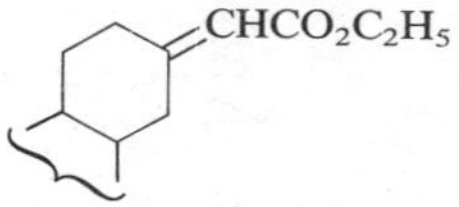 [structure] $=CHCO_2C_2H_5$ (82)	326
		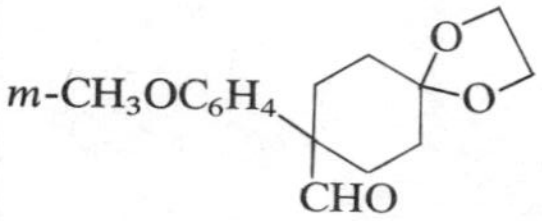 *m*-$CH_3OC_6H_4$ [structure] CHO	NaH/C_6H_6	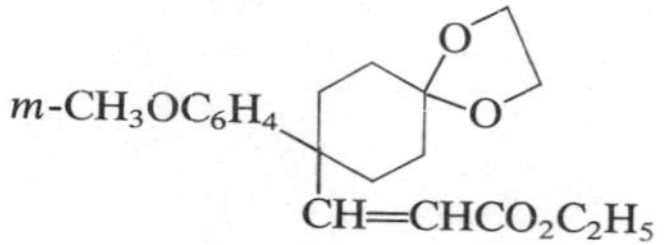 *m*-$CH_3OC_6H_4$ [structure] $CH{=}CHCO_2C_2H_5$ (–)	327
		p-$C_6H_5OC_6H_4(CH_2)_2COCH_3$	NaH/glyme	*p*-$C_6H_5OC_6H_4(CH_2)_2C(CH_3){=}CHCO_2C_2H_5$ (—)	412
		p-$C_6H_5SC_6H_4(CH_2)_2COCH_3$	NaH/glyme	*p*-$C_6H_5SC_6H_4(CH_2)_2C(CH_3){=}CHCO_2C_2H_5$ (—)	412
		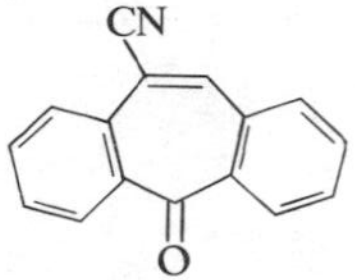 CN [structure] O	NaH/glyme	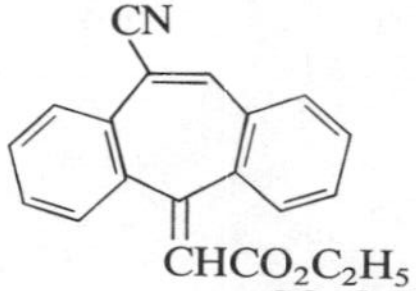 CN [structure] $CHCO_2C_2H_5$ (–)	262
		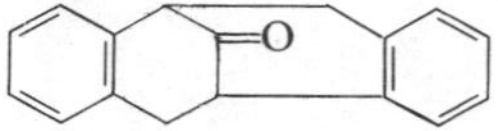 [structure] O	NaH/glyme	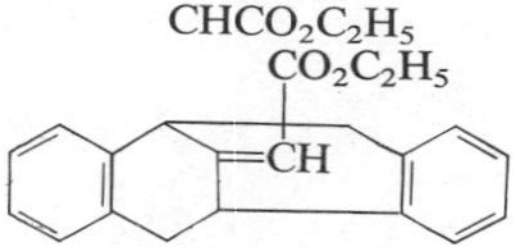 [structure] $=CH$ $CO_2C_2H_5$ (–)	262

Reactant	Conditions	Product(s) and Yield(s) (%)	Refs.
$(C_6H_5)_2$-cyclopropyl-$COCH_3$	NaH/DMF	$(C_6H_5)_2$-cyclopropyl-$C(CH_3){=}CHCO_2C_2H_5$ (93)	403
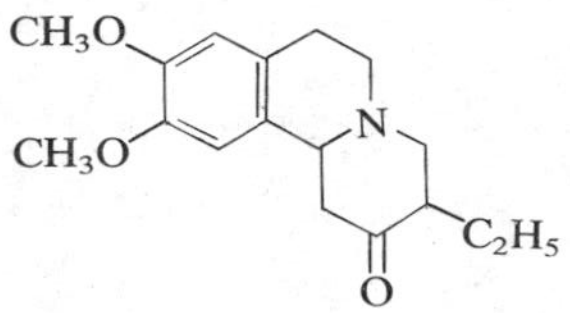 9,10-Dimethoxy-3-ethyl-2-oxo-benzo[a]quinolizidine (CH_3O, CH_3O, C_2H_5, O)	t-C_4H_9OK/DMF 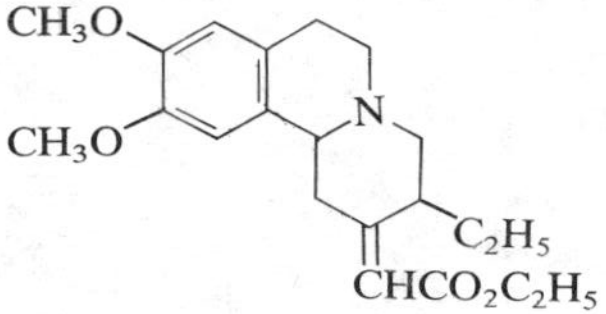	2-($CHCO_2C_2H_5$)-3-C_2H_5 analog (CH_3O, CH_3O) (–)	116
Farnesyl acetone ($COCH_3$)	NaH/glyme	Farnesyl-type ester ($CO_2C_2H_5$) (–)	126
CH_3O-benzazocinone, N-$SO_2C_6H_4CH_3$-p	NaH/glyme	$CHCO_2C_2H_5$ analog, N-$SO_2C_6H_4CH_3$-p (–)	328
17-Hydroxy-3-keto steroid (OH)	NaH/glyme	$C_2H_5O_2CCH{=}$ (3-ylidene) (–)	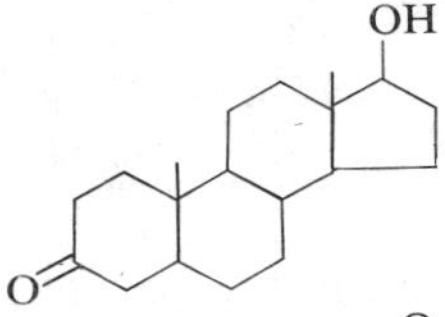 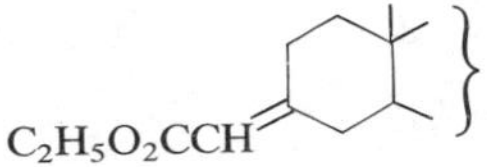 36, 268, 329, 330
CH_3CO_2-4,4-dimethyl steroid ketone	NaH/DMF	$=CHCO_2C_2H_5$ (82)	326
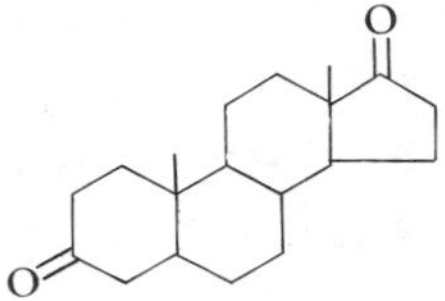 Androstane-3,17-dione	C_2H_5ONa/DMF	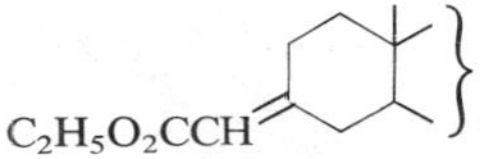 $C_2H_5O_2CCH{=}$ (3-ylidene) (51)	36, 86

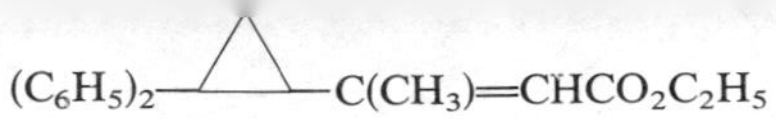

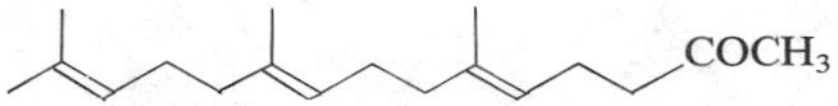

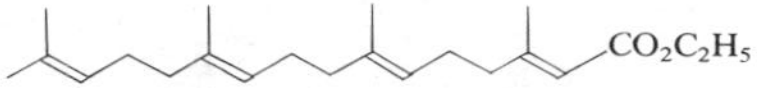

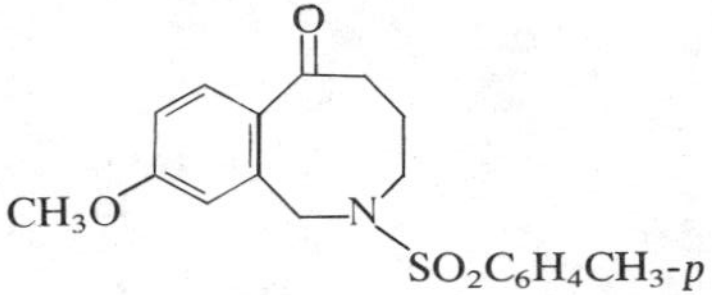

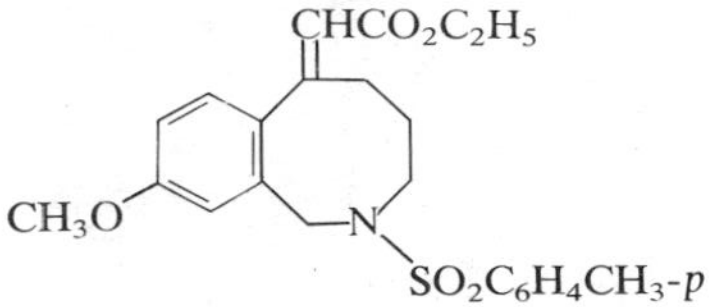

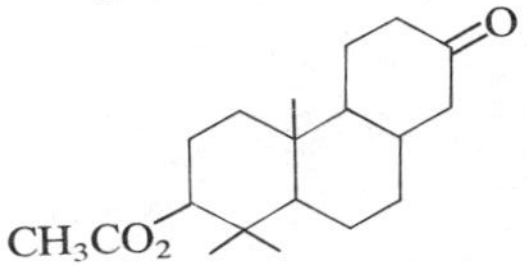

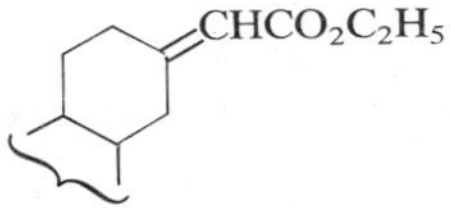

Note: References 228–415 are on pp. 250–253.

TABLE III. UNSATURATED ESTERS FROM PHOSPHORYL REAGENTS CONTAINING A CARBOXYLIC GROUP
$R_2P(O)CHR^1CO_2R^2 + R^3R^4CO \rightarrow R^3R^4C{=}CR^1CO_2R^2 + R_2PO_2^-$ (*Continued*)

No. of C Atoms	Reagent	Reactant	Base/Solvent	Product(s) and Yield(s) (%)	Refs.
C_8 (*Contd.*)	$(C_2H_5O)_2P(O)CH_2CO_2C_2H_5$ (*Contd.*)				
		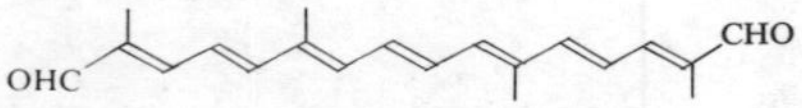 OHC–(polyene)–CHO	CH_3ONa/CH_3OH	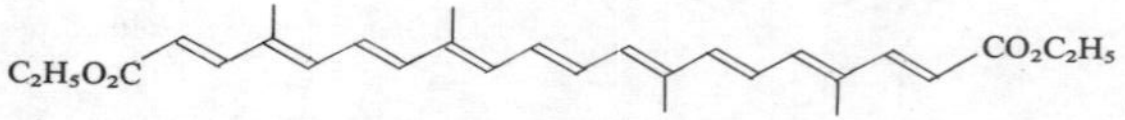 $C_2H_5O_2C$–(polyene)–$CO_2C_2H_5$ (—)	323
		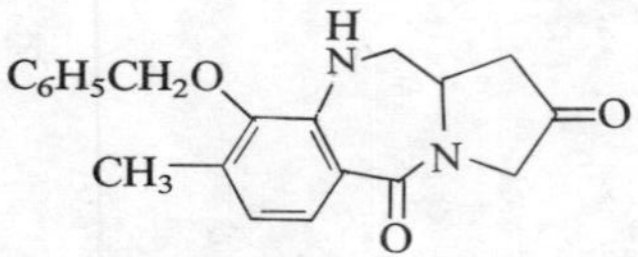 $C_6H_5CH_2O$, CH_3, H N, O, N, =O (tricyclic ketone)	NaH/glyme	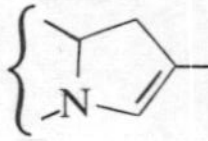 { N –$CH_2CO_2C_2H_5$ (74)	331
	$(C_2H_5O)_2P(O)CH(CH_3)CO_2CH_3$	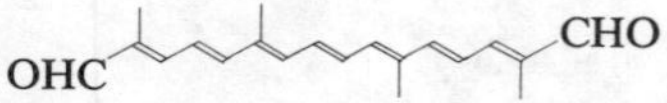 OHC–(polyene)–CHO	CH_3ONa/CH_3OH	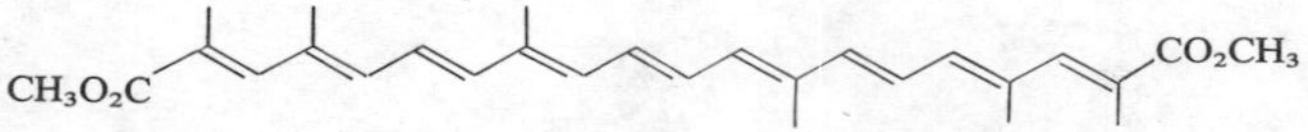 CH_3O_2C–(polyene)–CO_2CH_3 (—)	323
	$(C_2H_5O)_2P(O)CH_2CO_2C_2H_5$	n-$C_{10}H_{21}COCH_3$	NaH/glyme	n-$C_{10}H_{21}C(CH_3){=}CHCO_2C_2H_5$ (64)	289

Reactant	Conditions	Product (Yield, %)	Refs.
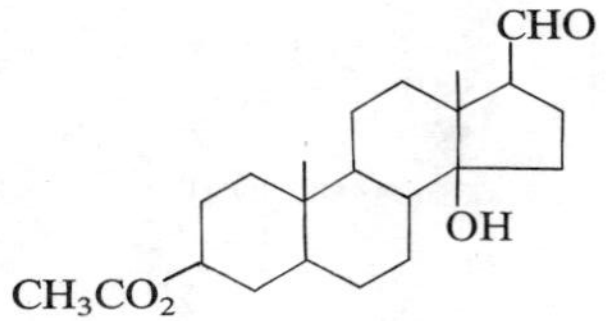 CH_3CO_2, CHO, OH (steroid)	NaH/glyme	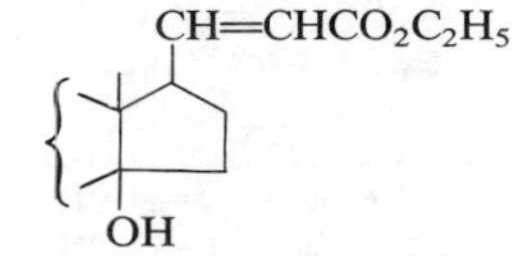 $CH=CHCO_2C_2H_5$, OH (−)	332
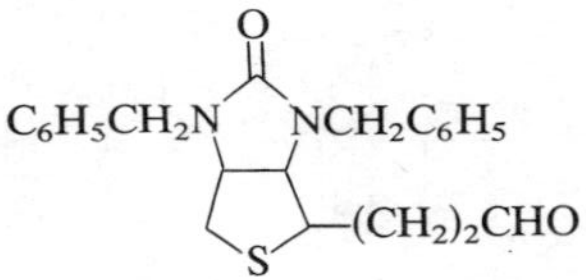 $C_6H_5CH_2N$, $NCH_2C_6H_5$, O, S, $(CH_2)_2CHO$	NaH/glyme	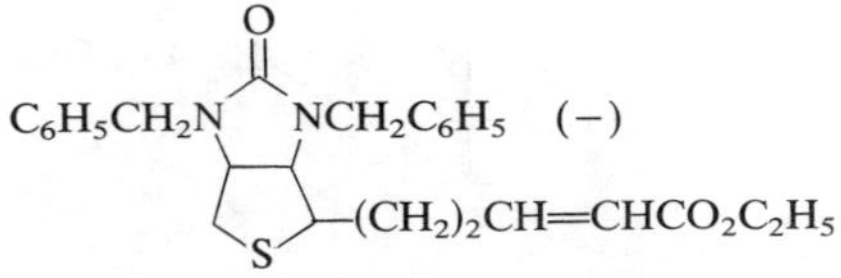 $C_6H_5CH_2N$, $NCH_2C_6H_5$, O, S, $(CH_2)_2CH=CHCO_2C_2H_5$ (−)	333
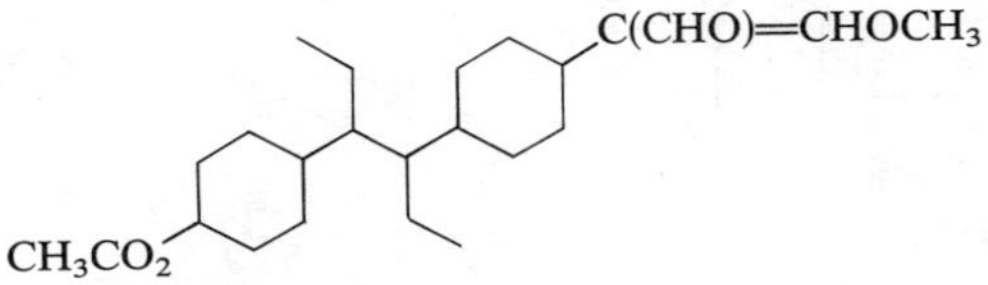 CH_3CO_2, $C(CHO)=CHOCH_3$	NaH/glyme	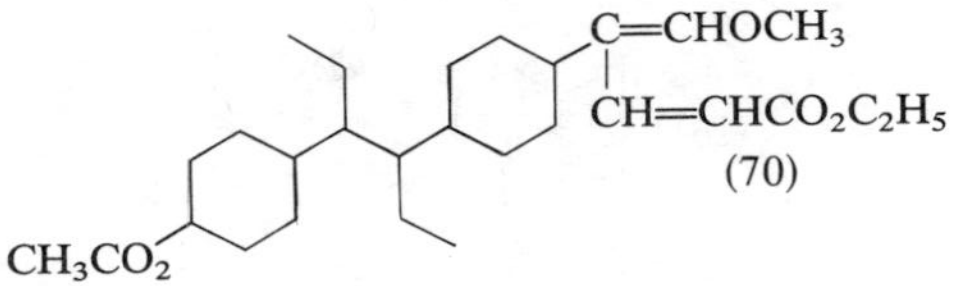 CH_3CO_2, $C=CHOCH_3$, $CH=CHCO_2C_2H_5$ (70)	365
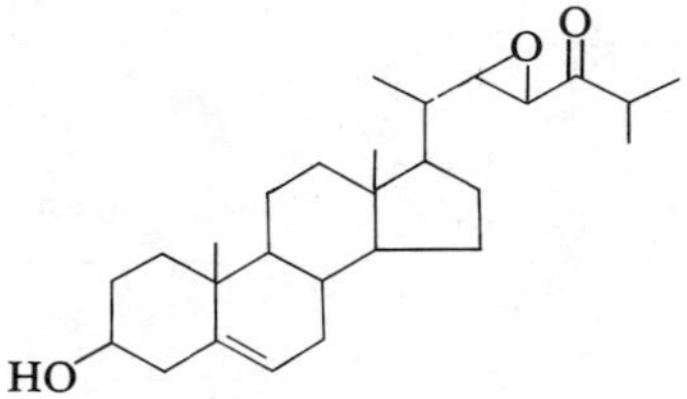 HO, O, O (steroid)	$NaNH_2$/THF	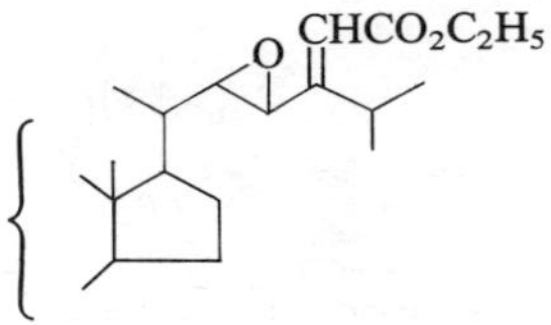 O, $CHCO_2C_2H_5$ (40)	375

Note: References 228–415 are on pp. 250–253.

TABLE III. UNSATURATED ESTERS FROM PHOSPHORYL REAGENTS CONTAINING A CARBOXYLIC GROUP

$R_2P(O)CHR^1CO_2R^2 + R^3R^4CO \rightarrow R^3R^4C{=}CR^1CO_2R^2 + R_2PO_2^-$ (*Continued*)

No. of C Atoms	Reagent	Reactant	Base/Solvent	Product(s) and Yield(s) (%)	Refs.
C_8 (*Contd.*)	$(C_2H_5O)_2P(O)CH_2CO_2C_2H_5$ (*Contd.*)	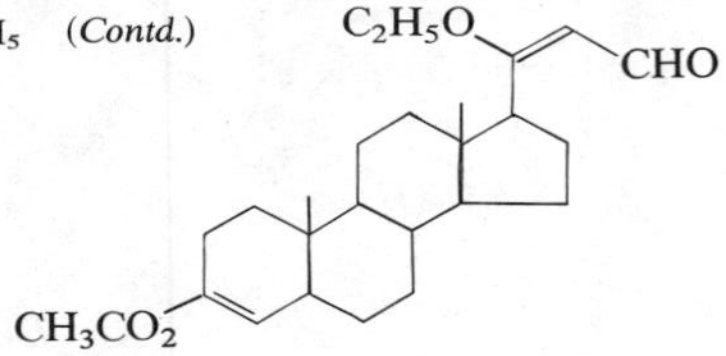 [structure: C_2H_5O, CHO, CH_3CO_2]	NaH/THF	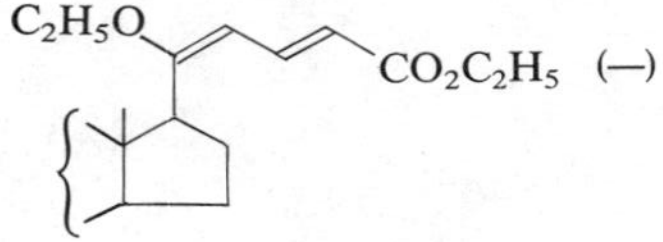 [structure: C_2H_5O, $CO_2C_2H_5$] (—)	267, 334, 335
		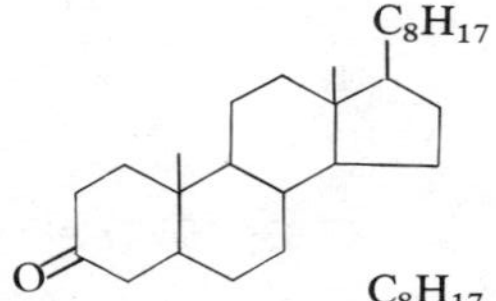 [structure: C_8H_{17}, O]	C_2H_5ONa/DMF	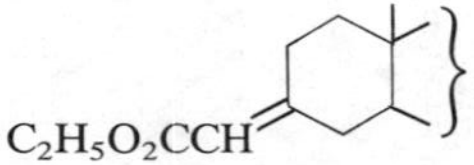 [structure: $C_2H_5O_2CCH$] (93)	36, 86
		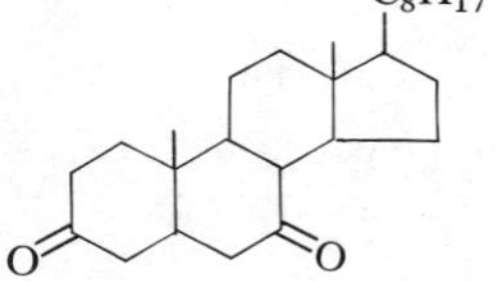 [structure: C_8H_{17}, O, O]	C_2H_5ONa/DMF	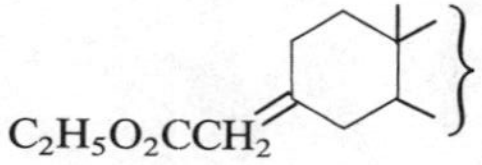 [structure: $C_2H_5O_2CCH_2$] (71)	36, 86
C_9	$(C_2H_5O)_2P(O)CH(CH_3)CO_2C_2H_5$	CH_3CHO	NaH/glyme	$CH_3CH{=}C(CH_3)CO_2C_2H_5$ (—)	20
		C_2H_5CHO	NaH/glyme	$C_2H_5CH{=}C(CH_3)CO_2C_2H_5$ (—)	20
		i-C_3H_7CHO	NaH/glyme	i-$C_3H_7CH{=}C(CH_3)CO_2C_2H_5$ (13)	20, 285, 288
		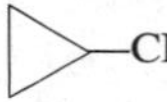 [cyclopropyl]—CHO	NaH/glyme	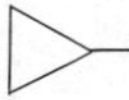 [cyclopropyl]—$CH{=}C(CH_3)CO_2C_2H_5$ (−)	288, 285

	n-C_3H_7CHO	NaH/glyme	n-$C_3H_7CH{=}C(CH_3)CO_2C_2H_5$ (69)	286
	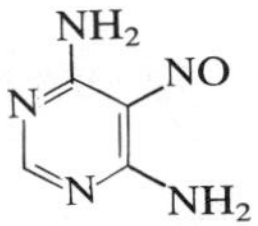 NH_2, NO, NH_2 (pyrimidine)	NaH/THF	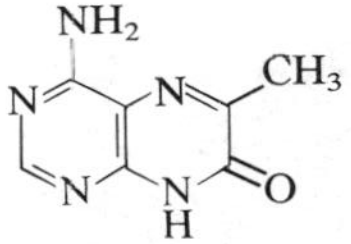 NH_2, CH_3, O, N–H (78)	107
	i-C_4H_9CHO	NaH/glyme	i-$C_4H_9CH{=}C(CH_3)CO_2C_2H_5$ (—)	20
	t-C_4H_9CHO	NaH/glyme	t-$C_4H_9CH{=}C(CH_3)CO_2C_2H_5$ (—)	20
	i-$C_3H_7COCH_3$	NaH/glyme	i-$C_3H_7C(CH_3){=}C(CH_3)CO_2C_2H_5$ (28)	288, 285
	n-$C_3H_7COCH_3$	NaH/glyme	n-$C_3H_7C(CH_3){=}C(CH_3)CO_2C_2H_5$ (92)	286
	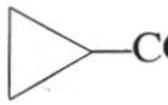 (cyclopropyl)–$COCH_3$	NaH/glyme	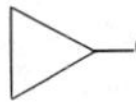 (cyclopropyl)–$C(CH_3){=}C(CH_3)CO_2C_2H_5$ (45)	285, 288
	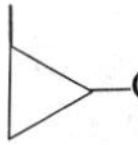 (methylcyclopropyl)–CHO	NaH/glyme	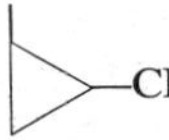 (methylcyclopropyl)–$CH{=}C(CH_3)CO_2C_2H_5$ (−)	285, 288
	$CH_2{=}C(CH_3)(CH_2)_2CHO$	NaH/glyme	$CH_2{=}C(CH_3)(CH_2)_2CH{=}C(CH_3)$-$CO_2C_2H_5$ (—)	127
$(C_2H_5O)_2P(O)CH_2CH{=}CHCO_2CH_3$	n-$C_5H_{11}CHO$	CH_3ONa/DMF	n-$C_5H_{11}(CH{=}CH)_2CO_2CH_3$ (25)	54
$(C_2H_5O)_2P(O)CH[N(CH_3)_2]CO_2CH_3$	C_6H_5NO	NaH/DMF	$C_6H_5N{=}C[N(CH_3)_2]CO_2CH_3$ (—)	41
$(C_2H_5O)_2P(O)CH(CH_3)CO_2C_2H_5$	C_6H_5CHO	NaH/THF	$C_6H_5CH{=}C(CH_3)CO_2C_2H_5$ (—)	15
	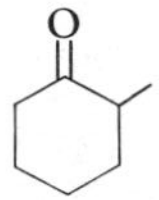 O (methylcyclohexanone)	NaH/glyme	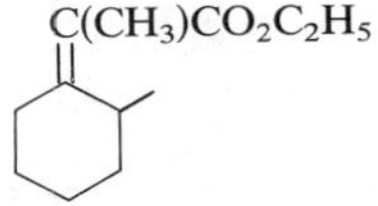 $C(CH_3)CO_2C_2H_5$ (methylcyclohexylidene) (−)	296
	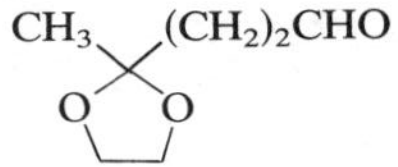 CH_3, $(CH_2)_2CHO$, O, O (dioxolane)	NaH/glyme	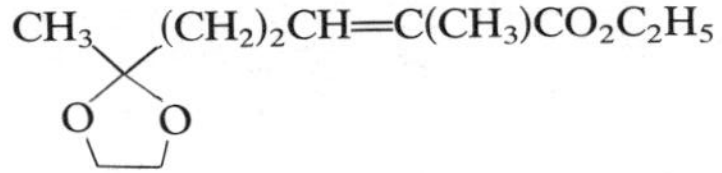 CH_3, $(CH_2)_2CH{=}C(CH_3)CO_2C_2H_5$, O, O (dioxolane) (−)	336
	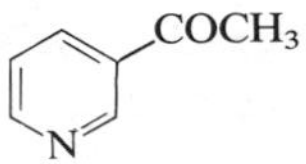 $COCH_3$ (pyridine, N)	C_2H_5ONa/C_2H_5OH	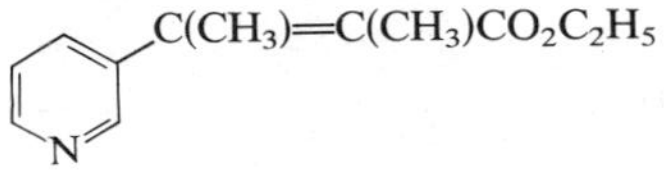 $C(CH_3){=}C(CH_3)CO_2C_2H_5$ (pyridine, N) (60)	337

Note: References 228–415 are on pp. 250–253.

TABLE III. Unsaturated Esters from Phosphoryl Reagents Containing a Carboxylic Group

$R_2P(O)CHR^1CO_2R^2 + R^3R^4CO \rightarrow R^3R^4C{=}CR^1CO_2R^2 + R_2PO_2^-$ (*Continued*)

No. of C Atoms	Reagent	Reactant	Base/Solvent	Product(s) and Yield(s) (%)	Refs.
	$(C_2H_5O)_2P(O)CH[N(CH_3)_2]$-$CO_2CH_3$	C_6H_5CHO	NaH/glyme	$C_6H_5CH{=}C[N(CH_3)_2]CO_2CH_3$ (—)	41
	$(C_2H_5O)_2P(O)CH(CH_3)CO_2C_2H_5$	$C_6H_5CH_2CHO$	NaH/glyme	$C_6H_5CH_2CH{=}C(CH_3)CO_2C_2H_5$ (67)	338
		$C_6H_5COCH_3$	C_2H_5ONa/C_2H_5OH	$C_6H_5C(CH_3){=}C(CH_3)CO_2C_2H_5$ (65)	337
		m-$ClC_6H_4COCH_3$	C_2H_5ONa/C_2H_5OH	m-$ClC_6H_4C(CH_3){=}C(CH_3)CO_2C_2H_5$ (82)	337
		m-$O_2NC_6H_4COCH_3$	C_2H_5ONa/C_2H_5OH	m-$O_2NC_6H_4C(CH_3){=}C(CH_3)CO_2C_2H_5$ (40)	337
		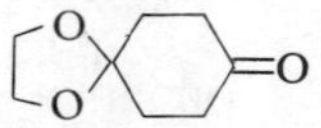 (structure: O, O, =O)	NaH/glyme	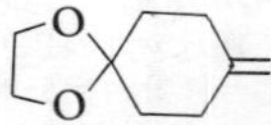 (structure: O, O) $=C(CH_3)CO_2C_2H_5$ (72)	124
	$(C_2H_5O)_2P(O)CH_2CH{=}CHCO_2CH_3$	n-$C_7H_{15}CHO$	CH_3ONa/DMF	n-$C_7H_{15}(CH{=}CH)_2CO_2CH_3$ (30)	54
	$(C_2H_5O)_2P(O)CH(CH_3)CO_2C_2H_5$	p-$CH_3C_6H_4CH_2CHO$	NaH/glyme	p-$CH_3C_6H_4CH_2CH{=}C(CH_3)CO_2C_2H_5$ (—)	338
		m-$CF_3C_6H_4COCH_3$	C_2H_5ONa/C_2H_5OH	m-$CF_3C_6H_4C(CH_3){=}C(CH_3)CO_2C_2H_5$ (75)	337
		m-$CH_3OC_6H_4COCH_3$	C_2H_5ONa/C_2H_5OH	m-$CH_3OC_6H_4C(CH_3){=}C(CH_3)$-$CO_2C_2H_5$ (52)	337
		p-$C_2H_5C_6H_4CH_2CHO$	NaH/glyme	p-$C_2H_5C_6H_4CH_2CH{=}C(CH_3)CO_2C_2H_5$ (—)	338
		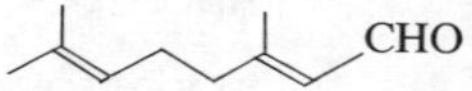 (structure) CHO	$NaNH_2$/THF	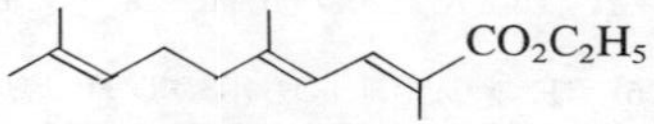 (structure) $CO_2C_2H_5$ (–)	315
		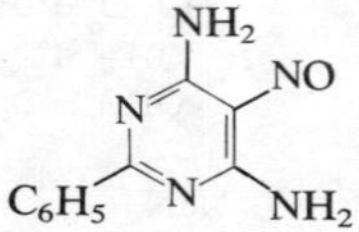 (structure: NH_2, NO, N, N, C_6H_5, NH_2)	NaH/THF	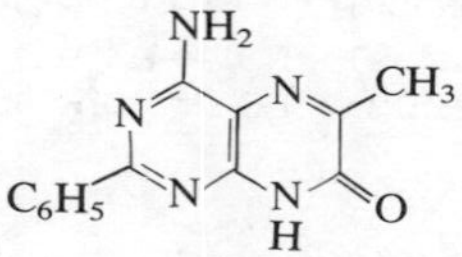 (structure: NH_2, N, N, CH_3, C_6H_5, N, N, H, O) (74)	107
	$(C_2H_5O)_2P(O)CH_2CH{=}CHCO_2CH_3$	n-$C_9H_{19}CHO$	CH_3ONa/DMF	n-$C_9H_{19}(CH{=}CH)_2CO_2CH_3$ (30)	54

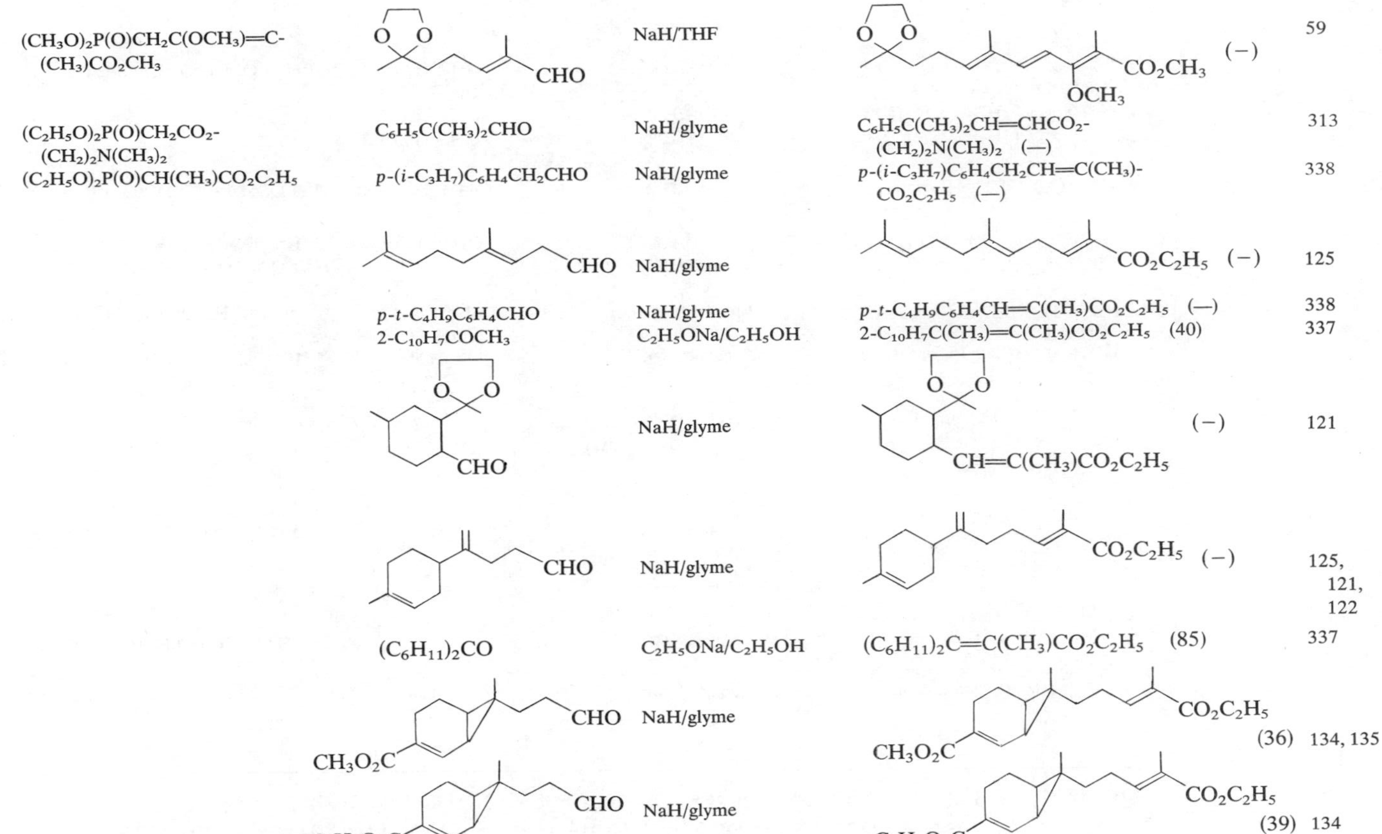

Reagent	Carbonyl Compound	Base/Solvent	Product (Yield, %)	Refs.
$(CH_3O)_2P(O)CH_2C(OCH_3){=}C(CH_3)CO_2CH_3$		NaH/THF	(–)	59
$(C_2H_5O)_2P(O)CH_2CO_2(CH_2)_2N(CH_3)_2$	$C_6H_5C(CH_3)_2CHO$	NaH/glyme	$C_6H_5C(CH_3)_2CH{=}CHCO_2(CH_2)_2N(CH_3)_2$ (—)	313
$(C_2H_5O)_2P(O)CH(CH_3)CO_2C_2H_5$	p-$(i$-$C_3H_7)C_6H_4CH_2CHO$	NaH/glyme	p-$(i$-$C_3H_7)C_6H_4CH_2CH{=}C(CH_3)CO_2C_2H_5$ (—)	338
		NaH/glyme	(–)	125
	p-t-$C_4H_9C_6H_4CHO$	NaH/glyme	p-t-$C_4H_9C_6H_4CH{=}C(CH_3)CO_2C_2H_5$ (—)	338
	2-$C_{10}H_7COCH_3$	C_2H_5ONa/C_2H_5OH	2-$C_{10}H_7C(CH_3){=}C(CH_3)CO_2C_2H_5$ (40)	337
		NaH/glyme	(–)	121
		NaH/glyme	(–)	125, 121, 122
	$(C_6H_{11})_2CO$	C_2H_5ONa/C_2H_5OH	$(C_6H_{11})_2C{=}C(CH_3)CO_2C_2H_5$ (85)	337
		NaH/glyme	(36)	134, 135
		NaH/glyme	(39)	134

Note: References 228–415 are on pp. 250–253.

TABLE III. Unsaturated Esters from Phosphoryl Reagents Containing a Carboxylic Group

$R_2P(O)CHR^1CO_2R^2 + R^3R^4CO \rightarrow R^3R^4C{=}CR^1CO_2R^2 + R_2PO_2^-$ (*Continued*)

No. of C Atoms	Reagent	Reactant	Base/Solvent	Product(s) and Yield(s) (%)	Refs
C_9 (*Contd.*)	$(C_2H_5O)_2P(O)CH(CH_3)CO_2C_2H_5$ (*Contd.*)	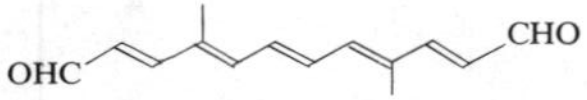 OHC … CHO	CH_3ONa/CH_3OH	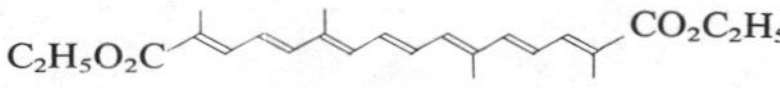 $C_2H_5O_2C$ … $CO_2C_2H_5$ (—)	323
	$(C_2H_5O)_2P(O)CH_2CH{=}CHCO_2CH_3$	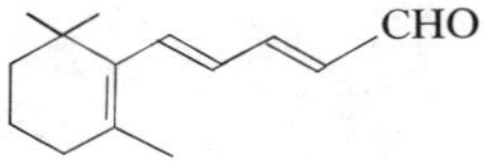 … CHO	$NaNH_2$/THF	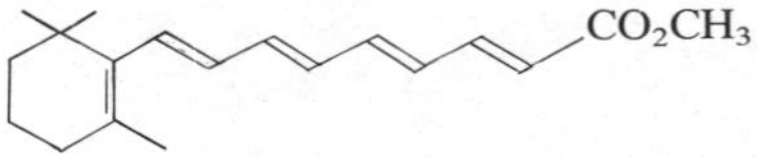 … CO_2CH_3 (80)	99
		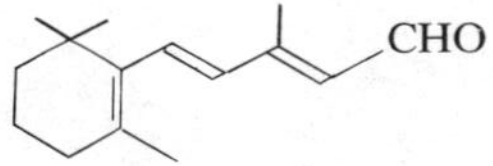 … CHO	$NaNH_2$/THF	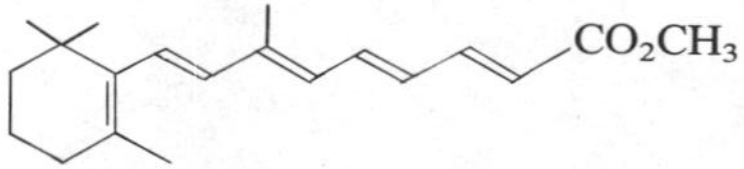 … CO_2CH_3 (80)	99
C_{10}	$(C_2H_5O)_2P(O)CH(C_2H_5)CO_2C_2H_5$	CH_3CHO	NaH/glyme	$CH_3CH{=}C(C_2H_5)CO_2C_2H_5$ (—)	20
		C_2H_5CHO	NaH/glyme	$C_2H_5CH{=}C(C_2H_5)CO_2C_2H_5$ (—)	20
	$(C_2H_5O)_2P(O)CH_2C(CH_3){=}CH$-$CO_2CH_3$	CH_3CH_2CHO	CH_3ONa/DMF	$CH_3CH_2CH{=}CHC(CH_3){=}CHCO_2CH_3$ (83)	55
	$(C_2H_5O)_2P(O)CH_2CH{=}CHCO_2C_2H_5$	$(CH_3)_2CO$	NaH/THF	$(CH_3)_2C{=}CHCH{=}CHCO_2C_2H_5$ (47)	227
	$(C_2H_5O)_2P(O)CH(C_2H_5)CO_2C_2H_5$	i-C_3H_7CHO	NaH/glyme	i-$C_3H_7CH{=}C(C_2H_5)CO_2C_2H_5$ (—)	20
	$(C_2H_5O)_2P(O)CH_2CH{=}CHCO_2C_2H_5$	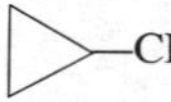 —CHO	NaH/glyme	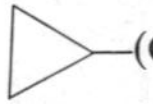 —$(CH{=}CH)_2CO_2C_2H_5$ (–)	285, 288
		i-C_3H_7CHO	NaH/glyme	i-$C_3H_7(CH{=}CH)_2CO_2C_2H_5$ (52)	227
		t-C_4H_9CHO	NaH/glyme	t-$C_4H_9(CH{=}CH)_2CO_2C_2H_5$ (50)	285, 288
		$CH_3CO(CH_2)_3CHO$	NaH/glyme	$CH_3CO(CH_2)_3(CH{=}CH)_2CO_2C_2H_5$ (—)	87
	$(C_2H_5O)_2P(O)CH(C_2H_5)CO_2C_2H_5$	t-C_4H_9CHO	NaH/glyme	t-$C_4H_9CH{=}C(C_2H_5)CO_2C_2H_5$ (—)	20

	O (cyclic ketone)	NaH/DMF	$C(C_2H_5)CO_2C_2H_5$ (10)	39
$(C_2H_5O)_2P(O)CH_2C(CH_3){=}CH$-$CO_2CH_3$	$CH_3CO_2CH_2C(CH_3)_2NO$	NaH/glyme	$CH_3CO_2CH_2C(CH_3)_2N{=}CHC(CH_3){=}CH$-$CO_2CH_3$ (38)	108
	n-$C_5H_{11}CHO$	$[(CH_3)_2CH]_2NLi$/THF	n-$C_5H_{11}CH{=}CHC(CH_3){=}CHCO_2CH_3$ (61)	57
$(C_2H_5O)_2P(O)CH_2(OC_2H_5)CO_2C_2H_5$	$C_2H_5O_2CCO_2C_2H_5$	NaH/dioxane	$C_2H_5O_2CC(OC_2H_5){=}C(OC_2H_5)CO_2C_2H_5$ (—)	100
	$CH_3(CH_2)_4CHO$	NaH/THF	$CH_3(CH_2)_4CH{=}C(OC_2H_5)CO_2C_2H_5$ (81)	380
$C_6H_5(CH_3O)P(O)CH_2CO_2CH_3$	O (cyclic ketone)	NaH/glyme	$CHCO_2CH_3$ (36)	373
$(C_2H_5O)_2P(O)CH(C_2H_5)CO_2C_2H_5$	C_6H_5CHO	NaH/DMF	$C_6H_5CH{=}C(C_2H_5)CO_2C_2H_5$ (30)	39
	p-ClC_6H_4CHO	NaH/DMF	p-$ClC_6H_4CH{=}C(C_2H_5)CO_2C_2H_5$ (75)	39
$(C_2H_5O)_2P(O)CH_2CH{=}CHCO_2C_2H_5$	C_6H_5CHO	NaH/THF	$C_6H_5(CH{=}CH)_2CO_2C_2H_5$ (75)	401, 227
	3,4-$Cl_2C_6H_3CHO$	NaH/glyme	3,4-$Cl_2C_6H_3(CH{=}CH)_2CO_2C_2H_5$ (79)	53
	p-BrC_6H_4CHO	NaH/glyme	p-$BrC_6H_4(CH{=}CH)_2CO_2C_2H_5$ (54)	53
	p-$O_2NC_6H_4CHO$	NaH/glyme	p-$O_2NC_6H_4(CH{=}CH)_2CO_2C_2H_5$ (56)	53
	$CH_3CO(CH_2)_4CHO$	NaH/glyme	$CH_3CO(CH_2)_4(CH{=}CH)_2CO_2C_2H_5$ (—)	87
$(C_2H_5O)_2P(O)CH_2C(CH_3){=}CHCO_2CH_3$	C_6H_5CHO	NaH/glyme	$C_6H_5CH{=}CHC(CH_3){=}CHCO_2CH_3$ (—)	56
$(C_2H_5O)_2P(O)CH_2CH{=}C(CH_3)CO_2CH_3$	C_6H_5CHO	NaH/glyme	$C_6H_5CH{=}CHCH{=}C(CH_3)CO_2CH_3$ (—)	56, 304, 339
$C_6H_5(CH_3O)P(O)CH_2CO_2CH_3$	O (cyclic ketone)	NaH/glyme	$CHCO_2CH_3$ (71)	373
$(C_2H_5O)_2P(O)CH(C_2H_5)CO_2C_2H_5$	p-$CH_3C_6H_4CHO$	NaH/DMF	p-$CH_3C_6H_4CH{=}C(C_2H_5)CO_2C_2H_5$ (48)	39
$(C_2H_5O)_2P(O)CH_2CH{=}CHCO_2C_2H_5$	p-$CH_3OC_6H_4CHO$	NaH/glyme	p-$CH_3OC_6H_4(CH{=}CH)_2CO_2C_2H_5$ (77)	53
	p-NCC_6H_4CHO	NaH/glyme	p-$NCC_6H_4(CH{=}CH)_2CO_2C_2H_5$ (79)	53
	p-$CH_3C_6H_4CHO$	NaH/glyme	p-$CH_3C_6H_4(CH{=}CH)_2CO_2C_2H_5$ (72)	53

Note: References 228–415 are on pp. 250–253.

TABLE III. Unsaturated Esters from Phosphoryl Reagents Containing a Carboxylic Group $R_2P(O)CHR^1CO_2R^2 + R^3R^4CO \rightarrow R^3R^4C{=}CR^1CO_2R^2 + R_2PO_2^+$ (*Continued*)

No. of C Atoms	Reagent	Reactant	Base/Solvent	Product(s) and Yield(s) (%)	Refs.
C_{10} (*Contd.*)	(*Contd.*)	$C_6H_5COCH_3$	NaH/THF	$C_6H_5C(CH_3){=}CHCH{=}CHCO_2C_2H_5$ (52)	227
		i-$C_3H_7(CH_2)_3COCH_3$	NaH/THF	i-$C_3H_7(CH_2)_3C(CH_3){=}CH$-$CH{=}CHCO_2C_2H_5$ (39)	227
		p-$(CH_3)_2NC_6H_4CHO$	NaH/glyme	p-$(CH_3)_2NC_6H_4(CH{=}CH)_2CO_2C_2H_5$ (70)	340
	$(C_2H_5O)_2P(O)CH_2{=}CHCO_2C_2H_5$	CHO	$NaNH_2$/THF	$CO_2C_2H_5$ (–)	315
		CHO	NaH/glyme	$CO_2C_2H_5$ (–)	341, 342
	$(C_2H_5O)_2P(O)CH_2CH{=}CHCO_2C_2H_5$	$CH_3CO(CH_2)_3CH(OC_2H_5)_2$	NaH/glyme	$(C_2H_5O)_2CH(CH_2)_3C(CH_3){=}CHCH{=}CH$-$CO_2C_2H_5$ (—)	87
	$(C_2H_5O)_2P(O)CH_2C(CH_3){=}CHCO_2CH_3$	CHO	NaH/glyme	CO_2CH_3 (70)	55, 56
		CHO	NaH/glyme	CO_2CH_3 (—)	383
		O CHO	NaH/glyme	O CO_2CH_3 (—)	415

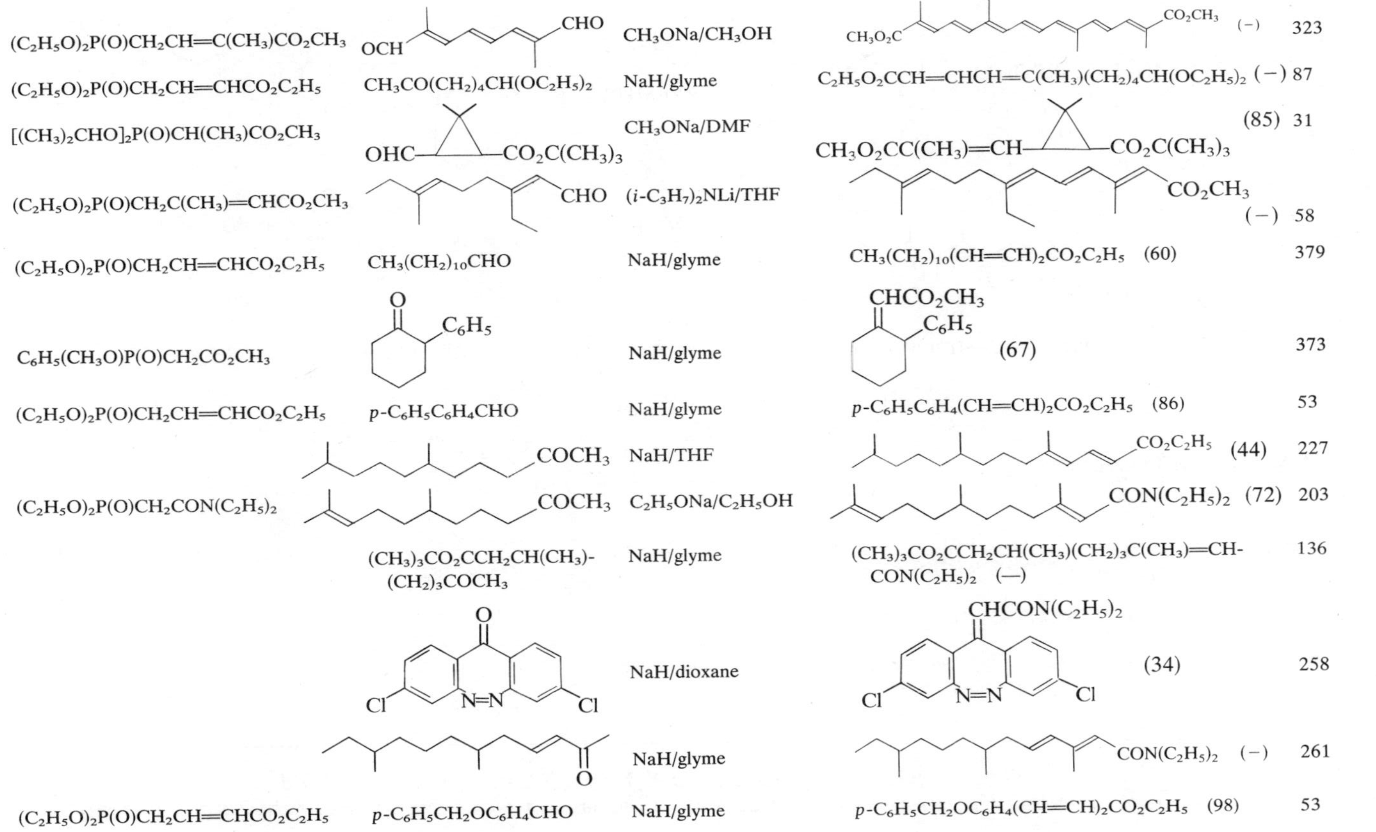

$(C_2H_5O)_2P(O)CH_2CH{=}C(CH_3)CO_2CH_3$		CH_3ONa/CH_3OH	(−)	323
$(C_2H_5O)_2P(O)CH_2CH{=}CHCO_2C_2H_5$	$CH_3CO(CH_2)_4CH(OC_2H_5)_2$	NaH/glyme	$C_2H_5O_2CCH{=}CHCH{=}C(CH_3)(CH_2)_4CH(OC_2H_5)_2$ (−)	87
$[(CH_3)_2CHO]_2P(O)CH(CH_3)CO_2CH_3$		CH_3ONa/DMF	(85)	31
$(C_2H_5O)_2P(O)CH_2C(CH_3){=}CHCO_2CH_3$		$(i\text{-}C_3H_7)_2NLi/THF$	(−)	58
$(C_2H_5O)_2P(O)CH_2CH{=}CHCO_2C_2H_5$	$CH_3(CH_2)_{10}CHO$	NaH/glyme	$CH_3(CH_2)_{10}(CH{=}CH)_2CO_2C_2H_5$ (60)	379
$C_6H_5(CH_3O)P(O)CH_2CO_2CH_3$		NaH/glyme	(67)	373
$(C_2H_5O)_2P(O)CH_2CH{=}CHCO_2C_2H_5$	$p\text{-}C_6H_5C_6H_4CHO$	NaH/glyme	$p\text{-}C_6H_5C_6H_4(CH{=}CH)_2CO_2C_2H_5$ (86)	53
		NaH/THF	(44)	227
$(C_2H_5O)_2P(O)CH_2CON(C_2H_5)_2$		C_2H_5ONa/C_2H_5OH	(72)	203
	$(CH_3)_3CO_2CCH_2CH(CH_3)(CH_2)_3COCH_3$	NaH/glyme	$(CH_3)_3CO_2CCH_2CH(CH_3)(CH_2)_3C(CH_3){=}CHCON(C_2H_5)_2$ (—)	136
		NaH/dioxane	(34)	258
		NaH/glyme	(−)	261
$(C_2H_5O)_2P(O)CH_2CH{=}CHCO_2C_2H_5$	$p\text{-}C_6H_5CH_2OC_6H_4CHO$	NaH/glyme	$p\text{-}C_6H_5CH_2OC_6H_4(CH{=}CH)_2CO_2C_2H_5$ (98)	53

Note: References 228–415 are on pp. 250–253.

TABLE III. Unsaturated Esters from Phosphoryl Reagents Containing a Carboxylic Group

$R_2P(O)CHR^1CO_2R^2 + R^3R^4CO \rightarrow R^3R^4C{=}CR^1CO_2R^2 + R_2PO_2^-$ (*Continued*)

No. of C Atoms	Reagent	Reactant	Base/Solvent	Product(s) and Yield(s) (%)	Refs.
C_{10}	$(C_2H_5O)_2P(O)CH_2CH{=}CHCO_2C_2H_5$ (*Contd.*)	$CH_3(CH_2)_{12}CHO$	NaH/glyme	$CH_3(CH_2)_{12}(CH{=}CH)_2CO_2C_2H_5$ (60)	379
	$(C_2H_5O)_2P(O)CH_2C(CH_3){=}CHCO_2CH_3$	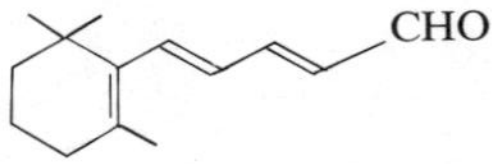 CHO	$NaNH_2$/THF	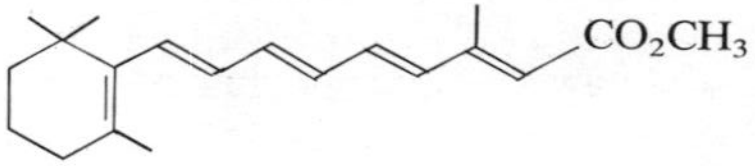 CO_2CH_3 (−)	99
		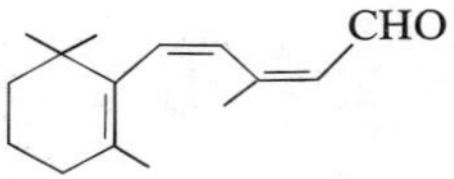 CHO	NaH/glyme	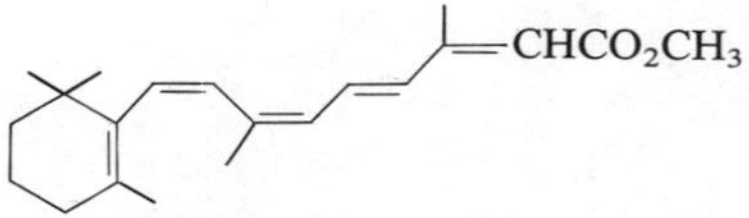 $=CHCO_2CH_3$ (71)	374
	$(C_2H_5O)_2P(O)CH_2CH{=}CHCO_2C_2H_5$	$CH_3(CH_2)_{14}CHO$	NaH/glyme	$CH_3(CH_2)_{14}(CH{=}CH)_2CO_2C_2H_5$ (60)	379
	$(C_2H_5O)_2P(O)CH(OC_2H_5)CO_2C_2H_5$	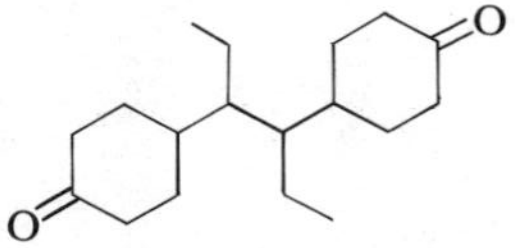 O, O	NaH/glyme	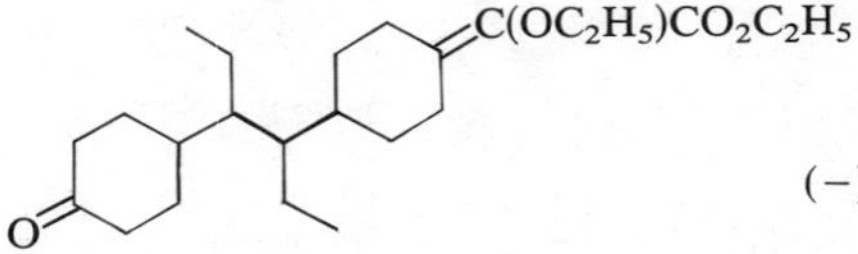 $C(OC_2H_5)CO_2C_2H_5$, O (−)	365
		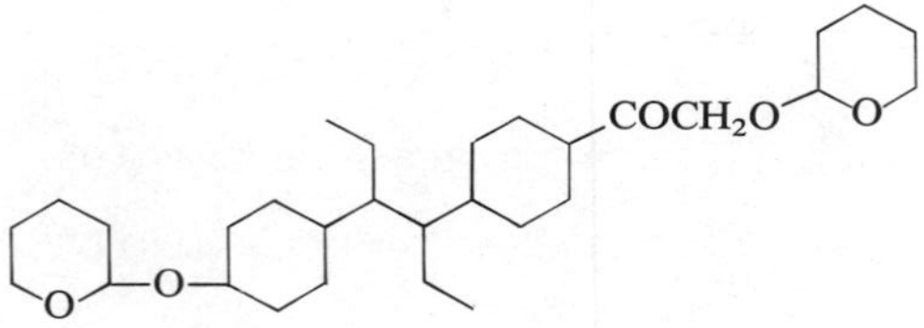 $COCH_2O$, O, O	NaH/glyme	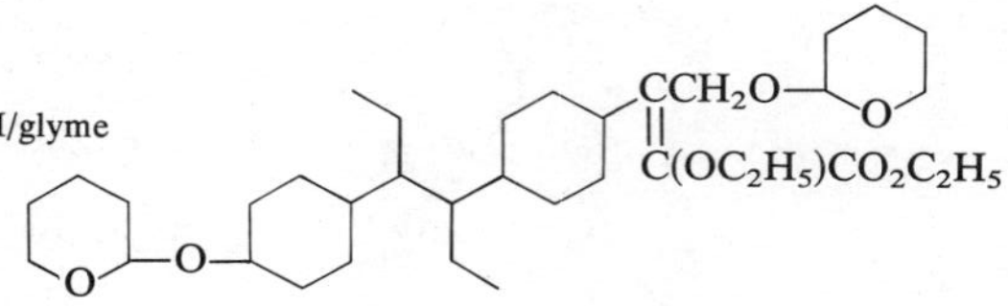 CCH_2O, $C(OC_2H_5)CO_2C_2H_5$, O, O (100)	365

	Phosphonate	Carbonyl Component	Base/Solvent	Product (Yield, %)	Refs.
	$(C_2H_5O)_2P(O)CH_2CO_2C(CH_3)_3$		NaH/glyme	(78)	399
C_{11}	$(C_2H_5O)_2P(O)CH_2(C_4H_9\text{-}n)CO_2C_2H_5$	CH_2O	NaH/glyme	$CH_2{=}C(C_4H_9\text{-}n)CO_2C_2H_5$ (60)	4
	$(C_2H_5O)_2P(O)CH_2C(CH_3){=}CHCO_2C_2H_5$	C_2H_5CHO	NaH/glyme	$C_2H_5CH{=}CHC(CH_3){=}CHCO_2C_2H_5$ (—)	56, 339
		$(CH_3)_2CO$	$NaNH_2$/ether	$(CH_3)_2C{=}CHC(CH_3){=}CHCO_2C_2H_5$ (70)	165
	$(C_2H_5O)_2P(O)CH_2CON$ (piperidine)	$C_2H_5O_2CCO_2C_2H_5$	NaH/ether	$C_2H_5O_2CC(OC_2H_5){=}CHCON$ (piperidine) (57)	100
	$(C_2H_5O)_2P(O)CH_2C(CH_3){=}CHCO_2C_2H_5$	C_6H_5CHO	NaH/glyme	$C_6H_5CH{=}CHC(CH_3){=}CHCO_2C_2H_5$ (—)	56, 339
	$(C_2H_5O)_2P(O)CH(CO_2C_2H_5)_2$	C_6H_5CHO	NaH/DMF	$C_6H_5CH{=}C(CO_2C_2H_5)_2$ (30)	39
			NaH/DMF	(−)	307
	$(C_2H_5O)_2P(O)CH_2C(CH_3){=}CH$-$CO_2C_2H_5$		NaH/glyme	(−)	56, 339
			C_2H_5ONa/DMF	(−)	343
			NaH/glyme	(−)	261

Note: References 228–415 are on pp. 250–253.

TABLE III. UNSATURATED ESTERS FROM PHOSPHORYL REAGENTS CONTAINING A CARBOXYLIC GROUP

$R_2P(O)CHR^1CO_2R^2 + R^3R^4CO \rightarrow R^3R^4C{=}CR^1CO_2R^2 + R_2PO_2^-$ (*Continued*)

No. of C Atoms	Reagent	Reactant	Base/Solvent	Product(s) and Yield(s) (%)	Refs.
C_{11} (*Contd.*)	$(C_2H_5O)_2P(O)CH_2COC_5H_{11}$-*n* (*Contd.*)	$(CH_3O)_2CHCH_2CH(NO_2)CH[(CH_2)_6CN]CHO$	NaH/glyme	$(CH_3O)_2CHCH_2CH(NO_2)CH[(CH_2)_6CN]CH{=}CHCOC_5H_{11}$-*n* (80)	151
	$(CH_3O)_2P(O)CH_2COCH_2OC_6H_5$	bicyclic lactone with *p*-$C_6H_5C_6H_4CO_2$ and CHO substituents	NaH/glyme	bicyclic lactone with *p*-$C_6H_5C_6H_4CO_2$ and $CH{=}CHCOCH_2OC_6H_5$ substituents (—)	344
C_{12}	$(C_2H_5O)_2P(O)CH(C_4H_9$-*n*$)$-$CO_2C_2H_5$	CH_2O	NaH/glyme	$CH_2{=}C(C_4H_9$-*n*$)CO_2C_2H_5$ (60)	4
	$(C_2H_5O)_2P(O)CH_2(CH{=}CH)_2$-$CO_2C_2H_5$	2-pyridinecarboxaldehyde (C_5H_4N-2-CHO)	$NaNH_2$/THF	2-$C_5H_4N(CH{=}CH)_3CO_2C_2H_5$ (41)	233
	$(C_2H_5O)_2P(O)CH_2CONHC_6H_5$	C_6H_5CHO	NaH/glyme	$C_6H_5CH{=}CHCONHC_6H_5$ (75)	39
	$(C_2H_5O)_2P(O)CH(NHCH_3)CON(CH_2)_5$ (piperidide)	C_6H_5CHO	NaH/glyme	$C_6H_5CH{=}C(NHCH_3)CON(CH_2)_5$ (piperidide) (−)	41
	$(C_2H_5O)_2P(O)CH(NC_5H_{10})CO_2CH_3$ (piperidino)	C_6H_5CHO	NaH/glyme	$C_6H_5CH{=}C(NC_5H_{10})CO_2CH_3$ (piperidino) (−)	41

	$(C_2H_5O)_2P(O)CH_2CONHC_6H_5$	$C_6H_5COCH_3$	NaH/glyme	$C_6H_5C(CH_3){=}CHCONHC_6H_5$ (60)	39
	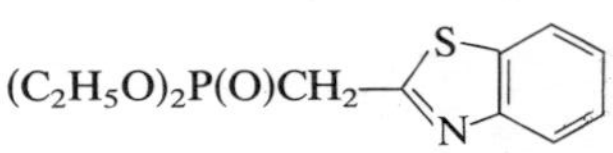 $(C_2H_5O)_2P(O)CH_2$–[benzothiazol-2-yl structure]	$C_2H_5O_2C(CH{=}CH)_2CHO$	$NaNH_2$/THF	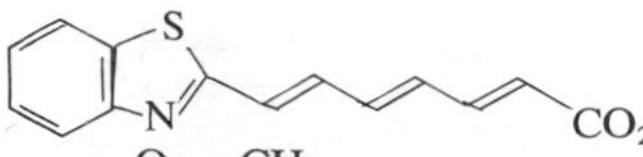 [benzothiazol-2-yl structure]$(CH{=}CH)_3CO_2C_2H_5$ (44)	233
	$(C_2H_5O)_2P(O)CH_2(CH{=}CH)_2CO_2C_2H_5$	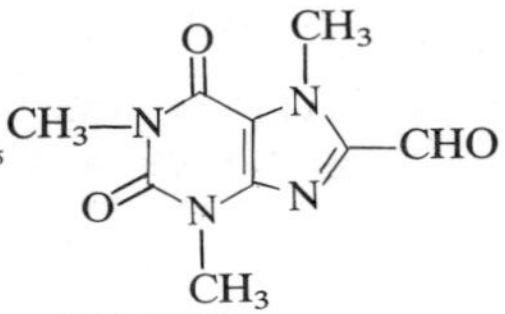 [structure: 1,3,7-trimethylxanthine ring with CH_3, O, N labels]–CHO	$NaNH_2$/THF	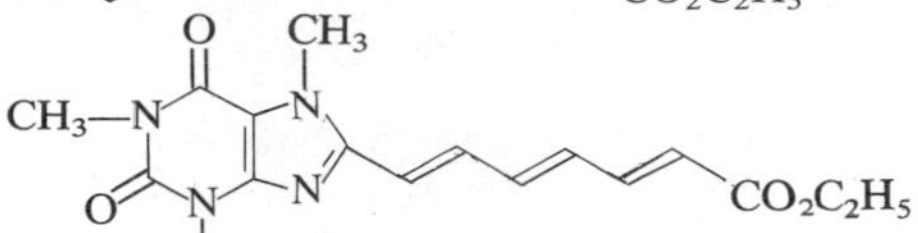 [structure: 1,3,7-trimethylxanthine ring]$(CH{=}CH)_3CO_2C_2H_5$ (6)	233
C_{13}	$(C_2H_5O)_2P(O)CH(CO_2C_2H_5)$-$(CH_2)_2CO_2C_2H_5$	C_2H_5CHO	NaH/THF	$C_2H_5CH{=}C(CO_2C_2H_5)(CH_2)_2$-$CO_2C_2H_5$ (53)	345
		$(CH_3)_2CO$	NaH/THF	$(CH_3)_2C{=}C(CO_2C_2H_5)(CH_2)_2CO_2C_2H_5$ (71)	345
	$(C_2H_5O)_2P(O)CH(NHC_6H_5)CO_2CH_3$	i-C_3H_7CHO	NaH/glyme	i-$C_3H_7CH{=}C(NHC_6H_5)CO_2CH_3$ (—)	41
		p-$O_2NC_6H_4CHO$	NaH/glyme	p-$O_2NC_6H_4CH{=}C(NHC_6H_5)CO_2CH_3$ (—)	41
	$(C_2H_5O)_2P(O)CH(NHCH_3)CONHC_6H_5$	C_6H_5CHO	NaH/glyme	$C_6H_5CH{=}C(NHCH_3)CONHC_6H_5$ (—)	41
	$(C_2H_5O)_2P(O)CH_2CON(CH_3)C_6H_5$	C_6H_5CHO	NaH/DMF	$C_6H_5CH{=}CHCON(CH_3)C_6H_5$ (60)	39
	$(C_2H_5O)_2P(O)CH(C_6H_5)CO_2CH_3$	C_6H_5CHO	NaH/DMF	$C_6H_5CH{=}C(C_6H_5)CO_2CH_3$ (81)	39
	$(C_2H_5O)_2P(O)CH(CO_2C_2H_5)$-$(CH_2)_2CO_2C_2H_5$	C_6H_5CHO	NaH/THF	$C_6H_5CH{=}C(CO_2C_2H_5)(CH_2)_2CO_2C_2H_5$ (84)	345
		p-$CH_3C_6H_4OCH_2CHO$	NaH/THF	p-$CH_3C_6H_4OCH_2CH{=}C(CO_2C_2H_5)$-$(CH_2)_2CO_2C_2H_5$ (70)	345
C_{14}	$(C_2H_5O)_2P(O)CH(CO_2C_2H_5)$-$(CH_2)_3CO_2C_2H_5$	C_2H_5CHO	NaH/THF	$C_2H_5CH{=}C(CO_2C_2H_5)(CH_2)_3CO_2C_2H_5$ (54)	345
		$(CH_3)_2CO$	NaH/THF	$(CH_3)_2C{=}C(CO_2C_2H_5)(CH_2)_3CO_2C_2H_5$ (10)	345
	$(C_2H_5O)_2P(O)CH(CO_2C_2H_5)$-$(CH_2)_3CO_2C_2H_5$	C_6H_5CHO	NaH/THF	$C_6H_5CH{=}C(CO_2C_2H_5)(CH_2)_3CO_2C_2H_5$ (86)	345
	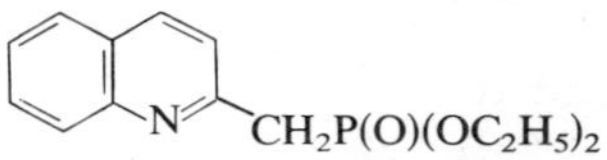 [quinolin-2-yl structure]$CH_2P(O)(OC_2H_5)_2$	$C_2H_5O_2C(CH{=}CH)_2CHO$	$NaNH_2$/THF	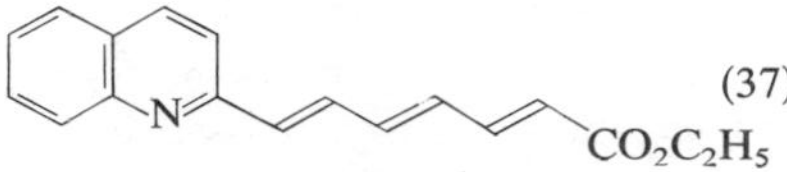 [quinolin-2-yl structure]$(CH{=}CH)_3CO_2C_2H_5$ (37)	233

Note: References 228–415 are on pp. 250–253.

TABLE III. UNSATURATED ESTERS FROM PHOSPHORYL REAGENTS CONTAINING A CARBOXYLIC GROUP
$R_2P(O)CHR^1CO_2R^2 + R^3R^4CO \rightarrow R^3R^4C{=}CR^1CO_2R^2 + R_2PO_2^-$ (*Continued*)

No. of C Atoms	Reagent	Reactant	Base/Solvent	Product(s) and Yield(s) (%)	Refs.
C_{14} (*Contd.*)	$(C_2H_5O)_2P(O)CH(CO_2C_2H_5)$-$(CH_2)_3CO_2C_2H_5$	p-$CH_3C_6H_4OCH_2CHO$	NaH/THF	p-$CH_3C_6H_4OCH_2CH{=}C(CO_2C_2H_5)$-$(CH_2)_3CO_2C_2H_5$ (77)	345
C_{15}	$(C_2H_5O)_2P(O)$–$CH(CO_2CH_2C_6H_4OCH_3$-$p)NHCSH$	$CH_3CO_2CH_2COCH_2Cl$	NaH/glyme	S, N; $CH_2O_2CCH_3$; $CO_2CH_2C_6H_4OCH_3$-p (—)	371
	$(C_2H_5O)_2P(O)CH_2CO_2(CH_2)_2$-$N(CH_3)C_6H_5$	$C_6H_5C(CH_3)_2CHO$	NaH/glyme	$C_6H_5C(CH_3)_2CH{=}CHCO_2(CH_2)_2$-$N(CH_3)C_6H_5$ (—)	313
C_{16}	$(C_6H_5CH_2O)_2P(O)CH_2CO_2H$	O	$(i$-$C_3H_7)_2NLi$/THF	$CHCO_2H$ (66)	192
		C_6H_5CHO	$(i$-$C_3H_7)_2NLi$/THF	$C_6H_5CH{=}CHCO_2H$ (70)	192
C_{18}	$(C_2H_5O)_2P(O)CH(CO_2C_2H_5)(CH_2)_2C_6H_5$	p-$O_2NC_6H_4CHO$	NaH/DMF	p-$O_2NC_6H_4CH{=}C(CO_2C_2H_5)(CH_2)_2C_6H_5$ (19)	158
	$(C_2H_5O)_2P(O)CH(CO_2C_2H_5)$-$(CH_2)_7CO_2C_2H_5$	CH_3CH_2CHO	NaH/THF	$CH_3CH_2CH{=}C(CO_2C_2H_5)(CH_2)_7CO_2C_2H_5$ (56)	345
		$(CH_3)_2CO$	NaH/THF	$(CH_3)_2C{=}C(CO_2C_2H_5)(CH_2)_7CO_2C_2H_5$ (35)	345
		C_6H_5CHO	NaH/THF	$C_6H_5CH{=}C(CO_2C_2H_5)(CH_2)_7CO_2C_2H_5$ (79)	345
		p-$CH_3C_6H_4OCH_2CHO$	NaH/THF	p-$CH_3C_6H_4OCH_2CH{=}C(CO_2C_2H_5)$-$(CH_2)_7CO_2C_2H_5$ (61)	345

	Reactant	Carbonyl Compound	Conditions	Product(s) and Yield(s) (%)	Refs.
C_{20}	CH_3, O, N, Cl, N, C_6H_4Cl-*o*, $-P(O)(OC_2H_5)_2$	CH_2O	NaH/glyme	CH_3, O, N, Cl, N, C_6H_4Cl-*o*, $=CH_2$ (52)	392
	O_2N, O, O, $P(O)(OC_6H_5)_2$	C_6H_5NO	NaH/DMF	O_2N, O, NC_6H_5, O (−)	346
		C_6H_5CHO	NaH/DMF	O_2N, O, O, CHC_6H_5 (−)	346
C_{22}	N_3, O, N, O, O_2CCH_3, $P(O)(OC_2H_5)_2$, $CO_2CH_2C_6H_5$	—	Na/glyme	N_3, O, N, O_2CCH_3, $CO_2CH_2C_6H_5$ (−)	68, 69
	$(C_2H_5O)_2P(O)CH(C_{14}H_{29}$-*n*$)CO_2C_2H_5$	O, $\overset{+}{N}$, O^-	NaH/glyme	$C(C_{14}H_{29}$-*n*$)CO_2C_2H_5$, $\overset{+}{N}$, O^- (25)	389

Note: References 228–415 are on pp. 250–253.

TABLE IV. Unsaturated Ketones from Phosphoryl Reagents Containing a Carbonyl Group

$$R_2P(O)CHR^1COR^2 + R^3R^4CO \rightarrow R^3R^4C{=}CR^1COR^2 + R_2PO_2^-$$

No. of C Atoms	Reagent	Reactant	Base/Solvent	Product(s) and Yield(s) (%)	Refs.
C_7	$(C_2H_5O)_2P(O)CH_2COCH_3$	O (cyclopentanone ring)	NaH/glyme	CHCOCH₃ on cyclopentane ring (16)	182
		$CH_2{=}CH(CH_2)_2CHO$	CH_3ONa/CH_3OH	$CH_2{=}CH(CH_2)_2CH{=}CHCOCH_3$ (—)	275
		$(CH_3O)_2P(O)OCH(CH_3)CHO$	NaH/glyme	$CH_3COCH{=}CHCH(CH_3)OP(O)(OCH_3)_2$ (72)	385
		$CH_3CO(CH_2)_3CHO$	NaH/glyme	$CH_3CO(CH_2)_3CH{=}CHCOCH_3$ (—)	87
		O (cyclohexanone ring)	NaH/glyme	CHCOCH₃ on cyclohexane ring (70)	182
		$(CH_3O)_2P(O)OCH(C_2H_5)CHO$	NaH/glyme	$CH_3COCH{=}CHCH(C_2H_5)OP(O)(OCH_3)_2$ (75)	385
		$CH_3CO(CH_2)_4CHO$	NaH/glyme	$CH_3CO(CH_2)_4CH{=}CHCOCH_3$ (—)	87
		n-$C_5H_{11}COCH_3$	NaH/glyme	n-$C_5H_{11}C(CH_3){=}CHCOCH_3$ (61)	182
		O (2-methylcyclohexanone ring)	NaH/glyme	CHCOCH₃ on 2-methylcyclohexane ring (40)	182
		C_6H_5CHO	$NaNH_2/THF$	$C_6H_5CH{=}CHCOCH_3$ (—)	250
		$(CH_3O)_2P(O)OCH(C_3H_7\text{-}n)CHO$	NaH/glyme	$CH_3COCH{=}CHCH(C_3H_7\text{-}n)OP(O)(OCH_3)_2$ (50)	385
		$C_6H_5COCH_3$	NaH/glyme	$C_6H_5C(CH_3){=}CHCOCH_3$ (46)	182
		$(CH_3)_2C{=}CH(CH_2)_2COCH_3$	NaH/glyme	$(CH_3)_2C{=}CH(CH_2)_2C(CH_3){=}CHCOCH_3$ (80)	182
		i-$C_5H_{11}CH(CH_3)CH_2CHO$	NaH/glyme	i-$C_5H_{11}CH(CH_3)CH_2CH{=}CHCOCH_3$ (—)	318

	Phosphonate	Carbonyl compound	Base/solvent	Product (yield, %)	Refs.
		$i\text{-}C_6H_{13}CH(CH_3)CH_2CHO$	NaOH/DMF	$i\text{-}C_6H_{13}CH(CH_3)CH_2CH{=}CHCOCH_3$ (—)	347
		$CH_3CO(CH_2)_3CH(OC_2H_5)_2$	NaH/glyme	$CH_3COCH{=}C(CH_3)(CH_2)_3CH(OC_2H_5)_2$ (—)	87
		$(CH_3)_2C{=}CH(CH_2)_2C(CH_3){=}CHCHO$	$NaNH_2$/THF	$(CH_3)_2C{=}CH(CH_2)_2C(CH_3){=}CHCH{=}CHCOCH_3$ (49)	18
		$CH_3CO(CH_2)_4CH(OC_2H_5)_2$	NaH/glyme	$CH_3COCH{=}C(CH_3)(CH_2)_4CH(OC_2H_5)_2$ (—)	87
		$OHCC(CH_3)_2CH_2OSO_2C_6H_4CH_3\text{-}p$	NaH/glyme	$CH_3COCH{=}CHC(CH_3)_2CH_2OSO_2C_6H_4CH_3\text{-}p$ (—)	348
	$(CH_3O)_2P(O)CH_2CO(CH_2)_2OCH_3$	$p\text{-}C_6H_5C_6H_4CO_2$– (lactone, O, =O), CHO	NaH/glyme	$p\text{-}C_6H_5C_6H_4CO_2$– (lactone, O, =O), $CH{=}CHCO(CH_2)_2OCH_3$ (—)	368
C_8	$(C_2H_5O)_2P(O)CH(CH_3)COCH_3$	cyclohexanone (O)	NaH/glyme	$C(CH_3)COCH_3$ (=cyclohexylidene) (60)	182
		$(CH_3)_2C{=}CH(CH_2)_2C(CH_3){=}CHCHO$	$NaNH_2$/THF	$(CH_3)_2C{=}CH(CH_2)_2C(CH_3){=}CHCH{=}C(CH_3)COCH_3$ (63)	18, 315
	$(C_2H_5O)_2P(O)CH_2COCH_2CH_3$	$(CH_3)_2C{=}CH(CH_2)_2C(CH_3){=}CHCHO$	$NaNH_2$/THF	$(CH_3)_2C{=}CH(CH_2)_2C(CH_3){=}CHCH{=}CHCOC_2H_5$ (75)	18
		cyclopentanone (O), $(CH_2)_4CO_2CH_3$, CHO	NaH/glyme	cyclopentanone (O), $(CH_2)_4CO_2CH_3$, $CH{=}CHCOCH_2CH_3$ (–)	149
		cyclopentanone (O), $CH_2CH{=}CH(CH_2)_2CO_2CH_3$, CHO	NaH/glyme	cyclopentanone (O), $CH_2CH{=}CH(CH_2)_2CO_2CH_3$, $CH{=}CHCOCH_2CH_3$ (–)	149

Note: References 228–415 are on pp. 250–253.

TABLE IV. Unsaturated Ketones from Phosphoryl Reagents Containing a Carbonyl Group

$$R_2P(O)CHR^1COR^2 + R^3R^4CO \rightarrow R^3R^4C{=}CR^1COR^2 + R_2PO_2^-$$ (*Continued*)

No. of C Atoms	Reagent	Reactant	Base/Solvent	Product(s) and Yield(s) (%)	Refs.
C_8 (*Contd.*)	$(C_2H_5O)_2P(O)CH_2COCH_2CH_3$ (*Contd.*)	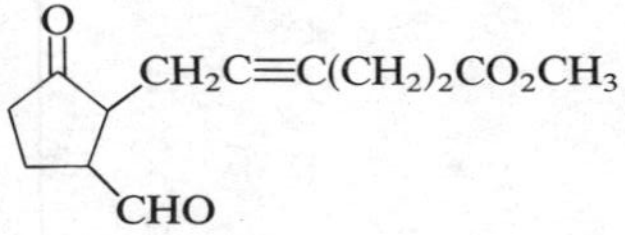 2-oxocyclopentane bearing $CH_2C{\equiv}C(CH_2)_2CO_2CH_3$ and CHO	NaH/glyme	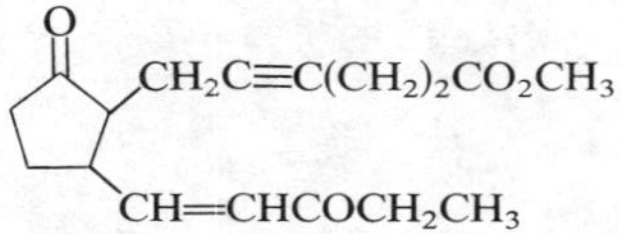 2-oxocyclopentane bearing $CH_2C{\equiv}C(CH_2)_2CO_2CH_3$ and $CH{=}CHCOCH_2CH_3$ (−)	149
	$(CH_3O)_2P(O)CH_2CO(CH_2)_3OCH_3$	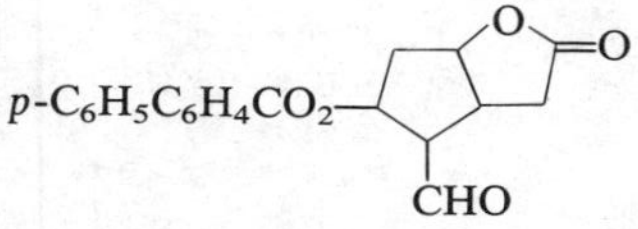 p-$C_6H_5C_6H_4CO_2$– bicyclic lactone (O, =O) bearing CHO	NaH/glyme	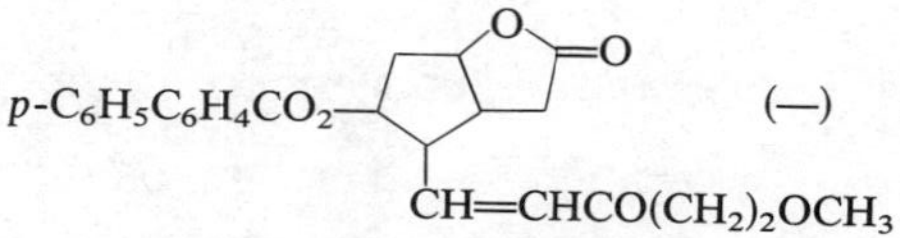 p-$C_6H_5C_6H_4CO_2$– bicyclic lactone (O, =O) bearing $CH{=}CHCO(CH_2)_2OCH_3$ (—)	368
C_9	$(C_2H_5O)_2P(O)CH(C_2H_5)COCH_3$	$(CH_3)_2C{=}CH(CH_2)_2C$-$(CH_3){=}CHCHO$	$NaNH_2$/DMF	$(CH_3)_2C{=}CH(CH_2)_2C(CH_3){=}CHCH{=}C$-$(C_2H_5)COCH_3$ (18)	315
	$(CH_3)_2P(O)CH_2CO(CH_2)_4CH_3$	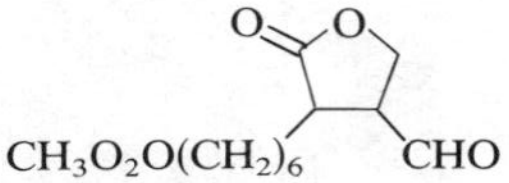 γ-lactone (O=, O) bearing $CH_3O_2O(CH_2)_6$ and CHO	NaH/glyme	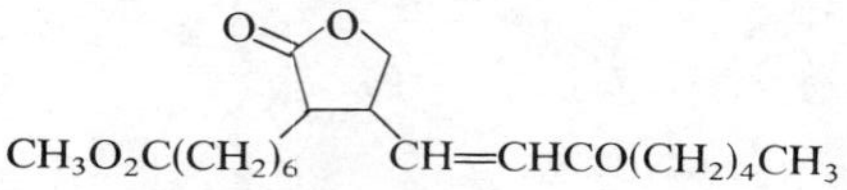 γ-lactone (O=, O) bearing $CH_3O_2C(CH_2)_6$ and $CH{=}CHCO(CH_2)_4CH_3$ (−)	150
C_{10}	$(CH_3O)_2P(O)CH_2COCH$-$(CH_3)C_4H_9$-n	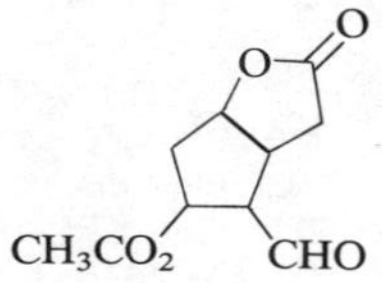 bicyclic lactone (O, =O) bearing CH_3CO_2 and CHO	NaH/glyme	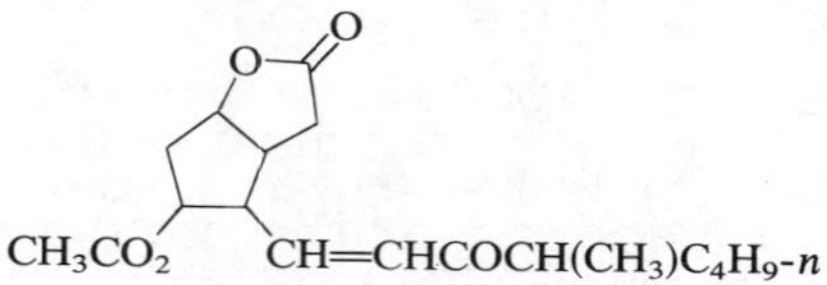 bicyclic lactone (O, =O) bearing CH_3CO_2 and $CH{=}CHCOCH(CH_3)C_4H_9$-n (−)	148

	Phosphonate	Carbonyl compound	Conditions	Product(s) and Yield(s) (%)	Refs.
		Bicyclic lactone with CH_3CO_2 and CHO substituents	NaH/THF	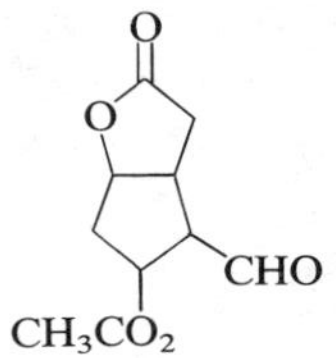 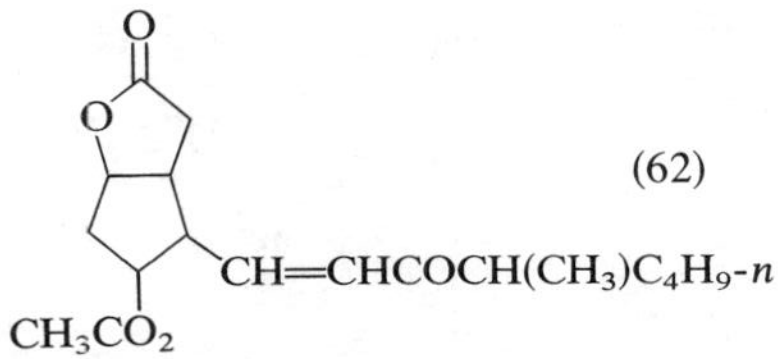 Bicyclic lactone with CH_3CO_2 and $CH{=}CHCOCH(CH_3)C_4H_9\text{-}n$ substituents (62)	158
C_{11}	$(CH_3O)_2P(O)CH_2CO(CH_2)_6CH_3$	$i\text{-}C_3H_7CHO$	NaH/glyme	$i\text{-}C_3H_7CH{=}CHCO(CH_2)_6CH_3$ (30)	405
	$(C_2H_5O)_2P(O)CH_2COC_5H_{11}\text{-}n$ 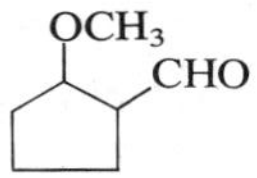 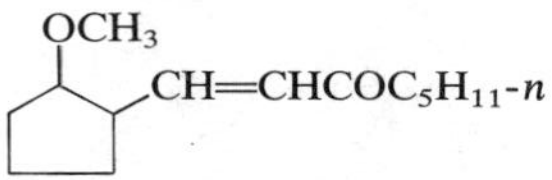	Cyclopentane with OCH_3 and CHO	NaH/DMSO	Cyclopentane with OCH_3 and $CH{=}CHCOC_5H_{11}\text{-}n$ (55)	349
	$(C_2H_5O)_2P(O)$-substituted 4-CH_3-cyclohexanone	Pyrimidine with NH_2, NO, NH_2, C_6H_5	NaH/THF	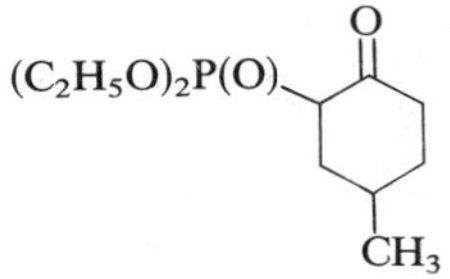 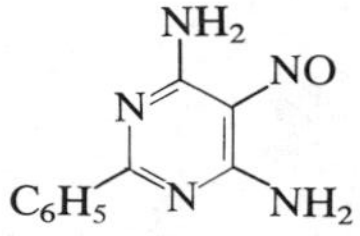 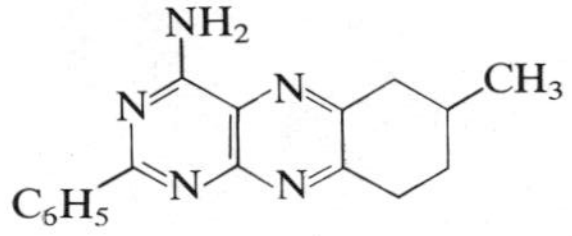 Tetrahydropteridine with NH_2, C_6H_5, CH_3 (32)	107, 254
	$(C_2H_5O)_2P(O)CH_2COC_5H_{11}\text{-}n$ 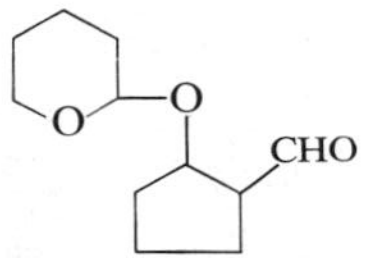 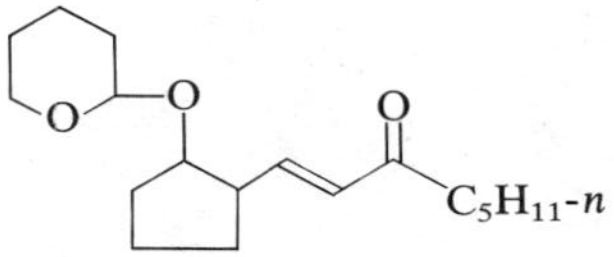	Cyclopentane with tetrahydropyranyloxy and CHO	NaH/DMSO	Cyclopentane with tetrahydropyranyloxy and $CH{=}CHCOC_5H_{11}\text{-}n$ (64)	349
		Cyclopentane ketal with OCH, $(CH_2)_6CO_2CH_3$, CH_3CO_2	NaH/glyme	Cyclopentane ketal with $CH_3(CH_2)_4COCH{=}CH$, $(CH_2)_6CO_2CH_3$, CH_3CO_2 (—)	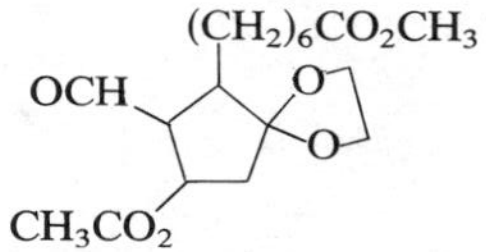 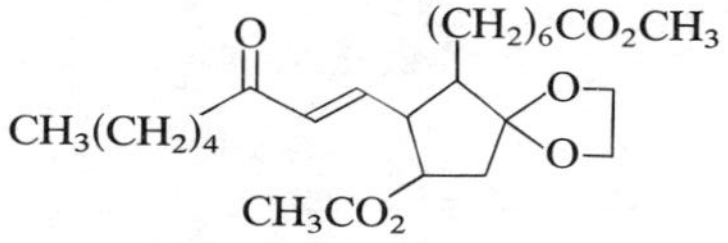 144, 145

Note: References 228–415 are on pp. 250–253.

TABLE IV. UNSATURATED KETONES FROM PHOSPHORYL REAGENTS CONTAINING A CARBONYL GROUP

$R_2P(O)CHR^1COR^2 + R^3R^4CO \rightarrow R^3R^4C{=}CR^1COR^2 + R_2PO_2^-$ (*Continued*)

No. of C Atoms	Reagent	Reactant	Base/Solvent	Product(s) and Yield(s) (%)	Refs.
C_{11} (*Contd.*)	$(CH_3O)_2P(O)CH_2COCH_2OC_6H_5$	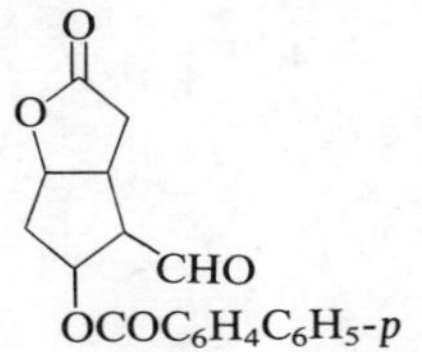		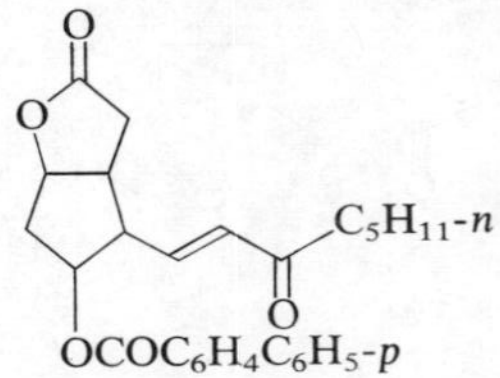	
		O, O, CHO, $OCOC_6H_4C_6H_5$-*p*	NaH/glyme	O, O, C_5H_{11}-*n*, O, $OCOC_6H_4C_6H_5$-*p* (−)	147
		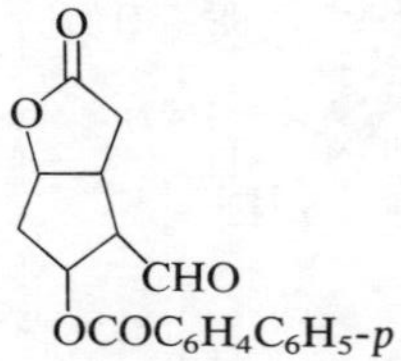 O, O, CHO, $OCOC_6H_4C_6H_5$-*p*	NaH/glyme	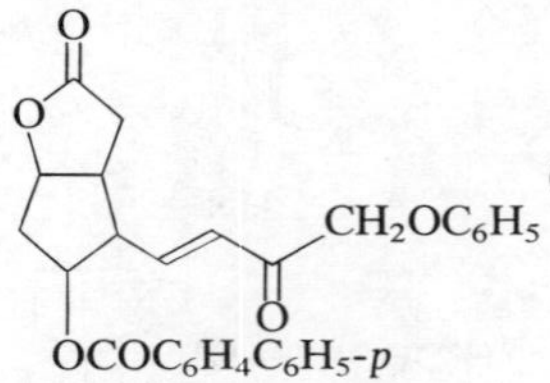 O, O, $CH_2OC_6H_5$, O, $OCOC_6H_4C_6H_5$-*p* (−)	344
C_{12}	$(C_2H_5O)_2P(O)CH_2COC_6H_5$	*p*-$ClC_6H_4N_3$	C_2H_5ONa/C_2H_5OH	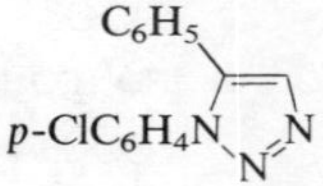 C_6H_5, *p*-ClC_6H_4N, N, N (36)	221
		p-$O_2NC_6H_4N_3$	C_2H_5ONa/C_2H_5OH	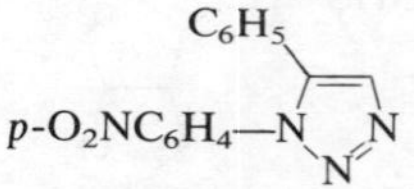 C_6H_5, *p*-$O_2NC_6H_4$—N, N, N (64)	221
		C_6H_5CHO	NaH/glyme	$C_6H_5CH{=}CHCOC_6H_5$ (61)	4

			NaH/THF	(75)	107, 254
			NaH/dioxane	(10)	258
C_{13}	$(C_2H_5O)_2P(O)CH_2COC_6H_4CH_3$-*p*	*p*-$C_6H_4(CHO)_2$	C_2H_5ONa/C_2H_5OH	*p*-$C_6H_4(CH{=}CHCOC_6H_4CH_3$-*p*$)_2$ (—)	88
		p-$C_6H_4(CH{=}CHCHO)_2$	C_2H_5ONa/C_2H_5OH	*p*-$C_6H_4[(CH{=}CH)_2COC_6H_4CH_3$-*p*$]_2$ (—)	88
C_{14}	$(CH_3O)_2P(O)CH_2$	CHO	NaH/glyme	(95)	350
C_{19}	$(CH_3O)_2P(O)$	CHO	NaH/glyme	(78)	405

Note: References 228–415 are on pp. 250–253.

TABLE V. UNSATURATED NITROGEN COMPOUNDS FROM PHOSPHORYL REAGENTS CONTAINING A NITROGEN GROUP

No. of C Atoms	Reagent	Reactant	Base/Solvent	Product(s) and Yield(s) (%)	Refs.
C_5	$(C_2H_5O)_2P(O)NHOCH_3$	C_6H_5CHO	NaH/glyme	$C_6H_5CH{=}NOCH_3$ (82)	351, 352
C_6	$(C_2H_5O)_2P(O)NHN(CH_3)_2$	CO_2	NaH/C_6H_6	$(CH_3)_2NNCO$ (32)	81
	$(C_2H_5O)_2P(O)CH_2NC$	t-C_4H_9CHO	n-C_4H_9Li/THF	$(CH_3)_3CCH{=}CHNC$ (72)	353
		$(CH_2)_5C{=}O$	n-C_4H_9Li/THF	$(CH_2)_5C{=}CHNC$ (84)	353
		C_6H_5CHO	n-C_4H_9Li/THF	$C_6H_5CH{=}CHNC$ (75)	353
		p-$CH_3OC_6H_4CHO$	n-C_4H_9Li/THF	p-$CH_3OC_6H_4CH{=}CNC$ (68)	353
C_8	$(C_2H_5O)_2P(O)NHC_4H_9$-n	CS_2	NaH/glyme	n-C_4H_9NCS (79)	351, 352
C_9	$(C_2H_5O)_2P(O)CH{=}CHCH_2N(CH_3)_2$	i-C_3H_7CHO	NaH/HMPT	$(CH_3)_2CH(CH{=}CH)_2N(CH_3)_2$ (83)	55
		n-C_4H_9CHO	NaH/HMPT	n-$C_4H_9(CH{=}CH)_2N(CH_3)_2$ (46)	55
		$(CH_2)_5C{=}O$	NaH/HMPT	$(CH_2)_5C{=}CHCH{=}CHN(CH_3)_2$ (60)	55
		C_6H_5CHO	NaH/HMPT	$C_6H_5(CH{=}CH)_2N(CH_3)_2$ (42)	55
C_{10}	$(C_2H_5O)_2P(O)NHC_6H_{11}$	CO_2	NaH/glyme	$C_6H_{11}NCO$ (24)	351, 352
		$C_6H_{11}NCO$	NaH/glyme	$C_6H_{11}N{=}C{=}NC_6H_{11}$ (53)	351, 352
		C_6H_5NCO	NaH/glyme	$C_6H_5N{=}C{=}NC_6H_{11}$ (60)	351, 352
		$C_6H_5(C_2H_5)C{=}CO$	NaH/glyme	$C_6H_5(C_2H_5)C{=}C{=}NC_6H_{11}$ (58)	351, 352
C_{11}	$(C_2H_5O)_2P(O)CH{=}C(CH_3)NHC_4H_9$-$n$	p-$CH_3OC_6H_4CHO$	NaH/THF	p-$CH_3OC_6H_4CH{=}CHC(CH_3){=}NC_4H_9$-$n$ (66)	198

C_{12}	$(C_2H_5O)_2P(O)CH{=}CHNHC_6H_{11}$	 O	NaH/THF	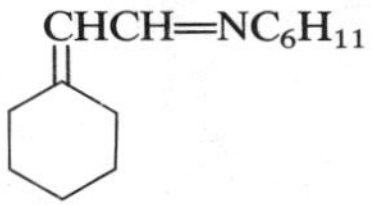 $CHCH{=}NC_6H_{11}$ (86)	45
		C_6H_5CHO	NaH/THF	$C_6H_5CH{=}CHCH{=}NC_6H_{11}$ (77)	45
		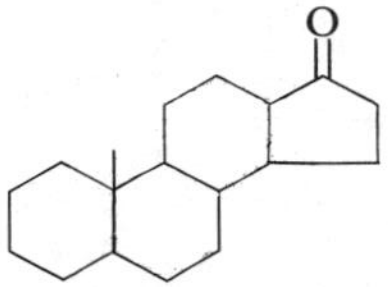 O	NaH/THF	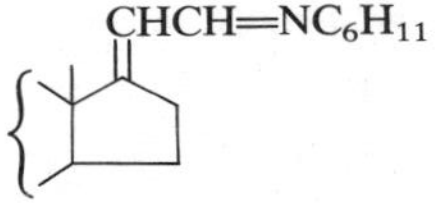 $CHCH{=}NC_6H_{11}$ (74)	45
		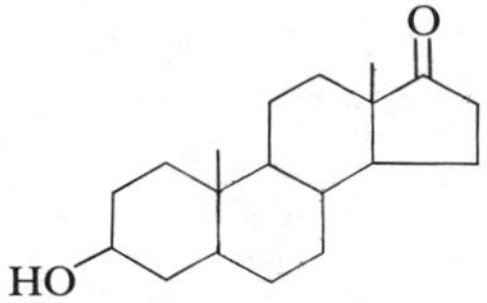 O, HO	NaH/THF	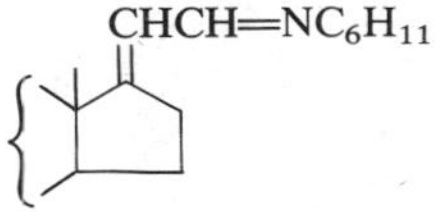 $CHCH{=}NC_6H_{11}$ (71)	45
		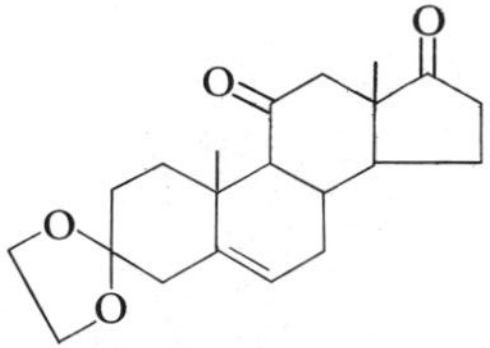 O, O, O, O	NaH/THF	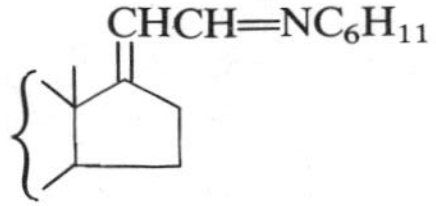 $CHCH{=}NC_6H_{11}$ (73)	45

Note: References 228–415 are on pp. 250–253.

TABLE V. UNSATURATED NITROGEN COMPOUNDS FROM PHOSPHORYL REAGENTS CONTAINING A NITROGEN GROUP

No. of C Atoms	Reagent	Reactant	Base/Solvent	Product(s) and Yield(s) (%)	Refs.
C_{12} (*Contd.*)	$(C_2H_5O)_2P(O)CH{=}CHNHC_6H_{11}$ (*Contd.*)	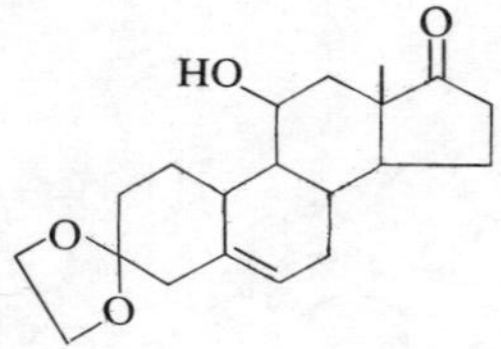 HO, O (structure)	NaH/THF	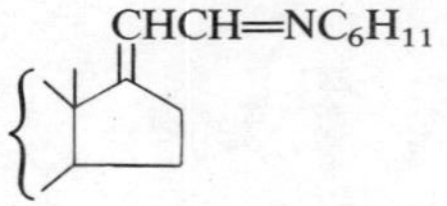 $CHCH{=}NC_6H_{11}$ (structure) (65)	45
		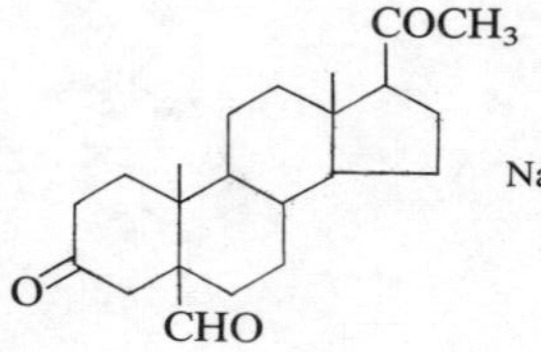 $COCH_3$, CHO (structure)	NaH/THF	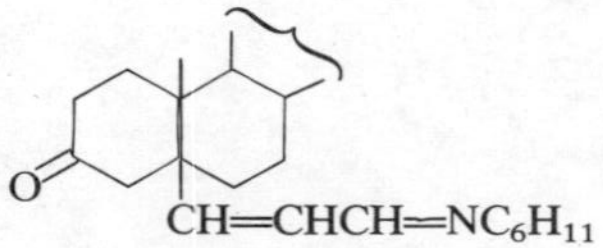 $CH{=}CHCH{=}NC_6H_{11}$ (structure) (83)	45
		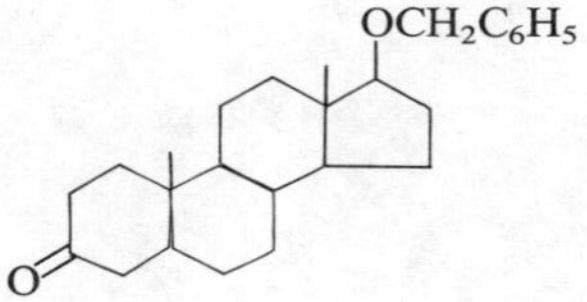 $OCH_2C_6H_5$ (structure)	NaH/THF	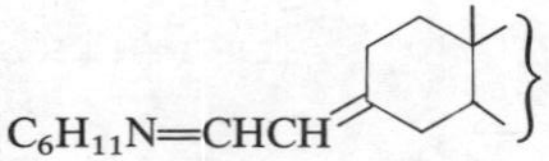 $C_6H_{11}N{=}CHCH$ (structure) (84)	45

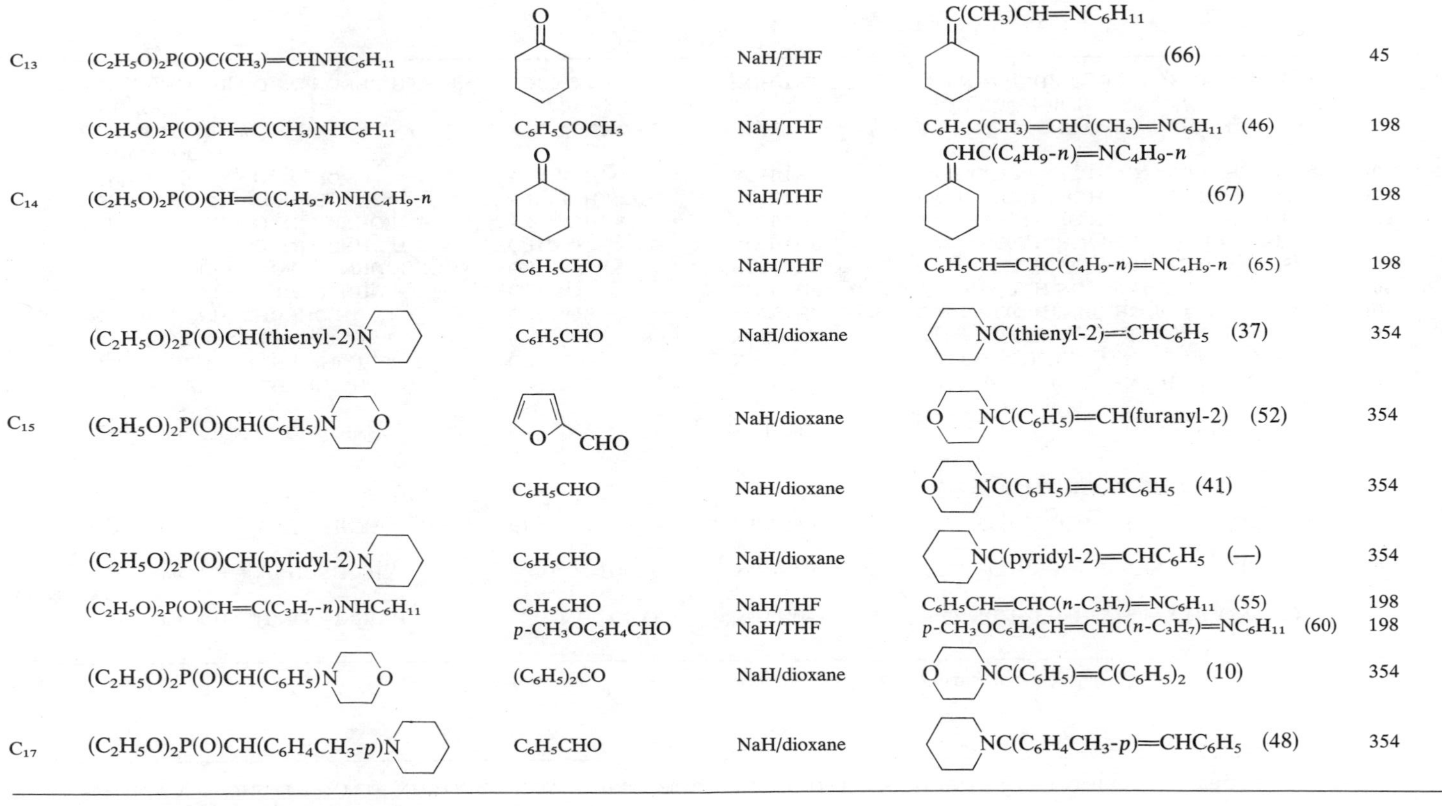

C_{13}	$(C_2H_5O)_2P(O)C(CH_3){=}CHNHC_6H_{11}$	O	NaH/THF	$C(CH_3)CH{=}NC_6H_{11}$ (66)	45
	$(C_2H_5O)_2P(O)CH{=}C(CH_3)NHC_6H_{11}$	$C_6H_5COCH_3$	NaH/THF	$C_6H_5C(CH_3){=}CHC(CH_3){=}NC_6H_{11}$ (46)	198
C_{14}	$(C_2H_5O)_2P(O)CH{=}C(C_4H_9\text{-}n)NHC_4H_9\text{-}n$	O	NaH/THF	$CHC(C_4H_9\text{-}n){=}NC_4H_9\text{-}n$ (67)	198
		C_6H_5CHO	NaH/THF	$C_6H_5CH{=}CHC(C_4H_9\text{-}n){=}NC_4H_9\text{-}n$ (65)	198
	$(C_2H_5O)_2P(O)CH(\text{thienyl-2})N$	C_6H_5CHO	NaH/dioxane	$NC(\text{thienyl-2}){=}CHC_6H_5$ (37)	354
C_{15}	$(C_2H_5O)_2P(O)CH(C_6H_5)N$ O	O CHO	NaH/dioxane	O $NC(C_6H_5){=}CH(\text{furanyl-2})$ (52)	354
		C_6H_5CHO	NaH/dioxane	O $NC(C_6H_5){=}CHC_6H_5$ (41)	354
	$(C_2H_5O)_2P(O)CH(\text{pyridyl-2})N$	C_6H_5CHO	NaH/dioxane	$NC(\text{pyridyl-2}){=}CHC_6H_5$ (—)	354
	$(C_2H_5O)_2P(O)CH{=}C(C_3H_7\text{-}n)NHC_6H_{11}$	C_6H_5CHO	NaH/THF	$C_6H_5CH{=}CHC(n\text{-}C_3H_7){=}NC_6H_{11}$ (55)	198
		$p\text{-}CH_3OC_6H_4CHO$	NaH/THF	$p\text{-}CH_3OC_6H_4CH{=}CHC(n\text{-}C_3H_7){=}NC_6H_{11}$ (60)	198
	$(C_2H_5O)_2P(O)CH(C_6H_5)N$ O	$(C_6H_5)_2CO$	NaH/dioxane	O $NC(C_6H_5){=}C(C_6H_5)_2$ (10)	354
C_{17}	$(C_2H_5O)_2P(O)CH(C_6H_4CH_3\text{-}p)N$	C_6H_5CHO	NaH/dioxane	$NC(C_6H_4CH_3\text{-}p){=}CHC_6H_5$ (48)	354

Note: References 228–415 are on pp. 250–253.

TABLE V. UNSATURATED NITROGEN COMPOUNDS FROM PHOSPHORYL REAGENTS CONTAINING A NITROGEN GROUP

No. of C Atoms	Reagent	Reactant	Base/Solvent	Product(s) and Yield(s) (%)	Refs.
C_{17} (*Contd.*)	$(C_2H_5O)_2P(O)CH(C_6H_4NO_2\text{-}p)NHC_6H_5$	$C_5H_{10}NCOCHO$	NaH/glyme	$p\text{-}O_2NC_6H_4C(NHC_6H_5){=}CHCONC_5H_{10}$ (−)	41
	$(C_2H_5O)_2P(O)CH{=}C(C_6H_5)NHC_6H_{11}$	$p\text{-}CH_3OC_6H_4CHO$	NaH/glyme	$p\text{-}CH_3OC_6H_4CH{=}CHC(C_6H_5){=}NC_6H_{11}$ (53)	198
C_{19}	$(C_2H_5O)_2P(O)CH[C_6H_4CH(CH_3)_2\text{-}p]NC_5H_{10}$	CH_2O	NaH/dioxane	$C_5H_{10}NC[C_6H_4CH(CH_3)_2\text{-}p]{=}CH_2$ (−)	354
	$(C_2H_5O)_2P(O)C(CH_2C_6H_5){=}CHNHC_6H_{11}$	$(CH_2)_5C{=}O$	NaH/THF	$(CH_2)_5C{=}C(CH_2C_6H_5)CH{=}NC_6H_{11}$ (63)	45
C_{24}	$(C_6H_5)_2P(O)CH{=}C(C_6H_5)NHC_4H_9\text{-}n$	C_6H_5CHO	NaH/THF	$C_6H_5CH{=}CHC(C_6H_5){=}NC_4H_9\text{-}n$ (—)	43
	$(C_6H_5)_2P(O)CH{=}C(C_6H_{11})NHC_4H_9\text{-}n$	C_6H_5CHO	NaH/THF	$C_6H_5CH{=}CHC(C_6H_{11}){=}NC_4H_9\text{-}n$ (—)	43
	$(C_6H_5)_2P(O)CH{=}C(C_6H_5)NHC_4H_9\text{-}n$	$(C_6H_5)_2CO$	NaH/THF	$(C_6H_5)_2C{=}CHC(C_6H_5){=}NC_4H_9\text{-}n$ (—)	43
C_{25}	$(C_6H_5O)_2P(O)CH(C_6H_5)NHC_6H_5$	C_6H_5NO	KOH/THF	$C_6H_5N{=}C(C_6H_5)NHC_6H_5$ (66)	105
	$(C_6H_5O)_2P(O)CH(p\text{-}O_2NC_6H_4)NHC_6H_5$	C_6H_5NO	KOH/THF	$C_6H_5N{=}C(C_6H_4NO_2\text{-}p)NHC_6H_5$ (97)	105
	$(C_6H_5O)_2P(O)CH(p\text{-}O_2NC_6H_4)NHC_6H_4NO_2\text{-}p$	C_6H_5NO	KOH/THF	$C_6H_5N{=}C(C_6H_4NO_2\text{-}p)NHC_6H_4NO_2\text{-}p$ (78)	105
	$(C_6H_5O)_2P(O)CH(p\text{-}O_2NC_6H_4)NHC_6H_4Cl\text{-}p$	C_6H_5NO	KOH/THF	$C_6H_5N{=}C(C_6H_4NO_2\text{-}p)NHC_6H_4Cl\text{-}p$ (68)	105
	$(C_6H_5O)_2P(O)CH(p\text{-}O_2NC_6H_4)NHC_6H_5$	$o\text{-}CH_3C_6H_4NO$	KOH/THF	$o\text{-}CH_3C_6H_4N{=}C(C_6H_4NO_2\text{-}p)NHC_6H_5$ (76)	105
	$(C_6H_5O)_2P(O)CH(C_6H_5)NHC_6H_4NO_2\text{-}p$	$o\text{-}CH_3C_6H_4NO$	KOH/THF	$o\text{-}CH_3C_6H_4N{=}C(C_6H_5)NHC_6H_4NO_2\text{-}p$ (45)	105
	$(C_6H_5O)_2P(O)CH(p\text{-}O_2NC_6H_4)\text{-}NHC_6H_4NO_2\text{-}p$	$o\text{-}CH_3C_6H_4NO$	KOH/THF	$o\text{-}CH_3C_6H_4N{=}C(C_6H_4NO_2\text{-}p)NHC_6H_4NO_2\text{-}p$ (67)	105
C_{26}	$(C_6H_5)_2P(O)CH{=}C(C_6H_5)NHCH_2\text{-}CH_2(OC_2H_5)_2$	CH_3CH_2CHO	NaH/THF	$CH_3CH_2CH{=}CHC(C_6H_5){=}NCH_2CH_2\text{-}(OC_2H_5)_2$ (60)	43
	$(C_6H_5O)_2P(O)CH(p\text{-}O_2NC_6H_4)NHC_6H_4OCH_3\text{-}p$	C_6H_5NO	KOH/THF	$C_6H_5N{=}C(C_6H_4NO_2\text{-}p)NHC_6H_4OCH_3\text{-}p$ (69)	105

TABLE VI. UNSATURATED SULFIDES, SULFONES, AND ETHERS FROM PHOSPHORYL REAGENTS

No. of C Atoms	Reagent	Reactant	Base/Solvent	Product(s) and Yield(s) (%)	Refs.
C_4	$(CH_3O)_2P(O)CH_2SO_2CH_3$	Cyclohexanone	NaH/C_6H_6	Cyclohexylidene=$CHSO_2CH_3$ (–)	49
		C_6H_5CHO	NaH/C_6H_6	$C_6H_5CH{=}CHSO_2CH_3$ (—)	49
		p-ClC_6H_4CHO	NaH/C_6H_6	p-$ClC_6H_4CH{=}CHSO_2CH_3$ (—)	49
		p-$CH_3OC_6H_4CHO$	NaH/C_6H_6	p-$CH_3OC_6H_4CH{=}CHSO_2CH_3$ (—)	49
		$C_6H_5COCH_3$	NaH/DMF	$C_6H_5C(CH_3){=}CHSO_2CH_3$ (—)	49
		$(C_6H_5)_2CO$	NaH/DMF	$(C_6H_5)_2C{=}CHSO_2CH_3$ (—)	49
C_6	$(C_2H_5O)_2P(O)CH_2SO_2CH_3$	Cyclohexanone	n-C_4H_9Li/THF	Cyclohexylidene=$CHSO_2CH_3$ (97)	355
		$CH_3(CH_2)_4COCH_3$	n-C_4H_9Li/THF	$CH_3(CH_2)_4C(CH_3){=}CHSO_2CH_3$ (86)	48
		$CH_3(CH_2)_5CHO$	n-C_4H_9Li/THF	$CH_3(CH_2)_5CH{=}CHSO_2CH_3$ (97)	48
	$(C_2H_5O)_2P(O)CH_2SCH_3$	C_6H_5CHO	NaH/glyme	$C_6H_5CH{=}CHSCH_3$ (65)	48, 408
	$(C_2H_5O)_2P(O)CH_2SO_2CH_3$	C_6H_5CHO	n-C_4H_9Li/THF	$C_6H_5CH{=}CHSO_2CH_3$ (87)	355, 408
		p-ClC_6H_4CHO	NaOH, H_2O/CH_2Cl_2, $[(C_2H_5)_3NCH_2C_6H_5]^+Cl^-$	p-$ClC_6H_4CH{=}CHSO_2CH_3$ (100)	408
	$(C_2H_5O)_2P(O)CH_2S(O)CH_3$	C_6H_5CHO	NaOH, H_2O/CH_2Cl_2, $[(C_2H_5)_3NCH_2C_6H_5]^+Cl^-$	$C_6H_5CH{=}CHS(O)CH_3$ (51)	408
	$(C_2H_5O)_2P(O)CH_2SCH_3$	$C_6H_5COCH_3$	NaH/glyme	$C_6H_5C(CH_3){=}CHSCH_3$ (43)	48
	$(C_2H_5O)_2P(O)CH_2SO_2CH_3$	p-$CH_3OC_6H_4CHO$	NaOH, H_2O/CH_2Cl_2, $[(C_2H_5)_3NCH_2C_6H_5]^+Cl^-$	p-$CH_3OC_6H_4CH{=}CHSO_2CH_3$ (100)	408
	$(C_2H_5O)_2P(O)CH_2SCH_3$	1-$C_{10}H_7CHO$	NaH/glyme	1-$C_{10}H_7CH{=}CHSCH_3$ (60)	48
		$(C_6H_5)_2CO$	NaH/glyme	$(C_6H_5)_2C{=}CHSCH_3$ (35)	48
C_7	$(C_2H_5O)_2P(O)CH_2SO_2C_2H_5$	CCl_3CHO	NaH/ether	$Cl_3CCH{=}CHSO_2C_2H_5$ (76)	50
		$CH_3SCH_2CH_2CHO$	NaH/ether	$CH_3SCH_2CH_2CH{=}CHSO_2C_2H_5$ (75)	50
		2-Thienyl-CHO	NaH/ether	2-Thienyl-$CH{=}CHSO_2C_2H_5$ (93)	50

Note: References 228–415 are on pp. 250–253.

TABLE VI. Unsaturated Sulfides, Sulfones, and Ethers from Phosphoryl Reagents (*Continued*)

No. of C Atoms	Reagent	Reactant	Base/Solvent	Product(s) and Yield(s) (%)	Refs.
C_7 (*Contd.*)	$(C_2H_5O)_2P(O)CH_2SO_2C_2H_5$ (*Contd.*)	2-furyl-CHO	NaH/glyme	2-furyl-$CH{=}CHSO_2C_2H_5$ (80)	50
		2-pyrrolyl-CHO	NaH/toluene	2-pyrrolyl-$CH{=}CHSO_2C_2H_5$ (90)	50
	$(C_2H_5O)_2P(O)CH(CH_3)SCH_3$	cyclopentanone	n-C_4H_9Li/THF	cyclopentylidene=$C(CH_3)SCH_3$ (10)	35
		cyclohexanone	n-C_4H_9Li/THF	cyclohexylidene=$C(CH_3)SCH_3$ (82)	35
		n-$C_5H_{11}CHO$	n-C_4H_9Li/THF	n-$C_5H_{11}CH{=}C(CH_3)SCH_3$ (67)	35
	$(C_2H_5O)_2P(O)CH_2SO_2C_2H_5$	2-pyridyl-CHO	NaH/ether	2-pyridyl-$CH{=}CHSO_2C_2H_5$ (70)	50
		3-pyridyl-CHO	$NaH/C_6H_5CH_3$	3-pyridyl-$CH{=}CHSO_2C_2H_5$ (89)	50
		4-pyridyl-CHO	$NaH/C_6H_5CH_3$	4-pyridyl-$CH{=}CHSO_2C_2H_5$ (88)	50

	$(C_2H_5O)_2P(O)CH_2SO_2C_2H_5$ (cont.)	C_6H_5CHO	NaH/glyme	$C_6H_5CH{=}CHSO_2C_2H_5$ (84)	50
		3,4-$Cl_2C_6H_3CHO$	NaH/glyme	3,4-$Cl_2C_6H_3CH{=}CHSO_2C_2H_5$ (91)	50
	$(C_2H_5O)_2P(O)CH(CH_3)SCH_3$	C_6H_5CHO	n-C_4H_9Li/THF	$C_6H_5CH{=}C(CH_3)SCH_3$ (80)	35
	$[(C_2H_5O)_2P(O)CH_2\overset{+}{S}(CH_3)_2]ClO_4^-$	C_6H_5CHO	NaH/THF	$[C_6H_5CH{=}CH\overset{+}{S}(CH_3)_2]ClO_4^-$ (64)	356
	$(C_2H_5O)_2P(O)CH_2SO_2C_2H_5$	p-$CH_3COC_6H_4CHO$	NaH/ether	p-$CH_3COC_6H_4CH{=}CHSO_2C_2H_5$ (96)	50
		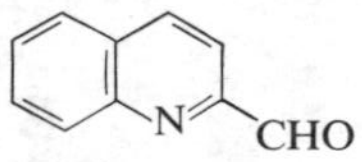 CHO	NaH/ether	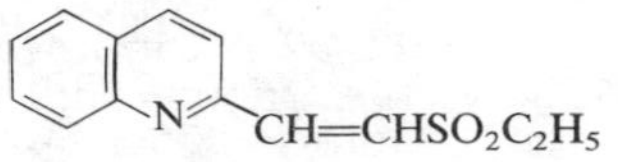 $CH{=}CHSO_2C_2H_5$ (77)	50
	$(C_2H_5O)_2P(O)CH(CH_3)SCH_3$	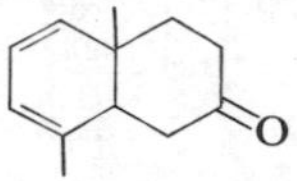 O	NaH/glyme	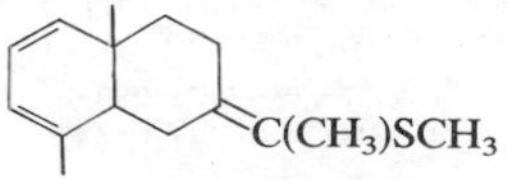 $C(CH_3)SCH_3$ (72)	357
		$(C_6H_5)_2CO$	n-C_4H_9Li/THF	$(C_6H_5)_2C{=}C(CH_3)SCH_3$ (84)	35
		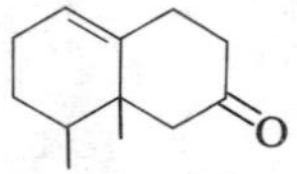 O	NaH/glyme	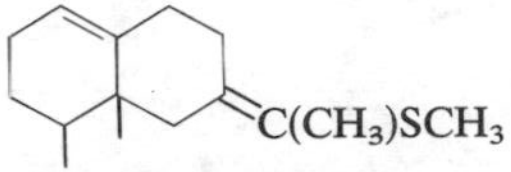 $C(CH_3)SCH_3$ (83)	396
C_8	$(C_2H_5O)_2P(O)CH_2SCH_2CH{=}CH_2$	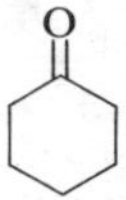 O	n-C_4H_9Li/THF	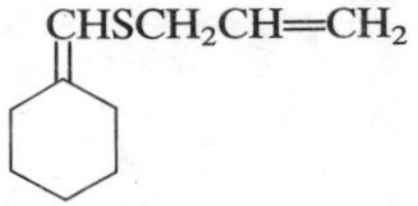 $CHSCH_2CH{=}CH_2$ (83)	51
		C_6H_5CHO	n-C_4H_9Li/THF	$C_6H_5CH{=}CHSCH_2CH{=}CH_2$ (83)	51
		Norbornanone	n-C_4H_9Li/THF	Norbornyl$=CHSCH_2CH{=}CH_2$ (65)	51
		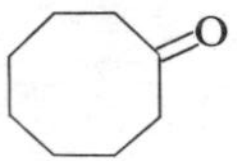 O	n-C_4H_9Li/THF	 $CHSCH_2CH{=}CH_2$ (19)	51
		Adamantanone	n-C_4H_9Li/THF	Adamantyl$=CHSCH_2CH{=}CH_2$ (72)	51
		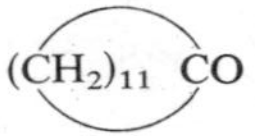 $(CH_2)_{11}$ CO	n-C_4H_9Li/THF	 $(CH_2)_{11}$ $C{=}CHSCH_2CH{=}CH_2$ (54)	51

Note: References 228–415 are on pp. 250–253.

TABLE VI. UNSATURATED SULFIDES, SULFONES, AND ETHERS FROM PHOSPHORYL REAGENTS (*Continued*)

No. of C Atoms	Reagent	Reactant	Base/Solvent	Product(s) and Yield(s) (%)	Refs.
C_8 (*Contd.*)	$(C_2H_5O)_2P(O)CH_2SCH_2CH{=}CH_2$ (*Contd.*)	(structure) O	n-C_4H_9Li/THF	(structure) $=CHSCH_2CH{=}CH_2$ (81)	51
C_9	$(C_2H_5O)_2P(O)CH_2CH{=}CHOC_2H_5$	$C_2H_5COCH_3$	NaH/glyme	$C_2H_5C(CH_3){=}CHCH{=}CHOC_2H_5$ (33)	55
		i-C_3H_7CHO	NaH/glyme	i-$C_3H_7CH{=}CHCH{=}CHOC_2H_5$ (57)	55
		O (structure)	NaH/glyme	$CHCH{=}CHOC_2H_5$ (structure) (84)	55
		C_6H_5CHO	NaH/HMPT	$C_6H_5CH{=}CHCH{=}CHOC_2H_5$ (66)	55
	$(C_2H_5O)_2P(O)CH_2CH{=}CHSC_2H_5$	C_6H_5CHO	NaH/glyme	$C_6H_5CH{=}CHCH{=}CHSC_2H_5$ (—)	55
	$(C_2H_5O)_2P(O)CH_2CH{=}CHOC_2H_5$	p-$CH_3OC_6H_4CHO$	NaH/HMPT	p-$CH_3OC_6H_4CH{=}CHCH{=}CHOC_2H_5$ (47)	55
		p-$(CH_3)_2NC_6H_4CHO$	NaH/glyme	p-$(CH_3)_2NC_6H_4CH{=}CHCH{=}CHOC_2H_5$ (80)	55
		$(C_6H_5)_2CO$	NaH/HMPT	$(C_6H_5)_2C{=}CHCH{=}CHOC_2H_5$ (63)	55
C_{10}	$(C_2H_5O)_2P(O)CH_2CH{=}C(CH_3)OC_2H_5$	i-C_3H_7CHO	NaH/HMPT	i-$C_3H_7CH{=}CHCH{=}C(CH_3)OC_2H_5$ (48)	55
		O (structure)	NaH/HMPT	$CHCH{=}C(CH_3)OC_2H_5$ (structure) (40)	55
		C_6H_5CHO	NaH/HMPT	$C_6H_5CH{=}CHCH{=}C(CH_3)OC_2H_5$ (60)	55

		p-$CH_3C_6H_4CHO$	NaH/HMPT	p-$CH_3C_6H_4CH{=}CHCH{=}C(CH_3)OC_2H_5$ (38)	55
		$(C_6H_5)_2CO$	NaH/HMPT	$(C_6H_5)_2C{=}CHCH{=}C(CH_3)OC_2H_5$ (47)	55
C_{11}	$(C_2H_5O)_2P(O)CH_2SO_2C_6H_4Cl$-p	C_6H_5CHO	n-C_4H_9Li/THF	$C_6H_5CH{=}CHSO_2C_6H_4Cl$-p (90)	355
		O	n-C_4H_9Li/THF	$CHSO_2C_6H_4Cl$-p (72)	355
	$(C_2H_5O)_2P(O)CH_2SC_6H_5$	C_6H_5CHO	NaOH, H_2O/CH_2CL_2, $[(C_2H_5)_3NCH_2C_6H_5]^+Cl^-$	$C_6H_5SCH{=}CHC_6H_5$ (81)	408
	$(C_2H_5O)_2P(O)CH_2S(O)C_6H_5$	C_6H_5CHO	NaOH, H_2O/CH_2Cl_2, $[(C_2H_5)_3NCH_2C_6H_5)^+Cl^-$	$C_6H_5CH{=}CHS(O)C_6H_5$ (54)	408
		p-ClC_6H_4CHO	NaOH, H_2O/CH_2Cl_2, $[(C_2H_5)_3NCH_2C_6H_5]^+Cl^-$	p-$ClC_6H_4CH{=}CHS(O)C_6H_5$ (57)	408
		n-$C_6H_{13}CHO$	n-C_4H_9Li/THF	n-$C_6H_{13}CH{=}CHSO_2C_6H_4Cl$-p (80)	355
	$(C_2H_5O)_2P(O)CH_2SO_2C_6H_4Br$-p	p-FC_6H_4CHO	NaH/glyme	p-$BrC_6H_4SO_2CH{=}CHC_6H_4F$-p (—)	47, 39
		m-$O_2NC_6H_4CHO$	NaH/glyme	p-$BrC_6H_4SO_2CH{=}CHC_6H_4NO_2$-m (—)	47, 39
		m-HOC_6H_4CHO	NaH/glyme	p-$BrC_6H_4SO_2CH{=}CHC_6H_4OH$-m (—)	47, 39
		o-$O_2NC_6H_4CHO$	NaH/glyme	p-$BrC_6H_4SO_2CH{=}CHC_6H_4NO_2$-o (—)	47, 39
		o-HOC_6H_4CHO	NaH/glyme	p-$BrC_6H_4SO_2CH{=}CHC_6H_4OH$-o (—)	47, 39
		C_6H_5CHO	NaH/glyme	p-$BrC_6H_4SO_2CH{=}CHC_6H_5$ (—)	47, 39
		p-ClC_6H_4CHO	NaH/glyme	p-$BrC_6H_4SO_2CH{=}CHC_6H_4Cl$-p (—)	47, 39
	$(C_2H_5O)_2P(O)CH_2SC_6H_5Br$-p	p-ClC_6H_4CHO	NaH/glyme	p-$BrC_6H_4SCH{=}CHC_6H_4Cl$-p (80)	47, 39
		C_6H_5CHO	NaH/glyme	p-$BrC_6H_4SCH{=}CHC_6H_5$ (75)	47, 39
		p-$CH_3OC_6H_4CHO$	NaH/glyme	p-$BrC_6H_4SCH{=}CHC_6H_4OCH_3$-p (93)	47, 39
	$(C_2H_5O)_2P(O)CH_2S(O)C_6H_5$	p-$CH_3OC_6H_4CHO$	NaOH, H_2O/CH_2Cl_2, $[(C_2H_5)_3NCH_2C_6H_5]^+Cl^-$	p-$CH_3OC_6H_4CH{=}CHS(O)C_6H_5$ (48)	408
	$(C_2H_5O)_2P(O)CH_2SO_2C_6H_4Br$-p	o-$CH_3OC_6H_4CHO$	NaH/glyme	p-$BrC_6H_4SO_2CH{=}CHC_6H_4OCH_3$-o (—)	47, 39
		p-$CH_3C_6H_4CHO$	NaH/glyme	p-$BrC_6H_4SO_2CH{=}CHC_6H_4CH_3$-p (—)	47, 39
		p-$CH_3OC_6H_4CHO$	NaH/glyme	p-$BrC_6H_4SO_2CH{=}CHC_6H_4OCH_3$-p (—)	47, 39
		p-$(CH_3)_2NC_6H_4CHO$	NaH/glyme	p-$BrC_6H_4SO_2CH{=}CHC_6H_4N(CH_3)_2$-p (—)	47, 39
		$C_6H_5CH{=}CHCHO$	NaH/glyme	p-$BrC_6H_4SO_2(CH{=}CH)_2C_6H_5$ (—)	47, 39
	$(C_2H_5O)_2P(O)CH_2SC_6H_5$	p-$(CH_3)_2NC_6H_4CHO$	NaOH, H_2O/CH_2Cl_2- $[(C_2H_5)_3NCH_2C_6H_5]^+Cl^-$	p-$(CH_3)_2NC_6H_4CH{=}CHSC_6H_5$ (40)	408

Note: References 228–415 are on pp. 250–253.

TABLE VI. Unsaturated Sulfides, Sulfones, and Ethers from Phosphoryl Reagents (*Continued*)

No. of C Atoms	Reagent	Reactant	Base/Solvent	Product(s) and Yield(s) (%)	Refs.
C_{11} (*Contd.*)	$(C_2H_5O)_2P(O)CH_2SC_6H_4Br$-*p*	*p*-$(CH_3)_2NC_6H_4CHO$	NaH/glyme	*p*-$BrC_6H_4SCH{=}CHC_6H_4N(CH_3)_2$-*p* (75)	47, 39
		$C_6H_5CH{=}CHCHO$	NaH/glyme	*p*-$BrC_6H_4S(CH{=}CH)_2C_6H_5$ (60)	47, 39
	$(C_2H_5O)_2P(O)CH_2SO_2C_6H_4Br$-*p*	2,3,4-$(CH_3)_3C_6H_2CHO$	NaH/glyme	*p*-$BrC_6H_4SO_2CH{=}CHC_6H_2$-$(OCH_3)_3$-2,3,4 (—)	47, 39
C_{12}	$[(C_2H_5O)_2P(O)CH_2\overset{+}{S}(CH_3)C_6H_5]ClO_4^-$	C_6H_5CHO	NaH/THF	$[C_6H_5CH{=}CH\overset{+}{S}(CH_3)C_6H_5]ClO_4^-$ (66)	356
C_{13}	$(C_2H_5O)_2P(O)CH_2CH{=}CHSC_6H_5$	$(CH_3)_2CO$	NaH/glyme	$(CH_3)_2C{=}CHCH{=}CHSC_6H_5$ (68)	55
		$C_2H_5COCH_3$	NaH/glyme	$C_2H_5C(CH_3){=}CHCH{=}CHSC_6H_5$ (75)	55
		i-C_3H_7CHO	NaH/glyme	*i*-$C_3H_7CH{=}CHCH{=}CHSC_6H_5$ (80)	55
		n-C_4H_9CHO	NaH/glyme	*n*-$C_4H_9CH{=}CHCH{=}CHSC_6H_5$ (70)	55
		[2-thienyl]CHO	NaH/glyme	[2-thienyl]$CH{=}CHCH{=}CHSC_6H_5$ (80)	55
		[cyclohexanone]	NaH/glyme	[cyclohexylidene]$CHCH{=}CHSC_6H_5$ (83)	55
		C_6H_5CHO	NaH/glyme	$C_6H_5CH{=}CHCH{=}CHSC_6H_5$ (80)	55
		$C_6H_5COCH_3$	NaH/glyme	$C_6H_5(CH_3)C{=}CHCH{=}CHSC_6H_5$ (72)	55
		$(C_6H_5)_2CO$	NaH/glyme	$(C_6H_5)_2C{=}CHCH{=}CHSC_6H_5$ (50)	55
C_{14}	$(C_2H_5O)_2P(O)CH[(CH_2)_7CH_3]SCH_3$	[cyclohexanone]	*n*-C_4H_9Li/THF	[cyclohexylidene]$C[(CH_2)_7CH_3]SCH_3$ (72)	35

Note: References 228–415 are on pp. 250–253.

TABLE VII. Polyenes from Difunctional Phosphoryl Reagents

No. of C Atoms	Reagent	Reactant	Base/Solvent	Product(s) and Yield(s) (%)	Refs.
C_{12}	$(C_2H_5O)_2P(O)CH_2CH{=}CHCH_2P(O)(OC_2H_5)_2$	$(C_2H_5O)_2CHCOCH_3$	CH_3ONa/CH_3OH	$(C_2H_5O)_2CHC(CH_3){=}CHCH{=}CH$-$CH{=}C(CH_3)CH(OC_2H_5)_2$ (—)	323
		$OCH(CH{=}CH)_3CHO$	t-C_4H_9OK/DMF	$\text{-}(CH{=}CH)_x\text{-}$ (—)	110
		$C_6H_5CH{=}CHCHO$	t-C_4H_9OK/C_6H_6	$C_6H_5(CH{=}CH)_5C_6H_5$ (70)	204
		$(C_6H_5)_2CO$	t-C_4H_9OK/C_6H_6	$(C_6H_5)_2C{=}CHCH{=}CHCH{=}C(C_6H_5)_2$ (61)	204
C_{14}	$(C_2H_5O)_2P(O)CH_2(CH{=}CH)_2CH_2P(O)$-$(OC_2H_5)_2$	$OHC(CH{=}CH)_3CHO$	t-C_4H_9OK/DMF	$\text{-}(CH{=}CH)_x\text{-}$ (—)	110
C_{16}	$(C_2H_5O)_2P(O)CH_2C_6H_4CH_2P(O)(OC_2H_5)_2$	cyclohexanone	t-$C_4H_9OK/C_6H_5CH_3$	$C_6H_4(CH{=}C_6H_{10})_2$-p (65)	249, 204
		3-pyridinecarboxaldehyde	t-C_4H_9OK/DMF	$C_6H_4(CH{=}CH$-3-pyridyl$)_2$-p (62)	256, 204
		C_6H_5CHO	t-C_4H_9OK/C_6H_6	$C_6H_4(CH{=}CHC_6H_5)_2$-p (—)	4, 204, 249
		p-$O_2NC_6H_4CHO$	t-C_4H_9OK/DMF	$C_6H_4(CH{=}CHC_6H_4NO_2$-$p)_2$-p (—)	256
	$C_6H_4[CH_2P(O)(OC_2H_5)_2]_2$-1,2	C_6H_5CHO	NaH/glyme	o-$C_6H_4(CH{=}CHC_6H_5)_2$ (—)	76
	$C_6H_4[CH_2P(O)(OC_2H_5)_2]_2$-1,4	p-$C_6H_4(CHO)_2$	t-C_4H_9OK/DMF	p-$(CH{=}CHC_6H_4)_2$ (—)	110
		$OCH(CH{=}CH)_3CHO$	t-C_4H_9OK/DMF	p-$C_6H_4\text{-}[(CH{=}CH)_5C_6H_4]_2$ (—)	110
		$C_6H_5CH{=}CHCHO$	NaH/glyme	p-$C_6H_4(CH{=}CHCH{=}CHC_6H_5)_2$ (84)	4, 204, 249

Note: References 228–415 are on pp. 250–253.

TABLE VII. Polyenes from Difunctional Phosphoryl Reagents (*Continued*)

No. of C Atoms	Reagent	Reactant	Base/Solvent	Product(s) and Yield(s) (%)	Refs.
C_{16} (*Contd.*)	$(C_6H_4[CH_2P(O)(O_2C_2H_5)_2]$-1,4 (*Contd.*)	*p*-$(CH_3)_2NC_6H_4CHO$	*t*-C_4H_9OK/DMF	*p*-$C_6H_4(CH{=}CHC_6H_4N(CH_3)_2$-*p*$)_2$ (67)	256, 204
		1-$C_{10}H_7CHO$	*t*-$C_4H_9OK/C_6H_5CH_3$	*p*-$C_6H_4(CH{=}CH$-1-$C_{10}H_7)_2$ (—)	249
		p-$C_6H_4(CH{=}CHCHO)_2$	*t*-C_4H_9OK/DMF	*p*-$(CH{=}CHCH{=}CHC_6H_4)_2$ (—)	110
		p-$C_6H_5C_6H_4CHO$	*t*-C_4H_9OK/DMF	*p*-$C_6H_4(CH{=}CHC_6H_4C_6H_5)_2$ (92)	204, 249 256
		$(C_6H_5)_2CO$	*t*-$C_4H_9OK/C_6H_5CH_3$	*p*-$C_6H_4[CH{=}C(C_6H_5)_2]_2$ (62)	204, 249
		$(C_6H_{11})_2CO$	*t*-C_4H_9OK/DMF	*p*-$C_6H_4[CH{=}C(C_6H_{11})_2]_2$ (—)	256, 204
		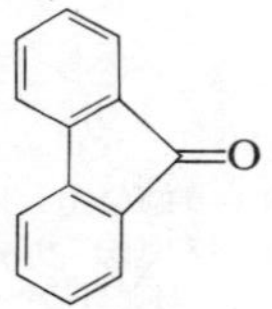 =O	*t*-C_4H_9OK/C_6H_6	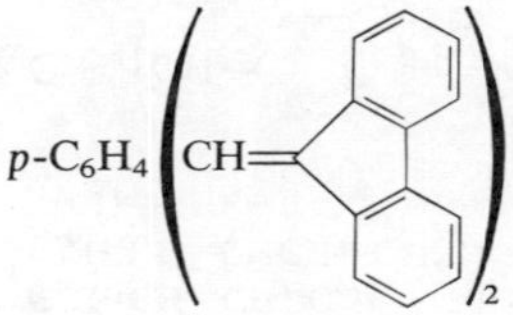 *p*-C_6H_4(CH=)$_2$ (70)	204
		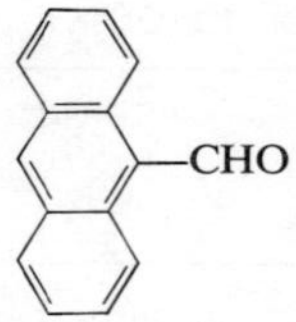 —CHO	*t*-C_4H_9OK/C_6H_6	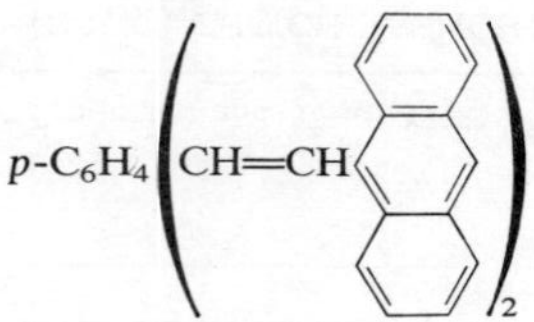 *p*-C_6H_4(CH=CH)$_2$ (76)	204, 249

		CHO (phenanthrene-9-carbaldehyde)	$t\text{-}C_4H_9OK/C_6H_6$	$p\text{-}C_6H_4(CH{=}CH\text{-phenanthryl})_2$ (65)	204, 249
C_{18}	$[(C_2H_5O)_2P(O)CH_2C(CH_3){=}CH{-}C{\equiv}]_2$	$C_6H_5CH{=}CHCHO$	CH_3ONa/DMF	$(C_6H_5$—polyene—$\equiv)_2$ (—)	358
		CHO (polyene aldehyde)	CH_3ONa/DMF	(polyene—$\equiv)_2$ (—)	323
C_{20}	$(C_2H_5O)_2P(O)CH_2$ $CH_2P(O)(OC_2H_5)_2$ (naphthalene-1,8-diyl)	C_6H_5CHO	CH_3ONa/DMF	$C_6H_5CH{=}CH$ $CH{=}CHC_6H_5$ (naphthalene-1,8-diyl) (40)	77
C_{26}	$(C_6H_5)_2P(O)CH_2CH_2P(O)(C_6H_5)_2$	$OHC(CH{=}CH)_3CHO$	$t\text{-}C_4H_9OK/DMF$	$-(CH{=}CH)_x-$ (−)	110
		$(C_6H_5)_2CO$	$t\text{-}C_4H_9OK/C_6H_5CH_3$	$(C_6H_5)_2C{=}CHCH{=}C(C_6H_5)_2$ (41)	249, 204
C_{28}	$(C_6H_5)_2P(O)(CH_2)_4P(O)(C_6H_5)_2$	$(C_6H_5)_2CO$	$t\text{-}C_4H_9OK/C_6H_6$	$(C_6H_5)_2C{=}CH(CH_2)_2CH{=}C(C_6H_5)_2$ (25)	204
C_{30}	$(C_6H_5)_2P(O)(CH_2)_6P(O)(C_6H_5)_2$	$(C_6H_5)_2CO$	$t\text{-}C_4H_9OK/C_6H_6$	$(C_6H_5)_2C{=}CH(CH_2)_4CH{=}C(C_6H_5)_2$ (25)	204, 249
C_{32}	$(C_6H_5)_2P(O)CH_2C_6H_4CH_2P(O)(C_6H_5)_2$	$C_6H_5CH{=}CHCHO$	$t\text{-}C_4H_9OK/C_6H_6$	$C_6H_4(CH{=}CHCH{=}CHC_6H_5)_2\text{-}p$ (27)	204

Note: References 228–415 are on pp. 250–253.

TABLE VIII. UNSATURATED PHOSPHONATES FROM METHYLENE BISPHOSPHONATES

No. of C Atoms	Reagent	Reactant	Base/Solvent	Product(s) and Yield(s) (%)	Refs.
C_9	$[(C_2H_5O)_2P(O)]_2CH_2$	C_2H_5CHO	t-C_4H_9OK/t-C_4H_9OH	$C_2H_5CH{=}CHP(O)(OC_2H_5)_2$ (—)	70
	$[(C_2H_5O)_2P(O)]_2CCl_2$	$(CH_3)_2CO$	n-C_4H_9Li/THF	$(CH_3)_2C{=}C(Cl)P(O)C(OC_2H_5)_2$ (85)	228
		t-C_4H_9CHO	n-C_4H_9Li/THF	t-$C_4H_9CH{=}C(Cl)P(O)(OC_2H_5)_2$ (83)	228
	$[(C_2H_5O)_2P(O)]_2CH_2$	(S ring) CHO	t-C_4H_9OK/t-C_4H_9OH	(S ring) $CH{=}CHP(O)(OC_2H_5)_2$ (–)	70
		(N ring) CHO	t-C_4H_9OK/t-C_4H_9OH	(N ring) $CH{=}CHP(O)(OC_2H_5)_2$ (–)	70
		(O, O ring) CHO	NaH/glyme	(O, O ring) $CH{=}CHP(O)(OC_2H_5)_2$ (—)	366
		(O, O ring) OH, CHO	NaH/glyme	(O, O ring) OH, $CH{=}CHP(O)(OC_2H_5)_2$ (—)	73
		C_6H_5CHO	NaH/glyme	$C_6H_5CH{=}CHP(O)(OC_2H_5)_2$ (67)	4, 70
		p-$C_6H_4(CHO)_2$	t-C_4H_9OK/t-C_4H_9OH	p-$C_6H_4[CH{=}CHP(O)(OC_2H_5)_2]_2$ (—)	70
		$C_6H_5CH{=}CHCHO$	t-C_4H_9OK/t-C_4H_9OH	$C_6H_5CH{=}CHCH{=}CHP(O)(OC_2H_5)_2$ (—)	70
		(O, O, N ring) CHO	NaH/C_6H_6	(O, O, N ring) $P(O)(OC_2H_5)_2$ (95)	72

Reagent	Carbonyl Compound	Base/Solvent	Product(s) and Yield(s) (%)	Refs.
	CH_2—CHCH—CHCHO (bis-isopropylidene acetal)	NaH/glyme	CH_2—CHCH—CHCH=CHP(O)(OC_2H_5)$_2$ (bis-isopropylidene acetal) (–)	73
	(diisopropylidene furanos-3-ulose)	n-C_4H_9Li/THF	$(C_2H_5O)_2(O)PCH$= (diisopropylidene furanose) (81)	74
	CHO—(CHO_2CCH_3)$_4$—$CH_2O_2CCH_3$	NaH/glyme	CH=CHP(O)(OC_2H_5)$_2$—(CHO_2CCH_3)$_4$—$CH_2O_2CCH_3$ (–)	73
$[(C_2H_5O)_2P(O)]_2CHN(CH_3)_2$	C_6H_5CHO	NaH/glyme	C_6H_5CH=C[N(CH_3)$_2$]P(O)(OC_2H_5)$_2$ (—)	71
$C_2H_5OP(O)[CH_2P(O)(OC_2H_5)_2]_2$	i-C_3H_7CHO	NaH/glyme	i-C_3H_7CH=CHP(O)(OC_2H_5)$_2$ (57)	32
	C_6H_5CHO	NaH/glyme	C_6H_5CH=CHP(O)(OC_2H_5)$_2$ (41)	32

Note: References 228–415 are on pp. 250–253.

TABLE IX. Acetylenes, Allenes, and Cyclopropanes from Phosphoryl Reagents

No. of C Atoms	Reagent	Reactant	Base/Solvent	Product(s) and Yield(s) (%)	Refs.
C_3	$(CH_3O)_2P(O)CHN_2$	$C_6H_5COCH_3$	n-C_4H_9Li/THF	$C_6H_5C{\equiv}CCH_3$ (16)	359
		$C_6H_5CH_2CHO$	n-C_4H_9Li/THF	$C_6H_5CH_2C{\equiv}CH$ (30)	359
C_7	$(C_2H_5O)_2P(O)CH_2COCH_3$	$CH_3CH{-}CH_2$ (epoxide, O bridging)	NaH/glyme	CH_3–cyclopropane–$COCH_3$ (49)	182
C_8	$(C_2H_5O)_2P(O)CH_2CO_2C_2H_5$	$(C_2H_5)_2C{=}CO$	NaH/glyme	$(C_2H_5)_2C{=}C{=}CHCO_2C_2H_5$ (—)	102
		$CH_3CH{-}CHCH_3$ (epoxide, O bridging)	NaH/glyme	CH_3, CH_3, $CO_2C_2H_5$-substituted cyclopropane (18)	395
		cyclohexene oxide	LiH/C_6H_6	bicyclo[4.1.0]heptane–$CO_2C_2H_5$ (58)	208
	$(C_2H_5O)_2P(O)CH(I)CO_2C_2H_5$	C_6H_5CHO	NaH/glyme	$C_6H_5C{\equiv}CCO_2C_2H_5$ (59)	4
	$(C_2H_5O)_2P(O)CH_2CO_2C_2H_5$	$C_6H_5CH{-}CH_2$ (epoxide, O bridging)	NaH/glyme	C_6H_5–cyclopropane–$CO_2C_2H_5$ (48)	4, 206
		$C_6H_5CH{=}CO$	NaH/glyme	$C_6H_5CH{=}C{=}CHCO_2C_2H_5$ (—)	102
		$C_6H_5(CH_3)C{=}CO$	NaH/glyme	$C_6H_5(CH_3)C{=}C{=}CHCO_2C_2H_5$ (—)	102
	$(C_2H_5O)_2P(O)$–(2-oxotetrahydrofuran-3-yl)	$C_6H_5C(CH_3){=}CO$	NaH/C_6H_6	$C_6H_5C(CH_3){=}C{=}$(2-oxotetrahydrofuran-3-ylidene) (81)	281
		$C_6H_5C(C_2H_5){=}CO$	NaH/C_6H_6	$C_6H_5C(C_2H_5){=}C{=}$(2-oxotetrahydrofuran-3-ylidene) (100)	281
	$(C_2H_5O)_2P(O)CH_2CO_2C_2H_5$	$C_6H_5(C_2H_5)C{=}CO$	NaH/glyme	$C_6H_5(C_2H_5)C{=}C{=}CHCO_2C_2H_5$ (—)	4

		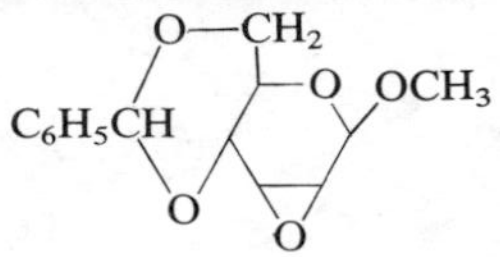 C_6H_5CH / O—CH_2 / O / OCH_3 / O / O	NaH/dioxane	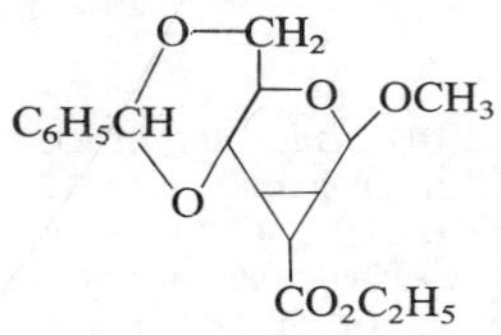 C_6H_5CH / O—CH_2 / O / OCH_3 / O / $CO_2C_2H_5$ (46)	360, 361
	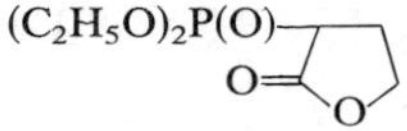 $(C_2H_5O)_2P(O)$— / O= / O	$(C_6H_5)_2C{=}CO$	NaH/C_6H_6	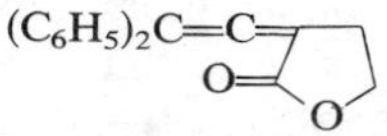 $(C_6H_5)_2C{=}C{=}$ / O= / O (100)	281
C_9	$(C_2H_5O)_2P(O)CH(CH_3)CO_2C_2H_5$	$C_6H_5(CH_3)C{=}CO$	NaH/glyme	$C_6H_5(CH_3)C{=}C{=}C(CH_3)CO_2C_2H_5$ (67)	102
		$C_6H_5(C_2H_5)C{=}CO$	NaH/glyme	$C_6H_5(C_2H_5)C{=}C{=}C(CH_3)CO_2C_2H_5$ (81)	102
C_{10}	$(C_2H_5O)_2P(O)CH(C_2H_5)CO_2C_2H_5$	$C_6H_5(CH_3)C{=}CO$	NaH/glyme	$C_6H_5(CH_3)C{=}C{=}C(C_2H_5)CO_2C_2H_5$ (75)	102
		$C_6H_5(C_2H_5)C{=}CO$	NaH/glyme	$C_6H_5(C_2H_5)C{=}C{=}C(C_2H_5)CO_2C_2H_5$ (80)	102
	$C_6H_5(CH_3O)P(O)CH_2CO_2CH_3$	$C_6H_5(C_2H_5)C{=}CO$	NaH/glyme	$C_6H_5(C_2H_5)C{=}C{=}CHCO_2CH_3$ (42)	373
C_{11}	$(C_2H_5O)_2P(O)CH(Br)C_6H_4NO_2$-*p*	C_6H_5CHO	C_2H_5ONa/C_2H_5OH	$C_6H_5C{\equiv}CC_6H_4NO_2$-*p* (—)	62
C_{14}	$(C_2H_5O)_2P(O)CH(C_6H_5)CO_2C_2H_5$	$CH_3(C_2H_5)C{=}CO$	NaH/glyme	$CH_3(C_2H_5)C{=}C{=}C(C_6H_5)CO_2C_2H_5$ (74)	102
		$(C_6H_5)_2C{=}CO$	NaH/glyme	$(C_6H_5)_2C{=}C{=}C(C_6H_5)CO_2C_2H_5$ (73)	102
C_{16}	$[(C_6H_5)_2P(O)C{=}C(CH_3)_2]MgBr$	C_6H_5CHO	NaH/ether	$C_6H_5CH{=}C{=}C(CH_3)_2$ (85)	103
C_{19}	$(C_6H_5)_2P(O)CH_2C_6H_5$	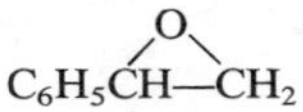 O / $C_6H_5CH{-}CH_2$	C_6H_5Li/ether	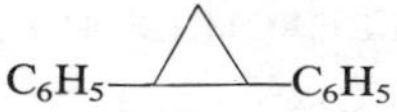 C_6H_5— —C_6H_5 (37)	362, 204
C_{21}	$[(C_6H_5)_2P(O)C{=}C(CH_3)C_6H_5]MgBr$	C_6H_5CHO	NaH/ether	$C_6H_5CH{=}C{=}C(CH_3)C_6H_5$ (85)	103
C_{22}	$(C_6H_5)_2P(O)CH{=}C(C_4H_9\text{-}t)NHC_4H_9$-*n*	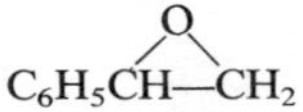 O / $C_6H_5CH{-}CH_2$	*n*-C_4H_9Li/THF	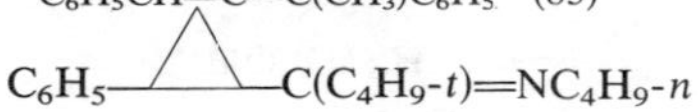 C_6H_5— —$C(C_4H_9\text{-}t){=}NC_4H_9$-*n* (70)	44
C_{23}	$(C_6H_5)_2P(O)CH_2C_{10}H_7$-2	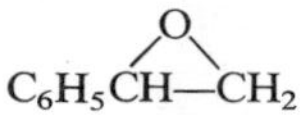 O / $C_6H_5CH{-}CH_2$	C_6H_5Li/ether	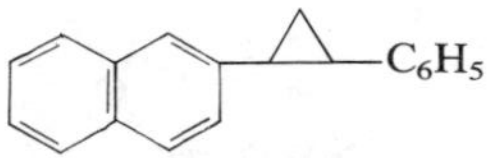 —C_6H_5 (63)	362
C_{24}	$(C_6H_5)_2P(O)CH{=}C(C_6H_5)NHC_4H_9$-*n*	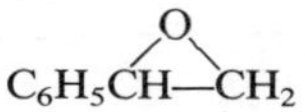 O / $C_6H_5CH{-}CH_2$	*n*-C_4H_9Li/THF	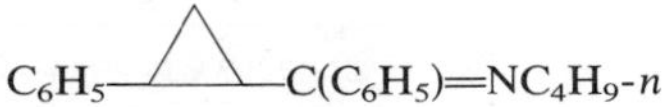 C_6H_5— —$C(C_6H_5){=}NC_4H_9$-*n* (56)	44
	$(C_6H_5)_2P(O)CH{=}C(C_6H_5)NHC_4H_9$-*t*	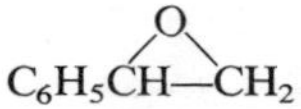 O / $C_6H_5CH{-}CH_2$	*n*-C_4H_9Li/THF	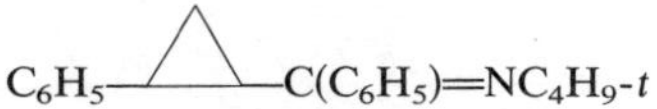 C_6H_5— —$C(C_6H_5){=}NC_4H_9$-*t* (60)	44
C_{26}	$[(C_6H_5)_2P(O)C{=}C(C_6H_5)_2]MgBr$	C_6H_5CHO	NaH/ether	$C_6H_5CH{=}C{=}C(C_6H_5)_2$ (85)	103

Note: References 228–415 are on pp. 250–253.

TABLE X. Reactions of P(O)-Stabilized Anions that Do Not Lead to Phosphate Elimination

No. of C Atoms	Reagent	Reactant	Base/Solvent	Product(s) and Yield(s) (%)	Refs.
C_5	$(C_2H_5O)_2P(O)CH_2Cl$	$(CH_3)_2CO$	NaH/DMSO	$\underbrace{(CH_3)_2C\text{—}CH}_{O}P(O)(OC_2H_5)_2$ (–)	220
		cyclohexanone	NaH/DMSO	$\underbrace{(CH_2)_5C\text{—}CH}_{O}P(O)(OC_2H_5)_2$ (–)	220
		C_6H_5CHO	NaH/DMSO	$\underbrace{C_6H_5CH\text{—}CH}_{O}P(O)(OC_2H_5)_2$ (–)	220
	$[(CH_3)_2N]_2P(O)CH_2Cl$	$p\text{-}ClC_6H_4CHO$	$n\text{-}C_4H_9Li$/THF	$p\text{-}ClC_6H_4\underbrace{CH\text{—}CH}_{O}P(O)[N(CH_3)_2]_2$ (85)	363
	$(C_2H_5O)_2P(O)CH_2Cl$	$C_6H_5COCH_3$	NaH/DMSO	$\underbrace{C_6H_5C(CH_3)\text{—}CH}_{O}P(O)(OC_2H_5)_2$ (–)	220
		$(C_6H_5)_2CO$	NaH/DMSO	$\underbrace{(C_6H_5)_2C\text{—}CH}_{O}P(O)(OC_2H_5)_2$ (–)	220
C_6	$(C_2H_5O)_2P(O)CH_2CN$	$(CH_3)_2CO$	Piperidine/C_2H_5OH	$(CH_3)_2C{=}C(CN)P(O)(OC_2H_5)_2$ (30)	218
		C_2H_5CHO	Piperidine/CH_3OH	$C_2H_5CH{=}C(CN)P(O)(OC_2H_5)_2$ (70)	213
		$CH_2{=}CHCHO$	Piperidine/C_6H_6	$CH_2{=}CHCH{=}C(CN)P(O)(OC_2H_5)_2$ (20)	213
		$n\text{-}C_3H_7CHO$	Piperidine/CH_3OH	$CH_3(CH_2)_2CH{=}C(CN)P(O)(OC_2H_5)_2$ (70)	213
		$i\text{-}C_3H_7CHO$	Piperidine/CH_3OH	$(CH_3)_2CHCH{=}C(CN)P(O)(OC_2H_5)_2$ (70)	213
		$C_2H_5O_2CCO_2C_2H_5$	NaH/ether	$C_2H_5O_2CCOCH(CN)P(O)(OC_2H_5)_2$ (13)	100

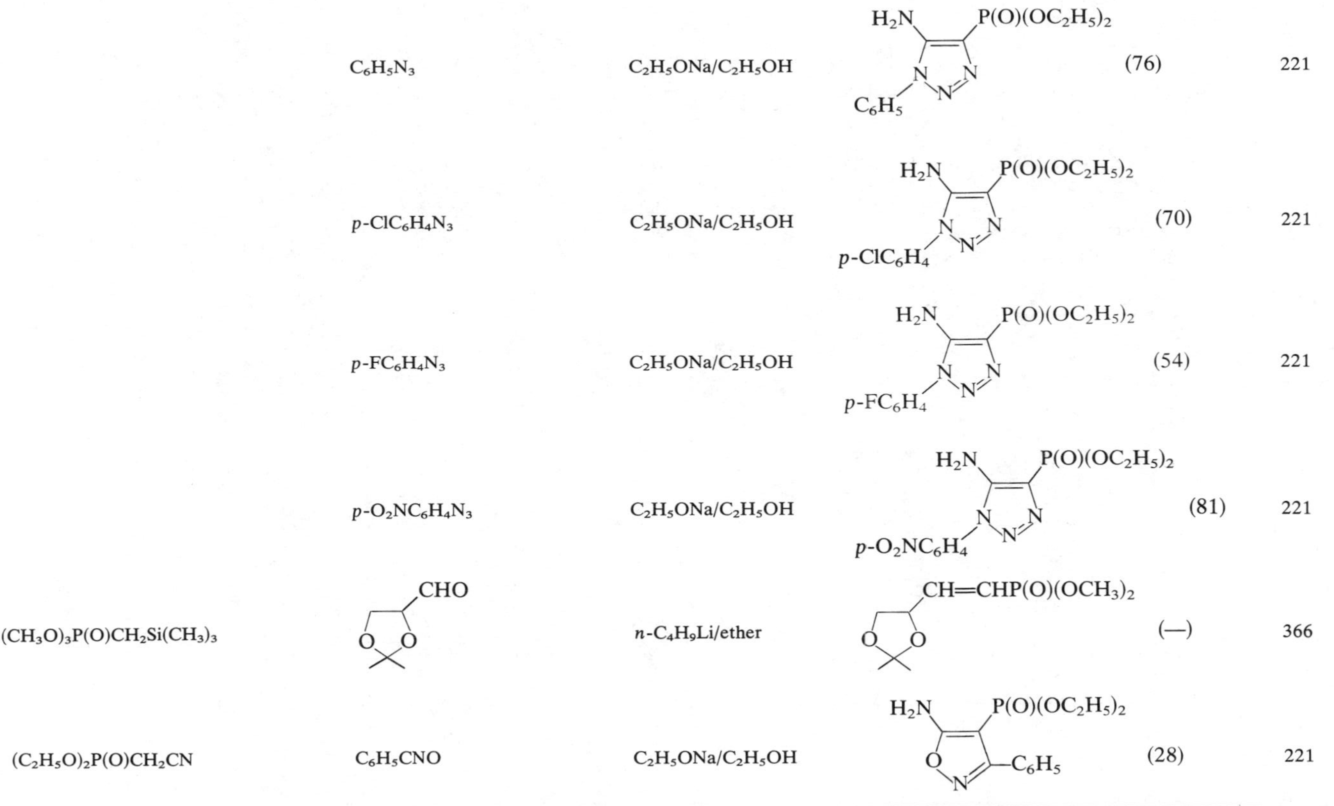

	$C_6H_5N_3$	C_2H_5ONa/C_2H_5OH	H_2N, $P(O)(OC_2H_5)_2$, N, N, N, C_6H_5	(76)	221
	p-$ClC_6H_4N_3$	C_2H_5ONa/C_2H_5OH	H_2N, $P(O)(OC_2H_5)_2$, N, N, N, p-ClC_6H_4	(70)	221
	p-$FC_6H_4N_3$	C_2H_5ONa/C_2H_5OH	H_2N, $P(O)(OC_2H_5)_2$, N, N, N, p-FC_6H_4	(54)	221
	p-$O_2NC_6H_4N_3$	C_2H_5ONa/C_2H_5OH	H_2N, $P(O)(OC_2H_5)_2$, N, N, N, p-$O_2NC_6H_4$	(81)	221
$(CH_3O)_3P(O)CH_2Si(CH_3)_3$	CHO	n-C_4H_9Li/ether	$CH{=}CHP(O)(OCH_3)_2$, O, O	(—)	366
$(C_2H_5O)_2P(O)CH_2CN$	C_6H_5CNO	C_2H_5ONa/C_2H_5OH	H_2N, $P(O)(OC_2H_5)_2$, O, N, C_6H_5	(28)	221

Note: References 228–415 are on pp. 250–253.

TABLE X. Reactions of P(O)-Stabilized Anions that Do Not Lead to Phosphate Elimination (*Continued*)

No. of C Atoms	Reagent	Reactant	Base/Solvent	Product(s) and Yield(s) (%)	Refs.
C_6 (*Contd.*)	$(C_2H_5O)_2P(O)CH_2CN$ (*Contd.*)	*p*-ClC_6H_4CNO	C_2H_5ONa/C_2H_5OH	5-amino-3-(C_6H_4Cl-*p*)-4-[$P(O)(OC_2H_5)_2$]isoxazole (H_2N, $P(O)(OC_2H_5)_2$, C_6H_4Cl-*p*) (31)	221
		p-FC_6H_4CNO	C_2H_5ONa/C_2H_5OH	5-amino-3-(C_6H_4F-*p*)-4-[$P(O)(OC_2H_5)_2$]isoxazole (H_2N, $P(O)(OC_2H_5)_2$, C_6H_4F-*p*) (–)	221
		p-$O_2NC_6H_4CNO$	C_2H_5ONa/C_2H_5OH	5-amino-3-($C_6H_4NO_2$-*p*)-4-[$P(O)(OC_2H_5)_2$]isoxazole (H_2N, $P(O)(OC_2H_5)_2$, $C_6H_4NO_2$-*p*) (23)	213, 162
		C_6H_5CHO	Piperidine/CH_3OH	$C_6H_5CH{=}C(CN)P(O)(OC_2H_5)_2$ (70)	213, 162
		p-ClC_6H_4CHO	Piperidine/CH_3OH	*p*-$ClC_6H_4CH{=}C(CN)P(O)(OC_2H_5)_2$ (—)	213
		p-$O_2NC_6H_4CHO$	Piperidine/CH_3OH	*p*-$O_2NC_6H_4CH{=}C(CN)P(O)(OC_2H_5)_2$ (61)	213
		m-$O_2NC_6H_4CHO$	Piperidine/C_6H_6	*m*-$O_2NC_6H_4CH{=}C(CN)P(O)(OC_2H_5)_2$ (—)	214
	$C_2H_5O(C_2H_5)P(O)CH_2CN$	C_6H_5CHO	Piperidine/C_6H_6	$C_6H_5CH{=}C(CN)P(O)(C_2H_5)OC_2H_5$ (60)	61
	$(CH_3O)_2P(O)CH_2CH{=}C(CH_3)Cl$	C_6H_5CHO	NaH/glyme	$C_6H_5\overset{O}{CH{-}C}(CH_3)CH{=}CHP(O)(OCH_3)_2$ (epoxide) (–)	61

		C_6H_5CHO	NaH/glyme	CH_3–(2,5-dihydrofuran)–$P(O)(OCH_3)_2$, C_6H_5 (–)	61
	$(C_2H_5O)_2P(O)CH_2CN$	$C_6H_5COCH_3$	Piperidine/C_6H_6	$C_6H_5C(CH_3){=}C(CN)P(O)(OC_2H_5)_2$ (—)	218
		p-$ClC_6H_4COCH_3$	Piperidine/C_6H_6	p-$ClC_6H_4C(CH_3){=}C(CN)P(O)(OC_2H_5)_2$ (—)	218
	$[(CH_3)_2N]_2P(O)CH_2CH_3$	$C_6H_5CO_2CH_3$	n-C_4H_9Li/THF	$C_6H_5COCH(CH_3)P(O)[N(CH_3)_2]_2$ (—)	34
	$(C_2H_5O)_2P(O)CH_2CN$	p-$(CH_3)_2NC_6H_4CHO$	Piperidine/C_6H_6	p-$(CH_3)_2NC_6H_4CH{=}C(CN)P(O)(OC_2H_5)_2$ (—)	213
		$C_6H_5CH{=}CHCHO$	Piperidine/CH_3OH	$C_6H_5CH{=}CHCH{=}C(CN)P(O)(OC_2H_5)_2$ (—)	213, 214
		p-$CH_3C_6H_4COCH_3$	Piperidine/C_6H_6	p-$CH_3C_6H_4C(CH_3){=}C(CN)P(O)(OC_2H_5)_2$ (—)	218
		2,4,6-$(CH_3)_3C_6H_2N_3$	C_2H_5ONa/C_2H_5OH	2,4,6-$(CH_3)_3C_6H_2$—N(triazole), H_2N, $P(O)(OC_2H_5)_2$ (–)	221
	$(C_2H_5)_2P(O)CH_2CN$	$C_6H_5CH{=}CHCHO$	Piperidine/C_6H_6	$C_6H_5CH{=}CHCH{=}C(CN)P(O)(C_2H_5)_2$ (56)	214
		2,4,6-$(CH_3)_3C_6H_2CNO$	C_2H_5ONa/C_2H_5OH	H_2N, $P(O)(C_2H_5)_2$ (isoxazole)—$C_6H_2(CH_3)_3$-2,4,6	221
C_7	$(C_2H_5O)_2P(O)CH_2COCH_3$	CH_2O	Piperidine/CH_3OH	$CH_2{=}C(COCH_3)P(O)(OC_2H_5)_2$ (11)	221
		CH_3CHO	Piperidine/C_6H_6	$CH_3CH{=}C(COCH_3)P(O)(OC_2H_5)_2$ (28)	221
		C_2H_5CHO	Piperidine/C_6H_6	$C_2H_5CH{=}C(COCH_3)P(O)(OC_2H_5)_2$ (40)	212
		$CH_2{=}CHCO_2CH_3$	C_2H_5ONa/C_2H_5OH	$CH_3COCH(CH_2CH_2CO_2CH_3)P(O)(OC_2H_5)_2$ (—)	92, 93, 94, 95
		C_6H_5CHO	Piperidine/C_6H_6	$C_6H_5CH{=}C(COCH_3)P(O)(OC_2H_5)_2$ (67)	215, 216
		p-ClC_6H_4CHO	Piperidine/C_6H_6	p-$ClC_6H_4CH{=}C(COCH_3)P(O)(OC_2H_5)_2$ (35)	213
		$C_6H_5CH{=}CHCHO$	Piperidine/C_6H_6	$C_6H_5CH{=}CHCH{=}C(COCH_3)P(O)(OC_2H_5)_2$ (72)	215
C_8	$(n$-$C_3H_7O)_2P(O)CH_2CN$	CH_2O	Piperidine/CH_3OH	$CH_2{=}C(CN)P(O)(OC_3H_7$-$n)_2$ (40)	213

Note: References 228–415 are on pp. 250–253.

TABLE X. Reactions of P(O)-Stabilized Anions that Do Not Lead to Phosphate Elimination (*Continued*)

No. of C Atoms	Reagent	Reactant	Base/Solvent	Product(s) and Yield(s) (%)	Refs.
C_8 (*Contd.*)	$(C_2H_5O)_2P(O)CH_2CO_2C_2H_5$	$ClCH_2CHO$	$O(CH_2CH_2)_2NCH_3$ + $TiCl_4$/THF	$ClCH_2CH{=}C(CO_2C_2H_5)P(O)(OC_2H_5)_2$ (58)	219
		$CH_2{=}CHCN$	C_2H_5ONa/C_2H_5OH	$C_2H_5O_2CC(CH_2CH_2CN)_2P(O)(OC_2H_5)_2$ (—)	92, 93
	$(C_2H_5O)_2P(O)CH_2Si(CH_3)_3$	$(CH_3)_2CO$	n-C_4H_9Li/THF	$(CH_3)_2C{=}CHP(O)(OC_2H_5)_2$ (55)	364
		i-C_3H_7CHO	n-C_4H_9Li/THF	i-$C_3H_7CH{=}CHP(O)(OC_2H_5)_2$ (92)	364
	$(C_2H_5O)_2P(O)CH_2CO_2C_2H_5$	n-C_3H_7CHO	$O(CH_2CH_2)_2NCH_3$ + $TiCl_4$/THF	n-$C_3H_7CH{=}C(CO_2C_2H_5)P(O)(OC_2H_5)_2$ (63)	219
		i-C_3H_7CHO	$O(CH_2CH_2)_2NCH_3$ + $TiCl_4$/THF	i-$C_3H_7CH{=}C(CO_2C_2H_5)P(O)(OC_2H_5)_2$ (85)	219
		$CH_3CH{=}CHCHO$	$O(CH_2CH_2)_2NCH_3$ + $TiCl_4$/THF	$CH_3CH{=}CHCH{=}C(CO_2C_2H_5)P(O)(OC_2H_5)_2$ (55)	219

$CF_3CO_2C_2H_5$	NaH/glyme	$CF_3COCH(CO_2C_2H_5)P(O)(OC_2H_5)_2$ (—)	100
$C_2H_5O_2CCCl_3$	NaH/glyme	$C_2H_5O_2CCH(Cl)P(O)(OC_2H_5)_2$ (52)	100
$C_2H_5O_2CCH_2Cl$	NaH/ether	$C_2H_5O_2CCH_2CH(CO_2C_2H_5)P(O)(OC_2H_5)_2$ (65)	100
$t\text{-}C_4H_9CHO$	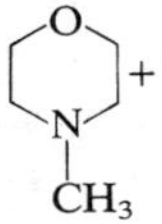 [N-methylmorpholine: O, N–CH_3 ring] + $TiCl_4$/THF	$t\text{-}C_4H_9CH{=}C(CO_2C_2H_5)P(O)(OC_2H_5)_2$ (70)	219
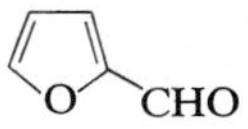 [2-furyl]CHO	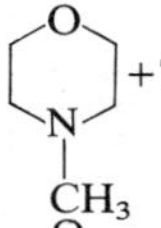 [N-methylmorpholine: O, N–CH_3 ring] + $TiCl_4$/THF	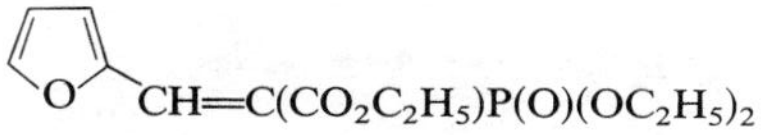 [2-furyl]$CH{=}C(CO_2C_2H_5)P(O)(OC_2H_5)_2$ (90)	219
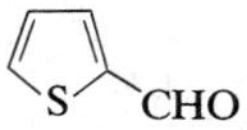 [2-thienyl]CHO	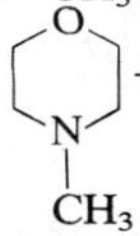 [N-methylmorpholine: O, N–CH_3 ring] + $TiCl_4$/THF	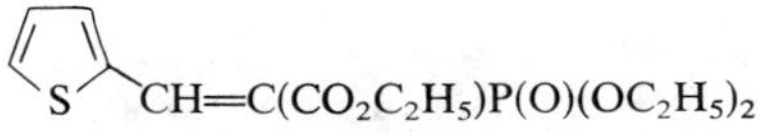 [2-thienyl]$CH{=}C(CO_2C_2H_5)P(O)(OC_2H_5)_2$ (94)	219
$Cl_3CCH{=}CHCO_2C_2H_5$	NaH/ether	 $C_2H_5O_2CCH$—$CHCH(CO_2C_2H_5)P(O)(OC_2H_5)_2$ [cyclopropane ring bridged by CCl_2] (50)	100
$p\text{-}ClC_6H_4N_3$	C_2H_5ONa/C_2H_5OH	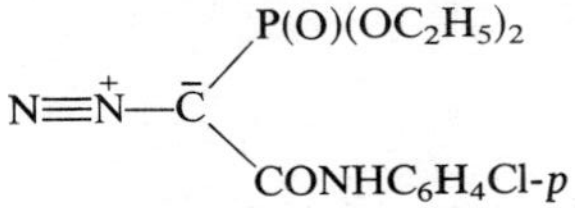 $N{\equiv}\overset{+}{N}{-}\bar{C}(P(O)(OC_2H_5)_2)CONHC_6H_4Cl\text{-}p$ (68)	221
$p\text{-}NO_2C_6H_4N_3$	C_2H_5ONa/C_2H_5OH	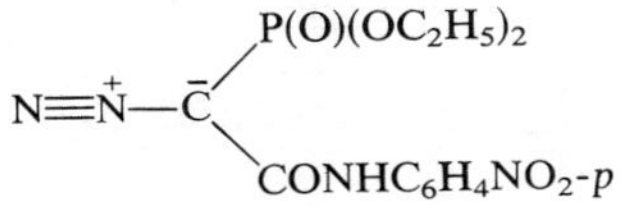 $N{\equiv}\overset{+}{N}{-}\bar{C}(P(O)(OC_2H_5)_2)CONHC_6H_4NO_2\text{-}p$ (65)	221

Note: References 228–415 are on pp. 250–253.

TABLE X. REACTIONS OF P(O)-STABILIZED ANIONS THAT DO NOT LEAD TO PHOSPHATE ELIMINATION (*Continued*)

No. of C Atoms	Reagent	Reactant	Base/Solvent	Product(s) and Yield(s) (%)	Refs.
C_8 (*Contd.*)	$(C_2H_5O)_2P(O)CH_2Si(CH_3)_3$	Cyclohexanone	n-C_4H_9Li/THF	1-[$CH_2P(O)(OC_2H_5)_2$]cyclohexene (65)	364
	$(C_2H_5O)_2P(O)CH_2CO_2C_2H_5$	C_6H_5CHO	N-Methylmorpholine + $TiCl_4$/THF	$C_6H_5CH{=}C(CO_2C_2H_5)P(O)(OC_2H_5)_2$ (94)	219
		o-ClC_6H_4CHO	N-Methylmorpholine + $TiCl_4$/THF	o-$ClC_6H_4CH{=}C(CO_2C_2H_5)P(O)(OC_2H_5)_2$ (89)	219
		p-ClC_6H_4CHO	N-Methylmorpholine + $TiCl_4$/THF	p-$ClC_6H_4CH{=}C(CO_2C_2H_5)P(O)(OC_2H_5)_2$ (86)	219
		p-$O_2NC_6H_4CHO$	N-Methylmorpholine + $TiCl_4$/THF	p-$O_2NC_6H_4CH{=}C(CO_2C_2H_5)P(O)(OC_2H_5)_2$ (96)	219
		C_6H_5CHO	Piperidine/C_6H_6	$C_6H_5CH{=}C(CO_2C_2H_5)P(O)(OC_2H_5)_2$ (63)	217
		p-ClC_6H_4CHO	Piperidine/C_6H_6	p-$ClC_6H_4CH{=}C(CO_2C_2H_5)P(O)(OC_2H_5)_2$ (26)	217, 213

	$o\text{-}HOC_6H_4CHO$	Piperidine/C_6H_6	$P(O)(OC_2H_5)_2$ (62)	217
$(C_2H_5O)_2P(O)CH_2Si(CH_3)_3$	C_6H_5CHO	$n\text{-}C_4H_9Li$/THF	$C_6H_5CH{=}CHP(O)(OC_2H_5)_2$ (63)	364
$(C_2H_5O)_2P(O)CH_2CO_2C_2H_5$	$p\text{-}CH_3OC_6H_4CHO$	+ $TiCl_4$/THF	$p\text{-}CH_3OC_6H_4CH{=}C(CO_2C_2H_5)P(O)(OC_2H_5)_2$ (92)	219
	$p\text{-}CH_3OC_6H_4CHO$	Piperidine/C_6H_6	$p\text{-}CH_3OC_6H_4CH{=}C(CO_2C_2H_5)P(O)(OC_2H_5)_2$ (69)	217
	$p\text{-}CH_3C_6H_4CHO$	Piperidine/C_6H_6	$p\text{-}CH_3C_6H_4CH{=}C(CO_2C_2H_5)P(O)(OC_2H_5)_2$ (36)	217
	$C_2H_5O_2CCH{=}CHCO_2C_2H_5$	NaH/ether	$C_2H_5O_2CCH_2CH(CO_2C_2H_5)CH(CO_2C_2H_5)P(O)\text{-}(OC_2H_5)_2$ (49)	100, 92
	$C_6H_5CH{=}CHCHO$	Piperidine/C_6H_6	$C_6H_5CH{=}CHCH{=}C(CO_2C_2H_5)P(O)(OC_2H_5)_2$ (—)	213
	$3,4\text{-}(CH_3O)_2C_6H_3CH{=}CHCHO$	Piperidine/C_6H_6	$3,4\text{-}(CH_3O)_2C_6H_3CH{=}CHCH{=}C(CO_2C_2H_5)\text{-}P(O)(OC_2H_5)_2$	217
	$p\text{-}(CH_3)_2NC_6H_4CHO$	+ $TiCl_4$/THF	$p\text{-}(CH_3)_2NC_6H_4CH{=}C(CO_2C_2H_5)P(O)\text{-}(OC_2H_5)_2$ (87)	219
$(C_2H_5O)_2P(O)CH_2Si(CH_3)_3$	$C_6H_5CON(CH_3)_2$	$n\text{-}C_4H_9Li$/THF	$C_6H_5C[N(CH_3)_2]{=}CHP(O)(OC_2H_5)_2$ (24)	364
$(C_2H_5O)_2P(O)CH_2CO_2C_2H_5$	$1\text{-}C_{10}H_7CHO$	Piperidine/C_6H_6	$1\text{-}C_{10}H_7CH{=}C(CO_2C_2H_5)P(O)(OC_2H_5)_2$ (28)	217
	$C_6H_5CH{=}CHCO_2C_2H_5$	C_2H_5ONa/C_2H_5OH	$C_2H_5O_2CCH[CH(C_6H_5)CH_2CO_2C_2H_5]P(O)\text{-}(OC_2H_5)_2$ (—)	92, 93
		+ $TiCl_4$/THF	$C(CO_2C_2H_5)P(O)(OC_2H_5)_2$ (70)	219

Note: References 228–415 are on pp. 250–253.

No. of C Atoms	Reagent	Reactant	Base/Solvent	Product(s) and Yield(s) (%)	Refs.
C_8 (*Contd.*)	$(C_2H_5O)_2P(O)CH_2Si(CH_3)_3$	$(C_6H_5)_2CO$	n-C_4H_9Li/THF	$(C_6H_5)_2C{=}CHP(O)(OC_2H_5)_2$ (83)	364
		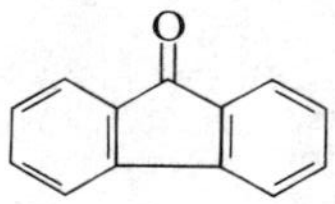 O	n-C_4H_9Li/THF	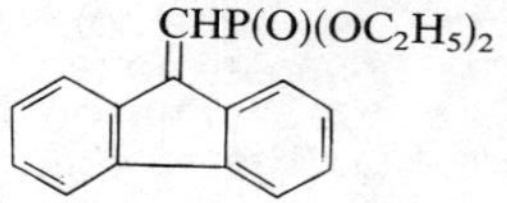 $CHP(O)(OC_2H_5)_2$ (42)	364
	$(C_2H_5O)_2P(O)CH_2CO_2C_2H_5$	$C_6H_5CH{=}CHCOC_6H_5$	C_2H_5ONa/C_6H_6	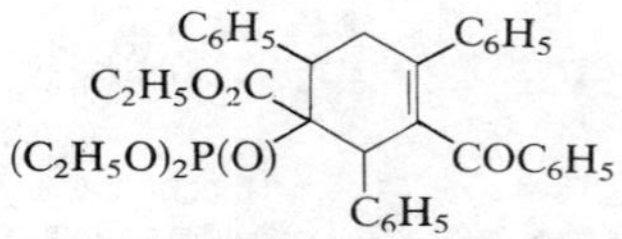 C_6H_5, C_6H_5, $C_2H_5O_2C$, $(C_2H_5O)_2P(O)$, COC_6H_5, C_6H_5 (–)	96
		$C_6H_5CH{=}CHCOC_6H_5$	$NaNH_2$/ether	$C_6H_5COCH_2CH(C_6H_5)CH(CO_2C_2H_5)P(O)(OC_2H_5)_2$ (—)	96
C_9	$(C_2H_5O)_2P(O)CH(CH_3)CO_2C_2H_5$	$CH_2{=}CHCN$	C_2H_5ONa/C_2H_5OH	$C_2H_5O_2CC(CH_3)(CH_2CH_2CN)P(O)(OC_2H_5)_2$ (—)	92–95
	$[(C_2H_5O)_2P(O)]_2CH_2$	$CH_2{=}CHCN$	Piperidine/xylene	$[(C_2H_5O)_2P(O)]_2C(CH_2CH_2CN)_2$ (54)	195
		$CH_2{=}CHCO_2CH_3$	Piperidine/xylene	$[(C_2H_5O)_2P(O)]_2CHCH_2CH_2CO_2CH_3$ (42)	195
		i-C_3H_7CHO	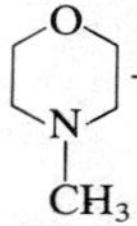 O, N, CH_3 + $TiCl_4$/THF	i-$C_3H_7CH{=}C[P(O)(OC_2H_5)_2]_2$ (50)	219
		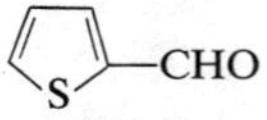 S, –CHO	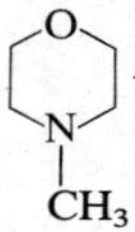 O, N, CH_3 + $TiCl_4$/THF	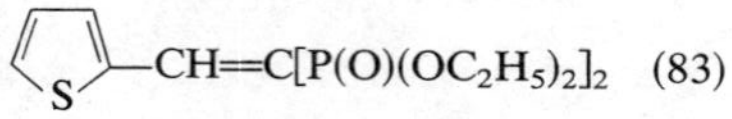 S, $-CH{=}C[P(O)(OC_2H_5)_2]_2$ (83)	219

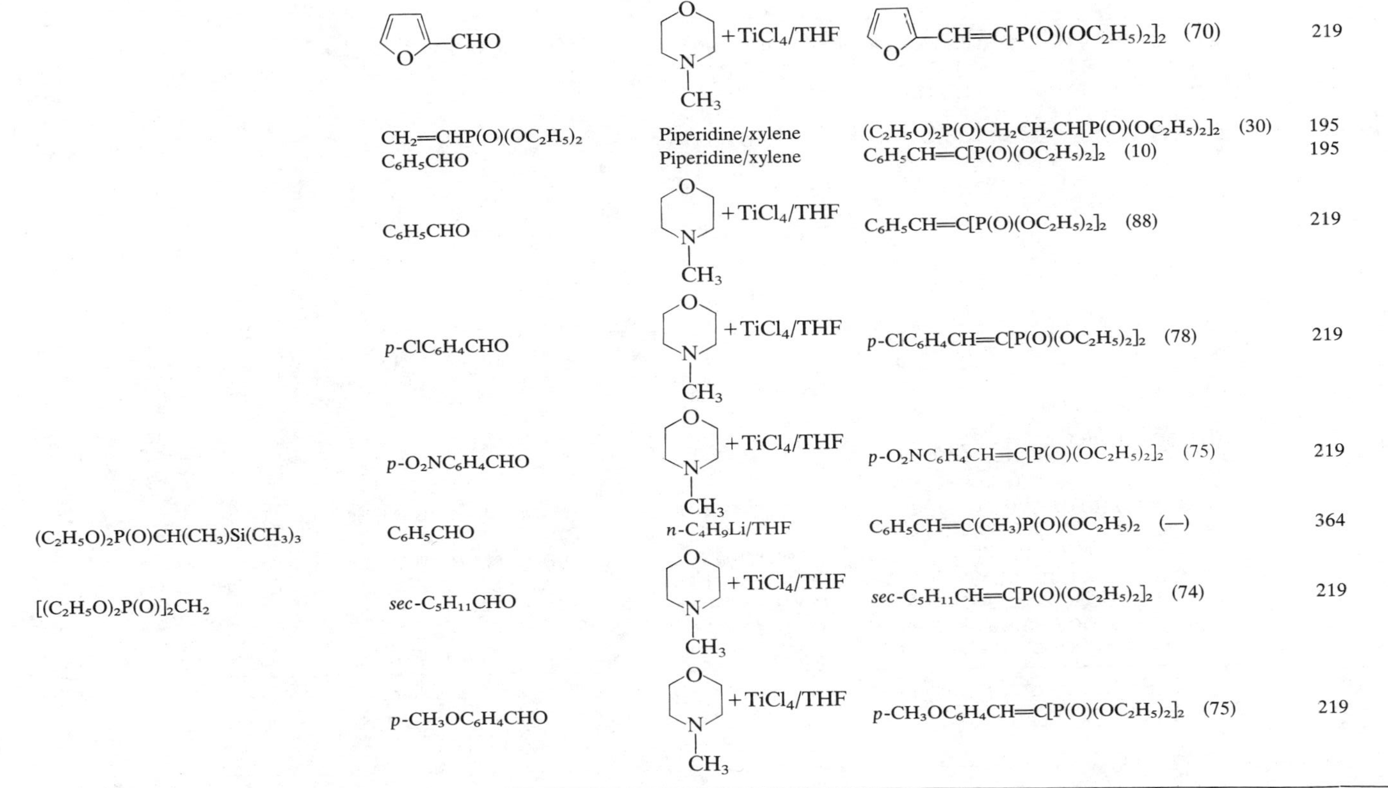

	—CHO	$+ TiCl_4$/THF	—$CH{=}C[P(O)(OC_2H_5)_2]_2$ (70)	219
	$CH_2{=}CHP(O)(OC_2H_5)_2$	Piperidine/xylene	$(C_2H_5O)_2P(O)CH_2CH_2CH[P(O)(OC_2H_5)_2]_2$ (30)	195
	C_6H_5CHO	Piperidine/xylene	$C_6H_5CH{=}C[P(O)(OC_2H_5)_2]_2$ (10)	195
	C_6H_5CHO	$+ TiCl_4$/THF	$C_6H_5CH{=}C[P(O)(OC_2H_5)_2]_2$ (88)	219
	p-ClC_6H_4CHO	$+ TiCl_4$/THF	p-$ClC_6H_4CH{=}C[P(O)(OC_2H_5)_2]_2$ (78)	219
	p-$O_2NC_6H_4CHO$	$+ TiCl_4$/THF	p-$O_2NC_6H_4CH{=}C[P(O)(OC_2H_5)_2]_2$ (75)	219
$(C_2H_5O)_2P(O)CH(CH_3)Si(CH_3)_3$	C_6H_5CHO	n-C_4H_9Li/THF	$C_6H_5CH{=}C(CH_3)P(O)(OC_2H_5)_2$ (—)	364
$[(C_2H_5O)_2P(O)]_2CH_2$	*sec*-$C_5H_{11}CHO$	$+ TiCl_4$/THF	*sec*-$C_5H_{11}CH{=}C[P(O)(OC_2H_5)_2]_2$ (74)	219
	p-$CH_3OC_6H_4CHO$	$+ TiCl_4$/THF	p-$CH_3OC_6H_4CH{=}C[P(O)(OC_2H_5)_2]_2$ (75)	219

Note: References 228–415 are on pp. 250–253.

TABLE X. Reactions of P(O)-Stabilized Anions that Do Not Lead to Phosphate Elimination (*Continued*)

No. of C Atoms	Reagent	Reactant	Base/Solvent	Product(s) and Yield(s) (%)	Refs.
C_{10}	$(n\text{-}C_4H_9O)_2P(O)CH_2CN$	CH_3CHO	Piperidine/CH_3OH	$CH_3CH{=}C(CN)P(O)(OC_4H_9\text{-}n)_2$ (70)	213
		C_2H_5CHO	Piperidine/CH_3OH	$C_2H_5CH{=}C(CN)P(O)(OC_4H_9\text{-}n)_2$ (70)	213
		$n\text{-}C_3H_7CHO$	Piperidine/CH_3OH	$n\text{-}C_3H_7CH{=}C(CN)P(O)(OC_4H_9\text{-}n)_2$ (70)	213
C_{11}	$(C_2H_5O)_2P(O)CH_2C_6H_5$	$C_2H_5O_2CCO_2C_2H_5$	NaH/glyme	$C_2H_5O_2CCOCH(C_6H_5)P(O)(OC_2H_5)_2$ (—)	100
		$C_6H_5CH{=}CHCOC_6H_5$	C_2H_5ONa/C_6H_6	$[C_6H_5COCH_2CH(C_6H_5)]_2C(C_6H_5)P(O)(OC_2H_5)_2$ (—)	96
		$C_6H_5CH{=}CHCOC_6H_5$	NaH_2/ether	$[C_6H_5COCH_2CH(C_6H_5)]CH(C_6H_5)P(O)(OC_2H_5)_2$ (—)	96
C_{12}	$(C_2H_5O)_2P(O)CH_2COC_6H_5$	$p\text{-}ClC_6H_4N_3$	C_2H_5ONa/C_2H_5OH	1-($p\text{-}ClC_6H_4$)-5-C_6H_5-1,2,3-triazol-4-yl $P(O)(OH)(OC_2H_5)$ (8)	221
		$p\text{-}O_2NC_6H_4N_3$	C_2H_5ONa/C_2H_5OH	1-($p\text{-}O_2NC_6H_4$)-5-C_6H_5-1,2,3-triazol-4-yl $P(O)(OH)(OC_2H_5)$ (24)	221
C_{13}	$(C_6H_5)_2P(O)CH_3$	CO_2	$n\text{-}C_4H_9Li$/ether	$HO_2CCH_2P(O)(C_6H_5)_2$ (63)	202, 190
	$(n\text{-}C_6H_{13})_2P(O)CH_3$	CO_2	$n\text{-}C_4H_9Li$/ether	$HO_2CCH_2P(O)(C_6H_{13}\text{-}n)_2$ (67)	202
		CH_3CHO	$n\text{-}C_4H_9Li$/ether	$CH_3CH(OH)CH_2P(O)(C_6H_{13}\text{-}n)_2$ (44)	202
		$CH_3OCO_2CH_3$	$n\text{-}C_4H_9Li$/ether	$CH_3O_2CCH_2P(O)(C_6H_{13}\text{-}n)_2$ (50)	202
		$n\text{-}C_3H_7CHO$	$n\text{-}C_4H_9Li$/ether	$n\text{-}C_3H_7CH(OH)CH_2P(O)(C_6H_{13}\text{-}n)_2$ (33)	202

	$(C_6H_5)_2P(O)CH_3$	$(C_2H_5)_2CO$	n-C_4H_9Li/ether	$(C_2H_5)_2C(OH)CH_2P(O)(C_6H_{13}$-$n)_2$ (46)	202
		$(C_2H_5)_2CO$	n-C_4H_9Li/ether	$(C_2H_5)_2C(OH)CH_2P(O)(C_6H_5)_2$ (61)	202
		cyclohexane epoxide (ring with O)	C_6H_5Li/ether	2-(OH)-cyclohexyl-$CH_2P(O)(C_6H_5)_2$ (92)	362
	$(n$-$C_6H_{13})_2P(O)CH_3$	C_6H_5CHO	n-C_4H_9Li/ether	$C_6H_5CH(OH)CH_2P(O)(C_6H_5)_2$ (31)	202
		C_6H_5CHO	n-C_4H_9Li/ether	$C_6H_5CH(OH)CH_2P(O)(C_6H_{13}$-$n)_2$ (50)	202
			n-C_4H_9Li/ether	$C_6H_5CH{=}CHP(O)(C_6H_{13}$-$n)_2$ (10)	202
		$C_6H_5COCH_3$	n-C_4H_9Li/ether	$CH_3(C_6H_5)C(OH)CH_2P(O)(C_6H_{13}$-$n)_2$ (51)	202
	$(C_6H_5)_2P(O)CH_3$	$C_6H_5CH\overset{O}{—}CH_2$	n-C_4H_9Li/ether	$HOCH_2CH(C_6H_5)CH_2P(O)(C_6H_5)_2$ (60)	202
		$C_6H_5CO_2C_2H_5$	n-C_4H_9Li/ether	$C_6H_5COCH_2P(O)(C_6H_5)_2$ (40)	202
		$(C_6H_5)_2P(O)Cl$	n-C_4H_9Li/ether	$(C_6H_5)_2P(O)CH_2P(O)(C_6H_5)_2$ (25)	202
		$(C_6H_5)_2CO$	n-C_4H_9Li/ether	$(C_6H_5)_2C(OH)CH_2P(O)(C_6H_5)_2$ (81)	202, 190
	$(n$-$C_6H_{13})_2P(O)CH_3$	$(C_6H_5)_2CO$	n-C_4H_9Li/ether	$(C_6H_5)_2C(OH)CH_2P(O)(C_6H_{13}$-$n)_2$ (49)	202
	$(C_6H_5)_2P(O)CH_3$	$(n$-$C_8H_{17})_2P(O)Cl$	n-C_4H_9Li/ether	$(n$-$C_8H_{17})_2P(O)CH_2P(O)(C_6H_5)_2$ (25)	202
C_{19}	$(C_6H_5)_2P(O)CH_2C_6H_5$	$CH_2\overset{O}{—}CH_2$	C_6H_5Li/ether	$HOCH_2CH_2CH(C_6H_5)P(O)(C_6H_5)_2$ (91)	362
		$C_6H_5CH\overset{O}{—}CH_2$	C_6H_5Li/ether	$HOCH_2CH(C_6H_5)CH(C_6H_5)P(O)(C_6H_5)_2$ (60)	362
		$C_6H_5CH\overset{O}{—}CHC_6H_5$	C_6H_5Li/ether	$C_6H_5CH(OH)CH(C_6H_5)CH(C_6H_5)P(O)$- $(C_6H_5)_2$ (85)	362
	$(C_6H_5)_2P(O)CH(CH_3)$—(cyclohexen-1-yl)	C_6H_5CHO	n-C_4H_9Li/ether	$(C_6H_5)_2P(O)C(CH_3)$(cyclohexenylidene)$CH(C_6H_5)OH$ (—)	377
C_{21}	$(C_6H_5)_2P(O)CH(C_6H_5)CH_2CH_2Cl$	—	t-C_4H_9OK/C_6H_6	(cyclopropyl)—$CH(C_6H_5)P(O)(C_6H_5)_2$ (71)	362

Note: References 228–415 are on pp. 250–253.

REFERENCES TO TABLES

[228] D. Seyferth and R. S. Marmor, *J. Organometal. Chem.*, **59,** 237 (1973).
[229] R. M. Kellogg, M. B. Groen, and H. Wynberg, *J. Org. Chem.*, **32,** 3093 (1967).
[230] B. Yom-Tov and S. Gronowitz, *Chem. Scr.*, **1973,** 37 [*C.A.*, **78,** 97519x (1973)].
[231] S. V. Krivun, S. N. Baranov, and O. F. Voziyanova, *Zh. Obshch. Khim.*, **43,** 359 (1973) [*C.A.*, **79,** 5217s (1973)].
[232] J. M. Bastian, A. Ebnother, E. Jucker, E. Rissi, and A. P. Stoll, *Helv. Chem. Acta*, **49,** 214 (1966).
[233] H. De Koning, G. N. Mallo, A. Springer-Fidder, K. E. C. Subramanian-Erhart, and H. O. Huisman, *Rec. Trav. Chim.*, **92,** 683 (1973).
[234] M. Dvolaitaky, H. B. Kagani, and I. Jacques, *Bull. Soc. Chim. Fr.*, **1961,** 598.
[235] H. De Koning, A. Springer-Fidder, M. J. Moslenaar, and H. O. Huisman, *Rec. Trav. Chim.*, **92,** 237 (1973).
[236] E. J. Sews and C. V. Wilson, *J. Org. Chem.*, **26,** 5243 (1961).
[237] R. Filler and Y. S. Rao, *J. Org. Chem.*, **39,** 3421 (1974).
[238] W. Sahm, E. Schinzel, and G. Roesch, Ger. Pat. 2,105,305 [*C.A.*, **78,** 31420b (1973)].
[239] F. W. Bachelor, A. A. Loman, and L. R. Snowdon, *Can. J. Chem.*, **48,** 1554 (1970).
[240] P. J. Hattersley, I. M. Lockhart, and M. Wright, *J. Chem. Soc., C.*, **1969,** 217.
[241] J. M. Bruce, D. Creed, and K. Dawes, *J. Chem. Soc., C,* **1971,** 3749.
[242] Farbwerke Hoechst A. G., Brit. Pat. 891,178 (1962) [*C.A.*, **60,** 11938e (1964)].
[243] L. Horner, H. Hoffmann, and H. G. Wippel, *Chem. Ber.*, **91,** 61 (1958).
[244] J. A. Eenkhoorn, S. Osmund, D. Silva, and V. Snieckus, *Can. J. Chem.*, **51,** 792 (1973).
[245] P. J. Nelson and A. F. A. Wallis, *Tappi*, **56,** 132 (1973). [*C.A.*, **80,** 14679e (1974)].
[246] L. Horner and H. Winkler, *Tetrahedron Lett.*, **1964,** 3265.
[247] A. Maercker, *Angew. Chem., Int. Ed. Engl.*, **6,** 557 (1967).
[248] D. Redmore, *J. Org. Chem.*, **34,** 1420 (1969).
[249] Farbwerke Hoechst A. G., Ger. Pat. 1,138,757 (1962) [*C.A.*, **58,** 9143a (1962)].
[250] H. Takahashi, K. Fwjiwara, and M. Ohta, *Bull. Chem. Soc. Jpn.*, **35,** 1498 (1962).
[251] W. A. Nasutavicus, S. W. Tobey, and F. Johnson, *J. Org. Chem.*, **32,** 3325 (1967).
[252] J. W. McFarland, L. H. Conover, H. L. Howes, J. E. Lynch, D. R. Chisholm, W. C. Austin, R. L. Cornwell, J. C. Danilewicz, W. Courtney, and D. H. Morgan, *J. Med. Chem.*, **12,** 1066 (1969).
[253] J. A. Babler, D. O. Olsen, and W. H. Arnold, *J. Org. Chem.*, **39,** 1656 (1974).
[254] E. C. Taylor and K. Lenard, *Ann.*, **726,** 100 (1969).
[255] J. A. Marshall and G. M. Cohen, *J. Org. Chem.*, **36,** 877 (1971).
[256] W. Stilz and H. Pommer, Ger. Pat. 1,108,208 (1961) [*C.A.*, **56,** 11422e (1962)].
[257] V. Ramamurthy, G. Tustin, C. C. Yaw, and R. S. H. Liw, *Tetrahedron*, **31,** 193 (1975).
[258] F. D. Popp, R. J. Dubois, and A. C. Casey, *J. Heterocycl. Chem.*, **6,** 285 (1969).
[259] S. Bien and U. Michael, *J. Chem. Soc., C,* **1968,** 2151.
[260] G. W. Stacy, D. L. Eck, and T. E. Wollner, *J. Org. Chem.*, **35,** 3495 (1970).
[261] C. A. Henrick, Ger. Pat. 2,202,016 (1972) [*C.A.*, **78,** 110633b (1973).]
[262] G. N. Walker and A. R. Engle, *J. Org. Chem.*, **37,** 4294 (1972).
[263] V. Ramamurthy and R. S. H. Liu, *Tetrahedron*, **31,** 201 (1975).
[264] T. Harayama, M. Ohtani, M. Oki, and Y. Inubushi, *Chem. Commun.*, **1974,** 827.
[265] Ferbwerke Hoechst A. G., Fr. Pat. 2,013,358 (1970) [*C.A.*, **74,** 3797d (1971)].
[266] J. Heider, W. Eberlein, W. Kobinger, and W. Diederen, Ger. Pat. 2,126,518 (1972) [*C.A.*, **78,** 72443c (1973)].
[267] G. R. Pettit, J. C. Knight, and C. L. Herald, *J. Org. Chem.*, **35,** 1393 (1970).
[268] A. K. Bose and R. M. Ramer, *Steroids*, **11,** 27 (1968).
[269] D. E. McGreer and N. W. K. Chiu, *Can. J. Chem.*, **46,** 2225 (1968).
[270] A. F. Thomas, B. Willhalm, and R. Muller, *Org. Mass. Spectrom.*, **2,** 223 (1969).
[271] J. M. Forrester and T. Money, *Can. J. Chem.*, **50,** 3310 (1972).
[272] G. W. K. Cavill and P. J. Williams, *Aust. J. Chem.*, **22,** 1737 (1969).
[273] K. Mori, T. Mitsui, J. Fukami, and T. Ohtaki, *Agri. Biol. Chem.*, **35,** 1116 (1971).
[274] A. G. Gonzalez, J. D. Martin, and M. L. Rodriguez, *Tetrahedron Lett.*, **1973,** 3657.

[275] J. K. Crandall and C. F. Mayer, *J. Org. Chem.*, **35,** 3049 (1970).
[276] W. S. Johnson, D. H. Rich, and P. Loew, Ger. Pat. 2,129,785 (1972) [*C.A.*, **76,** 85402z (1972)].
[277] F. Eloy and A. Deryckere, *Helv. Chim. Acta*, **53,** 645 (1970).
[278] K. Mori and M. Matsui, *Tetrahedron*, **22,** 2883 (1966).
[279] W. Fritsch, U. Stache, W. Haede, K. Radscheit, and H. Ruschig, *Ann.*, **721,** 168 (1969).
[280] L. A. Yanovskaya and V. F. Kucherov, *Izv. Akad. Nauk SSSR, Ser. Khim.*, **1964** 1341 [*C.A.*, **61,** 11887d (1964)].
[281] T. Minami, I. Niki, and T. Agawa, *J. Org. Chem.*, **39,** 3236 (1974).
[282] S. Bory, D. J. Lin, and M. Fetizon, *Bull. Soc. Chim. Fr.*, **1971,** 1298.
[283] E. Wenkert, K. G. Dave, F. Haglid, R. G. Lewis, T. Oishi, R. V. Stevens, and M. Terashima, *J. Org. Chem.*, **33,** 747 (1968).
[284] A. F. Tolochko and A. V. Dombrovskii, *Ukr. Khim. Zh.*, **31,** 220 (1965) [*C.A.*, **63,** 1727e (1965)].
[285] M. J. Jorgenson and T. Leung, *J. Amer. Chem. Soc.*, **90,** 3769 (1968).
[286] C. E. Moppett and J. K. Sutherland, *J. Chem. Soc., C.*, **1968,** 3040.
[287] R. S. Marmor, *J. Org. Chem.*, **37,** 2901 (1972).
[288] M. J. Jorgenson, *J. Amer. Chem. Soc.*, **91,** 6432 (1969).
[289] S. Trippett and D. M. Walker, *Chem. Ind.* (London) **990** (1961).
[290] P. Baret, H. Buffet, and J. Pierre, *Bull. Soc. Chim. Fr.*, **1972,** 2493.
[291] J. W. Wilson and V. S. Stubblefield, *J. Amer. Chem. Soc.*, **90,** 3423 (1968).
[292] J. A. Edwards, J. S. Mills, J. Sundeen, and J. H. Fried, *J. Amer. Chem. Soc.*, **91,** 1248 (1969).
[293] L. D. Quin and R. C. Stocks, *J. Org. Chem.*, **39,** 686 (1974).
[294] R. A. Moss and C. B. Mallon, *Tetrahedron Lett.*, **1973,** 4481.
[295] J. J. Gajewski and L. T. Burka, *J. Amer. Chem. Soc.*, **94,** 8865 (1971).
[296] L. Mamlok and L. Lacombe, *Bull. Soc. Chim. Fr.*, **1973,** 1524.
[297] R. C. Cookson and N. W. Hughes, *J. Chem. Soc., Perkin I*, **1973,** 2738.
[298] H. Gerlach, *Helv. Chem. Acta*, **49,** 1291 (1966).
[299] R. S. Bly, R. K. Bly, A. O. Bedenbaugh, and O. R. Vail, *J. Amer. Chem. Soc.*, **89,** 880 (1967).
[300] Y. Sein, *Union Burma J. Sci. Technol.*, **1969,** 417 [*C.A.*, **75,** 151482n (1971)].
[301] L. N. Yakhontov, L. T. Mastafanova, and A. V. Rubtsov, *Zh. Obshch. Khim.*, **33,** 3211 (1963) [*C.A.*, **60,** 4109e (1964)].
[302] A. A. Khalaf and R. M. Roberts, *J. Org. Chem.*, **36,** 1040 (1971).
[303] B. G. Kovalev, L. A. Yanovskaya, and V. F. Kucherov, *Izv. Akad. Nauk SSSR, Otd. Khim. Nauk*, **1962,** 1876 [*C.A.*, **58,** 9148d (1963)].
[304] H. Kjoesen and L. Jensen, *Acta Chem. Scand.* **24,** 2259 (1970).
[305] S. T. Joung, J. R. Turner, and D. S. Tarbell, *J. Org. Chem.*, **28,** 928 (1963).
[306] H. Machleidt and W. Grell, *Ann.*, **690,** 79 (1965).
[307] S. Otsuka and M. Kawakami, *Angew. Chem.*, **75,** 858 (1963).
[308] A. G. Andrews, S. Hetzberg, L. Jensen, and M. P. Starr, *Acta Chem. Scand.*, **27,** 2383 (1973).
[309] S. G. Boots, M. R. Boots, and K. E. Guyer, *J. Pharm. Sci.*, **60,** 614 (1971).
[310] H. Kjoesen and L. Jensen, *Acta Chem. Scand.*, **26,** 4121 (1972).
[311] E. Van Heyningen, C. N. Brown, F. Jose, J. K. Henderson, and P. Stark, *J. Med. Chem.*, **9,** 675 (1966).
[312] T. Kubota, T. Matsuura, T. Tsutsui, S. Uyeo, H. Irie, A. Numiata, T. Fujita, and T. Suzuki, *Tetrahedron*, **22,** 1659 (1966).
[313] D. Dekeukeleire, E. C. Sanford, and G. S. Hammond, *J. Amer. Chem. Soc.*, **95,** 7904 (1973).
[314] R. J. Liedtke, A. I. Gerrard, J. Diekman, and C. Djerassi, *J. Org. Chem.*, **37,** 776 (1972).
[315] K. Sasaki, *Bull. Chem. Soc. Jpn.*, **40,** 2967 (1967).
[316] D. L. Adams and W. R. Vaughan, *J. Org. Chem.*, **37,** 3906 (1972).
[317] H. P. Schelling and F. Schaub, Ger. Pat. 2,226,523 (1972) [*CA.*, **78,** 71483f (1973)].
[318] C. A. Henrick, Fr. Pat. 2,124,279 (1972) [*CA.*, **78,** 136467a (1973)].
[319] R. Azerad and M. O. Cyrot, *Bull. Soc. Chim. Fr.*, **65,** 3740, (1965).
[320] A. G. Andrews and L. Jensen, *Acta Chem. Scand.*, **27,** 1401 (1973).
[321] S. Nozoc and K. Hirai, *Tetrahedron*, **27,** 6073 (1971).
[322] A. K. Bose, R. T. Dahill, and J. Noboken, *Angew. Chem.*, **76,** 796 (1964).

[323] H. Pommer, *Angew. Chem.*, **72,** 911 (1960).
[324] A. A. Khalaf and R. M. Roberts, *J. Org. Chem.*, **37,** 4227 (1972).
[325] A. J. Birch and J. J. Wright, *Chem. Commun.*, **1969,** 788.
[326] H. Hauth and D. Stauffacher, *Helv. Chem. Acta*, **54,** 1278 (1971).
[327] S. Uyco, H. Shirai, A. Koshiro, T. Yashiro, and K. Kagei, *Chem. Pharm. Bull.* (Tokyo), **14,** 1033, (1966) [*C.A.*, **66,** 18661p (1967)].
[328] A. Melean and G. R. Proctor, *J. Chem. Soc., Perkin I*, **1973,** 1084.
[329] A. K. Bose, M. S. Manhas, and R. M. Ramer, *J. Chem. Soc., C*, **1969,** 2728.
[330] H. Kaneko and M. O. Kazaki, *Tetrahedron Lett.*, **1966,** 219.
[331] W. Leimgruber, V. Stefanovic, F. Schenker, A. Karr, and J. Berger, *J. Amer. Chem. Soc.*, **90,** 5641 (1968).
[332] J. S. Boutagy and R. E. Thomas, *Aust. J. Chem.*, **24,** 2723 (1971).
[333] G. F. Field, W. J. Zally, and L. H. Sternbach, *J. Amer. Chem. Soc.*, **92,** 3520 (1970).
[334] J. C. Knight, G. R. Pettit, and C. L. Herald, *Chem. Commun.*, **1967,** 445.
[335] J. P. Dusza, J. P. Joseph, and S. Bernstein, U.S. Pat. 3,351,638 (1967) [*C.A.*, **68,** 114864u (1968)].
[336] O. P. Vig, B. S. Bhatt, J. Kaur, and J. C. Kapur, *J. Indian Chem. Soc.*, **50,** 139 (1973).
[337] G. Gallagher, Jr., and R. L. Webb, *Synthesis*, **1974,** 122.
[338] L. J. Dolby and G. N. Riddle, *J. Org. Chem.*, **32,** 3481 (1967).
[339] G. Pattenden, *J. Chem. Soc., C*, **1970,** 1404.
[340] D. J. Martin, M. Gordon, and C. E. Griffin, *Tetrahedron*, **23,** 1831 (1967).
[341] K. Sasaki, *Bull. Chem. Soc. Jpn.*, **41,** 1252 (1968).
[342] L. A. Yanovskaya and V. F. Kucherov, *Izv. Akad. Nauk SSSR., Ser. Khim.*, **1965,** 1504. [*C.A.*, **63,** 16387d (1965)].
[343] C. A. Henrick and J. B. Siddall, Ger. Pat. 2,202,031 (1972) [*C.A.*, **78,** 110626b (1973)].
[344] D. Binder, J. Bowler, E. D. Brown, N. S. Crossley, J. Hutton, M. Senior, L. Slater, P. Wilkinson, and N. C. A. Wright, *Prostaglandins*, **1974,** 87.
[345] K. Sato, M. Hirayama, T. Inoue, and S. K. Kuchi, *Bull. Chem. Soc. Jpn.*, **42,** 250 (1969).
[346] A. Yamaguchi and M. Okazaki, *Nippon Kagaku Kaishi*, **1973,** 110 [*C.A.*, **78,** 84494k (1973)].
[347] C. A. Henrick, W. E. Willy, D. R. McKean, E. Baggiolini, and J. B. Siddall, *J. Org. Chem.*, **40,** 8 (1975).
[348] H. Marschall, K. Tantau, and P. Weyerstahl, *Chem. Ber.*, **107,** 887 (1974).
[349] N. Finch, J. J. Fitt, and I. H. S. Hsu, *J. Org. Chem.*, **40,** 206 (1975).
[350] W. G. Dauben, G. H. Beasley, M. D. Broadhurst, B. Muller, D. J. Peppard. P. Pesnelle, and C. Suter, *J. Amer. Chem. Soc.*, **96,** 4724 (1974).
[351] W. S. Wadsworth, Jr., and W. D. Emmons, *J. Amer. Chem. Soc.*, **84,** 1316 (1962).
[352] W. S. Wadsworth, Jr., and W. D. Emmons, *J. Org. Chem.*, **29,** 2816 (1964).
[353] U. Schollkopf and R. Schroder, *Tetrahedron Lett.*, **1973,** 633.
[354] H. Boehme, M. Haake, and G. Auterhoff, *Arch. Pharm. (Weinheim)*, **305,** 88 (1972). [*C.A.*, **76,** 113024y (1972)].
[355] G. H. Posner and D. J. Brunelle, *J. Org. Chem.*, **37,** 3547 (1972).
[356] K. Kondo and D. Tunemoto, *Chem. Commun.*, **1972,** 952.
[357] D. S. Watt and E. J. Corey, *Tetrahedron Lett.*, **1972,** 4655.
[358] W. Stilz and H. Pommer, Ger. Pat. 1,092,472 (1961) [*C.A.*, **56,** 413b (1962)].
[359] E. W. Colvin and B. J. Hamill, *Chem. Commun.*, **1973,** 151.
[360] W. M. Reckendorf and U. Kamprath-Scholtz, *Angew. Chem., Int. Ed. Engl.*, **7,** 142 (1968).
[361] W. M. Reckendorf and U. Kamprath-Scholtz, *Chem. Ber.*, **105,** 673 (1972).
[362] L. Horner, H. Hoffmann, and V. G. Toscano, *Chem. Ber.*, **95,** 536 (1962).
[363] G. Lavielle, M. Carpentier, and P. Sagvignac, *Tetrahedron Lett.*, **1973,** 173.
[364] F. A. Carey and A. S. Court, *J. Org. Chem.*, **37,** 939 (1972).
[365] W. Kreiser and G. Neef, *Ann.*, **1974,** 1279.
[366] H. Paulsen and W. Bartsch, *Chem. Ber.*, **108,** 1732 (1975).
[367] C. Piechucki, *Synthesis*, **1974,** 869.
[368] Pfizer, Inc., Ger. Pat. 2,344,829 [*C. A.*, **81,** 25182w (1974)].
[369] P. Savignac, J. Petrova, M. Dreux, and P. Coutrot, *Synthesis*, **1975,** 535.
[370] E. D'Incan and J. Seyden-Penne, *Synthesis*, **1975,** 516.

[371] R. W. Ratcliffe and B. G. Christensen, *Tetrahedron Lett.*, **1973,** 4649.
[372] H. J. Trede, E. F. Jenny, and K. Heusler, *Tetrahedron Lett.*, **1973,** 3425.
[373] S. Musierowicz, A. Wroblewski, and H. Krawczyk, *Tetrahedron Lett.*, **1975,** 437.
[374] A. E. Asato and R. S. Liu, *J. Amer. Chem. Soc.*, **97,** 4128 (1975).
[375] C. R. Popplestone and A. M. Unrau, *Can. J. Chem.*, **51,** 1223 (1973).
[376] F. Camps, J. Coll, A. Messequer, and A. Roca, *Tetrahedron Lett.*, **1976,** 791.
[377] A. H. Davidson, P. K. G. Hodgson, D. Howells, and S. Warren, *Chem. Ind.* (London), **1975,** 455.
[378] E. J. Corey and D. S. Watt, *J. Amer. Chem. Soc.*, **95,** 2303 (1973).
[379] O. P. Vig, V. D. Ahuja, M. L. Sharma, and S. D. Sharma, *Indian J. Chem.*, **13,** 1358 (1975).
[380] R. E. Ireland, R. H. Mueller, and A. K. Willard, *J. Org. Chem.*, **41,** 986 (1976).
[381] O. P. Vig, B. Ram, J. C. Kapur, and J. Kaur, *Indian J. Chem.*, **11,** 857 (1973).
[382] C. Piechucki, *Synthesis,* **1976,** 187.
[383] J. E. Johnson and S. Liaaen-Jensen, *Tetrahedron Lett.*, **1976,** 955.
[384] D. Danion and R. Carrie, *Bull. Soc. Chim. Fr.*, **1974,** 2065.
[385] J. L. Kraus and G. Sturtz, *Bull. Soc. Chim. Fr.*, **1974,** 943.
[386] D. A. Evans, G. C. Andrews, T. T. Fujimoto, and D. Wells, *Tetrahedron Lett.*, **1973,** 1389.
[387] M. L. Raggio and D. S. Watt, *J. Org. Chem.* **41,** 1873 (1976).
[388] D. K. Manh, M. Fetizon, and M. Kone, *Tetrahedron,* **31,** 1903 (1975).
[389] M. Dvolaitzky, C. Taupin, and F. Poldy, *Tetrahedron Lett.*, **1975,** 1469.
[390] B. M. Trost and L. S. Melvin, Jr., *J. Amer. Chem. Soc.*, **98,** 1204 (1976).
[391] N. Bensol, H. Marschall, and P. Weyerstahl, *Tetrahedron Lett.*, **1976,** 2293.
[392] J. H. Sellstedt, *J. Org. Chem.*, **40,** 1508 (1975).
[393] J. A. Marshall, C. P. Hagan, and G. A. Flynn, *J. Org. Chem.*, **40,** 1162 (1975).
[394] R. K. Hill and D. W. Ladner, *Tetrahedron Lett.*, **1975,** 989.
[395] R. A. Tzydore and R. G. Ghirardelli, *J. Org. Chem.*, **38,** 1790 (1973).
[396] H. M. McGuire, H. C. Odom, and A. R. Pinder, *J. Chem. Soc., Perkin I,* **1974,** 1879.
[397] D. N. Brattesani and C. H. Heathcock, *J. Org. Chem.*, **40,** 2165 (1975).
[398] S. M. Kupchan, R. W. Britton, J. A. Lacadie, M. F. Ziegler, and C. W. Sigel, *J. Org. Chem.*, **40,** 648 (1975).
[399] G. F. Griffiths, G. W. Kenner, S. W. McCombie, K. M. Smith, and M. J. Sutton, *Tetrahedron,* **32,** 275 (1976).
[400] U. Horn, F. Mutterer, and C. D. Weis, *Helv. Chem. Acta,* **59,** 190 (1976).
[401] O. P. Vig, V. D. Ahuja, V. K. Segal, and A. K. Vig, *Indian J. Chem.*, **13,** 1129 (1975).
[402] O. P. Vig, J. Chander, and B. Ram, *Indian J. Chem.*, **12,** 1156 (1974).
[403] H. E. Zimmerman and C. J. Samuel, *J. Amer. Chem. Soc.*, **97,** 4025 (1975).
[404] D. L. Walker and B. Fraser-Reid, *J. Amer. Chem. Soc.*, **97,** 6251 (1975).
[405] W. G. Dauben, G. H. Beasley, M. D. Broadhurst, B. Muller, D. J. Peppard, D. Pesnelle, and C. Suter, *J. Amer. Chem. Soc.*, **97,** 4973 (1975).
[406] E. Piers and W. M. Phillips-Johnson, *Can. J. Chem.*, **53,** 1281 (1975).
[407] R. A. Moss and C. B. Mallon, *J. Org. Chem.*, **40,** 1368 (1975).
[408] M. Mikolajczyk, Crzeiszezak, W. Midura, and A. Zatorski, *Synthesis,* **1975,** 278.
[409] V. Jarolim, K. Slama, and F. Sorm, Czech. Pat. 155,364 [*C. A.*, **82,** 139400s (1975).]
[410] V. Jarolim, K. Slama, and F. Sorm, Czech. Pat. 155,393 [*C. A.*, **82,** 139391q (1975).]
[411] A. Franke, G. Mattern, and W. Traber, *Helv. Chim. Acta.*, **58,** 293 (1975).
[412] A. Franke, G. Mattern, and W. Traber, *Helv. Chim. Acta.*, **58,** 283 (1975).
[413] A. Franke, G. Mattern, and W. Traber, *Helv. Chim. Acta.*, **58,** 268 (1975).
[414] L. Streinz, M. Romanuk, and F. Sorm, *Collect. Czech. Chem. Commun.*, **1974,** 1898 [*C.A.*, **81,** 151611y (1974)].
[415] V. Jarolim, F. Sehnal, and F. Sorm, U.S. Pat. 3, 799, 957 [*C.A.*, **80,** 132832j (1974)].

CHAPTER 3

HYDROCYANATION OF CONJUGATED CARBONYL COMPOUNDS

Wataru Nagata and Mitsuru Yoshioka*

Shionogi Research Laboratory, Shionogi & Co., Ltd.
Fukushima-ku, Osaka, Japan

CONTENTS

* We gratefully thank Mrs. Carolyn F. Sidor of E. I. du Pont de Nemours and Company for a literature search which unveiled several papers we had overlooked. We acknowledge Miss Michiko Katayama of Shionogi Research Laboratory for typing the manuscript with skill and forbearance. We would like to express our appreciation to Professor Dwight S. Fullerton, Oregon State University, who read a major part of this manuscript and made many helpful suggestions.

INTRODUCTION

Hydrogen cyanide adds to a multiple bond in the presence of an appropriate catalyst. The process is called hydrocyanation. The addition to an α,β-unsaturated carbonyl compound **1** gives a 1,2 adduct, α-cyanohydrin **2**, and a 1,4 adduct, β-cyano ketone **3**. Both 1,2 and 1,4 additions are reversible in principle, but in the presence of a base the reverse 1,2 addition is much faster than the reverse 1,4 reaction, leading to the 1,4 adduct as a major (usually sole) product (Eq. 1).

$$\underset{\mathbf{1}}{R^1R^2\overset{4}{C}{=}\overset{3}{C}(R^3){-}\overset{2}{C}(R^4){=}\overset{1}{O}} + HCN \;\begin{cases} \xrightleftharpoons{\text{1,2 Addition}} \underset{\mathbf{2}}{R^1R^2C{=}C(R^3){-}C(R^4)(CN){-}OH} \\ \xrightarrow{\text{1,4 Addition}} \underset{\mathbf{3}}{R^1R^2\overset{\beta}{C}(CN){-}\overset{\alpha}{C}H(R^3){-}C(R^4){=}O} \end{cases} \qquad \text{(Eq. 1)}$$

In this sense conjugate hydrocyanation is a special case of the Michael reaction.[1] Predominance of 1,2 or 1,4 addition depends on the substrate structure, the reagent, and the reaction conditions. In this chapter 1,2 addition is discussed only when necessary. We deal chiefly with the 1,4 addition and, more generally, with conjugate addition including that on conjugated polyethylenic compounds giving 1,6 or more extended conjugated adducts. Carbonyl substrates include ketones, aldehydes, carboxylic acid derivatives, carbonitriles, and carboimines. Also included are β-functionalized carbonyl compounds, **4**, and π-allylpalladium complexes, **5**, that generate α,β-unsaturated carbonyl compounds, **1**, under hydrocyanation conditions. Although

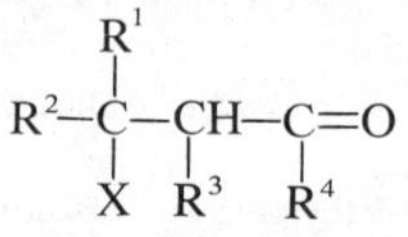

$$R^2{-}\underset{X}{\overset{R^1}{C}}{-}\underset{R^3}{CH}{-}\underset{R^4}{C}{=}O$$

4

X = OH, OCH_3, NR_2, $\overset{+}{N}R_3$, Cl, SO_3Na

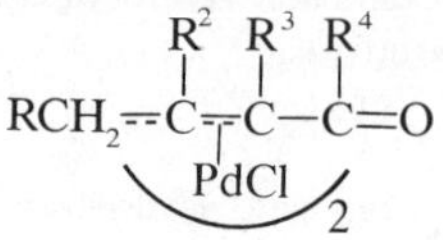

$$\left(RCH_2{\cdots}\overset{R^2}{C}{\cdots}\overset{R^3}{C}{-}\overset{R^4}{C}{=}O \quad PdCl\right)_2$$

5

hydrogen cyanide adds to α,β-unsaturated compounds activated by other groups such as nitro ($\geq C{=}C{-}NO_2$)[2–4] and sulfone ($\geq C{=}C{-}SO_2R$),[2] these reactions are not considered in this chapter.

The first example of 1,4 addition of hydrogen cyanide to α,β-unsaturated carbonyl derivatives appeared more than a century ago, in 1873, when Claus prepared tricarballylic acid (1,2,3-propanetricarboxylic acid) by heating an ethanolic solution of 2,3-dichloropropene and potassium cyanide in a sealed tube.[5] Five years later he obtained methylsuccinic acid by a similar treatment of allyl iodide.[6] Apparently α,β-unsaturated carbonitriles are intermediates in these reactions. Addition of hydrogen cyanide to a carbon-carbon double bond activated by a carbonyl group was first observed by Bredt and Kallen who obtained phenylsuccinic acid by treatment of ethyl benzalmalonate with potassium cyanide and sulfuric acid followed by hydrolysis and decarboxylation.[7]

The discovery in 1903 that the reactive species in hydrocyanation is

[1] E. D. Bergmann, D. Ginsberg, and R. Pappo, *Org. React.*, **10,** 179 (1959).

[2] P. Kurtz, *Ann. Chem.*, **572,** 23 (1951).

[3] O. von Schichh, H. G. Padeken, and A. Segnitz in *Methoden der Organischen Chemie*, Vol. X/1, p. 415, George Thieme, Stuttgart, 1971.

[4] J. McKenna, J. M. McKenna, and P. B. Smith, *Tetrahedron*, **21,** 2983 (1965).

[5] A. Claus, *Ann. Chem.*, **170,** 126 (1873).

[6] A. Claus, *Ann. Chem.*, **191,** 33 (1878).

[7] J. Bredt and J. Kallen, *Ann. Chem.*, **293,** 338 (1896).

the cyanide anion, CN^-, and that the rate of the reaction is proportional to the concentration of CN^-[8], formed the foundation for active study of hydrocyanation chemistry. A fundamental procedure for conjugate hydrocyanation was developed which used 2 molar equivalents of potassium cyanide and 1 molar equivalent of acetic acid in aqueous alcohol.[9] The reaction mode was clarified by Ingold[10] and Michael[11] in the 1930s. Many chemists began extensive application of the reaction to organic syntheses, and attempts were made to explore improved procedures with higher efficiency and fewer side reactions.[2,12–16] Detailed reviews cover developments of conjugate hydrocyanations up to the 1950s.[17–21]

In recent years conjugate hydrocyanation has been used in total syntheses of complex natural products such as terpenes, steroids, and alkaloids. However, conventional procedures using hydrogen cyanide in the presence of a base or an alkali or alkaline-earth metal cyanide were not efficient. Furthermore, the efficiency could not be improved without increasing side reactions. The difficulty was overcome in 1962 when methods using organoaluminum compounds were discovered by Nagata and co-workers.[22] They developed two new hydrocyanation methods, one involving a combination of hydrogen cyanide and an alkylaluminum (HCN–AlR_3) and the other an alkylaluminum cyanide (R_2AlCN), both in an aprotic solvent.[22–26] Applications of the new methods have led to total syntheses of complex natural products having a methyl, a functionalized methyl, or a carbon bridge at an

[8] A. Lapworth, *J. Chem. Soc.*, **83,** 995 (1903).

[9] A. C. O. Hann and A. Lapworth, *J. Chem. Soc.*, **85,** 1335 (1904).

[10] D. A. Duff and C. K. Ingold, *J. Chem. Soc.*, **1934,** 87.

[11] A. Michael and N. Weiner, *J. Amer. Chem. Soc.*, **59,** 744 (1937).

[12] E. Knoevenagel, *Ber.*, **37,** 4065 (1904).

[13] E. Knoevenagel and E. Lange, *Ber.*, **37** 4059 (1904).

[14] I. N. Nazarov and S. I. Zav'yalov, *J. Gen. Chem. USSR* (*Engl. Transl.*), **24,** 475 (1954) [*C.A.*, **49,** 6139f (1955)].

[15] W. Nagata, S. Hirai, H. Itazaki, and K. Takeda, *J. Org. Chem.*, **26,** 2413 (1961).

[16] N. Mayer and G. E. Gantert, Ger. Pat. 1,085,871 (1960) [*C.A.*, **56,** 4639f (1962)].

[17] F. Adickes and H. du Mont in *Hanbuch der Katalyse*, G.-M. Schwab, Ed., Band 7, p. 356, Springer, Wien, 1943.

[18] V. Migradichian, *The Chemistry of Organic Cyanogen Compounds*, pp. 219–225, Reinhold Publishing Corp., New York, 1947.

[19] D. T. Mowry, *Chem. Rev.*, **42,** 189–244 (1948).

[20] P. Kurz in *Methoden der Organischen Chemie*, Vol. VIII, pp. 265–302, George Thieme, Stuttgart, 1952.

[21] H. Gilman, *Organic Chemistry*, Vol. I, 2nd ed., pp. 678–682, Wiley, New York, 1943.

[22] W. Nagata, M. Yoshioka, and S. Hirai, *Tetrahedron Lett.*, **1962,** 461.

[23] W. Nagata and M. Yoshioka, *Tetrahedron Lett.*, **1966,** 1913.

[24] W. Nagata and M. Yoshioka, *Proc. Int. Congr. Horm. Steroids*, 2nd, **1966,** 327 (1967).

[25] W. Nagata, M. Yoshioka, and S. Hirai, *J. Amer. Chem. Soc.*, **94,** 4635 (1972).

[26] W. Nagata, T. Okumura, and M. Yoshioka, *J. Chem. Soc., C*, **1970,** 2347.

angular position.[27] Short reviews on recent development in hydrocyanation have appeared.[28–30]

MECHANISM

Base-Catalyzed Addition of Hydrogen Cyanide

Hydrogen cyanide, a weak acid of pK_a 9.21, generally does not add by itself to an activated carbon-carbon double bond.[8,9,31] On the other hand, CN^-, like SH^- and I^-, has a nucleophilicity great enough to add to an electrophilic carbon atom of a conjugated carbonyl system.[32] The observation that the addition rate is proportional to CN^- concentration,[8] as well as later kinetic studies,[33,34] disproved the view that nascent hydrogen cyanide is the reactive species.[7] Addition of cyanide ion is reversible[18,35] and, with an α,β-ethylenic ketone as the substrate, 1,2 and 1,4 additions occur competitively. Thus, in principle, the overall hydrocyanation process can be described by Eq. 2. Cyanide ion adds to the electrophilic position 2 or 4 in the conjugated carbonyl system of substrate **A** to give cyanohydrin anion **B** or enolate anion **D**. Reversible protonation of **B** and **D** with hydrogen cyanide or a protic solvent, if any, leads to cyanohydrin **C** and β-cyano ketone **E**. The keto function of **E** further undergoes reversible cyanide addition to give β-cyano ketone cyanohydrin **G** via the anion **F**. All the steps are reversible. With many substrates the reversible 1,2 addition $\mathbf{A} \rightleftharpoons \mathbf{B}$ is faster than the reversible 1,4 addition $\mathbf{A} \rightleftharpoons \mathbf{D}$, and the reverse 1,4 addition $\mathbf{D} \rightarrow \mathbf{A}$ is much slower than the forward one. Therefore the latter step, $\mathbf{A} \rightarrow \mathbf{D}$, is rate determining in practice. Apparently the cyanohydrin is unstable in basic medium (1,2 adducts **C** and **G** are stabilized in acid medium). Thus the final products are β-cyano ketone, **E**, and/or its cyanohydrin, **G**, formation of the latter being negligible in strongly basic medium. Cyanohydrins **C** and **G** can be isolated in the reaction with anhydrous hydrogen cyanide

[27] W. Nagata, *Nippon Kagaku Zasshi*, **90,** 837 (1969).

[28] H. O. House, *Modern Synthetic Reactions*, 2nd ed., pp. 623–628, Benjamin, Menlo Park, Calif., 1972.

[29] W. Nagata and M. Yoshioka, *Yuki Gosei Kagaku Kyokai Shi*, **26,** 2 (1968) [*C.A.*, **68,** 77327y (1968)].

[30] K. Friedlich and K. Wallenfels, in *The Chemistry of the Cyano Group*, Z. Rappoport, Ed., pp. 67–110, Interscience, New York, 1970.

[31] A. Lapworth, *J. Chem. Soc.*, **85,** 1214 (1904).

[32] P. R. Wells, *Chem. Rev.*, **63,** 212 (1963).

[33] W. J. Jones, *J. Chem. Soc.*, **105,** 1547 (1914).

[34] R. V. Serebryakov, M. A. Dalin, and A. G. Konoval'chukov, *Dokl. Akad. Nauk Azerb. SSR*, **19,** 31 (1963) [*C.A.*, **60,** 15700e (1964)].

[35] W. Nagata, M. Yoshioka, and M. Murakami, *J. Amer. Chem. Soc.*, **94,** 4644 (1972).

$$\underset{\mathbf{A}}{(R^1)(R^2)\overset{4}{C}{=}\overset{3}{C}(R^3){-}\overset{2}{C}(R^4){=}\overset{1}{O}} \;\underset{-CN^-}{\overset{+CN^-}{\rightleftharpoons}}\; \underset{\mathbf{B}}{(R^1)(R^2)C{=}C(R^3){-}C(O^-)(R^4){-}CN} \;\underset{-H^+}{\overset{+H^+}{\rightleftharpoons}}\; \underset{\mathbf{C}}{(R^1)(R^2)C{=}C(R^3){-}C(OH)(R^4){-}CN} \quad \text{(1,2 Addition)}$$

$$\mathbf{A} \;\underset{-CN^-}{\overset{+CN^-}{\rightleftharpoons}}\; \underset{\mathbf{D}}{R^2{-}C(R^1)(CN){-}C(R^3){=}C(O^-)(R^4)} \;\underset{-H^+}{\overset{+H^+}{\rightleftharpoons}}\; \underset{\mathbf{E}}{R^2{-}\overset{\beta}{C}(R^1)(CN){-}\overset{\alpha}{C}H(R^3){-}C(R^4){=}O} \quad \text{(1,4 Addition)}$$

$$\mathbf{E} \;\underset{-CN^-}{\overset{+CN^-}{\rightleftharpoons}}\; \underset{\mathbf{F}}{R^2{-}C(R^1)(CN){-}CH(R^3){-}C(O^-)(R^4){-}CN} \;\underset{-H^+}{\overset{+H^+}{\rightleftharpoons}}\; \underset{\mathbf{G}}{R^2{-}C(R^1)(CN){-}CH(R^3){-}C(OH)(R^4){-}CN} \quad \text{(Eq. 2)}$$

or concentrated hydrocyanic acid in the presence of a basic catalyst, since the proton concentration is large enough to stabilize 1,2 adducts. While reverse 1,2 and 1,4 additions are both more favorable at higher temperatures than the forward reactions,[36,37] the 1,2 reversal is generally more complete.[38] Thus the reactions at higher temperatures favor the conversions **C** → **B** → **A** and **G** → **F** → **E** and afford only β-cyano ketone, **E**. This situation is well illustrated by the hydrocyanation of methyl vinyl ketone (see p. 269). In summary, hydrocyanation of an α,β-ethylenic ketone with hydrogen cyanide in the presence of a basic catalyst selectively gives 1,2 adduct **C** at lower temperatures and 1,4 adduct **E** at higher temperatures. Reaction involving an alkali or alkaline-earth metal cyanide affords only 1,4 adduct **E** in most cases owing to the highly basic reaction medium.

Addition of Organoaluminum Cyanides

The two new hydrocyanation methods using organoaluminum compounds are referred to as Method A [which employs a combination of

[36] E. O. Leupold and H. Vollmann, U.S. Pat. 2,166,600 (1939) [*C.A.*, **32,** 6668 (1938)].

[37] M. Tanaka and N. Murata, *Kogyo Kagaku Zasshi*, **60,** 433 (1957) [*C.A.*, **53,** 9043f (1959)].

[38] Ref. 19, p. 231.

hydrogen cyanide and an alkylaluminum (HCN—AlR_3) in tetrahydrofuran (THF)] and Method B [using an alkylaluminum cyanide (R_2AlCN) in an inert organic solvent]. Mechanisms of the new methods are fundamentally different from those of conventional methods in that the carbonyl function of the α,β-ethylenic carbonyl system is activated through coordination with the Lewis-acidic organoaluminum species.[25] The mechanisms of these methods were investigated in detail[35] and are reviewed briefly here.

Method A. Since alkylaluminums in tetrahydrofuran show appreciable electric conductivities, the dissociation shown in Eq. 3 seems likely.[39] An increase in conductivity by addition of hydrogen cyanide indicates dissociation to a proton and a cyanoaluminate anion, R_3AlCN^- (Eq. 4). Presumably, the enone is activated by H^+ or R_2Al^+,

$$2\,R_3Al \underset{\text{THF}}{\rightleftharpoons} R_2Al^+ + R_4Al^- \qquad \text{(Eq. 3)}$$

$$R_3Al + HCN \underset{\text{THF}}{\rightleftharpoons} H^+ + R_3AlCN^- \qquad \text{(Eq. 4)}$$

and the anion R_3AlCN^- acts as the cyanating species. Substitution of chlorine for an alkyl in AlR_3 increases the activation of the enone system but decreases the cyanating power of R_3AlCN^-. Thus the reactivity is roughly in the order $HCN\text{-}Al(C_2H_5)_3 \approx HCN\text{-}(C_2H_5)_2AlCl > HCN\text{-}C_2H_5AlCl_2 \gg HCN\text{-}AlCl_3$. Addition of a small amount of water to the $HCN\text{-}Al(C_2H_5)_3$ combination markedly increases the 1,4-addition rate. The acceleration is considered to arise from a catalytic action of diethylaluminum hydroxide or bis(diethylaluminum) oxide formed by the action of water on triethylaluminum. The combination reagents eventually form alkylaluminum cyanides, the Method B reagents, according to Eq. 5. Thus Method A hydrocyanations are partly effected by the initial $HCN\text{-}AlR_3$ reagent and partly

$$HCN + AlR_3 \rightarrow R_2AlCN + RH \qquad \text{(Eq. 5)}$$

by the secondarily formed R_2AlCN by a reaction mechanism discussed later (see Eq. 8, p. 264). The participation of Method B reagent (R_2AlCN) has proved minimal in the $HCN\text{-}C_2H_5AlCl_2$ and $HCN\text{-}Al(C_2H_5)_3$—(H_2O)* combinations, thus rendering these combinations ideal Method A reagents. Ideal Method A hydrocyanation does not involve formation of 1,2-adduct but involves irreversible 1,4 addition and catalysis by AlR_3. These facts and the product analysis led to the

* See footnote on p. 266.

[39] E. Bonitz, *Chem. Ber.*, **88**, 742 (1955).

mechanism shown in Eq. 6 for ideal Method A hydrocyanation of an α,β-unsaturated ketone. According to this mechanism, enone **A** is activated with a cation L^+ (H^+ or R_2Al^+), the activated species **H** undergoes rate-determining, nucleophilic attack of the cyanoaluminate anion R_3AlCN^- at position 4 to afford, via transition state **I**, the initial 1,4 adduct **J** and AlR_3. Adduct **J** is irreversibly transformed to β-cyano ketone **E** which reacts with any excess reagent to give dinitrile **K**. The dinitrile can be reconverted into the desired β-cyano ketone on basic workup.

$$\underset{\mathbf{A}}{R^1R^2\overset{4}{C}{=}\overset{3}{C}(R^3){-}\overset{2}{C}(R^4){=}\overset{1}{O}} \underset{}{\overset{L^+}{\rightleftharpoons}} \left[\underset{\mathbf{H}}{R^1R^2\overset{\delta+}{C}{\cdots}C(R^3){\cdots}\overset{\delta+}{C}(R^4){\cdots}OL} \xrightarrow[\text{Slow}]{(R_3AlCN)^-} \underset{\mathbf{I}}{R^1{\cdots}C(R^2)(CN{\cdots}AlR_3){\cdots}C(R^3){\cdots}C(R^4){\cdots}OL}\right]$$

$$\longrightarrow \underset{\mathbf{J}}{R^1{-}C(R^2)(CN){-}C(R^3){=}C(R^4){-}OL} + AlR_3 \xrightarrow{H^+\ (\text{when } L = R_2Al)}$$

$$\underset{\mathbf{E}}{R^1{-}C(R^2)(CN){-}CH(R^3){-}C(R^4){=}O} \overset{HCN\text{-}AlR_3}{\rightleftharpoons} \underset{\mathbf{K}}{R_1{-}C(R^2)(CN){-}CH(R^3){-}C(OL)(R^4){-}CN} \qquad \text{(Eq. 6)}$$

$(L = H \text{ or } AlR_2)$

Method B. In hydrocyanation of α,β-unsaturated ketones using dialkylaluminum cyanide the reacting species has proved to be the nonionic R_2AlCN itself which activates the enone system and also acts as the nucleophilic cyanating agent. Ebullioscopic molecular-weight determination has indicated that the degree of association of dialkylaluminum cyanide is 4–5 in boiling benzene and that of diethylaluminum cyanide is 1.86 in boiling tetrahydrofuran. It is reasonable to assume that a monomeric solvated species "$S{\cdot}R_2AlCN$" is present, although in low concentration, according to Eq. 7. Kinetic studies using

$$(R_2AlCN)_n \underset{\text{Fast}}{\overset{S}{\rightleftharpoons}} n\text{"}S{\cdot}R_2AlCN\text{"} \qquad \text{(Eq. 7)}$$

(S, solvent; n, 2–5)

diethylaluminum cyanide have revealed the following features of Method B hydrocyanation: (1) the monomeric "$S{\cdot}(C_2H_5)_2AlCN$" acts as the reacting species, that is, the activating and cyanating species; (2)

the reaction consists of a very rapid, reversible 1,2 addition (pre-equilibrium) followed by a slow, reversible 1,4 addition; (3) 1,2 addition is first order in both enone and "S·R_2AlCN" and reverse 1,2 addition is first order in 1,2 adduct and zero order in the reagent; (4) 1,4 addition is first order in enone and second order in "S·$(C_2H_5)_2$AlCN," the rate-determining step being the attack of "S·$(C_2H_5)_2$AlCN" at position 4 of the activated enone; (5) reverse 1,4 addition is first order in 1,4 adduct and zero order in the reagent; (6) dinitrile **K** is not formed. Equation 8 shows a mechanism of Method B hydrocyanation of an α,β-unsaturated ketone in tetrahydrofuran consistent with the features mentioned. According to this mechanism, substrate **A** is converted with "S·R_2AlCN" into an activated species **M** which is subjected to 1,2 addition (**M** ⇌ **O**) and 1,4 addition (**M** → **Q**). Fast 1,2 addition through the four-centered, low-energy transition state **N** precedes slower 1,4 addition. The 1,2 adduct **O** predominant in the early stages of the reaction is reconverted into the activated enone **M**, which is attacked by another molecule of "S·R_2AlCN" at position 4, leading to 1,4-adduct **Q** through transition state **P**. The 1,4 adduct **Q** is reconverted directly into the starting enone **A**. A more likely

$$\mathbf{A}:\ R^1R^2C{=}C(R^3){-}C(R^4){=}O \xrightleftharpoons[\text{Fast}]{\text{"S·}R_2\text{AlCN"}} \mathbf{M}:\ R^1R^2C{=}C(R^3){-}C(R^4){=}O\cdots NCAlR_2 \xrightleftharpoons{\text{Fast}} \mathbf{N}\ \left(R^1R^2C{=}C(R^3){-}C(R^4){\cdots}O\ /\ NC{\cdots}AlR_2\right) \xrightleftharpoons{\text{Fast}} \mathbf{O}:\ R^1R^2C{=}C(R^3){-}C(R^4)(CN){-}OAlR_2\ \text{(1,2 Addition)}$$

$$\mathbf{M} \xrightarrow[\text{Slow}]{\text{"S·}R_2\text{AlCN"}} \mathbf{P}\ \left(R^1{-}C(R^2)(CN{\cdots}R_2Al{\cdots}S){\cdots}C(R^3){\cdots}C(R^4){\cdots}O,\ R_2AlCN\right) \longrightarrow \mathbf{Q}:\ R^1{-}C(R^2)(CN){-}C(R^3){=}C(R^4){-}O{-}AlR_2 + \text{"S·}R_2\text{AlCN"}\ \text{(1,4 Addition)}$$

$$\mathbf{Q} \xrightarrow{\text{Very slow}} \mathbf{A} \qquad \text{(Eq. 8)}$$

reverse reaction, **Q** → **P** → **M** → **A**, is not consistent with the kinetic evidence that the reverse 1,4 addition is zero order in "S·R_2AlCN." In the last stage of the reaction an equilibrium is established as indicated in Eq. 9. With no proton source present in the Method B

$$\text{Enone } \mathbf{A}\ \text{(negligible)} \rightleftharpoons \text{1,2 Adduct } \mathbf{O}\ \text{(minor)}; \qquad \text{Enone } \mathbf{A}\ \text{(negligible)} \rightleftharpoons \text{1,4 Adduct } \mathbf{Q}\ \text{(major)} \qquad \text{(Eq. 9)}$$

reaction medium the 1,2- and 1,4-adducts are present before workup in the form of a cyanohydrin dialkylaluminum alkoxide **O** and a dialkylaluminum enoxide **Q**, respectively.

Since the reaction with organoaluminum cyanides is highly efficient and has relatively small steric requirements compared with the conventional procedures, one can imagine that the true cyanating agent might be a dialkylaluminum isocyanide in equilibrium with the dialkylaluminum cyanide (Eq. 10). However, there is no unequivocal evidence of this isocyanide structure.[40]

$$R_2Al-C\equiv N \rightleftharpoons R_2Al-\overset{+}{N}\equiv\overset{-}{C} \qquad \text{(Eq. 10)}$$

Factors Controlling Conjugate or Nonconjugate Addition

It is important from the synthetic viewpoint to be able to predict the mode of addition in the hydrocyanation of a conjugated carbonyl system, *i.e.*, 1,2; 1,4; 1,6, or more extended conjugated addition. The addition modes are governed by electronic, entropic, and steric factors related to the substrate structure, the nature of the reagent, and the reaction conditions. An understanding of these relationships is essential for prediction of the addition mode.

A driving force for conjugate (Michael-type) addition is thought to arise from "better electron delocalization."* The transition state of 1,2 addition is unfavorable relative to the 1,4-transition state $\mathbf{T_4}$ and the 1,6-transition state $\mathbf{T_6}$ in the hydrocyanation of α,β-ethylenic and

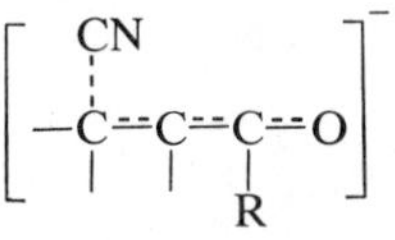

$$\left[-\overset{\overset{\displaystyle CN}{\vdots}}{C}{=}{-}C{=}{-}\underset{\underset{\displaystyle R}{|}}{C}{=}{-}O\right]^- \qquad \left[-\overset{\overset{\displaystyle CN}{\vdots}}{C}{=}{-}C{=}{-}C{=}{-}C{=}{-}\underset{\underset{\displaystyle R}{|}}{C}{=}{-}O\right]^-$$

$\mathbf{T_4}$ 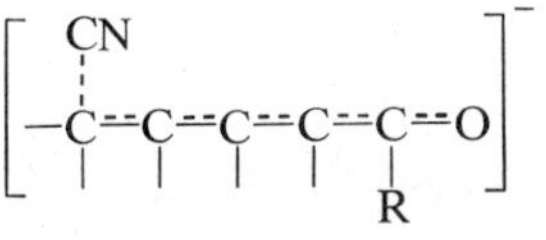 $\qquad$ $\mathbf{T_6}$

$\alpha,\beta,\gamma,\delta$-diethylenic carbonyl compounds. This concept has experimental support in the hydrocyanation of most conjugated esters and ketones but *not* in that of most conjugated aldehydes (which give 1,2 adducts with little Michael-type addition). Thus the orientation of the cyanide addition is greatly affected by the carbonyl substituent R (O-alkyl, alkyl, or hydrogen) and other carbon substituents and cannot be explained only by the "better electron delocalization" concept. Another important electronic factor is the *electron density of the carbon atoms in the conjugated carbonyl system, where cyanide anion attacks*

* This concept is also referred to as "better neutralization."

[40] W. Nagata, *Proc. R. A. Welch Foundation on Chem. Res.*, **XVII**, 185 (1973).

the most electron-deficient carbon atom(s). In general, cyanide ion attacks the carbon atom(s) indicated by arrows in formulas **a–d**.

$$\overset{}{>}C{=}C{-}\overset{\downarrow}{C}{=}O \;(\text{C}{-}\text{H}) \quad \textbf{a}$$

$$>\overset{\downarrow}{C}{=}C{-}\overset{\downarrow}{C}(R){=}O \quad \textbf{b}$$

$$>\overset{\downarrow}{C}{=}C{-}C(OR){=}O \quad \textbf{c}$$

$$>\overset{\downarrow}{C}{=}C{-}C{=}C{-}C{=}O \quad \textbf{d}$$

This general trend may be altered by changes in the electron distribution caused by substituents. For example, hydrocyanation of vinyl crotonate (**6**) with hydrogen cyanide in the presence of a catalytic amount of potassium cyanide gives the unexpected products **7** and **8**. They can be explained by the intermediacy of a 1,2 adduct (Eq. 11).[2]

$$\underset{\mathbf{6}}{CH_3CH{=}CH{-}\overset{\downarrow}{C}({=}O){-}O{-}CH{=}CH_2} \xrightarrow{\text{HCN-(KCN)}^*} \underset{\mathbf{7}\ (7\%)}{CH_3CH{=}CH{-}C({=}O){-}O{-}CH(CN)CH_3} + \underset{\mathbf{8}\ (15\%)}{CH_3CH(CN)CH_2{-}C({=}O){-}O{-}CH(CN)CH_3}$$

$$\mathbf{6} \xrightarrow{\text{1,2 Addition}} \underset{\mathbf{9}}{CH_3CH{=}CH{-}C(CN)(OH){-}OCH{=}CH_2} \longrightarrow \underset{\mathbf{10a}}{CH_3CH{=}CHCOCN} + CH_3CHO$$

$$\mathbf{10a} \longrightarrow \underset{\mathbf{10b}}{CH_3CH(CN)CHCOCN}$$

$$CH_3CHO \xrightarrow{CN^-} CH_3CH(O^-)(CN) \xrightarrow{\mathbf{6}\text{ or }\mathbf{10a}} \mathbf{7} \xrightarrow{HCN} \mathbf{8}; \quad CH_3CH(O^-)(CN) \xrightarrow{\mathbf{10b}} \mathbf{8} \qquad \text{(Eq. 11)}$$

This unusual 1,2-adduct formation is considered to result from activation of the carbonyl carbon by the vinyloxy group. Another example of a substituent changing the orientation of the cyanide attack is the hydrocyanation of dimethyl cinnamylidenemalonate (**11**), which gives the 1,4 adduct **12** rather than the 1,6 adduct.[10,41] This anomalous 1,4

* In this chapter a metal cyanide or a base used in a catalytic amount is shown in parentheses to differentiate it from that used as the reagent. Also, a small amount of water or an alcohol used with hydrogen cyanide and trialkylaluminum is shown in parentheses.

[41] J. Thiele and J. Meisenheimer, *Ann. Chem.*, **306,** 247 (1899).

addition is explained as an electronic effect of the phenyl group at position δ of **11** which makes the β-carbon atom the most electron-deficient carbon.[10]

$$C_6H_5CH{=}CHCH{=}C(CO_2CH_3)_2 \xrightarrow{KCN} C_6H_5CH{=}CHCH(CN)CH(CO_2CH_3)_2$$

11 **12**

Steric hindrance is an important factor affecting the 1,2- or 1,4-addition mode. Hydrocyanation of steroidal enones **13** and **15** with hydrogen cyanide and triethylaluminum exclusively affords 1,4 adduct **14** and 1,2 adduct **16**, respectively.[22,42,43] In both reactions, cyanide

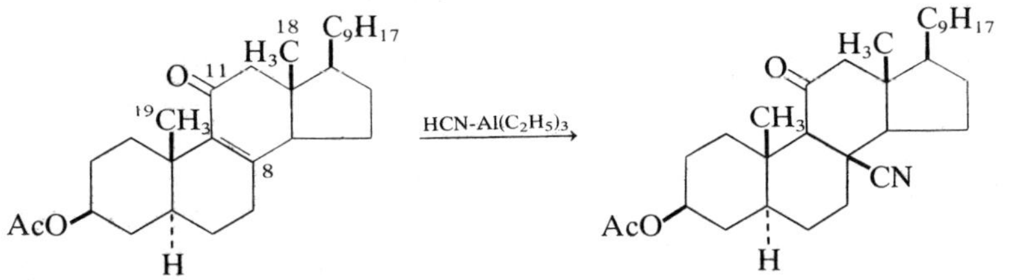

HCN-Al(C_2H_5)$_3$

13 **14** (86%)

HCN-Al(C_2H_5)$_3$
X = H, OH, OAc

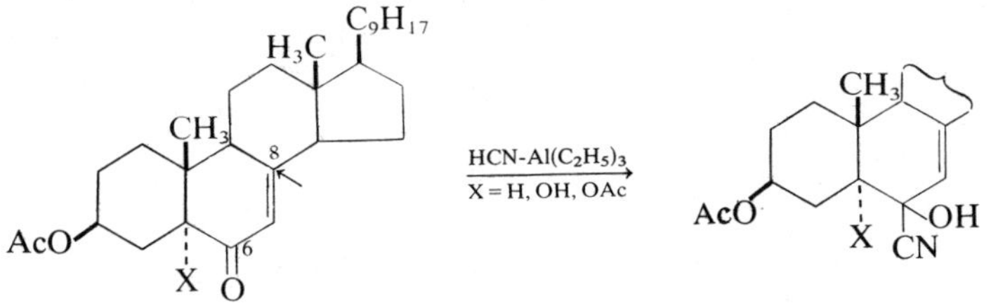

15 **16** (Quantitative)

anion can attack carbon 8 only from the β side of the molecule, *i.e.*, the stereoelectronically controlled direction (see p. 310). The β-carbon atom of both enones is severely sterically hindered by the 18- and 19-methyl groups, especially the former. Molecular models show that the steric interaction between the 18-methyl group and the entering cyano group at position 8 in the transition state is greater in the Δ^7-6-one **15** than in the Δ^8-11-one **13**. Furthermore, the 11-keto function is greatly hindered by the two methyl groups, whereas the 6-keto group is relatively open. Thus steric considerations can explain the divergent addition pathways for enones **13** and **15**.[43]

[42] W. Nagata, H. Itazaki, T. Tomita, F. Mukawa, T. Aoki, T. Okada, M. Murakami, N. Haga, and S. Hirai, *Shionogi Kenkyusho Nenpo*, **23,** 47 (1973).

[43] W. Nagata, M. Yoshioka, and M. Murakami, *J. Amer. Chem. Soc.*, **94,** 4654 (1972).

Hydrocyanation of transoid enones with diethylaluminum cyanide proceeds via the four-centered transition state $\mathbf{T}_f$ to favor 1,2 addition in an early stage of the reactions, as shown in Eq. 8. Presumably the six-centered transition state $\mathbf{T}_s$ is entropically advantageous for cisoid enones, though so far there is no experimental evidence to support this hypothesis.

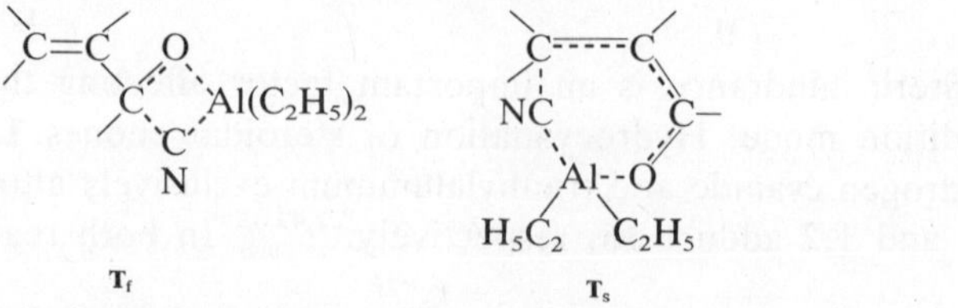

C=C O C Al$(C_2H_5)_2$ C N NC C Al O H_5C_2 C_2H_5

$\mathbf{T}_f$ $\mathbf{T}_s$

In addition to the substrate structural factors mentioned above, the properties of the hydrocyanating reagents can also affect addition modes. In general, 1,4 addition is favored in hydrocyanation with an alkali or alkaline-earth metal cyanide and, to a lesser extent, with hydrogen cyanide in the presence of a basic catalyst, where the cyanating species is the completely dissociated cyanide ion. In hydrocyanation with hydrogen cyanide and trialkylaluminum 1,4 addition is also favored, because the reacting species is most likely the "ate" anion R_3AlCN^- as discussed earlier (see Eq. 4, p. 262). On the other

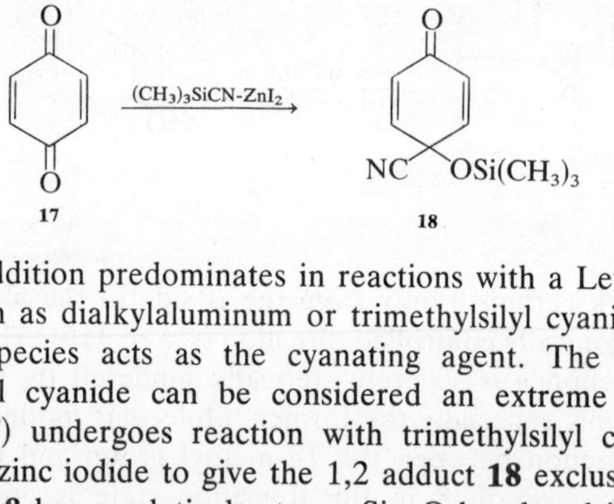

$$\xrightarrow{(CH_3)_3SiCN\text{-}ZnI_2}$$

NC OSi$(CH_3)_3$

17 **18**

hand, 1,2 addition predominates in reactions with a Lewis-acid metal cyanide such as dialkylaluminum or trimethylsilyl cyanide in which a molecular species acts as the cyanating agent. The reaction with trimethylsilyl cyanide can be considered an extreme case. Benzoquinone (**17**) undergoes reaction with trimethylsilyl cyanide in the presence of zinc iodide to give the 1,2 adduct **18** exclusively.[44,45] The 1,2 adduct **18** has a relatively strong Si—O bond and is stable under the reaction conditions. Thus the silyl reagent is suitable for preparing 1,2 adducts of conjugated carbonyl compounds of higher reactivity.

The final important factors to influence addition modes are the reaction conditions. As discussed earlier, 1,2 addition is reversible

[44] D. A. Evans, L. K. Truesdale, and G. L. Carroll, *J. Chem. Soc., Chem. Commun.*, **1973,** 55.

[45] D. A. Evans, J. M. Hoffmann, and L. K. Truesdale, *J. Amer. Chem. Soc.*, **95,** 5822 (1973).

except for special cases, and the reverse 1,2 addition is favored by higher basicity of the reaction medium and higher reaction temperature. Therefore 1,2 adducts can rarely be isolated in hydrocyanations with an alkali or alkaline-earth metal cyanide, except for α,β-ethylenic aldehydes, whereas 1,2 adducts can be isolated in the early stages of hydrocyanation with hydrogen cyanide in the presence of a basic catalyst. For example, hydrocyanation of methyl vinyl ketone (**19**) with hydrogen cyanide in the presence of a basic catalyst such as potassium cyanide, potassium carbonate, or pyridine gives the 1,2 adduct **20** as a major product at temperatures below 15° (preferably below −5°) and the 1,4 adduct **21** above 20°, the latter increasing with time even at −5°.[36,37,46] The 1,2 and 1,4 adducts **20** and **21** are kinetically and thermodynamically controlled products, respectively, with regard to the addition mode. Selective preparation of 1,2 and 1,4 adducts is

$$CH_2{=}CHCOCH_3 \;(\mathbf{19}) \xrightarrow{<15^\circ} CH_2{=}CHC(OH)(CN)CH_3 \;(\mathbf{20})$$

$$CH_2{=}CHCOCH_3 \;(\mathbf{19}) \xrightarrow{>20^\circ} NCCH_2CH_2COCH_3 \;(\mathbf{21})$$

possible in the hydrocyanation of α,β-ethylenic ketones and aldehydes with diethylaluminum cyanide if the reaction conditions are controlled. Cholestenone (**22**) reacts with diethylaluminum cyanide in tetrahydrofuran to give the 1,2 adduct **23** after 15 minutes at −60° and the 1,4 adduct **24** at room temperature, both in high yield.

C_8H_{17}

22

$(C_2H_5)_2AlCN/THF$, Room temp, 4 hr →

24 (quantitative)

$(C_2H_5)_2AlCN/THF$, −60°, 15 min (Acid workup) ↓

23 (92%)

[46] H. B. Dykstra, U.S. Pat. 2,188,340 (1940) [*C.A.*, **34,** 3764[8] (1940)].

SCOPE AND LIMITATIONS

Methods

Hydrogen Cyanide Catalyzed by a Base

As pointed out by Lapworth in 1903, hydrogen cyanide is not reactive by itself and it requires a basic catalyst which generates cyanide ion.[8] The catalyst can be selected from cyanides [*e.g.*, KCN, $K_4Ni(CN)_4$, and $CaK_2Ni(CN)_4$], inorganic bases (*e.g.*, potassium carbonate and sodium hydroxide), or organic bases (*e.g.*, triethylamine, pyridine, and piperidine). The reaction is carried out in the presence or absence of a solvent such as benzene, ethanol, acetonitrile, or dimethylformamide.

In a typical procedure a substrate is heated with an equimolar amount of liquid hydrogen cyanide in the presence of 1–2% of potassium cyanide at 40–80°. Although acrylonitrile (**25a**) and methyl acrylate (**27a**) could be converted into succinonitrile (**26a**) and methyl β-cyanopropionate (**28a**) in high yields by this hydrogen cyanide-(potassium cyanide) procedure, less reactive substrates such as methacrylonitrile (**25b**), methyl methacrylate (**27b**), crotononitrile (**25c**), and methyl crotonate (**27c**) gave the corresponding 1,4 adducts **26b**, **28b**, **26c**, and **28c** in yields of less than 5% together with the unchanged starting materials and hydrogen cyanide polymers.[2]

$$RCH{=}C(R')CN \longrightarrow RCH(CN)CH(R')CN$$

25a, R = R′ = H → **26a**, R = R′ = H

25b, R = H, R′ = CH_3 → **26b**, R = H, R′ = CH_3,

25c, R = CH_3 R′ = H, → **26c**, R = CH_3 R′ = H

$$RCH{=}C(R')CO_2CH_3 \longrightarrow RCH(CN)CH(R')CO_2CH_3$$

27a, R = R′ = H → **28a**, R = R′ = H

27b, R = H, R′ = CH_3 → **28b**, R = H, R′ = CH_3

27c, R = CH_3, R′ = H → **28c**, R = CH_3, R′ = H

Inorganic and organic bases are as effective as potassium cyanide as the catalyst,[36,37,46,47] whereas potassium nickel cyanide and calcium nickel cyanide seem to be less effective.[48]

Because of the quite low reactivity of hydrogen cyanide-(base)

[47] Asahi Kasei Kabushiki Kaisha, Brit. Pat. 1,055,060 (1967) [*C.A.*, **64**, 6504b (1966)].

[48] G. R. Coraor and W. Z. Heldt, U.S. Pat. 2,904,581 (1959) [*C.A.*, **54**, 4393f (1960)].

combinations, hydrocyanation of less reactive substrates such as β,β-dialkylated α-enones is possible only at high temperature when an autoclave or a high-boiling solvent is used. In particular, use of aprotic polar solvents such as dimethylformamide (DMF), dimethylacetamide (DMA), or N-methylpyrrolidone is claimed to be advantageous. Thus mesityl oxide (**29**) could be converted into cyano ketone **30** in 86–92% yield by treatment with hydrogen cyanide-(potassium cyanide) at 135–140° in an autoclave.[49] Hydrocyanation of 3,5,5-trimethylcyclohex-2-en-1-one (**31**) into nitrile **32** (66% yield) was effected with hydrogen cyanide-(potassium carbonate) in dimethylacetamide at 175° at atmospheric pressure.[16]

$$CH_3COCH{=}C(CH_3)_2 \longrightarrow CH_3COCH_2C(CN)(CH_3)_2$$

29 → **30**

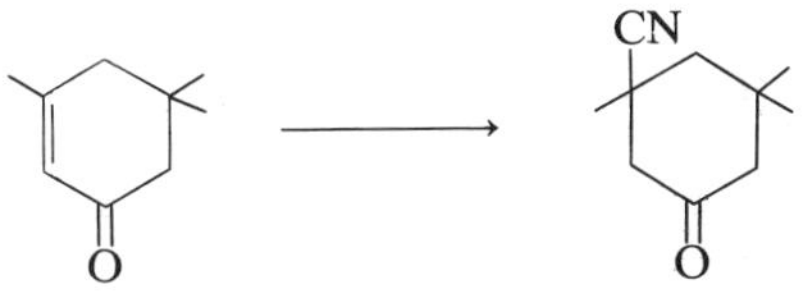

31 → **32**

An advantageous feature of the hydrogen cyanide-(base) procedure is the lack of alkali-induced side reactions. Although its low reactivity has made this procedure unsuitable for laboratory preparation of complex molecules, it has found use in the industrial preparation of simple cyano compounds.

Acetone Cyanohydrin Catalyzed by a Base

Acetone cyanohydrin can regenerate cyanide ion in the presence of a base and is therefore useful for hydrocyanation. A hydrocyanation method was reported by Nazarov: α,β-unsaturated ketones were treated with excess acetone cyanohydrin in the presence of a small amount of potassium carbonate or triethylamine in refluxing aqueous methanol for several hours.[14] By this method, 2-methylcyclohexenone (**33**) and substituted chalcones **35** can be converted into the corresponding 1,4 adducts **34**[14] and **36**[50] in moderate to good yields. That the efficiency of the catalyst is parallel to its basicity is shown in hydrocyanation of dienone **37** where bisadduct **38** is formed in 43% yield with potassium carbonate as the catalyst and monoadduct **39** in

[49] Rohm & Haas Company, Brit. Pat. 887,411 (1962) [*C.A.*, **57,** 3294e (1962)].

[50] B. E. Betts and W. Davey, *J. Chem. Soc.*, **1958,** 4193.

only 23% yield with triethylamine.[50] This method does not involve

$$\textbf{33} \xrightarrow[\text{Reflux}]{(CH_3)_2C(OH)CN\text{-}(Na_2CO_3)/aq\ CH_3OH} \textbf{34}\ (73\%)$$

$$XC_6H_4CH{=}CHCOC_6H_4Y \longrightarrow XC_6H_4CH(CN)CH_2COC_6H_4Y$$

35 **36**

X, Y = H CH_3, CH_3O, $N(CH_3)_2$, halogen

$CO(CH{=}CHC_6H_5)_2$ **37**

$CO(CH_2CH(CN)C_6H_5)_2$ **38** (43%)

$C_6H_5CH{=}CHCOCH_2CH(CN)C_6H_5$ **39** (23%)

significant side reactions and can be conveniently used in the laboratory. However, as in the previously mentioned procedure using hydrogen cyanide and a basic catalyst, the reactivity of the Nazarov method is low because of the low cyanide concentration in the reaction medium. The reactivity can be increased to some extent by using a large excess of acetone cyanohydrin without solvent and with saturated methanolic potassium hydroxide as the catalyst.[51a] Recently crown ethers were found to be effective in increasing the reactivity.[51b] This modification is discussed in the next section (p. 283).

Alkali or Alkaline-Earth Metal Cyanide (MCN)

Since cyanide anion is the cyanating species in conjugate hydrocyanation, one can predict that alkali or alkaline-earth metal cyanides (MCN) such as sodium, potassium, or calcium cyanide, which provide this ion in high concentration in a protic solvent, should be very reactive. In fact, these cyanides are the most powerful of the conventional hydrocyanating reagents. For example, diester **40**, which is resistant to hydrogen cyanide in the presence of a potassium cyanide catalyst at 50°,[2] undergoes hydrocyanation with potassium cyanide, subsequent hydrolysis and decarboxylation, giving nitrile **42** in excellent yield.[7,52]

[51a] G. R. Ames and W. Davey, *J. Chem. Soc.*, **1957,** 3480.

[51b] C. L. Liotta, A. M. Dabdoub, and L. H. Zalkow, *Tetrahedron Lett.*, **1977,** 1117.

[52] C. F. H. Allen and H. B. Johnson, *Org. Synth.*, *Coll. Vol.* **IV,** 804 (1963).

$$\underset{\mathbf{40}}{C_6H_5CH{=}C(CO_2C_2H_5)_2} \xrightarrow[60^\circ]{KCN/aq\ C_2H_5OH} \underset{\mathbf{41}}{\left[C_6H_5\underset{\displaystyle CN}{\underset{|}{C}}HCH(CO_2C_2H_5)_2\right]} \longrightarrow \underset{\substack{\mathbf{42}\ (94\%) \\ R = C_2H_5,\ H}}{C_6H_5\underset{\displaystyle CN}{\underset{|}{C}}HCH_2CO_2R}$$

An alkali cyanide is considerably basic in a protic solvent, and the basicity of the reaction medium increases as cyanide ion is consumed to liberate a strong base (*e.g.*, potassium hydroxide). Thus side reactions such as hydrolysis, decarboxylation, and condensation of substrate and product can take place. These side reactions can be prevented by neutralizing the alkaline reaction medium, although the neutralization also lowers the hydrocyanating potency. Schemes for neutralizing the basic medium include: (1) use of a substrate having a functional group that can neutralize alkali; (2) addition of a neutralizing agent; (3) use of an aprotic, polar solvent; and (4) a combination of these means.

Without Neutralization. As discussed above, the method using an alkali cyanide in a protic solvent is the most powerful of the conventional methods although it frequently promotes side reactions. The method is useful when (1) substrates are reactive enough to undergo hydrocyanation before the side reactions occur and (2) products are relatively stable to alkali. For example, isopropenyl *p*-bromophenyl ketone (**43**)[53] is smoothly hydrocyanated by potassium cyanide in

$$\underset{\mathbf{43}}{p\text{-}BrC_6H_4COC(CH_3){=}CH_2} \xrightarrow[\text{Room temp}]{KCN/CH_3OH} \underset{\mathbf{44}\ (96\%)}{p\text{-}BrC_6H_4COCH(CH_3)CH_2CN}$$

methanol. Hydrocyanation of α,β-unsaturated dicarbonyl compounds such as **45** proceeds readily, because the activation of the β-carbon atom is doubled by the two carbonyl groups and the initial adduct **46** is

$$\underset{\mathbf{45}}{(R^1)(R^2)\overset{\beta}{C}{=}\overset{\alpha}{C}(COR^3)(COR^4)} \xrightarrow{KCN} \underset{\mathbf{46}}{R^1{-}\underset{\displaystyle CN}{\underset{|}{\overset{\displaystyle R^2}{\overset{|}{C}}}}{-}\bar{C}(C(R^3){=}O)(C(R^4){=}O)}$$

[53] F. J. McEvoy and G. R. Allen, Jr., *J. Org. Chem.*, **38**, 4044 (1973).

a relatively weak base and does not react to a large extent with protic solvents. In other words, the 1,3-dicarbonyl grouping neutralizes the potassium cyanide. The compounds in this category include diketones, keto esters, keto nitriles, cyano esters, and dinitriles.

Quinones such as naphthoquinone **47** are reactive and readily hydrocyanated by potassium cyanide without neutralization to give stable aromatic nitriles such as **48**. The weakly basic phenoxide grouping is generated in the addition process.[54]

$\xrightarrow[\text{Room temp}]{\text{NaCN/aq } C_2H_5OH\text{-dioxane}}$

47 **48** (64–84%)

Of the conventional procedures only the potassium cyanide method can effect hydrocyanation of less reactive (hindered) polyalkylated α,β-unsaturated ketones. Since α-alkylated cyano products are less reactive toward condensation with the substrate in the alkaline reaction medium, hydrocyanation of these enones with potassium cyanide in aqueous alcohol followed by alkaline hydrolysis provides a good method for preparing polyalkylated γ-keto acids. The method is exemplified in the reaction of the α,β-dialkylcyclohexenones **49**.[55,56]

$\xrightarrow[\text{2. Aq KOH, reflux}]{\text{1. KCN/aq } CH_3OH\text{, reflux}}$

$R = CH_3, C_2H_5, CH_2CO_2H, CH_2CH_2CO_2H$

49 **50** (Good)

Use of a nonaqueous solvent in the hydrocyanation of less reactive enones also prevents most hydrolysis of the introduced cyano group. For example, negligible amounts of hydrolysis products are formed in the hydrocyanation of the $\Delta^{5(10)}$-6-keto steroid **51** with potassium cyanide in refluxing ethylene glycol[57] and of Δ^4-3-ones **53** with calcium cyanide in N-methylpyrrollidone at room temperature.[58] Therefore the

[54] A. N. Griner, A. P. Klyagina, and A. P. Terentev, *Zh. Obshch. Khim.*, **29**, 2773 (1959) [*C.A.*, **54**, 10970c (1960)].

[55] E. Adlerová, L. Novák, and M. Protiva, *Collect. Czech. Chem. Commun.*, **23**, 681 (1958).

[56] H. H. Inhoffen and E. Prinz, *Chem. Ber.*, **87**, 684 (1954).

[57] J. Fishman and H. Guzik, *Tetrahedron Lett.*, **1966**, 1483.

[58] H. B. Henbest and W. R. Jackson, *J. Chem. Soc., C*, **1967**, 2465.

angular cyano steroids **52** and **54** are produced in high yields. Lithium cyanide dissolved in warm tetrahydrofuran was found to effect hydrocyanation of steroidal Δ^4-3-ones and Δ^8-11-ones, but the use of this reagent has not been studied extensively.[25,59]

OH

KCN/ethylene glycol, Reflux

HO, O, NC, H

51 **52** (80–85%)

R

$Ca(CN)_2$ / N-methylpyrrolidone, Room temp, 65 hr

O, CN

R = H, OH, $COCH_3$, C_8H_{17}

53 **54** (78–86%)

When hydrocyanation-hydrolysis (decarboxylation) products are desired, the alkali cyanide method in aqueous solvents can be used successfully, usually without isolating the hydrocyanation products. Succinic acids and levulinic acids can be prepared in this manner. For example, ethyl crotonate (**55**)[60] upon treatment with potassium cyanide in refluxing aqueous ethanol followed by hydrolysis with aqueous barium oxide yields the corresponding succinic acid **56**. Similarly, α-phenyllevulinic acid (**58**) is obtained in 83% overall yield from ethyl benzalacetoacetate (**57**); see reference 252.

$$CH_3CH{=}CHCO_2C_2H_5 \xrightarrow[\text{2. Aq Ba(OH)}_2\text{, reflux}]{\text{1. KCN/aq C}_2\text{H}_5\text{OH, reflux}} CH_3CH(CO_2H)CH_2CO_2H$$

55 **56** (66–70%)

$$C_6H_5CH{=}C(COCH_3)(CO_2C_2H_5) \xrightarrow{\text{1. NaCN/aq C}_2\text{H}_5\text{OH, room temp; 2. HCl, reflux; 3. 80\% AcOH, reflux}} C_6H_5CH(CO_2H)CH_2COCH_3$$

57 **58** (83%)

[59] W. Nagata *et al.*, unpublished observations.

[60] G. B. Brown, *Org. Synth., Coll. Vol.* **III,** 615 (1955).

In summary, hydrocyanation with an alkali cyanide without neutralization is not a generally applicable method and gives good results only when: (1) substrates are very reactive; (2) cyano enolates formed are stabilized (neutralization by product); (3) substrates and products are stable in alkaline medium; (4) a nonaqueous solvent is used; and (5) the desired compounds are actually hydrolysis (decarboxylation) derivatives of the hydrocyanation products.

With Partial Neutralization. The side reactions arising from an alkaline reaction medium in hydrocyanations with alkali or alkaline-earth metal cyanides can be minimized by partial neutralization of the medium with an acid or its equivalent. The neutralization can be effected by an acid-liberating group in the substrate (internal neutralization) or by addition of an acidic co-reagent. The neutralization agent is selected from organic acids, mineral acids, esters, or acidic salts. The amount of the agent should be controlled to neutralize an alkali cyanide partially while maintaining the cyanide concentration high enough to effect the addition and the basicity of the reaction medium low enough to prevent side reactions. Usually the neutralization agent is added in a 0.5 to 0.75 molar equivalent to the alkali cyanide (usually employed in excess), but the amount is subject to the substrate reactivity and the stability of the substrate and the product.

Internal Partial Neutralization. Internal neutralization can be effected when the enone substrate contains a carboxyl group or an easily hydrolyzable carboalkoxy group or both and when the substrate has an acid residue (leaving group X) β to the carbonyl group.

The first condition is exemplified by smooth hydrocyanation of β-benzoylacrylic acid (**59**) (room temperature in water)[61] and the steroidal 3-acetoxy-Δ^{16}-20-one **60** (reflux in aqueous methanol)[62] with potassium cyanide.

$C_6H_5COCH{=}CHCO_2H$

59

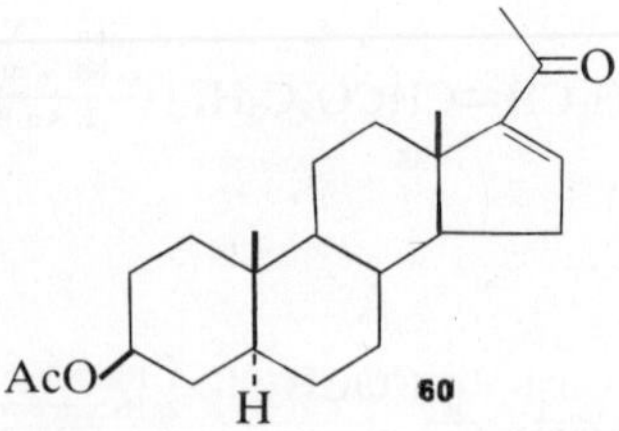

AcO H

60

An example of the second condition is the reaction of β-functionalized carbonyl compounds **4** (masked α,β-unsaturated

[61] J. Bougault, *Ann. Chim.* (Paris), [8] **15,** 504 (1908).

[62] P. Crabbé, M. Pérez, and G. Vera, *Can. J. Chem.*, **41,** 156 (1963).

ketones) with an alkali cyanide (MCN) to regenerate enones **1** and hydrogen cyanide. Here hydrocyanation proceeds as in the hydrogen cyanide-(alkali cyanide) or hydrogen cyanide-alkali cyanide method (Eq. 12). Similarly, hydrocyanation of β-functionalized enones **61** proceeds by an addition-elimination mechanism to effect internal neutralization (Eq. 13).

$$\underset{\mathbf{4}}{\text{RCO—CH—C—X}} \xrightarrow[(\beta\text{ Elimination})]{\text{MCN}} \underset{\mathbf{1}}{\text{RCO—C=C}} + \text{HCN} + \text{MX}$$

$$\mathbf{1} \xrightarrow{\text{HCN-(MCN) or HCN-MCN}} \underset{\mathbf{3}}{\text{RCO—CH—C—CN}} \qquad \text{(Eq. 12)}$$

$$\underset{\mathbf{61}}{\text{RCO—C=C—X}} \xrightarrow{\text{MCN}} \text{RC(O—M)=C—C(CN)—X} \longrightarrow \underset{\mathbf{62}}{\text{RCO—C=C—CN}} + \text{MX} \qquad \text{(Eq. 13)}$$

Examples of masked α,β-unsaturated ketones are shown in formulas **63–66**.

$CO[CH_2C(Cl)(CH_3)_2]_2$ **63**

3-Oxo-5-R-cyclohexyl with CH_3 and SO_3Na on C-1 **64**

$ArCOCH_2CH_2N(CH_3)_2\cdot HCl$ **65**

$ArCOCH(R)CH_2\overset{+}{N}(CH_3)_3I^-$ **66**

Because the chloride **63**[63] and the sulfite **64**[64] (Knoevenagel method[13]) must be prepared from the corresponding enones, their reaction with potassium cyanide or sodium cyanide in refluxing aqueous ethanol or water has little advantage over direct hydrocyanation of the enones by external-neutralization procedures (*e.g.*, potassium cyanide-acetic acid, potassium cyanide-ammonium chloride discussed

[63] R. Anschütz, *Ber.*, **26,** 827 (1893).

[64] W. F. Whitmore and C. W. Roberts, *J. Org. Chem.*, **13,** 31 (1948).

in the next section). The Mannich salts **65**[65] and **66**[53] react with 2 molar equivalents of potassium cyanide in water or aqueous methanol at reflux (for **65**) or room temperature (for **66**). One molar equivalent of potassium cyanide is neutralized with hydrogen chloride or iodide to liberate dimethylamine (removed by distillation) or trimethylamine. Yields of cyano ketones depend on the aromatic group. This procedure is useful when α,β-unsaturated ketones are to be prepared via the Mannich bases.

A typical method for preparing β-cyano enones by addition-elimination (Eq. 13) is seen in the conversion of β-chlorovinyl ketones **67** into quaternary salts **68** followed by treatment with potassium cyanide (1.5 molar equivalents) and trimethylamine hydrochloride (0.5 molar equivalent) in a mixture of water and benzene.[66] Since both substrates **68** and products **69** are alkali sensitive, hydrocyanation with potassium cyanide should be carried out with the addition of trimethylamine hydrochloride to buffer the medium (see potassium

$$\underset{\mathbf{67}}{\mathrm{ArCOCH{=}CHCl}} \xrightarrow{\mathrm{N(CH_3)_3}}$$

$$\underset{\mathbf{68}}{\mathrm{ArCOCH{=}CH\overset{+}{N}(CH_3)_3Cl^-}} \xrightarrow[\mathrm{H_2O\text{-}C_6H_6,\ cooling}]{\mathrm{KCN,\ N(CH_3)_3.HCl}} \underset{\mathbf{69}\ (77\text{–}97\%)}{\mathrm{ArCOCH{=}CHCN}}$$

cyanide-ammonium chloride method, p. 280) and with cooling and stirring in the two-layer system to transfer the product to the organic layer. The direct conversion is occasionally possible, *e.g.*, the reaction of chloranil (**70**) with potassium cyanide giving cyano compound **71**. In this reaction hydrolysis of the vinyl chloride groupings also takes place.[67]

$$\text{Cl, Cl, Cl, Cl-quinone } \mathbf{70} \xrightarrow{\text{2 Eq KCN/aq } \mathrm{CH_3OH}} \text{HO, CN, NC, OH-quinone } \mathbf{71}$$

70 **71**

As the foregoing examples show, hydrocyanation with an alkali cyanide with internal neutralization is limited to special substrates and is often less effective than the external-neutralization procedures discussed in next.

[65] E. B. Knott, *J. Chem. Soc.*, **1947,** 1190.

[66] A. N. Nesmeyanov and M. I. Rybinskaya, *Doklady Akad. Nauk SSSR*, **115,** 315 (1957) [*C.A.*, **52,** 7158g (1958)].

[67] M. M. Richter, *Ber.*, **44,** 3469 (1911).

External Partial Neutralization. (a) With Ethyl Acetate. In 1911 a procedure was reported that used a slight excess of potassium cyanide in aqueous ethanol in the presence of ethyl acetate.[68] Studies of the hydrocyanation of (−)-carvone showed that neutralization of potassium hydroxide (from potassium cyanide) with the ethyl acetate is slow and is accompanied by alkali-induced epimerization of a cyano product (see p. 309).[69]

Thus the potassium or sodium cyanide-ethyl acetate procedure is considered to be less effective in preventing side reactions than the potassium cyanide-acetic acid process and so has not been utilized often.

(b) With Acetic Acid (KCN—AcOH: The Lapworth Procedure).[9,70] A typical example is the addition of 2 molar equivalents of aqueous potassium cyanide to an alcoholic solution of benzalacetophenone (**72**) and 1 molar equivalent of acetic acid, followed by stirring at 35° for 3 hours to afford α-phenyl-β-benzoylpropionitrile (**73**) in 93–96% yield.[71] Here the amount of acetic acid used is important; if too much is used, hydrocyanation will not take place; with too little, side reactions will occur. Potassium cyanide can

$$\underset{\textbf{72}}{C_6H_5CH{=}CHCOC_6H_5} \xrightarrow[35°]{\text{KCN-AcOH/aq } C_2H_5OH} \underset{\textbf{73}\ (93\text{–}96\%)}{C_6H_5\underset{\displaystyle CN}{\underset{|}{C}}HCH_2COC_6H_5}$$

be replaced by sodium cyanide.

This procedure can be successfully applied to hydrocyanation of α,β-unsaturated carbonyl substrates having high to moderate reactivity. Diethyl benzalmalonate (**40**) is converted into cyano ester **41** by this procedure,[72] whereas with the potassium cyanide method (p. 272) the decarboxylated product **42** is obtained.[7] This example illustrates

$$\underset{\textbf{40}}{C_6H_5CH{=}C(CO_2C_2H_5)_2} \xrightarrow[4°]{\text{NaCN-AcOH/aq } C_2H_5OH} \underset{\textbf{41}\ (90\text{–}94\%)}{C_6H_5\underset{\displaystyle CN}{\underset{|}{C}}HCH(CO_2C_2H_5)_2}$$

$$\underset{\textbf{40}}{C_6H_5CH{=}C(CO_2C_2H_5)_2} \xrightarrow[60°]{\text{KCN/aq } C_2H_5OH} \underset{\textbf{42}\ (94\%)}{C_6H_5\underset{\displaystyle CN}{\underset{|}{C}}HCH_2CO_2R}$$

$R = C_2H_5$ (major), H

[68] A. Lapworth and V. Steele, *J. Chem. Soc.*, **99,** 1877 (1911).

[69] C. Djerassi, R. A. Schneider, H. Vorbrueggen, and N. L. Allinger, *J. Org. Chem.*, **28,** 1632 (1963).

[70] A. Lapworth and E. Wechsler, *J. Chem. Soc.*, **97,** 38 (1910).

[71] C. F. H. Allen and R. K. Kimball, *Org. Synth., Coll. Vol.* **II,** 498 (1941).

[72] C. F. Koelsch and C. H. Stratton, *J. Amer. Chem. Soc.*, **66,** 1883 (1944).

the utility of acetic acid in preventing side reactions. In the hydrocyanation of cycloheptenone **74** the use of tetrahydrofuran and water (two-layer system) instead of aqueous methanol prevented conversion of nitrile **75a** to the lactam **75b**.[73]

KCN-AcOH

74 **75a** **75b**

Aq CH_3OH, room temp: **75a** (minor), **75b** (43%)
THF-H_2O, reflux: **75a** (47%)

Hydrocyanation of the moderately reactive substrate **76** gives the cyano product **77** in fair yield accompanied by starting material.[74]

KCN-AcOH/aq C_2H_5OH
Room temp

(*cis* and *trans*)

76 **77** (76%)

In summary, the potassium cyanide-acetic acid procedure provides a general, convenient method for hydrocyanation of highly to moderately reactive α,β-unsaturated carbonyl substrates, usually with fewer side reactions. However, the efficiency is not high enough to effect hydrocyanation of less reactive substrates such as hindered polycyclic enones, and side reactions are not always avoidable.

(c) *With Mineral Acid.* Replacement of acetic acid in the Lapworth procedure by a mineral acid like sulfuric acid or hydrochloric acid would not essentially change the features of the method except that the reaction medium is less basic (potassium acetate *vs.* potassium sulfate). The hydrocyanation of substituted chalcones (ArCOCH=CHAr′) has been claimed to proceed more smoothly with potassium cyanide and sulfuric acid than with potassium cyanide and acetic acid.[75]

(d) *With Ammonium Chloride.* The conventional hydrocyanation procedure is markedly improved by using ammonium chloride as a co-reagent in dimethylformamide (DMF).[15,76] In the standardized procedure a solution of a substrate, 2.0 molar equivalents of potassium

[73] H. Newman and T. L. Fields, *Tetrahedron*, **28,** 4051 (1972).
[74] D. K. Banerjee and V. B. Angadi, *Tetrahedron*, **21,** 281 (1965).
[75] W. Davey and D. J. Tivey, *J. Chem. Soc.*, **1958,** 1230.
[76] W. Nagata, S. Hirai, H. Itazaki, and K. Takeda, *Ann. Chem.*, **641,** 184 (1961).

cyanide, and 1.5 molar equivalents of ammonium chloride in aqueous dimethylformamide (usually 90%) is heated on a water bath.[15] Potassium cyanide can be replaced by sodium cyanide.[77] Replacement of dimethylformamide by tetrahydrofuran, dioxane, methanol, or ethanol is not advisable (see later discussion). The reaction temperature can be lowered or raised as required by the reactivity of the substrate. In this procedure, ammonium chloride serves as a neutralizing agent to liberate ammonia. The relative amounts of potassium cyanide and ammonium chloride are critical, since the use of 2 molar equivalents of both potassium cyanide and ammonium chloride resulted in a lowered yield of the cyano product and recovery of the starting material.[78,79] The use of dimethylformamide, an aprotic, polar, basic, high-boiling solvent, is advantageous for three reasons: (1) dimethylformamide with a small amount of water can dissolve the reagents and a variety of substrates; (2) the rather basic dimethylformamide helps to expel ammonia from the reaction system, thus lowering its alkalinity and preventing hydrolysis and amino nitrile formation; and (3) the reaction temperature can be raised to facilitate hydrocyanation of hindered substrates.

Of the conventional hydrocyanation methods using an alkali cyanide, the potassium cyanide-ammonium chloride procedure is the most effective. It has the fewest side reactions and can be widely applied to most substrates, even those of quite low reactivity. The efficiency of this procedure is illustrated by hydrocyanation of methyloctalinone **76** at room temperature into cyano ketones **77** in quantitative yield (compared with a 76% conversion by the Lapworth procedure, p. 280); and by smooth angular hydrocyanation of less reactive polycyclic enones such as cholestenone (**22**) (84% conversion at 100°),[15] steroidal D-homo-12-en-11-one **78** (90%),[80] 13(17)-en-20-one **79** (79%),[81] and the tricyclic enone **80a** (75%).[43] High-temperature hydrocyanation of nitrogen-containing enone **81** into epimeric nitriles **82**[82] illustrates the effect of the reaction temperature (for use of diethylaluminum cyanide see p. 312).

However, even this highly effective conventional procedure is not sufficient to effect appreciable hydrocyanation at 100° of highly hindered substrates of very low reactivity such as the tricyclic enone **83**

[77] D. H. R. Barton, E. F. Lier, and J. F. McGhie, *J. Chem. Soc., C*, **1968,** 1031.

[78] W. Nagata, *Tetrahedron*, **13,** 278 (1961).

[79] S. Hirai, Ph.D. Dissertation, Kyoto University, 1961.

[80] W. Nagata, M. Narisada, and T. Sugasawa, *J. Chem. Soc., C*, **1967,** 648.

[81] W. Nagata, I. Kikkawa, and K. Takeda, *Chem. Pharm. Bull.* (Tokyo), **9,** 79 (1961).

[82] Y. Inubushi, T. Kikuchi, T. Ibuka, K. Tanaka, I. Saji, and K. Tokane, *Chem. Pharm. Bull.* (Tokyo), **22,** 349 (1974); *J. Chem. Soc., Chem. Commun.*, **1972,** 1252.

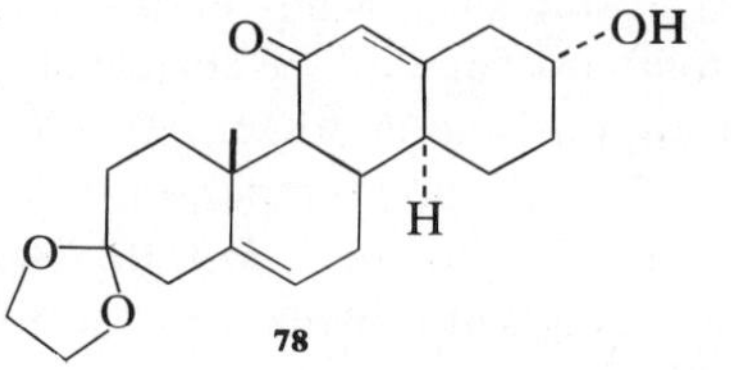

78 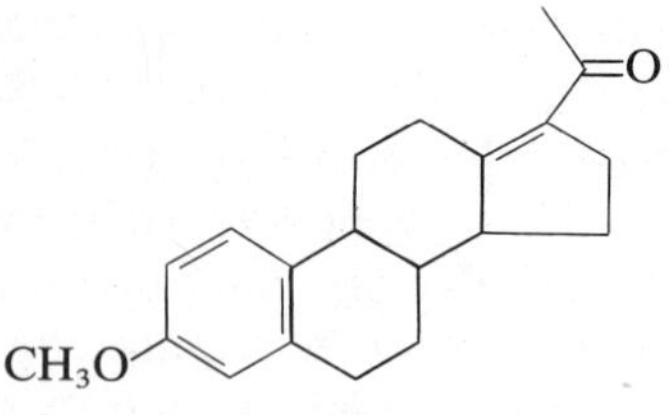 79

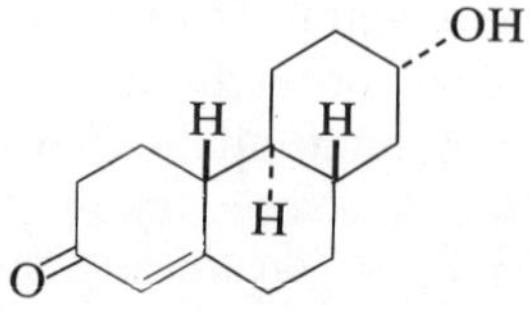

80a

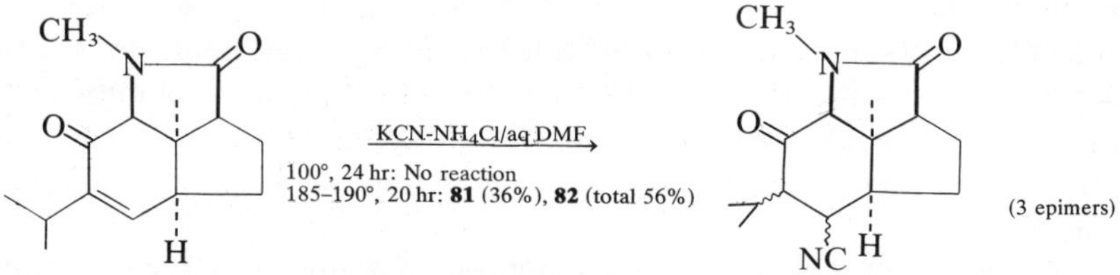

KCN-NH_4Cl/aq DMF

100°, 24 hr: No reaction
185–190°, 20 hr: **81** (36%), **82** (total 56%)

(3 epimers)

81 82

(14% conversion)[43,83] and steroidal 8-en-11-ones **84**[25,84] (no reaction). Higher reaction temperature might facilitate hydrocyanation of these substrates to some extent as anticipated from the reaction of enone **81**. Hydrocyanation of the electronically deactivated enone **85** failed also.[85] Poor yields obtained in hydrocyanation of β-methoxy enones **86** to β-cyano enones **87**[86] may arise because of deactivation by the

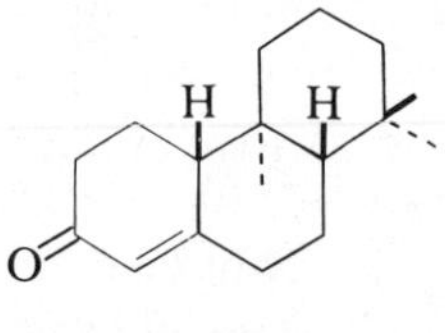

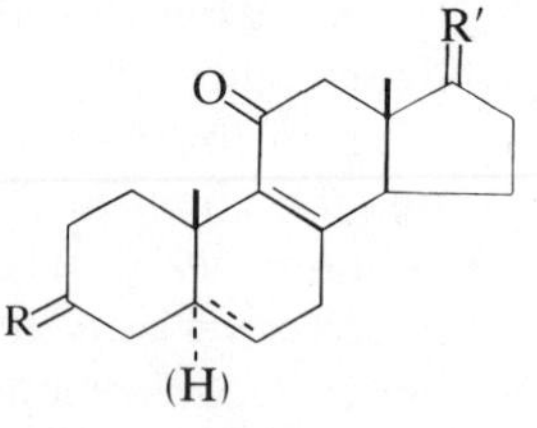

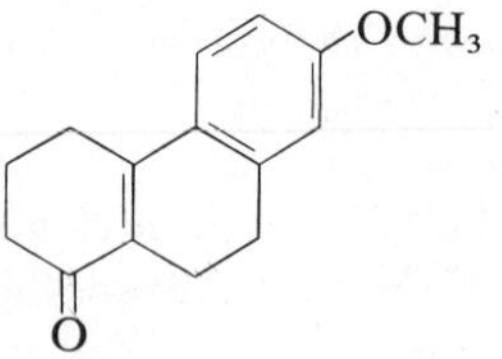

83 84 85

[83] M. Narisada, Ph.D. Dissertation, Kyoto University, 1964.

[84] W. Nagata, U.S. Pat. 3,231,566 (1966) [*C.A.*, **63,** 10035c (1965)].

[85] W. Nagata, T. Sugasawa, M. Narisada, T. Wakabayashi, and Y. Hayase, *J. Amer. Chem. Soc.*, **89,** 1483 (1967).

[86] J. Katsube and M. Matsui, *Agr. Biol. Chem.* (Tokyo), **35,** 401 (1971) [*C.A.*, **74,** 141052h (1971)].

R″, R, R′, OCH_3 (86) $\xrightarrow[\text{Reflux}]{KCN\text{-}NH_4Cl/aq\ THF}$ R″, R, R′, CN (87 (0–27%))

$R = CH_3$, n-Pr, n-C_5H_{11}, $(CH_2)_6CO_2CH_3$; $R' = R'' = H$, OH

methoxy group (see Reactivity-Structure Relationships, p. 300) as well as substitution of tetrahydrofuran for dimethylformamide. Recovery of cholestenone (**22**) in the reaction with potassium cyanide and ammonium chloride in aqueous dioxane[79] compared with the 84% conversion into the cyano ketones **24** in aqueous dimethylformamide illustrates the importance of the solvent in this procedure. Use of an aqueous alcohol is not favorable, because side reactions such as hydrolysis (p. 292) and Strecker amino nitrile formation occur more easily. It should be noted that the hydrocyanation of many simple enones is successful at room temperature, and consequently the reaction conditions should be selected according to the reactivity of the substrate and stability of the cyano product.

Use of Phase-Transfer Catalysts and Crown Ethers. Since phase-transfer catalysts and crown ethers have been known to catalyze the substitution reactions with potassium cyanide, application of these reagents to the conjugate hydrocyanation is interesting.

No papers have appeared concerning conjugate hydrocyanation of conjugated carbonyl compounds under phase-transfer conditions. Our preliminary experiments indicated that cholestenone (**22**) could be hydrocyanated with an excess of potassium cyanide and an equimolar amount of tetrabutylammonium bromide in benzene-water at 80° to give 5-cyano-3-oxo compounds **24** (predominantly the *cis* isomer) with little formation of by-products.[59] A small amount of the starting enone remained unchanged even after 24 hours at 80° and the reaction did not proceed at room temperature. Presumably the cyano enolates formed in the organic phase were slowly trapped with water at the interface. Further investigation of this process may be necessary.

Recently the use of crown ethers has been studied in the conjugate hydrocyanation of cholestenone (**22**).[51b] Although potassium cyanide and 18-crown-6 did not hydrocyanate the enone in refluxing benzene, the reaction proceeded in the presence of acetone cyanohydrin. In a typical procedure, enone **22** (0.16 *M*) and 18-crown-6 (0.19 *M*) in refluxing benzene are vigorously stirred with 1.2 molar equivalents of acetone cyanohydrin and a catalytic amount of potassium cyanide for 5 hours; the 1,4 adducts **24** are formed (with the *cis* isomer predominant) in high yield with no by-products. No reaction takes place in the absence of the crown ether. The following mechanism has been suggested.[51b] Clearly acetone cyanohydrin acts as a cyanide and proton source. Because the crown ether increases the

+ "naked" CN^- ⇌ CN

+ NC $C(CH_3)_2$ ⟶ CN
H—O

+ $(CH_3)_2C{=}O$ + CN^- (recycled)

effectiveness of the Nazarov method which is inherently free of side reactions, this crown ether procedure will be useful in the conjugate hydrocyanation of moderately reactive substrates.

Organoaluminum Cyanides

Aluminum alkoxides were the first organoaluminum compounds used in hydrocyanation. Cholestenone (**22**) reacted with hydrogen cyanide and aluminum isopropoxide and with hydrogen cyanide and aluminum *tert*-butoxide, giving cyano derivatives **88** and **24**, respectively.[87] The complex cyanide, lithium aluminum cyanide, was also effective, leading to adduct **24** in 47% yield.[87] The observed effectiveness of these aluminum compounds has led to the important discovery of two new hydrocyanation methods using aluminum alkyls.

C_8H_{17}

22

HCN-Al(OC_3H_7-*i*)$_3$/C_6H_6
Reflux, 6 hr

HO CN (4 isomers)

88 (52%)

HCN-Al(OC_4H_9-*t*)$_3$/C_6H_6
Reflux, 6 hr

O CN (2 isomers)

24 (50%)

[87] S. Hirai, *Chem. Pharm. Bull.* (Tokyo), **9,** 837 (1961).

Two methods[25] recently developed use a combination of anhydrous hydrogen cyanide and an alkylaluminum (HCN-AlR_3) in tetrahydrofuran (referred to as Method A) or an alkylaluminum cyanide (R_2AlCN) in an inert organic solvent (referred to as Method B). These new methods are different from the conventional procedures, and they give superior hydrocyanation results in laboratory use. Major advantages include high hydrocyanation efficiency even with the least reactive substrates, high uniformity (few side reactions), high selectivity (other functional groups are usually unaffected, see p. 297), and high stereoselectivity (see Stereochemistry, p. 313). The mechanistic features of these new methods are discussed on pp. 261–265. It is appropriate here to summarize their characteristics from a preparative viewpoint.[25,35] (1) Dialkylaluminum cyanide is more powerful than hydrogen cyanide-aluminum alkyl. (2) Hydrogen cyanide-triethylaluminum [actually containing a small amount of water; see (5) below] and diethylaluminum cyanide are the reagents most often used. (3) The reactivity order of Method A reagents is hydrogen cyanide-ethylaluminum dichloride < hydrogen cyanide-trimethylaluminum < hydrogen cyanide-triethylaluminum < hydrogen cyanide-diethylaluminum chloride < hydrogen cyanide-triethylaluminum-(water).* (4) The reactivity of diethylaluminum cyanide increases with decreasing basicity of the solvent, *i.e.*, tetrahydrofuran < dioxane < diisopropyl ether < benzene, toluene. (5) The small amounts of water normally present in the reaction medium accelerate the hydrocyanation with hydrogen cyanide-triethylaluminum. (6) In the initial stage of hydrocyanation the reversible 1,2 addition is predominant with dialkylaluminum cyanide, decreasing in the order hydrogen cyanide-triethylaluminum > hydrogen cyanide-diethylaluminum chloride > hydrogen cyanide-triethylaluminum-(water)* > hydrogen cyanide-ethylaluminum dichloride. (7) The 1,4 addition is irreversible with hydrogen cyanide-aluminum alkyl but reversible with dialkylaluminum cyanide. Thus there is some recovery of starting material in many reactions. (8) Stereochemically, Method A gives kinetic products, whereas Method B can lead to thermodynamic products when long reaction times are employed.

It should be noted that sodium triethylaluminum cyanide (**89**)[88] is a poor hydrocyanation reagent.[40]

$$Na[(C_2H_5)_3AlCN]$$

89

* See footnote on p. 266.

[88] K. Ziegler, R. Köster, H. Lehmkuhl, and K. Reinert, *Ann. Chem.*, **629**, 33 (1960).

The advantages and characteristics of Methods A and B are illustrated by the accompanying examples and discussion.

Hydrogen Cyanide-Alkylaluminum (Method A). In a typical example, treatment of 3β-acetoxycholest-5-en-7-one (**90c**) with a combination of hydrogen cyanide (3 molar equivalents) and triethylaluminum (5 molar equivalents) in the presence of 1 molar equivalent of water in tetrahydrofuran at room temperature for 7 hours gave the 5α-cyano-7-one **91c** in 92–93% yield.[43,89] The related enone **90a** could be hydrocyanated in the same way to nitrile **91a** in 93% yield.[43] These yields are compared to 51% and 43% of 5α-cyano products **91a** and **91b** by the potassium cyanide-ammonium chloride method.[43,90] The molar ratio of reagent to substrate may be changed depending on

C_8H_{17}

R O → R O CN

90a, R = H **91a,** R = H

90b, R = OH **91b,** R = OH

90c, R = OAc **91c,** R = OAc

the reactivity of the substrate (see p. 358). Triethylaluminum, although catalytic in principle, is used in a 1–2 molar excess over the amount of hydrogen cyanide to prevent possible polymerization of the latter (indicated by reddening of the reaction mixture). Addition of water (usually a 10–20% molar amount relative to triethylaluminum; see p. 262) is not necessary for more reactive substrates. Because rigorous exclusion of water from hydrogen cyanide and tetrahydrofuran, as well as during the hydrocyanation, is difficult and, moreover, unnecessary, the water contaminating the reaction medium is usually sufficient to accelerate the reaction. Therefore addition of water is not usually specified in Method A. When present in substrates, hydroxyl groups serve as rate-accelerating proton sources. The high efficiency of Method A compared with conventional procedures is demonstrated by the successful hydrocyanation of tricyclic enone **83**,[83] steroidal 8-en-11-ones **84**,[25,84] and β-methoxy enone **94**.[86] (Also compare the reaction of similar substrates **86** with potassium cyanide and ammonium chloride, p. 283.)

[89] W. Nagata and M. Yoshioka, *Org. Synth.*, **52,** 100 (1972).

[90] G. Snatzke and K. Kinsky, *Tetrahedron*, **28,** 295 (1972).

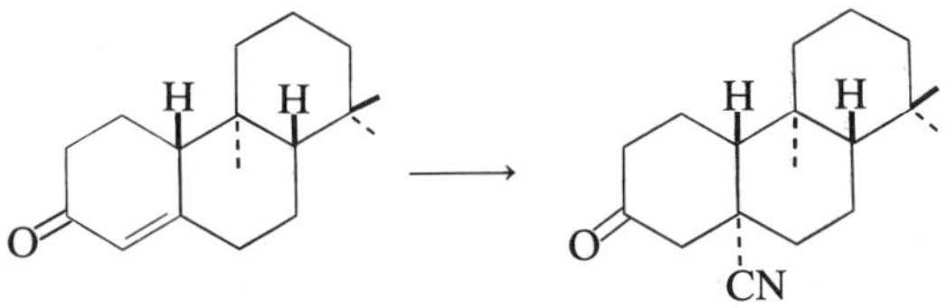

3 → 92

KCN-NH_4Cl/aq DMF, 100°, 54 hr: **92** (14%)
HCN-$Al(C_2H_5)_3$/THF, room temp, 27 hr: **92** (72%)

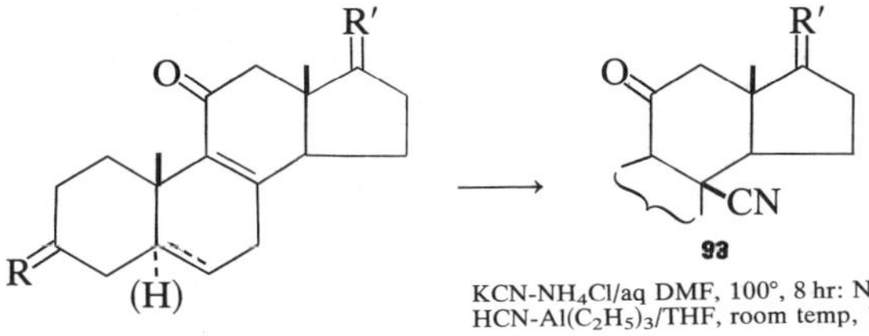

84 → 93

KCN-NH_4Cl/aq DMF, 100°, 8 hr: No reaction
HCN-$Al(C_2H_5)_3$/THF, room temp, 18–25 hr: 70–74%

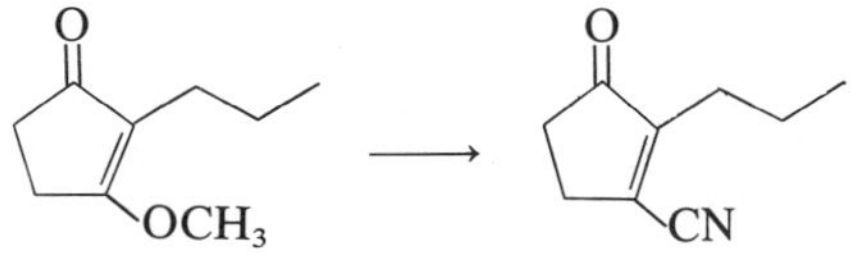

94 → 95

$(CH_3)_2C(OH)CN$-$N(C_2H_5)_3$, reflux, 45 min: 8%
HCN-$Al(C_2H_5)_3$/THF, room temp, 3.5 hr: 86%

The high uniformity of Method A is illustrated by the hydrocyanation of steroidal 21-acetoxy-16-en-20-one **96**. Reaction with sodium cyanide gives the deacetylated cyano ketone **97a** in only 12% yield,[91] whereas Method A hydrocyanation affords the acetoxy cyano ketone **97b** in 72% yield.[43] Apparently the base sensitivity of the 21-acetoxy-20-keto or the 21-hydroxy-20-keto grouping is responsible for the poor yield in the conventional procedure.

Replacement of the hydrogen cyanide-triethylaluminum combination by hydrogen cyanide-diethylaluminum chloride or hydrogen cyanide-ethylaluminum dichloride is effective when 1,2 addition or reverse 1,4 addition or both are predominant. The effect results from better activation of the enone system by the chloro reagents, which are stronger Lewis acids. For example, hydrocyanation of aromatic enone **85** with hydrogen cyanide and triethylaluminum gave the *trans*-cyano ketone **98t** in only 11% yield, whereas with hydrogen cyanide and

[91] J. Romo, *Tetrahedron*, **3**, 37 (1958).

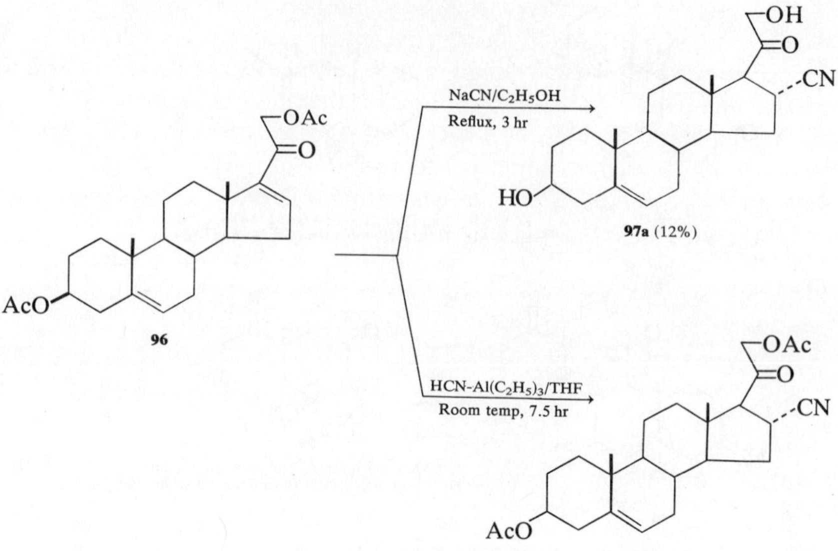

NaCN/C_2H_5OH
Reflux, 3 hr

HCN-$Al(C_2H_5)_3$/THF
Room temp, 7.5 hr

96

97a (12%)

97b (72%)

diethylaluminum chloride the total yield of *cis*- and *trans*-cyano ketones **98t** and **98c** was 75%.[85] Also, the α,β-unsaturated imine **99** reacted with hydrogen cyanide and triethylaluminum to give only the 1,2 adduct **100**, whereas with hydrogen cyanide and ethylaluminum dichloride the 1,4 adduct **101** was obtained in reasonable yield after hydrolysis[92] (see p. 344).

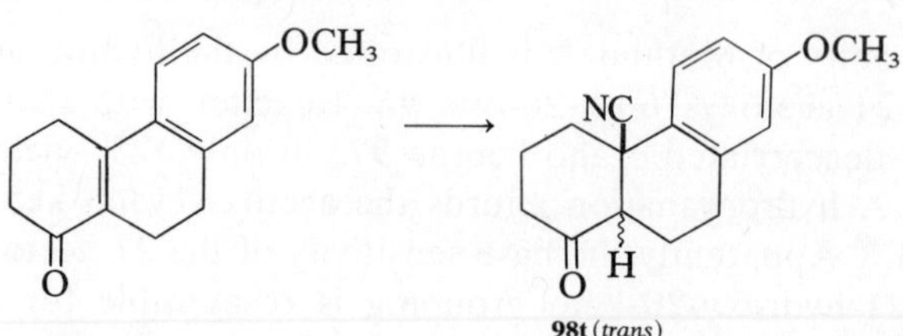

85

98t (*trans*)
98c (*cis*)

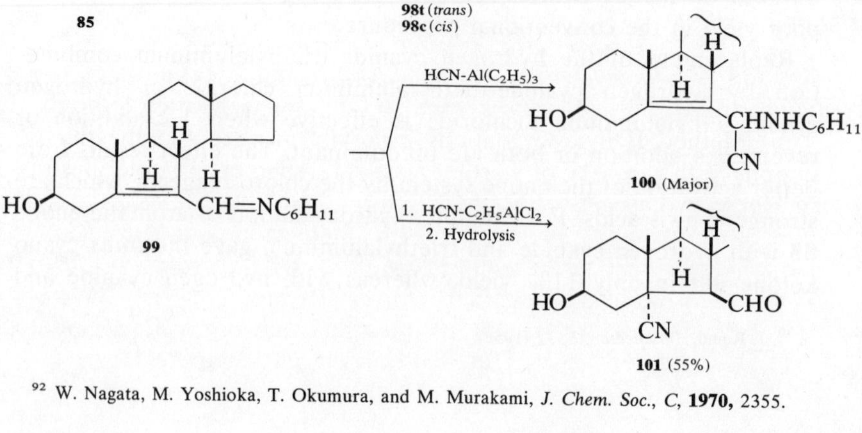

HCN-$Al(C_2H_5)_3$

1. HCN-$C_2H_5AlCl_2$
2. Hydrolysis

99

100 (Major)

101 (55%)

[92] W. Nagata, M. Yoshioka, T. Okumura, and M. Murakami, *J. Chem. Soc., C*, **1970,** 2355.

Replacement of tetrahydrofuran by another ether solvent such as dioxane, diethyl ether, or diisopropyl ether, if necessary, should be limited to reactions at low temperature; hydrogen cyanide reacts with triethylaluminum rapidly in solvents other than tetrahydrofuran to give diethylaluminum cyanide and the advantages of Method A are thus lost. The only such example is the reaction of cyclohex-2-en-1-one with hydrogen cyanide and triethylaluminum in diethyl ether at −15° giving 3-oxocyclohexane-1-carbonitrile in 80% yield (p. 291). In this reaction the use of the low-boiling ether was preferred for separation of the product. Hydrocarbon solvents such as benzene, toluene, and hexane are not suitable because of the instantaneous formation of diethylaluminum cyanide.

Dialkylaluminum Cyanide (Method B). Owing to its easier preparation and handling, diethylaluminum cyanide is the only Method B reagent commonly used, although other alkylaluminum cyanides such as diisobutylaluminum cyanide, dimethylaluminum cyanide, and ethyl(chloro)aluminum cyanide have been prepared[25] and their reactivities tested.[35] Ethyl(chloro)aluminum cyanide was found to be sparingly soluble in organic solvents and not very reactive.[59] In a typical Method B hydrocyanation the steroidal 8-en-11-one **102** was treated with 6 molar equivalents of diethylaluminum cyanide in benzene-toluene at 0° for 10 minutes to give the 8β-cyano ketone **103** in 92% yield after retreatment of recovered starting material.[25] In comparison,

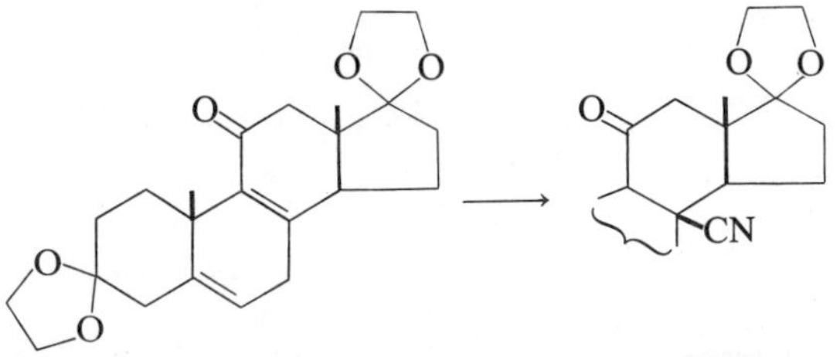

102 **103** (92%)

Method A hydrocyanation of enone **102** in tetrahydrofuran at room temperature afforded cyano ketone **103** in 74% yield after 25 hours.[25] Thus Method B hydrocyanations generally proceed rapidly but frequently with recovery of starting material. Another example illustrating these two aspects of Method B is the hydrocyanation of the rigid nitrogen-containing enone **81** with diethylaluminum cyanide in benzene-toluene at room temperature. The cyanated products **82** were obtained in 67% yield together with unchanged enone in 17% yield (for formulas and the result by the potassium cyanide-ammonium chloride procedure, see p. 282). The high efficiency of Method B is illustrated by the reaction of the 14-en-16-one **104** (no reaction with potassium cyanide and ammonium chloride)[93] and of the less reactive 6β-substituted B-nor-4-en-3-ones **106a–106c**.[43] The reluctance of the 6β-ethyl enone **106b** to react even at 70° and the poor yield of the 6β-vinyl nitrile **107c** produced by Method A indicate the low reactivity of this type of B-nor enone (see Reactivity-Structure Relationships, p. 306). Use of a neutral reagent, diethylaluminum cyanide, is the key to successful conversion of the unstable 6β-vinyl enone **106c** which rapidly isomerizes to the corresponding conjugated dienone under basic conditions.

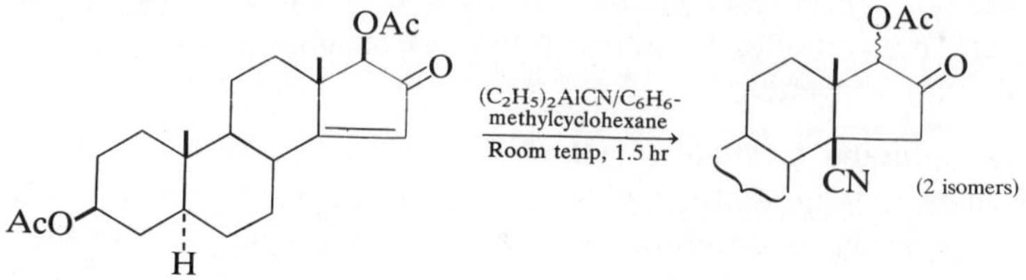

$(C_2H_5)_2AlCN/C_6H_6$-methylcyclohexane, Room temp, 1.5 hr

104 → **105** (58%) (2 isomers)

$(C_2H_5)_2AlCN$

106a, R = CH_3
106b, R = C_2H_5
106c, R = CH=CH_2

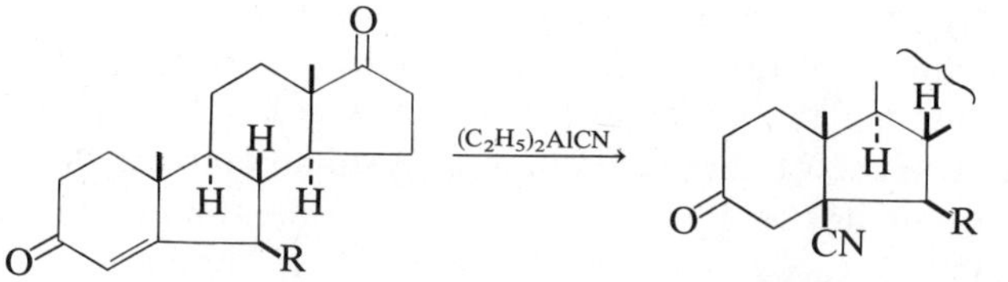

107a, R = CH_3
107b, R = C_2H_5
107c, R = CH=CH_2

THF-C_6H_6, room temp, 2 hr: **107a** (80%)
THF-toluene, 70°, 1 hr: **107b** (13%)
C_6H_6-CH_2Cl_2, room temp, 3 hr: **107c** (61%)
[HCN-Al$(C_2H_5)_3$/THF, room temp, 6 hr: **107c** (16%)]

When a substrate is moderately reactive but its 1,4 adduct is susceptible to reversion or its 1,2 adduct is highly stable, Method B hydrocyanation is less suitable than the irreversible Method A reaction (especially with hydrogen cyanide and diethylaluminum chloride). This situation is exemplified by reactions of cyclohexenone **108**[43] and the bicyclic dienone **110**.[94] In the presence of a proton source, diethylaluminum cyanide reacts like a Method A reagent, effecting conjugate

[93] A. C. Campbell, W. Lawrie, and J. McLean, *J. Chem. Soc., C*, **1969,** 554.
[94] H. Minato and T. Nagasaki, *J. Chem. Soc., C*, **1966,** 1866; *Chem. Commun.*, **1965,** 377. See footnote on p. 291.

addition of the α,β-ethylenic imine **112**. This modification is attractive because the diethylaluminum cyanide-alcohol combination will have both high efficiency and irreversibility. Further investigations would be desirable.

108 → **109** (CN)

Method A: **109** (80%)

Method B: **108** (13%), **109** (57%)

110 → **111***

Method A: **111** (quantitative)

Method B: **110** (major), **111** (0–5%)

$C_6H_{11}N{=}CH$ ... $OCOC_6H_5$ **112**

$(C_2H_5)_2AlCN$ → $C_6H_{11}NH$, CN, CH ... H **113**

1. $HCN\text{-}Al(C_2H_5)_3\text{-}(H_2O)$ or $(C_2H_5)_2AlCN\text{-}(i\text{-}C_3H_7OH)$
2. Hydrolysis

→ OHC ... CN H **114**

Generally, Method B is preferred for 1,6 addition to $\alpha,\beta,\gamma,\delta$-unsaturated ketones, as is seen from the reaction of steroidal dienone **115**.[43,95,96] The enolate intermediate **118** formed in Method B prevents the formation of secondary products from the δ-cyano ketone **116** observed in conventional procedures and in Method A (here the

* On the basis of the infrared absorption at 1723 cm^{-1}, structure **111′** was assigned.[94] The authors have agreed to revise it to the more likely structure **111**.

CN **111′**

[95] G. H. Rasmusson, A. Chen, and G. E. Arth, *J. Org. Chem.*, **38,** 3670 (1973).

[96] R. G. Christiansen and W. S. Johnson, *Steroids*, **1,** 620 (1963).

reaction was stopped earlier to prevent the formation of the 5,7-dinitrile).[96] The success of Method A with the bicyclic dienone **110** may be ascribed to the stability of the unconjugated structure of the product **111** (see p. 339).

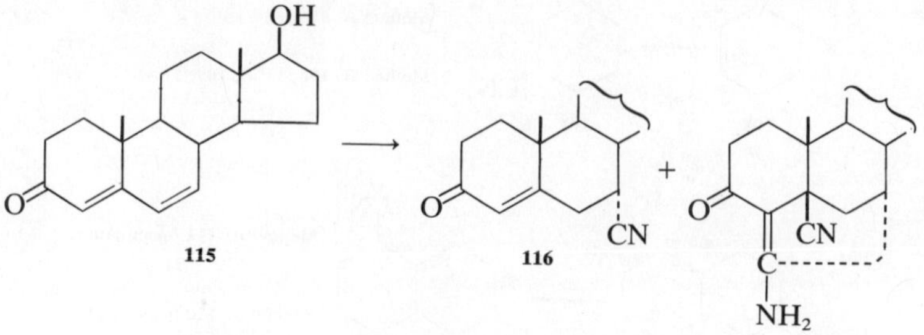

115 **116** **117**

KCN-$AcOC_2H_5$ or KCN-NH_4Cl: **116** (6–55%), **117** (10–32%)
HCN-$Al(C_2H_5)_3$, 0°: **115** (32%), **116** (32%)
$(C_2H_5)_2AlCN$, room temp: **116** (92%)

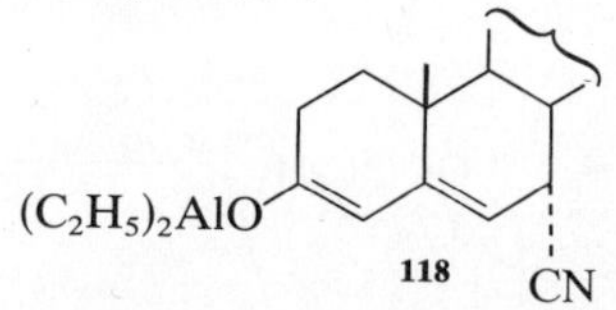

$(C_2H_5)_2AlO$ ... CN

118

Hydrocyanation of the sensitive acid derivatives **119** is effected by Method B which uses neutral diethylaluminum cyanide having poor

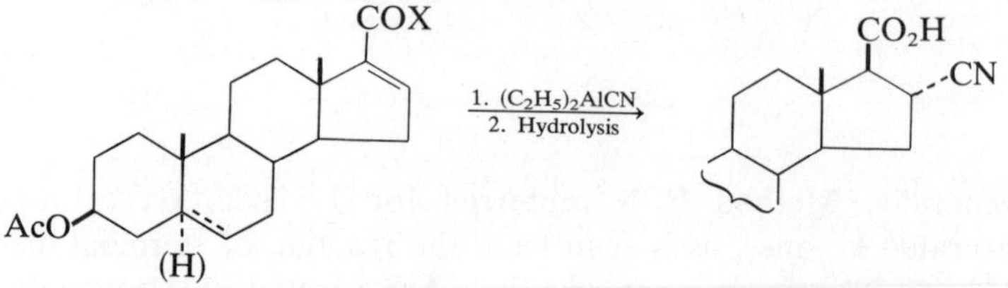

1. $(C_2H_5)_2AlCN$
2. Hydrolysis

119a, X = Cl, 5α-H
119b, X = CN, 5-ene

120a, 5α-H
120b, 5-ene

alkylating ability. An aprotic ether solvent would be preferable to avoid formation of the ethyl ketone.

Side Reactions and Incompatible Groups

Here we define "side reactions" as reactions other than hydrocyanation occurring at the conjugated system. Other isolated functional groups that react with the hydrocyanation reagents are defined as incompatible groups; their reactions are described on pp. 297–299.

Side Reactions

The less efficient conventional methods using anhydrous hydrogen cyanide or acetone cyanohydrin in the presence of a basic catalyst are almost free of side reactions. In the more efficient conventional methods using an alkali cyanide, side reactions arise from the reaction of an alkaline species formed from an alkali cyanide in aqueous medium. The presence of an alcoholic solvent facilitates the side reactions, because an alkoxide ion is more nucleophilic than a hydroxide ion. Common side reactions are discussed below.

Hydrolysis. The side reaction occurring most frequently is alkaline hydrolysis of ester and cyano groups present in the substrate and of the newly introduced cyano group. Hydrolysis may be followed by decarboxylation (*e.g.*, **42**), cyclization (*e.g.*, **122** and **125a**), or elimination (*e.g.*, **124**). Sometimes hydrogen cyanide (*e.g.*, **125b**) or solvent (*e.g.*, **75b**, p. 280) may add to products, as shown in the accompanying examples. The intermediates for **125b** and **75b** might be imines such as **126**.

$$C_6H_5CH{=}CH(CO_2C_2H_5)_2 \xrightarrow[60^\circ]{KCN/aq\ C_2H_5OH} C_6H_5CH(CN)CH_2CO_2R$$

40 → **42**[7]

$R = C_2H_5$ (80%), $R = H$ (14%)

121 $\xrightarrow[75^\circ]{KCN/aq\ CH_3OH}$ **122** (50%)[97] (2 isomers)

123 $\xrightarrow[Reflux]{KCN\text{-}AcOH/aq\ C_2H_5OH}$ **124** (86%)[98]

$$CH_3COCH{=}C(CH_3)_2 \xrightarrow[Reflux]{KCN/aq\ C_2H_5OH}$$

29

125a, R = OH[99]
125b, R = CN

126

Because hydrolysis is so common in hydrocyanations of α,β-unsaturated carboxylic acid derivatives with an alkali cyanide, a deliberate hydrolysis step is usually included in such cases to give dicarboxylic acids such as **128**.[52,100]

$$C_6H_5CH{=}C(R)CO_2C_2H_5 \xrightarrow[\text{2. HCl, } H_2O]{\text{1. KCN}} C_6H_5CH(CO_2H)CH_2CO_2H$$

127 ($R = CN, CO_2C_2H_5$) → **128**

Addition of Solvent. When dry methanol is used as the hydrocyanation solvent at room temperature, competitive 1,4 addition of the solvent occurs. For example, the reaction of the quaternary salt **66a**[53] gives the β-methoxy carbonyl derivative **130** as well as the cyano product **129**. Conjugate addition of an alkoxide ion to α,β-ethylenic

$$ArCOCH(CH_3)CH_2\overset{+}{N}(CH_3)_3I^- \xrightarrow[\text{Room temp, 3 hr}]{KCN/CH_3OH} ArCOCH(CH_3)CH_2CN + ArCOCH(CH_3)CH_2OCH_3$$

66a ($Ar = p{-}BrC_6H_4$); **129** (62%); **130** (*ca.* 20%)

substrates usually is not serious since the reaction is reversible and the cyano adduct is therefore eventually formed. On the other hand, competitive addition of an alkoxide ion to α,β-acetylenic substrates gives enol ether **132a** which is not reconverted into substrate **131** under the reaction conditions.[11] When aqueous alcohol is used as the solvent, the above-mentioned conjugate addition of an alkoxide decreases greatly or does not occur at all.

$$C_6H_5C{\equiv}CCO_2CH_3 \xrightarrow[\text{2. Dry HCl (neutralization)}]{\text{1. KCN/dry } CH_3OH\text{, reflux}} C_6H_5C(OCH_3){=}CHCO_2CH_3 + C_6H_5C({=}O)CH_2CO_2CH_3 + C_6H_5CH(CN)CH(CN)CO_2CH_3$$

131 → **132a** (48%) + **132b** (20%) + **133** (19%)

Condensation. Intermolecular Michael-type condensation of a cyanated product (or its hydrolysis product) and the α,β-ethylenic substrate is a common side reaction in conventional hydrocyanation

[97] W. L. Meyer and N. G. Schnautz, *J. Org. Chem.*, **27,** 2011 (1962).

[98] R. W. L. Clarke and A. Lapworth, *J. Chem. Soc.*, **89,** 1869 (1906).

[99] R. V. Stevens, C. G. Christensen, W. L. Edmonson, M. Kaplan, E. B. Reid, and M. P. Wentland, *J. Amer. Chem. Soc.*, **93,** 6629 (1971); R. V. Stevens and M. Kaplan, *Chem. Commun.*, **1970,** 822.

[100] A. Lapworth and W. Baker, *Org. Synth., Coll. Vol.* **I,** 451 (1941).

methods. A typical example is the formation of the tricyclic enamino ketone **134** in 53% yield during hydrocyanation of cyclohex-2-en-1-one (**108**) with potassium cyanide in refluxing methanol.[101] The Michael adduct **135** is isolated in 18% yield from the reaction of 4(10)-octalin-3-one with potassium cyanide and ammonium chloride.[102] Similar cyano lactam dimers were isolated from the hydro-

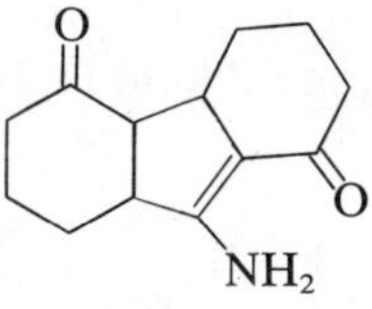

134

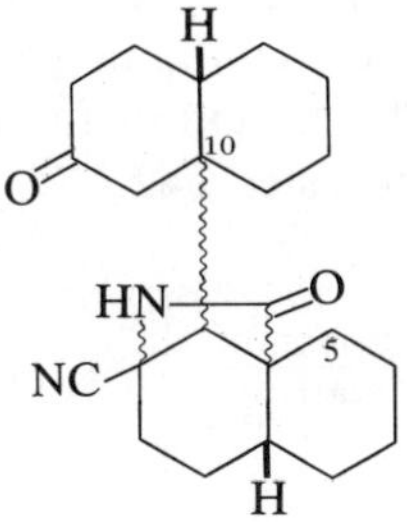

135

(or C_{10}—C_5 dimer)

cyanation of cholest-4-en-3-one (**22**)[15] and of D-homo-3β-hydroxy-18-norandrost-13(17a)-en-17-one[103] with potassium cyanide in refluxing methanol. This fact indicates that even steroidal compounds of rather low reactivity are not free from intermolecular condensation in reactions with an alkali cyanide. Formation of two different condensation products, **139** and **140**, in anhydrous and aqueous methanol from the reaction of methyl fumarate (**136**) with potassium cyanide has been reported.[11] The reaction no doubt proceeds through the Michael adduct **138** of the substrate and the cyano product **137**. As expected,

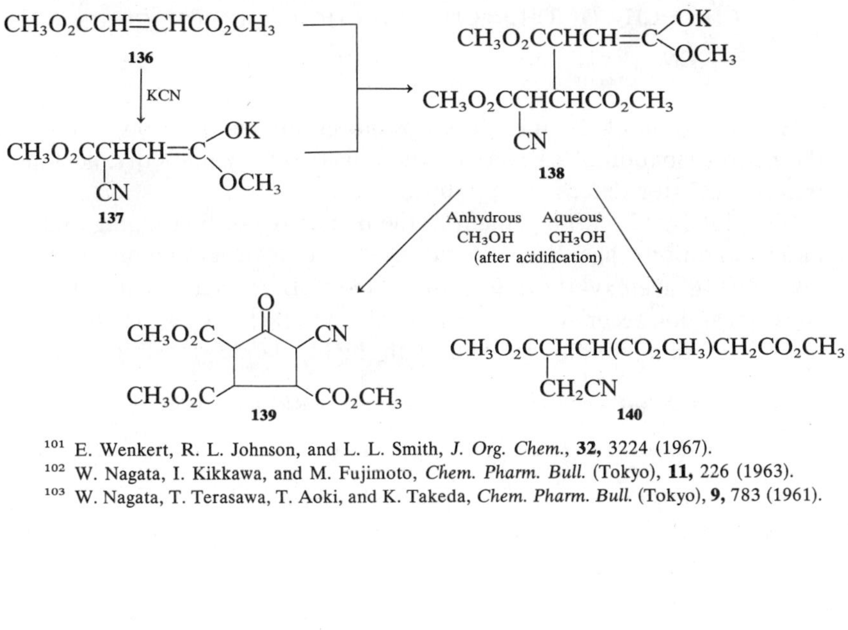

$CH_3O_2CCH{=}CHCO_2CH_3$ **136**

↓ KCN

$CH_3O_2CCH(CN)CH{=}C(OK)OCH_3$ **137**

→ $CH_3O_2CCHCH{=}C(OK)OCH_3$ / $CH_3O_2CCHCHCO_2CH_3$ (CN) **138**

Anhydrous CH_3OH (after acidification) → **139**

Aqueous CH_3OH → $CH_3O_2CCHCH(CO_2CH_3)CH_2CO_2CH_3$ (CH_2CN) **140**

[101] E. Wenkert, R. L. Johnson, and L. L. Smith, *J. Org. Chem.*, **32**, 3224 (1967).

[102] W. Nagata, I. Kikkawa, and M. Fujimoto, *Chem. Pharm. Bull.* (Tokyo), **11**, 226 (1963).

[103] W. Nagata, T. Terasawa, T. Aoki, and K. Takeda, *Chem. Pharm. Bull.* (Tokyo), **9**, 783 (1961).

the Michael-type condensation is greatly suppressed in reactions of α-alkyl substrates. This situation is illustrated by reaction of the α-alkyl-β-methylcyclohexenones **49** with potassium cyanide in refluxing aqueous methanol to give, after hydrolysis, β-carboxy ketones **50** in high yields without formation of dimeric products[55] (for formulas, see p. 274).

Another alkali-induced side reaction is the intramolecular condensation of cyanated products containing a reactive functional group. A typical example is the formation of the bicyclic cyano diketone **143** in high yield via the cyano ester **142**.[104]

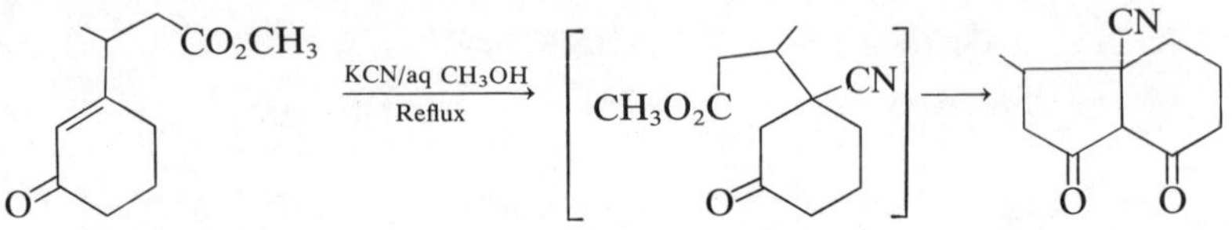

CO_2CH_3 (**141**) $\xrightarrow[\text{Reflux}]{\text{KCN/aq } CH_3OH}$ [CH_3O_2C … CN (**142**)] $\longrightarrow$ CN … O O (**143**)

Isomerization, Epimerization, and Rearrangement. Base-catalyzed double-bond isomerization of a β,γ-unsaturated substrate can also occur in hydrocyanation reactions; it is followed by addition to the isomerized double bond as exemplified by the reaction of the $\alpha,\beta,\beta',\gamma'$-unsaturated ketone **144**.[105]

$$\underset{\textbf{144}}{CH_2{=}CHCH_2COC(CH_3){=}CH_2} \xrightarrow[45\text{–}50^\circ]{\text{KCN-AcOH}} \underset{\textbf{145}\ (14\%)}{CH_2{=}CHCH_2COCH(CH_3)CH_2CN} +$$

$$\underset{\textbf{146}\ (18\%)}{CH_3\underset{\displaystyle CN}{\underset{|}{C}}HCH_2COC(CH_3){=}CH_2} + \underset{\textbf{147}\ (3\%)}{CH_3\underset{\displaystyle CN}{\underset{|}{C}}HCH_2COCH(CH_3)CH_2CN}$$

Epimerization of the introduced cyano group has been observed in the hydrocyanation of carvone by the potassium cyanide-ethyl acetate procedure[69] (for details see p. 309).

Base-catalyzed rearrangement of the α-acetoxy keto grouping under basic conditions has been reported in the hydrocyanation of the steroidal 3β-acetoxy-1(10)-en-2-one **148**.[106] Base-induced side reactions generally occur with substrates of low reactivity, which require severe reaction conditions, and with highly labile substrates. The

[104] A. J. Birch, J. E. T. Corrie, P. L. Macdonald, and G. S. Rao, *J. Chem. Soc., Perkin I*, **1972,** 1186.

[105] I. N. Nazarov and M. V. Kuvarzina, *Izv. Akad. Nauk SSSR, Ser. Khim.*, **1949,** 386 [*C.A.*, **44,** 3458d (1950)].

[106] J. Fishman, M. Torigoe, and J. A. Settepani, *Tetrahedron*, **21,** 3677 (1965).

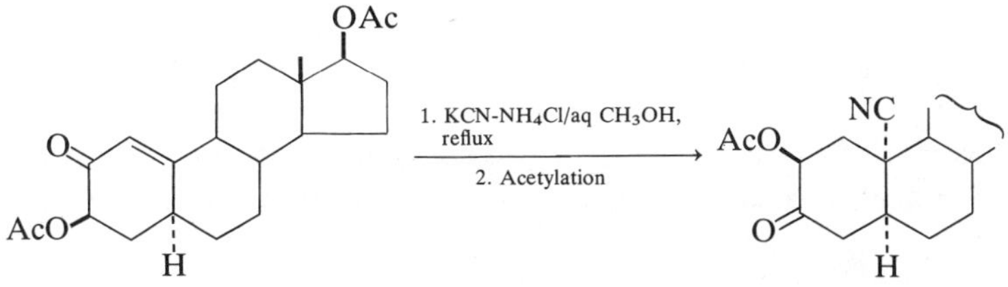

1. $KCN\text{-}NH_4Cl$/aq CH_3OH, reflux
2. Acetylation

148 → **149** (60%)

recently developed procedures using phase-transfer catalysts and crown ethers may be effective in preventing these side reactions.

Hydrocyanation methods using alkylaluminum compounds (Method A and B) are completely free of base-induced side reactions. Reactive acid chlorides such as **150** may be converted into the corresponding ethyl ketone **152** under some conditions. It should be noted that this ethylation did not occur in an ether solvent.[26]

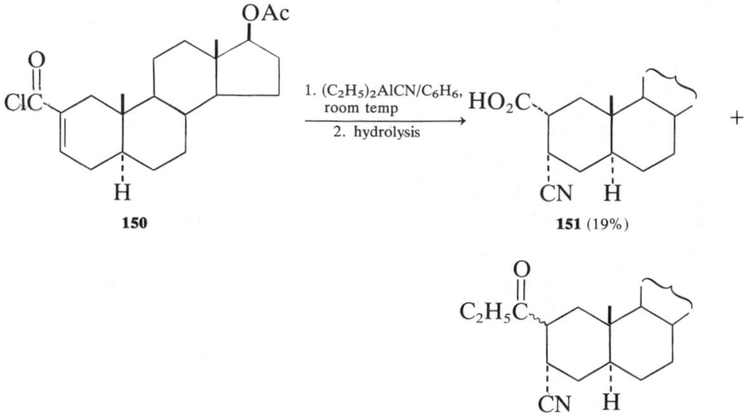

1. $(C_2H_5)_2AlCN/C_6H_6$, room temp
2. hydrolysis

150 → **151** (19%) + **152** (4%)

Incompatible Groups

Because of its basic nature and high nucleophilicity, an alkali cyanide can attack alkali-sensitive or cyanide-exchangeable functionalities or both. Examples of isolated functional groups that can be hydrolyzed are acetoxyl,[103] alkoxycarbonyl, and cyano (see the preceding section). A steroidal 21-hydroxy-20-oxo grouping (α-ketol) was not stable to sodium cyanide, as shown in the reaction of 21-acetoxy-16-en-20-one **96** (see p. 287). An alkali cyanide might also be expected to destroy other alkali-sensitive functions such as lactone,

β-ketol, and α- and β-dicarbonyl compounds. Examples of cyanide-exchangeable groups are halo (except for fluoro), tosyloxy, and mesyloxy.[19,30] Such alkali cyanide-sensitive functionalities would be compatible with other hydrocyanating reagents in the order: potassium cyanide-acid (potassium cyanide-ammonium chloride)[107] $<$ hydrogen cyanide or acetone cyanohydrin-(base)[2,108] $\ll$ hydrogen cyanide-alkylaluminum $\approx$ diethylaluminum cyanide. With the organoaluminum reagents, attempted substitution of these functionalities resulted in recovery of the starting material.[59]

Cleavage of an epoxy group to form a β-cyanohydrin depends on the reactivity of the substrate and can be controlled by choice of the reagent. The organoaluminum reagents can cleave all epoxy derivatives including the least reactive steroidal $9\alpha,11\alpha$-epoxides.[109] Simple aliphatic epoxides react with hydrogen cyanide-(base) reagents.[19,20] A steroidal $5\alpha,6\alpha$-epoxide is cleaved, although in low yield, with potassium cyanide in ethylene glycol above 90°[110] and in the presence of magnesium sulfate (to generate hydrogen cyanide) in refluxing ethanol.[111] The potassium cyanide-ammonium chloride reagent failed to react with a steroidal $5\beta,6\beta$-epoxide.[109] These data suggest that selective hydrocyanation of a conjugated carbonyl system in the presence of an epoxy group may be possible with potassium cyanide or potassium cyanide-ammonium chloride. β-Halohydrins (precursors of epoxides) react with potassium cyanide[20] and the new organoaluminum reagents.[109]

The formyl group (aldehydes) usually reacts with all the hydrocyanating reagents giving α-cyanohydrins.[30,92] Most saturated ketones form α-cyanohydrins with the organoaluminum reagents,[25,59] but usually not with the conventional reagents whose basicity destabilizes the ketone cyanohydrins. The formation of these α-cyanohydrins is not a serious problem, since they can be reconverted into aldehydes and ketones by alkaline treatment upon workup. Unlike Grignard reagents, the organoaluminum reagents do not enolize ketones.[25] The aldimine system ($\rangle$C=N—) reacts to give amino nitriles[19,30] whose reconversion into imine would be difficult. However, they can be converted into the parent ketones or aldehydes by acid hydrolysis.[92]

Examples of groups compatible to hydrocyanation are ethylenedioxy (ketals), bismethylenedioxy, isolated acetylenic, isolated ethylenic, N-

[107] W. S. Johnson, J. A. Marshall, J. F. W. Keana, R. W. Franck, D. G. Martin, and V. J. Bauer, *Tetrahedron*, Suppl. 8, Part II, 541 (1966).

[108] M. P. L. Caton, T. Parker, and G. L. Watkins, *Tetrahedron Lett.*, **1972,** 3341.

[109] W. Nagata, M. Yoshioka, and T. Okumura, *J. Chem. Soc., C*, **1970,** 2365.

[110] A. Bowers, E. Denot, M. B. Sánchez, L. M. Sánchez-Hidalgo, and H. Ringold, *J. Amer. Chem. Soc.*, **81,** 5233 (1959).

[111] J. C. Jacquesy and J. Levisalles, *Bull. Soc. Chim. Fr.*, **1965,** 1538.

acyl, N-sulfonyl, alkyloxy (ethers), and hydroxyl.[43,57,59,112] The vinyloxy group (vinyl ethers) might be compatible, but not the vinyl ester (see p. 266). The favorable effect of the hydroxyl group in the organoaluminum methods has been discussed on pp. 286 and 291.

The accompanying examples illustrate special cases in which usually compatible groups are attacked by the organoaluminum reagents. The reaction of 4-chlorocepham **159** (probably via imine **161**) also shows that the alkali-sensitive β-lactam ring can survive the reaction with diethylaluminum cyanide.

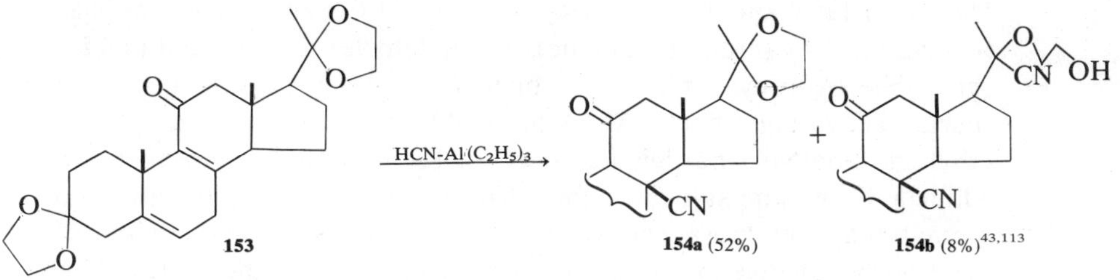

HCN-Al(C_2H_5)$_3$

153 **154a** (52%) **154b** (8%)[43,113]

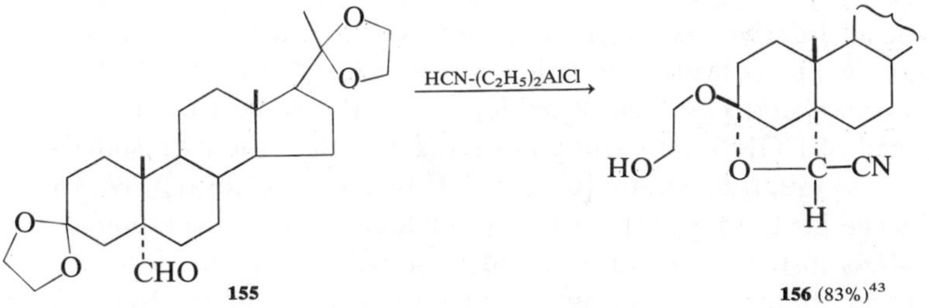

HCN-(C_2H_5)$_2$AlCl

155 **156** (83%)[43]

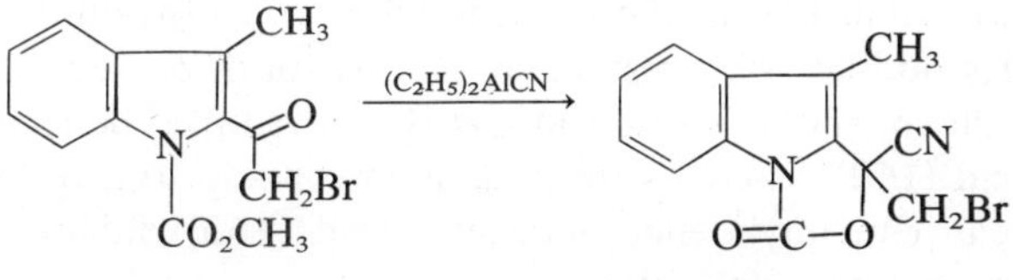

(C_2H_5)$_2$AlCN

157 **158** (65%)[40]

(C_2H_5)$_2$AlCN

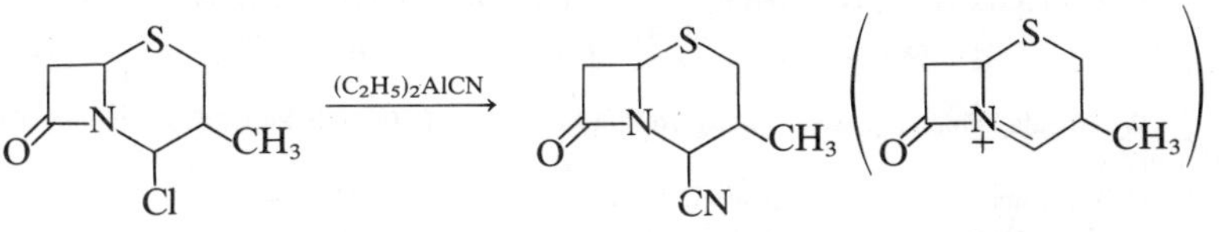

159 **160** (72%)[114] **161**

[112] K. Wiesner, E. W. K. Jay, T. Y. R. Tsai, C. Demerson, L. Jay, T. Kanno, J. Křepinský, A. Vilím, and C. S. Wu, *Can. J. Chem.*, **50,** 1925 (1972).

Survey of Structure-Reactivity Relationships

Classes of Reactive Carbonyl Substrates

Reactivities of α,β-ethylenic carbonyl substrates **1** to conjugate hydrocyanation are primarily affected by the carbonyl substituent R^4. The reactivity increases with decreasing electron-donating ability (in-

$$\mathrm{(R^1)(R^2)C{=}C(R^3){-}C(R^4){=}O}$$

1

ductive or resonance) of R^4: NR_2 (amide) < OR (ester) < alkyl (aliphatic ketone) < Ar (aromatic ketone) < H (aldehyde). Conjugated amides cannot be hydrocyanated unless one of the other substituents has a special activating effect. Although highly reactive in nature, α,β-ethylenic aldehydes participate poorly in 1,4 addition because their 1,2 adducts form almost irreversibly. Conjugated enones (aliphatic and aromatic) are the most reactive carbonyl substrates. The reactivities of the poorly reactive α,β-ethylenic esters increase in the order $R^4 =$ OH[115] < O-alkyl < O⌒⟍R$'^2$ ≪ OAr. The reactivities of α,β-ethylenic carboxylic acid derivatives are greatly improved by replacement of the OR group by a less electron-donating group such as SR (thio ester), Cl (acid chloride), and CN (acid cyanide).[26] α,β-Ethylenic carbonitriles (CN instead of COR^4) are slightly more reactive than the corresponding esters.[2,116] Imino substrates [$C(R^4){=}NR$ instead of COR^4] possess a reactivity similar to that of the carbonyl analog.[92] Iminium compounds [$C(R^4){=}\overset{+}{N}R_2$ instead of COR^4] should be more reactive.[43]

The susceptibility of α,β,γ,δ-unsaturated carbonyl compounds to 1,6 addition is lower than that of the corresponding α,β-ethylenic derivatives toward 1,4 addition,[117] presumably because of dispersed carbonyl activation of the δ and β carbon atoms. For example, benzylidenecyanoacetic acid (**162**) reacted with potassium cyanide to give the 1,4 adduct, whereas cinnamylidenecyanoacetic acid (**163**) did not react under the same reaction conditions.[118]

$$C_6H_5CH{=}C(CN)CO_2H \qquad C_6H_5CH{=}CHCH{=}C(CN)CO_2H$$

162 **163**

[113] W. Nagata, *Proc. Int. Symp. Drug Res.*, **1967,** 188; H. Itazaki, Ph.D. Dissertation, Tokyo University, 1969.

[114] H. W. Schnabel, D. Grimm, and H. Jensen, *Ann. Chem.*, **1974,** 484.

[115] L. Higginbothan and A. Lapworth, *J. Chem. Soc.*, **121,** 49 (1922).

[116] W. Franke and H. Weber, Ger. Pat. 808,835 (1951) [*C.A.*, **47,** 5427d (1953)].

[117] O. R. Rodig and N. J. Johnston, *J. Org. Chem.*, **34,** 1949 (1969).

[118] A. Lapworth and J. A. McRae, *J. Chem. Soc.*, **121,** 1699 (1922).

α,β-Acetylenic carbonyl substrates would be expected to be more reactive than the ethylenic analogs.[119] However, the yields of 1,4 adducts are actually low because of the side reactions which these reactive products undergo.[2,11,120,121]

Cyclic α,β-ethylenic carbonyl compounds generally are more reactive than the corresponding acyclic compounds. For example, acyclic α-phenyl-β-methoxy-α,β-unsaturated ketones **164** are inert to hydrocyanation by the Nazarov procedure (acetone cyanohydrin catalyzed by a base), whereas cyclic enones **165** and **166** are successfully hydrocyanated under similar conditions to the corresponding addition-elimination products.[122–124] Coumarin (**168**) was more reactive than the corresponding cinnamate **167**.[7]

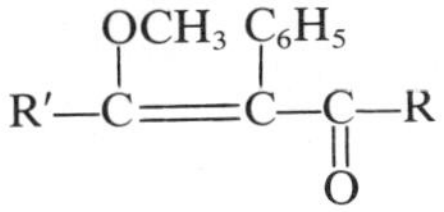

$R'—C(OCH_3){=}C(C_6H_5)—C(=O)—R$

164 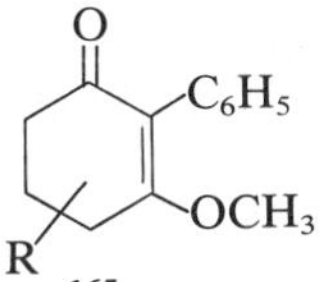 **165** 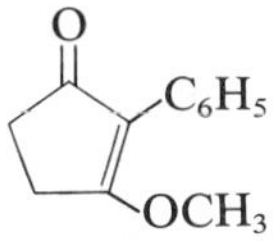 **166**

$C_6H_5CH{=}CHCO_2C_2H_5$

167 **168**

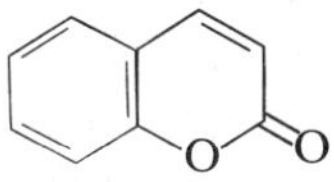

The ring size of cyclic enones also affects their reactivities in the order: 7- or 8-membered < 6-membered < 5-membered rings. A higher reactivity of 5-membered cyclic enones is illustrated by the inertness of β-methoxycyclohexenone **169** to the Nazarov procedure[123] compared with the conversion of the cyclopentenone substrate **94** under similar conditions into its cyanated product, although in poor yield.[86] Another example is the quantitative conjugate hydrocyanation of dimethylcyclopentenone **170**[125] with diethylaluminum cyanide compared with a 57% conversion of cyclohex-2-en-1-one (**108**).[43] The

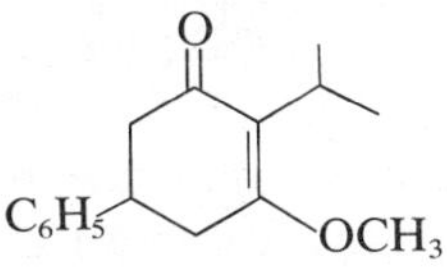

169 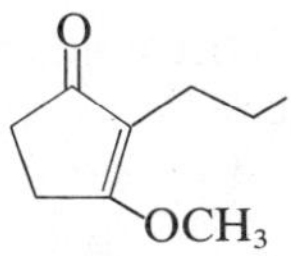 **94** 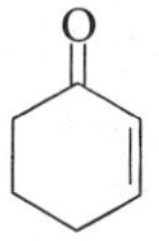 **108** 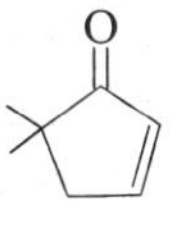 **170**

[119] G. H. Posner, *Org. Reactions*, **19,** 1 (1972).
[120] P. H. Cobb, *Amer. Chem. J.*, **45,** 604 (1911).
[121] H. P. L. Gitsels and J. P. Wilbaut, *Rec. Trav. Chim. Pays-Bas*, **59,** 1093 (1940).
[122] B. E. Betts and W. Davey, *J. Chem. Soc.*, **1961,** 1683.
[123] B. E. Betts and W. Davey, *J. Chem. Soc.*, **1961,** 3333.
[124] B. E. Betts and W. Davey, *J. Chem. Soc.*, **1961,** 3340.
[125] W. C. Agosta and A. B. Smith, III, *J. Amer. Chem. Soc.*, **93,** 5513 (1971).

higher reactivity of the 5-membered cyclic enones can be explained in terms of larger relief of ring strain, smaller steric hindrance, and lower irreversibility of the 1,4 addition than of the 6-membered cyclic enones. Appropriate examples to show the lower reactivity of 7- or 8-membered cyclic enones are not available.

Effect of Substituents and Ring System

Effect of Substituents on the Double Bond. The substituents on the α- and β-carbon atoms (R^1, R^2, and R^3) greatly affect the reactivities of α,β-ethylenic carbonyl compounds **1** toward nucleophilic hydrocyanation. The effects are summarized as follows: (1) electron-donating groups retard conjugate hydrocyánation; (2) electron-

$$\underset{R^2}{\overset{R^1}{>}}\overset{\beta}{C}=\overset{\alpha}{\underset{R^3}{C}}-\underset{R^4}{C}=O$$

1

withdrawing groups accelerate the 1,4 addition; (3) substitution of phenyl for hydrogen at the β carbon atom (R^1 or $R^2 = C_6H_5$) decreases reactivity, whereas the same substitution at the α carbon atom ($R^3 = C_6H_5$) increases it; and (4) the Hammett equation is correlated to conjugate hydrocyanation of substituted benzalmalononitriles. Examples of the four cases follow.

Case 1. Acrylonitrile ($CH_2{=}CHCN$) and acrylates ($CH_2{=}CH$-CO_2R) are converted into their 1,4 adducts in 73–93% yields by the hydrogen cyanide-(potassium cyanide) process, whereas methacrylonitrile [$CH_2{=}C(CH_3)CN$], methyl methacrylate [$CH_2{=}C(CH_3)$-CO_2CH_3], crotononitrile ($CH_3CH{=}CHCN$), and methyl crotonate ($CH_3CH{=}CHCO_2CH_3$) under similar or more vigorous reaction conditions afford only 2–5% of their 1,4 adducts (see p. 270).[2] Low reactivities of β,β-dialkylated substrates are exemplified by the 30% conversion of ethyl β-methylcrotonate [$(CH_3)_2C{=}CHCO_2C_2H_5$] by the potassium cyanide process[115] as opposed to the 66–70% conversion of ethyl crotonate[60] and by selective hydrocyanation of vinyl isobutylidene ketone **171** into the β-cyanoethyl ketone **172**. (See reference 277.)

$$\underset{\textbf{171}}{(CH_3)_2C{=}CHCOCH{=}CH_2} \xrightarrow{\text{KCN-AcOH}} \underset{\textbf{172}\ (77\%)}{(CH_3)_2C{=}CHCOCH_2CH_2CN}$$

The rate retardation by a δ-methyl group in $\alpha,\beta,\gamma,\delta$-diethylenic substrates is exemplified by the inertness of sorbonitrile (**173**) to the hydrogen cyanide-(potassium cyanide) reaction compared with a low-yield conversion of β-vinylacrylonitrile (**174**) into its 1,6 adduct.[2]

$CH_3CH{=}CHCH{=}CHCN$ (**173**) $CH_2{=}CHCH{=}CHCN$ (**174**)

Substitution of a methoxy group at the β-carbon atom of α,β-ethylenic ketones greatly reduces the reactivity.[86] For example, 2-methylcyclopent-2-en-1-one (**175**) was smoothly hydrocyanated by the Nazarov process,[14] whereas the β-methoxy analog **176** was converted into the cyanated product in poor yield even by more effective procedures.[86]

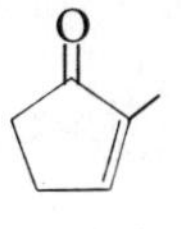

175 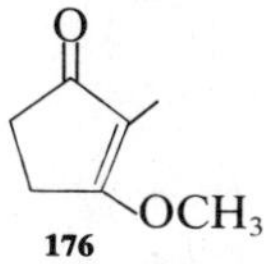   **176** (OCH_3)

Case 2. Substitution of an electron-withdrawing group such as COR, CO_2R, or CN at the α position of α,β-ethylenic substrates results in the greatest rate increase owing to "double activation" of the β-carbon atom. Thus α,β-ethylenic dicarbonyl compounds **177** are the

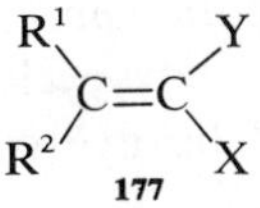

$(R^1)(R^2)C{=}C(Y)(X)$ X, Y = COR, CO_2R, CN

177

most reactive substrates and can be successfully hydrocyanated irrespective of the β-substituents, R^1 and R^2. The 1,4-addition reaction usually proceeds well under mild conditions (often exothermic) by the potassium cyanide and the potassium cyanide-acid procedures. The yields of cyano products are high because of the stabilization of the products by internal neutralization (see p. 273).

An electron-withdrawing group substituted on a β-carbon atom also increases the reactivity, although the high reactivity often leads to side reactions (see p. 295). For example, fumaronitrile ($NCCH{=}CHCN$) reacted violently with hydrogen cyanide-(potassium cyanide) but did not give an isolable product.[2] The Michael adduct $\{NC[CH(CO_2CH_3)]_3CH_2CO_2CH_3\}$ was the only cyano product isolated in the reaction of methyl maleate ($CH_3O_2CCH{=}CHCO_2CH_3$).[2] In contrast, the hydrocyanation of quinone substrates usually gives cyano products in high yields because of the high substrate reactivity and stabilization of the products as phenolic derivatives. This situation is illustrated by the smooth hydrocyanation of 2-arylnaphthoquinones **47** with sodium cyanide at room temperature (p. 274). The reactivities of the substrates are increased by substitution of chloro or $\overset{+}{N}R_3$ at the

β-carbon atom. This type of reaction is illustrated by conversion of the quaternary salts **68** into the β-cyano enones **69** under very mild conditions (p. 278).

Case 3. The opposite effects of substitution of a phenyl group for a hydrogen atom at the α- vs. β-carbon atoms can be interpreted in terms of stabilization of the enolate product **178** by the α-phenyl group and by neutralization of the positive β-carbon atom of substrate **179** by the β-phenyl group. A typical example of the favorable effect

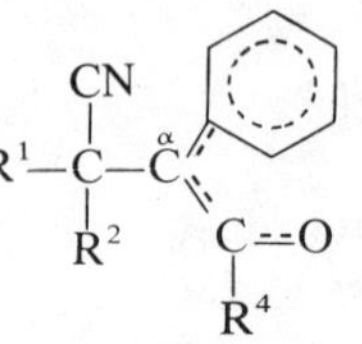

178 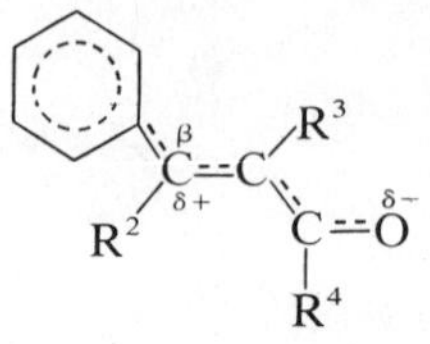 **179**

of α-phenyl substitution is the smooth hydrocyanation (87% conversion) of α,β-diphenylacrylonitrile **180** with potassium cyanide.[118] On the other hand, β-phenylacrylonitrile **181** is inert to the hydrogen cyanide-(potassium cyanide) and the potassium cyanide procedures.[2,126] Another example is the β-methoxycyclohexenones **182** where only the α-phenyl derivative could be hydrocyanated by the Nazarov procedure.[123]

$C_6H_5CH{=}C(C_6H_5)CN$ **180**

$C_6H_5CH{=}CHCN$ **181**

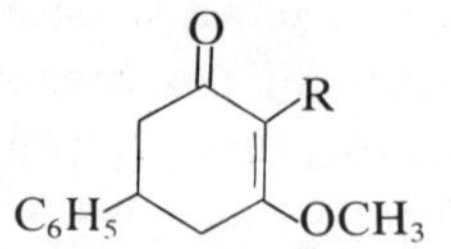

182

$R = C_6H_5$: 95% conversion;

$R = H$, C_3H_5-*i*: no reaction

The retarding effect of a β-phenyl group is apparent from the poor reactivity of β-phenylacrylonitrile (**181**). Retardation by a β-phenyl group was suggested to be equivalent to that of a β,β-dimethyl.[115] The anomalous 1,4 hydrocyanation of dimethyl cinnamylidenemalonate (**11**) (deactivation by the δ-phenyl group) has been noted (p. 267).

Case 4. Kinetic studies on conjugate hydrocyanation of substituted benzalmalononitriles **183** have led to the following Hammett correlation for this reaction (X and Y are *meta, para* substituents).[127]

[126] D. T. Mowry, *J. Amer. Chem. Soc.*, **68,** 2108 (1946).

[127] R. B. Pritchard, C. E. Lough, J. B. Reesor, H. L. Holmes, and D. J. Currie, *Can. J. Chem.*, **45,** 775 (1967).

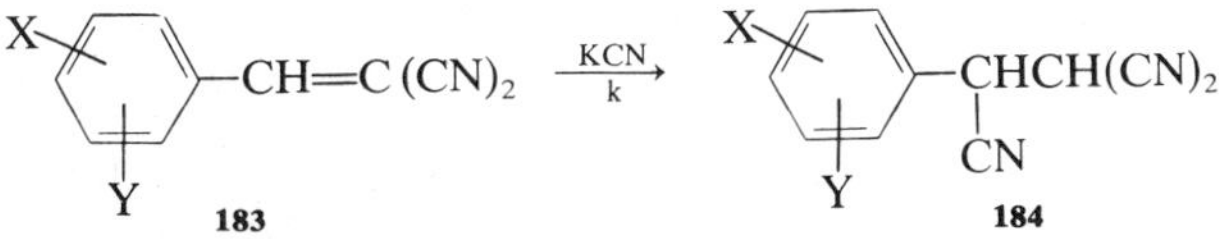

$$\text{X,Y-}C_6H_3\text{-}CH{=}C(CN)_2 \xrightarrow[k]{KCN} \text{X,Y-}C_6H_3\text{-}CH(CN)CH(CN)_2$$

183 → **184**

$\log k = 1.35\sigma - 0.33$ (σ = substituent constant)

Before this quantitative study was made, it was pointed out that: (1) electron-withdrawing groups such as NO_2, CN, and Cl on the benzene ring, irrespective of the position, accelerate the 1,4 addition above; (2) electron-donating groups such as OH, OCH_3, and NR_2 placed at the *ortho* or *para* position retard the reaction; and (3) CH_3, AcO, and AcNH groups exert little effect on the conjugate hydrocyanation.[128] Similar effects of substituents on the hydrocyanation of substituted chalcones **35** (p. 272) have been reported.[50,75,129,130] A marked decrease in reactivity was noted when the *p*-MeO group was present.

Steric Effects. Although the steric requirement of the linear cyano group is not large, as the small conformational energy (**A** value = 0.2 kcal/mol) shows,[131] cyanide addition is not free from steric hindrance. Steric factors in the reactions of flexible acyclic systems have received little study. Hydrocyanation of the β-mesityl derivative **185a** by the potassium cyanide-acetic acid procedure proceeds in 92% conversion.[132] This result shows that the mesityl group alone is not bulky enough to prevent the attack of cyanide ion. It has been suggested that hydrocyanation of the β-methyl homolog **185b** would be difficult.[132] Here steric interactions between two *ortho* methyl groups and the β-methyl group would favor the non-coplanar conformation.

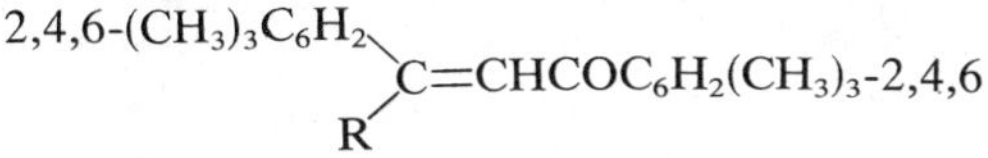

$$2,4,6\text{-}(CH_3)_3C_6H_2(R)C{=}CHCOC_6H_2(CH_3)_3\text{-}2,4,6$$

185a, R = H

185b, R = CH_3

CH_3, CH_3, H, $COC_6H_2(CH_3)_3$-2,4,6, H_3C, CH_3

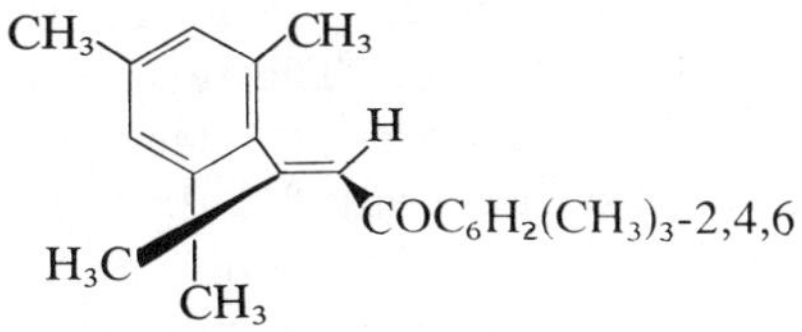

[128] L. Horner and K. Klüpfel, *Ann. Chem.*, **591,** 69 (1955).

[129] R. B. Davis, *J. Org. Chem.*, **24,** 879 (1959).

[130] E. P. Kohler and L. Leers, *J. Amer. Chem. Soc.*, **56,** 981 (1934).

[131] E. L. Eliel, N. L. Allinger, S. J. Angyal, and G. A. Morrison, *Conformational Analysis*, pp. 436–442, Interscience, New York, 1965.

[132] R. C. Fuson and R. G. Bannister, *J. Amer. Chem. Soc.*, **74,** 1631 (1952).

A substituent *syn*-diaxial to the entering cyano group in the cyclohexenone system greatly retards the reaction. For example, hydrocyanation of 3-methyl-cyclohex-2-en-1-one (**186**) by the potassium cyanide-acetic acid procedure at 0° overnight gave its 1,4 adduct in 73% yield,[133] whereas the 5,5-dimethyl analog **31** could be hydrocyanated at room temperature after 1 week in only 60% yield by the same procedure.[64]

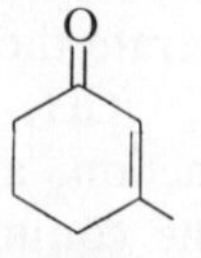

186 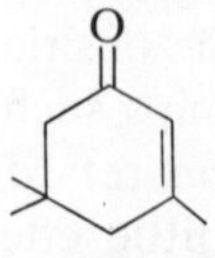 **31**

The steric hindrance of a *syn*-diaxial substituent becomes more significant in rigid polycyclic systems. For example, 7α-isopropenyl-4,10-dimethyl compound **187b** could not be hydrocyanated by conventional methods which are effective for the reaction of the 7β isomer **187a**.[134] Another example is the reluctance of the 9α-methyl tricyclic enone **83** and steroidal 8-en-11-ones **84** having two *syn*-diaxial methyl groups to undergo hydrocyanation by the potassium cyanide-ammonium chloride method (see p. 287).

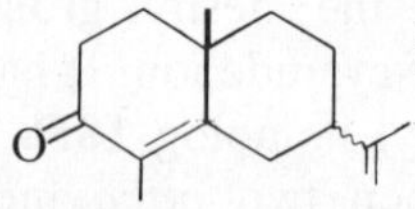

187a, 7β-C_3H_5
187b, 7α-C_3H_5

Effect of Strain in Fused Ring Systems (Rigidity). In conjugate hydrocyanation of α,β-ethylenic carbonyl compounds the cyanide ion attacks the β-carbon atom via a transition state having maximal orbital overlap (stereoelectronic control, see p. 317). This requirement results in deformation of the ring system in the reaction of cyclic substrates. The ring-strain energy arising from this deformation becomes greater when the enone system is part of a rigid polycyclic system and thus retards the conjugate hydrocyanation. The relative rates of conjugate hydrocyanation of polycyclic enones with diethylaluminum cyanide are: **121**, 10; **76**, 1.7; **22**, 1.0 (standard); **90a** 0.26; **106d**, 0.13.[43] These data indicate that the rate is reduced by introduction of an angular methyl group at the γ-position, an increase in the number of rings, the presence of the enone system in the B or C ring of steroids, and the fusion of a five-membered ring. These factors bring about an increase

[133] N. K. Chakravarty and D. K. Banerjee, *J. Indian Chem. Soc.*, **23**, 377 (1946) [*C.A.*, **41**, 6536a (1947)].

[134] D. W. Theobald, *Tetrahedron*, **21**, 791 (1965).

in the ring-strain energy (rigidity) in the transition state to retard the rate

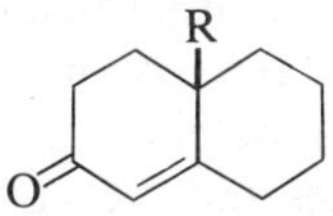

121, R = H
76, R = CH_3

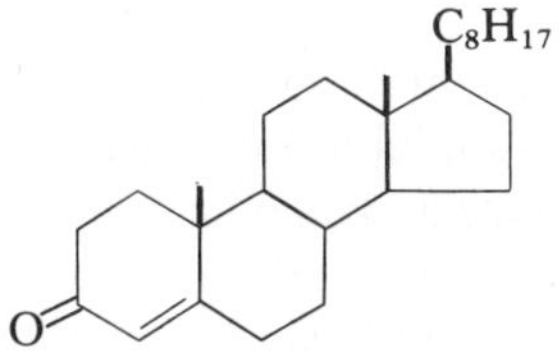

22

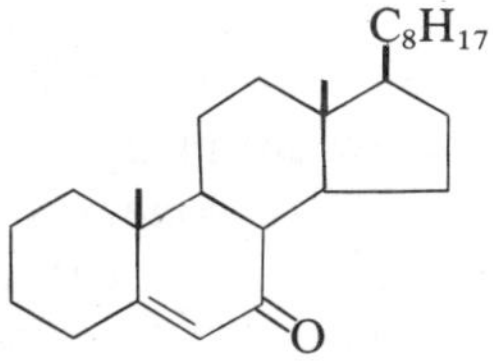

90a

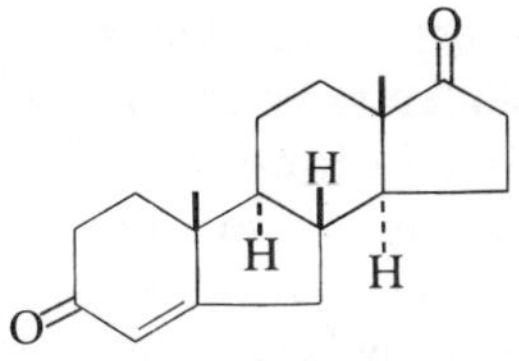

106d

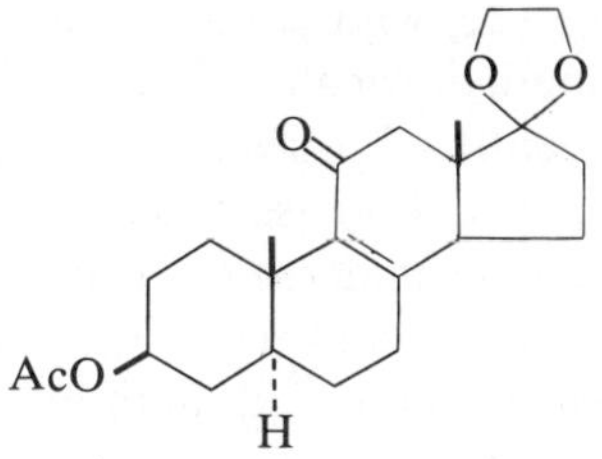

188

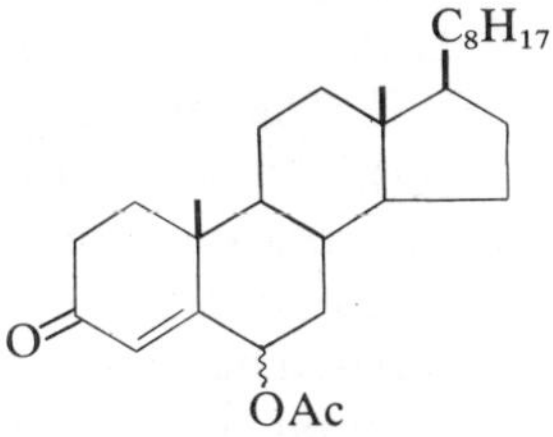

189a, 6α-OAc
189b, 6β-OAc

Miscellaneous Effects. 1,2 Addition precedes 1,4 addition in hydrocyanation with diethylaluminum cyanide. The ease of forming the 1,4 adduct decreases with increasing stability of the 1,2 adduct. The hydrocyanation of α,β-unsaturated aldehydes illustrates this point.[92] In steroidal 8-en-11-ones the extreme instability of the 1,2 adducts favors the 1,4 addition and is an important factor in their successful hydrocyanation despite the hindrance by the two angular groups (the relative rate of **188** is 0.17).[43] The rates of 6α- and 6β-acetoxycholestenones (**189a** and **189b**) relative to the parent enone **22** are 0.12 and 0.016, respectively. The retardation stems from steric hindrance and participation of the neighboring acetoxyl group to deactivate the reaction center, C_5. A greater electronic participation of the axial acetoxyl group accounts for the eightfold retardation of the 6β compared with the 6α isomer.[43]

Stereochemistry

General Aspects

No reports on asymmetric conjugate hydrocyanation and on stereochemistry of the reaction of acyclic substrates have appeared. Therefore the following discussion is limited to stereochemistry of the reaction of cyclic substrates.

As discussed in the mechanism section, conjugate hydrocyanation can be reversible, and the stereoisomeric ratio of the cyano products is subject to the reversibility. The reversibility depends on the structure of the substrate, the method, and the reaction conditions. In discussing the stereochemistry it is essential to know whether the hydrocyanation procedure is kinetic or thermodynamic. Only a few reports on the reversibility of conventional procedures have appeared. 17β-Hydroxy-3-oxo-5α- and 5β-androstane-5-carbonitriles (**54**, R = OH, p. 275) were not interconvertible under hydrocyanation conditions using calcium cyanide in N-methylpyrrollidone (19°, 68 hours).[58] Hydrocyanation of cholestenone (**22**) with potassium cyanide and ammonium chloride in aqueous dimethylformamide at 100° is essentially kinetic.[135] The following conventional procedures are assumed to be essentially kinetic: hydrogen cyanide-(base), acetone cyanohydrin-(base), and potassium or sodium cyanide-acid. The potassium or sodium cyanide (without neutralization) and the potassium cyanide-ethyl acetate methods would be *partly* controlled thermodynamically because of the strongly alkaline medium.

It has been clearly established that the hydrogen cyanide-alkylaluminum process (Method A) is completely kinetic and the diethylaluminum cyanide process (Method B) is thermodynamic in nature.[25,135] It should be noted that the equilibration with diethylaluminum cyanide is very slow in a basic solvent such as tetrahydrofuran. Therefore the reaction must be carried out in a hydrocarbon solvent such as benzene for more than a few hours in order to obtain thermodynamically controlled products.

It is appropriate here to list methods for assigning the configuration of the introduced cyano group before discussing the stereochemistry of conjugate hydrocyanation. Readers should refer to the papers cited for accounts of the principles and techniques.

Physical Methods. (1) Comparison of the C≡N intensities in the infrared (the most convenient method when two epimers are available);[27,85,136–139] (2) interpretation of the optical rotatory dispersion or circular dichroism curve (applicable to a single epimer, if optically active) (Refs. 15, 69, 76, 90, 140, 141); (3) measurement of the dipole moment (applicable to epimers having another polar substituent) (Refs. 85, 117, 137, 138, 142); (4) interpretation of the nmr

[135] W. Nagata, M. Yoshioka, and T. Terasawa, *J. Amer. Chem. Soc.*, **94**, 4672 (1972).

[136] W. Nagata, M. Yoshioka, M. Narisada, and H. Watanabe, *Tetrahedron Lett.*, **1964**, 3133.

[137] W. Nagata, T. Sugasawa, M. Narisada, T. Wakabayashi, and Y. Hayase, *J. Amer. Chem. Soc.*, **85**, 2342 (1963).

[138] O. R. Rodig and N. J. Johnston, *J. Org. Chem.*, **34**, 1942 (1969).

[139] R. E. Ireland and S. C. Welch, *J. Amer. Chem. Soc.*, **92**, 7232 (1970).

spectrum (convenient for secondary nitriles, often ambiguous for angular cyano compounds) (Refs. 26, 42, 117, 138, 139, 141, 143–146).

Chemical Methods. (1) Comparison of the hydrolysis or reduction rates of epimeric nitriles (tedious but useful when other methods are not available);[15,85,137] (2) epimerization to the more stable isomer (applicable to secondary cyano epimers);[147–149] (3) transformation into a known compound (the most tedious but unambiguous) (Refs. 78, 80, 81, 113, 146, 150–152).

Assignment Based upon Substrate Structure. An empirical structure-selectivity relationship has been established (see the following discussion),[135] which often makes possible an assignment of configuration to a cyano product if it is the sole or a major product.

Monocyclic Compounds

In the conjugate hydrocyanation of (−)-carvone (**190**) in aqueous ethanol the formation of the axial cyano epimer **191a** as a major product in the kinetic potassium cyanide-acetic acid process indicates

	190 →	**191a** (e) CN (a) +	**191b** (e) CN (e) +	**191c** (a) CN (e)
KCN-AcOH, room temp		Major	Trace	Minor
$KCN\text{-}AcOC_2H_5$, reflux		1 : 1		Trace

that the cyanide ion attacks the β-carbon atom from an axial direction.[69] The equatorial cyano epimer **191c** is considered to be formed from **191a** by epimerization of the 2-methyl group followed by conformational ring inversion. The increase of the all-equatorial epimer **191b**

[140] A. Bowers, *J. Org. Chem.*, **26,** 2043 (1961).

[141] A. D. Cross and I. T. Harrison, *J. Amer. Chem. Soc.*, **85,** 3223 (1963).

[142] W. Nagata, T. Wakabayashi, M. Narisada, Y. Hayase, and S. Kamata, *J. Amer. Chem. Soc.*, **93,** 5740 (1971); **92,** 3202 (1970).

[143] J. C. Jacquesy, J.-M. Lehn, and J. Levisalles, *Bull. Soc. Chim. Fr.*, **1961,** 2444.

[144] R. F. Zürcher, *Helv. Chim. Acta*, **46,** 2054 (1963).

[145] M. Torigoe and J. Fishman, *Tetrahedron*, **21,** 3669 (1965).

[146] P. N. Rao and J. E. Burdett, Jr., *J. Org. Chem.*, **34,** 1090 (1969).

[147] E. Wenkert and D. P. Strike, *J. Amer. Chem. Soc.*, **86,** 2044 (1964).

[148] A. T. Glen, W. Lawrie, and J. McLean, *J. Chem. Soc., C*, **1966,** 661; A. T. Glen and J. McLean, *Tetrahedron Lett.*, **1964,** 1387.

[149] E. W. Cantrall, R. Littell, and S. Bernstein, *J. Org. Chem.*, **29,** 64 (1964).

[150] W. Nagata, M. Narisada, and T. Sugasawa, *Tetrahedron Lett.*, **1962,** 1041.

[151] W. Nagata, T. Terasawa, and T. Aoki, *Tetrahedron Lett.*, **1963,** 865.

[152] W. Nagata, T. Terasawa, and T. Aoki, *Chem. Pharm. Bull.* (Tokyo), **11,** 820 (1963).

by the potassium cyanide-ethyl acetate process is accounted for by base-induced epimerization of the cyano group.

The preferred axial attack of cyanide anion is more apparent in the more rigid substrate **192**, which gave exclusively the axial cyano compounds **193a** and **193b**.[153] The acetyl groups in the two epimers are interconvertible under the reaction conditions. These results are in accord with the view that the cyanide anion always enters from an axial direction to give an initial product which can be epimerized in the basic reaction medium.[154a]

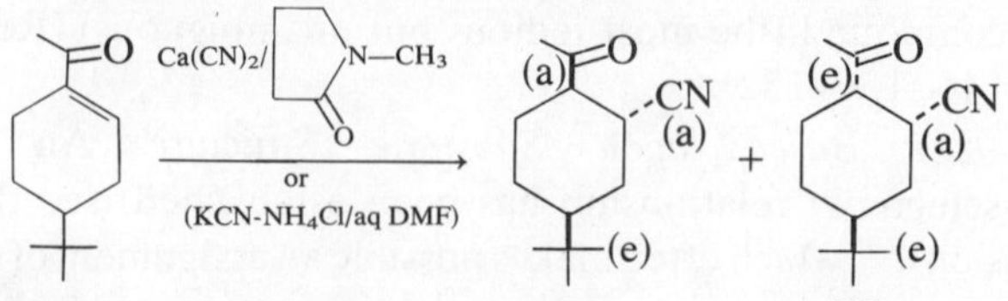

No stereochemical study of cyclopentenone hydrocyanations has been reported.

In the conjugate hydrocyanation of the substituted α-pyrone **194** with sodium cyanide, the formation of olefinic isomers **195a** and **195b** depends on the solvent.[154b] In aqueous acetone, **195a** was obtained in 90% yield, whereas isomer **195b** was isolated in 70% yield in dry dimethylformamide. This difference has been explained in terms of the stability of two conformations, **a** and **b**, in protic and aprotic solvents. In a protic solvent the cyano group becomes bulkier than the methyl group because of hydration and avoids a skew interaction with a large ethoxycarbonyl group by taking conformation **a**, giving the retention product **195a**. In an aprotic solvent, conformation **b** becomes favorable, giving the inversion product **195b**.

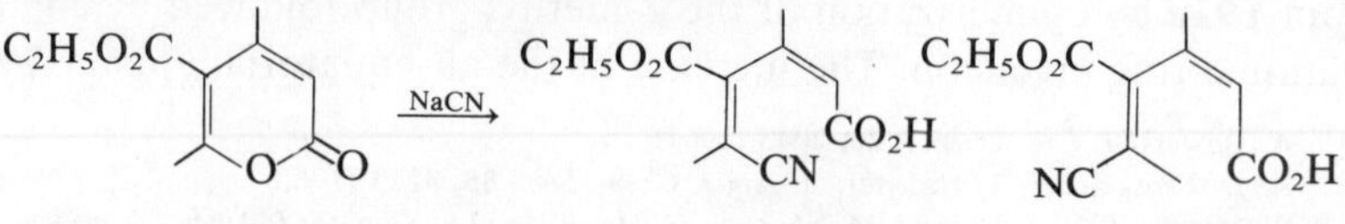

[153] C. W. Alexander and W. R. Jackson, *J. Chem. Soc., Perkin II,* **1972,** 1601.

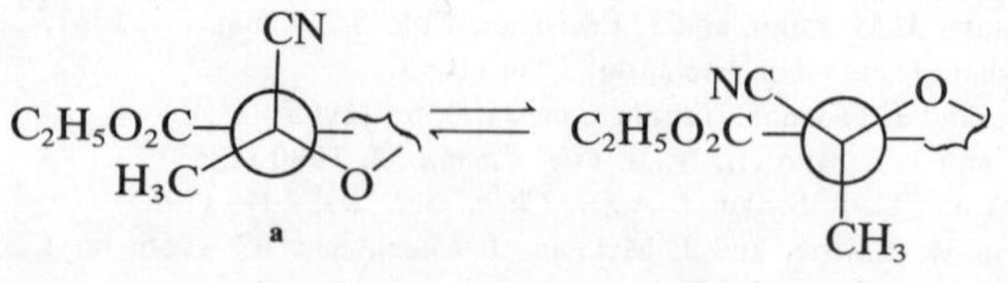

[154a] E. Toromanoff, *Bull. Soc. Chim. Fr.,* **1962,** 708.

[154b] G. Vogel, *J. Org. Chem.,* **30,** 203 (1965); *Chem. Ind.* (London), **1962,** 1829.

[154c] W. L. Meyer and K. K. Maheshwari, *Tetrahedron Lett.,* **1964,** 2175.

Polycyclic Compounds

The stereochemistry of conjugate hydrocyanation of rigid polycyclic substrates has been studied in more detail than that of monocyclic enones. The reactions are nonangular and angular cyanations.

Nonangular Cyanation. When the cyano group is introduced at a β-carbon atom that does not occupy a bridgehead position of a polycyclic system, it tends to prefer an axial or a quasi-axial (in the cyclopentane system) orientation unless the reaction conditions are sufficiently basic to cause epimerization. Typical 1,4 and 1,6 products and the hydrocyanation method(s) used are shown in the accompanying formulas.

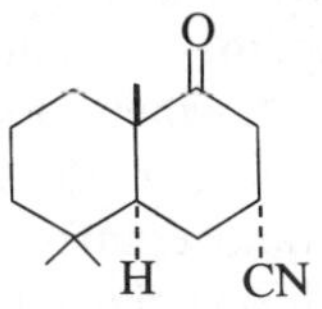

196[147] (KCN-NH_4Cl/aq DMF)

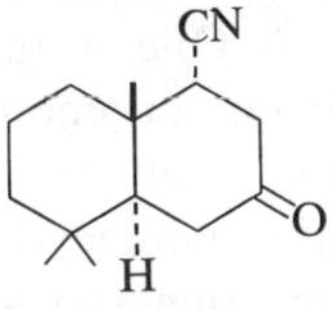

197[147] (KCN-NH_4Cl/aq DMF)

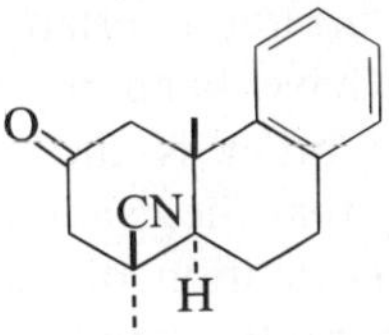

278[154c] (—)

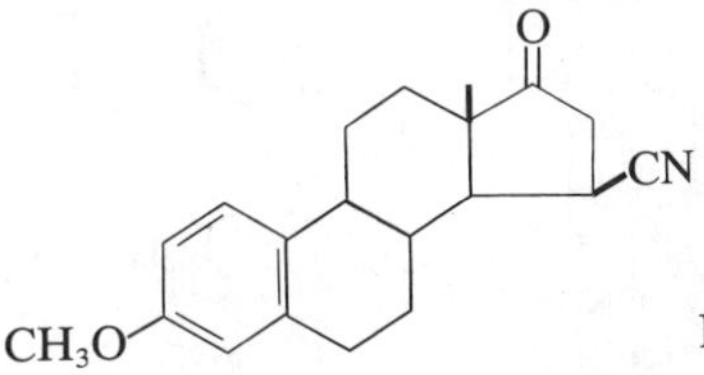

199[149] (NaCN/aq THF)

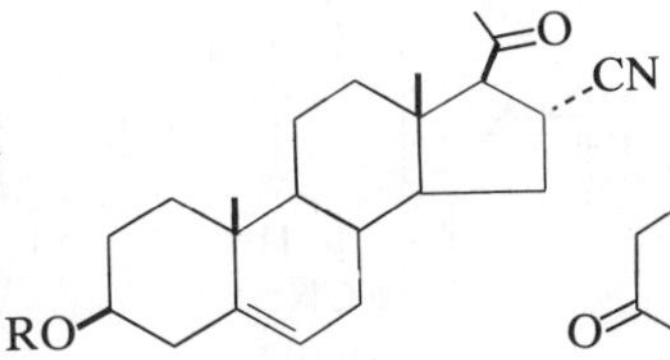

200 (KCN-$AcOC_2H_5$)[155] [HCN-$Al(C_2H_5)_3$][43]

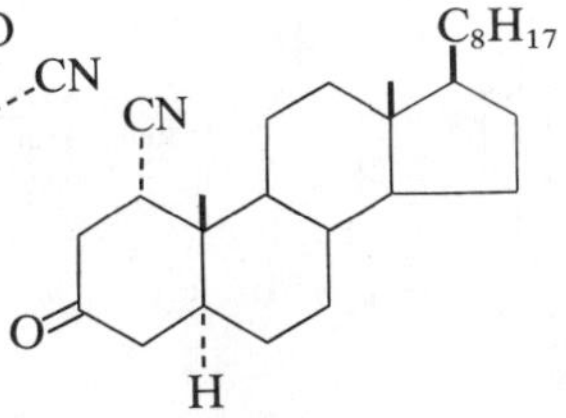

201 [$(CH_3)_2C(OH)CN$-(Na_2CO_3)][156] (KCN-NH_4Cl/aq DMF)[148] [HCN-$Al(C_2H_5)_3$][43]

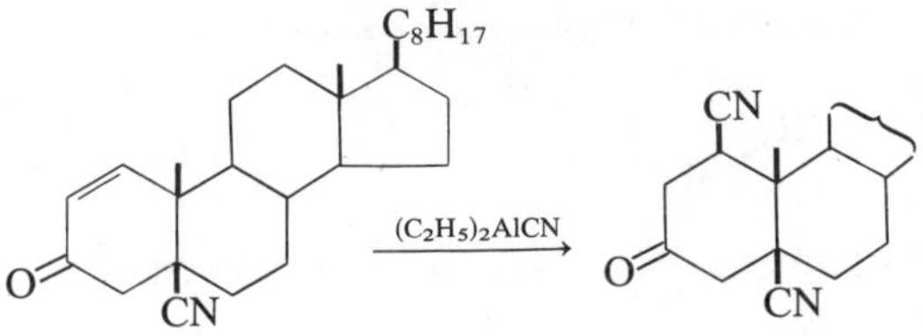

202 $\xrightarrow{(C_2H_5)_2AlCN}$ **203**[43]

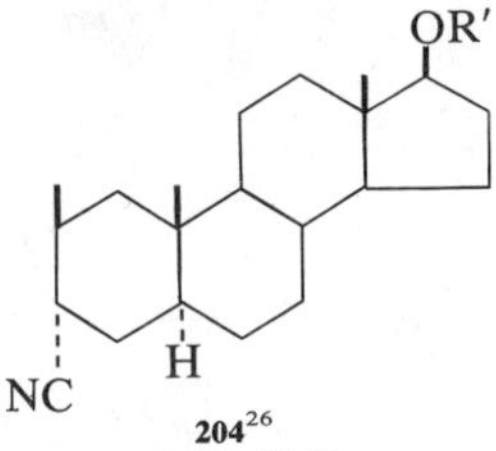

204[26] X = OCH_3, SC_2H_5 [$(C_2H_5)_2AlCN$]

[155] R. H. Mazur and J. A. Cella, *Tetrahedron*, **7,** 130 (1959).
[156] S. Julia, H. Linarès, and P. Simon, *Bull. Soc. Chim. Fr.*, **1963,** 2471.

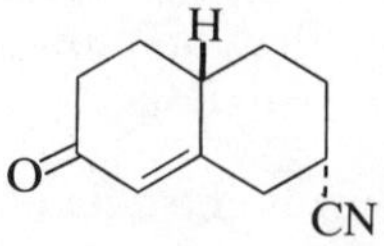

205
$[HCN—Al(C_2H_5)_3]$[157]
$[(C_2H_5)_2AlCN]$[158]

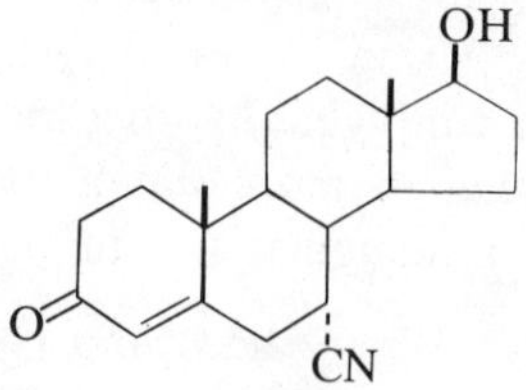

116[43]
$[(C_2H_5)_2AlCN]$

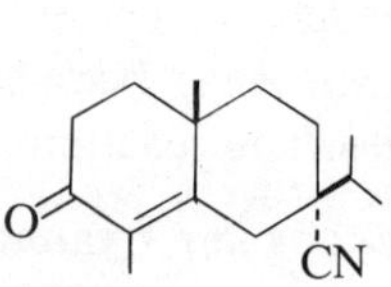

206[134]
$(KCN—NH_4Cl/aq\ C_2H_5OH)$

When the six-membered enone system is fixed with two or more fused rings, the enone ring is in a planar conformation and the cyanide anion can attack from either α or β directions while keeping the maximal orbital overlap to give both epimeric nitriles. Two such examples have been reported. One is the hydrocyanation of enone **81**.[82] The other is the reaction of enone **207**, where there is the possibility that the cyano group is epimerized under the forcing conditions.[159] Formation of the two pentacyclic cyano epimers **209a** and **209b** apparently results from epimerization during the long reaction time (90 hours).[77]

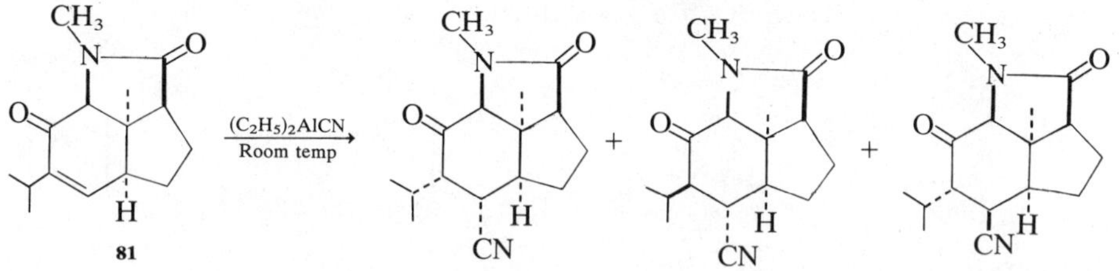

81 $\xrightarrow{(C_2H_5)_2AlCN,\ \text{Room temp}}$ **82a** (18%) + **82b** (20%) + **82c** (29%)

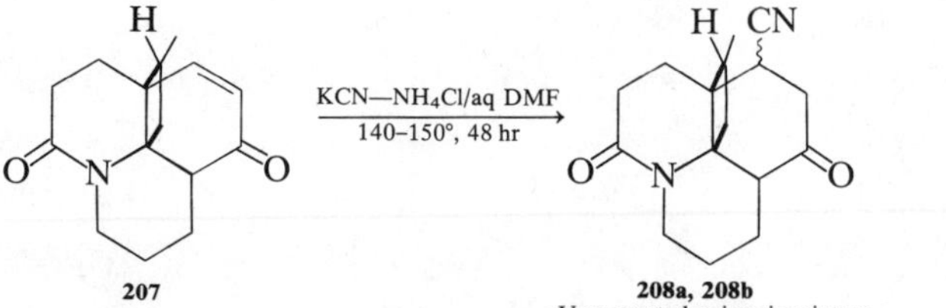

207 $\xrightarrow{KCN—NH_4Cl/aq\ DMF,\ 140–150°,\ 48\ hr}$ **208a, 208b**
Unseparated epimeric mixture

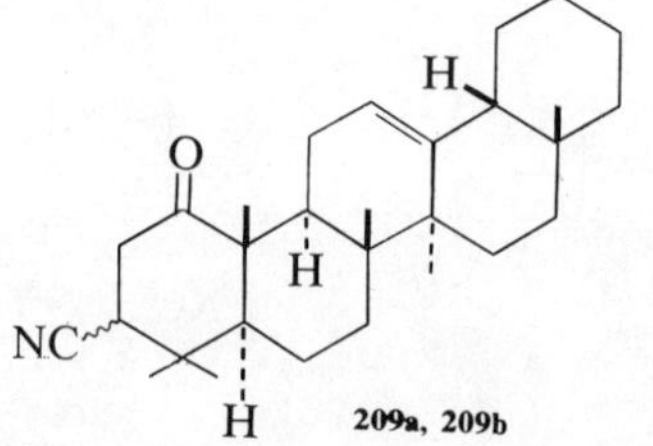

209a, 209b
Two unidentified epimers
$(NaCN—NH_4Cl/aq\ DMF,\ 100°,\ 90\ hr)$

[157] R. A. Finnegan and P. L. Bachman, *J. Org. Chem.*, **36,** 3196 (1971).

[158] E. Fujita, Kyoto University, Sakyo-ku, Kyoto, Japan, personal communication.

[159] K. Wiesner, L. Poon, I. Jirkovský, and M. Fishman, *Can. J. Chem.*, **47,** 433 (1969); K. Wiesner and I. Jirkovsky, *Tetrahedron Lett.*, **1967,** 2077.

Angular Cyanation. The conjugate hydrocyanation of polycyclic α,β-ethylenic ketones in which the β-carbon atom occupies a bridgehead position introduces an angular *trans* or *cis* cyano group. The stereochemical direction of this process is important for the synthesis of isoprenoid natural products. The *trans* to *cis* ratios of the nitrile products are sensitive to the hydrocyanation method and the reaction conditions (Table A).[25,43,135]

The predominant formation of the *trans* isomer by Method A [HCN-Al(C_2H_5)$_3$] in all the examples in Table A should be compared with the poor stereoselectivity in the most effective conventional Method C (KCN-NH_4Cl). The difference is striking in the reaction of the steroidal 17(13)-en-20-one **79**. The *trans* isomer of **213**, which was the desired intermediate in the total synthesis of steroids by Nagata's group, was isolated in 67% yield by Method A. Since both methods are kinetically controlled, the poor selectivity in Method C can be ascribed to solvation of the cyanide ion to increase its steric bulk. Since the number of *syn*-axial CN-H interactions is larger in the *trans* than in the *cis* transition state, the bulkier cyanating species in Method C results in a decrease of the *trans* isomers.

In hydrocyanations of methyloctalinone **76** and cholestenone (**22**) by the recently reported Method D, the predominant formation of the thermodynamically favorable *cis* isomers has been claimed.[51b] In the presence of the crown ether, the cyanating species is considered to be "naked" cyanide which should favor the formation of the *trans* isomer. *Cis-trans* equilibration of the cyano ketone does not take place under the reaction conditions.[51b,59] Because the hydrocyanation does not proceed in the absence of a proton source (acetone cyanohydrin), the stereochemical result of Method D can be interpreted by assuming that the rapid and reversible addition of cyanide ion allows equilibration to take place to a considerable extent before protonation of the cyano enolate anion.

O ⇌ (CN^-, rapid) O CN → ($(CH_3)_2C(OH)CN$, slow) O CN

Stereoselectivity in the thermodynamically controlled Method B [(C_2H_5)$_2$AlCN] needs detailed discussion. The thermodynamic nature of this method in benzene is demonstrated in Fig. 1.[25,135] Except for octalinone **121**, for which the kinetic and thermodynamic product ratios are similar, the decrease of the *trans* isomers in the reaction of the other substrates is large and equilibrium is reached in several hours. The product ratios in the early stage of the reaction are approximately the same as those in kinetic Method A. Thus it is clear that formation of the *trans* or the *cis* isomer can often be controlled by selection of

TABLE A. EFFECT OF PROCEDURE AND CONDITIONS ON STEREOCHEMISTRY OF CONJUGATE ANGULAR HYDROCYANATION OF POLYCYCLIC α,β-ETHYLENIC KETONES

Substrate	Product	Method,[a] Conditions[b]	Yield (%)[c] *trans*	Yield (%)[c] *cis*
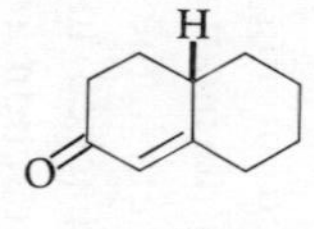 121	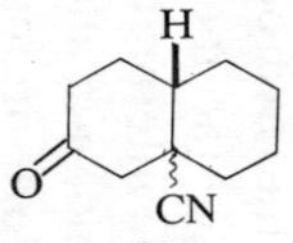 210	A(THF), 25°, 6 hr (K)	89	11
		B(C_6H_6), 25°, 2 min (K)	87	13
		B(C_6H_6), 25°, 6 hr (T)	84	16
		C(aq DMF), 20°, 29 hr (K)	45	22
		D(C_6H_6), room temp, 3 hr (K)	66	19
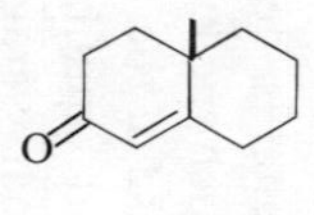 76	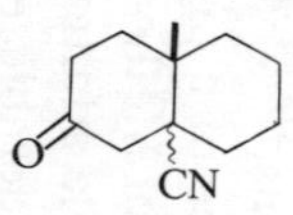 77	A(THF), 25°, 8 hr (K)	71	29
		B(C_6H_6), 25°, 2 min (K)	64	36
		B(C_6H_6), 25°, 20 hr (T)	13	87
		C(aq DMF), room temp, 21 hr (K)	57	43
		D(C_6H_6), room temp, 15 hr (K)	17	68
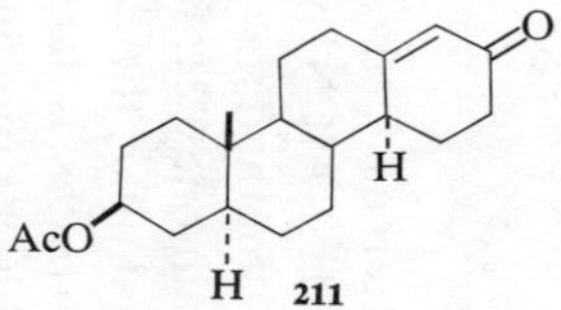 211	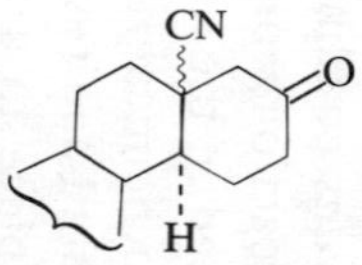 212	A(THF), room temp, 2.5 hr (K)	68	13
		C(aq CH_3OH), reflux, 1.5 hr (K)	41	25

Substrate	Product	Conditions		
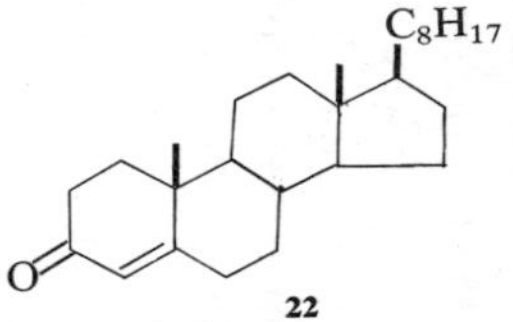 C_8H_{17} **22**	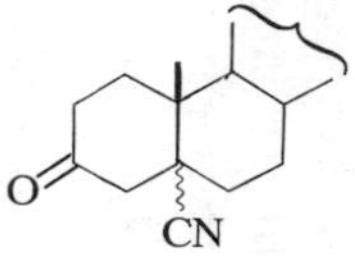 CN **24**	A(THF), room temp, 3.5 hr (K)	49	42
		B(THF), 15°, 4 hr (K)	45	42
		$B(C_6H_6)$, 25°, 10 hr (T)	10	90
		C(aq DMF), 100°, 8 hr (K)	33	51
		$D(C_6H_6)$, room temp, 15 hr (K)	7	70
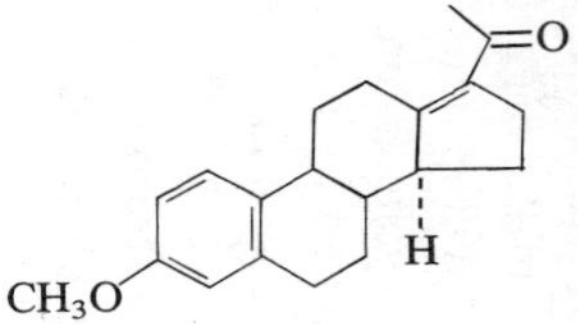 CH_3O H **79**	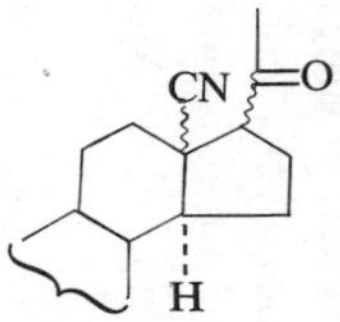 CN H **213**	A(THF), room temp, 12 hr (K)	67	9
		$B(C_6H_6)$, 25°, 3 min (K)	69	31
		$B(C_6H_6)$, 25°, 10 hr (T)	46	54
		C(aq DMF), 100°, 8 hr (K)	22	57

[a] A, HCN-$Al(C_2H_5)_3$; B, $(C_2H_5)_2AlCN$; C, KCN-NH_4Cl; D, $(CH_3)_2C(OH)CN$-KCN-18-crown-6.
[b] K, kinetically controlled; T, thermodynamically controlled.
[c] Isolated yields, unless the total is 100%; then the product ratio was estimated from gas–liquid chromatography.

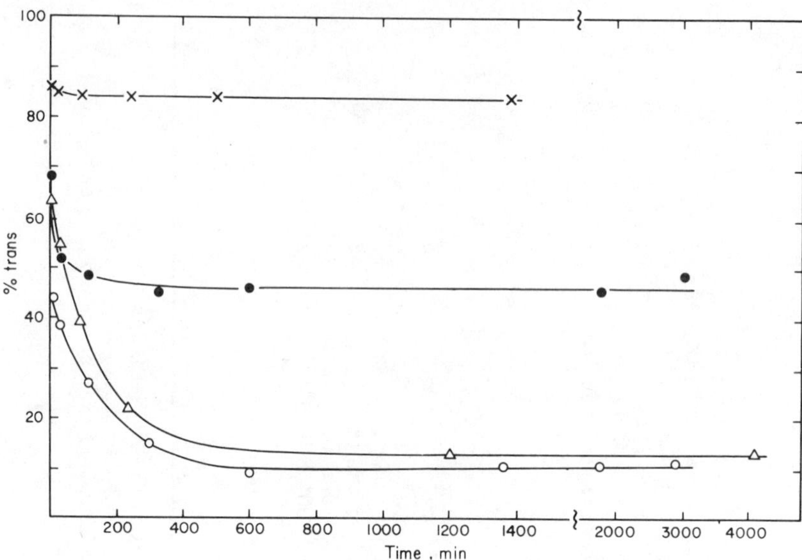

Fig. 1. Plots of the percentages of the *trans* isomers as a function of time in Method B hydrocyanation of $\Delta^{4(10)}$-octalin-3-one (**121**) (—×—), acetylhydrindene **79** (—●—), 9-methyl-$\Delta^{4(10)}$-octalin-3-one (**76**) (—Δ—), and cholestenone (**22**) (—O—) (each 0.1 *M*) with $(C_2H_5)_2AlCN$ (0.3 *M*) in benzene at 25°.

Method A or B and the reaction time in Method B. An extreme case is the hydrocyanation of polycyclic enone **214**.[139] Quite striking is the exclusive formation of the *trans* isomer **215t** (a key intermediate in the total synthesis of alunusenone) by Method A, and the *cis* isomer **215c** by Method B.

A smaller steric requirement of the unsolvated cyanating species (AlC≡N or $AlN^+{\equiv}C^-$) is one of the important features of Methods A and B with organoaluminum compounds in aprotic solvents compared with the requirements of Method C and other conventional methods. Thus the stereochemistry of the organoaluminum methods is considered to be subject to product-development control rather than steric approach control. For example, tricyclic enone **83** and the steroidal 8-en-11-one **84** (see p. 287) are attacked by the cyanide exclusively from the more hindered sides. The kinetic and thermodynamic hydrocyanation product ratios of octalinone **121** and its methyl homolog **76** have been explained using the total strain energies of the cyano products.[135] The details of the calculation need not be presented here, but the following factors are taken into account: (1) Toromanoff's concept of stereo-

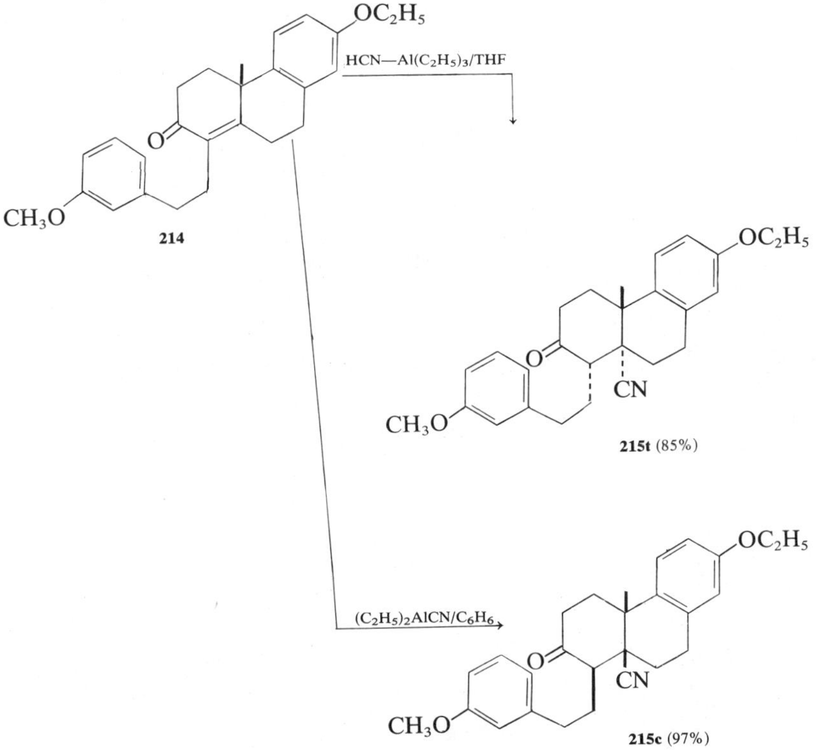

OC_2H_5

HCN—$Al(C_2H_5)_3$/THF

CH_3O

214

OC_2H_5

CH_3O

CN

215t (85%)

OC_2H_5

$(C_2H_5)_2AlCN/C_6H_6$

CH_3O

CN

215c (97%)

chemical pathways is adopted.[160] (2) The thermodynamic product ratios are calculated from the total strain energies of *trans* and *cis* stable product conformers as enolates. (3) The kinetic ratios are estimated from the qualitative energy difference between two primary products (postulated to be similar to the energy difference between the transition states). The conformation used in the estimation is illustrated with cyanocyclohexene in the accompanying figures A and B (where

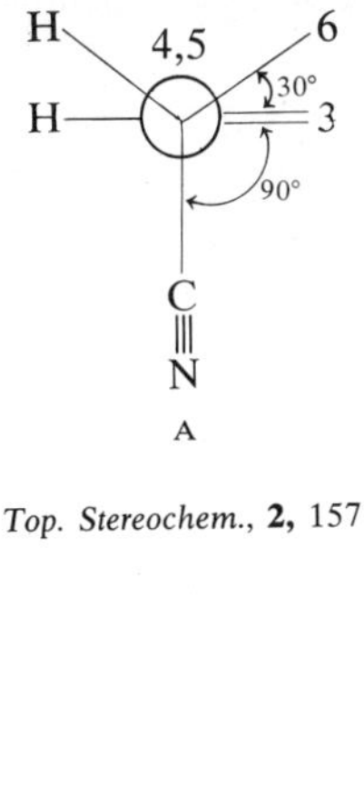

A

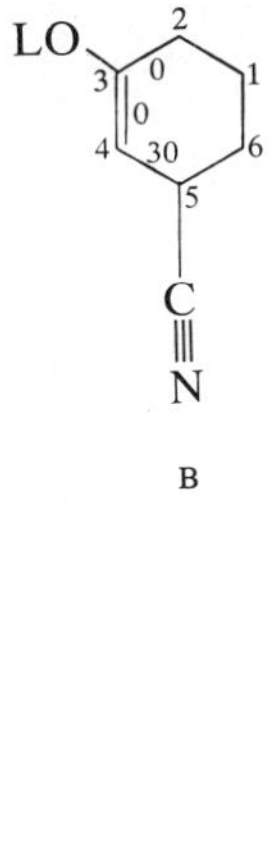

B

[160] E. Toromanoff, *Top. Stereochem.*, **2**, 157 (1969).

L = ligand). (4) The total strain energy is the sum of the ring-strain energies (Pitzer and Baeyer) calculated by the Bucourt and Hainaut method[161] and the nonbonded interaction energies evaluated from the empirical conformational energies of substituents (**A** values). Qualitative estimation of these factors makes it possible to explain stereochemical results obtained with various types of α,β-ethylenic ketones.

These considerations led to several generalizations regarding the stereochemistry of angular conjugate hydrocyanation under kinetic conditions.[135] As illustrated in the accompanying formulations, substrates are classified into five types with indication of the stereochemical preference of the cyano products by Method A or B under kinetic conditions. The proportion of *cis*-isomers increases under thermodynamic conditions and by the potassium cyanide-ammonium chloride procedure (see Table A) except for Type III (only *trans*, irrespective of the methods). Typical substrates are shown (for details see Table VI).

Type I (Octalinone type) → *trans* > *cis*

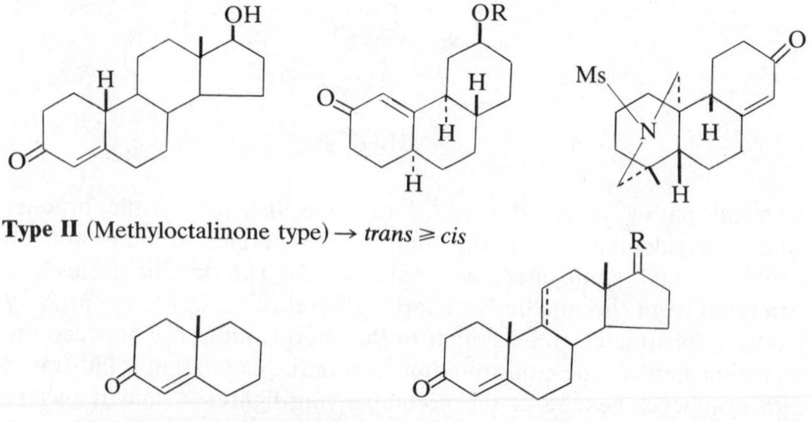

Type II (Methyloctalinone type) → *trans* ≥ *cis*

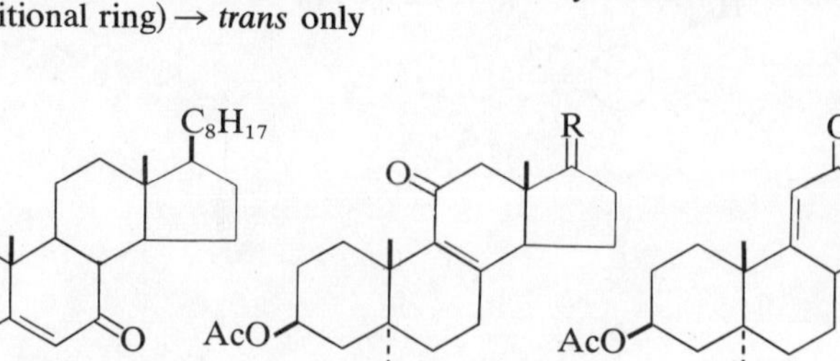

Type III (Octalinone or hydrindenone whose cyclohexenone is *trans*-fused to an additional ring) → *trans* only

[161] R. Bucourt and D. Hainaut, *Bull. Soc. Chim. Fr.*, **1965,** 1366.

Type IV (Acetylhydrindenone type)→ *trans* ≫ *cis*

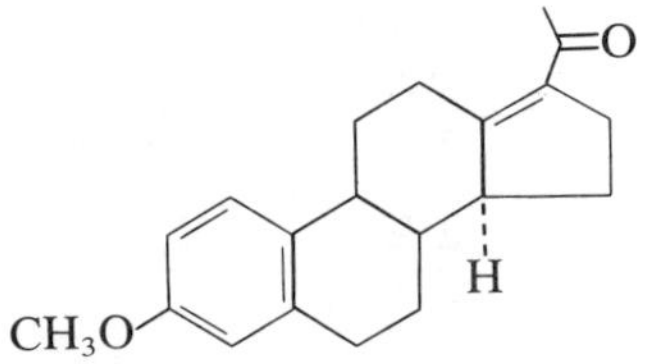

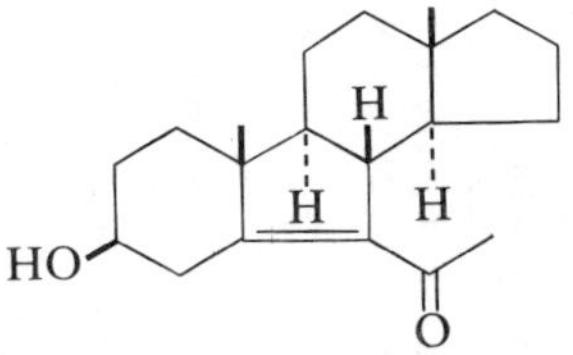

Type V (Hydrindenone type)→ *cis* ≫ *trans*

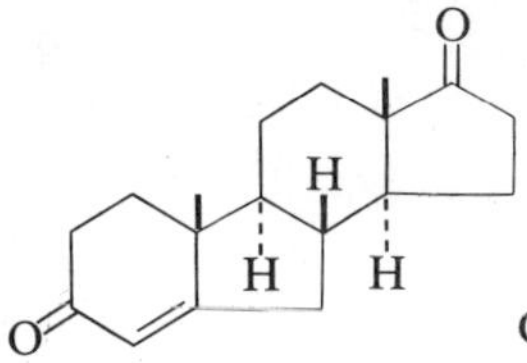

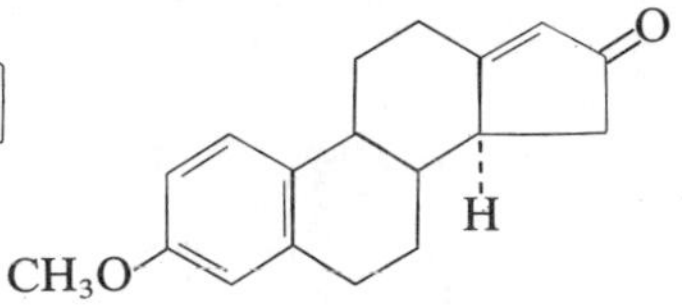

Selection of Method

Selection of the hydrocyanation method is a function of the efficiency of the method, reaction conditions, reactivity of the substrate, stability of the product, structure of the desired product, side reactions, stereochemistry, and so on. These features have been discussed in preceding sections. Table B summarizes the selection of method in terms of the substrate and the desired product. The substrates (subdivided) are placed in the same order as in the next. The reported methods are chosen from the successful examples included in Tables I–IX. When more than one method is listed, the best method is set in boldface. Promising methods are selected from procedures which the authors feel will work better than the previously cited methods. For reaction conditions the reader should examine the next section and Tables I–IX.

The potassium cyanide-phase-transfer catalyst process and the acetone cyanohydrin-potassium cyanide-crown ether procedure are not included in this section, because the scope and limitations of these methods are not yet clarified.

α,β-Unsaturated Carbonyl Substrates

In this section the scope and limitations of conjugate hydrocyanation are discussed in terms of substrates in the same order as in the Tabular Survey.

Acyclic α,β-Ethylenic Carboxylic Acid Derivatives

Acrylates and acrylonitrile, the simplest substrates in this class of compounds, are sufficiently reactive to undergo smooth hydrocyanation by the hydrogen cyanide-(base) process at 60–80° on an

TABLE B. SELECTION OF METHOD

Substrate Type[a]	Desired Product	Reported Method(s)	Promising Method(s)
$CH_2{=}CHX$ ($X = CN, CO_2R$)	$NCCH_2CH_2X$	HCN-(base)	KCN-NH_4Cl, $(C_2H_5)_2AlCN$
$RCH{=}CHX$ or $CH_2{=}C(R)X$ ($X = CN, CO_2R$)	$NCCH(R)CH_2X$ or $NCCH_2CH(R)X$	HCN-(base) (poor results)	KCN-NH_4Cl, $(C_2H_5)_2AlCN$
	$HO_2CCH(R)CH_2CO_2H$ or $HO_2CCH_2CH(R)CO_2H$	1. KCN. 2. Hydrolysis	None
$ArCH{=}CHX$ ($X = CN, CO_2R$)	$NCCH(Ar)CH_2X$	No example	KCN-NH_4Cl (high temp)
	$HO_2CCH(Ar)CH_2CO_2H$	1. KCN. 2. Hydrolysis (poor results)	None
$R'CH{=}C(Ar)CO_2R$ ($R' = H, R, Ar$)	$NCCH(R')CH(Ar)CO_2R$	No example	KCN-NH_4Cl, $(C_2H_5)_2AlCN$
$R'CH{=}C(Ar)CN$ ($R' = H, R, Ar$)	$NCCH(R')CH(Ar)CN$	HCN-(base), KCN, **KCN-NH_4Cl**	$(C_2H_5)_2AlCN$
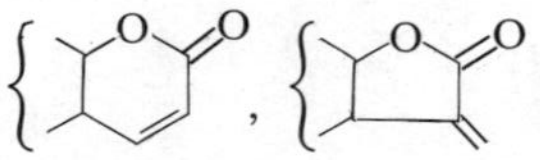	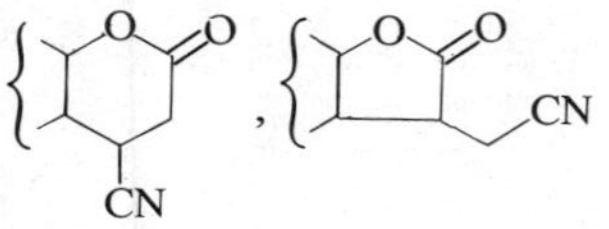	KCN	KCN-AcOH, KCN-NH_4Cl, HCN-AlR_3, $(C_2H_5)_2AlCN$
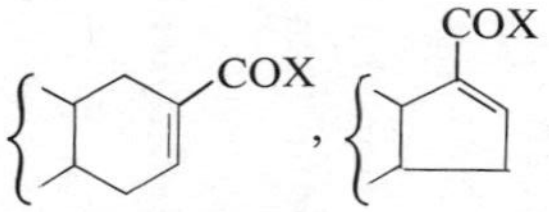	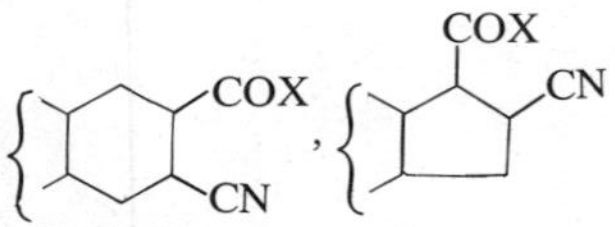		
($X = OR$)	($X = OR$)	$(C_2H_5)_2AlCN$ (poor results)	None
($X = Cl, CN, SR$)	($X = OH, SR$)	1. $(C_2H_5)_2AlCN$. 2. (Hydrolysis)	HCN-organic base

$R'R''C{=}C(X)CO_2R$ (R', $R'' = H$, R, Ar) ($X = CO_2R$, CN, COR)	$R'R''C(CN)CH(X)CO_2R$	HCN-(base), KCN, **KCN-Acid**	Unnecessary
	$R'R''C(CO_2H)CH_2CO_2H$	1. KCN. 2. Hydrolysis	Unnecessary
$R'R''C{=}C(X)X'$ (R', $R'' = H$, R, Ar) (X, $X' = CN$, COR)	$R'R''C(CN)CH(X)X'$	KCN	Unnecessary
$XCH{=}CHX'$ (X, $X' = CN$, CO_2R)	$XCH(CN)CH_2X'$	None (side reactions)	$KCN\text{-}NH_4Cl$, $HCN\text{-}AlR_3$, $(C_2H_5)_2AlCN$
$CH_2{=}CR'COR''$ ($R' = H$, R) ($R'' = R$, Ar)	$NCCH_2CH(R')COR''$	HCN-(base) ($R'' = R$); KCN, **NaCN-AcOH** ($R'' = Ar$)	$KCN\text{-}NH_4Cl$, $HCN\text{-}AlR_3$
$ArCH{=}CHCOR'$ ($R' = R$, Ar)	$NCCH(Ar)CH_2COR'$	$(CH_3)_2C(OH)CN$-(base), **KCN-acid**	$KCN\text{-}NH_4Cl$, $HCN\text{-}AlR_3$ Unnecessary
$RR'C{=}CHCOR$ ($R' = H$, R)	$RR'C(CN)CH_2COR$	HCN-(base), $KCN\text{-}NH_4Cl$, **$HCN\text{-}AlR_3$**	
$R'COCH{=}CH\overset{+}{N}R_3Cl^-$ ($R' = R$, Ar)	$R'COCH{=}CHCN$	$KCN\text{-}NR_3.HCl$ (two layers)	$(C_2H_5)_2AlCN$
$ArCOCH(R')CH_2\overset{+}{N}R_3Cl^-$	$ArCOCH(R')CH_2CN$	KCN	$KCN\text{-}NH_4Cl$
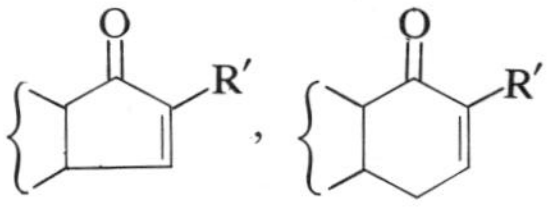 ($R' = H$, R)	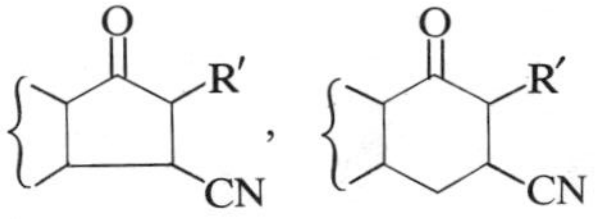	$(CH_3)_2C(OH)CN$-(base), KCN, KCN-acid, $KCN\text{-}NH_4Cl$, **$HCN\text{-}AlR_3$,** $(C_2H_5)_2AlCN$	Unnecessary

TABLE B. SELECTION OF METHOD (*Continued*)

Substrate Type[a]	Desired Product	Reported Method(s)	Promising Method(s)
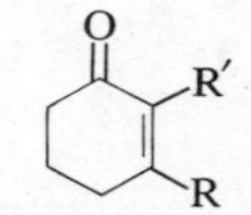 (R′ = H, R)	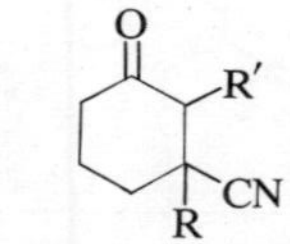	NaCN on $NaHSO_3$ adduct, KCN-acid, **$KCN\text{-}NH_4Cl$**	$HCN\text{-}AlR_3$, $(C_2H_5)_2AlCN$
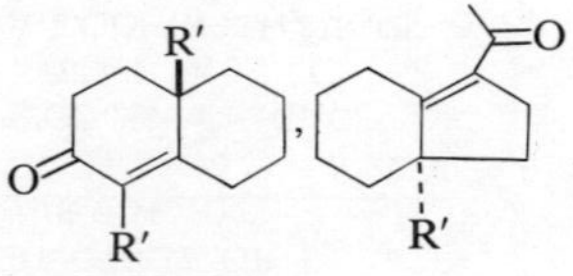 (R′ = H, R) (Including polycyclic)	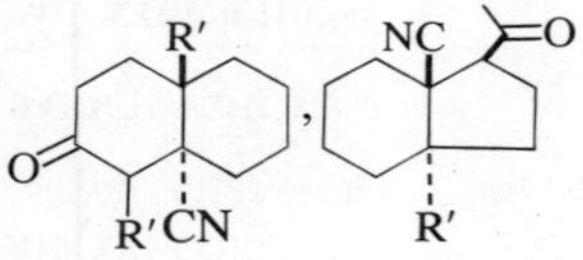	$KCN\text{-}NH_4Cl$, **$HCN\text{-}AlR_3$**, $(C_2H_5)_2AlCN$ (kinetic)	Unnecessary
	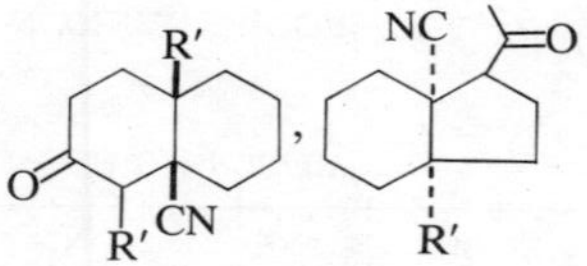	$KCN\text{-}NH_4Cl$, $HCN\text{-}AlR_3$, **$(C_2H_5)_2AlCN$ (thermodynamic)**	Unnecessary
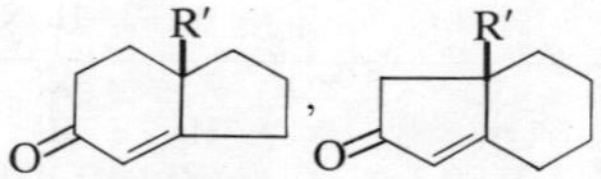 (R′ = H, R) (Polycyclic)	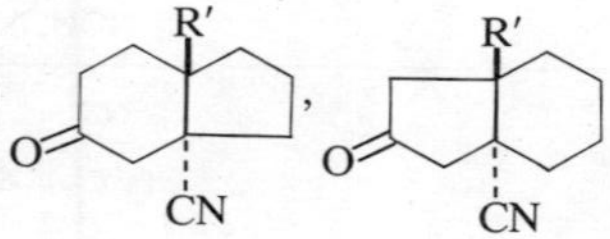	$HCN\text{-}AlR_3$ (poor result)	None
	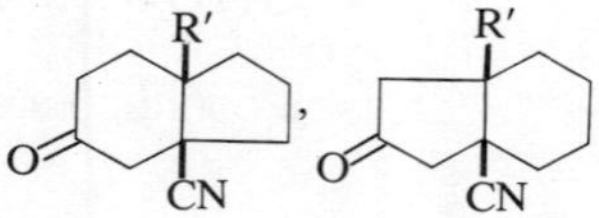	$KCN\text{-}NH_4Cl$, **$(C_2H_5)_2AlCN$**, $HCN\text{-}AlR_3$	Unnecessary

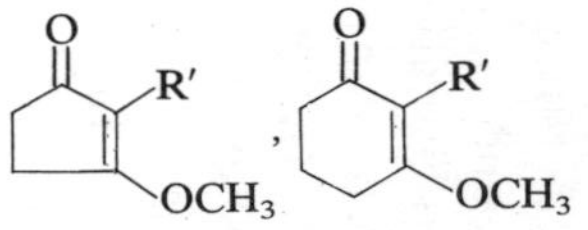 [structures: 2-R'-3-methoxycyclopent-2-enone, 2-R'-3-methoxycyclohex-2-enone] (R' = H, R, Ar)	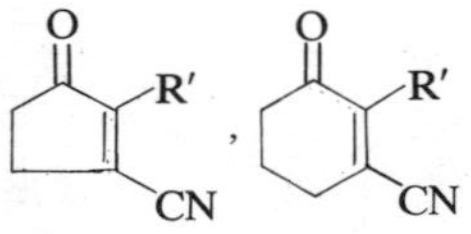 [structures: 2-R'-3-cyanocyclopent-2-enone, 2-R'-3-cyanocyclohex-2-enone]	KCN-NH_4Cl (poor results), HCN-AlR_3, **$(C_2H_5)_2AlCN$**	Unnecessary
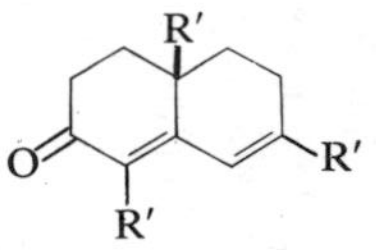 [structure: polycyclic dienone with R' substituents] (R' = H, R) (Polycyclic)	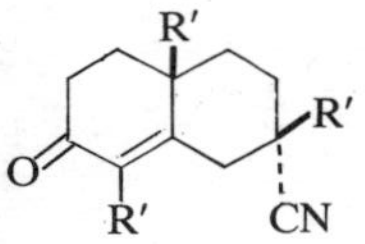 [structure: polycyclic enone with R' and CN]	KCN-NH_4Cl (poor results), HCN-AlR_3, **$(C_2H_5)_2AlCN$**	Unnecessary
$R'C{\equiv}CCO_2R$ (R' = H, R)	$NCC(R'){=}CHCO_2R$	HCN-(base), KCN (no good method)	None
$RR'C{=}CHCHO$ (Including cyclic) (R' = H, R)	$RR'C(CN)CH_2CHO$	None	None
$RR'C{=}CHCH{=}NR$ (Including cyclic) (R' = H, R)	$RR'C(CN)CH_2CHO$	1. **HCN-AlR_3** or $(C_2H_5)_2AlCN$-(ROH). 2. Hydrolysis	None

[a] R = alkyl; Ar = substituted or unsubstituted phenyl.

industrial scale to give β-cyanopropionates and succinonitrile in high yields.[2,162] This simplest process is not effective with less reactive alkylated substrates such as crotonates **27c** and **55**, methacrylates **27′b**, crotononitrile (**25c**), and methacrylonitrile (**25b**). They can be hydrocyanated with potassium or sodium cyanide (without neutralization) in refluxing ethanol. This reaction is accompanied by hydrolysis of the cyano group and the ester grouping. For example, ethyl crotonate (**55**) underwent hydrocyanation followed by treatment with aqueous barium hydroxide to complete the hydrolysis to give methylsuccinic acid (**56**) in 60–70% yields.[60,115] Ethyl cinnamate (**167**), deactivated by its β-phenyl substituent, was converted into phenylsuccinic acid (**128**) in 18% yield by the same treatment. Cinnamonitrile (**181**) would be hydrocyanated similarly. Preparation of monosubstituted succinic acids is easier by hydrocyanation of more reactive alkylidene- or benzylidene-malonates or cyanoacetates followed by hydrolysis. This conversion is discussed later (p. 327). Hydrocyanation of reactive α-aryl-substituted substrates (see p. 304) such as α-phenylcinnamonitrile (**180**) is smooth and gives the 1,4 adducts such as **217** by most

$RCH{=}C(R')CO_2R''$

27′a, $R = R' = H$
27′b, $R = H, R' = CH_3$
55, $R = CH_3, R' = H, R'' = C_2H_5$
27c, $R = R'' = CH_3, R' = H$
167, $R = C_6H_5, R' = H, R'' = C_2H_5$

$RCH{=}C(R')CN$

25a, $R = R' = H$
25b, $R = H, R' = CH_3$
25c, $R = CH_3, R' = H$
181, $R = C_6H_5, R' = H$
180, $R = R' = C_6H_5$
218, $R = R' = Ar$

$NCCH_2CH_2CO_2R$

28′a

$RCH(CN)CH(R')CN$

26a, $R = R' = H$
216, $R = C_6H_5, R' = H$
217, $R = R' = C_6H_5$
219, $R = R' = Ar$

$RCH(CO_2H)CH_2CO_2H$

56, $R = CH_3$
128, $R = C_6H_5$

conventional methods.[2,118,163] A one-step synthesis of diphenylsuccinonitrile (**217**) in 72–77% yield was reported which treated benzaldehyde and phenylacetonitrile with sodium cyanide (condensation to **180** followed by hydrocyanation).[164] The hydrocyanation of β-aryl-α-phenylacrylonitriles **218** ($R = Ar$, $R' = C_6H_5$) by conventional procedures was successful on the substrate with a p-CH_3O or m-O_2N substituent, whereas the reaction on the p-O_2N compound **218** ($R = p$-$O_2NC_6H_4$, $R' = C_6H_5$) gave unchanged starting material.[165] We do not understand this retardation by the p-O_2N substituent, for it should

[162] R. L. Heider and H. M. Walker, U.S. Pat. 2,698,337 (1954) [*C.A.*, **50,** 1896i, (1954)].
[163] J. A. McRae and R. A. B. Bannard, *Org. Synth., Coll. Vol.* **IV,** 393 (1963).
[164] R. B. Davis and J. A. Ward, Jr., *Org. Synth., Coll. Vol.* **IV,** 392 (1963).
[165] K. Brand and O. Loehr, *J. Prakt. Chem.*, [2], **109,** 359 (1925).

enhance the nucleophilic cyanide addition, as discussed earlier (p. 305). Hydrocyanation of *o*-nitrophenyl-substituted substrates **218a** and **218b** results in intramolecular condensation of the initial cyanated product **219a** under the basic reaction conditions to give bicyclic nitriles **220** and **221** as major products.[166] A similar condensation takes place in the hydrocyanation of *o*-nitrobenzylidenemalonates and related compounds (for the results and mechanisms see p. 328).

CN, $C{=}CHC_6H_5$, NO_2 — **218a** or $CH{=}CC_6H_5$, CN, NO_2 — **218b** →(KCN) [CN, $CHCHC_6H_5$, CN, NO_2 — **219a**] → CN, C_6H_5, N, OH — **220** + CN, C_6H_5, NH_2, ^+N, ^-O — **221**

The organoaluminum reagents have not been tested on acyclic α,β-ethylenic carboxylic acid derivatives. The reluctance of these reagents to react with steroidal α,β-ethylenic carboxylates and carbonitriles[26] suggests that a significant improvement in yields cannot be expected with unactivated substrates. Presumably, the reaction of diethylaluminum cyanide with activated carboxylic acid derivatives (acid chloride, acid cyanide, and thio ester) would be an effective route to β-cyanopropionic acid derivatives (see the following discussion).

Cyclic α,β-Ethylenic Carboxylic Acid Derivatives

Only a few reports have appeared on the hydrocyanation of α,β-ethylenic lactones and lactams. Coumarin (**168**) and alantolactone (**168′**) undergo smooth hydrocyanation with potassium cyanide.[7] Practically all the procedures should be effective for this type of substrate. Recently the preparation of 5- and 6-cyanouracil derivatives was reported.[167] The initial 6-cyano product **223** obtained from 5-bromo- or 6-chloro-1,3-dimethyluracil (**222a** or **222b**) under mild conditions

[166] J. D. Loudon and G. Tennant, *J. Chem. Soc.*, **1960,** 3466.
[167] S. Senda, K. Hirota, and T. Asao, *J. Org. Chem.*, **40,** 353 (1975).

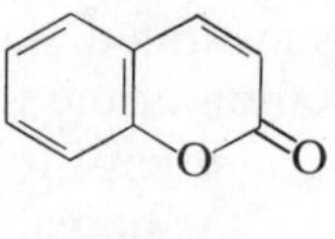

168 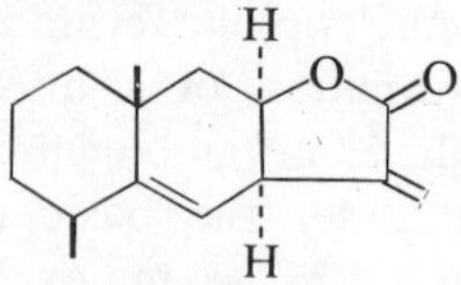 168′

was found to be isomerized to the 5-cyano isomer **224** by heating with a catalytic amount of sodium cyanide. Thus the 5-cyano product **224** was the sole isolated product in the reaction of excess sodium cyanide in dimethylformamide or dimethyl sulfoxide with heating.[168] These conversions are explained by 1,4 addition of the cyanide ion, proton transfer if necessary, and elimination of halogen or cyanide.

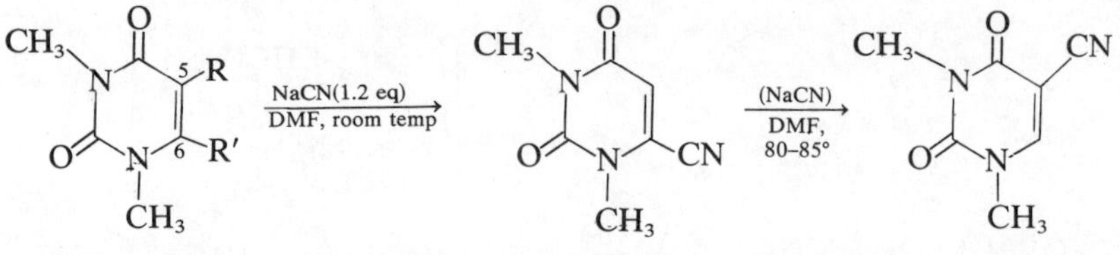

NaCN(1.2 eq) DMF, room temp → (NaCN) DMF, 80–85°

222a, R = Br, R′ = H
222b, R = H, R′ = Cl

223 **224**

Steroidal α,β-ethylenic carboxylates and carbonitriles react sluggishly with the organoaluminum reagents; therefore a general, alternative process using activated substrates for preparing β-cyano carboxylic acid derivatives has been devised.[26] For example, ester **119d** was treated repeatedly with diethylaluminum cyanide in toluene at 120° to give the 16α-cyano compound **120d** in only 36% yield. Nitrile **119e** was inert to the hydrogen cyanide-triethylaluminum reagent.[59] On the other hand, hydrocyanation of acid chloride **119a**, acid cyanide **119b**, and thio ester **119c** with diethylaluminum cyanide proceeded smoothly

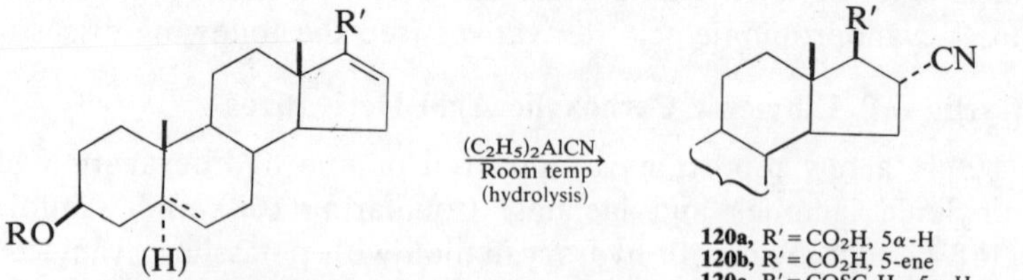

$(C_2H_5)_2AlCN$, Room temp (hydrolysis)

R = H or Ac
5α-H or 5-ene

119a, R′ = COCl, 5α-H
119b, R′ = COCN, 5-ene
119c, R′ = $COSC_2H_5$, 5α-H
119d, R′ = CO_2CH_3, 5-ene
119e, R′ = CN, 5α-H: no reaction

120a, R′ = CO_2H, 5α-H
120b, R′ = CO_2H, 5-ene
120c, R′ = $COSC_2H_5$, 5α-H
120d, R′ = CO_2CH_3, 5-ene

[168] W. Liebenow and H. Liedtke, *Chem. Ber.*, **105**, 2095 (1972).

to give the 16α-cyano-17-carboxylic acid derivatives **120a–120c** in high yields.

α,β-Ethylenic α,α-Dicarbonyl Derivatives

Because of their higher reactivities and product stabilities (see p. 303), hydrocyanations of these doubly activated substrates usually are successful by conventional methods. The newer methods would be effective but usually unnecessary. The reactions of the six representative substrates, **162**, **40**, **225**, **183a**, **57**, and **226** (in the order of increasing reactivity), are illustrative of the scope and limitations. Diethyl benzalmalonate (**40**), which is of moderate reactivity, did not react with hydrogen cyanide-potassium cyanide at 45–50°, but was smoothly converted into the 1,4 adduct **41** by treatment with sodium cyanide and acetic acid at 4°.[72] The use of potassium cyanide in aqueous ethanol at a higher temperature (>60°) resulted in hydrolysis and decarboxylation of the product **41** giving cyano ester **42a** and cyano acid **42b**.[7] Substitution of cyano for ethoxycarbonyl increases the reactivity. Although somewhat less reactive than the diester **40**, the sodium salt of **162** reacted with potassium cyanide and acetic acid at room temperature to give the 1,4 adduct **227**, which was then hydrolyzed to phenylsuccinic acid (**128**).[118] Reaction without acetic acid resulted in formation of the Michael condensation product **229**.[169] The reactive ester **225** reacted more smoothly with sodium cyanide (2 equivalents) in refluxing aqueous ethanol for 2 minutes to give adduct **227b** in quantitative yield. The adduct was then hydrolyzed to phenylsuccinic acid (**128**) in 91–95% overall yields.[100] Since the adduct **227b** is acidic enough to neutralize the alkali, use of excess sodium cyanide does not lead to side reactions. The highly reactive substrates, **183a**, **57**, and **226**, are smoothly hydrocyanated with potassium cyanide at room temperature to give 1,4 adducts, **184a**, **58** (after hydrolysis), and **228**, respectively, in high yields. The effect of phenyl substituents on the reactivity of hydrocyanation is discussed on p. 305. A number of α,α-dialkylsuccinic acid derivatives, including intermediates for steroid syntheses (see p. 345) were prepared by hydrocyanation of the corresponding alkylidenecyanoacetates followed by acid hydrolysis.

$C_6H_5CH{=}C(X)Y$

162, X = CN, Y = CO_2H
40, X = Y = $CO_2C_2H_5$
225, X = CN, Y = $CO_2C_2H_5$
183a, X = Y = CN
57, X = $CO_2C_2H_5$, Y = $COCH_3$
226, X = Y = $COCH_3$

$C_6H_5CH(CN)CH(X)Y$

227a, X = CN, Y = CO_2H
41, X = Y = $CO_2C_2H_5$
42a, X = H, Y = $CO_2C_2H_5$
42b, X = H, Y = CO_2H
227b, X = CN, Y = $CO_2C_2H_5$
184a, X = Y = CN
228, X = Y = $COCH_3$

[169] M. Henze, *J. Prakt. Chem.*, [2], **119,** 157 (1928).

$C_6H_5CH(CO_2H)CH_2X$

128, $X = CO_2H$
58, $X = COCH_3$

$C_6H_5CH(CN)CH(CN)CH(C_6H_5)CH(CN)CO_2H$

229

In an interesting synthesis of spiro compounds by intramolecular alkylation of the cyano product, treatment of γ-bromo-α,β-unsaturated dicarbonyl substrates **230** with potassium cyanide in 78% methanol at 35° for 2 hours gave the cyclopropanes **231** in high yields.[170]

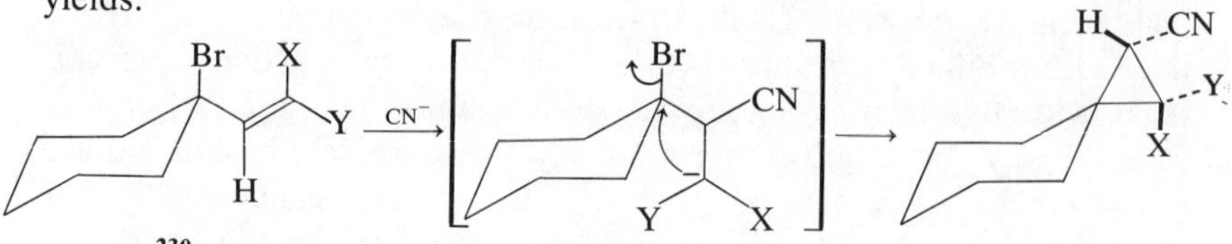

230 $\xrightarrow{CN^-}$ [intermediate] $\longrightarrow$ **231**

(X, Y = CN, CO_2CH_3)

Hydrocyanation of *o*-nitrobenzalmalonate **232** is complex.[171] Under mild conditions the reaction gave the normal 1,4 adduct **233**, whereas with potassium cyanide in refluxing aqueous ethanol the major product was the hydroxyquinolone **234**. This product was formed by action of a strong base (KOH) on the 1,4 adduct **233**, whereas with a weak base (Na_2CO_3) the indole derivative **235** was formed. The indole **235** can be produced by way of the aldol-type condensation product **i**, and formation of the quinolone **234** must involve reduction by an unspecified external agent* (for an analogous conversion, see p. 325).[171] This reaction has been applied to other acyl analogs **236**, giving the quinoline derivatives **237** or their hydrolysis products.[172]

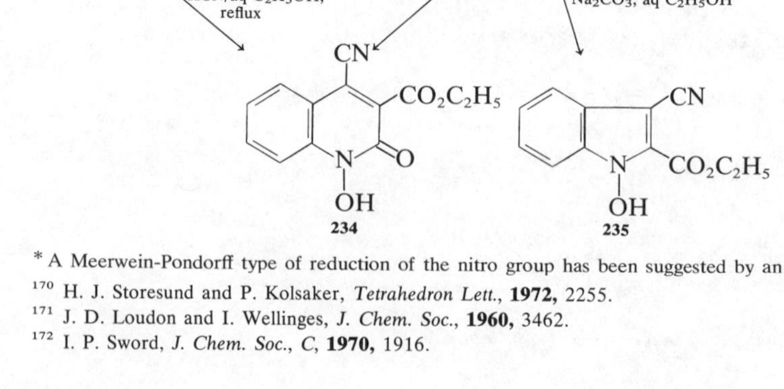

232 $\xrightarrow[\text{Room temp}]{\text{KCN–AcOH}}$ **233**

232 $\xrightarrow{\text{KCN/aq } C_2H_5OH\text{, reflux}}$ **234**

233 $\xrightarrow{KOH/C_2H_5OH}$ **234**

233 $\xrightarrow{Na_2CO_3\text{, aq } C_2H_5OH}$ **235**

* A Meerwein-Pondorff type of reduction of the nitro group has been suggested by an editor.

[170] H. J. Storesund and P. Kolsaker, *Tetrahedron Lett.*, **1972,** 2255.

[171] J. D. Loudon and I. Wellinges, *J. Chem. Soc.*, **1960,** 3462.

[172] I. P. Sword, *J. Chem. Soc., C,* **1970,** 1916.

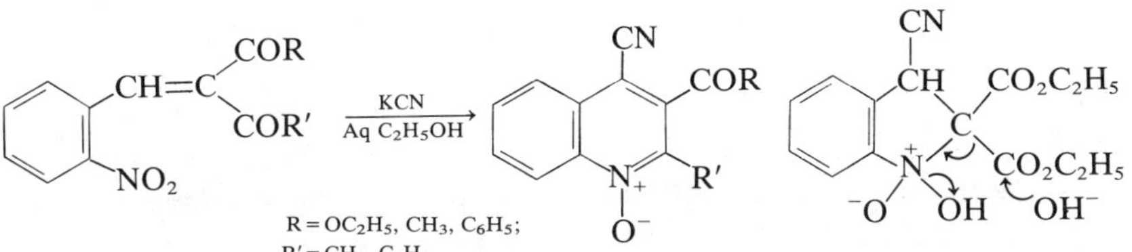

$R = OC_2H_5, CH_3, C_6H_5$; $R' = CH_3, C_6H_5$

236 $\xrightarrow[\text{Aq } C_2H_5OH]{\text{KCN}}$ 237

i

α,β-Ethylenic α,β-Dicarbonyl Derivatives

A high reactivity of substrates such as dimethyl maleate and fumarate ($CH_3O_2CCH{=}CHCO_2CH_3$) and fumaronitrile (NCCH=CHCN) and the tendency of their products to form Michael adducts or polymers have been noted (pp. 295, 303). Successful hydrocyanation of this class of substrates is limited to cases in which the product is stable or stabilized by internal neutralization. This situation is illustrated by the formation of the acidic, stable (aromatized) cyano compounds **238**[173] and **48**[54] by hydrocyanation of quinone substrates **17a**, **17b**, and **47** with potassium cyanide-sulfuric acid and sodium cyanide, respectively. The reaction to give **238** would proceed by way of **17c**. Thus autoxidation must have taken place in the reaction of **17a**. The 1,4 adduct **239** (enolate form) obtained from benzoylacrylic acid **59** seems to be stabilized by the phenyl group.[174]

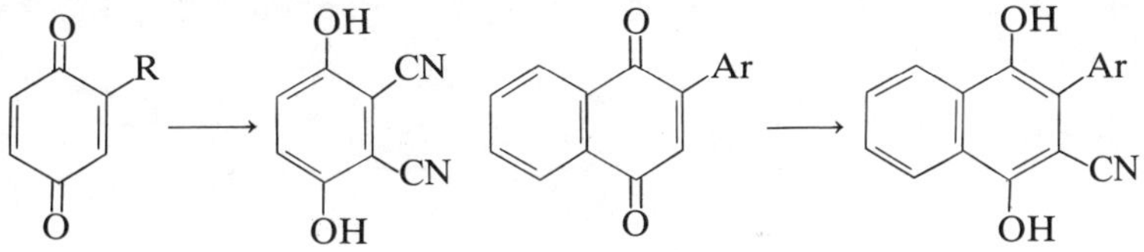

17a, R = H
17b, R = Cl
17c, R = CN

→ **238**

47 → **48**

$$C_6H_5COCH{=}CHCO_2H \xrightarrow[\text{Room temp}]{KCN/H_2O} C_6H_5COCH_2CH(CN)CO_2H$$

59 **239**

Acyclic α,β-Ethylenic Ketones

Acyclic enones are highly to moderately reactive depending on the substituents. The simplest type of enones, such as methyl vinyl ketone and phenyl vinyl ketone, can be smoothly hydrocyanated with hydrogen cyanide-(base) and sodium cyanide-acetic acid into levulinonitrile[2] and β-benzoylpropionitrile, respectively.[175] Hydrocyanation of masked aryl vinyl ketones **66** with potassium cyanide

[173] B. Helferich, *Ber.*, **54,** 155 (1921).

[174] J. Bougault, *C.R. Acad. Sci., Paris*, **146,** 936 (1908).

[175] C. F. H. Allen, M. R. Gilbert, and D. M. Young, *J. Org. Chem.*, **2,** 231 (1937).

occurs in good yields to give nitriles **129**, as discussed on p. 277.[53] substitution of the poorer leaving group, $N(CH_3)_2$.HCl, for the quaternary ammonium group made it necessary to run the reaction of **65** with potassium cyanide in boiling water.[65] Highly reactive substrates such as **68′** require a buffered two-layer system of preserve the reactive products **69′** (see p. 278).[66]

$$\mathrm{ArCOCH(R)CH_2\overset{+}{N}(CH_3)_3I^-} \xrightarrow{} \mathrm{ArCOCH(R)CH_2CN}$$

66 R = H, alkyl, C_6H_5 **129**

$$\mathrm{ArCOCH_2CH_2N(NH_3)_2{\cdot}HCl}$$

65

$$\mathrm{RCOCH{=}CH\overset{+}{N}(CH_3)_3Cl^-} \xrightarrow{} \mathrm{RCOCH{=}CHCN}$$

68′ R = Alkyl, Ar **69′**

The hydrocyanation of chalcones **35** has been intensively studied. Cyano ketones **36** were obtained in high yields, irrespective of the substituent(s), by treatment with hydrogen cyanide-potassium cyanide (not catalytic),[129] Nazarov reagent [$(CH_3)_2C(OH)CN$-(base)],[50] potassium cyanide-acetic acid,[176,177] or potassium cyanide-sulfuric acid.[75] The potassium cyanide-ammonium chloride and hydrogen cyanide-alkylaluminum procedures would undoubtedly be effective as well. The *o*-nitrophenyl-substituted enone **240** gave quinoline and indole derivatives **241** and **242**, as did the dicarbonyl compounds **232** and **236** (p. 328).[172]

$$\mathrm{ArCOCH{=}CHAr'} \longrightarrow \mathrm{ArCOCH_2CH(CN)Ar'}$$

35 **36**

o-$O_2NC_6H_4CH{=}CHCOC_6H_5$ (**240**) $\xrightarrow{\text{KCN}}$ 4-cyano-2-phenylquinoline N-oxide (C_6H_5, CN; $\overset{+}{N}$–O^-) (**241**) + 3-cyano-2-benzoyl-1-hydroxyindole (CN, COC_6H_5, N–OH) (**242**)

Reactivities of β-methylated enones are low, and forcing conditions are necessary with the older procedures. For example, mesityl oxide **29** could be converted into its 1,4 adduct **30** in 86–92% yield with hydrogen cyanide-(potassium cyanide) at 135–140° under pressure.[49] The use of potassium cyanide in refluxing aqueous ethanol resulted in

[176] C. F. H. Allen and G. F. Frame, *Can. J. Res.*, **6**, 605 (1932) [*C.A.*, **26**, 5086 (1932)].
[177] F. G. Baddar and S. Sherif, *J. Chem. Soc.*, **1960**, 2309.

formation of hydrolyzed products **125a** and **125b**, as discussed on p. 293. This difficulty is easily overcome by the use of hydrogen cyanide-alkylaluminum[43] and diethylaluminum cyanide,[99] which give the nitrile **30** in 86–88% and 55% yields, respectively. The lower yield by the latter method can be ascribed to reversibility of the cyano enolate.

$CH_3COCH{=}C(CH_3)_2 \longrightarrow CH_3COCH_2C(CH_3)_2(CN)$

29 **30**

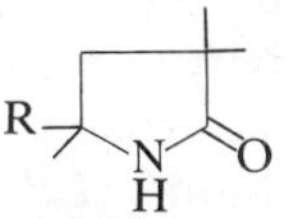

125a, R = OH
125b, R = CN

Cyclic α,β-Ethylenic Ketones

The reaction of this important but complex class of substrates had not been studied extensively until the potassium cyanide-ammonium chloride and the more recent organoaluminum methods were introduced. In preceding sections we have discussed the effect of ring size, ring rigidity, steric hindrance, and 1,2-adduct stability on the reactivity of the substrates and the stereochemistry of the reaction. The scope and limitations of this class of substrates are discussed in terms of the ring systems (monocyclic, polycyclic) and the cyanating position (angular or nonangular).

Monocyclic Enones. Unsubstituted (R^1, R^2, R^3 = H) and α-substituted (R^1 = alkyl) cyclopentenones and cyclohexenones are highly reactive and can be smoothly hydrocyanated by conventional methods unless the β-carbon atom is highly hindered. Side reactions such as hydrolysis and Michael condensation are common when basic conditions (potassium cyanide without neutralization) are used. The usefulness of the hydrogen cyanide-alkylaluminum method is demonstrated by the 80% conversion of cyclohex-2-en-1-one (**245**, R^1, R^2, R^3 = H)

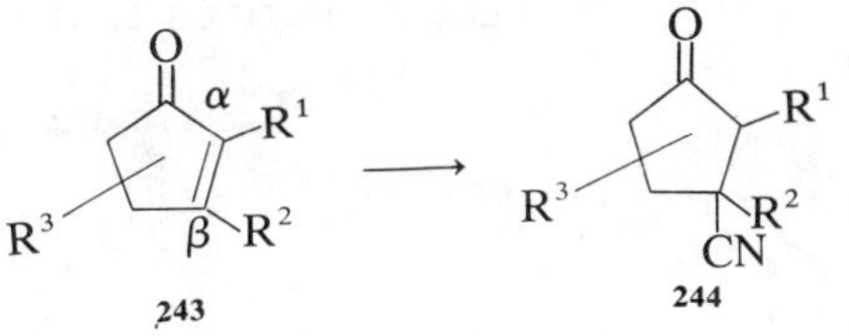

243 → **244**

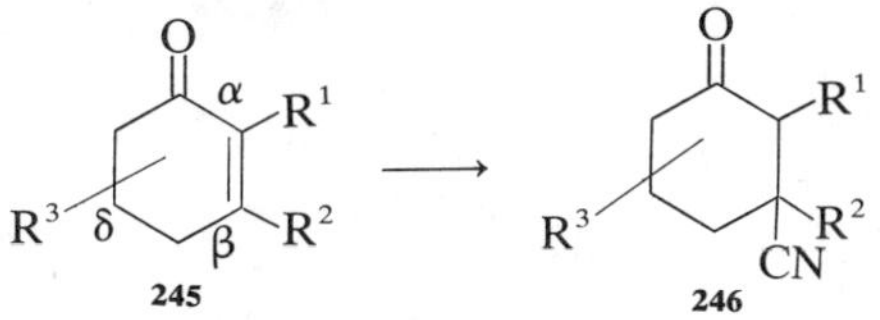

245 → **246**

at −15° into the 1,4 adduct.[43] In contrast, conventional methods proceed in low yields or produce Michael condensation side products (see p. 295). The lower yield (57%) obtained by the diethylaluminum cyanide method is ascribed to the reversible nature of the enolate process.

Substitution of an alkyl group at the β position decreases the reactivity of the cyclic enone. The retardation becomes apparent when an axial group is also present at the δ position, as in **245**. For example, isophorone (**245**, $R^1 = H$, $R^2 = CH_3$, $R^3 = \delta$-dimethyl) could not be hydrocyanated by the Knoevenagel process (sodium cyanide on sodium bisulfite adduct).[178] The sodium cyanide-acetic acid process gave a considerable amount of the starting material;[64,178] the hydrogen cyanide-(potassium carbonate) process required forcing conditions (175° in dimethylacetamide) to effect the reaction. By analogy, the hydrogen cyanide-alkylaluminum method may be successful.

The higher reactivity of five-membered cyclic enones **243** than that of six-membered ones **245** is noted (p. 301).

Only the organoaluminum methods are effective for hydrocyanation of less reactive, cyclic β-methoxy enones **86** and **247** whose adducts are also reactive and subject to side reactions. α-Alkyl substrates (**86**, $R^1 = CH_3$, n-C_3H_7, n-C_5H_{11}, $(CH_2)_6CO_2CH_3$; $R^2 = H$ or γ-OH) react slowly and incompletely with the Nazarov reagent and with potassium cyanide-ammonium chloride, whereas the organoaluminum reagents afford the cyano products **87** in high yield.[86] Hydrocyanation of β-methoxycyclohexenones **247** by the Nazarov process was successful only when the reactivity was increased by introduction of an α-phenyl group ($R^1 = C_6H_5$). The efficiency of the diethylaluminum cyanide method has been shown by the 55% conversion of the unsubstituted substrate (**247**, $R^1 = R^2 = H$) into the adduct **248**; see reference 284. Failure of the Nazarov process on an α-nitro compound (**247**, $R^1 = NO_2$, $R^2 = 5,5-(CH_3)_2$)[123] might be ascribed to the stabilization of its enol form **249**.

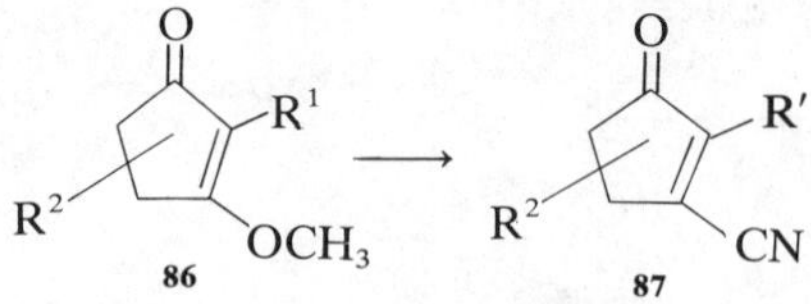

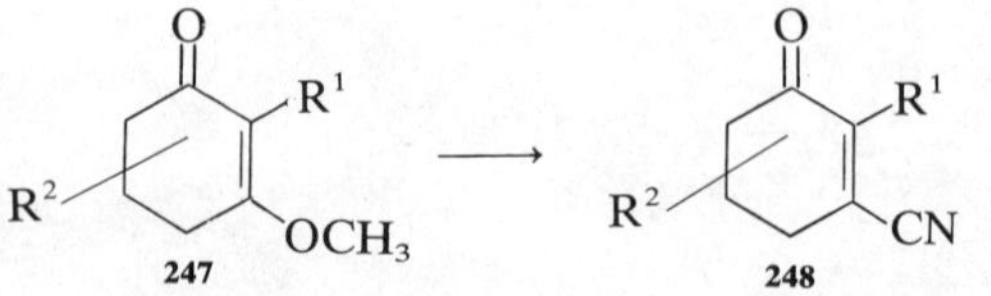

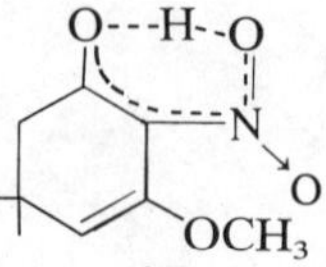

[178] M. S. Ziegler and R. M. Herbst, *J. Org. Chem.*, **16**, 920 (1951).

Hydrocyanation of spirocyclopentenone **250** by the potassium cyanide-ammonium chloride method gives a 1:2 mixture of **251a** and **251b** in 91% yield.[179]

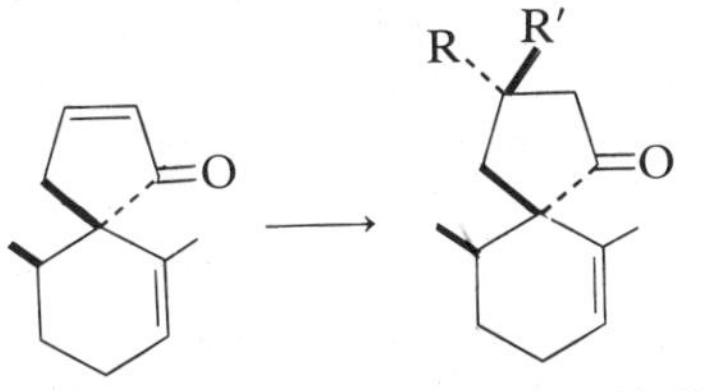

250

251a, R = CN, R′ = H
251b, R = H, R′ = CN

The stereochemistry of hydrocyanation of six-membered enones has been discussed on p. 309.

Polycyclic Enones. *Nonangular Cyanation.* Reactivities of polycyclic enones are much lower than those of monocyclic substrates. However, the conventional methods often are useful for nonangular cyanation of less hindered polycyclic enones, such as steroidal 1-en-3-ones, 15-en-17-ones, and 16-en-20-ones. The potassium cyanide-ammonium chloride method requires forcing conditions for highly hindered substrates. For example, the reactions of bridged substrates **81**[82] and **207**[159] (for formula see p. 312) were effected at 185–190° and 140–150°, respectively. The organoaluminum methods are more generally useful, as shown by the facile hydrocyanation of the highly hindered substrate **81** with diethylaluminum cyanide.[82]

The stereochemistry of nonangular cyanation is discussed on p. 311.

An unusual conversion of 1,2-naphthoquinone derivatives **252** into α-aminonaphthol **253** has been reported without any rational mechanism.[180]

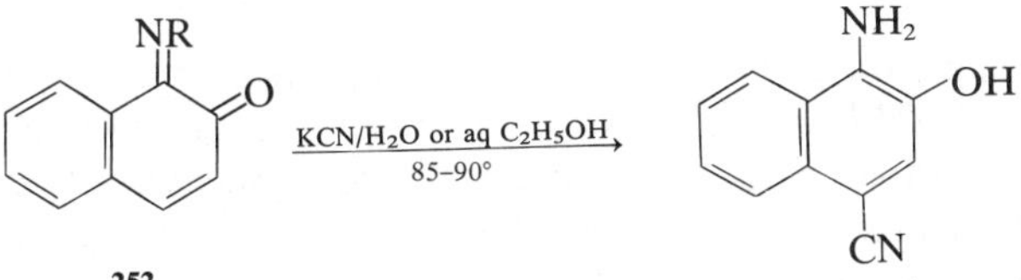

$\xrightarrow[85\text{–}90^\circ]{KCN/H_2O \text{ or aq } C_2H_5OH}$

252

R = OH, NHC_6H_5, $NHC_6H_4SO_3Na$-*p*

253 (49–69%)

Angular Cyanation. The potassium cyanide-ammonium chloride, hydrogen cyanide-alkylaluminum, and diethylaluminum cyanide methods are widely used for angular cyanation of polycyclic enones whose reactivities are usually low. The stereochemistry of kinetic and

[179] B. M. Trost, M. Preckel, and L. M. Leichter, *J. Amer. Chem. Soc.*, **97,** 2227 (1975).
[180] W. Bradley and R. Robinson, *J. Chem. Soc.*, **1934,** 1484.

thermodynamic hydrocyanations has been discussed in detail (see p. 313).

Bicyclic enones have moderate reactivities. The reaction of the simplest substrate, octalinone **121**, with many reagents has been examined. The conventional reagents, hydrogen cyanide-(piperidine),[181] potassium cyanide-acetic acid,[182] and potassium cyanide-ammonium chloride (aqueous dimethylformamide, room temperature),[102] gave the 1,4 adduct **210** in 57–68% yields. Hydrolysis and other side reactions occurred with potassium cyanide[97] or potassium cyanide-ammonium chloride[102] at high temperature. The organoaluminum methods were quite successful.[43,102,135] The reaction of methyloctalinone **76** with potassium cyanide-acetate acid gave the product **77** in 76% yield.[74] This conversion was smoother in the potassium cyanide-ammonium chloride process and with the organoaluminum methods.[135] Hydrolysis products (lactamols) were the major or only products in hydrocyanation of dimethyloctalinone **254**[138] with potassium cyanide and of the 6β-isopropenyl compound **187a** with potassium cyanide-ammonium chloride in refluxing aqueous ethanol. The 6α-alkyl compounds **187b** and **255** were inert to potassium cyanide or potassium cyanide-acetic acid.[134] By analogy, the organoaluminum methods may be effective.

121, $R^1 = R^2 = R^3 = H$
76, $R^1 = CH_3$, $R^2 = R^3 = H$
254, $R^1 = R^2 = CH_3$, $R^3 = H$
187a, $R^1 = R^2 = CH_3$, $R^3 = \beta\text{-}C_3H_5\text{-}i$
187b, $R^1 = R^2 = CH_3$, $R^3 = \alpha\text{-}C_3H_5\text{-}i$
255, $R^1 = R^2 = CH_3$, $R^3 = \alpha\text{-}C_3H_7\text{-}i$

210, $R^1 = R^2 = R^3 = H$
77, $R^1 = CH_3$, $R^2 = R^3 = H$

(*trans* and *cis* isomers)

An interesting stereochemical example has been found in the total synthesis of shionone. Treatment of the hydroxy enone **256** with either hydrogen cyanide-triethylaluminum or diethylaluminum cyanide (thermodynamic conditions) gave only the *trans* isomer **257** in good yields.[183,184] Since the predominant formation of *cis* isomers under the thermodynamic conditions (diethylaluminum cyanide/benzene, several hours) has been established for hydroxyl-free

[181] R. D. Haworth, B. G. Hutley, R. G. Leach, and G. Rodgers, *J. Chem. Soc.*, **1962,** 2720.

[182] N. G. Kundu and P. C. Dutta, *J. Chem. Soc.*, **1962,** 533.

[183] R. E. Ireland, C. A. Lipinski, C. J. Kowalski, J. W. Tilley, and D. M. Walba, *J. Amer. Chem. Soc.*, **96,** 3333 (1974).

[184] R. E. Ireland, M. I. Dawson, C. J. Kowalski, C. A. Lipinski, D. R. Marshall, J. W. Tilley, J. Bordner, and B. L. Trus, *J. Org. Chem.*, **40,** 973 (1975).

substrates such as the tricyclic homolog **214** (see p. 317) and methyloctalinone **76** (see p. 314), formation of the *trans* isomer **257** may be ascribed to the presence of the hydroxyl group (a proton source), which will make the diethylaluminum cyanide process kinetic (see p. 291). Experimental evidence clarifying this point has not been obtained.[184]

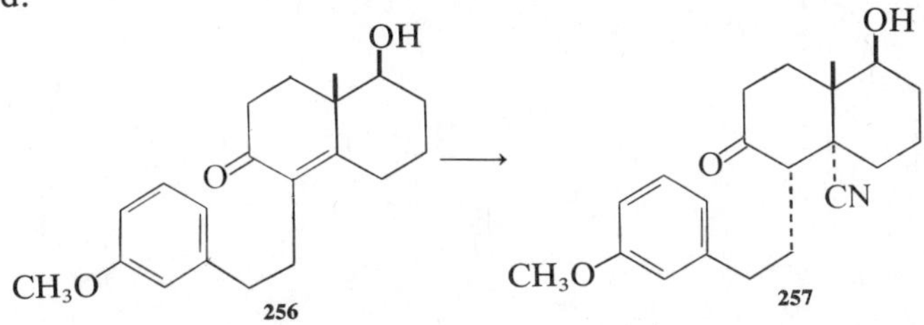

OH OH CN CH_3O CH_3O **256** **257**

Hydrocyanation of acetylhydrindene **258**, another simple bicyclic substrate, with potassium cyanide-ammonium chloride gave the *cis* isomer **259c** in 49% yield and the *trans* product **259t** in 30% yield.[185]

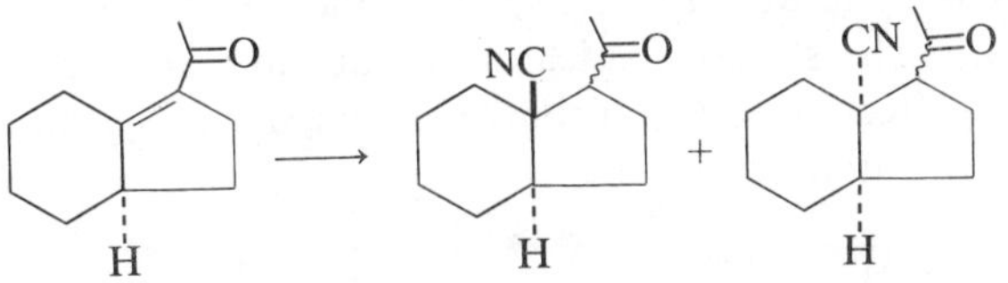

O NC O CN O H H H **258** **259t** **259c**

(Two isomers) (Two isomers)

Effective methods are necessary for the angular cyanation of tricyclic enones whose poor reactivities are close to those of tetracyclic substrates. Resistance of the hindered substrate **83** to the potassium cyanide-ammonium chloride method (14% conversion) and successful conversion (72%) with hydrogen cyanide-triethylaluminum are noted (pp. 287 and 306). Introduction of the cyano group from the highly hindered α side has been discussed (p. 316). A similar example is hydrocyanation of bridged substrate **260** with hydrogen cyanide-triethylaluminum.[85] The *trans* nitrile **261** was obtained in 60% yield together with a trace of its *cis* isomer. The reaction of the electronically deactivated substrate **85** was successful only with hydrogen cyanide-diethylaluminum chloride. Tetrahydrofluorenone **262** is reactive enough to be converted into 1,4 adduct **263** (71%) by the sodium cyanide-ethyl acetate process.[186] The high reactivity can be explained by the anti-aromatic nature of the indenone system.[187] The position of

[185] W. L. Meyer and J. F. Wolfe, *J. Org. Chem.*, **29,** 170 (1964).

[186] W. E. Parham and L. J. Czuba, *J. Org. Chem.*, **34,** 1899 (1969).

[187] A. Streitwieser, Jr., *Molecular Orbital Theory for Organic Chemists*, Wiley, New York, 1961, p. 271.

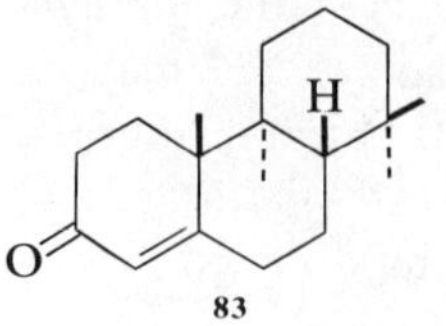

83 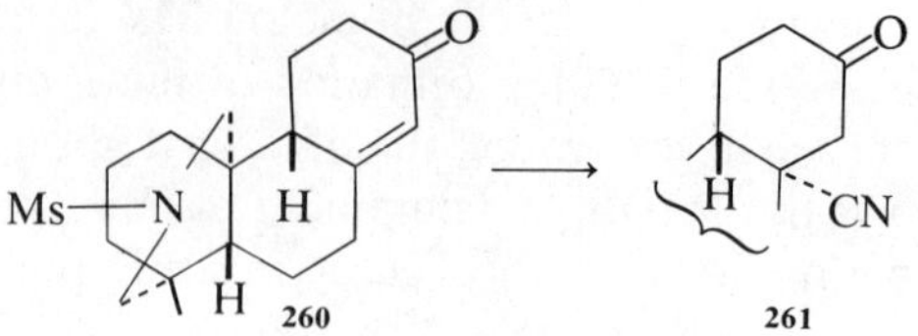 260 261

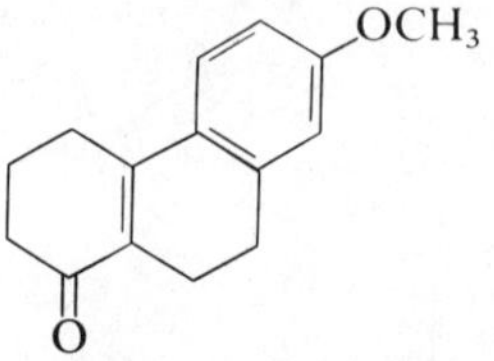

85 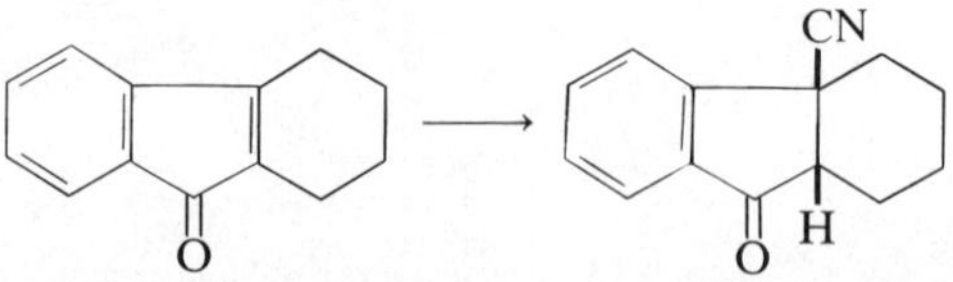 262 263

the third ring relative to the enone system affects the stereochemistry. Thus the exclusive formation of the *trans* isomer **265** from the linear tricyclic enone **264**[188] should be compared with the increasing formation of the *cis* isomers **266c** (10%) and **268c** (30%) from the nonlinear tricyclic 4-en-3-one **80b** and 1(10)-en-2-one **267** (steroid numbering), respectively.[43] The stereochemical difference may be accounted for by an increasing steric interaction between the third ring and the enone ring in the transition state.[135]

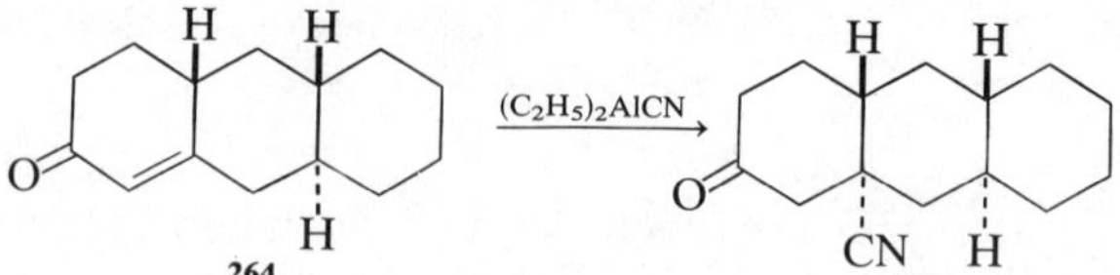

264 $\xrightarrow{(C_2H_5)_2AlCN}$ 265

80b $\xrightarrow{HCN—Al(C_2H_5)_3}$ **266t,** α-CN (65%) **266c,** β-CN (10%)

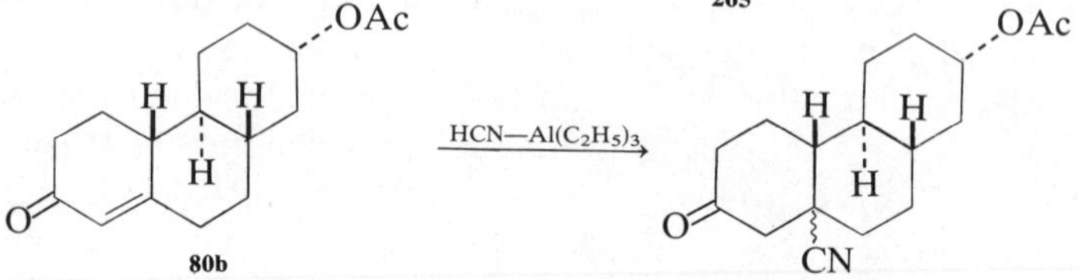

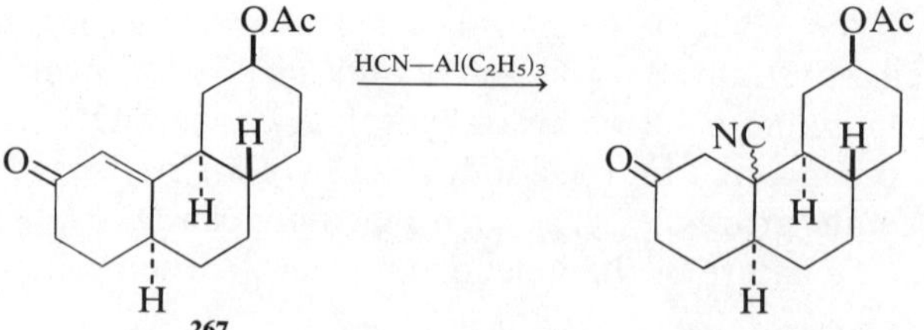

267 $\xrightarrow{HCN—Al(C_2H_5)_3}$ **268t,** β-CN (50%) **268c,** α-CN (30%)

[188] G. Stork and R. E. Boeckman, *J. Amer. Chem. Soc.*, **95,** 2016 (1973).

Angular cyanation of tetracyclic enones has been extensively studied in connection with syntheses of steroids and modified steroids. The cyano group has been introduced into all the angular positions. The accompanying (partial) formulas show various types of steroidal angular nitriles that have been prepared. References are given at the end of the formulas. When *trans* and *cis* isomers are formed, the predominant

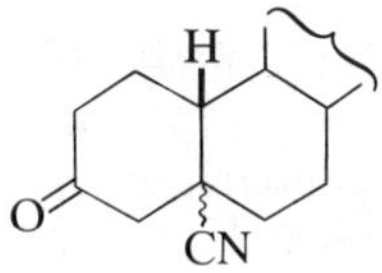

a (*trans*) 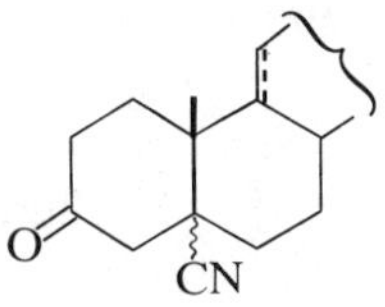 **b** (*trans*) 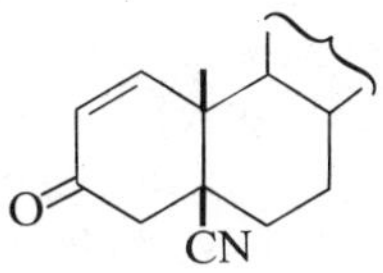 **c** 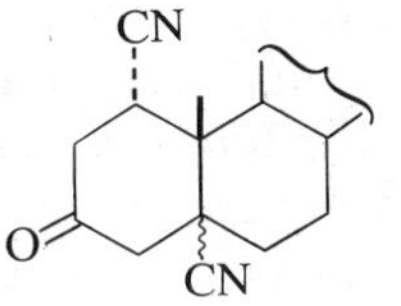 **d** (*cis*)

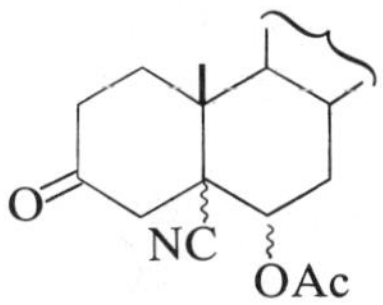

e 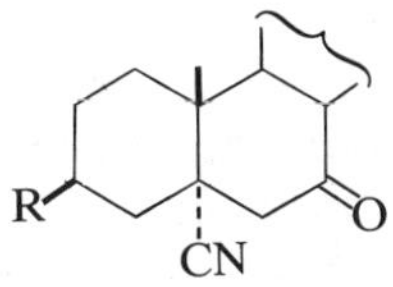 **f** 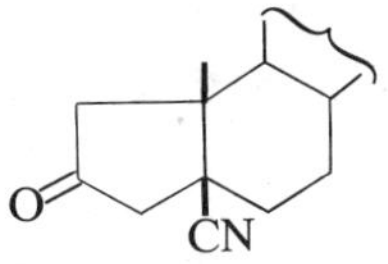 **g** 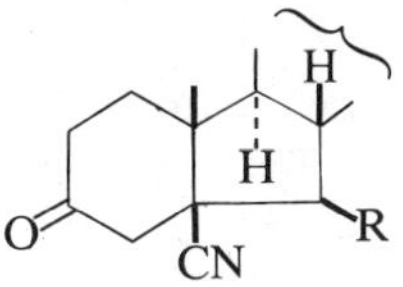 **h**

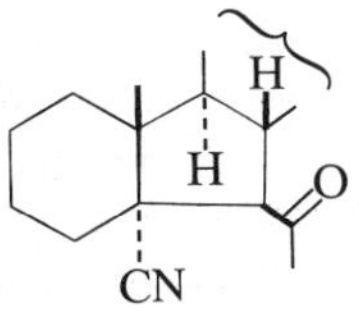

i 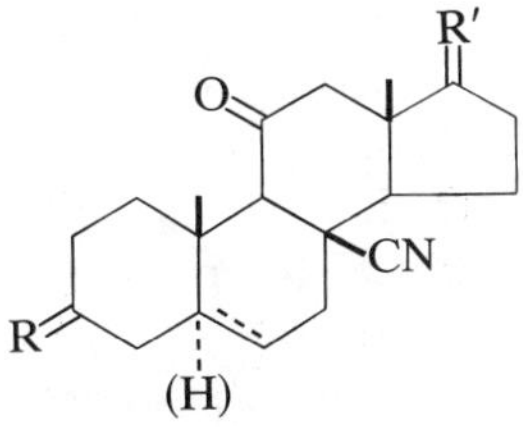 **j** 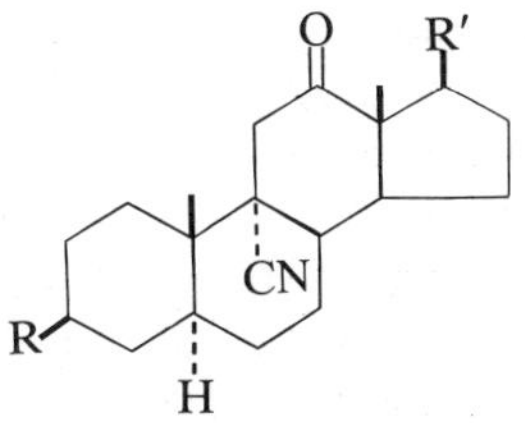 **k**

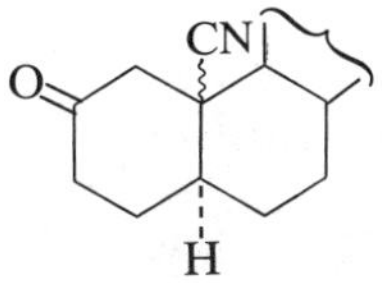

l (*cis*) 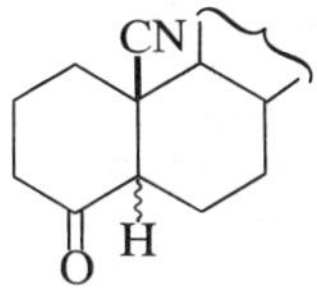 **m** 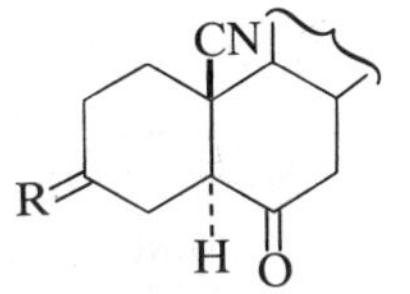 **n** 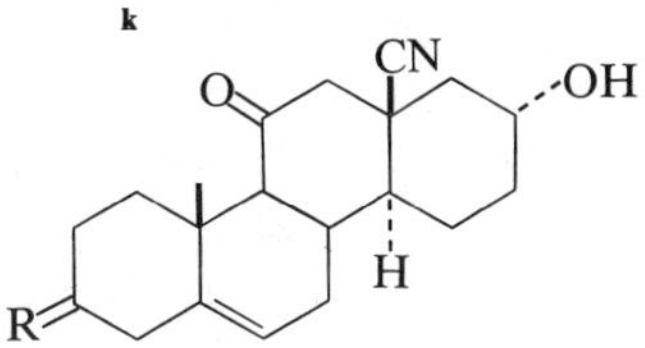 **o**

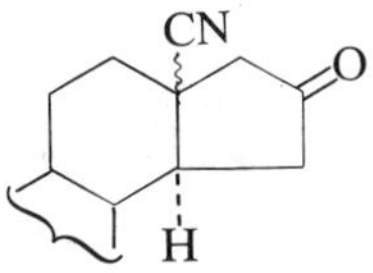

p (*cis*) 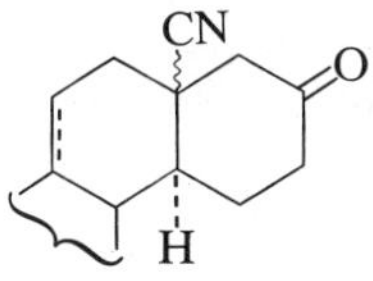 **q** (*trans*) 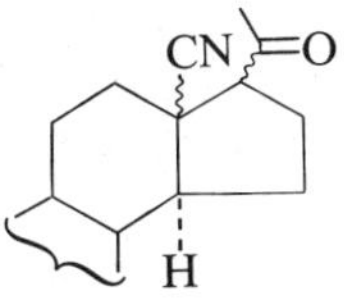 **r** (*trans*) 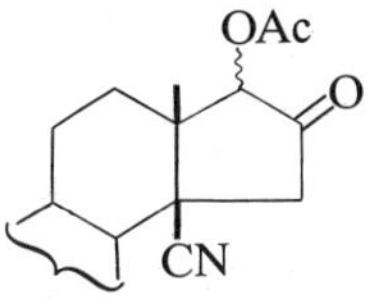 **s**

References for formulas **a–s**: **a**: 43, 87, 15, 189, 25, 190, 58, 141, 140, 79; **b–e**: 191; **f**: 43, 89, 90; **g**: 192–194; **h**: 195, 43; **i**: 43; **j**: 22, 84, 113, 25, 43, 42; **k**: 43; **l**: 196, 145; **m**: 197; **n**: 57; **o**: 80; **p**: 43; **q**: 78, 103, 43, 152; **r**: 81, 25, 151, 43, 107; **s**: 93. (References on p. 338)

isomer produced by the hydrogen cyanide-alkylaluminum or diethylaluminum cyanide method under kinetic conditions (if tested) is shown in parentheses. For details see Table VI. The conventional methods usually gave poor results except the potassium cyanide-ammonium chloride process which has limited use for hydrocyanation of steroidal enones with medium reactivities. The use of calcium cyanide in N-methylpyrrolidone was successful for the reaction of 4-en-3-ones.[58] Preparation of 10β-cyano steroids from 5(10)-en-7-ones by treatment with potassium cyanide in refluxing ethylene glycol is an exceptional case where the product is stable (see p. 275).[145] The organoaluminum methods are now the widely applied, successful means for the angular cyanation of steroids. The superiority of these methods over the most effective conventional process, potassium cyanide-ammonium chloride, has been discussed (see p. 287).

The stereochemistry of angular cyanation has been discussed in detail (p. 313). Some other interesting stereochemical results are discussed here. The predominance of the *trans* isomers in the acetylhydrindene system (types **i** and **r**) contrasts with the reverse results in the hydrindenone system (types **g**, **h**, **p**, and **s**). These results indicate that the stereochemical course can be changed by selecting the substrate type. The *trans*:*cis* ratios in types **b** and **q** are increased by the presence of a double bond at the 9,11 position as a result of disappearance of one syn-axial CN-H interaction. The 1α-cyano group retards the cyanide attack at C_5 from the α side (1,3-diaxial interaction) and favors the formation of the *cis* isomer (type **d** as compared with **b**).

Conjugated Polyenic Carbonyl Derivatives

In preceding sections we have discussed the lower reactivity of $\alpha,\beta,\gamma,\delta$-dienic carbonyl compounds compared with α,β-ethylenic substrates (see p. 300), the predominance of 1,6 addition over 1,4 addition except for the reaction of cinnamylidenemalonate (**11**) (p. 267), and the stereochemistry (see p. 312). The effect of the conjugating

[189] K. Takeda, K. Igarashi, and M. Narisada, *Steroids*, **4,** 305 (1964).

[190] J. Fishman and M. Torigoe, *Steroids*, **5,** 599 (1965).

[191] W. Nagata, M. Narisada, T. Wakabayashi, and T. Sugasawa, *J. Amer. Chem. Soc.*, **86,** 929 (1964).

[192] S. D. Levine, *Steroids*, **7,** 477 (1966).

[193] S. D. Levine, U.S. Pat. 3,271,437 (1966) [*C.A.*, **66,** 2702p (1967)].

[194] S. D. Levine, U.S. Pat. 3,526,642; 3,660,457 (1970) [*C.A.*, **72,** 90746m (1970)].

[195] W. Nagata, M. Narisada, T. Wakabayashi, Y. Hayase, and M. Murakami, *Chem. Pharm. Bull.* (Tokyo), **19,** 1567 (1971).

[196] M. Torigoe and J. Fishman, *Tetrahedron Lett.*, **1963,** 1251.

[197] D. P. Strike, D. Herbst, and H. Smith, *J. Med. Chem.*, **10,** 446 (1967).

$$-\overset{\delta}{C}=\overset{\gamma}{C}-\overset{\beta}{C}=\overset{\alpha}{C}-C=O \xrightarrow{CN^-} -\underset{CN}{C}-\overset{\gamma}{C}=C-\overset{\alpha}{C}=C-O^-$$

$$\xrightarrow{\alpha\ \text{Protonation}} -\underset{CN}{C}-C=C-CH-C=O$$

$$\xrightarrow{\gamma\ \text{Protonation}} -\underset{CN}{C}-CH-C=C-C=O \qquad \text{(Eq. 14)}$$

electron-withdrawing group ($CO_2R < CN < COR$) and of cyclic systems on the reactivity is the same as for α,β-ethylenic carbonyl systems. The complex results often observed arise from possible protonation of the initial 1,6 adduct at position α or γ as indicated in Eq. 14. The protonated products may undergo further hydrocyanation and other side reactions. Thus diethylaluminum cyanide in the absence of a proton source usually gives the cleanest results. This situation is demonstrated by the reaction of the steroidal dienone **115** (see also p. 292). Potassium cyanide in aqueous solvents gave the secondary product **117** (36% yield), although its formation was decreased to 10% in the less basic potassium cyanide-ammonium chloride process. The hydrogen cyanide-alkylaluminum method also gave poor results, since the reaction should be stopped at an early stage (recovery of **115** in 32% yield) to prevent the formation of the 5,7-dicyano product. Only the diethylaluminum cyanide method was satisfactory (92% of **116**).

OH

O CN O CN C NH$_2$

116 **117**

The reaction of two types of bicyclic conjugated dienenones has been examined. The reaction of **269** was smoother with diethylaluminum cyanide (high yield of **205**)[158] than with hydrogen cyanide-triethylaluminum (48% of **205**)[157] (compare the steroidal substrate

115). The protonation occurs at position γ. On the other hand, α protonation occurs in the reaction of dienone **110** with hydrogen cyanide-triethylaluminum to give the unconjugated product **111** in quantitative yield.[94] This favorable result is due to the inertness of the product **111** to further hydrocyanation. The failure of the diethylaluminum cyanide process in the reaction of **110** has been noted (see p. 291) and contrasts with the 53% conversion of the steroidal substrate **270** into 1,6 adduct **271** by the diethylaluminum cyanide method.[198] This difference may arise because the 17-hydroxyl group would block the reverse 1,6 addition [essentially the $(C_2H_5)_2AlCN$-(ROH) process]. (This process would be very effective when the reverse 1,6 addition and the α protonation are prone to occur.) On the other hand, the organoaluminum methods failed on the tetracyclic substrate **272,** possibly because of an increase in the ring strain.[198]

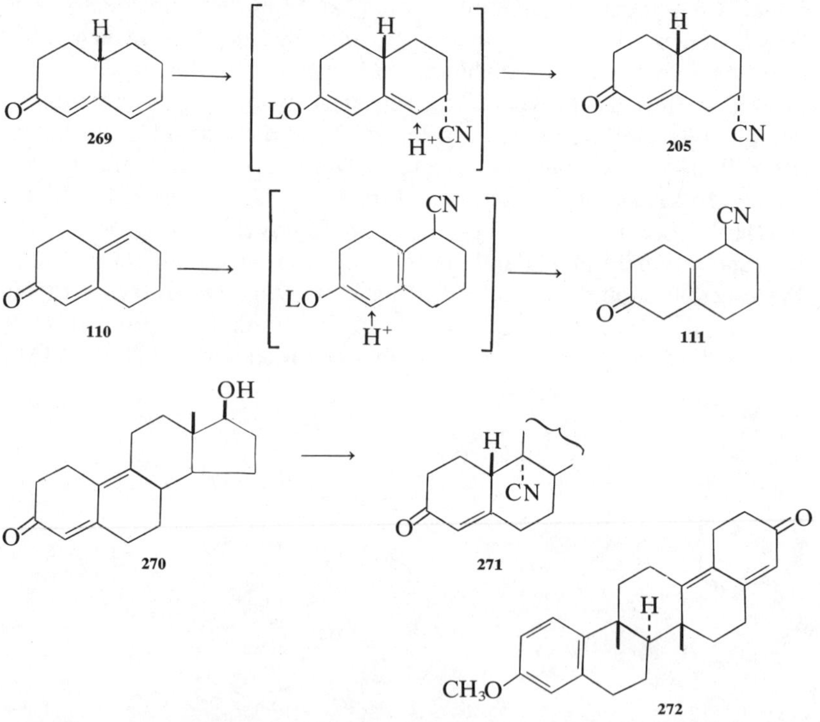

269 → [LO, H⁺, CN] → 205

110 → [LO, CN, H⁺] → 111

270 → 271

272

α,β-Acetylenic Carbonyl Derivatives

α,β-Acetylenic carbonyl compounds are expected to be more reactive than the corresponding α,β-ethylenic substrates.[119] However,

[198] R. E. Ireland, G. Pfister, D. J. Dawson, D. Dennis, and R. H. Stanford, *Synth. Commun.*, **2,** 175 (1972).

there are not sufficient examples to illustrate the difference in reactivity. The reaction of only two acetylenic esters has been reported. Methyl propiolate (**273**) reacted with hydrogen cyanide-(potassium cyanide) (slightly exothermic) at 45° to give the 1,4 adduct **274** in 24% yield, accompanied by a large amount of starting material.[2] The reaction of β-phenyl substrates **131** and **275** gave only the secondary products **133** and **216**.[11,120] Addition of methanol in the reaction of methyl ester **131** has been noted (see p. 294). No papers dealing with the reaction of α,β-acetylenic carbonyl derivatives with the organoaluminum reagents have appeared. Preliminary experiments have shown the inertness of hydrogen cyanide-triethylaluminum or diethylaluminum cyanide to ynoate **275** and the formation of complex products in the reaction of acetylenic ketones **276** and **277**.[59] These complexities may arise from susceptibility of the primary products **278a–278c** to further hydrocyanation and other side reactions.

$$\underset{\textbf{273}}{HC{\equiv}CCO_2CH_3} \longrightarrow \underset{\textbf{274}}{NCCH{=}CHCO_2CH_3}$$

$$C_6H_5C{\equiv}CCO_2R \xrightarrow{KCN} \left[C_6H_5\underset{CN}{\underset{|}{C}}{=}CHCO_2R\right] \longrightarrow C_6H_5\underset{CN}{\underset{|}{C}}H\underset{CN}{\underset{|}{C}}HCO_2R$$

131, $R = CH_3$
275, $R = C_2H_5$

133, $R = CH_3$

$$\longrightarrow C_6H_5\underset{CN}{\underset{|}{C}}HCH_2CN$$

216

$$RC{\equiv}CCOR' \xrightarrow{CN^-} \left[R\underset{CN}{\underset{|}{C}}{=}C{=}\underset{R'}{\underset{|}{C}}{-}O^- \rightleftharpoons R\underset{CN}{\underset{|}{C}}{=}\underline{C}COR'\right]$$

276, $R = H$, $R' = C_3H_7\text{-}n$
277, $R = n\text{-}C_5H_{11}$, $R' = CH_3$

278a **278b**

$$\xrightarrow{H^+} R\underset{CN}{\underset{|}{C}}{=}CHCOR'$$

278c

Miscellaneous Substrates

α,β-Unsaturated Aldehydes. Hydrocyanation of α,β-ethylenic aldehydes usually gives only 1,2 adducts. A few successful examples of conjugate hydrocyanation include conversion of simple substrates **279** or their cyanohydrins **280** into 1,4 adducts **281** or their cyanohydrins **282** by the hydrogen cyanide-(base) process[47,199] and the formation of

[199] T. Warner and O. A. Moe, U.S. Pat. 2,565,537 (1951) [*C.A.*, **46,** 2565c (1952)].

the 16α-cyano aldehyde **284** (58%) from 16-ene-17-carboxyaldehyde **283** (the 1,2 adduct is unstable) by the diethylaluminum cyanide method.[92] No 1,4 adducts were isolated by treatment of cyclohexylideneacetaldehyde and several steroidal α-enals with the organoaluminum reagents.[92] A general, advantageous method for preparing β-cyano aldehydes via α,β-unsaturated imino derivatives is described in the following paragraph.

$$\mathrm{RCH{=}C(R')CHO}\ (\mathbf{279}) \rightleftharpoons \mathrm{RCH{=}C(R')CH(OH)CN}\ (\mathbf{280})$$

$$\mathrm{RCH{=}C(R')CHO}\ (\mathbf{279}) \longrightarrow \mathrm{RCH(CN)CH(R')CHO}\ (\mathbf{281}) \rightleftharpoons \mathrm{RCH(CN)CH(R')CH(OH)CN}\ (\mathbf{282})$$

R, R′ = H, CH_3

CHO
AcO
283

→

CHO
CN
284

α,β-Unsaturated Imino Derivatives. Only a few reports have appeared on hydrocyanation of this class of substrates (usually base-sensitive) by conventional methods. No cyanated products were isolated in the reaction of cinnamylidene imines **285a**.[200] The vinyloxazine derivative **289** was converted into 1,4 adduct **290** in 45% yield under special conditions (hydrogen cyanide-m-dinitrobenzene in refluxing acetic acid).[201] Hydrocyanation of naphthoquinone diimine **291** gave dinitrile **292**.[202] The product is formed by simultaneous 1,2 and 1,4 additions of hydrogen cyanide followed by elimination of benzenesulfonamide. Quite recently, the preparation of 3-indolylacetonitriles **295** by the action of potassium cyanide on unstable quaternary imino compounds **294** was reported.[203] The imino compounds were formed *in situ* by [2, 3], [3, 3] double sigmatropic rearrangement of propargyl N-oxides **293**.

[200] J. S. Walia, D. H. Rao, and M. Singh, *Indian J. Chem.*, **2,** 437 (1964).
[201] A. I. Meyers, *J. Org. Chem.*, **25,** 145 (1960).
[202] R. Adams and W. Moje, *J. Amer. Chem. Soc.*, **74,** 5562 (1952).
[203] Y. Makisumi and S. Takada, *Chem. Pharm. Bull.* (Tokyo), **24,** 770 (1976).

$C_6H_5CH{=}CHCH{=}NR \xrightarrow[\text{Reflux}]{KCN/CH_3OH} C_6H_5CH_2CH{=}C(OCH_3)NHR$ **286**

285a, R = CH_3, C_6H_5
285b, R = C_4H_9-*t*

+ $C_6H_5(CH_2)_2CONHR$ **287**

$\xrightarrow[\text{2. Hydrolysis}]{\text{1. } HCN\text{-}(C_2H_5)_2AlCl}$ $C_6H_5CH(CN)CH_2CHO$ **288** (54%)

289 → **290** (CN)

291 ($NSO_2C_6H_5$, $NSO_2C_6H_5$) $\xrightarrow[C_6H_6,\ 0°]{HCN\text{-}[N(C_2H_5)_3]}$ **292** (85%) (CN, CN, $NHSO_2C_6H_5$)

293 (R′, C≡C, CH_2, $\overset{+}{N}$, ^{-}O, $CH_2C_6H_5$) → [**294** (R, R′, +N, C_6H_5)] $\xrightarrow[\text{Aq DMF}]{KCN}$ **295** (R, CN, N, R′, C_6H_5)

(R = H, Cl, OCH_3; R′ = H, CH_3)

A general method has been devised for preparing β-cyano aldehydes **298** by hydrocyanation of α,β-ethylenic aldimines **296** by the organoaluminum methods and subsequent acid hydrolysis of the resulting 1,3-dicyanopropylamines **297**.[92] In some reactions, 2-iminopyrrolidines **299** are formed as by-products. Of the two alkyl groups used as R′ of imines **296**, the *t*-butyl group proved superior to the cyclohexyl group in effecting the 1,4 addition (although slower) and in reducing the formation of the by-products **299**. A proton source is important in this hydrocyanation, as illustrated in the reaction of **99** (p. 288), **112** (p. 291), and **285b** (compare the failure by the potassium cyanide method; see p. 343).

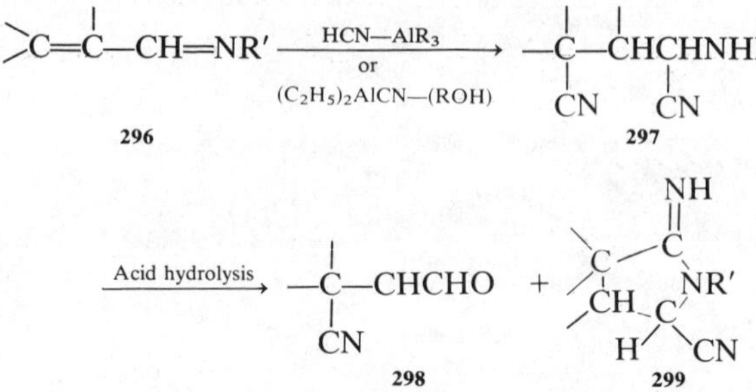

$$\text{>C=C}(\text{—})\text{—CH=NR}' \xrightarrow[\text{(C}_2\text{H}_5)_2\text{AlCN—(ROH)}]{\text{HCN—AlR}_3 \text{ or}} \text{—C(CN)—CH(CN)NHR}'$$

296 → **297**

$$\xrightarrow{\text{Acid hydrolysis}} \text{—C(CN)—CHCHO} + \text{299}$$

298 **299**

Successful conjugate hydrocyanation of the steroidal dienamine **300a** and its perchlorate **300b** into nitrile **301** has been reported.[43]

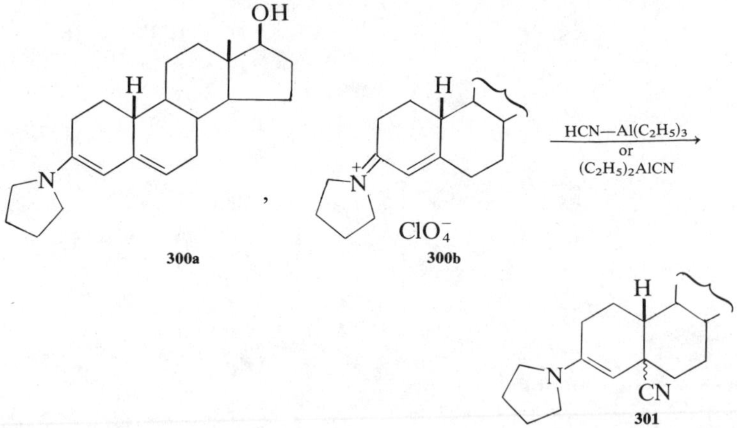

300a, **300b** (ClO_4^-) $\xrightarrow[\text{(C}_2\text{H}_5)_2\text{AlCN}]{\text{HCN—Al(C}_2\text{H}_5)_3 \text{ or}}$ **301** (2 isomers)

SYNTHETIC UTILITY AND COMPARISON WITH CONJUGATE ADDITION OF ORGANOCOPPER REAGENTS

Application of conjugate hydrocyanation before the early twentieth century was limited to preparation of various derivatives of succinic acid and γ-ketobutyric acid which were industrial chemicals themselves and important as synthetic intermediates.[204]

[204] O. Bayer, *Angew. Chem.*, **61,** 230 (1949).

The wide synthetic utility of hydrocyanation has since been recognized, and β-cyano carbonyl compounds have been used as important intermediates for syntheses of natural products such as terpenes, terpene alkaloids, steroids, steroidal alkaloids, prostaglandins, antibiotics, and corrinoids.

The first hydrocyanation applied to the synthesis of degradation products from bile acids and sex hormones[133,205,206] consisted of constructing the steroidal D-ring by introducing a carbon-skeleton at C_{13} via conjugate hydrocyanation (**i**→**ii**→**iii**). This route led to the synthesis of deoxyequilenin,[207] and equilenin.[208] Further development of this approach,[209] its modification,[210] and its application to the synthesis of hydroaromatic steroids[211] have been reported. In the total synthesis of vitamin D the sequence **iv**→**v**→**iii** for construction of the C,D-ring system was used.[55,56] This sequence also involves the formation of a quaternary carbon at C_{13} by conjugate hydrocyanation.

CH_3 CO_2R' R CN (**i**) —1. HCN; 2. CH_2=CHCN→ (**ii**) R, CH_3, CO_2R', O → (**iii**) CH_3, 13, R″, H

(**iv**) CH_3, CO_2R, O —1. HCN; 2. Hydrolysis→ (**v**) CH_3, CO_2R', CO_2R, O → **iii**

R″ = CO_2R'/H, O

Several total syntheses of steroids are based on new approaches. They include conversion by conjugate hydrocyanation of polycyclic α-enones into angular nitriles, **q**, **o**, and **r** (see p. 337), which are key intermediates for construction of the functionalized C,D-ring systems, **q**′, **o**′, and **r**′. The usefulness of these approaches for synthesis of isoprenoids with an angular (functionalized) methyl group has been

[205] D. K. Banerjee, *J. Indian Chem. Soc.*, **17**, 453 (1940) [*C.A.*, **35**, 2152[1] (1941)].

[206] D. K. Banerjee and S. K. Das Gupta, *J. Amer. Chem. Soc.*, **74**, 1318 (1952).

[207] R. A. Barnes and R. Miller, *J. Amer. Chem. Soc.*, **82**, 4960 (1960).

[208] D. K. Banerjee, H. N. Khastgir, J. Dutta, E. J. Jacob, W. S. Johnson, C. F. Allen, B. K. Bhattacharyya, J. C. Collins, Jr., A. L. McCloskey, W. T. Tsatsos, W. A. Vredenburg, and K. L. Williamson, *Tetrahedron Lett.*, **1961**, 76.

[209] D. K. Banerjee, *J. Indian Chem. Soc.*, **47**, 1 (1970).

[210] L. Novak, M. Borovička, and M. Protiva, *Collect. Czech. Chem. Commun.*, **27**, 1261 (1962).

[211] M. Chaykovsky and R. E. Ireland, *J. Org. Chem.*, **28**, 748 (1963).

demonstrated by successful syntheses of *dl*-3β-hydroxy-5α-pregn-16-en-20-one,[78,212] *dl*-3α-acetoxy-5α-pregna-9(11),16-dien-20-one,[152] *dl*-3-hydroxy-19-norpregna-1,3,5(10)-trien-20-one,[81] *dl*-3β-hydroxy-5α-pregnan-20-one,[151] *dl*-latifolin,[213] and aldosterone.[80] The high efficiency and the high *trans* selectivity of the new hydrocyanation methods are of great significance in these steroid syntheses.

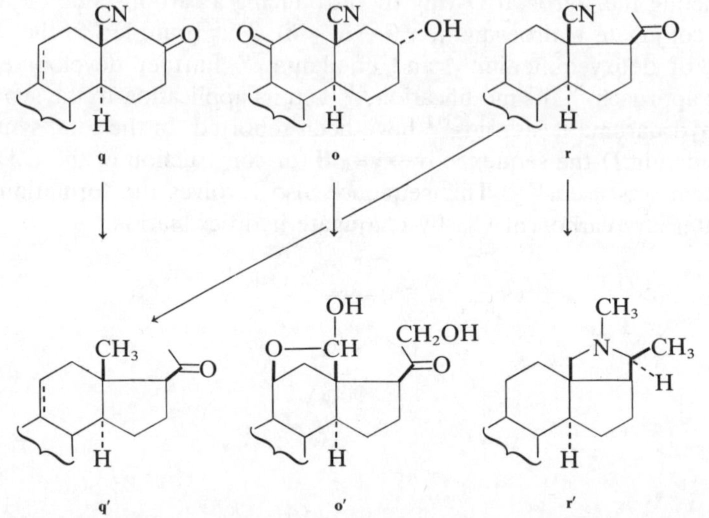

q o r

q′ o′ r′

These approaches have been applied to total syntheses of progesterone, conessine, and cholesterol.[107] Angular cyanation has been applied also to introduction of the C_{19}-methyl group via the 10β-cyano intermediate (type **m** or **n** on p. 337). This approach was also used in the total syntheses of testosterone and 18-methyltestosterone.[197] The usefulness of the 10β-cyano (^{14}CN) intermediate for the synthesis of ^{14}C-labeled steroids has been demonstrated.[57]

Conversion of the angular cyano group of β-cyano ketones into a methyl group or other carbon-functional groups is important in syntheses of natural products. The conversions have been examined with 5α-cyano-3-oxo steroids,[76] 8β-cyano-11-oxo steroids,[42,113] 13β-cyano-11-oxo-D-homo steroids,[80] 10α-cyanodecahydronaphthalen-3-ones,[102] and 13β-cyano-20-oxo steroids.[81,151,213] The conversion is summarized in partial structures in the two accompanying schemes.[27] It should be noted that conversion of the angular cyano group generally is much more difficult than that of the nonangular

[212] W. Nagata, T. Terasawa, S. Hirai, and K. Takeda, *Tetrahedron Lett.*, **1960, No. 17,** 27.
[213] W. Nagata, T. Terasawa, and T. Aoki, *Tetrahedron Lett.*, **1963,** 869.

group but is often assisted by participation of the β-carbonyl or β-axial hydroxyl group. The unusual CN → OH conversion with retention has been reported to occur in lithium aluminum hydride reduction of the angular cyano group when tetrahydrofuran is used as a solvent.[59,94,156,214] Conversion of secondary cyano groups into aminomethyl and alkoxycarbonyl groups has been achieved most effectively by catalytic hydrogenation[94] and by acid-catalyzed solvolysis,[159,179] respectively.

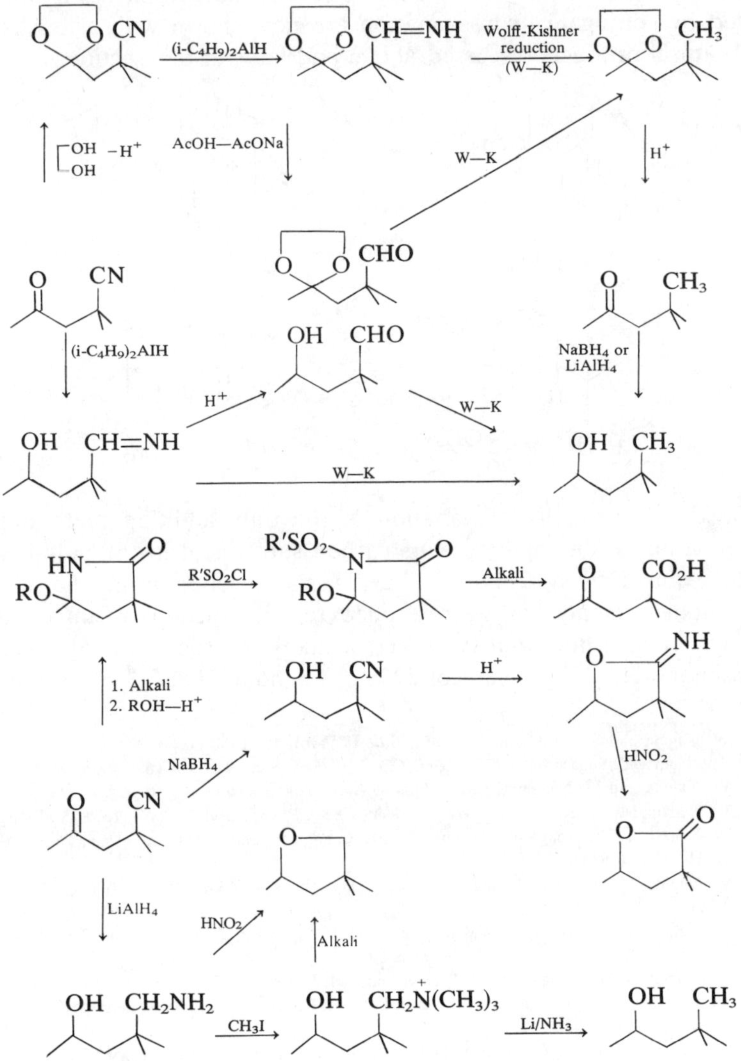

[214] M. M. Janot, X. Lusinchi, and R. Goutarel, *C.R. Acad. Sci.*, **258,** 4780 (1964).

Extensive studies have been made on conversion of the β-cyano keto grouping into bridged cyclic systems such as pyrrolidine,[215] piperidine,[216] philocladene-[181,217] and kaurene-type[157,218,219] bicyclo[3.2.1]octanes, and bicyclo[2.2.2]octane.[220] Using these conversions, total syntheses have been achieved of representative diterpene alkaloids such as atisine **302**,[85] veatchine **303**,[221] and garryine **304**,[221] as well as a diterpene, gibberellin A_{15} **305**.[142] In these syntheses angular cyanation by organoaluminium methods has been utilized twice for construction of the ring systems. The carbon atoms introduced by conjugate hydrocyanation are now shown with asterisks in the natural products discussed above and later in this section.

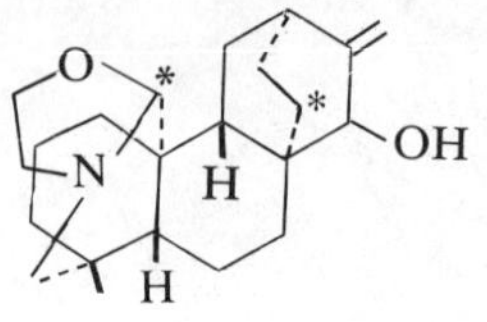

302

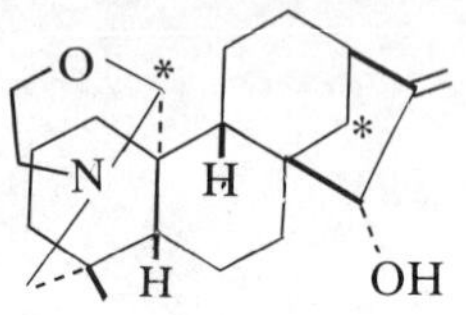

303

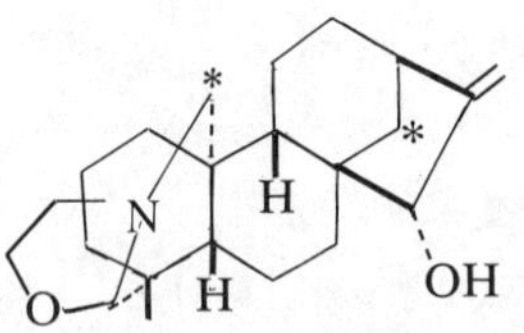

304

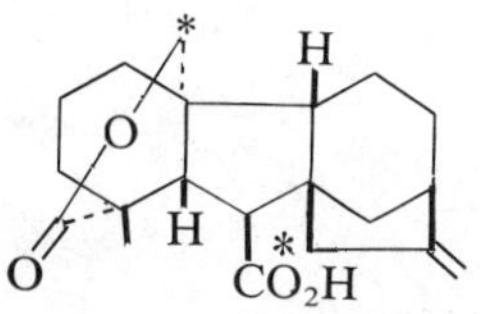

305

Since conjugate hydrocyanation is the only addition method for introduction of an angular, *trans* carbon substituent in the poly-fused cyclohexane ring system (see p. 351), the reaction using the alkylaluminum reagents has been applied to construction of an angular methyl group in the total syntheses of many triterpenes. They include alnusenone **306**,[139] germanicol **307**,[222] shionone **308**,[183,184] and lupeol **309**.[223]

[215] W. Nagata and S. Hirai, *Chem. Pharm. Bull.* (Tokyo), **16** 1550 (1968).

[216] W. Nagata, T. Sugasawa, and T. Aoki, *Chem. Pharm. Bull.* (Tokyo), **16,** 1556 (1968).

[217] W. Nagata and M. Narisada, *Chem. Pharm. Bull.* (Tokyo), **16,** 867 (1968).

[218] W. Nagata, M. Narisada, and T. Wakabayashi, *Chem. Pharm. Bull.* (Tokyo), **16,** 875 (1968).

[219] W. Nagata, T. Wakabayashi, M. Narisada, M. Yamaguchi, and Y. Hayase, *Chem. Pharm. Bull.* (Tokyo), **19,** 1582 (1971).

[220] W. Nagata, M. Narisada, T. Sugasawa, and T. Wakabayashi, *Chem. Pharm. Bull.* (Tokyo), **16,** 885 (1968).

[221] W. Nagata, M. Narisada, T. Wakabayashi, and T. Sugasawa, *J. Amer. Chem. Soc.*, **89,** 1499 (1967).

[222] R. E. Ireland, S. W. Baldwin, D. J. Dawson, M. I. Dawson, J. E. Dolfini, J. Newbould, W. S. Johnson, M. Brown, R. J. Crawford, P. F. Hudrlik, G. H. Rasmussen, and K. K. Schmiegel, *J. Amer. Chem. Soc.*, **92,** 5743 (1970).

[223] G. Stork, S. Uyeo, T. Wakamatsu, P. Grieco, and J. Labovitz, *J. Amer. Chem. Soc.*, **93,** 4945 (1971).

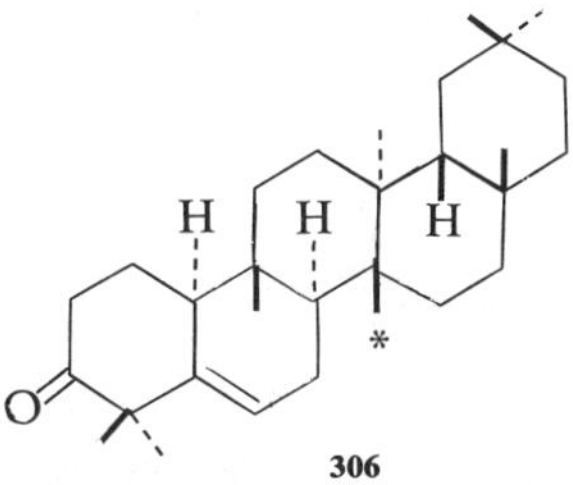

306 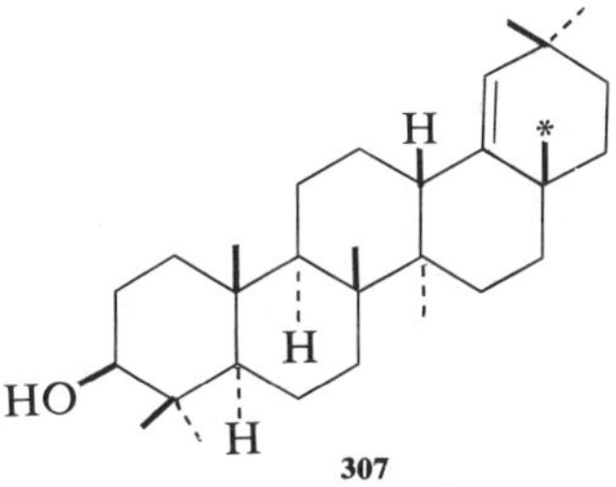 307

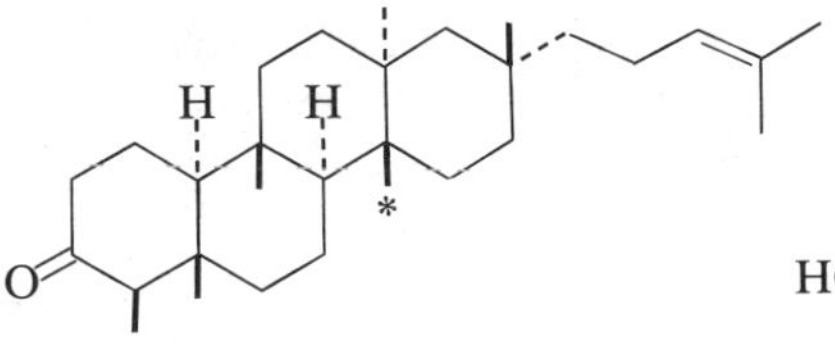

308 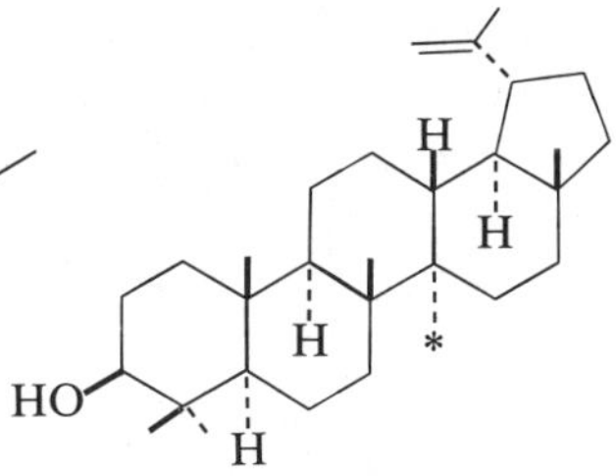 309

Nonangular cyanation also has been widely applied to total syntheses of natural products. The terpenes thus synthesized are podocarpic acid **310**,[154c] atractylon **311**,[94] epizizanoic acid **312**,[224] hinesol **313**,[179] and drimenin **314**.[147] Other sesquiterpene and diterpene syntheses include mirestrol[225] and an approach to eudesmane sesquiterpenes,[74,117,134,138] and to gibberellins.[226] Hydrocyanation has been applied to the construction of the bridged lactone ring of the sesquiterpene alkaloids, annotinine **315**[159] and dendrobine **316**.[82] Other interesting natural products synthesized by the application of

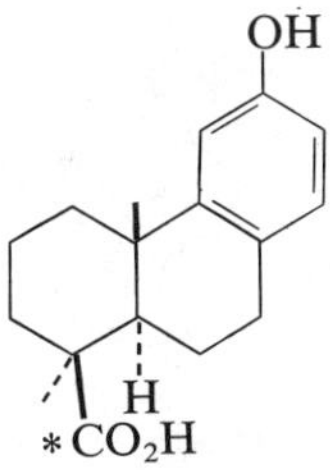

310 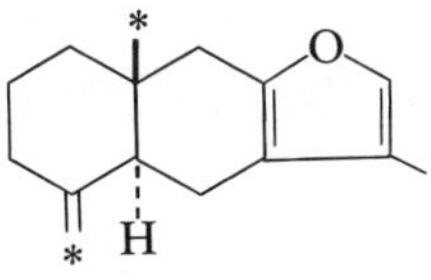 311 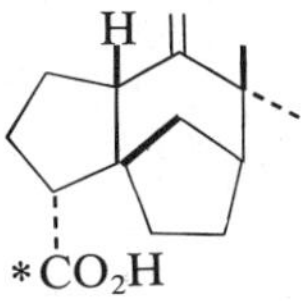 312 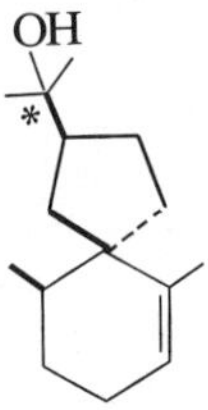 313

[224] F. Kido, H. Uda, and A. Yoshikoshi, *J. Chem. Soc., Perkin I*, **1972,** 1755; *Chem. Commun.*, **1969,** 1335.

[225] M. Miyano and C. R. Dorn, *J. Org. Chem.*, **37,** 268 (1972).

[226] K. Mori, M. Matsui, and Y. Sumiki, *Agr. Biol. Chem.* (Tokyo), **28,** 243 (1964) [*C.A.*, **61,** 9539e (1964)].

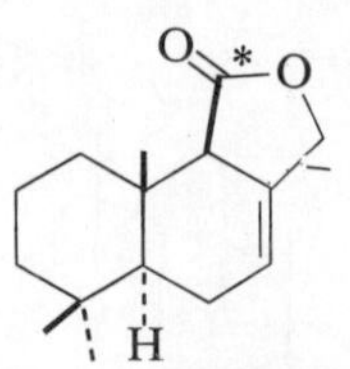

314 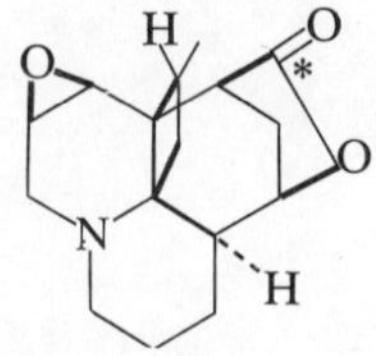 315 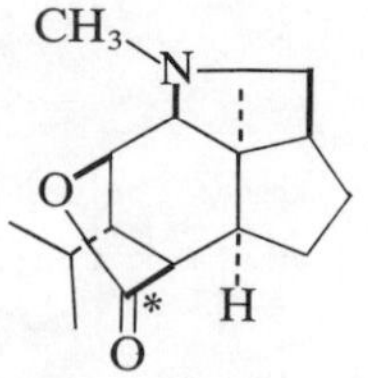 316

conjugate hydrocyanation are **317**[112] and **318**[227] (degradation products of diterpene alkaloids such as delphinine and songorine), isoajmaline **319**,[228] metacycloprodigiosin **320** (a tripyrrole pigment from *Streptomyces*),[229] and prostaglandin F_1 **321**.[86] Application of hydrocyanation has been extended to total syntheses of 11-deoxyprostaglandins,[108,230,231] the synthetic study of semicorrins,[99] and syntheses of various medicines.[53,232–236]

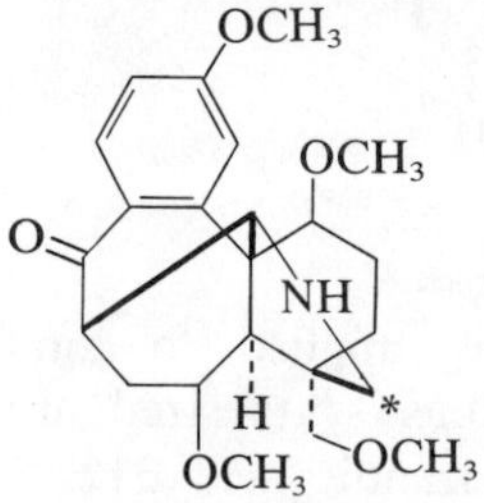

317 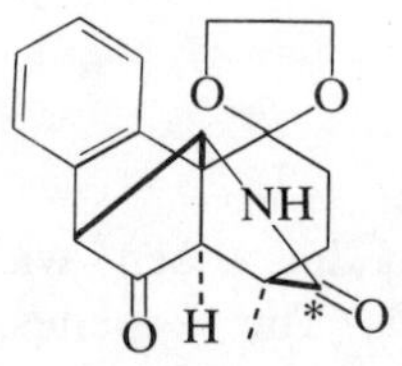 318 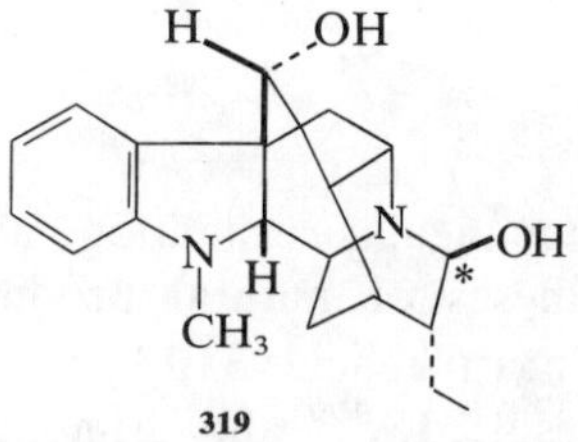 319

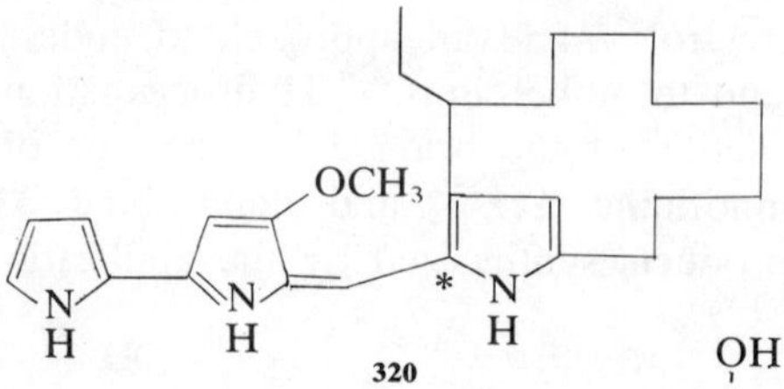

320

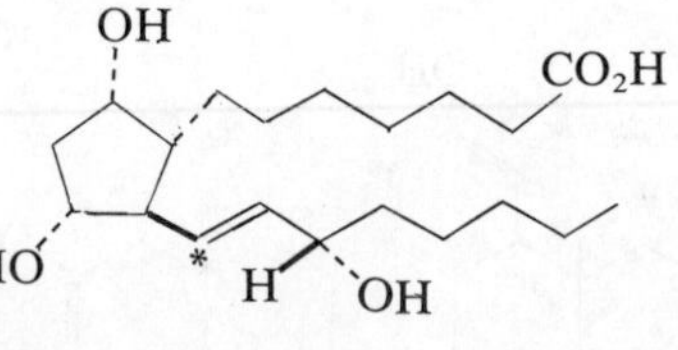

321

[227] K. Wiesner, A. Philipp, and P. Ho, *Tetrahedron Lett.*, **1968,** 1209.

[228] K. Mashimo and Y. Sato, *Tetrahedron,* **26,** 803 (1970); *Tetrahedron Lett.*, **1969,** 901.

[229] H. H. Wasserman, D. D. Keith, and J. Nadelson, *J. Amer. Chem. Soc.*, **91,** 1264 (1969).

[230] J. F. Bagli, T. Bogri, R. Deghenghi, and K. Wiesner, *Tetrahedron Lett.*, **1966,** 465.

[231] M. P. L. Caton, E. C. J. Coffee, and G. L. Watkins, *Tetrahedron Lett.*, **1972,** 773.

[232] K. Lundahl, J. Schut, J. L. M. A. Schlatmann, G. B. Paerels, and A. Peters, *J. Med. Chem.*, **15,** 129 (1972).

[233] T. Petrzilka, M. Demuth, and W. G. Lusuardi, *Helv. Chim. Acta*, **56,** 519 (1973).

[234] S. M. McElvain and R. E. McMahon, *J. Amer. Chem. Soc.*, **71,** 901 (1949).

[235] M. P. Mertes, A. A. Ramsey, P. E. Hanna, and D. D. Miller, *J. Med. Chem.*, **13,** 789 (1970).

[236] R. M. Weier and L. M. Hofmann, *J. Med. Chem.*, **18,** 817 (1975).

The broad synthetic utility of conjugate hydrocyanation, especially when the hydrogen cyanide-alkylaluminum and diethylaluminum cyanide reagents are used, is apparent from the discussion above. Carbon-carbon bond formation has long been a constant challenge to those engaged in synthetic organic chemistry. Another useful method for meeting this objective is conjugate addition of organocopper reagents.[119] It is appropriate and important to compare organocopper conjugate addition with conjugate hydrocyanation using organoaluminum reagents.

First, organocopper reagents are superior to the cyanide reagents in placing hydrocarbon groups such as alkyl, alkenyl, and aryl at a position β to a carbonyl function. Conversely, the organoaluminum cyanide reagents are better for introduction of functionalized carbon groups such as formyl, carboxyl, hydroxymethyl, carbamoyl, and various bridged ring systems.

Secondly, the steric requirement of organocopper reagents is much greater than that of the cyanide reagents. With increasing steric bulk the efficiency and stereoselectivity of a reagent generally decrease. For example, lithium dimethylcuprate, $(CH_3)_2CuLi$, is inert to moderately hindered bicyclic enones **322** and **187b**[119] whose conjugate hydrocyanation could be easily effected by the new cyanide reagents (see pp. 334 and 335). Conjugate addition of the cuprate reagent to highly hindered 8-en-11-oxo steroids would be impossible. In contrast, successful hydrocyanations of these steroids are noted often.

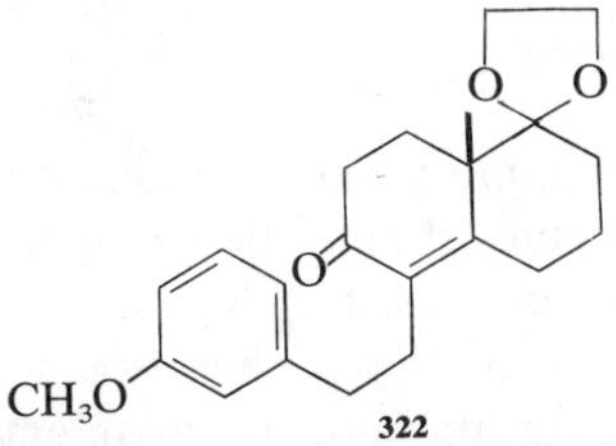

322 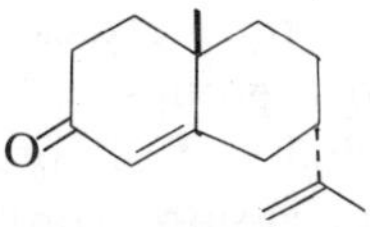 **187b**

The difference in stereoselectivity can be seen from the exclusive formation of the *cis* isomers **323a**, **323b**, and **325** by the reaction of the corresponding conjugated ketones **121**, **76**, and **324** with lithium dimethylcuprate or methylmagnesium iodide and cuprous acetate.[119] The predominant formation of the *trans* nitriles in hydrocyanation of these substrates with the cyanide reagents under kinetic conditions has been shown (p. 314). Thus *kinetic conjugate hydrocyanation using the organoaluminum reagents is the only addition method for introduction of a trans-carbon substituent at an angular position in six-membered polycyclic systems.* Another stereochemical advantage of hydrocyanation is that either *trans* or *cis* nitriles can be obtained from certain

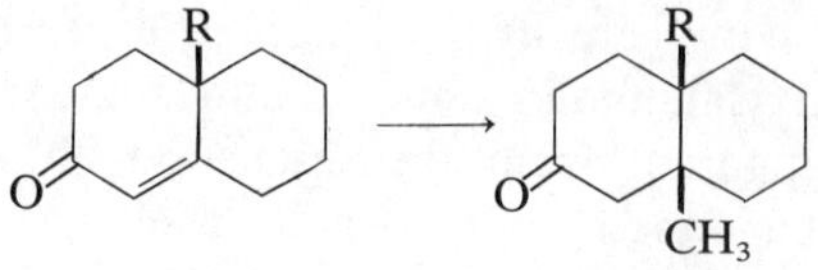

121, R=H
76, R=CH_3

323a, R=H
323b, R=CH_3

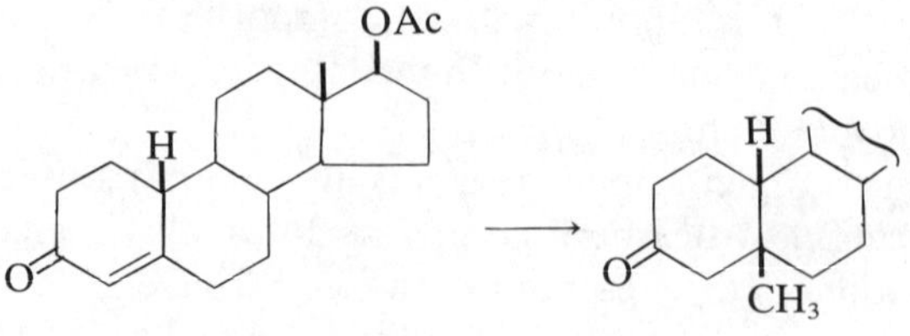

324

325

types of substrates by selecting the reagent and reaction conditions (see p. 313). A stereochemical disadvantage of the organocopper reagents is that nonangular conjugate addition does not always give the axial 1,4 adduct. For example, α,α-dimethyloctalone **326** gives 1,4 adduct **327** as a 54:46 mixture of isopropyl epimers.[119,237] The axial addition rule is followed in conjugate hydrocyanation with the organoaluminum reagents (p. 311).

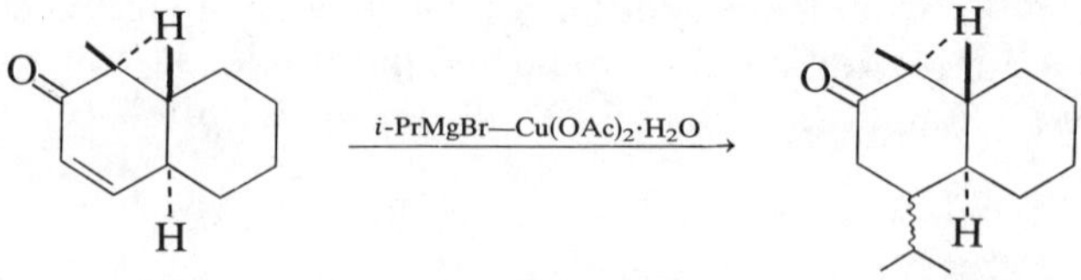

i-PrMgBr—$Cu(OAc)_2 \cdot H_2O$

326

327

Finally, the inertness of organoaluminum cyanide reagents to various functional groups, including halogen, should be compared to the susceptibility of the organocopper reagents to halogen.

In conclusion, conjugate hydrocyanation using organoaluminum reagents provides wide synthetic utility and is complementary to organocopper conjugate addition.

EXPERIMENTAL FACTORS

Preparation and Handling of Reagents

Caution! All hydrocyanation reagents are toxic and should be handled carefully in a well-ventilated hood. Hydrogen cyanide is low boiling and especially hazardous. However, with reasonable care and precaution these reagents present little danger.

[237] J. A. Marshall and N. H. Andersen, *J. Org. Chem.*, **31**, 667 (1966).

Alkali Cyanides. Commercially available (reagent-grade or purer) sodium cyanide, potassium cyanide, and calcium cyanide can be used without purification.

Acetone Cyanohydrin. This reagent is commercially available and is easily prepared by adding acetone to an aqueous solution of sodium or potassium cyanide and treating with sulfuric acid. An *Organic Syntheses* procedure has been published.[238] Freshly distilled material is recommended, since aged material reportedly led to formation of tars.[104]

Hydrogen Cyanide. Although used for more than a century, hydrogen cyanide, because of its high volatility (bp 26°), still scares most organic chemists. It is essential to handle this reagent carefully in a good hood. However, the reagent, neat or in a solution, is quite safely transferred by cold syringes (conveniently cooled in a plastic bag placed in dry ice for 10 minutes). Hydrogen cyanide has a distinctive almondlike odor.

Although details of the preparation have appeared in *Organic Syntheses*,[239] evolution of gaseous hydrogen cyanide using the *Organic Syntheses* apparatus is often not smooth and can be done better by the following procedure described by Bauer[240] and partly modified by us. A mixture of 1 kg of concentrated sulfuric acid, 400 ml of water, and 20 g of ferrous sulfate is heated at 90° on a water bath in a 5-l., round-bottomed, long-necked flask containing several boiling chips and fitted with a dropping funnel and a water-cooled condenser. The condenser leads to a drying apparatus. A solution of 1 kg of sodium cyanide in 1.2 l of water is added dropwise for 2–3 hours; a large volume of hydrogen cyanide is released. The water bath is then brought to boiling for an additional 30 minutes to drive off the hydrogen cyanide from the reaction flask. The hydrogen cyanide is dried by passing through 20 ml of 4 *N* sulfuric acid and through 200 g of anhydrous calcium chloride on a layer of glass wool, each placed in a bottle fitted with an inlet tube reaching to the bottom. The drying apparatus is immersed in a water bath kept at 50°. The gaseous hydrogen cyanide is further dried by passage through a U-tube filled with anhydrous calcium chloride (heated at 50°) and condensed in a 1-1, long-necked flask fitted with an inlet tube reaching to an inch below the neck and with an outlet leading to a coil condenser (a second

[238] R. F. B. Cox and R. T. Stormont, *Org. Synth., Coll. Vol.* **II,** 7 (1943).

[239] K. Ziegler, *Org. Synth., Coll. Vol.* **I,** 314 (1941).

[240] B. Brauer, *Handbook of Preparatory Inorganic Chemistry*, Vol. 1, Academic Press, New York, 1963, p. 658.

trap). The flask and the condenser should be well cooled with ice. The yield is nearly quantitative. The reader should familiarize himself also with the precautions described in *Organic Syntheses.*[239]

Neat hydrogen cyanide stabilized by addition of anhydrous calcium chloride can be stored as a liquid or, better, frozen in a refrigerator for months. However, we recommend keeping the reagent as a 10–30% solution in tetrahydrofuran, benzene, or toluene. The solution is stored in a tightly stoppered bottle or, preferably, in strong ampoules. The solution in ampoules seems to be stable for several years when kept in a freezer.

Alkylaluminums. Reagents such as triethylaluminum, diethylaluminum chloride, and ethylaluminum dichloride are commercially available in lecture bottles ($\frac{1}{2}$ lb) from Ethyl Corporation, Texas Alkyls, Inc., Toyo Ethyl Co. (Japan), and other chemical reagent firms. Handling procedures for these *pyrophoric* alkylaluminums have appeared in *Organic Syntheses.*[89]

Recently, standardized (15 or 25%) solutions of all the alkylaluminums in various hydrocarbon solvents stored in glass bottles have become available from Stauffer Chemical Company, New York, and some solutions are available from Tokyo Kasei Kogyo Co., Ltd., Tokyo. These *less pyrophoric*, easily handled solutions eliminate tedious handling of pyrophoric alkylaluminums. Instructions for using the solution are available from the suppliers. We are confident that the standardized solutions can replace neat alkylaluminums in Method A and B hydrocyanation procedures, provided the alkylaluminum solution is mixed first with a twofold or greater volume of the reaction solvent such as tetrahydrofuran or toluene. Exchange of the solvent by evaporation-addition under nitrogen is not difficult.

It should be noted that diethylaluminum chloride and ethylaluminum dichloride react slowly with tetrahydrofuran and so cannot be kept in a tetrahydrofuran-containing solvent mixture. Triethylaluminum is stable in tetrahydrofuran over a long period.

Alkylaluminum solutions, though less pyrophoric, react with air and moisture. Although rigorous exclusion of oxygen or moisture, as with a dry box, is unnecessary, the solutions are best transferred by dry hypodermic syringes and reactions should be run in an atmosphere of dry nitrogen or argon.

Diethylaluminum Cyanide. Details of the preparation of this reagent have appeared in *Organic Syntheses.*[241] A commercially available

[241] W. Nagata and M. Yoshioka, *Org. Synth.*, **52,** 90 (1972).

25% solution of triethylaluminum in benzene or heptane facilitates the preparation. When using a triethylaluminum solution in an aliphatic hydrocarbon, one should first mix the solution with a twofold or larger volume of benzene, toluene, or diisopropyl ether and then add a hydrogen cyanide solution in the same solvent, preferably at around −10°, since hydrogen cyanide is not miscible with aliphatic hydrocarbons. The crude diethylaluminum cyanide solution thus prepared can be used for preparative hydrocyanation without purification by distillation, as described in *Organic Syntheses.*[241] However, care must be taken not to let any unchanged triethylaluminum remain (use 1.1 molar equivalents of hydrogen cyanide). It has been found that a small amount of triethylaluminum retards hydrocyanation of 8-en-11-oxo steroids in benzene or toluene (not in tetrahydrofuran).[25] Also, the triethylaluminum in a hydrocarbon solvent works as an ethylating agent, except in ether solvents, to yield by-products such as ethylcarbinols.[59]

Diethylaluminum cyanide solutions in hydrocarbon solvents or diisopropyl ether stored in ampoules are stable at room temperature for many years. The reagent reacts slowly with tetrahydrofuran; thus this solvent is not suitable in preparation or storage of the reagent.

Recently, a 1–2 *M* solution of diethylaluminum cyanide in benzene has become commercially available from Alfa Products, Ventron Corporation, Danvers, Massachusetts.

Diethylaluminum cyanide is nonpyrophoric even as pure liquid. However, the reagent solutions are not inert to air and moisture and should be handled in the way described above for alkylaluminums (dry hypodermic syringes and a nitrogen atmosphere).

Reaction Conditions

Selection of suitable reaction conditions such as concentration, reagent ratio, solvent, temperature, and time is as important as selection of the hydrocyanation reagent in preparing cyano compounds in good yields and with minimal by-products. Generally, the concentrations of substrate and reagent(s) depend on their solubilities, and the reaction time depends on reactivity of the substrate. Effects of these two factors are not discussed here except for special cases. In this section we summarize effects of such reaction variables as reagent ratio, solvent, and temperature in the representative hydrocyanation procedures that have been partly discussed under "Scope and Limitations." Special emphasis is accorded the procedures using alkylaluminum compounds.

Hydrogen Cyanide-Base Catalyst. Having the lowest reactivity, this combination of reagents usually needs high temperatures to effect conjugate hydrocyanation. In the absence of a solvent (the substrate should be miscible with hydrogen cyanide), the use of a pressure vessel is inevitable; therefore this procedure is applicable to only an industrial preparation of simple cyano compounds. Use of a polar, high-boiling solvent such as dimethylformamide is the other means of keeping the reaction temperature high and also dissolving the substrate. To prevent the formation of hydrogen cyanide polymers it is essential to utilize just one or preferably less than one molar equivalent of hydrogen cyanide and a minimal amount of a basic catalyst, irrespective of the presence or absence of a solvent.

Acetone Cyanohydrin-Base Catalyst. This Nazarov procedure is usually carried out in refluxing aqueous alcohol. A polar high-boiling solvent may be used at a high temperature for less reactive substrates. Acetone cyanohydrin, which is resistant to polymerization, can be used in a small or large excess, even as the reaction solvent. The amount and kind of basic catalyst are variable also.

Alkali Cyanide without Neutralization. An aqueous alcohol which can dissolve potassium or sodium cyanide is the most common solvent. Side reactions resulting from use of anhydrous methanol have been noted (p. 294). When the substrate is poorly soluble in an aqueous alcohol, tetrahydrofuran[242] or dimethylformamide[243] containing water in an amount sufficient to dissolve the reagent can be used. Use of calcium cyanide in dry N-methylpyrrolidone[58] is unique. Two molar equivalents of an alkali cyanide are generally used, although a large excess of the reagent is needed for less reactive substrates. The reaction temperature should be kept as low as possible in this powerful but side-reaction-prone procedure. Temperatures between room temperature and 0° for a long period or warming for a short time is preferable for reactive substrates to maximize unhydrolyzed cyano products. The reaction with potassium cyanide in refluxing ethylene glycol has been reported successful,[57] but this combination should be limited to formation of products stable to base.

Alkali Cyanide with Partial Neutralization. Of the many internal and external neutralization procedures we discuss only the two generally applicable ones.

[242] I. M. Clark, W. A. Denny, E. R. H. Jones, G. D. Meakins, A. Pendlebury, and J. T. Pinhey, *J. Chem. Soc., Perkin I*, **1972,** 2765.

[243] O. Kovacs, A. F. Aboulez, and B. Matkovics, *Acta Chem. Acad. Sci. Hung.*, **48,** 241 (1966) [*C.A.*, **65,** 18647d (1966)].

(*a*) *Alkali Cyanide-Acetic Acid.* The importance of the substrate:cyanide:acid ratio (1:2:1) has been noted (p. 279). As in the foregoing nonneutralization procedure, an aqueous alcohol is the common solvent although other polar solvents may be utilized. It has been reported that the use of tetrahydrofuran and water in a two-layer system was effective in preventing side reactions.[73] The reaction temperature can be varied from 0° to reflux according to the reactivity of the substrate.

(*b*) *Alkali Cyanide-Ammonium Chloride.* The standardized conditions [potassium or sodium cyanide (2 molar equivalents)-ammonium chloride (1.5 molar equivalents) in about 90% dimethylformamide at 100°][15] are designed for hydrocyanation of substrates having medium reactivities. The temperature should be lowered to room temperature for reactive substrates or raised to 200° (in a sealed tube) for less reactive substrates. Additional reagent mixture is necessary when the reaction is not completed after 10 hours at temperatures higher than 100°, since little cyanide anion remains at this stage.[59] The disadvantage of substituting another polar solvent such as tetrahydrofuran or methanol for the dimethylformamide has been noted (p. 281).

Hydrogen Cyanide-Alkylaluminum (*Method A*). This newer procedure has usually been carried out at ambient temperature in purified tetrahydrofuran with exclusion of moisture and air (rigorous exclusion is unnecessary; see p. 286). For reactive substrates the reaction can be carried out at a lower temperature. However, higher reaction temperatures should be avoided, since the formation of an alkylaluminum cyanide (Method B reagent) from the combination reagent ($HCN\text{-}AlR_3$) is accelerated with loss of Method A features. Tetrahydrofuran is the only solvent known to slow this formation of the alkylaluminum cyanide and it has been utilized exclusively. Dioxane or dimethoxymethane might be usable, but neither has been tested. Another advantage of tetrahydrofuran is that it dissolves a wide range of substrates. When a commercially available alkylaluminum solution in a hydrocarbon solvent is used, a large amount of tetrahydrofuran should be added. When a low-boiling solvent is needed for separation of volatile products, diethyl ether can replace tetrahydrofuran but only at a temperature lower than −10° with the hydrogen cyanide-triethylaluminum combination. The temperature might be raised a little when the triethylaluminum is replaced by diethylaluminum chloride or ethylaluminum dichloride (which reacts more slowly with hydrogen cyanide). A hydrogen cyanide solution should be added to a cold solution of a 1 or 2 molar excess of an alkylaluminum to prevent polymerization of hydrogen cyanide. According to the reactivity of the substrate, substrate to hydrogen cyanide to triethylaluminum molar ratios of 1:2:3, 1:3:5, and 1:5:7 have been used. The favorable effect of a small amount of water is noted (pp. 262 and 286). Since the concentration, not the ratio, of the combination reagent affects the hydrocyanation rate, care should be taken not to lower the concentration of the substrate by using a large amount of a diluted reagent solution. The usual concentrations of the substrate are 0.15–0.25 *M*.

Diethylaluminum Cyanide (*Method B*). This most powerful procedure can be carried out in various inert, dry solvents over a wide range of temperatures under nitrogen. The inert solvents include tetrahydrofuran, dioxane, diisopropyl or diethyl ether, dichloromethane, benzene, toluene, and hexane. The reactivity of the reagent increases with decreasing basicity of the solvent.[35] Except in tetrahydrofuran or its solvent mixture, Method-B hydrocyanation usually is complete in several minutes at ambient temperature, giving an equilibrium mixture of the 1,4 adduct and the 1,2 adduct (and in some cases the substrate). One could be misled by the presence of the 1,2 adduct or the substrate as an indication of incomplete reaction and thus prolong the reaction time; often deterioration of the substrate and the product(s) results. When the amount of the 1,2 adduct or the substrate is not decreased with time, one should repeat the reaction on the recovered substrate. When the temperature is raised to accelerate the reaction, the time should be shortened to minimize side reactions. A tetrahydrofuran-containing solvent is preferred in the higher temperature reactions. Generally, 3–5 molar equivalents of the reagent and an 0.1–0.2 *M* solution of the substrate are used.

Workup Procedure

A special workup procedure is unnecessary for the conventional methods. Neutralization is the only common, but often neglected, step before the usual workup.

The workup and separation procedures may be critical in the non-basic, newer methods using alkylaluminums. In the hydrocyanation of unsaturated ketones the usual products in the reaction solution are 1,4 adducts **E** and **Q** and di-adducts **Ka** and **Kb** in Method A (HCN-AlR_3), and 1,2 adduct **O** (or substrate **A** in some cases) and 1,4 adduct **Q** in Method B [$(C_2H_5)_2AlCN$] (see Mechanism, pp. 263, 264). Therefore a basic treatment is necessary to isolate the product as **E** and the substrate as **A**. With a careless basic treatment the product might suffer hydrolysis, epimerization, and other base-induced reactions. Decomposition of alkylaluminum compounds is exothermic and should

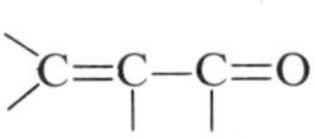

$>C{=}C{-}C{=}O$ (**A**) $-C(CN){-}CH{-}C{=}O$ (**E**) $-C(CN){-}CH{-}C(OH){-}CN$ (**Ka**) $-C(CN){-}CH{-}C(OAlR_2){-}CN$ (**Kb**)

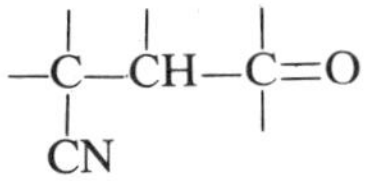

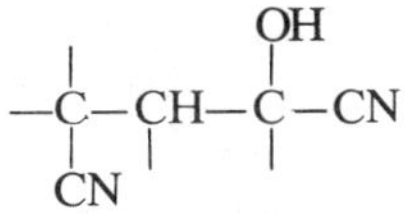

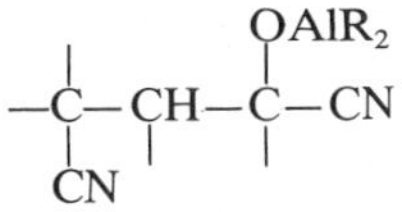

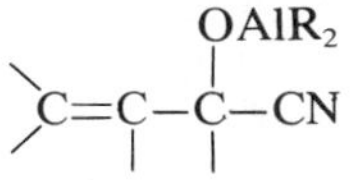

$>C{=}C{-}C(OAlR_2){-}CN$ (**O**) $-C(CN){-}C{=}C{-}OAlR_2$ (**Q**)

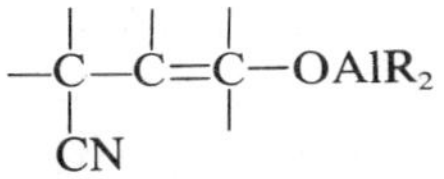

be carried out with good cooling. A recommended workup procedure for both Method A and Method B follows.[89] The reaction solution is poured slowly, with good stirring and cooling, into a tenfold volume of ice water containing hydrochloric acid (the acid is necessary to dissolve aluminum hydroxide) in an amount equimolar to the alkylaluminum. After it has been stirred for 20 minutes with cooling, the mixture is extracted three times with an appropriate solvent (mixture). The extracts are washed two or three times with cold 2 *N* aqueous sodium hydroxide, twice with cold water, and once with saturated aqueous sodium chloride, then dried, combined, and evaporated under reduced pressure. If the product is susceptible to aqueous sodium hydroxide, aqueous sodium carbonate may be substituted. Alternatively, the extracts are not washed with an alkaline solution, but passed through a short column packed with neutral alumina or silica gel. If the product is relatively stable to base and susceptible to acid, the hydrochloric acid should be replaced by sodium hydroxide (2 molar equivalents relative to alkylaluminum).

Special treatments are necessary for the products from α,β-unsaturated aldehydes,[92] imines,[92] carbonyl chlorides,[26] and carbonyl cyanides[26] (see the original papers).

When the product is a mixture of epimeric nitriles or contains the unchanged substrate, separation is usually difficult and often lowers the total yield of nitriles. Chromatography on basic or neutral alumina or silica gel often causes partial hydrolysis of nitriles. Better results would be obtained by using alumina or silica gel deactivated with 10% aqueous acetic acid (5–10% of the weight of the adsorbent).[58] Ketalization of cyano ketone products is sometimes effective for preventing hydrolysis and for the separation.

EXPERIMENTAL PROCEDURES

In this section, examples are chosen to illustrate experimental procedures of all conventional and newer hydrocyanation methods and to cover various types of substrates.

Succinonitrile [HCN-(KCN)/no solvent].[2] To 300 g (5.65 mol) of acrylonitrile containing 3 g (0.054 mol) of potassium cyanide was added at once 50 g (1.85 mol) of liquid hydrogen cyanide. When reaction did not start, the mixture was warmed to 30° for a short period. Soon an exothermic reaction began, and the reaction temperature was maintained at 55–60° with ice cooling. After the exothermic reaction subsided, 103 g (3.8 mol) of liquid hydrogen cyanide was added dropwise with cooling. After it had been heated at 60–70° for 2 hours, the resulting dark-brown reaction mixture was distilled under reduced pressure to give 420 g (93%) of succinonitrile as a colorless oil, bp 136° (10 mm), which solidified (mp 55°) on cooling.

Ethyl 4-Cyanocyclohex-1-ene-1-carboxylate [HCN-[$N(C_2H_5)_3$]/DMF, 1,6 addition].[244] A mixture of 15.4 g (102 mmol) of ethyl cyclohexa-1,3-dienyl-1-carboxylate, 2 g (74 mmol) of liquid hydrogen cyanide, 0.2 g (2 mmol) of triethylamine, and 50 ml of dimethylformamide was heated at 140° for 4 hours in a sealed vessel. After the solvent was removed by distillation, the residue was distilled under reduced pressure to give 10.1 g (56% or 76% based on substrate or HCN) of colorless oily product, bp 120–125° (4 mm).

3-Oxo-2,5-diphenylcyclohex-1-ene-1-carbonitrile [$(CH_3)_2C(OH)CN$-(KOH); Nazarov procedure, addition-elimination].[51a] A solution of 1 g (3.6 mmol) of 3-methoxy-2,5-diphenylcyclohex-2-en-1-one in 10 ml (110 mmol) of acetone cyanohydrin, containing 0.2 ml of saturated methanolic potassium hydroxide, was allowed to stand overnight and was then poured into 100 ml of water. The precipitate was recrystallized from benzene-petroleum ether to give 0.93 g (95%) of 3-oxo-2,5-diphenylcyclohex-1-ene-1-carbonitrile as prisms, mp 148–149°.

Ethyl 2,3-Dicyano-3,3-diphenylpropionate (KCN/aq C_2H_5OH;[245] see also an *Organic Syntheses* procedure[163]). Ethyl diphenylmethylenecyanoacetate (118.5 g, 0.428 mol) was dissolved in 180 ml of warm ethanol and treated with a solution of 58.5 g (0.9 mol) of potassium cyanide in 180 ml of water. The resulting clear yellow solution was refluxed for 15 minutes, cooled, and acidified with excess concentrated hydrochloric acid. The product separated, nearly quantitatively, as a viscous oil which solidified on cooling overnight. Recrystallization from aqueous ethanol gave 116.5 g (90%) of cyano product as white crystals, mp 89–91°.

[244] G. Inoue, F. Hukumi, and K. Sato, Jap. Pat. 7,111,648 (1971) [*C.A.*, **74**, 141092w (1971)].
[245] E. J. Cragoe, Jr., C. M. Robb, and J. M. Sprague, *J. Org. Chem.*, **15**, 381 (1950).

2-Ethyl-1-methyl-3-oxocyclohexane-1-carboxylic Acid (KCN/aq CH_3OH, then hydrolysis;[55] see also *Organic Syntheses* procedures[52,60,100]). To a solution of 47 g (0.34 mol) of 2-ethyl-3-methylcyclohex-2-en-1-one in 340 ml of methanol was added a solution of 47 g (0.72 mol) of potassium cyanide in 275 ml of water; the resulting mixture was refluxed for 3 hours. A solution of 42.5 g (0.77 mol) of potassium hydroxide in 640 ml of water was added. The mixture was heated at 100° for 32 hours in a flask fitted with an air condenser, then cooled, slowly acidified with hydrochloric acid, saturated with sodium chloride, and extracted with ether. Removal of the ether from the extract left a crystalline residue, which was filtered and washed with ether to give 40.6 g (65%) of 2-ethyl-1-methyl-3-oxocyclohexane-1-carboxylic acid, mp 137–138°, after recrystallization from ether-methanol.

3-Oxo-5α- and -5β-cholestane-5-carbonitriles [$Ca(CN)_2$/N-methylpyrrolidone].[58] A mixture of 1 g (2.6 mmol) of cholest-4-en-3-one, 1 g (11 mmol) of calcium cyanide, and 100 ml of purified N-methylpyrrolidone was stirred at 20° for 65 hours with exclusion of moisture. The reaction mixture was poured into water and extracted with ethyl acetate. The extract was washed with dilute hydrochloric acid and water, dried, and evaporated to give 1.2 g of a neutral product, which was chromatographed on 100 g of alumina deactivated with 10% acetic acid (5% of the weight of alumina). Elution with petroleum ether-benzene gave 0.425 g (40%) of 3-oxo-5α-cholestane-5-carbonitrile, mp 181–183° (from ethanol), $[\alpha]_D + 46°$ (chloroform). Elution with benzene-ether (49:1 and 19:1) gave 0.41 g (38%) of the 5β epimer, mp 126–128° (from ethanol), $[\alpha]_D + 26°$ (chloroform). A mixed fraction (19 mg) and 57 mg of a dimeric by-product, mp 196–197° [elution with benzene-ether (1:1)] were also obtained. Silica-gel chromatography would probably give better separation. Separation of the two epimers is also possible by repeated fractional crystallization carried out by seeding with crystals of the predominant epimer.[25] The seeds of the epimers can be obtained by preparative layer chromatography.

β-(4-Acetamidobenzoyl)-β-methylpropionitrile [KCN/aq CH_3OH on a masked enone (internal neutralization)].[53] To a stirred solution of 88.0 g (1.35 mol) of potassium cyanide in 1.76 l. of water was added a solution of 220 g (0.565 mol) of [2-(4-acetamidobenzoyl)propyl]trimethylammonium iodide (**66**, $R = CH_3$, $Ar = p\text{-}AcNHC_6H_4$, p. 277) in 440 ml of methanol and 2.2 l. of water. An oil separated and gradually solidified on stirring for 4 hours. The solid was collected and washed with water, giving 119 g (92%) of nitrile, mp 127–130°. Recrystallization from acetone-petroleum ether afforded a pure sample, mp 131–132°.

Diethyl (1-Cyano)propylmalonate (NaCN-AcOH/aq C_2H_5OH, Lapworth procedure;[72] see also an *Organic Syntheses* procedure).[71] To a solution of 59 g (1.2 mol) of sodium cyanide in 150 ml of water and 300 ml of ethanol was added 66 g (1.1 mol) of acetic acid below 10°. The resulting solution was added to 200 g (1 mol) of diethyl propylidenemalonate. After stirring for 1 hour, the reaction mixture was diluted with water, acidified, and extracted with ether. Distillation of the residue from the extract gave 141 g (62%) of diethyl (1-cyano)-propylmalonate, bp 130–140° (2.5 mm).

3,3-Ethylenedioxy-17α-hydroxy-11-oxo-18-nor-D-homoandrost-5-ene-13-carbonitrile ($KCN-NH_4Cl$/aq DMF).[80] To a solution of 1.452 g (4.22 mmol) of 3,3-ethylenedioxy-17α-hydroxy-18-nor-D-homoandrosta-5,12-dien-11-one (**78**, p. 282) in 73 ml of dimethylformamide was added a solution of 0.548 g (8.4 mmol) of potassium cyanide and 0.338 g (6.3 mmol) of ammonium chloride in 7.3 ml of water. The resulting mixture was heated at 100° for 7 hours, cooled, treated with 0.113 g (2.1 mmol) of ammonium chloride, and evaporated under reduced pressure to dryness. The residue was diluted with water and extracted with chloroform. The extract was washed with water, dried, and evaporated. The crystalline product was recrystallized from methanol to give 1.438 g (92%) of cyano product, mp 234–236°. Further recrystallization from methanol gave a pure sample as thick plates, mp 236.5–238.5°, and from acetone-ether as dimorphic crystals, mp 239–241°, $[\alpha]_D^{21}$ −5.0° (chloroform).

***dl*-2-Oxo-4b,8,8-trimethyl-1,10aα,2,3,4,4aβ,4bα,5,6,7,8,8aβ,9,-10-tetradecahydrophenanthrene-10a-carbonitrile** [HCN-$Al(C_2H_5)_3$/THF, Method A].[83] To an ice-cooled solution of 695 mg (6.1 mmol) of triethylaluminum in 6.9 ml of dry tetrahydrofuran were added successively a solution of 110 mg (4.1 mmol) of hydrogen cyanide in dry tetrahydrofuran and a solution of 500 mg (2.03 mmol) of *dl*-4b,8,8-trimethyl-2,3,4,4aβ,4bα,5,6,7,8,8aβ,9,10-dodecahydrophenanthren-2-one (**83**,* p. 287) in 5 ml of dry tetrahydrofuran. The reaction solution was allowed to stand overnight (27 hours) under nitrogen, carefully poured into a mixture of 2 *N* sodium hydroxide and ice with good stirring, and extracted with ether-chloroform (3:1). The extract was washed with ice water, cold 2 *N* hydrochloric acid, and ice water, dried, and evapo-

* The structure is inverted for convenience.

rated. Crystallization of the residue from ether gave 336 mg of cyano product (**92,*** p. 287), mp 140.5–141.5°. Chromatography of the residue from the mother liquor on neutral alumina (activity II; elution with 1:1 petroleum ether-benzene and benzene) afforded 21 mg of additional product, mp 142–144°. A pure sample melted at 144–144.5°. The total yield of the product was 72%.

3-Oxocyclohexane-1-carbonitrile [HCN-Al(C_2H_5)$_3$/ethyl ether, Method A].[43] To 6.03 g (62.3 mmol) of cyclohex-2-en-1-one in 50 ml of dry ethyl ether was added at −15° a reagent solution prepared by adding a solution of 4.87 g (180 mmol) of hydrogen cyanide in 56 ml of dry ethyl ether to a solution of 34.2 g (300 mmol) of triethylaluminum in 200 ml of dry ethyl ether with ice cooling. The resulting mixture was kept overnight (22 hours) at −15° under nitrogen, poured into a mixture of 2 *N* hydrochloric acid and ice, and extracted with dichloromethane. The extract was washed with cold 8.5% aqueous sodium carbonate and water, dried, and concentrated at atmospheric pressure to give 7.9 g of an oil, which was distilled to afford 6.16 g (80%) of 3-oxocyclohexane-1-carbonitrile, bp 126–127° (7 mm). The semicarbazone derivative melted at 169–172°.

17β-Benzoyloxy-3β-formyl-5α-androstane-2β-carbonitrile [HCN-Al(C_2H_5)$_3$—H_2O/THF;[92] see also an *Organic Syntheses* procedure:[89] Method A on α,β-ethylenic imine, then hydrolysis]. To a solution of 366 mg (0.793 mmol) of 3-N-*t*-butyliminomethyl-5α-androst-2-en-17β-ol 17-benzoate and 14 mg (0.79 mmol) of water in 4 ml of dry tetrahydrofuran was added a reagent solution prepared by adding 0.34 ml (2.4 mmol) of a 7 *M* tetrahydrofuran solution of hydrogen cyanide to 1.0 ml (4 mmol) of a 4 *M* tetrahydrofuran solution of triethylaluminum at 0°. The resulting solution was kept at room temperature for 2 hours under a nitrogen atmosphere, poured into a mixture of 2 *N* hydrochloric acid and ice, and extracted with dichloromethane. A mixture of the iminomethyl nitrile thus obtained, 8 ml each of 5% aqueous oxalic acid, ethanol, and tetrahydrofuran was refluxed for 1 hour, poured into a mixture of 2 *N* sodium bicarbonate and ice, and extracted with dichloromethane to give 370 mg of a crystalline product. Recrystallization from dichloromethane-ether afforded 222 mg (64%) of 17β-benzoyloxy-3β-formyl-5α-androstane-2β-carbonitrile, mp 231.5–233°. An analytical sample had mp 248–252° and $[\alpha]_D^{23}$ +54° (chloroform).

***dl*-7-Methoxy-1-oxo-1,2,3,4,4aα,9,10,10aβ-octahydrophenanthrene-4-carbonitrile** [HCN-(C_2H_5)$_2$AlCl/THF, Method A].[85] To a solution

* The structure is inverted for convenience.

of 43.7 g (0.192 mol) of 7-methoxy-1,2,3,4,9,10-hexahydrophenanthren-1-one (**85**,* p. 288) in 280 ml of dry tetrahydrofuran was added slowly a reagent solution prepared by addition of a solution of 36.2 ml (0.958 mol) of hydrogen cyanide in 250 ml of dry tetrahydrofuran to a solution of 161.5 g (1.34 mol) of diethylaluminum chloride in 376 ml of dry tetrahydrofuran with ice cooling under a nitrogen atmosphere. The resulting solution was allowed to stand at room temperature for 45 hours in a flask fitted with a mercury bubbler, then gradually poured, with vigorous stirring and ice cooling, into 10.5 kg of ice water containing 148 g (3.7 mol) of sodium hydroxide, and extracted with ether-chloroform (3:1). The extract was washed with water, dried, and evaporated to give 53 g of a residue, which yielded 23.9 g of a mixture of *cis*- and *trans*-cyano products (**98c** and **98t**,* p. 288) on crystallization from acetone. The mother liquor was treated with semicarbazide hydrochloride and sodium acetate in the usual way to give 21 g of a crystalline mixture of semicarbazones of the *cis* and *trans* products, which was refluxed with 790 ml of 3 *N* hydrochloric acid in 206 ml of benzene with vigorous stirring for 2 hours. Extraction with chloroform and recrystallization of a crystalline residue gave an additional crop of mixed cyano product. The combined crystalline mixture of *cis*- and *trans*-cyano ketones (*ca.* 1:1.5, 36.3 g, 75%) was recrystallized from acetone to give 9.3 g of *trans* product, mp 151.5–154.5°. Additional crops (24.2 g, mp 149–155°) of *trans* product were obtained by repeated epimerization of the mother liquor with 1 *N* hydrochloric acid in refluxing acetone for 20 minutes followed by crystallization from acetone. The total yield was 69%.

6β-Formyl-3β-hydroxy-B-nor-5α-androstane-5-carbonitrile (HCN-$C_2H_5AlCl_2$/THF, Method A, on α,β-ethylenic imine, then hydrolysis).[92] To 0.50 g (1.35 mmol) of 6-N-cyclohexyliminomethyl-B-norandrost-5-en-3β-ol (**99**, p. 288) placed in an ampoule was added a reagent solution prepared by addition of 0.97 ml (6.8 mmol) of a 7 *M* tetrahydrofuran solution of hydrogen cyanide to a solution of 1.21 g (9.5 mmol) of ethylaluminum dichloride in 4.5 ml of dry tetrahydrofuran with ice cooling. The ampoule was sealed and kept at room temperature for 20 hours. The reaction mixture was poured into a mixture of 2 *N* hydrochloric acid and ice and extracted with dichloromethane. The extract was washed with water, dried, and evaporated to give 0.58 g of imino nitrile. A mixture of the product, 8 ml each of 5% aqueous oxalic acid, tetrahydrofuran, and ethanol was refluxed for 1 hour, poured into ice water, neutralized with sodium bicarbonate, and extracted with dichloromethane to give 433 mg of product.

* The structure is inverted for convenience.

Chromatography on acid alumina (activity II: elution with benzene-dichloromethane and dichloromethane) followed by crystallization from ether-pentane and preparative layer chromatography of the mother liquor afforded 233 mg (55%) of 5α-cyano-6β-formyl product (**101,** p. 288), mp 141.5–143.5°, $[\alpha]_D^{23}$ −71° (chloroform).

3,3:17,17-Bisethylenedioxy-11-oxoandrost-5-ene-8-carbonitrile [$(C_2H_5)_2AlCN/C_6H_6—C_6H_5CH_3$, Method B].[25] To a solution of 2.000 g (5.175 mmol) of 3,3:17,17-bisethylenedioxyandrosta-5,8-dien-11-one (**102,** p. 289) in 32 ml of benzene and 19 ml of toluene was added 26 ml (31 mmol) of a 1.2 *M* benzene solution of diethylaluminum cyanide with ice cooling under nitrogen. After it had been kept at 0° for 10 minutes, the reaction solution was poured into 500 ml of ice water containing 20 g (0.5 mol) of sodium hydroxide and extracted with chloroform in the usual way (washing with water). Crystallization of the residue from methanol gave 1.567 g (73.3%) of cyano product (**103,** p. 289), mp 198–200°, $[\alpha]^{24}$ +61° (99:1 chloroform-ether). The residue from the mother liquor was hydrocyanated again with 4 molar equivalents of diethylaluminum cyanide in the same way as described above. Crystallization of the second product and neutral-alumina (activity II) chromatography of the residue from the mother liquor afforded an additional 0.391 g (18.3%) of the cyano product.

3,17-Dioxo-6β-vinyl-B-nor-5β-androstane-5-carbonitrile [$C_2H_5)_2$-$AlCN/CH_2Cl_2—C_6H_6$, Method B].[195] To a solution of 600 mg (2.01 mmol) of 6β-vinyl-B-norandrost-4-ene-3,17-dione (**106c,** p. 290) in 6 ml of dry dichloromethane was added 6 ml (7.6 mmol) of a 1.26 *M* benzene solution of diethylaluminum cyanide at room temperature under nitrogen. The resulting solution was kept for 2 hours, poured into a mixture of 2 *N* sodium hydroxide and ice, and extracted with ether-dichloromethane (3:1) in the usual way (washing with water). The residue (688 mg) was crystallized from dichloromethane-ether to give 184 mg of cyano product (**107c,** p. 290), mp 185–189°. The residue from the mother liquor was hydrocyanated again with 3.8 mmol of diethylaluminum cyanide as described above. Alumina chromatography (neutral activity II, elution with benzene and 9:1 benzene-chloroform) followed by crystallization afforded an additional 217 mg of the product, mp 189–191°. The total yield was 61%. An analytical sample had mp 189.5–191° and $[\alpha]_D^{22}$ +144° (chloroform).

17β-Hydroxy-3-oxoandrost-4-ene-7α-carbonitrile [$(C_2H_5)_2AlCN$/THF-toluene, Method B, 1,6 addition].[43] To a solution of 1.089 g (3.81 mmol) of 17β-hydroxyandrosta-4,6-dien-3-one (**115,**

p. 292) in 20 ml of dry tetrahydrofuran was added 12 ml (19 mmol) of a 1.63 *M* toluene solution of diethylaluminum cyanide. After it had been kept at room temperature under a nitrogen atmosphere for 45 minutes, the reaction solution was poured into a mixture of 2 *N* sodium hydroxide and ice and extracted with dichloromethane in the usual way (washing with water). Recrystallization of a crystalline residue from methanol gave 0.881 g of cyano product (**116,** p. 292), mp 289–292°. The residue (0.31 g) from the mother liquor was chromatographed on neutral alumina (activity II). Elution with benzene-dichloromethane (4:1 and 2:1) afforded an additional 41 mg of the cyano product. Fractions (0.24 g) eluted with benzene-dichloromethane (9:1) were treated with 4 mmol of diethylaluminum cyanide in the way described above. Recrystallization of the second product gave 0.149 g of the cyano product, mp 282–286°. Repeated hydrocyanation of the residue (85 mg) from the mother liquor afforded an additional 31 mg of the product, mp 279–284°. The total yield of the 7α-cyano product was 92%.

3β-Acetoxy-16α-cyano-5α-androstane-17β-carboxyic Acid [$(C_2H_5)_2AlCN$/THF-$(i\text{-}C_3H_7)_2O$, Method B, on acid chloride followed by hydrolysis].[26] A solution of 1.00 g (2.77 mmol) of 3β-acetoxy-5α-androst-16-ene-17-carboxylic acid in 10 ml of dry benzene was treated with 13 ml of thionyl chloride for 3.5 hours at room temperature. The resulting solution was evaporated to dryness. The crude acid chloride (**119a,** p. 292) was dissolved in 15 ml of dry tetrahydrofuran and mixed with 7.5 ml (13.7 mmol) of a 1.83 *M* solution of diethylaluminum cyanide in isopropyl ether. The reaction solution was kept at room temperature for 17 hours under nitrogen, poured into a mixture of 2 *N* hydrochloric acid and ice, and extracted with methanol-free dichloromethane. A solution of the residue (1.21 g) in 40 ml of 90% aqueous pyridine was kept at room temperature overnight to complete hydrolysis of the acyl chloride, concentrated to *ca.* 10 ml, and poured into a mixture of 2 *N* hydrochloric acid and ice. The crystals which formed were filtered, washed with water, dried, and recrystallized from dichloromethane-methanol to give 0.628 g of cyano product (**120a,** p. 292), mp 259–262°. Since the mother liquor contained a small amount of 3-hydroxy-16α-cyano-5α-androstane-17β-carboxylic acid, the residue was treated with 6 ml of pyridine and 3 ml of acetic anhydride overnight. The acetylation solution was treated with 1 ml of water at 0°, poured into ice water, and extracted with dichloromethane. A solution of the residue in 5 ml of 90% pyridine was heated at 100° for 1 hour to effect hydrolysis of the mixed anhydride, and concentrated.

The residue was chromatographed on acid alumina (activity II). Elution with dichloromethane-methanol-ethyl acetate (1:1:1) containing acetic acid (0.2–2%) followed by recrystallization from methanol afforded an additional crop (0.141 g) of the cyano product, mp 256–259°; total yield, 72%. An analytical sample had mp 260–262° and $[\alpha]_D^{24}$ + 1.1° (chloroform).

17β-Benzoyloxy-2α-formyl-5α-androstane-3α-carbonitrile [$(C_2H_5)_2$-AlCN-(i-C_3H_7OH)/THF-C_6H_6, Method B, in the presence of a proton source].[92] To a solution of 0.488 g (1.0 mmol) of 2-cyclohexyliminomethyl-5α-androst-2-en-17β-ol 17-benzoate (**112,** p. 291) and 60 mg (1.0 mmol) of isopropyl alcohol in 5.7 ml of dry tetrahydrofuran was added 4.1 ml (5 mmol) of a 1.22 *M* benzene solution of diethylaluminum cyanide under nitrogen. After 30 minutes at 25° the reaction solution was poured into a mixture of 2 *N* hydrochloric acid and ice and extracted with ether-dichloromethane (3:1). A mixture of the residue (0.5 g), 8 ml each of 5% aqueous oxalic acid, tetrahydrofuran, and ethanol was refluxed for 40 minutes, poured into ice water, and neutralized with sodium bicarbonate in the presence of dichloromethane with stirring. The organic layer was separated and the aqueous layer was extracted with dichloromethane. The organic layer and the extract were combined, washed with water, dried, and evaporated to yield 0.43 g of crude product, which was crystallized from dichloromethane-acetone to give 0.253 g of the 2α-formyl-3α-cyano product (**114,** p. 291), mp 236–239°. Chromatography of the residue from the mother liquor on acid alumina (activity II: elution with benzene and benzene-dichloromethane) afforded an additional 45 mg of product, mp 236–238°; total yield, 69%. An analytical sample had mp 237–240° and $[\alpha]_D^{23}$ +101° (chloroform).

TABULAR SURVEY

Although an attempt has been made to include in Tables I–IX all conjugate hydrocyanations of conjugated carbonyl compounds reported through the summer of 1975, lack of a systematic method of searching the literature for the reaction makes it likely that some examples were overlooked.

Tables I and II list conjugate hydrocyanation of acyclic and cyclic α,β-ethylenic monocarboxylic acid derivatives including nitriles. Tables III and IV cover the reaction of α,β-ethylenic α,α- and α,β-dicarbonyl derivatives (acyclic and cyclic), respectively, whose carbonyl functions are carboxylic acid, ester, nitrile, and ketone. Tables V and

VI refer to the reaction of acyclic and cyclic α,β-ethylenic monoketones, respectively. [Cyclic substrates (Tables II and VI) are molecules in which the double bond or the carbonyl function or both constitute a part of a ring.] Table VII includes the 1,6 or 1,8 addition of di- or tri-enic carbonyl compounds (acyclic and cyclic, mono- and di-carbonyl). Hydrocyanation of α,β-acetylenic carbonyl substrates are listed in Table VIII. Table IX covers the reaction of α,β-ethylenic aldehydes and imines.

Within each table the substrates are arranged in order of increasing number of carbon atoms. Carbon atom(s) of *R* or *COR* in the following functional groups are not counted: —CO_2R, —O*COR*, >N*COR*, —O*R*, and >N—*R* (—$\overset{+}{N}R_3$). However, the carbon atoms which constitute a ring containing the conjugated carbonyl system are counted: *e.g.*, C_4 for compound **328.** Compounds with the same

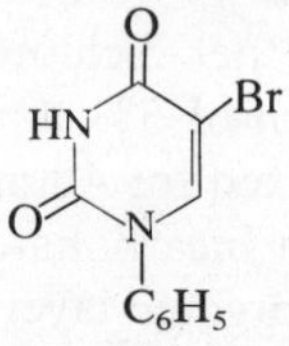

328

number of carbon atoms are listed in order of increasing complexity of the substrates (*e.g.*, aliphatic before aromatic, five-membered before six-membered cyclic, monocyclic before bicyclic, hydrocarbon before heteroatom-containing compound). Enantiomers are not differentiated. Unless otherwise indicated, the ring junctions BC and CD of steroids are *trans.* Thus, 8β-, 9α-, and 14α-hydrogens are not shown in steroidal formulas.

When more than one hydrocyanation reagent has been used for the same substrate, the reagents are arranged in the order: hydrogen cyanide-(base), acetone cyanohydrin-(base) (the parentheses indicate a catalyst used in a small amount), alkali cyanide (sodium cyanide, potassium cyanide, calcium cyanide), potassium or sodium cyanide-acid, potassium or sodium cyanide-ammonium chloride, hydrogen cyanide-alkylaluminum, diethylaluminum cyanide. Also shown in parentheses is a small amount of water or an alcohol used with the organoaluminum reagent to accelerate the conjugate hydrocyanation. In addition to the usual chemical symbols, the following abbreviations are used: THF, tetrahydrofuran; DMF, dimethylformamide; AcO, CH_3CO_2; pip, piperidine; and pyr, pyridine.

The reaction time is given only for the diethylaluminum cyanide method to indicate whether the condition is kinetic (within 10 minutes) or thermodynamic (more than a few hours, but kinetic in THF).

When *trans*- and *cis*-fused cyano compounds are formed in the angular cyanation of a polycyclic substrate, the *trans* nitrile is placed before the *cis* isomer. Hydrolysis products, if formed, are listed after nitriles.

A dash (—) in the reaction conditions or yield column means that no information is given in the literature. When there is more than one reference, the experimental data are taken from the first one, and the remaining references are listed in numerical order.

TABLE I. Conjugate Hydrocyanation of Acyclic α,β-Ethylenic Carboxylic Acid Derivatives

	Substrate	Reagent(s)	Reaction Conditions	Product(s) and Yield(s) (%)	Refs.
C_3	$CH_2{=}CHCO_2R$ (I)			$NCCH_2CH_2CO_2R$ (II)	
	I, $R = CH_3$	HCN-(KCN)	No solvent, 70–80°	II, $R = CH_3$ (73)	2
	I, $R = C_2H_5$	HCN-(KCN)	No solvent, 70–80°	II, $R = C_2H_5$ (80)	2
		HCN-[$K_4Ni(CN)_4$]	" , 150°	II, $R = C_2H_5$ (79)	48
		HCN-[$Na_4Ni(CN)_4$]	" "	II, $R = C_2H_5$ (90)	48
	I, $R = n\text{-}C_4H_9$	HCN-(KCN)	No solvent, 70–80°	$R = n\text{-}C_4H_9$, I (41), II (45)	2
	$CH_2{=}CHCN$	HCN-[$N(C_2H_5)_3$]	DMF, 80–90°	$NCCH_2CH_2CN$ (I, 91)	162
		HCN-(KCN)	No solvent, 60–70°	I (93)	2, 246
		HCN-pyr	" "	I (86)	2
		HCN-[$K_4Ni(CN)_4$]	" , 100°	I (93)	48
		HCN-[$CaK_2Ni(CN)_4$]	" , 150°	I (25)	48
		NaCN-AcOH	H_2O, 70°	I (—), succinimide (—)	247
		1. $NaCN\text{-}H_2SO_4$ or H_3PO_4 2. Hydrolysis	H_2O, 50°	$HO_2CCH_2CH_2CO_2H$ (88–90)	247
	$CH_2{=}C(Cl)CH_2Cl$ [via $CH_2{=}C(CN)CH_2CN$?]	KCN	Aq C_2H_5OH, reflux	$HO_2CCH_2CH(CO_2H)CH_2CO_2H$ (—)	5
C_4	$CH_3CH{=}CHCO_2R$ (I)			$CH_3CH(CN)CH_2CO_2R$ (II), $CH_3CH(CO_2H)CH_2CO_2H$ (III)	

	I, $R = Na$	1. KCN 2. NaOH	H_2O, reflux Hydrolysis	III (9)	115
	I, $R = CH_3$	HCN-(KCN)	No solvent, 50–60°	$R = CH_3$, I (Major), II (2)	2
		KCN	Dry CH_3OH, reflux	I, $R = CH_3$ (22), 2 dimers (17 wt %)	11
	I, $R = C_2H_5$	1. KCN 2. Aq $Ba(OH)_2$	Aq C_2H_5OH, reflux Hydrolysis	III (66–70)	60, 115
	I, $R = CH{=}CH_2$	HCN-(KCN)	No solvent, 60–70°	$R = CH(CN)CH_3$, I (7), II (15)	2
	I, $R = n\text{-}C_4H_9$	HCN-(KCN)	No solvent, 150°	$R = n\text{-}C_4H_9$, I (10), II (76)	116
	$CH_2{=}C(CH_3)CO_2CH_3$ (I)	HCN-(KCN)	No solvent, 35–40°	I (90), $NCCH_2CH(CH_3)CO_2CH_3$ (II, 5)	2
		HCN-[$Na_4Ni(CN)_4$]	,, , 150°	II (22)	48
		HCN-[$CaK_2Ni(CN)_4$]	,, ,,	II (14)	48
	$CH_2{=}CHCH_2I$ (via $CH_3CH{=}CHCN$)	1. KCN 2. KOH	C_2H_5OH, high temp Hydrolysis	$CH_3CH(CO_2H)CHCO_2H$ (—)	6
	$CH_3CH{=}CH_2CN$ (I) (*cis* and *trans* mixture)	HCN-(KCN)	No solvent, 110°	$NCCH_2CH(CH_3)CN$ (II, 5)	2
	$CH_2{=}CHCH_2CN$ (via I)	HCN-(aq KCN)	No solvent, 110°	II (26)	2
		KCN	Dry CH_3OH, reflux	No reaction	11
	$CH_2{=}C(CH_3)CN$ (I)	HCN-(KCN)	No solvent, 40°	I (—), $NCCH_2CH(CH_3)CN$ (1.8)	2
	$RCH_2CH{=}CHCN$ (I)				2
	I, $R = OCH_3$	HCN-(KCN)	No solvent, 65°	I, $R = OCH_3$ (60), $CH_3OCH_2CH(CN)CH_2CN$ (28)	
	I, $R = OCOCH_3$	HCN-(KCN)	No solvent, 35–40°	No reaction	
	I, $R = N(CH_3)_2$	HCN	No solvent, 30–35°	No reaction	
C_5	$(CH_3)_2C{=}CHCO_2C_2H_5$	1. KCN 2. NaOH	Aq C_2H_5OH, reflux Hydrolysis	$(CH_3)_2C(CO_2H)CH_2CO_2H$ (30)	115
	$NCCH{=}CHCH_2CN$	HCN-(KCN)	No solvent, 30–35°	Decomposed product	2

TABLE I. Conjugate Hydrocyanation of Acyclic α,β-Ethylenic Carboxylic Acid Derivatives (*Continued*)

	Substrate	Reagent(s)	Reaction Conditions	Product(s) and Yield(s) (%)	Refs.
C_5	(*Contd.*)				
	$CH_2{=}C(CO_2CH_3)CH_2CO_2CH_3$	KCN	Dry CH_3OH, room temp	O, HN, O (ring), $-CH_2CO_2CH_3$ (I, 27–38)	11
				$CH_3OCH_2CH(CO_2CH_3)CH_2CO_2CH_3$ (II, 37–53)	
			Dry CH_3OH, reflux	I (41)	
	$CH_3O_2CCH_2CH{=}CHCO_2CH_3$ (I)	HCN-(KCN)	No solvent, 40°	I (72), $CH_3O_2CCH_2CH(CN)CH_2CO_2CH_3$ (9)	2
	(ring) $NCH_2CH{=}CHCH_2CN$ (I) (via (ring) $N(CH_2)_2CH{=}CHCN$)	HCN	No solvent, 80°	I (15), (ring) $N(CH_2)_2CH(CN)CH_2CN$ (51)	2
C_6	$NC(CH_2)_2CH{=}CHCN$ (I)	HCN-(KCN-pyr)	No solvent, 35°	I (18), $NC(CH_2)_2CH(CN)CH_2CN$ (53)	248
C_9	n-$C_6H_{13}CH{=}CHCO_2C_2H_5$	1. KCN 2. NaOH	Aq C_2H_5OH, reflux Hydrolysis	n-$C_6H_{13}CH(CONH_2)CH_2CO_2H$ (80)	115
	$C_6H_5CH{=}CHCO_2C_2H_5$	HCN-(KCN)	No solvent, 35–40°	No reaction	2
		1. KCN or NaCN 2. HCl or $Ba(OH)_2$	Aq C_2H_5OH, reflux Hydrolysis	$C_6H_5CH(CO_2H)CH_2CO_2H$ (18–20)	115, 249
	$C_6H_5CH{=}CHCN$ *cis* and *trans* mixture	HCN-(KCN)	No solvent, 35–40°	No reaction	2
	trans	NaCN	—, —	No reaction	126
C_{15}	$ArCH{=}C(C_6H_5)CN$ (I)			$ArCH(CN)CH(C_6H_5)CN$ (II)	

I, $Ar = C_6H_5$	HCN	No solvent, cold	No reaction	12
	HCN	C_2H_5OH, 100°	II, $Ar = C_6H_5$ (45)	12
	HCN-[$N(C_2H_5)_3$]	,, , ,,	II, $Ar = C_6H_5$ (64)	12
	HCN-(KCN)	No solvent, 40–45°	I, $Ar = C_6H_5$ (38), II (58)	2
	KCN	CH_3OH—$(C_2H_5)_2O$—H_2O, reflux	II, $Ar = C_6H_5$ (92–94)	163, 118, 250
	KCN-HCN (2:1)	Aq C_2H_5OH, 100°	II, $Ar = C_6H_5$ (57)	8
	KCN-NH_4Cl	,, , reflux	II, $Ar = C_6H_5$ (—)	251
I, $Ar = p\text{-}CH_3OC_6H_4$	KCN-NH_4Cl	Aq C_2H_5OH, reflux	II, $Ar = p\text{-}CH_3OC_6H_4$ (—)	165
I, $Ar = o\text{-}O_2NC_6H_4$	HCN-(pip)	C_2H_5OH, room temp	No reaction	165
	KCN			
	KCN	Aq C_2H_5OH, warm	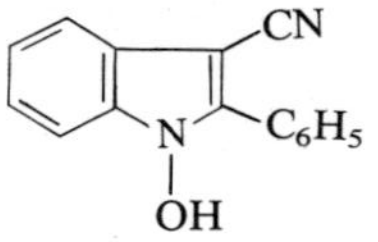 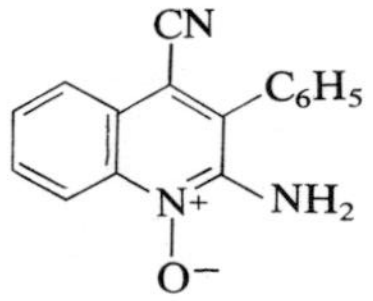 1-hydroxy-2-phenylindole-3-carbonitrile (III, 24), 2-amino-3-phenylquinoline-4-carbonitrile 1-oxide (IV, 46)	166
	KCN-NH_4Cl-(pip)	Aq C_2H_5OH, reflux	III (30), IV (43)	166, 165
I, $Ar = p\text{-}O_2NC_6H_4$	HCN-(pip)	C_2H_5OH, room temp	No reaction	165
I, $Ar = m\text{-}O_2NC_6H_4$	KCN-NH_4Cl-(pip)	Aq C_2H_5OH, reflux	No reaction	165
	HCN-(pip)	C_2H_5OH, 100°	II, $Ar = m\text{-}O_2NC_6H_4$ (40)	165
	KCN-NH_4Cl-(pip)	Aq C_2H_5OH, reflux	II, $Ar = m\text{-}O_2NC_6H_4$ (91)	165

TABLE I. CONJUGATE HYDROCYANATION OF ACYCLIC α,β-ETHYLENIC CARBOXYLIC ACID DERIVATIVES (*Continued*)

Substrate	Reagent(s)	Reaction Conditions	Product(s) and Yield(s) (%)	Refs.
C_{15} (*Contd.*)				
o-$O_2NC_6H_4C(CN){=}CHC_6H_5$ (I)			3-cyano-2-phenyl-1-hydroxyindole (II), 4-cyano-3-phenyl-2-aminoquinoline 1-oxide (III)	166
	KCN	Aq C_2H_5OH, 18°	I (81), II (15)	
	,,	,, , warm	II (60), III (20)	

Note: References 246–305 are on pp. 475–476.

TABLE II. CONJUGATE HYDROCYANATION OF CYCLIC α,β-ETHYLENIC CARBOXYLIC ACID DERIVATIVES

	Substrate	Reagent(s)	Reaction Conditions	Product(s) and Yield(s) (%)	Refs.
C_4	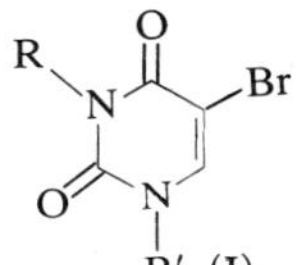 1-R'-3-R-5-bromouracil (I)	NaCN		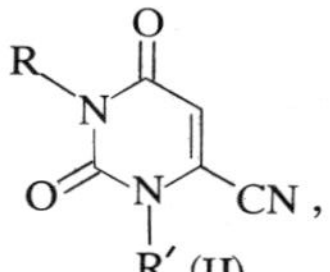 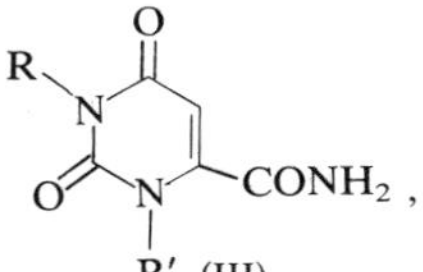 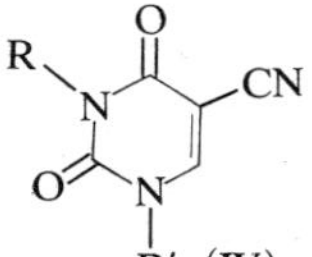 1-R'-3-R-6-cyanouracil (II), 1-R'-3-R-uracil-6-carboxamide ($CONH_2$) (III), 1-R'-3-R-5-cyanouracil (IV)	167
	I, R = H, R' = CH_3		Aq C_2H_5OH, reflux	II, R = H, R' = CH_3 (93)	
			Aq DMF, room temp	II, R = H, R' = CH_3 (53)	
	I, R = R' = CH_3		Aq C_2H_5OH, reflux	III, R = R' = CH_3 (86)	
			Aq DMF, room temp	II, R = R' = CH_3 (95)	
			DMF, 80–90°	IV, R = R' = CH_3 (92)	
	I, R = H, R' = C_6H_5		Aq DMF, room temp	II, R = H, R' = C_6H_5 (69)	
	I, R = CH_3, R' = cyclohexyl		Aq C_2H_5OH, reflux	III, R = CH_3, R' = cyclohexyl (93)	
			Aq DMF, room temp	II, R = CH_3, R' = cyclohexyl (94)	
	I, R = CH_3, R' = C_6H_5		Aq C_2H_5OH, reflux	III, R = CH_3, R' = C_6H_5 (90)	
			Aq DMF, room temp	II, R = CH_3, R' = C_6H_5 (86)	

TABLE II. Conjugate Hydrocyanation of Cyclic α,β-Ethylenic Carboxylic Acid Derivatives (*Continued*)

Substrate	Reagent(s)	Reaction Conditions	Product(s) and Yield(s) (%)	Refs.
C_4 (*Contd.*) 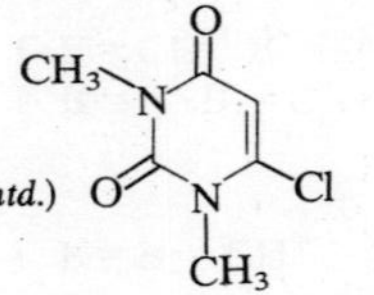 [structure]			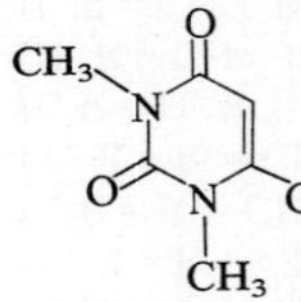 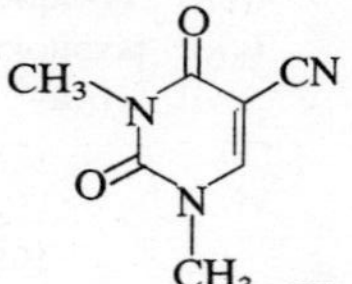 CH_3, O, N, CN, CH_3 (I); CH_3, O, N, CN, CH_3 (II)	
	NaCN	Dry DMF, room temp	I (88)	167
	,,	DMSO, 90–120°	II (73)	168
C_5 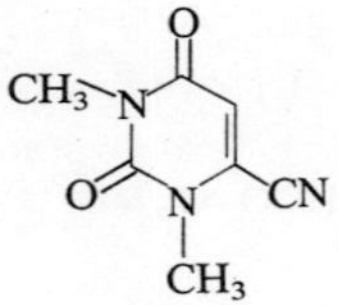 [structure]	NaCN	DMF, 80–85°	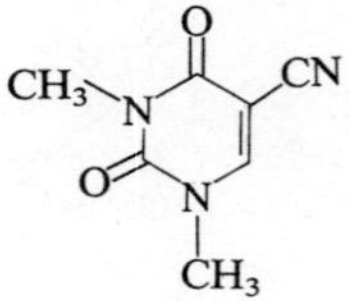 CH_3, O, N, CN, CH_3 (68)	167
C_9 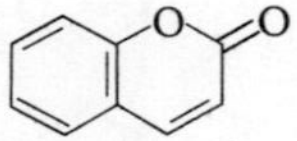 [structure]	1. KCN 2. KOH	Aq C_2H_5OH, reflux Hydrolysis	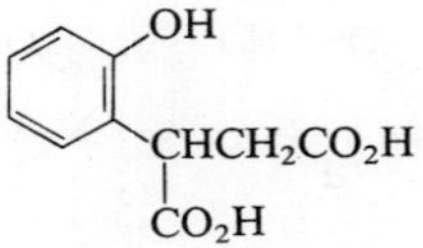 OH, $CHCH_2CO_2H$, CO_2H (—)	7
C_{15} 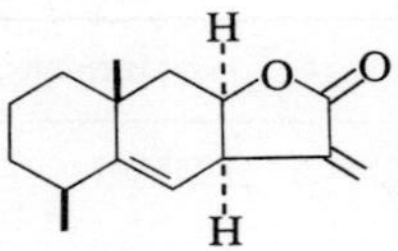 [structure]	KCN	Aq C_2H_5OH, reflux	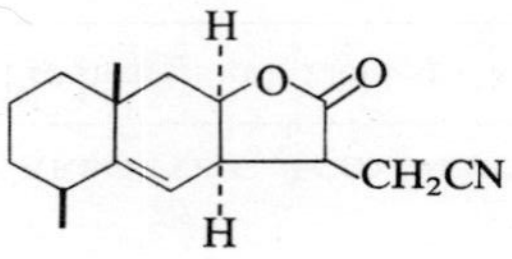 H, O, CH_2CN, H (—)	7

	Substrate	Reagent	Conditions	Product(s) and Yield(s) (%)	Refs.
C_{20}	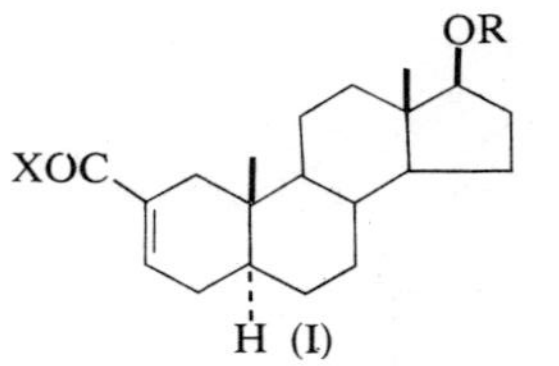 (I)			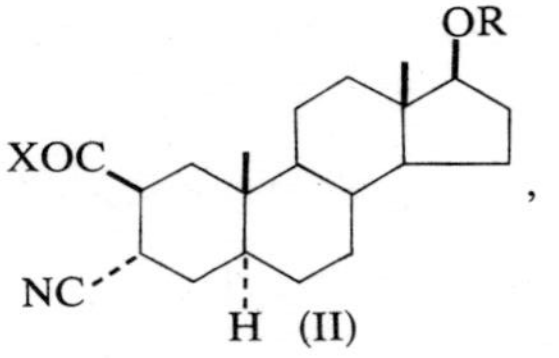 (II), 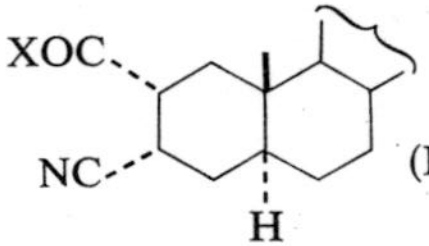 (III)	26
	I, R = H, X = OCH_3	$(C_2H_5)_2AlCN$	Toluene, 90°, 10 min[a]	II, R = H, X = OCH_3 (50)	
	I, R = Ac, X = SC_2H_5	$(C_2H_5)_2AlCN$	C_6H_6, room temp, 3 hr	R = Ac, X = SC_2H_5, II (55), III (6)	
	I, R = Ac, X = Cl	1. $(C_2H_5)_2AlCN$ 2. 2 *N* NaOH	C_6H_6, room temp, 3 hr Hydrolysis	R = Ac, X = OH, II (0.1), III (19); R = Ac, X = C_2H_5, II (2), III (2)	
		1. $(C_2H_5)_2AlCN$ 2. 90% pyr	$(i\text{-}C_3H_7)_2O—C_6H_6$, room temp, 3.5 hr Hydrolysis	III, R = Ac, X = OH (73)	
	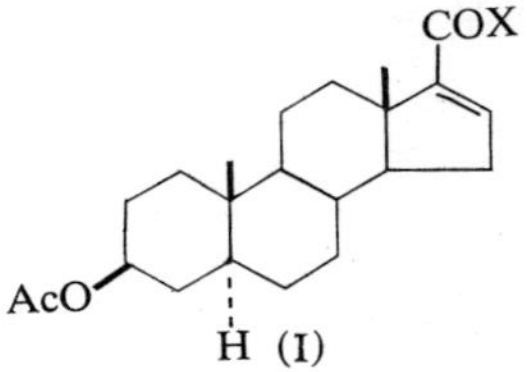 (I)			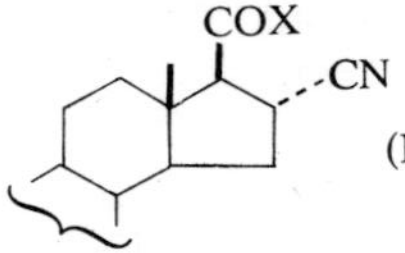 (II), 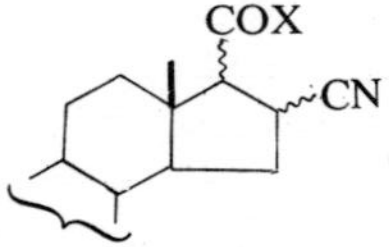 (III)	26

TABLE II. Conjugate Hydrocyanation of Cyclic α,β-Ethylenic Carboxylic Acid Derivatives (*Continued*)

Substrate	Reagent(s)	Reaction Conditions	Product(s) and Yield(s) (%)	Refs.
C_{20} (*Contd.*)				
I, $X=SC_2H_5$	$(C_2H_5)_2AlCN$	C_6H_6, room temp, 1.8 hr	$X=SC_2H_5$, II (87), III (2)	
I, $X=SC_6H_5$	$(C_2H_5)_2AlCN$	C_6H_6, room temp, 2 hr	$X=SC_6H_5$, II (85), III (3)	
I, $X=Cl$	1. $(C_2H_5)_2AlCN$ 2. 90% pyr	$THF-(i-C_3H_7)_2O$, room temp Hydrolysis	II, $X=OH$ (72)	
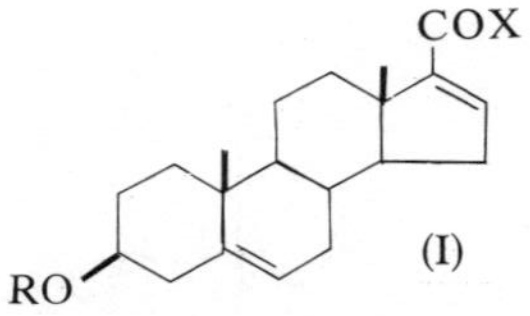 (I) [structure: RO, COX]			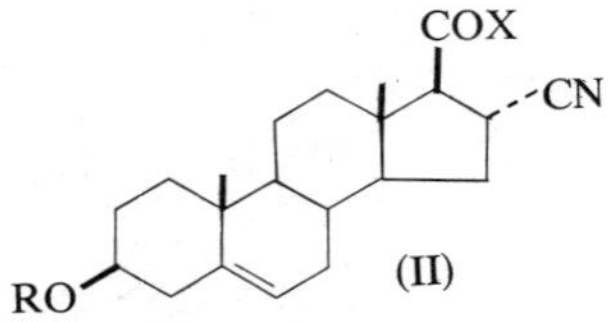 (II) [structure: RO, COX, CN]	26
I, $R=H$, $X=OCH_3$	$(C_2H_5)_2AlCN$	Toluene, 120°, 2 min[a]	$R=H$, $X=OCH_3$, I (33), II (36)	

Substrate	Reagent	Conditions	Product(s) (Yield, %)	Refs.
I, R = Ac, X = CN	1. $(C_2H_5)_2AlCN$ 2. CH_2N_2	Toluene, room temp, 10 min	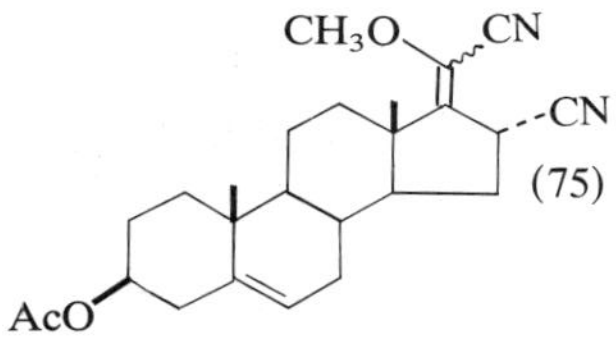 AcO, CH_3O, CN, CN (75)	
	1. $(C_2H_5)_2AlCN$ 2. 70% AcOH	Toluene, room temp, 5 min Hydrolysis	II, R = Ac, X = OH (71)	
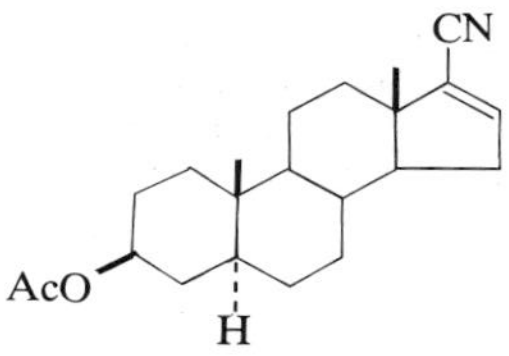 AcO, CN, H	$HCN\text{-}Al(C_2H_5)_3$	THF, room temp	No reaction	59

[a] The reaction was repeated on the recovered substrate.

Note: References 246–305 are on pp. 475–476.

TABLE III. CONJUGATE HYDROCYANATION OF α,β-ETHYLENIC α,α-DICARBONYL DERIVATIVES

	Substrate	Reagent(s)	Reaction Conditions	Product(s) and Yield(s) (%)	Refs.
C_5	$CH_3CH{=}C(CO_2C_2H_5)_2$	HCN-(KCN)	No solvent, exothermic	$CH_3CHCH(CO_2C_2H_5)_2$ (92) $\vert$ CN	2
		KCN	95% C_2H_5OH, reflux	$CH_3CHCH_2CO_2C_2H_5$ (Crude 71) $\vert$ CN	7
C_6	$CH_3CH{=}C(CO_2C_2H_5)COCH_3$	1. KCN 2. HCl	Aq C_2H_5OH, room temp Reflux	$CH_3CHCH_2COCH_3$ (67) $\vert$ CO_2H	252
	$C_2H_5CH{=}C(CO_2C_2H_5)_2$	NaCN-AcOH	Aq C_2H_5OH, 10°	$C_2H_5CHCH(CO_2C_2H_5)_2$ (62) $\vert$ CN	72
C_7	$C_2H_5O_2CCH{=}C(CH_3)$-$CH(CN)CO_2C_2H_5$ (via $C_2H_5O_2CCH_2$-$C(CH_3){=}C(CN)CO_2C_2H_5$)	KCN-HCl	Aq C_2H_5OH, room temp	$C_2H_5O_2CCH_2C(CH_3)CHCN$ (83) $\vert$ $\vert$ CN $CO_2C_2H_5$	253
C_8	n-$C_3H_7CH{=}C(CO_2C_2H_5)COCH_3$	1. KCN 2. HCl 3. 80% AcOH	Aq C_2H_5OH, room temp Reflux Reflux	n-$C_3H_7CHCH_2COCH_3$ (48), $\vert$ CO_2H [lactone structure] (23)	252
	[2-furyl]$CH{=}C(CO_2C_2H_5)_2$	1. KCN 2. KOH	Aq C_2H_5OH, reflux Hydrolysis	[2-furyl]$CHCH_2CO_2H$ (78) $\vert$ CO_2H	254

	Reactant	Reagent	Conditions	Product (Yield, %)	Ref.
	$C_2H_5O_2CC(CH_3){=}C(CH_3)$-$CH(CN)CO_2C_2H_5$ [via $C_2H_5O_2CH(CH_3)$-$C(CH_3){=}C(CN)CO_2C_2H_5$]	KCN-HCl	Aq C_2H_5OH, room temp	$C_2H_5O_2CCH(CH_3)$-$C(CH_3)(CN)CH(CN)CO_2C_2H_5$ (75)	253
	C_6H_5CON (piperidine)$={C(CN)CO_2C_2H_5}$	KCN	H_2O, room temp	C_6H_5CON (piperidine)-4-CN-4-$CH(CN)CO_2C_2H_5$ (I, 93)	234
		KCN-HCl	Aq C_2H_5OH, room temp	I (96)	
C_9	$C_2H_5O_2C(CH_2)_3$-$C(CH_3){=}C(CN)CO_2C_2H_5$	KCN	—	$C_2H_5O_2C(CH_2)_3$-$C(CH_3)(CN)CH(CN)CO_2C_2H_5$ (—)	205
C_{10}	2-acetyl-5,5-dimethyl-3-R-cyclohex-2-en-1-one (I)	KCN-NH$_4$Cl	DMF, room temp	2-acetyl-3-cyano-5,5-dimethyl-cyclohexadienol (OH, CN) (II)	255
	I, $R = OCH_3$			II (60)	
	I, $R = Cl$			II (60)	
	1-Br-1-[$CH{=}C(R)R'$]cyclohexane (I)	KCN	78% CH_3OH, 30°	spiro[2.5]octane, 1-H, 1-CN, 2-R, 2-R' (II)	170
	I, $R = R' = CO_2CH_3$			II, $R = R' = CO_2CH_3$ (High)	
	I, $R = CN$, $R' = CO_2CH_3$			II, $R = CN$, $R' = CO_2CH_3$ (High)	
	I, $R = R' = CN$			II, $R = R' = CN$ (40),	

TABLE III. Conjugate Hydrocyanation of α,β-Ethylenic α,α-Dicarbonyl Derivatives (*Continued*)

Substrate	Reagent(s)	Reaction Conditions	Product(s) and Yield(s) (%)	Refs.
C_{10} (*Contd.*)			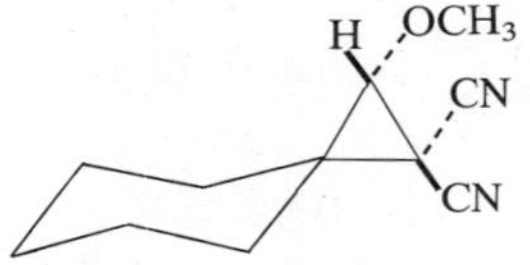 H, OCH_3, CN, CN (60)	
$ArCH{=}C(CO_2R)_2$ (I)			$ArCH(CN)CH(CO_2R)_2$ (II), $ArCH(CN)CH_2CO_2R$ (III)	
I, $Ar = C_6H_5$, $R = CH_3$	KCN	Dry CH_3OH, reflux	$[C_6H_5CH(CN)C(CO_2CH_3)_2]^-K^+$ (100) [on acidification: II, $Ar = C_6H_5$, $R = CH_3$ (100)]	11
		Aq CH_3OH, room temp	II, $Ar = C_6H_5$, $R = CH_3$ (50)	11
I, $Ar = C_6H_5$, $R = C_2H_5$	HCN-(KCN)	No solvent, 45–50°	No reaction	2
	1. KCN 2. $CaCO_3$	Aq C_2H_5OH, reflux	III, $Ar = C_6H_5$, $R = \frac{1}{2}Ca$ (51)	256, 7
	1. KCN 2. HCl	Aq C_2H_5OH, 65–75° Hydrolysis	$C_6H_5CH(CO_2H)CH_2CO_2H$ (67–70)	52
	KCN	Aq C_2H_5OH, 60°	III, $Ar = C_6H_5$, $R = C_2H_5$ (80), III, $Ar = C_6H_5$, $R = H$ (14)	7
	KCN-HCl	,, , room temp	II, $Ar = C_6H_5$, $R = C_2H_5$ (—)	7
	NaCN-AcOH	,, , 4°	II, $Ar = C_6H_5$, $R = C_2H_5$ (90–94)	72

I, Ar = m-$O_2NC_6H_4$, R = C_2H_5	KCN	Aq C_2H_5OH, 60°	III, Ar = m-$O_2NC_6H_4$, R = C_2H_5 (83)	257
I, Ar = o-$O_2NC_6H_4$, R = CH_3	KCN	Aq CH_3OH, —	II, Ar = o-$O_2NC_6H_4$, R = CH_3 (—)	128
I, Ar = o-$O_2NC_6H_4$, R = C_2H_5	KCN	Aq C_2H_5OH, room temp	II, Ar = o-$O_2NC_6H_4$, R = C_2H_5 (77)	171
	KCN	Aq C_2H_5OH, reflux	4-cyano-3-($CO_2C_2H_5$)-1-hydroxy-2-quinolone (N–OH, C=O, CN) (23-27)	171
$RC(CH_3){=}C(CN)CO_2R'$ (I)				
I, R = $CH_3C({=}CH_2)(CH_2)_2$, R′ = t-C_4H_9	KCN-HCl	Aq C_2H_5OH, 0°	$CH_3C({=}CH_2)(CH_2)_2C(CH_3)(CN)$-$CH(CN)CO_2C_4H_9$-$t$ (High)	211
I, R = $C_2H_5O_2C(CH_2)_2CH$-$CO_2C_2H_5$, R′ = C_2H_5	1. KCN-HCl 2. HCl 3. CH_3OH-H_2SO_4	Aq C_2H_5OH, — Hydrolysis Esterification	$CH_3O_2C(CH_2)_2CH(CO_2CH_3)$-$C(CH_3)(CO_2CH_3)CH_2CO_2CH_3$ (56)	206
$RCH{=}C(CN)CO_2R'$ (I)			$RCH(CN)CH(CN)CO_2R'$ (II), $RCH(CO_2H)CH_2CO_2H$ (III)	
I, R = n-C_6H_{13}, R′ = Na	1. KCN-AcOH 2. HCl	H_2O, room temp Hydrolysis	III, R = n-C_6H_{13} (60)	258
I, R = n-C_6H_{13}, R′ = C_2H_5	1. KCN 2. HCl	Aq C_2H_5OH, room temp Hydrolysis	III, R = n-C_6H_{13} (85)	258
I, R = cyclohexyl, R′ = C_2H_5	1. KCN 2. HCl	Aq C_2H_5OH, room temp Hydrolysis	III, R = cyclohexyl (88)	258

TABLE III. CONJUGATE HYDROCYANATION OF α,β-ETHYLENIC α,α-DICARBONYL DERIVATIVES (*Continued*)

Substrate	Reagent(s)	Reaction Conditions	Product(s) and Yield(s) (%)	Refs.
C_{10} (*Contd.*)				
I, R = C_6H_5, R′ = Na	1. KCN 2. Decarboxylation	H_2O, 0°	$C_6H_5[CH(CN)]_2CH(C_6H_5)$-$CH(CN)CO_2H$ (—)	169
	1. KCN-AcOH 2. HCl	H_2O, room temp Hydrolysis	III, R = C_6H_5 (85)	118
I, R = C_6H_5, R′ = CH_3	KCN	Dry CH_3OH, warm	$[C_6H_5CH(CN)C(CN)CO_2CH_3]^-K^+$ (100) [on acidification: II, R = C_6H_5, R′ = CH_3 (100)]	11
I, R = C_6H_5, R′ = C_2H_5	KCN	Aq C_2H_5OH, reflux	II, R = C_6H_5, R′ = C_2H_5 (64)	126, 249, 259
	1. KCN 2. HCl	Aq C_2H_5OH, reflux Hydrolysis	III, R = C_6H_5 (91–95)	100
I, R = *p*-$CH_3OC_6H_4$, R′ = C_2H_5	1. NaCN 2. HCl	Aq C_2H_5OH, reflux Hydrolysis	III, R = *p*-$CH_3OC_6H_4$ (70)	249
I, R = 4-HO-2-$CH_3OC_6H_3$, R′ = C_2H_5	1. NaCN 2. HCl	H_2O, reflux Hydrolysis	III, R = 4-HO-2-$CH_3OC_6H_3$ (45)	249
I, R = 3,4-methylenedioxyphenyl, R′ = Na	1. KCN-AcOH 2. HCl	H_2O, 40° Hydrolysis	III, R = 3,4-methylenedioxyphenyl (75)	118
I, R = *o*-$O_2NC_6H_4$, R′ = CH_3	KCN	Aq C_2H_5OH, 25–35°	II, R = *o*-$O_2NC_6H_4$, R′ = CH_3 (73)	128
I, R = *m*-$O_2NC_6H_4$, R′ = H	1. HCN addition 2. Hydrolysis	Unknown	*m*-$O_2NC_6H_4CH(CO_2H)CHCO_2H$ (—)	260

Reactant	Reagent	Conditions	Product (Yield, %)	Refs.
$ArCH{=}C(CN)_2$ (I)	KCN	Aq CH_3OH or aq C_2H_5OH, room temp or heat for a few min	$ArCH(CN)CH(CN)_2$ (II)	
I, $Ar = C_6H_5$			II, $Ar = C_6H_5$ (90)	261, 128, 262
I, $Ar = o\text{-}ClC_6H_4$			II, $Ar = o\text{-}ClC_6H_4$ (High)	128
I, $Ar = p\text{-}ClC_6H_4$			II, $Ar = p\text{-}ClC_6H_4$ (High)	128
I, $Ar = p\text{-}HOC_6H_4$			II, $Ar = p\text{-}HOC_6H_4$ (High)	128
I, $Ar = o\text{-}CH_3OC_6H_4$			II, $Ar = o\text{-}CH_3OC_6H_4$ (95)	261
I, $Ar = p\text{-}CH_3OC_6H_4$			II, $Ar = p\text{-}CH_3OC_6H_4$ (95)	261, 128
I, $Ar = p\text{-}(CH_3)_2NC_6H_4$			II, $Ar = p\text{-}(CH_3)_2NC_6H_4$ (96)	128
I, Ar = 3,4-(methylenedioxy)phenyl			II, Ar = 3,4-(methylenedioxy)phenyl (80)	261
2,4-dimethylpyrrol-3-yl–$CH{=}C(CN)_2$ (N–H)	KCN	Aq C_2H_5OH, mixed hot	2,4-dimethylpyrrol-3-yl–$CH(CN)CH(CN)_2$ (N–H) (Crude, 98)	263
3-($C_2H_5O_2C$)cyclohexylidene=$C(CN)CO_2C_2H_5$	KCN	Aq C_2H_5OH, —	3-($C_2H_5O_2C$)cyclohexane-1-(CN)-1-(CH_2CN) (68)	264
2,6-dimethyl-1-CH_3-piperidin-4-ylidene=$C(CN)_2$ ·HCl	1. KCN 2. HCl	H_2O, cold	4-(NC)-4-($CH(CN)_2$)-2,6-dimethyl-1-CH_3-piperidine ·HCl (82)	265

TABLE III. CONJUGATE HYDROCYANATION OF α,β-ETHYLENIC α,α-DICARBONYL DERIVATIVES (*Continued*)

	Substrate	Reagent(s)	Reaction Conditions	Product(s) and Yield(s) (%)	Refs.
C_{11}	$C_6H_5CH{=}CH(CO_2C_2H_5)COCH_3$	1. HCN 2. 33% NaOH	No solvent, room temp Reflux	$C_6H_5CH(CO_2H)CH_2CO_2H$ (I, 11)	12
		1. HCN-(pip) 2. 33% NaOH	No solvent, room temp Hydrolysis	I (22)	12
		1. NaCN 2. HCl 3. 80% AcOH	Aq C_2H_5OH, room temp Reflux	$C_6H_5CH(CO_2H)CH_2COCH_3$ (83)	252
		KCN	Aq C_2H_5OH, room temp	$C_6H_5CH(CN)CH_2CH(COCH_3)CO_2C_2H_5$ (—)	266
	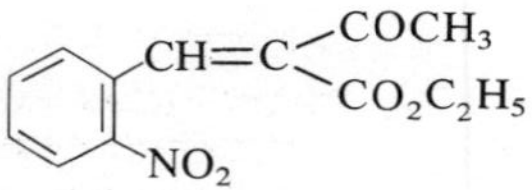 $o\text{-}O_2NC_6H_4CH{=}C(COCH_3)CO_2C_2H_5$	KCN	Aq C_2H_5OH, warm	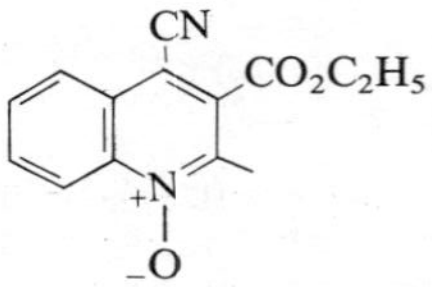 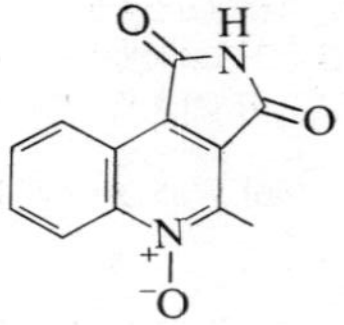 4-CN, 3-$CO_2C_2H_5$, 2-CH_3 quinoline N^{+}–O^{-} (3), pyrrolo-dione (O, NH, O) fused 2-CH_3 quinoline N^{+}–O^{-} (57)	172

	Substrate	Reagent	Conditions	Product (yield %)	Ref.
	$ArC(CH_3){=}C(CN)CO_2C_2H_5$ (I)	KCN	Aq C_2H_5OH, reflux	$ArC(CH_3)(CN)CH(CN)CO_2C_2H_5$ (II)	259
	I, $Ar = C_6H_5$			II, $Ar = C_6H_5$ (—)	
	I, $Ar = p\text{-}ClC_6H_4$			II, $Ar = p\text{-}ClC_6H_4$ (—)	
	I, $Ar = p\text{-}CH_3OC_6H_4$			II, $Ar = p\text{-}CH_3OC_6H_4$ (—)	
	I, $Ar = p\text{-}O_2NC_6H_4$			II, $Ar = p\text{-}O_2NC_6H_4$ (—)	
	5-RO_2C-4-methyl-2-R′-pyrrole-3-$CH{=}C(R'')CN$ (I)	KCN	Aq C_2H_5OH	5-RO_2C-4-methyl-2-R′-pyrrole-3-$CH(CN)CH(R'')CN$ (II)	263
	I, $R = H$, $R' = CH_3$, $R'' = CO_2C_2H_5$		Mixed warm	II, $R = H$, $R' = CH_3$, $R'' = CO_2C_2H_5$ (—)	
	I, $R = C_2H_5$, $R' = CH_3$, $R'' = CO_2C_2H_5$		75°	II, $R = C_2H_5$, $R' = CH_3$, $R'' = CO_2C_2H_5$ (50)	
	I, $R = C_2H_5$, $R' = CH_3$, $R'' = CN$		Room temp	II, $R = C_2H_5$, $R' = CH_3$, $R'' = CN$ (90)	
	I, $R = C_2H_5$, $R' = CO_2CH_3$, $R'' = CN$		Room temp	II, $R = C_2H_5$, $R' = CO_2CH_3$, $R'' = CN$ (Crude, quant)	
	$Ar(CH_3)C{=}C(CN)_2$ (I)			$Ar(CH_3)C(CN)CH(CN)_2$ (II)	262
	I, $R = C_6H_5$	KCN	Aq C_2H_5OH, reflux	II, $Ar = C_6H_5$ (Quant)	
	I, $R = p\text{-}CH_3OC_6H_4$	KCN	Aq C_2H_5OH, reflux	II, $Ar = p\text{-}CH_3OC_6H_4$ (Quant)	
	I, $R = p\text{-}O_2NC_6H_4$	KCN	Aq THF, cold	II, $Ar = p\text{-}O_2NC_6H_4$ (Quant)	
C_{12}	$C_6H_5CH{=}C(COCH_3)_2$	KCN	Aq C_2H_5OH, room temp	$C_6H_5CH(CN)CH(COCH_3)_2$ (Quant)	266
	$C_6H_5(CH_2)_2CH{=}C(CO_2CH_3)_2$	1. KCN 2. 10% KOH	Aq C_2H_5OH, reflux Hydrolysis	$C_6H_5(CH_2)_2CH(CO_2H)CH_2CO_2H$ (—)	41

TABLE III. CONJUGATE HYDROCYANATION OF α,β-ETHYLENIC α,α-DICARBONYL DERIVATIVES (*Continued*)

	Substrate	Reagent(s)	Reaction Conditions	Product(s) and Yield(s) (%)	Refs.
C_{12} (*Contd.*)	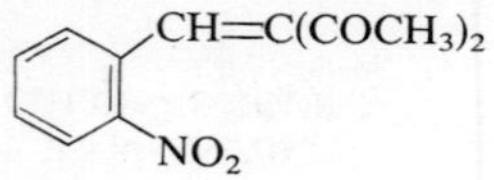 o-$O_2NC_6H_4CH{=}C(COCH_3)_2$ (structure)	KCN with or without $KHCO_3$	Aq C_2H_5OH, room temp	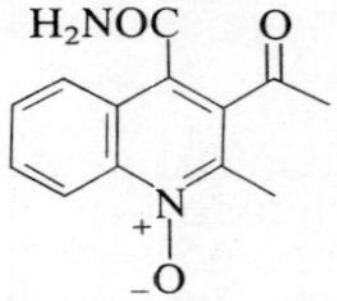 4-carbamoyl-3-acetyl-2-methylquinoline 1-oxide (structure: H_2NOC, O, N⁺–O⁻) (41–57)	172
	m-$O_2NC_6H_4CH{=}C(COCH_3)_2$	KCN	Aq C_2H_5OH, room temp	m-$O_2NC_6H_4CH(CN)CH(COCH_3)_2$ (—)	172
	p-$O_2NC_6H_4CH{=}C(COCH_3)_2$	KCN	Aq C_2H_5OH, room temp	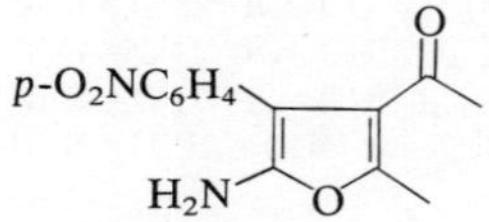 2-amino-3-(p-$O_2NC_6H_4$)-4-acetyl-5-methylfuran (structure: p-$O_2NC_6H_4$, H_2N, O) (76)	172
	$C_6H_5C(C_2H_5){=}C(CN)CO_2C_2H_5$	1. KCN 2. Hydrolysis	Aq C_2H_5OH, reflux	$C_6H_5C(C_2H_5)(CO_2H)CH_2CO_2H$ (77)	245
	p-$CH_3C_6H_4C(CH_3){=}C(CN)CO_2C_2H_5$	KCN	Aq C_2H_5OH, reflux	p-$CH_3C_6H_4C(CH_3)(CN)CH(CN)CO_2C_2H_5$ (—)	259
C_{13}	2,4,6-$(CH_3)_3C_6H_2CH{=}C(CO_2C_2H_5)_2$	KCN	Aq C_2H_5OH, 65–70°	2,4,6-$(CH_3)_3C_6H_2CH(CO_2H)CH_2CO_2H$ (2)	132
	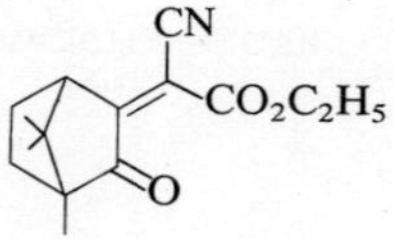 bicyclic ketone with $=C(CN)CO_2C_2H_5$ (structure: CN, $CO_2C_2H_5$, O)	1. KCN 2. AcOH	Aq C_2H_5OH, 40–50° Room temp	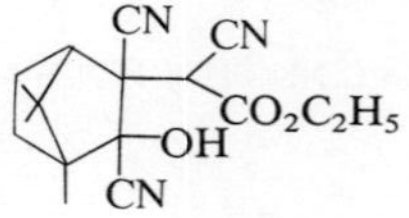 bicyclic product (structure: CN, CN, $CO_2C_2H_5$, OH, CN) (Quant)	267

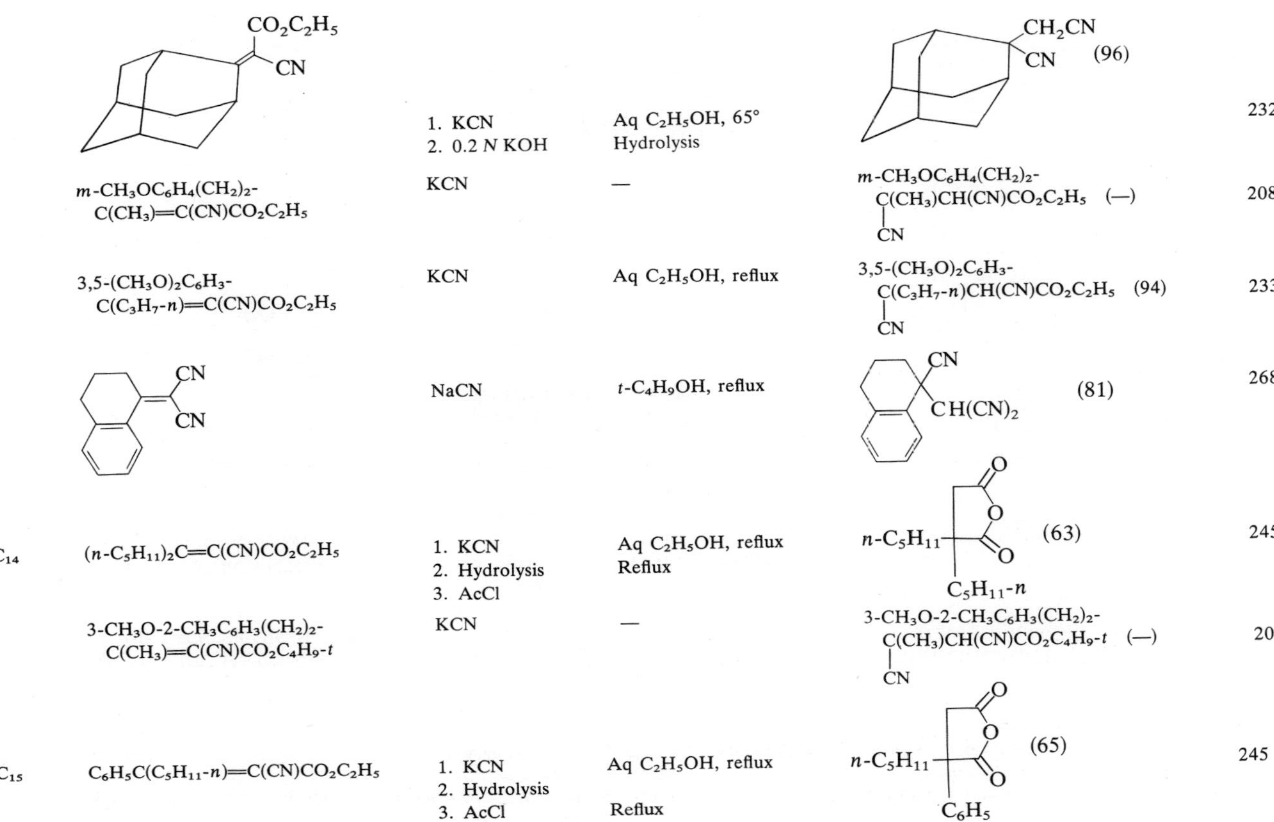

	Reactant	Reagents	Conditions	Product(s) (Yield, %)	Refs.
	(adamantylidene, $CO_2C_2H_5$, CN)	1. KCN 2. 0.2 *N* KOH	Aq C_2H_5OH, 65° Hydrolysis	(adamantane, CH_2CN, CN) (96)	232
	$m\text{-}CH_3OC_6H_4(CH_2)_2\text{-}$ $C(CH_3){=}C(CN)CO_2C_2H_5$	KCN	—	$m\text{-}CH_3OC_6H_4(CH_2)_2\text{-}$ $C(CH_3)CH(CN)CO_2C_2H_5$ (—) CN	208
	$3,5\text{-}(CH_3O)_2C_6H_3\text{-}$ $C(C_3H_7\text{-}n){=}C(CN)CO_2C_2H_5$	KCN	Aq C_2H_5OH, reflux	$3,5\text{-}(CH_3O)_2C_6H_3\text{-}$ $C(C_3H_7\text{-}n)CH(CN)CO_2C_2H_5$ (94) CN	233
	(tetralinylidene, CN, CN)	NaCN	$t\text{-}C_4H_9OH$, reflux	(tetralin, CN, $CH(CN)_2$) (81)	268
C_{14}	$(n\text{-}C_5H_{11})_2C{=}C(CN)CO_2C_2H_5$	1. KCN 2. Hydrolysis 3. AcCl	Aq C_2H_5OH, reflux Reflux	$n\text{-}C_5H_{11}$, $C_5H_{11}\text{-}n$ (anhydride) (63)	245
	$3\text{-}CH_3O\text{-}2\text{-}CH_3C_6H_3(CH_2)_2\text{-}$ $C(CH_3){=}C(CN)CO_2C_4H_9\text{-}t$	KCN	—	$3\text{-}CH_3O\text{-}2\text{-}CH_3C_6H_3(CH_2)_2\text{-}$ $C(CH_3)CH(CN)CO_2C_4H_9\text{-}t$ (—) CN	208
C_{15}	$C_6H_5C(C_5H_{11}\text{-}n){=}C(CN)CO_2C_2H_5$	1. KCN 2. Hydrolysis 3. AcCl	Aq C_2H_5OH, reflux Reflux	$n\text{-}C_5H_{11}$, C_6H_5 (anhydride) (65)	245

TABLE III. CONJUGATE HYDROCYANATION OF α,β-ETHYLENIC α,α-DICARBONYL DERIVATIVES (Continued)

	Substrate	Reagent(s)	Reaction Conditions	Product(s) and Yield(s) (%)	Refs.
C_{15} (*Contd.*)	$C_2H_5O_2CCH(C_6H_5)(CH_2)_2$-$C(CH_3)=C(CN)CO_2C_2H_5$	KCN	Aq C_2H_5OH, —	$C_2H_5O_2CCH(C_6H_5)CH_2CH_2$-$C(CH_3)(CN)CH(CN)CO_2C_2H_5$ (—)	205
	p-$CH_3OC_6H_4CH(CN)(CH_2)_2$-$C(CH_3)=C(CN)CO_2C_2H_5$	1. KCN-HCl 2. HCl-AcOH 3. CH_3OH-HCl	Aq C_2H_5OH, room temp Hydrolysis Esterification	p-$CH_3OC_6H_4CH(CO_2CH_3)$-$(CH_2)_2C(CH_3)(CO_2CH_3)CH_2CO_2CH_3$ (50)[a]	210
C_{16}	$CH_3(CH_2)_9C(C_2H_5)=C(CN)$-$CO_2C_2H_5$	1. KCN-HCl 2. Aq H_2SO_4 3. Aq KOH 4. C_2H_5OH-H_2SO_4	Aq C_2H_5OH, cold Reflux (hydrolysis) Reflux (decarboxylation) Esterification	$CH_3(CH_2)_9C(C_2H_5)(CO_2C_2H_5)CH_2CO_2C_2H_5$ (76)	269
	o-$NO_2C_6H_4CH=C(COC_6H_5)CO_2C_2H_5$	KCN	Aq C_2H_5OH, reflux	4-cyano-3-$CO_2C_2H_5$-2-C_6H_5-quinoline N-oxide (8), 4-C_6H_5-pyrrolo[3,4-c]quinoline-1,3-dione 5-oxide (—)	172

	Reactant	Reagent	Conditions	Product (Yield, %)	Refs.
	p-$CH_3OC_6H_4(R)C(CO_2C_2H_5)$-$(CH_2)_2C(CH_3){=}C(CN)CO_2C_2H_5$ (I)				210
	I, R = $CO_2C_2H_5$	KCN	Aq C_2H_5OH, reflux	p-$CH_3OC_6H_4C(CO_2C_2H_5)_2$-$(CH_2)_2C(CN)(CH_3)CH_2CN$ (78)	
	I, R = CN	1. KCN 2. NaOH 3. CH_3OH-HCl	Aq C_2H_5OH, 35° Hydrolysis Esterification	p-$CH_3OC_6H_4CH(CO_2CH_3)(CH_2)_2$-$C(CH_3){=}CHCO_2CH_3$ (34) (No hydrocyanation)	
	$ArC(Ar'){=}C(CN)CO_2C_2H_5$ (I)			$ArC(Ar')(CO_2H)CH_2CO_2H$ (II)	
	I, Ar = Ar' = C_6H_5	KCN	Aq C_2H_5OH, reflux	$C_6H_5C(C_6H_5)(CN)CH(CN)CO_2C_2H_5$ (80)	245, 259
		1. KCN 2. Hydrolysis	” , ”	II, Ar = Ar' = C_6H_5 (80)	245
	I, Ar = C_6H_5, Ar' = p-ClC_6H_4	1. KCN 2. Hydrolysis	Aq C_2H_5OH, reflux	II, Ar = C_6H_5, Ar' = p-ClC_6H_4 (82)	245
	I, Ar = p-ClC_6H_4, Ar' = o-ClC_6H_4	1. KCN 2. Hydrolysis	Aq C_2H_5OH, reflux	II, Ar = p-ClC_6H_4, Ar' = o-ClC_6H_4 (47)	245
	I, Ar = Ar' = p-ClC_6H_4	1. KCN 2. Hydrolysis	Aq C_2H_5OH, reflux	II, Ar = Ar' = p-ClC_6H_4 (47)	245
C_{17}	n-$C_7H_{15}C(C_6H_5){=}C(CN)CO_2C_2H_5$	1. KCN 2. Hydrolysis 3. AcCl	Aq C_2H_5OH, reflux Reflux	3-n-C_7H_{15}-3-C_6H_5-succinic anhydride (56)	245

TABLE III. CONJUGATE HYDROCYANATION OF α,β-ETHYLENIC α,α-DICARBONYL DERIVATIVES (*Continued*)

	Substrate	Reagent(s)	Reaction Conditions	Product(s) and Yield(s) (%)	Refs.
C_{17} (*Contd.*)	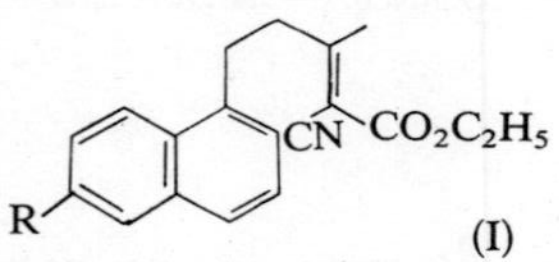 [structure: R-naphthyl–$CH_2CH_2C(CH_3)$=C(CN)$CO_2C_2H_5$] (I)			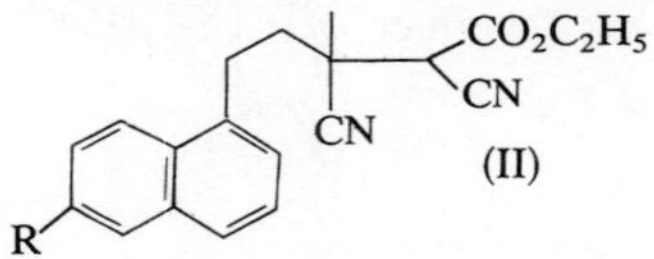 [structure: R-naphthyl–$CH_2CH_2C(CH_3)(CN)CH(CN)CO_2C_2H_5$] (II)	
	I, R = H	NaCN	50% C_2H_5OH, exothermic	II, R = H (Crude 98)	207
	I, R = OCH_3	KCN	—	II, R = OCH_3 (—)	208
C_{18}	$RC(R'){=}C(CN)CO_2C_2H_5$ (I)				
	I, R 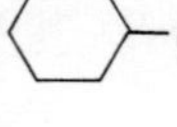 = cyclohexyl–$(CH_2)_2$, R′ = C_6H_5	1. KCN 2. Hydrolysis 3. AcCl	Aq C_2H_5OH, reflux Reflux	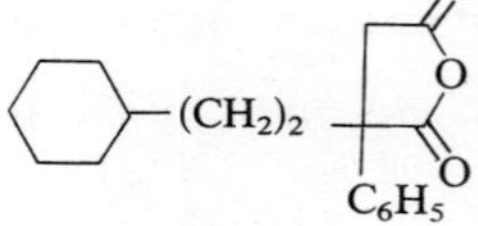 cyclohexyl–$(CH_2)_2$–C(C_6H_5) succinic anhydride (52)	245
	I, R = R′ = $CH_2C_6H_5$	KCN	Aq C_2H_5OH, reflux	$(C_6H_5CH_2)_2C(CN)CH(CN)CO_2C_2H_5$ (50)	270, 259
C_{20}	n-$C_4H_9C_6H_4C(C_6H_{13}$-$n){=}C(CN)$-$CO_2C_2H_5$	1. KCN 2. Hydrolysis 3. AcCl	Aq C_2H_5OH, reflux Reflux	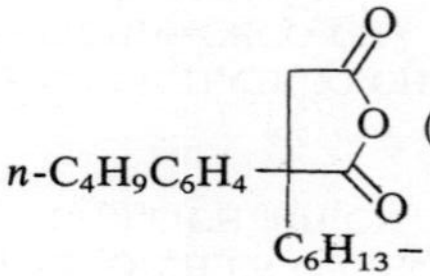 n-$C_4H_9C_6H_4$–C(C_6H_{13}-n) succinic anhydride (54)	245
	$(C_6H_5CH_2CH_2)_2C{=}C(CN)CO_2C_2H_5$	1. KCN 2. Hydrolysis	Aq C_2H_5OH, reflux	$(C_6H_5CH_2CH_2)_2C(CO_2H)CH_2CO_2H$ (16)	245

C_{21}	cyclohexyl-$(CH_2)_5C(C_6H_5)=C(CN)CO_2C_2H_5$	1. KCN 2. Hydrolysis 3. AcCl	Aq C_2H_5OH, reflux; Reflux	2-(5-cyclohexylpentyl)-2-phenylsuccinic anhydride [cyclohexyl-$(CH_2)_5$, C_6H_5] (45)	245
C_{22}	$(n\text{-}C_9H_{19})_2C=C(CN)CO_2C_2H_5$	1. HCN 2. Hydrolysis 3. AcCl	Aq C_2H_5OH, reflux; Reflux	2,2-di($n\text{-}C_9H_{19}$)succinic anhydride [$n\text{-}C_9H_{19}$, $C_9H_{19}\text{-}n$] (41)	245
	$o\text{-}O_2NC_6H_4CH=C(COC_6H_5)_2$	KCN	Aq C_2H_5OH, reflux	4-$CONH_2$-3-COC_6H_5-2-C_6H_5-quinoline N-oxide (—), $o\text{-}O_2NC_6H_4CH(CO_2H)CH_2COC_6H_5$ (10)	172
C_{23}	$n\text{-}C_{13}H_{27}C(C_6H_5)=C(CN)CO_2C_2H_5$	1. KCN 2. Hydrolysis 3. AcCl	Aq C_2H_5OH, reflux; Reflux	2-($n\text{-}C_{13}H_{27}$)-2-C_6H_5-succinic anhydride (48)	245

[a] The ethyl ester is prepared in lower yield.

Note: References 246–305 are on pp. 475-476.

TABLE IV. Conjugate Hydrocyanation of α,β-Ethylenic α,β-Dicarbonyl Derivatives

	Substrate	Reagent(s)	Reaction Conditions	Product(s) and Yield(s) (%)	Refs.
C_4	*cis*-$CH_3O_2CCH{=}CHCO_2CH_3$	HCN-(KCN)	No solvent, 45°	*cis*- and *trans*- $CH_3O_2CCH{=}CHCO_2CH_3$ (50), $CH_3O_2CC(CN)HCH(CO_2CH_3)$–$CH_3O_2CCHCH_2CO_2CH_3$ (18)	2
	trans-$CH_3O_2CCH{=}CHCO_2CH_3$	KCN	Dry CH_3OH, room temp	$CH_3O_2CCH(OCH_3)CH_2CO_2CH_3$ (I, 17–31), 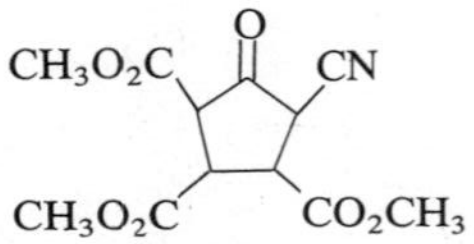 cyclopentanone bearing CH_3O_2C, CN, CH_3O_2C, CO_2CH_3 (II, 38–47)	11
			Dry CH_3OH, reflux	II (76)	11
			Aq CH_3OH, room temp	I (11), *trans*-$HO_2CCH{=}CHCO_2CH_3$ (60), $NCCH_2[CH(CO_2CH_3)]_2$-$CH_2CO_2CH_3$ (Crude 39)	11
	$NCCH{=}CHCN$	HCN-(KCN)	—	No uniform product	2
C_6	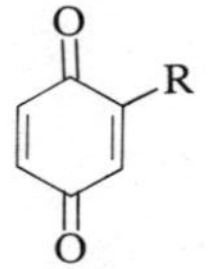 2-R-1,4-benzoquinone (O, R, O) (I)			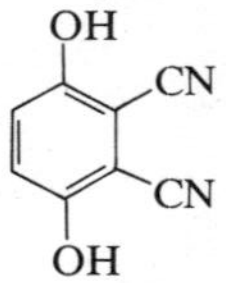 hydroquinone with OH, CN, CN, OH (II)	
	I, R = H	$KCN\text{-}H_2SO_4$	H_2O, mixed hot	II (65)	173
		,,	C_2H_5OH, room temp	II (70)	271, 272
	I, R = Cl	$KCN\text{-}H_2SO_4$	C_2H_5OH, room temp	II (Good)	273

	Reactant	Reagent	Conditions	Product (Yield, %)	Refs.
	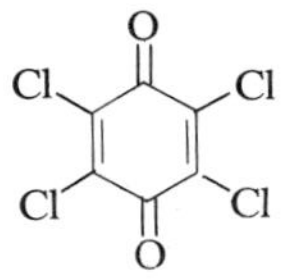 Cl, Cl, Cl, Cl, O, O (tetrachloro-p-benzoquinone structure)	KCN	Aq CH_3OH, reflux	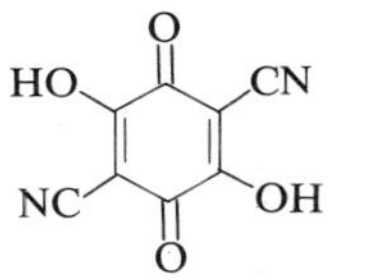 HO, CN, NC, OH, O, O (structure) (—)	67
C_{10}	$C_6H_5COCH{=}CHCO_2H$	KCN	H_2O, room temp	$C_6H_5COCH_2CH(CN)CO_2H$ (—)	174, 61
C_{13}	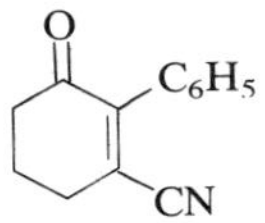 O, C_6H_5, CN (structure)	$(CH_3)_2C(OH)CN$- $[N(C_2H_5)_3]$	No solvent, 80–90°	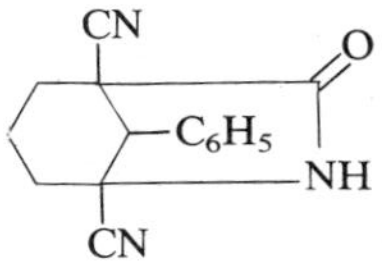 CN, O, C_6H_5, NH, CN (structure) (10)	123
C_{16}	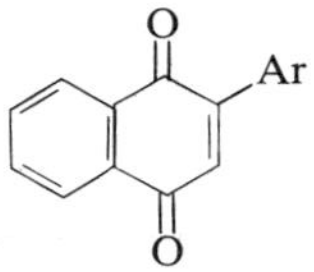 O, Ar, O (structure) (I)	NaCN	Dioxane-aq C_2H_5OH, room temp	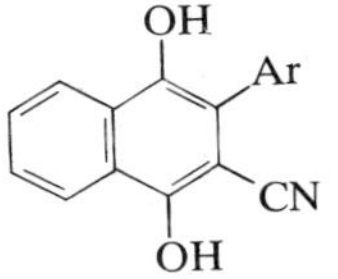 OH, Ar, CN, OH (structure) (II)	54
	I, $Ar = C_6H_5$ I, $Ar = p\text{-}CH_3OC_6H_4$ I, $Ar = p\text{-}O_2NC_6H_4$			II, $Ar = C_6H_5$ (78) II, $Ar = p\text{-}CH_3OC_6H_4$ (64) II, $Ar = p\text{-}O_2NC_6H_4$ (82)	
C_{17}	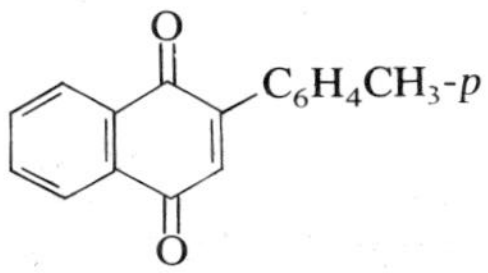 O, $C_6H_4CH_3\text{-}p$, O (structure)	NaCN	Dioxane-aq C_2H_5OH, room temp	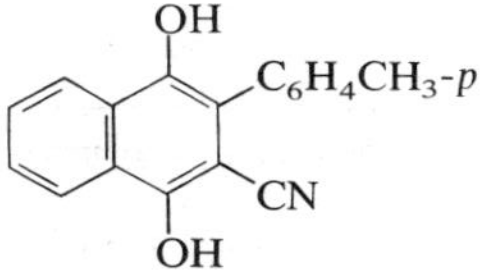 OH, $C_6H_4CH_3\text{-}p$, CN, OH (structure) (84)	54

TABLE IV. CONJUGATE HYDROCYANATION OF α,β-ETHYLENIC α,β-DICARBONYL DERIVATIVES (*Continued*)

	Substrate	Reagent(s)	Reaction Conditions	Product(s) and Yield(s) (%)	Refs.
C_{19}	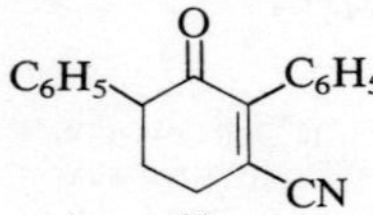 C_6H_5, O, C_6H_5, CN (I)	$(CH_3)_2C(OH)CN$-$[N(C_2H_5)_3]$	No solvent, reflux	I(40), 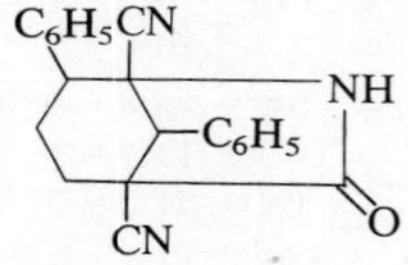 C_6H_5 CN, NH, C_6H_5, CN, O (19)	124

Note: References 246–305 are on pp. 475–476.

TABLE V. Conjugate Hydrocyanation of Acyclic α,β-Ethylenic Ketones

	Substrate	Reagent(s)	Reaction Conditions	Product(s) and Yield(s) (%)	Refs.
C_4	$CH_3COCH{=}CH_2$	HCN-(K_2CO_3)	No solvent, 80°	$CH_3COCH_2CH_2CN$ (I, 90)	36, 37
		HCN-(KCN)	C_6H_6, 30°	I (84)	46
	$CH_3(HO)C(CN)CH{=}CH_2$	HCN-(pyr)	No solvent, 70°	I (92)	47
	$CH_3COCH{=}CH\overset{+}{N}(CH_3)_3Cl^-$	KCN-$N(CH_3)_3{\cdot}HCl$	H_2O—C_6H_6, 50–60°	$CH_3COCH{=}CHCN$ (40)	66
C_5	$CH_3CH_2COCH{=}CH_2$	1. KCN-H_2SO_4 2. HCl 3. C_2H_5OH—H_2SO_4	Aq C_2H_5OH, room temp Hydrolysis Esterification	$CH_3CH_2COCH_2CH_2CO_2C_2H_5$ (—)	274
	$C_2H_5COCH{=}CH\overset{+}{N}(CH_3)_3Cl^-$	KCN-$N(CH_3)_3{\cdot}HCl$	H_2O—C_6H_6, 50–60°	$C_2H_5COCH{=}CHCN$ (22)	66
C_6	$CH_3COCH{=}C(CH_3)_2$	HCN	No solvent, room temp	$CH_3COCH_2C(CH_3)_2CN$ (I, —)	12
		HCN-(KCN)	No solvent, 135–140°	I (86–92)	49
		HCN-[$K_4Ni(CN)_4$]	C_6H_6, N_2 (100 atm), 60°	I (13)	48
		KCN	H_2O, —	I (—), $CH_3COCH_2C(CH_3)_2CO_2H$ 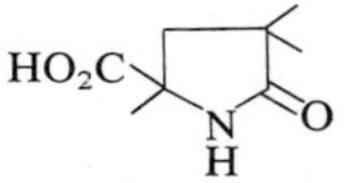 (—), HO_2C … N H … O (—)	275
		KCN	Aq C_2H_5OH, reflux	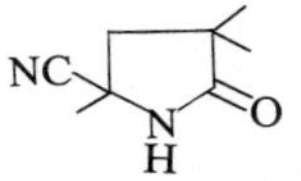 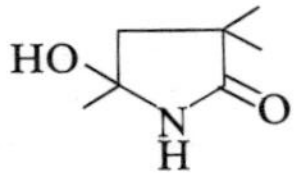 NC … N H … O (—), HO … N H … O (—)	99, 31
		HCN-$Al(C_2H_5)_3$	THF, room temp	I (88)	43

TABLE V. CONJUGATE HYDROCYANATION OF ACYCLIC α,β-ETHYLENIC KETONES (Continued)

	Substrate	Reagent(s)	Reaction Conditions	Product(s) and Yield(s) (%)	Refs.
C_6 (Contd.)	$CH_3COCH{=}C(CH_3)_2$				
		HCN-$(C_2H_5)_2AlCl$	THF, room temp	I (86)	43
		$(C_2H_5)_2AlCN$	C_6H_6-toluene, 0°, 3 hr	I (55)	99
	$CH_3COC(CH_3){=}CHCH_3$	KCN	Aq C_2H_5OH, 75°	$CH_3COCH(CH_3)CH(CONH_2)CH_3$ (60)	276
	n-$C_3H_7COCH{=}CH\overset{+}{N}(CH_3)_3Cl^-$	KCN-$N(CH_3)_3$·HCl	H_2O—C_6H_6, 50–60°	n-$C_3H_7COCH{=}CHCN$ (55)	66
C_7	$CH_2{=}CHCH_2COC(CH_3){=}CH_2$			$CH_2{=}CHCH_2COCH(CH_3)CH_2CN$ (I), $CH_3CH(CN)CH_2COC(CH_3){=}CH_2$ (II), $CH_3CH(CN)CH_2COCH(CH_3)CH_2CN$ (III), 2,4-dimethylcyclopent-2-en-1-one (IV)	105
		KCN-AcOH	Aq C_2H_5OH, 35°	I+III (15–18), II (36), IV (19)	
		KCN-AcOH	” , 45–50°	I (14), II (18), III (3), IV (13)	
	$(CH_3)_2C{=}CHCOCH{=}CH_2$	KCN-AcOH	Aq C_2H_5OH, 35–37°	$(CH_3)_2C{=}CHCOCH_2CH_2CN$ (77)	277
	$(CH_3)_2C(Cl)CH_2COCH{=}CH_2$	KCN	Aq C_2H_5OH, 42°	$(CH_3)_2C{=}CHCOCH{=}CH_2$ (25), $(CH_3)_2C{=}CHCOCH_2CH_2CN$ (49)	277
	$(CH_3)_2C{=}CHCOCH_2CH_2OCH_3$	KCN-AcOH	C_2H_5OH, —	Tars	277
	2-($COCH_2CH_2N(CH_3)_2$·HCl)-furan/thiophene (ring with X) (I)	KCN	H_2O, reflux	2-($COCH_2CH_2CN$)-furan/thiophene (ring with X) (II)	65
	I, X = O			II, X = O (57)	
	I, X = S			II, X = S (67)	

C_9	$(CH_2)_5C{=}CHCOCH_3$	HCN-(KCN)	No solvent, 150–160°	$CH_3COCH_2C(CN)(CH_2)_5$ (70–75)	16
	$CO[CH{=}C(CH_3)_2]_2$	KCN	C_2H_5OH, warm	$(CH_3)_2C(CN)CH_2COCH_2C(CN)(CH_3)_2$ (I, —)	250
		KCN-HCN (1:1)	Aq C_2H_5OH, 100°	I (43)	8
	{via $CO[CH_2CCl(CH_3)_2]_2$}	1. Dry HCl 2. KCN 3. Conc HCl	 Aq C_2H_5OH, reflux Hydrolysis	$(CH_3)_2C(CO_2H)CH_2COCH_2C(CO_2H)(CH_3)_2$ (78)	63
	$C_6H_5COCH{=}CH_2$	NaCN-AcOH	Aq C_2H_5OH, room temp	$C_6H_5COCH_2CH_2CN$ (65)	175
	$ArCOCH_2CH_2N(CH_3)_2{\cdot}HCl$ (I)	KCN	H_2O, reflux	$ArCOCH_2CH_2CN$ (II)	65
	I, Ar $= C_6H_5$			II, Ar $= C_6H_5$ (67)	
	I, Ar $= p\text{-}ClC_6H_4$			II, Ar $= p\text{-}ClC_6H_4$ (17)	
	I, Ar $= p\text{-}BrC_6H_4$			II, Ar $= p\text{-}BrC_6H_4$ (63)	
	I, Ar $= m\text{-}HOC_6H_4$			II, Ar $= m\text{-}HOC_6H_4$ (—)	
	I, Ar $= p\text{-}HOC_6H_4$			II, Ar $= p\text{-}HOC_6H_4$ (59)	
	I, Ar $= m\text{-}CH_3OC_6H_4$			II, Ar $= m\text{-}CH_3OC_6H_4$ (73)	
	I, Ar $= p\text{-}CH_3OC_6H_4$			II, Ar $= p\text{-}CH_3OC_6H_4$ (71)	
	I, Ar $= 3,4\text{-}(CH_3O)_2C_6H_3$			II, Ar $= 3,4\text{-}(CH_3O)_2C_6H_3$ (85)	
	$o\text{-}O_2NC_6H_4COCH_2CH_2\overset{+}{N}(CH_3)_3Cl^-$	KCN	CH_3OH, room temp	$o\text{-}O_2NC_6H_4COCH_2CH_2CN$ (59)	53
	$ArCOCH{=}CH\overset{+}{N}(CH_3)_3Cl^-$ (I)	$KCN\text{-}N(CH_3)_3.HCl$	$H_2O—C_6H_6$, under cooling	$ArCOCH{=}CHCN$ (II)	66
	I, Ar $= C_6H_5$			II, Ar $= C_6H_5$ (94)	
	I, Ar $= o\text{-}ClC_6H_4$			II, Ar $= o\text{-}ClC_6H_4$ (97)	
	I, Ar $= p\text{-}BrC_6H_4$			II, Ar $= p\text{-}BrC_6H_4$ (—)	
	I, Ar $= p\text{-}O_2NC_6H_4$			II, Ar $= p\text{-}O_2NC_6H_4$ (77)	
C_{10}	$CH_3COCH{=}CHC_6H_5$	HCN	No solvent, —	No reaction	12
		HCN-(base)	No solvent, —	Unidentified syrup	12
		$(CH_3)_2C(OH)CN$-(Na_2CO_3)	Aq CH_3OH, reflux	$CH_3COCH_2CH(CN)C_6H_5$ (53)	50

TABLE V. Conjugate Hydrocyanation of Acyclic α,β-Ethylenic Ketones (*Continued*)

	Substrate	Reagent(s)	Reaction Conditions	Product(s) and Yield(s) (%)	Refs.
C_{10} (*Contd.*)					
	p-$BrC_6H_4COC(CH_3){=}CH_2$	KCN	CH_3OH, room temp	p-$BrC_6H_4COCH(CH_3)CH_2CN$ (96)	53
	p-$CH_3C_6H_4COCH_2CH_2N(CH_3)_2$.HCl	KCN	H_2O, reflux	p-$CH_3C_6H_4COCH_2CH_2CN$ (52)	65
	$ArCOCH(CH_3)CH_2\overset{+}{N}(CH_3)_3Cl^-$ (I)	KCN	Room temp	$ArCOCH(CH_3)CH_2CN$ (II)	53
	I, Ar = m-$O_2NC_6H_4$		CH_3OH	II, Ar = m-$O_2NC_6H_4$ (92)	
	I, Ar = p-$AcNHC_6H_4$		Aq CH_3OH	II, Ar = p-$AcNHC_6H_4$ (92)	
	I, Ar = p-BrC_6H_4		H_2O	II, Ar = p-BrC_6H_4 (37), p-$BrC_6H_4COC(CH_3){=}CH_2$ (63)	
			CH_3OH	II, Ar = p-BrC_6H_4 (62), p-$BrC_6H_4COCH(CH_3)CH_2OCH_3$ (—)	
C_{11}	$CH_3COC[C_6H_3(NO_2)_2$-3,4]$=C(OCH_3)CH_3$	$(CH_3)_2C(OH)CN$-(base)	—	No reaction	123
	p-$AcNHC_6H_4COCH(C_2H_5)$-$CH_2\overset{+}{N}(CH_3)_3Cl^-$	KCN	H_2O, room temp	p-$AcNHC_6H_4COCH(C_2H_5)$-CH_2CN (>57)	53
	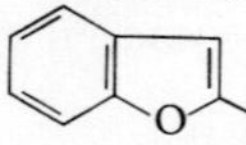 2-Benzofuryl-$COCH_2CH_2$-$N(CH_3)_2\cdot$HCl	KCN	H_2O, reflux	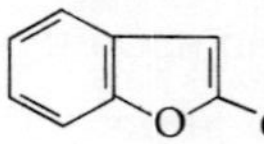 2-Benzofuryl-$COCH_2CH_2CN$ (21)	65
C_{12}	p-$AcNHC_6H_4COCH(C_3H_7$-$n)$-$CH_2\overset{+}{N}(CH_3)_3Cl^-$	KCN	H_2O, room temp	p-$AcNHC_6H_4COCH(C_3H_7$-$n)CH_2CN$ (>40)	53
C_{13}	$CH_3COCH{=}CHC_6H_2(CH_3)_3$-2,4,6	NaCN-AcOH	Aq C_2H_5OH, 35°	$CH_3COCH_2CH(CN)C_6H_2(CH_3)_3$-2,4,6 (63)	132
	$C_{10}H_7COCH_2CH_2N(CH_3)_2\cdot$HCl (I)	KCN	H_2O, reflux	$C_{10}H_7COCH_2CH_2CN$ (II)	65
	I, 1-Naphthyl			II, 1-Naphthyl (43)	
	I, 2-Naphthyl			II, 2-Naphthyl (38)	

	Reactant		Reagent	Conditions	Product(s) (Yield, %)	Refs.
C_{14}	[structure]		$KCN-NH_4Cl$	Aq DMF, 76–81°	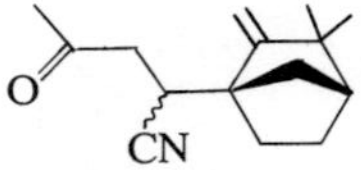 [structure] (High)	224
C_{15}	$C_6H_5COCH{=}CHC_6H_5$		HCN	C_2H_5OH, reflux	$C_6H_5COCH_2CH(CN)C_6H_5$ (I, Low)	278
			HCN-KCN	CH_3OH, 50–55°	I (77–82)	129
			$(CH_3)_2C(OH)CN$-(Na_2CO_3)	Aq CH_3OH, reflux	I (92)	50
			$(CH_3)_2C(OH)CN$-(methanolic KOH)	C_2H_5OH, ”	$C_6H_5COCH_2CH(C_6H_5)$-$C(CN)(C_6H_5)CH_2COC_6H_5$ (II, 66)	50
			KCN	Aq CH_3OH, reflux	II (76)	11
			KCN-AcOH	Aq C_2H_5OH, 35°	I (93–96)	71, 9, 70,
			$KCN-H_2SO_4$	” , reflux	I (74)	75
	$ArCOCH{=}CHAr'$				$ArCOCH_2CH(CN)Ar'$ (I), $ArCOCH_2CH(CO_2H)Ar'$ (II)	
	Ar:	Ar′:				
	C_6H_5	o-ClC_6H_4	HCN-KCN	CH_3OH, 50–55°	I (76)	129
	C_6H_5	p-ClC_6H_4	HCN-KCN	CH_3OH, 50–55°	I (89)	129
	C_6H_5	m-IC_6H_4	$(CH_3)_2C(OH)CN$-(Na_2CO_3)	Aq CH_3OH, reflux	I (65)	50
	C_6H_5	p-IC_6H_4	$(CH_3)_2C(OH)CN$-(Na_2CO_3)	Aq CH_3OH, reflux	I (83)	50
	C_6H_5	p-HOC_6H_4	$KCN-H_2SO_4$	Aq C_2H_5OH, reflux	I (68)	75
	C_6H_5	o-$CH_3OC_6H_4$	$(CH_3)_2C(OH)CN$-(Na_2CO_3)	Aq CH_3OH, reflux	I (82)	50
	C_6H_5	m-$CH_3OC_6H_4$	$(CH_3)_2C(OH)CN$-(Na_2CO_3)	Aq CH_3OH, reflux	I (83)	50

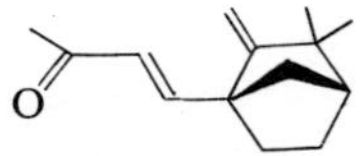

TABLE V. CONJUGATE HYDROCYANATION OF ACYCLIC α,β-ETHYLENIC KETONES (*Continued*)

Substrate			Reagent(s)	Reaction Conditions	Product(s) and Yield(s) (%)	Refs.
C_{15} (*Contd.*)	C_6H_5	p-$CH_3OC_6H_4$	HCN-KCN	CH_3OH, 50–55°	I (72)	129
			$(CH_3)_2C(OH)CN$-(Na_2CO_3)	Aq CH_3OH, reflux	I (81)	50
	C_6H_5	p-$(CH_3)_2NC_6H_4$	$(CH_3)_2C(OH)CN$-(Na_2CO_3)	Aq CH_3OH, reflux	I (70)	50
	m-ClC_6H_4	C_6H_5	$(CH_3)_2C(OH)CN$-(Na_2CO_3)	Aq CH_3OH, reflux	I (22)	50
	p-ClC_6H_4	C_6H_5	$(CH_3)_2C(OH)CN$-(Na_2CO_3)	Aq CH_3OH, reflux	I (86)	50
			KCN-AcOH	Aq C_2H_5OH, 35° or reflux	I (87–96)	176, 177
	m-BrC_6H_4	C_6H_5	$(CH_3)_2C(OH)CN$-(Na_2CO_3)	Aq CH_3OH, reflux	I (64)	50
	p-BrC_6H_4	C_6H_5	$(CH_3)_2C(OH)CN$-(Na_2CO_3)	Aq CH_3OH, reflux	I (89)	50
			KCN-AcOH	Aq C_2H_5OH, 35°	I (84)	176
	m-IC_6H_4	C_6H_5	$(CH_3)_2C(OH)CN$-(Na_2CO_3)	Aq CH_3OH, reflux	I (87)	50
	p-IC_6H_4	C_6H_5	$(CH_3)_2C(OH)CN$-(Na_2CO_3)	Aq CH_3OH, reflux	I (90)	50
	p-$CH_3OC_6H_4$	C_6H_5	KCN-H_2SO_4	Aq C_2H_5OH, reflux	I (86)	75
			KCN-AcOH	″ , 35° or reflux	I (60–65)	130, 176
	m-$O_2NC_6H_4$	C_6H_5	$(CH_3)_2C(OH)CN$-(Na_2CO_3)	Aq CH_3OH, reflux	I (62)	50
	p-$CH_3OC_6H_4$	o-ClC_6H_4	HCN-KCN	CH_3OH, 50–55°	I (85)	129
	p-$CH_3OC_6H_4$	p-$CH_3OC_6H_4$	KCN-H_2SO_4	Aq C_2H_5OH, reflux	I (66)	75
	p-$CH_3OC_6H_4$	2,4-$(CH_3O)_2C_6H_3$	KCN-H_2SO_4	Aq C_2H_5OH, reflux	I (83)	75
			1. KCN-AcOH 2. Aq H_2SO_4	″ , ″ Hydrolysis	II (—)	279

	p-$CH_3OC_6H_4$ 3,4-$(CH_3O)_2C_6H_3$	KCN-H_2SO_4	Aq C_2H_5OH, reflux	I (79)	75
		1. KCN-AcOH 2. Aq H_2SO_4	" " Hydrolysis	II (—)	279
	o-$NO_2C_6H_4CH{=}CHCOC_6H_5$	KCN	Aq DMF, 100°	4-cyano-2-C_6H_5-quinoline N-oxide (—), 1-hydroxy-3-cyano-2-COC_6H_5-indole (—)	172
	$C_6H_5COCH_2CH(Cl)C_6H_5$	KCN	Aq C_2H_5OH, mixed hot	$C_6H_5COCH_2CH(CN)C_6H_5$ (I, 7–9), $C_6H_5COCH{=}CHC_6H_5$ (II, 92)	278
		KCN	Aq C_2H_5OH, reflux	I (92–96), II (Minor)	278
		KCN	" , 60°	I (Quant)	280
	$ArCOCH(C_6H_5)CH_2\overset{+}{N}(CH_3)_3I^-$ (I) I, Ar = p-FC_6H_4 I, Ar = C_6H_5	KCN	MeOH, room temp	$ArCOCH(C_6H_5)CH_2CN$ (II) II, Ar = p-FC_6H_4 (91) II, Ar = C_6H_5 (97)	53
C_{16}	$ArCOCH{=}CHAr'$			$ArCOCH_2CH(CN)Ar'$ (I), $ArCOCH_2CH(CO_2H)Ar'$ (II)	
	Ar: C_6H_5 Ar′: p-$CH_3C_6H_4$	$(CH_3)_2C(OH)CN$-(Na_2CO_3)	Aq CH_3OH, reflux	I (99)	50

TABLE V. Conjugate Hydrocyanation of Acyclic α,β-Ethylenic Ketones (*Continued*)

	Substrate		Reagent(s)	Reaction Conditions	Product(s) and Yield(s) (%)	Refs.
C_{16} (*Contd.*)						
	o-$CH_3C_6H_4$	C_6H_5	$(CH_3)_2C(OH)CN$-(Na_2CO_3)	Aq CH_3OH, reflux	I (94)	50
	p-$CH_3C_6H_4$	C_6H_5	HCN-KCN	CH_3OH, 50–55°	I (51)	129
	p-$CH_3C_6H_4$	p-$CH_3OC_6H_4$	HCN-KCN	CH_3OH, 50–55°	I (67)	129
	p-$CH_3OC_6H_4$	2-CH_3-4-$CH_3OC_6H_3$	1. KCN-AcOH 2. Aq H_2SO_4	Aq C_2H_5OH, reflux Hydrolysis	II (—)	279
	p-$CH_3OC_6H_4$	3-CH_3-4-$CH_3OC_6H_3$	1. KCN-AcOH 2. Aq H_2SO_4	Aq C_2H_5OH, reflux Hydrolysis	II (—)	279
	p-$CH_3OC_6H_4$	3-CH_3-2-$CH_3OC_6H_3$	1. KCN-AcOH 2. Aq H_2SO_4	Aq C_2H_5OH, reflux Hydrolysis	II (—)	279
	2-CH_3-4-$CH_3OC_6H_3$	o-$CH_3OC_6H_4$	1. KCN-AcOH 2. Aq H_2SO_4	Aq C_2H_5OH, reflux Hydrolysis	II (—)	281
	3-CH_3-4-$CH_3OC_6H_3$	o-$CH_3OC_6H_4$	KCN-AcOH	—	No reaction	281
C_{17}	$CO(CH{=}CHC_6H_5)_2$		$(CH_3)_2C(OH)CN$-(Na_2CO_3)	Aq CH_3OH, reflux	$CO(CH_2CH(CN)C_6H_5)_2$ (43)	50
			$(CH_3)_2C(OH)CN$-$[N(C_2H_5)_3]$	CH_3OH, ”	$C_6H_5CH{=}CHCOCH_2CH(CN)C_6H_5$ (23)	50
C_{18}	$C_6H_5COCH{=}CHC_6H_2(CH_3)_3$-2,4,6		KCN-AcOH	Aq C_2H_5OH, 45–50°	$C_6H_5COCH_2CH(CN)$-$C_6H_2(CH_3)_3$-2,4,6 (84)	132
C_{21}	2,4,6-$(CH_3)_3C_6H_2COCH{=}CH$-$C_6H_2(CH_3)_3$-2,4,6		KCN-AcOH	Aq C_2H_5OH, 45°	2,4,6-$(CH_3)_3C_6H_2COCH_2$-$CH(CN)C_6H_2(CH_3)_3$-2,4,6 (92)	132

C_{22}	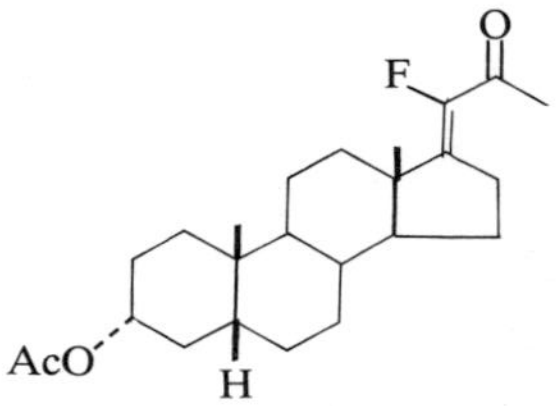	KCN-NH_4Cl	—	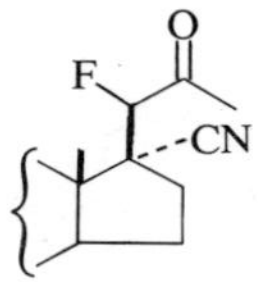 (—), 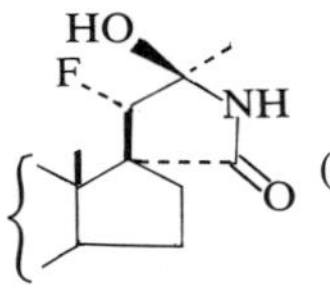 (—), 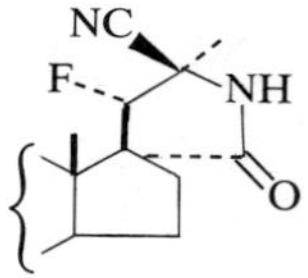 (—), 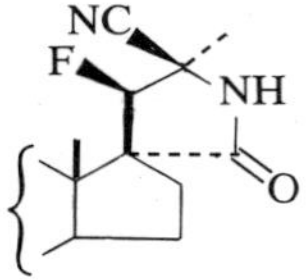 (—),	282
C_{23}	$C_6H_5CH_2COC(C_6H_5)$= $C(OCH_3)CH_2C_6H_5$	$(CH_3)_2C(OH)CN$- (base)		No reaction	123
C_{24}	$C_6H_4(CH{=}CHCOC_6H_5)_2$ (I)	$(CH_3)_2C(OH)CN$- (Na_2CO_3)	Aq CH_3OH, reflux	$C_6H_4(CH(CN)CH_2COC_6H_5)_2$ (II)	50
	m-I			*m*-II (70)	
	p-I			*p*-II (72)	

Note: References 246–305 are on pp. 475-476.

TABLE VI. Conjugate Hydrocyanation of Cyclic α,β-Ethylenic Ketones

	Substrate	Reagent(s)	Reaction Conditions	Product(s) and Yield(s) (%)	Refs.
C_6	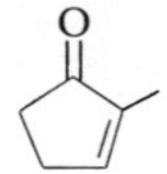 O	$(CH_3)_2C(OH)CN$-(Na_2CO_3)	Aq CH_3OH, reflux	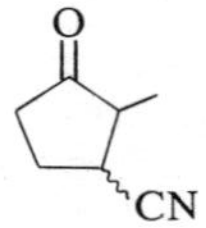 O, CN (2 isomers, 77)	14
	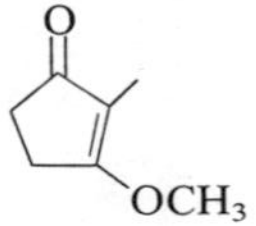 O, OCH_3	KCN-AcOH	95% C_2H_5OH, reflux	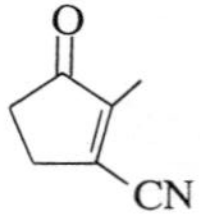 O, CN (I, poor)	86
		KCN-NH_4Cl	Aq THF, reflux	I (*ca.* 50)	86
	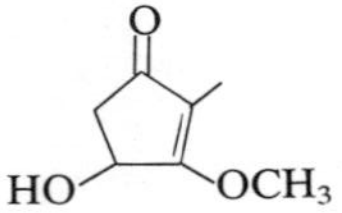 O, HO, OCH_3	KCN-NH_4Cl	Aq THF, reflux	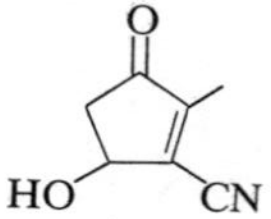 O, HO, CN (10)	86
	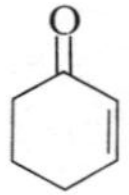 O (I)	$(CH_3)_2C(OH)CN$-(Na_2CO_3)	Aq CH_3OH, reflux	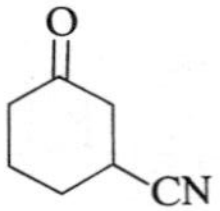 O, CN (II, 50)	235

	KCN	Aq C_2H_5OH, reflux	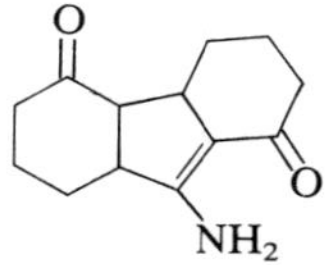 (53)	101
	NaCN-AcOH	Aq C_2H_5OH, —	II (44)	283
	$KCN-NH_4Cl$	Aq DMF, room temp	II (22)	235
	$HCN-Al(C_2H_5)_3$	$(C_2H_5)_2O$, $-15°$	II (80)	43
	$(C_2H_5)_2AlCN$	C_6H_6-hexane, $-15°$, 1 hr	I (13), II (57)	43
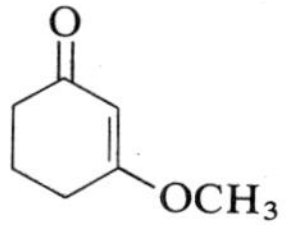	$(C_2H_5)_2AlCN$	C_6H_6-toluene, 0–25°, 1 hr	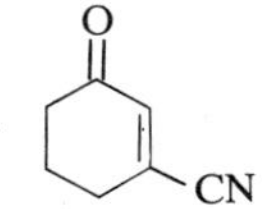 (55)	284
C_7 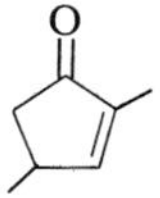	$(CH_3)_2C(OH)CN$-(Na_2CO_3)	Aq CH_3OH, reflux	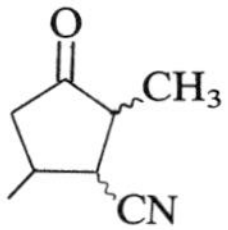 (mixed isomers, total 55)	14
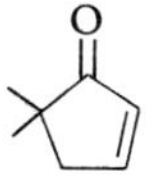	$(C_2H_5)_2AlCN$	C_6H_6, room temp, 3 hr	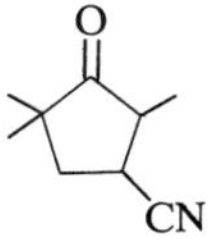 (100)	125

TABLE VI. CONJUGATE HYDROCYANATION OF CYCLIC α,β-ETHYLENIC KETONES (*Continued*)

Substrate	Reagent(s)	Reaction Conditions	Product(s) and Yield(s) (%)	Refs.
C_7 (*Contd.*)				
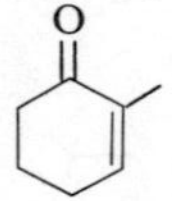	$(CH_3)_2C(OH)CN$-(Na_2CO_3)	Aq CH_3OH, reflux	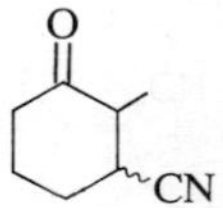 (2 isomers, total 73)	14
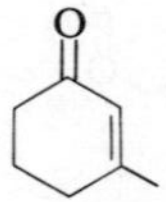	NaCN on $NaHSO_3$ adduct	H_2O, 100°	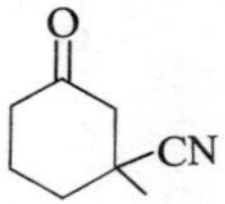 (I, 32)	64
	KCN-AcOH	C_2H_5OH, 0°	I (73)	133
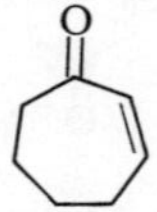	KCN-AcOH	Aq C_2H_5OH, 0°	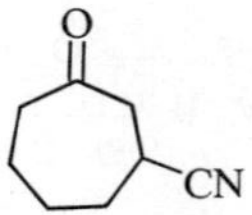 (I, 38)	285
	"	Aq CH_3OH, room temp	I (minor), 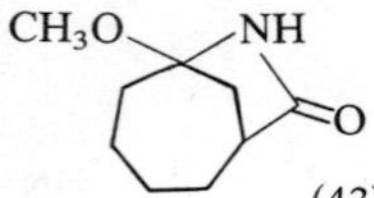 CH_3O … NH … O (43)	73
	"	THF-H_2O (2 layers), reflux	I (47)	73
C_8 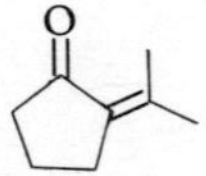	$(C_2H_5)_2AlCN$	Toluene, room temp, 2.5 hr	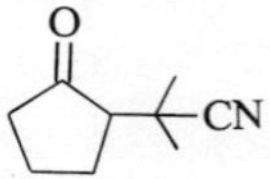 CN (46)	99

Substrate	Reagent	Conditions	Product(s) (Yield, %)	Refs.
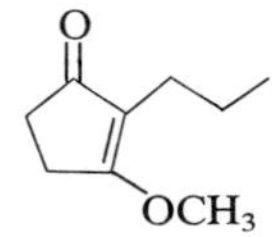 O, OCH$_3$	$(CH_3)_2C(OH)CN$-$N(C_2H_5)_3$	No solvent, room temp	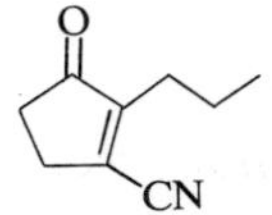 O, CN (I, 17)	86
	”	No solvent, reflux	I (8)	86
	HCN-$Al(C_2H_5)_3$	THF, room temp	I (86)	86
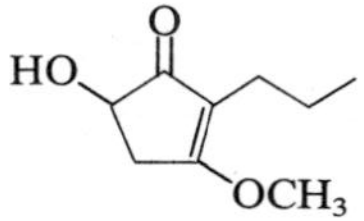 HO, O, OCH$_3$	KCN-NH_4Cl	Aq THF, reflux	No reaction	86
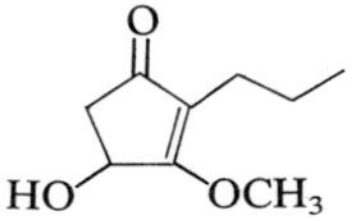 O, HO, OCH$_3$	KCN-NH_4Cl	Aq THF, reflux	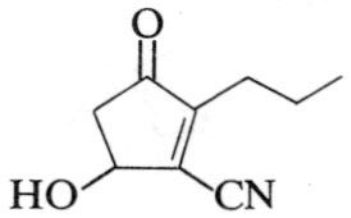 O, HO, CN (27)	86
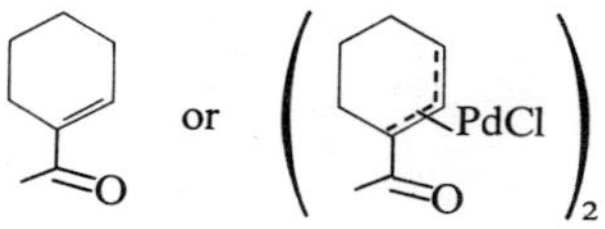 O or (PdCl, O)$_2$	$Ca(CN)_2$	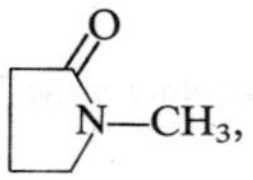 O, N—CH_3, room temp	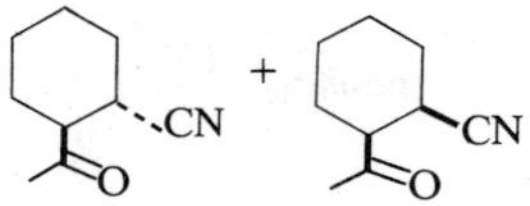 O, CN + O, CN (1:1, total 65)	153

TABLE VI. Conjugate Hydrocyanation of Cyclic α,β-Ethylenic Ketones (*Continued*)

Substrate	Reagent(s)	Reaction Conditions	Product(s) and Yield(s) (%)	Refs.
C_8 (*Contd.*) 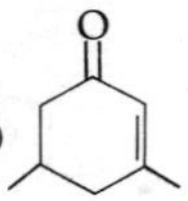 O	KCN-HCl	H_2O, room temp	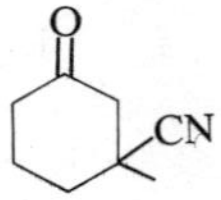 O CN (I, good)	12
	NaCN on $NaHSO_3$ adduct	H_2O, 100°, 30 min	I (61–79)	13, 64, 178
	NaCN on $NaHSO_3$ adduct	", ", 4–6 hr	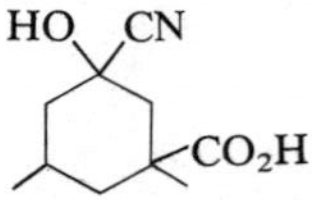 HO CN CO_2H (—)	13
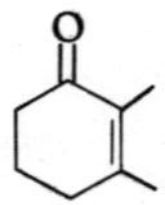 O	1. KCN 2. KOH	Aq CH_3OH, reflux Hydrolysis	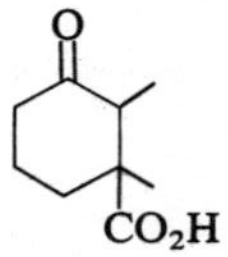 O CO_2H (High)	55
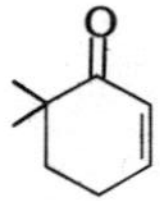 O	$KCN\text{-}NH_4Cl$	Aq DMF, 100°	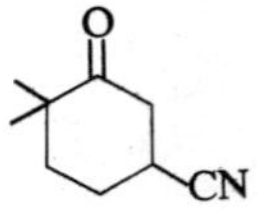 O CN (48)	225
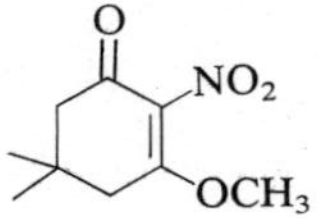 O NO_2 OCH_3	$(CH_3)_2C(OH)CN$- [methanolic KOH or $N(C_2H_5)_3$]	No solvent, room temp, −90°	No reaction	123

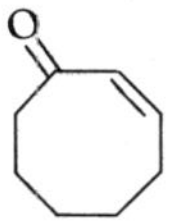

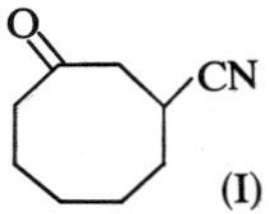

Carbon	Substrate	Reagent	Conditions	Product(s) (Yield, %)	Refs.
	[2-cyclooctenone]			(I) [3-cyanocyclooctanone], (II) [methoxy bicyclic lactam]	
		KCN-AcOH	Aq CH_3OH, room temp	I:II = 65:35 (Total —)	73
		KCN-AcOH	$THF\text{-}H_2O$ (2 layers), reflux	I (39)	73
C_9 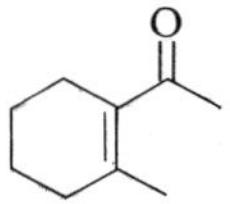	[1-acetyl-2-methylcyclohexene]	HCN-NaCN	DMF, 145–155°	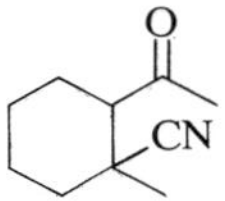 [2-acetyl-1-methylcyclohexanecarbonitrile] (76)	16
	[5-ethyl-3-methyl-2-cyclohexenone]	NaCN on $NaHSO_3$ adduct	H_2O, 100°	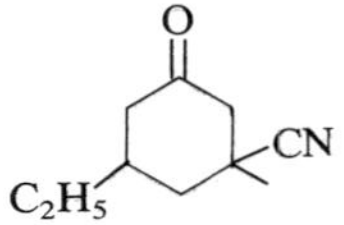 [5-ethyl-3-methyl-3-cyanocyclohexanone] (30)	64, 178
	[isophorone] (I)	HCN	—	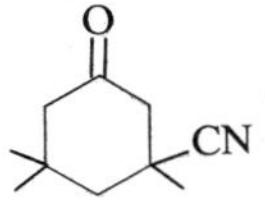 [3-cyano-3,5,5-trimethylcyclohexanone] (II, 18–20)	286
		HCN-(K_2CO_3)	$CH_3CON(CH_3)_2$, 175°	II (66)	16
		NaCN-AcOH	Aq CH_3OH, room temp	I (28), I+II (14), II (43)	64, 178
		NaCN on $NaHSO_3$ adduct	H_2O, 90°	No reaction	178

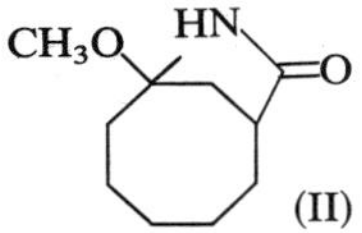

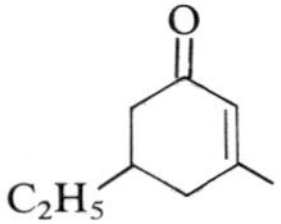

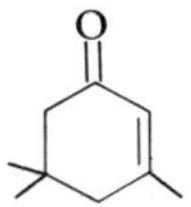

TABLE VI. CONJUGATE HYDROCYANATION OF CYCLIC α,β-ETHYLENIC KETONES (*Continued*)

	Substrate	Reagent(s)	Reaction Conditions	Product(s) and Yield(s) (%)	Refs.
C_9 (*Contd.*) 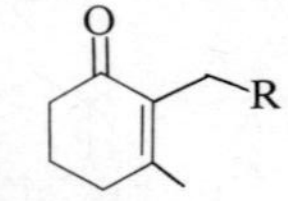	2-(RCH$_2$)-3-methylcyclohex-2-enone (I) I, R = CH_3 I, R = CO_2H	1. KCN 2. KOH	Aq CH_3OH, reflux " , "	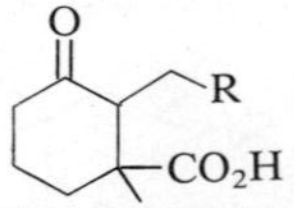 2-(RCH$_2$)-3-methyl-3-carboxycyclohexanone (II) II, R = CH_3 (65) II, R = CO_2H (65)	55
C_{10} 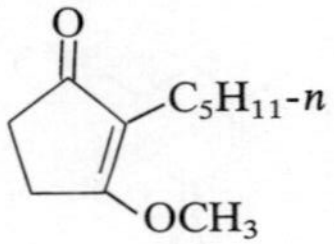	2-C_5H_{11}-n-3-OCH_3-cyclopent-2-enone	KCN-NH_4Cl	Aq THF, reflux	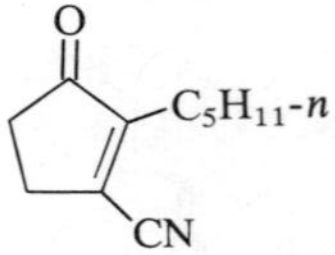 2-C_5H_{11}-n-3-CN-cyclopent-2-enone (13)	86
	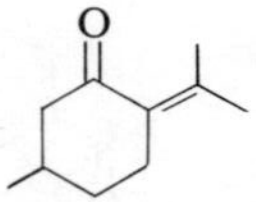 5-methyl-2-isopropylidenecyclohexanone (I)	HCN-(KCN)	—, cold	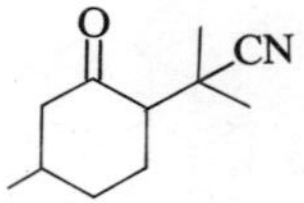 2-(1-cyano-1-methylethyl)-5-methylcyclohexanone (II, —)	287
			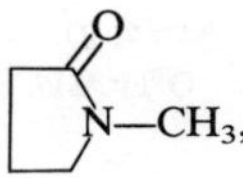 N-methylpyrrolidone (N—CH_3), 180°	II (Crude, 91)	16
		KCN-AcOH	Aq C_2H_5OH, reflux	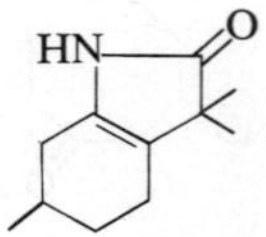 3,3,6-trimethyl-tetrahydroindol-2-one (HN, C=O) (86)	98

Substrate	Reagent	Conditions	Product(s) and Yield(s) (%)	Refs.
I or (PdCl)$_2$ 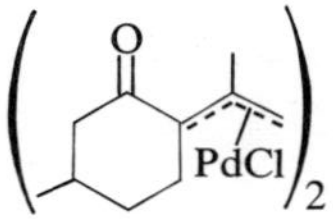	$KCN-NH_4Cl$	Aq DMF, 90–100°	II + unknown (9:1) (total 56)	153
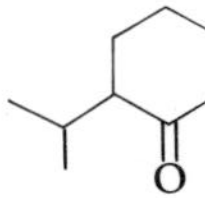 or 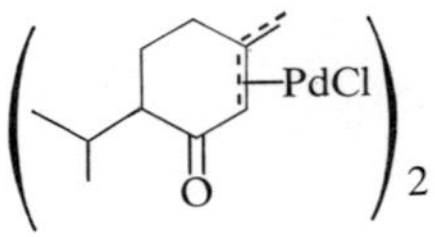 (PdCl)$_2$	$KCN-NH_4Cl$	Aq DMF, 90–100°	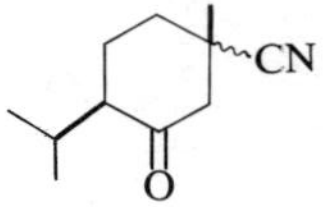 CN (7:3 mixture, total 88)	153
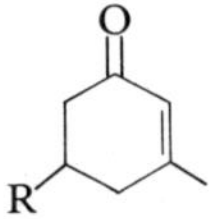 R (I) I, R = n-C_3H_7 I, R = i-C_3H_7	NaCN on $NaHSO_3$ adduct	H_2O, 100°	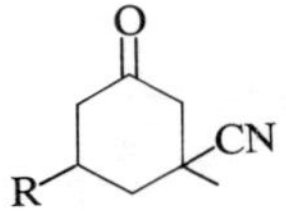 CN R (II) II, R = n-C_3H_7 (67) II, R = i-C_3H_7 (32)	64, 178
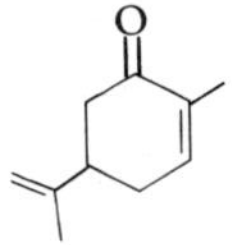			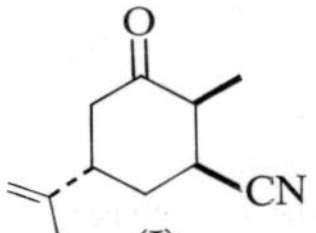 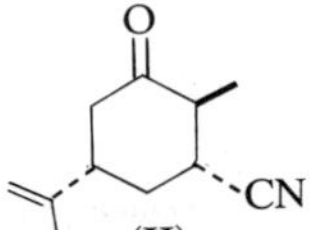 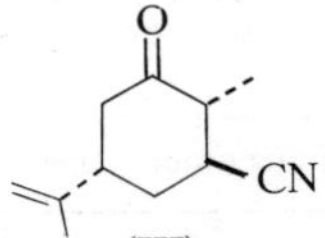 CN (I), CN (II), CN (III)	

TABLE VI. Conjugate Hydrocyanation of Cyclic α,β-Ethylenic Ketones (*Continued*)

Substrate	Reagent(s)	Reaction Conditions	Product(s) and Yield(s) (%)	Refs.
C_{10} (*Contd.*) 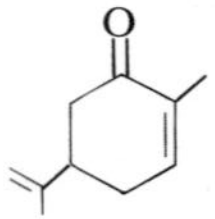 O	HCN-(KCN)	—, cold	I, II, and/or III (Quant)	287
	KCN-$AcOC_2H_5$	Aq C_2H_5OH, reflux	I+II (1:1, Major), III (trace) (total 96)	69, 68
	KCN-AcOH	,, , room temp	I (Major), II (trace), III (minor) (total 74)	69, 288
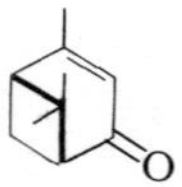 O	HCN	No solvent, 150°	 CN O (—)	289
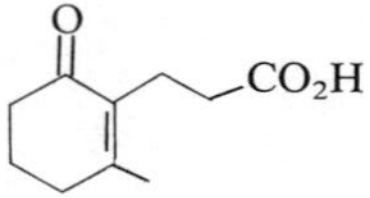 O CO_2H	1. KCN 2. KOH	Aq CH_3OH, reflux ,, , ,,	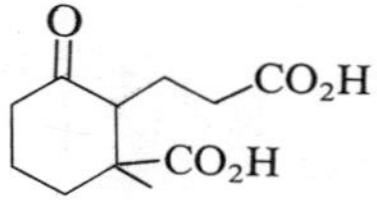 O CO_2H CO_2H (80)	56, 55
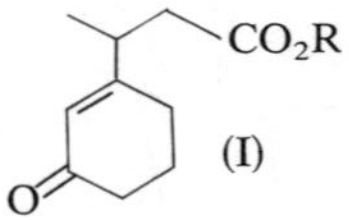 CO_2R O (I)				
I, R = H	KCN-NH_4Cl	Aq DMF, 97°	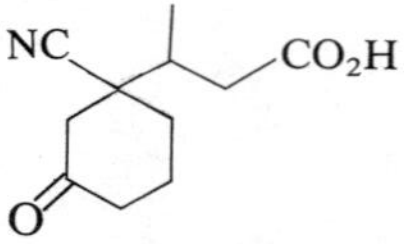 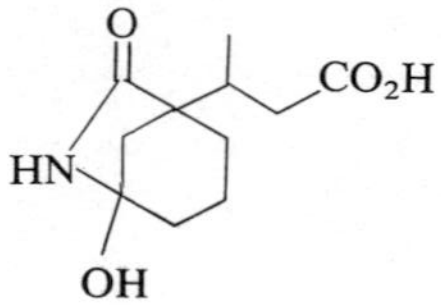 NC CO_2H O (17), O CO_2H HN OH (Major)	104

I, $R = CH_3$	KCN	Aq CH_3OH, reflux	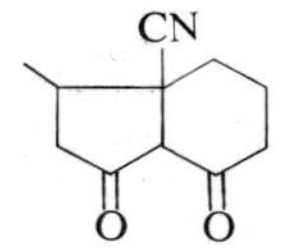 CN (High)	
H, H; (I)	$(CH_3)_2C(OH)CN$-(Na_2CO_3)	Aq CH_3OH-THF	CN, H, H (27); H, CN (II), H, CN (III), HO, H, HN, O (IV), HO, H, HN, O (V)	235

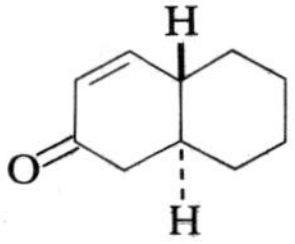

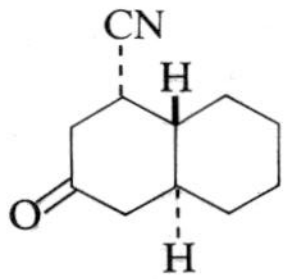

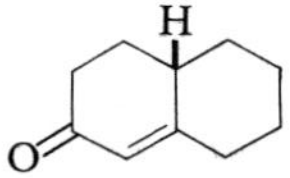

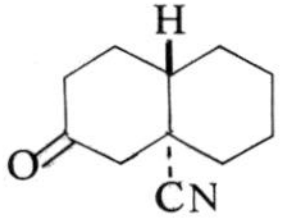

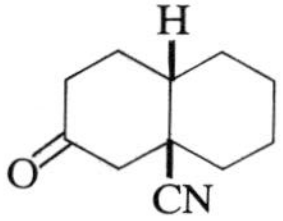

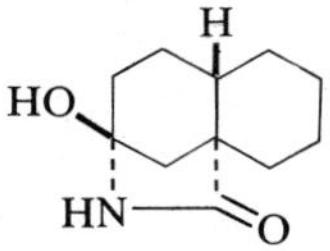

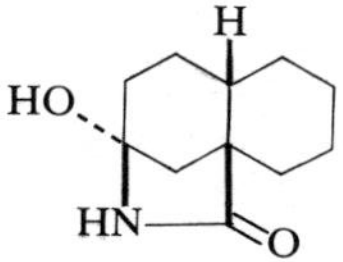

TABLE VI. CONJUGATE HYDROCYANATION OF CYCLIC α,β-ETHYLENIC KETONES (*Continued*)

Substrate	Reagent(s)	Reaction Conditions	Product(s) and Yield(s) (%)	Refs.
C_{10} (*Contd.*) 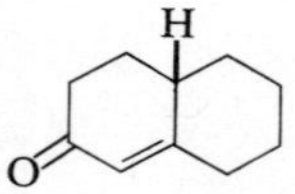 (I)	HCN-(pip)	C_2H_5OH, room temp	I (15), II (42)	181
	KCN	Aq C_2H_5OH, room temp	IV (23)	181
	”	Aq CH_3OH, 75°	IV (36), V (14)	97
	$(CH_3)_2C(OH)CN$-KCN-18-crown-6	C_6H_6, room temp	II + III (3.4:1, total 85), dinitrile (15)	51b
	$(CH_3)_2C(OH)CN$-KCN-18-crown-6	CH_3CN, room temp	I (30), II + III (2.7:1, total 35), dinitrile (35)	51b
	$(CH_3)_2C(OH)CN$-KCN-18-crown-6	” , reflux	II + III (2.6:1, total 64), dinitrile (36)	51b
	KCN-AcOH	Aq C_2H_5OH, 0°	II + III (1:1, total 68)	182
	KCN-NH_4Cl	Aq CH_3OH, reflux	II (17), III (8), 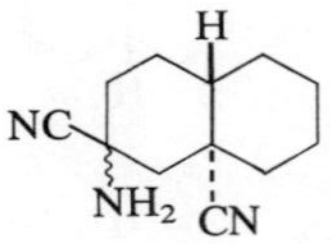 NC, NH_2, CN, H (22)	102
	”	Aq DMF, room temp	II (44), III (22)	102, 290, 291
	”	” , 100°	II (33), III (1), dimer ($C_{22}H_{30}O_2N_2$, 18)	102
	HCN-$Al(C_2H_5)_3$	THF, room temp	II (72), III (4)	102, 84
	”	” , 25°	II:III (89:11, total quant)	135
	HCN-$(C_2H_5)_2AlCl$	” , room temp	II (70), III (9)	43
	$(C_2H_5)_2AlCN$	C_6H_6, 25°, 2 min	II:III (87:13, total quant)	135
	”	” , ” , 6 hr	II:III (84:16, total quant)	135

	Reactant	Reagent	Conditions	Product (Yield, %)	Ref.
	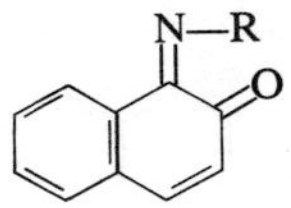 (I)	KCN		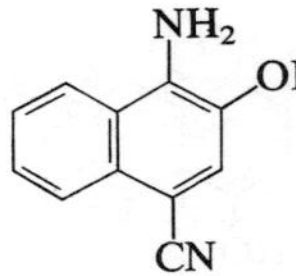 (II)	180
	I, R = OH		H_2O, 80–90°	II (69)	
	I, R = NHC_6H_5		Aq C_2H_5OH, reflux	II (69)	
	I, R=NHC_6H_4—SO_3Na-*p*		H_2O, reflux	II (49)	
			,, , 85–90°	II (63)	
C_{11}	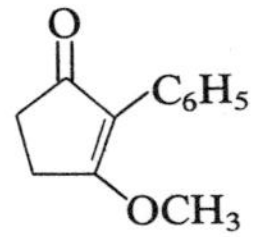	$(CH_3)_2C(OH)CN$- [methanolic KOH or $N(C_2H_5)_3$]	No solvent, room temp –90°	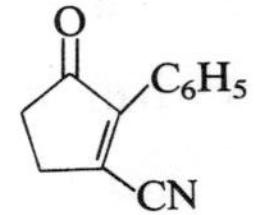 (76)	123
	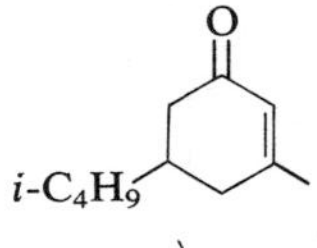	NaCN on $NaHSO_3$ adduct	H_2O, 100°	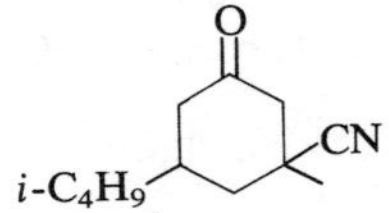 (56)	64
		KCN-NH_4Cl	Aq DMF, 100°	(mixed isomers, 30),	185
				(mixed isomers, 49)	

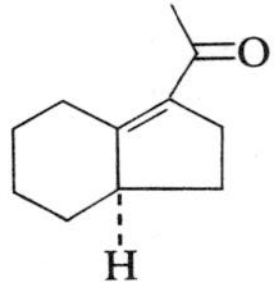

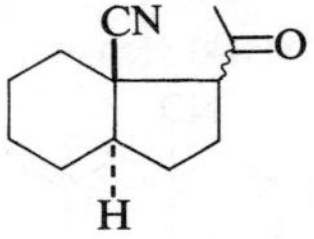

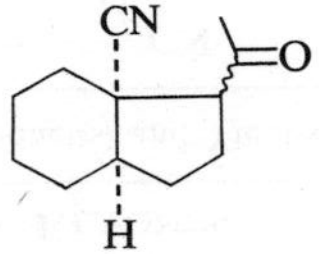

TABLE VI. CONJUGATE HYDROCYANATION OF CYCLIC α,β-ETHYLENIC KETONES (*Continued*)

Substrate	Reagent(s)	Reaction Conditions	Product(s) and Yield s) (%)	Refs.
C_{11} (*Contd.*) 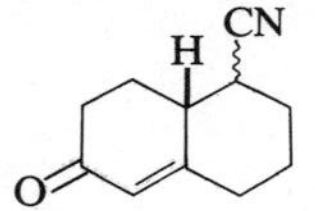 (I)	$HCN-Al(C_2H_5)_3$	THF, room temp	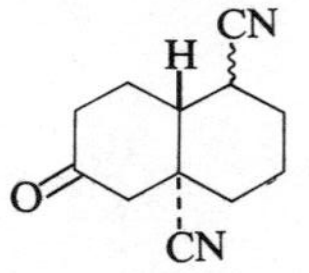 (II)	94
I, α-CN			II, α-CN (86)	
I, β-CN			II, β-CN (84)	
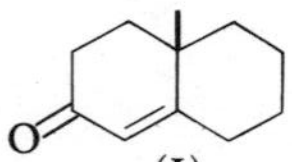 (I)			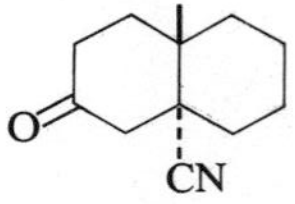 (II), 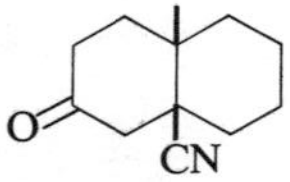 (III)	
	$(CH_3)_2C(OH)CN-KCN-$ 18-crown-6	C_6H_6, room temp	II + III (1:4, total 85)	51b
	KCN-AcOH	Aq C_2H_5OH, room temp	I (8), II (28), III (2) [II + III (76)]	74
	$KCN-NH_4Cl$	Aq DMF, room temp	II:III (57:43, total quant)	135
	$HCN-Al(C_2H_5)_3$	THF, 25°	II:III (71:29, total quant)	135, 292
	$(C_2H_5)_2AlCN$	C_6H_6, 25°, 2 min	II:III (64:36, total quant)	135
	”	” , ” , 20 hr	II:III (13:87, total quant)	135
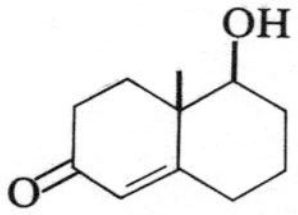	KCN-AcOH	Aq C_2H_5OH, room temp	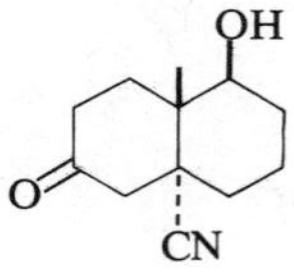 +	146, 74

	Substrate	Cyanating Agent	Conditions	Product(s) and Yield(s) (%)	Refs.
				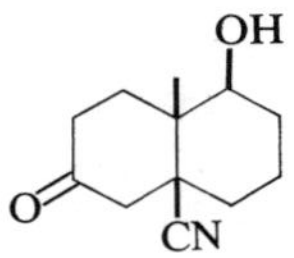 O; OH; CN — 2:1 (Total —)	
C_{12}	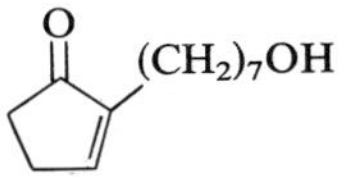 O; $(CH_2)_7OH$	$(CH_3)_2C(OH)CN$-(Na_2CO_3)	Aq CH_3OH, — 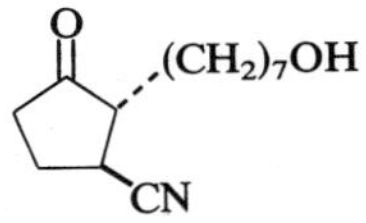	O; $(CH_2)_7OH$; CN (—)	231
	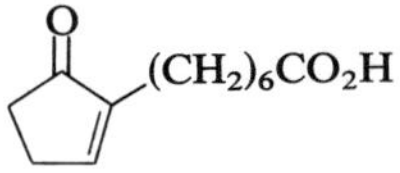 O; $(CH_2)_6CO_2H$	$(CH_3)_2C(OH)CN$-(Na_2CO_3)	Aq CH_3OH, — 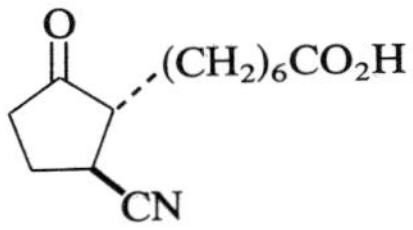	O; $(CH_2)_6CO_2H$; CN (—)	230
	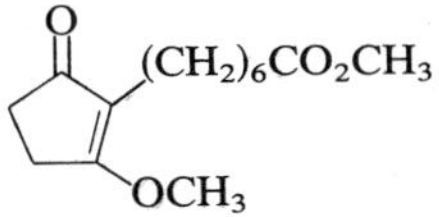 O; $(CH_2)_6CO_2CH_3$; OCH_3	HCN-$Al(C_2H_5)_3$	THF, room temp 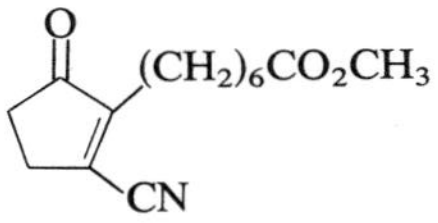	O; $(CH_2)_6CO_2CH_3$; CN (61)	86
	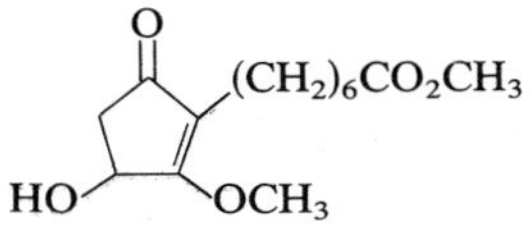 O; $(CH_2)_6CO_2CH_3$; HO; OCH_3	KCN-NH_4Cl	Aq THF, reflux	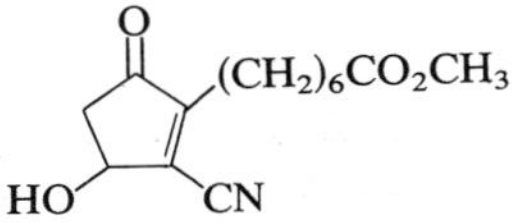 O; $(CH_2)_6CO_2CH_3$; HO; CN (I, 15)	86
		$(C_2H_5)_2AlCN$	C_6H_6-toluene, room temp, 2 hr	I (88)	86

TABLE VI. Conjugate Hydrocyanation of Cyclic α,β-Ethylenic Ketones (*Continued*)

Substrate	Reagent(s)	Reaction Conditions	Product(s) and Yield(s) (%)	Refs.
C_{12} (*Contd.*) [structure]	$KCN-NH_4Cl$	Aq DMF, 100°	[structure] CN ... O (1:2 $\alpha:\beta$ mixture, total 91)	179
[structure] or ([structure] PdCl)$_2$			[structure] CN, O (I) + [structure] CN, O (II)	153
	$Ca(CN)_2$	[structure] N—CH_3, room temp	I:II = 1:2 (total 37)	
	$KCN-NH_4Cl$	Aq DMF, 90–100°	I:II = 6:4 (total 72)	
[structure] O, Ar, OR	$(CH_3)_2C(OH)CN$-(base)	No solvent	[structure] O, Ar, CN (I)	123

Ar: C_6H_5 R: CH_3	Base Methanolic KOH or $N(C_2H_5)_3$	Room temp	I, Ar = C_6H_5 (37)	
C_6H_5 C_2H_5	Methanolic KOH or $N(C_2H_5)_3$	Room temp	I, Ar = C_6H_5 (30–37)	
C_6H_5 $CH_2C_6H_5$	Methanolic KOH or $N(C_2H_5)_3$	Room temp	I, Ar = C_6H_5 (30–40)	
p-$CH_3OC_6H_4$ CH_3	Methanolic KOH	Room temp	I, Ar = p-$CH_3OC_6H_4$ (76)	
	$N(C_2H_5)_3$	,, ,, or 80–90°	I, Ar = p-$CH_3OC_6H_4$ (71–76)	
2,4-$(O_2N)_2C_6H_3$ CH_3	$N(C_2H_5)_3$	80–90°	I, Ar = 2,4-$(O_2N)_2C_6H_3$ (13)	
	Methanolic KOH or $N(C_2H_5)_3$	Room temp	I, Ar = 2,4-$(O_2N)_2C_6H_3$ (25)	
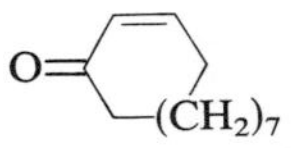 O= ... $(CH_2)_7$	KCN-AcOH	THF-H_2O (2 layers), reflux	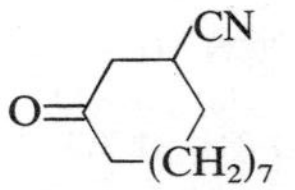 CN, O= ... $(CH_2)_7$ (52)	73
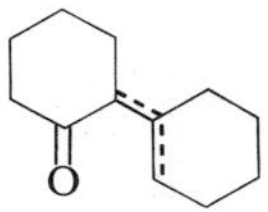 O, Isomeric	HCN-(KCN)	DMF, 160°	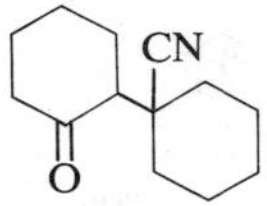 CN, O (83)	16
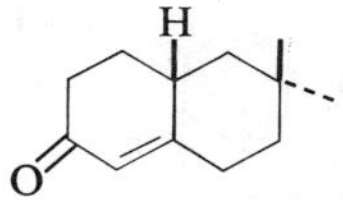 H, O			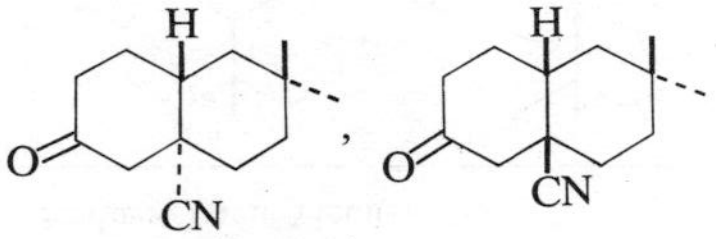 H, O, CN , H, O, CN	293
	KCN-NH_4Cl	Aq DMF, —	5 : 2 (total —)	
	HCN-$Al(C_2H_5)_3$	THF, —	17 : 1 (total —)	

TABLE VI. Conjugate Hydrocyanation of Cyclic α,β-Ethylenic Ketones (*Continued*)

Substrate	Reagent(s)	Reaction Conditions	Product(s) and Yield(s) (%)	Refs.
C_{12} (*Contd.*) 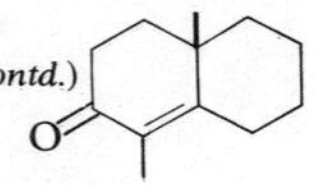 (I)			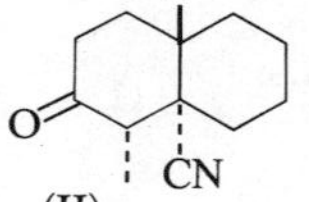 (II), 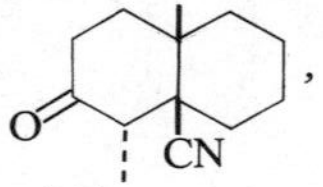 (III), 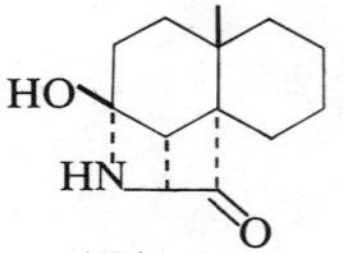 (IV), 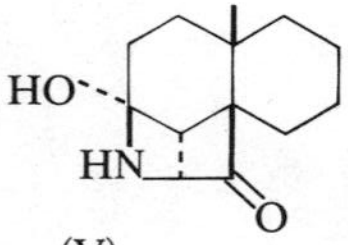 (V)	138
	KCN	Aq C_2H_5OH, reflux	I (3), II (1), IV (20), V (55), IV+V (8)	
	KCN-NH_4Cl	" , "	I (6), II (42), III (18), IV+V (4)	
C_{13} 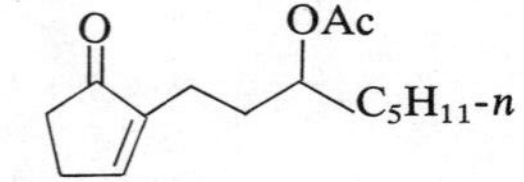 OAc, C_5H_{11}-*n*	$(CH_3)_2C(OH)CN$-(Na_2CO_3)	Aq CH_3OH, —	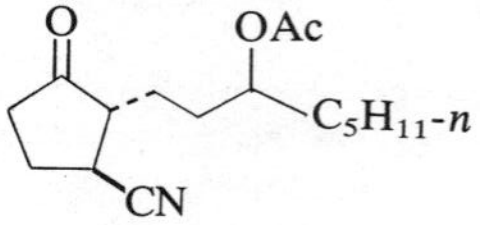 OAc, CN, C_5H_{11}-*n* (—)	108
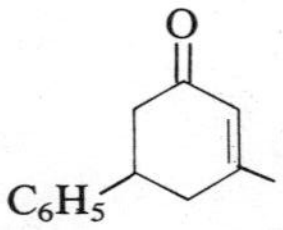 C_6H_5	NaCN on $NaHSO_3$ adduct	H_2O, 100°	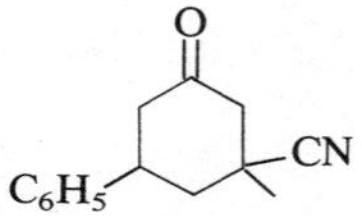 C_6H_5, CN (57)	64, 178

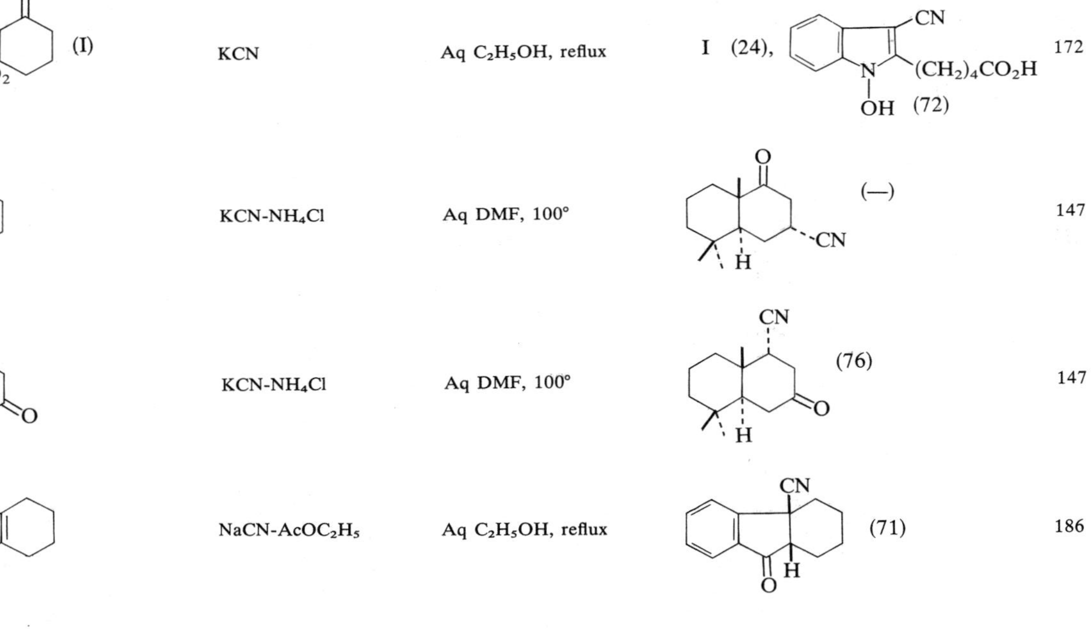

Substrate	Reagent	Conditions	Product (Yield, %)	Ref.
(I)	KCN	Aq C_2H_5OH, reflux	I (24), $(CH_2)_4CO_2H$, OH (72)	172
	KCN-NH_4Cl	Aq DMF, 100°	(—)	147
	KCN-NH_4Cl	Aq DMF, 100°	(76)	147
	NaCN-$AcOC_2H_5$	Aq C_2H_5OH, reflux	(71)	186
C_{14}	KCN-NH_4Cl	Aq DMF, 105°	(42)	229

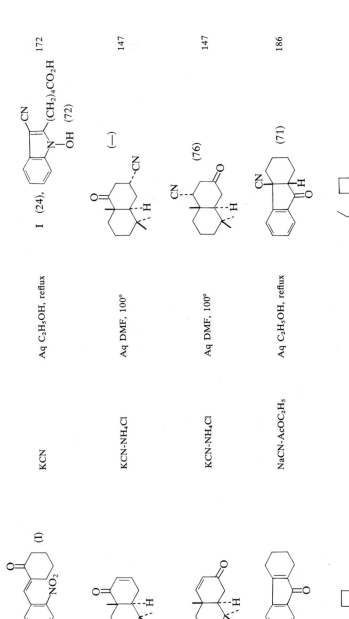

TABLE VI. Conjugate Hydrocyanation of Cyclic α,β-Ethylenic Ketones (*Continued*)

	Substrate	Reagent(s)	Reaction Conditions	Product(s) and Yield(s) (%)	Refs.
C_{14} (*Contd.*)	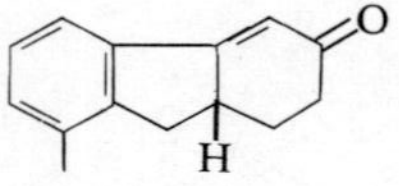 (I)	$HCN\text{-}Al(C_2H_5)_3$	THF, room temp	I (>10), 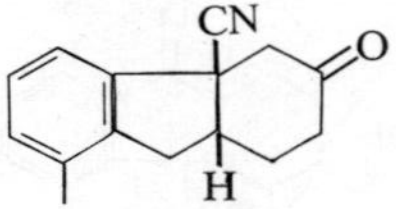 (>10)[a]	226
	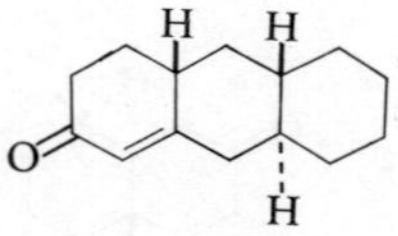	$(C_2H_5)_2AlCN$	—	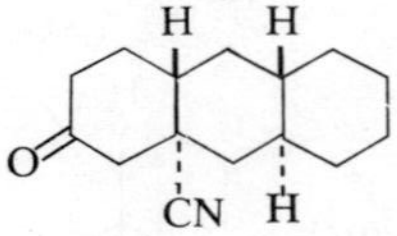 (—)	188
	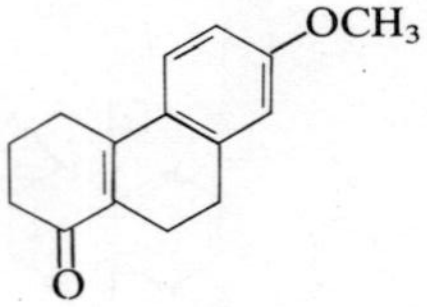			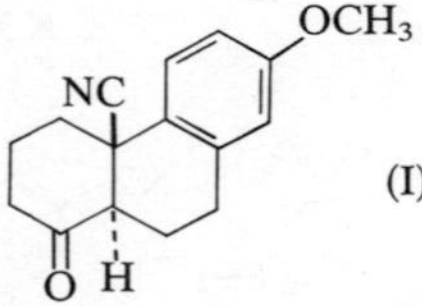 (I), 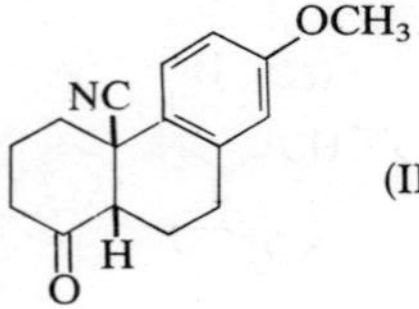 (II)	

	KCN-NH_4Cl	Aq DMF, 100°	No reaction	85
	HCN-$Al(C_2H_5)_3$	THF, room temp	I (11)	85
	HCN- $C_2H_5)_2AlCl$	,, , ,,	I (45), II (30)	85, 84
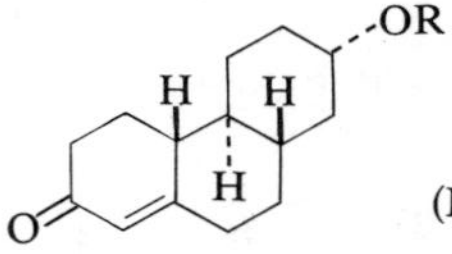 (I)			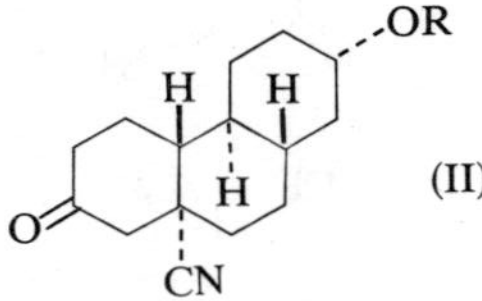 (II) ,	
			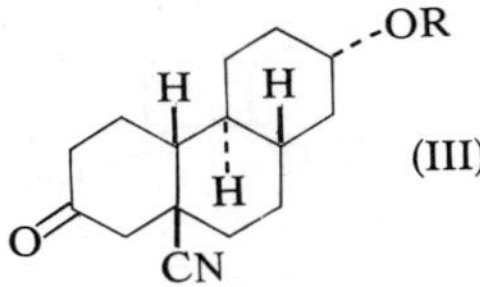 (III)	
I, R = H	KCN-NH_4Cl	Aq DMF, 100°	R = H, II (43), III (32)	43
I, R = Ac	HCN-$Al(C_2H_5)_3$	THF, room temp	R = Ac, II (65), III (10)	43, 84
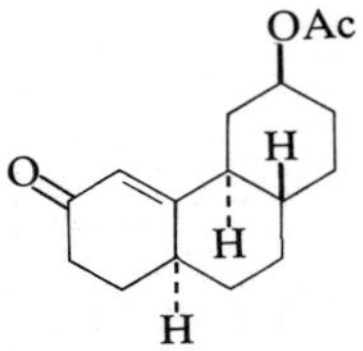			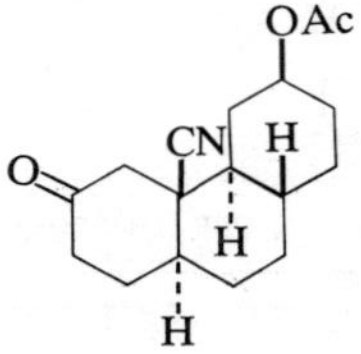 (I),	43, 135

TABLE VI. CONJUGATE HYDROCYANATION OF CYCLIC α,β-ETHYLENIC KETONES (*Continued*)

Substrate	Reagent(s)	Reaction Conditions	Product(s) and Yield(s) (%)	Refs.
C_{14} (*Contd.*)			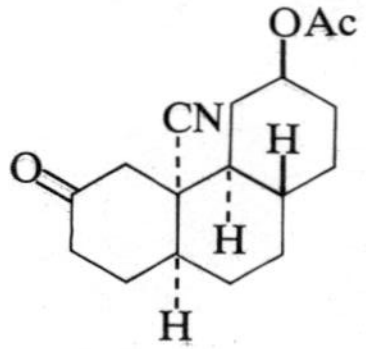 (II)	
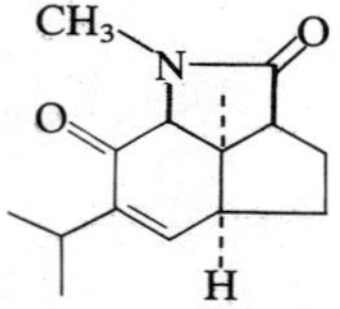 (I)	$HCN-Al(C_2H_5)_3$ $HCN-Al(C_2H_5)_2Cl$	THF, room temp " , "	I (50), II (30) I (41), II (27) 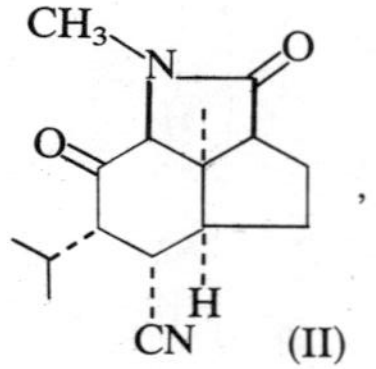 (II), 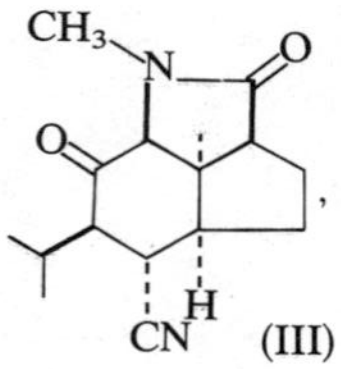 (III), 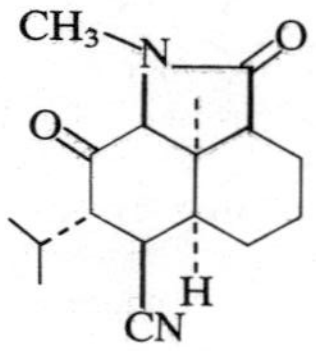 (IV)	82

	Substrate	Reagent	Conditions	Product(s) and Yield(s) (%)	Refs.
		KCN-NH_4Cl	Aq DMF, 100°	No reaction	
		,,	,, , 185–190°	I (36), II (28), III (8), IV (20)	
		$(C_2H_5)_2AlCN$	C_6H_6-toluene, room temp, 3 hr	I (17), II (18), III (20), IV (29)	
C_{15} 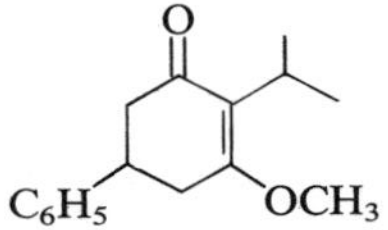		$(CH_3)_2C(OH)CN$-[methanolic KOH or $N(C_2H_5)_3$]	No solvent, room temp −90°	No reaction	123
		KCN-NH_4Cl	Aq DMF, 140–150°	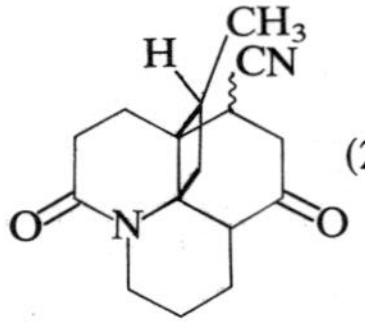 (2 isomers, total 82)	159
	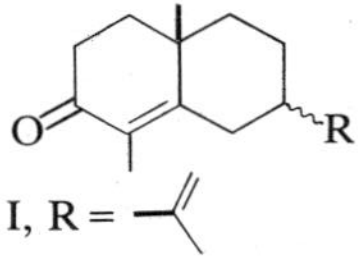 (I) I, R =	KCN-NH_4Cl	Aq C_2H_5OH, reflux	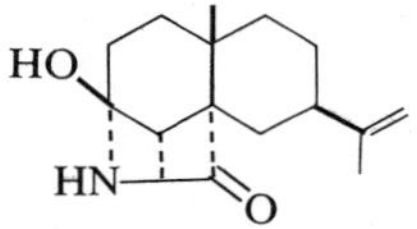 (53), 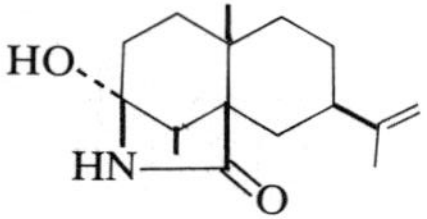 (15)	134
	I, R = or	KCN	Aq CH_3OH or aq C_2H_5OH, reflux	No reaction	134
		KCN-AcOH	Aq C_2H_5OH, 35°	No reaction	134

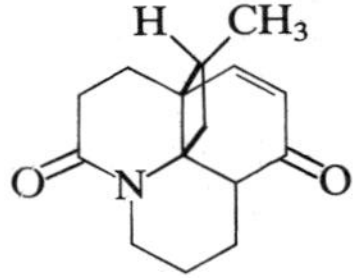

TABLE VI. Conjugate Hydrocyanation of Cyclic α,β-Ethylenic Ketones (*Continued*)

	Substrate	Reagent(s)	Reaction Conditions	Product(s) and Yield(s) (%)	Refs.
C_{16}	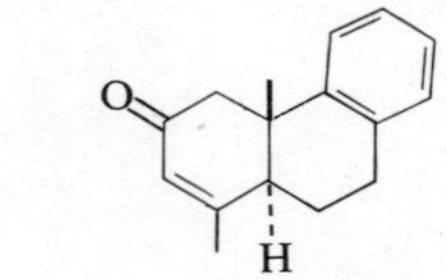 H	—	—	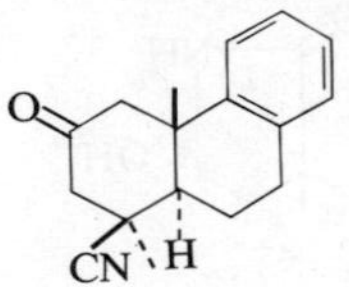 CN H (—)	154c
	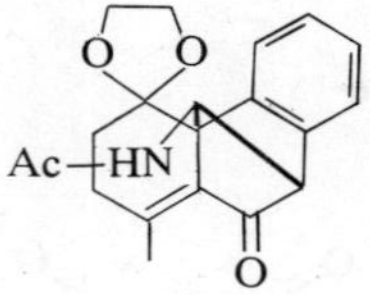 Ac—HN (I)	$KCN-NH_4Cl$	Aq DMF, 100°	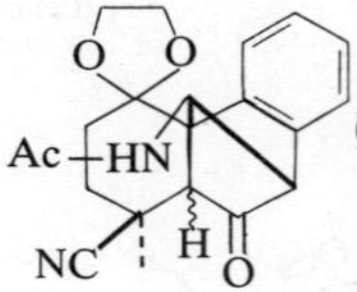 Ac—HN NC H (Major isomer, 69), (Minor isomer, 11), I (15)	227
C_{17}	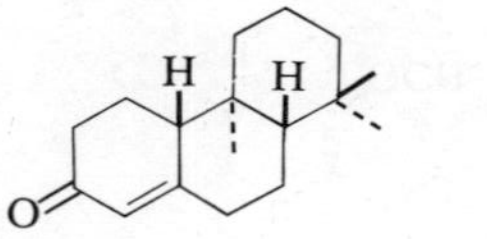 H H			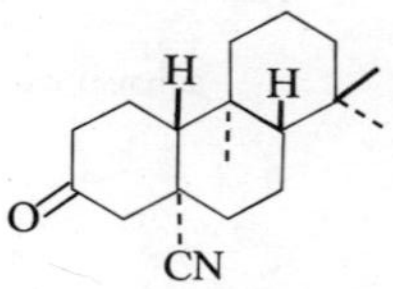 H H CN (I)	
		$KCN-NH_4Cl$	Aq DMF, 100°	I (14)	83, 43
		$HCN-Al(C_2H_5)_3$	THF, room temp	I (72)	83, 84

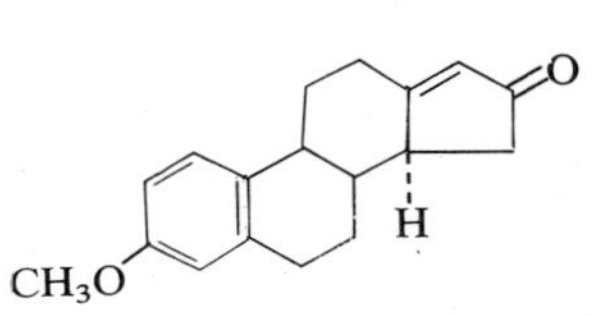	HCN-Al$(C_2H_5)_3$	THF, room temp	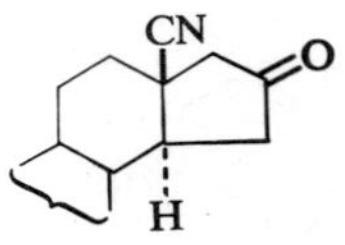 (16), 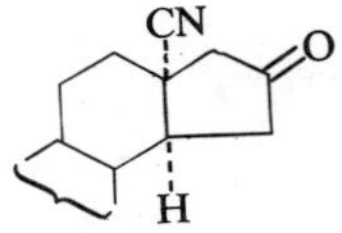 (65)	43
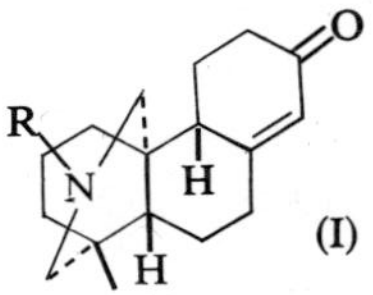 (I) I, R = Ac II, R = SO_2CH_3	HCN-Al$(C_2H_5)_3$	THF, room temp	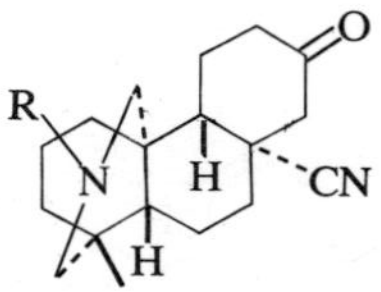 (II), 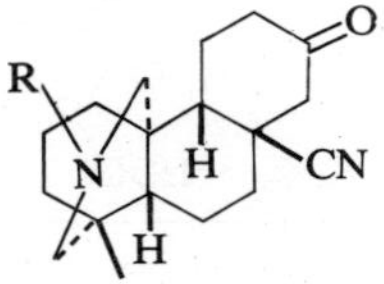 (III) II, R = Ac (28) R = SO_2CH_3, II (60), III (1)	85

TABLE VI. CONJUGATE HYDROCYANATION OF CYCLIC α,β-ETHYLENIC KETONES (*Continued*)

Substrate	Reagent(s)	Reaction Conditions	Product(s) and Yield(s) (%)	Refs.
C_{17} (*Contd.*) 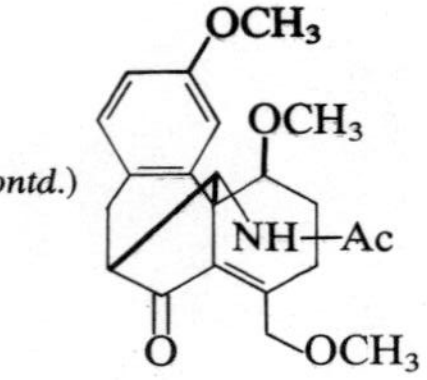 [structure: OCH_3, OCH_3, NH—Ac, OCH_3, O] [b]	KCN	Aq C_2H_5OH, reflux	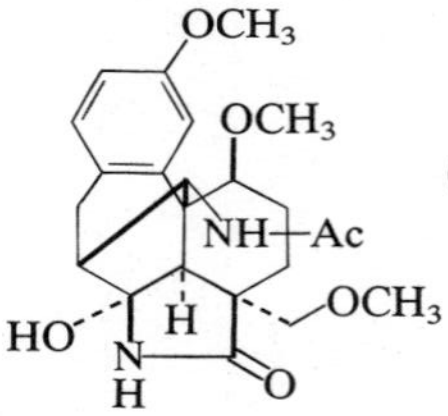 [structure: OCH_3, OCH_3, NH—Ac, HO, H, OCH_3, NH, O] (86)	112
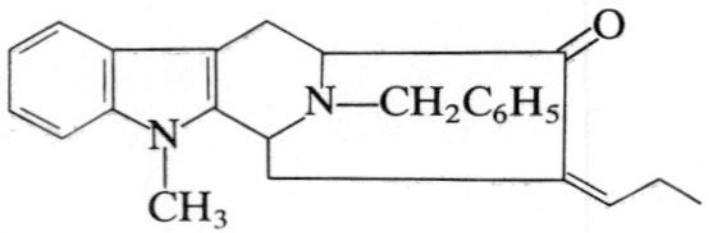 [structure: N—CH_3, N—$CH_2C_6H_5$, O]	KCN-NH_4Cl	Aq DMF, 100°	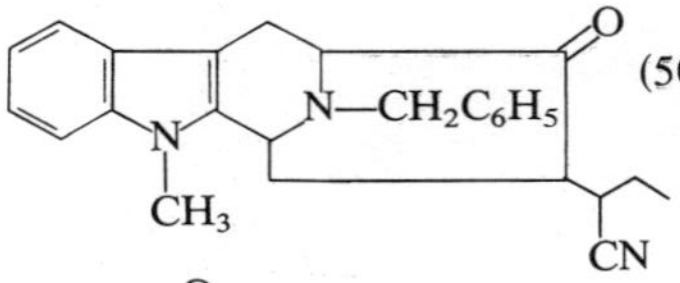 [structure: N—CH_3, N—$CH_2C_6H_5$, O, CN] (50)	228
C_{18} 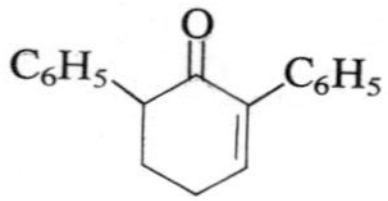 [structure: C_6H_5, O, C_6H_5]	$(CH_3)_2C(OH)CN$-$[N(C_2H_5)_3]$	No solvent, room temp	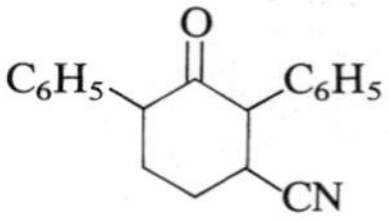 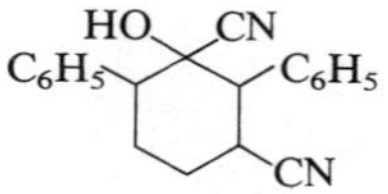 [structure: C_6H_5, O, C_6H_5, CN] (4), [structure: HO, CN, C_6H_5, C_6H_5, CN] (21)	124

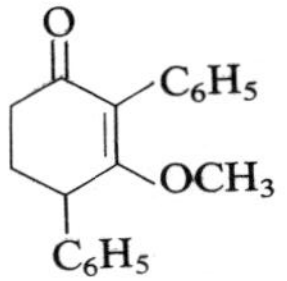	$(CH_3)_2C(OH)CN$- [methanolic KOH or $N(C_2H_5)_3$]	No solvent, room temp	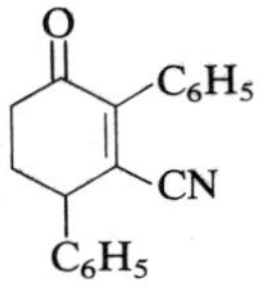 (27–33), 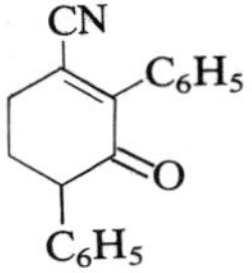 (24–27)	124
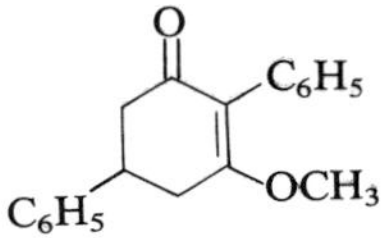			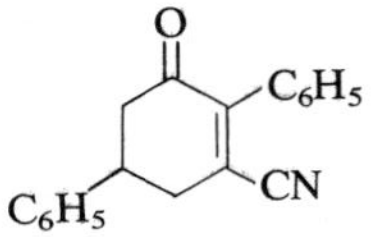 (I), 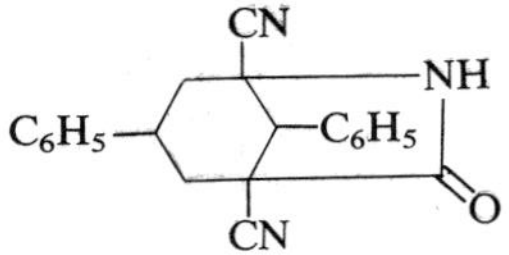 (II)	
	$(CH_3)_2C(OH)CN$- (methanolic KOH)	No solvent, room temp	I (95)	51a, 122
	$(CH_3)_2C(OH)CN$- [$N(C_2H_5)_3$]	No solvent, room temp	I (67), II (3)	122
	$(CH_3)_2C(OH)CN$- [$N(C_2H_5)_3$]	No solvent, 80–90°	II (50)	122

TABLE VI. CONJUGATE HYDROCYANATION OF CYCLIC α,β-ETHYLENIC KETONES (*Continued*)

Substrate	Reagent(s)	Reaction Conditions	Product(s) and Yield(s) (%)	Refs.
C_{18} (*Contd.*) 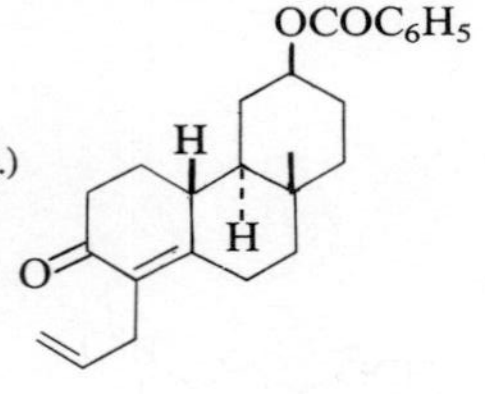 $OCOC_6H_5$, H, H, O	$(C_2H_5)_2AlCN$	C_6H_6-toluene, 45°, 1 hr	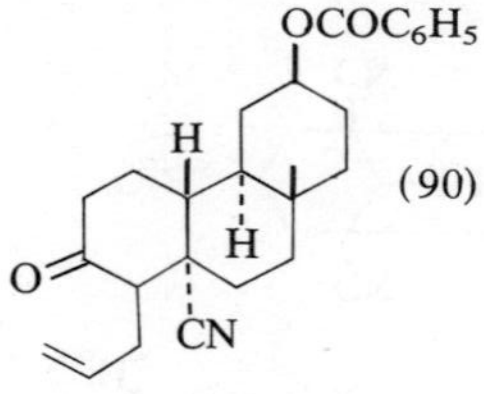 $OCOC_6H_5$, H, H, O, CN (90)	223
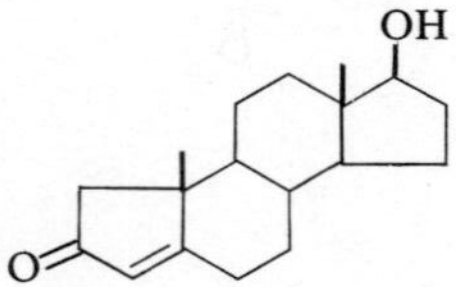 OH, O	KCN-NH_4Cl	Aq CH_3OH, reflux	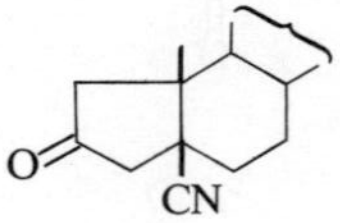 O, CN (53)	192, 193
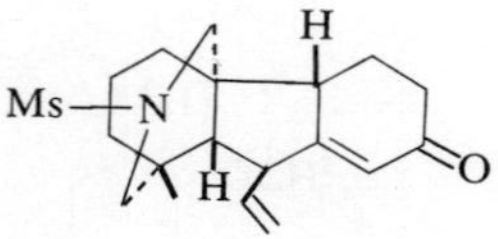 Ms—N, H, H, O	$(C_2H_5)_2AlCN$	CH_2Cl_2, room temp	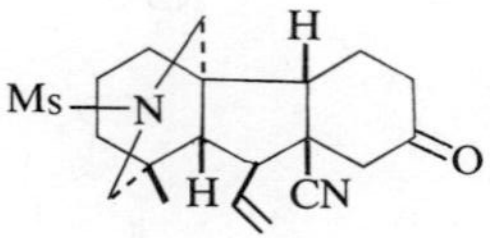 Ms—N, H, H, CN, O (87–90)	142

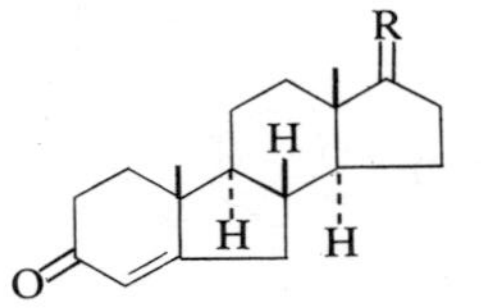 (I) I, R = H_2 I, R = O	$(C_2H_5)_2AlCN$	THF-C_6H_6 or toluene, room temp, 2 hr	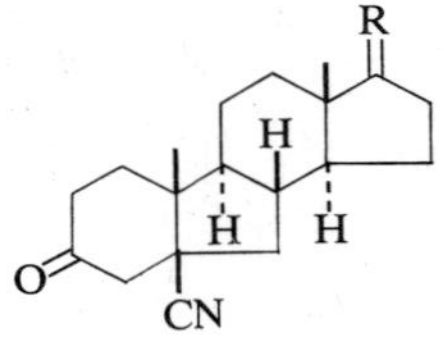 (II) II, R = H_2 (72) II, R = O (85)	43
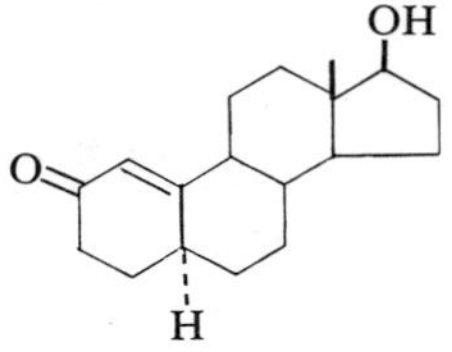	KCN-NH_4Cl HCN-$Al(C_2H_5)_3$	Aq CH_3OH, reflux THF, room temp	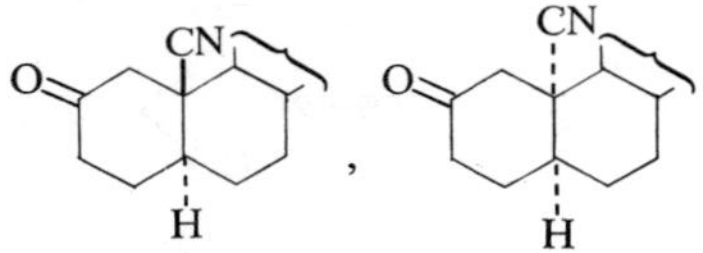 (I), (II) I (21), II (49) I (25), II (40)	145
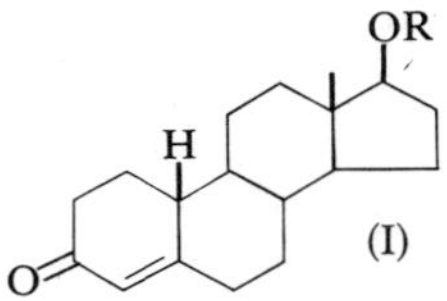 (I) I, R = H I, R = H I, R = Ac	KCN-NH_4Cl HCN-$Al(C_2H_5)_3$ HCN-$Al(C_2H_5)_3$	Aq CH_3OH, reflux THF, room temp " , "	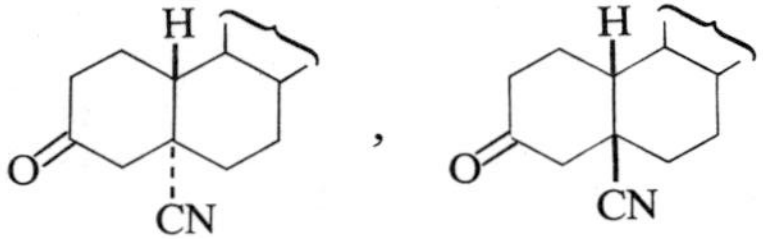 (II), (III) R = H, II (48), III (25) R = H, II (67), III (18) R = Ac, II (65), III (Minor)	190 43 84

TABLE VI. CONJUGATE HYDROCYANATION OF CYCLIC α,β-ETHYLENIC KETONES (*Continued*)

	Substrate	Reagent(s)	Reaction Conditions	Product(s) and Yield(s) (%)	Refs.
C_{18} (*Contd.*)	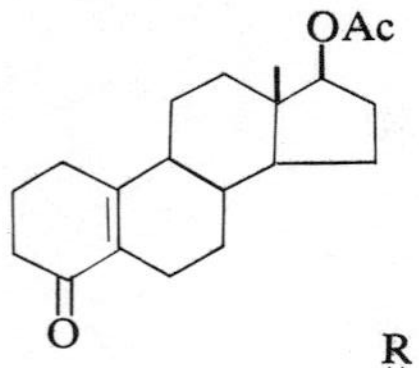 OAc	$HCN-(C_2H_5)_2AlBr$	THF-heptane room temp	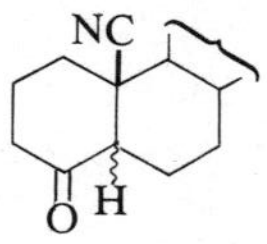 NC, H (73)	197
	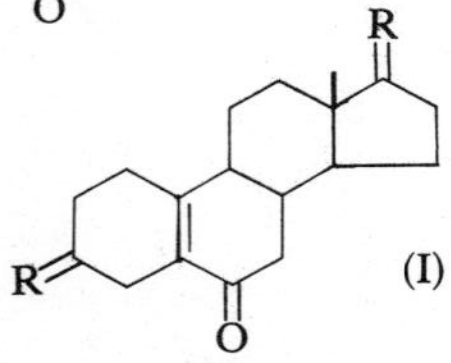 R (I)	KCN	OH, OH, reflux	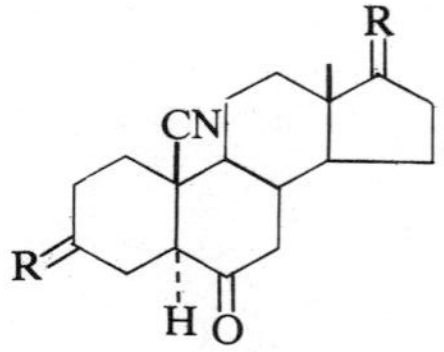 R, CN (II)	57
	I, R 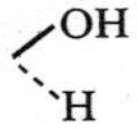 = OH, H			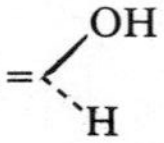 II, R = OH, H (80–85)	
	I, R 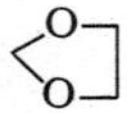 = O, O			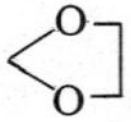 II, R = O, O (High)	
	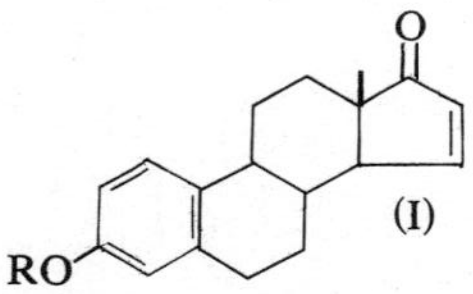 RO (I)			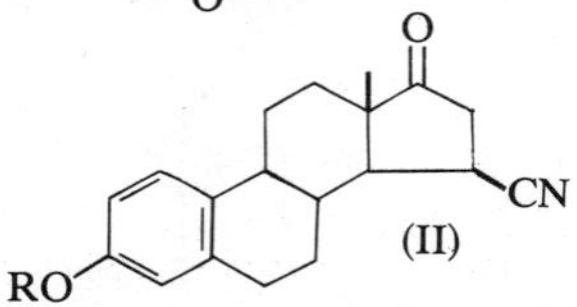 RO, CN (II)	
	I, R = H	NaCN	Aq THF, reflux	II, R = H (66)	294
	I, R = CH_3	”	” ”, ”	II, R = CH_3 (94)	149

	Reactant	Reagent	Conditions	Product(s) (Yield, %)	Refs.
	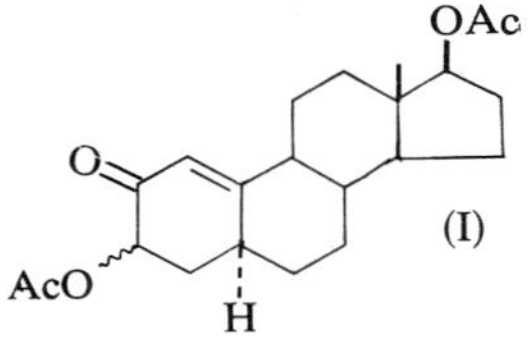 (I)	KCN-NH_4Cl	Aq CH_3OH, reflux	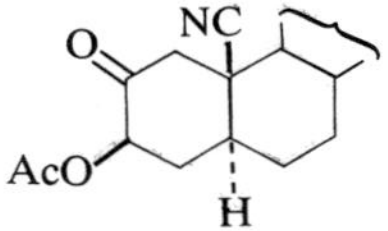 (II), 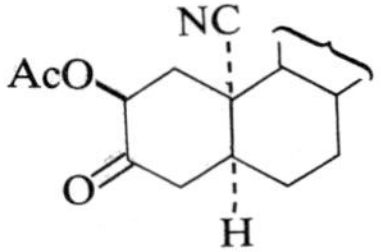 (III)	106
	I, 3α-OAc			III (60)[c]	
	I, 3β-OAc			I (29), II (20), III (21)[c]	
C_{19} 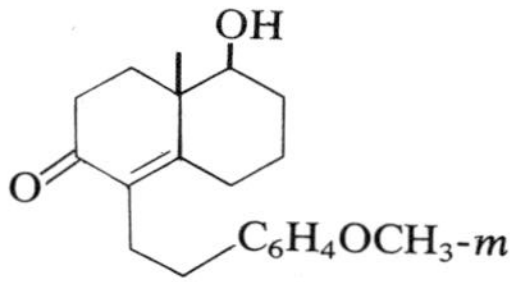	$C_6H_4OCH_3$-*m*	HCN-$Al(C_2H_5)_3$	THF, —	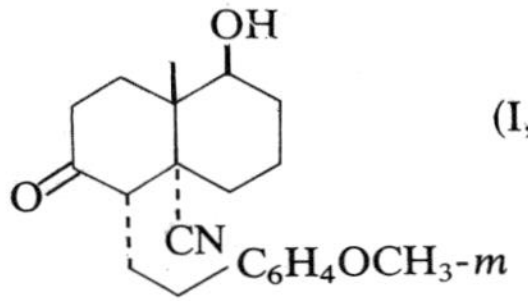 $C_6H_4OCH_3$-*m* (I, High)	183
	CH_3O	$(C_2H_5)_2AlCN$	C_6H_6, 0°, 2.5 hr	I (82) (I), (II)	184

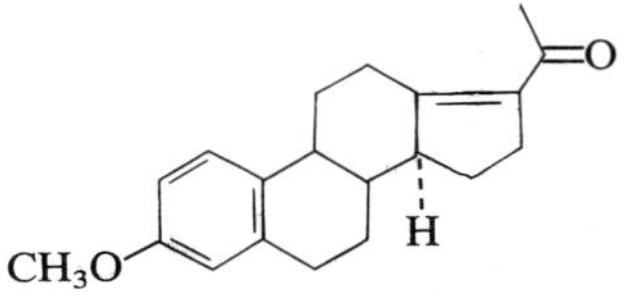

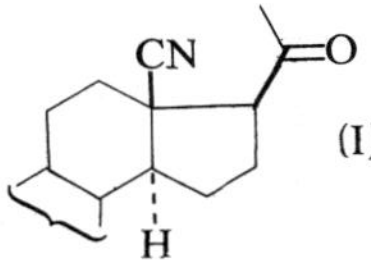

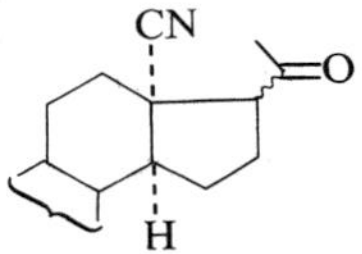

TABLE VI. CONJUGATE HYDROCYANATION OF CYCLIC α,β-ETHYLENIC KETONES (*Continued*)

Substrate	Reagent(s)	Reaction Conditions	Product(s) and Yield(s) (%)	Refs.
C_{19} (*Contd.*)	$KCN-NH_4Cl$	Aq DMF, 100°	I (22), II (57)	81
	$HCN-Al(C_2H_5)_3$	THF, room temp	I (67), II (7), I+II (8)	25, 84
	$(C_2H_5)_2AlCN$	C_6H_6, 25°, 3 min	I+II (69:31, total quant)	25
	"	" , " , 10 hr	I+II (46:54, total quant)	25
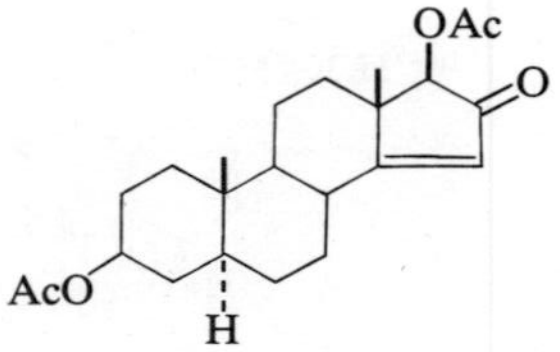 AcO, OAc, O, H	$KCN-NH_4Cl$	Aq DMF, 100°	Ineffective	93
	$(C_2H_5)_2AlCN$	C_6H_6-methylcyclo-hexane, room temp, 1.5 hr	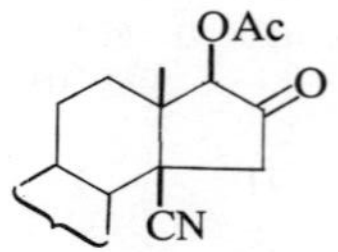 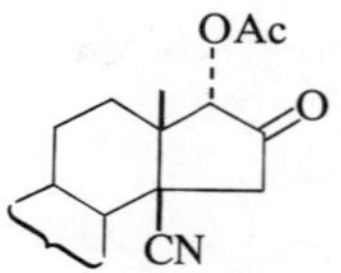 OAc, O, CN (55), OAc, O, CN (3)	93

Substrate	Reagent	Conditions	Product(s) (yield, %)	Ref.
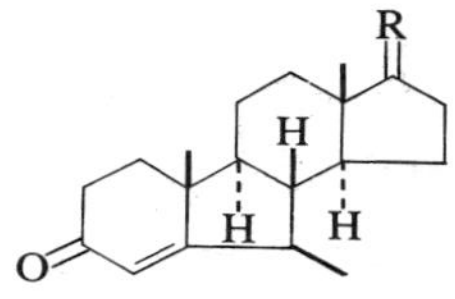 (I), I, $R = H_2$; I, $R = O$	$(C_2H_5)_2AlCN$	THF-C_6H_6 or toluene, room temp, 2–2.5 hr	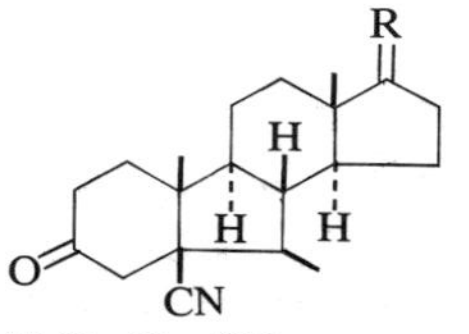 (II); II, $R = H_2$ (54); II, $R = O$ (80)	43
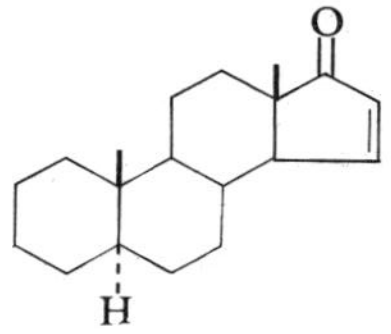	NaCN	Aq THF, reflux	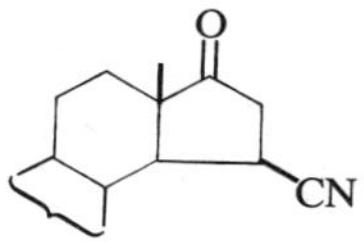 (77)	242
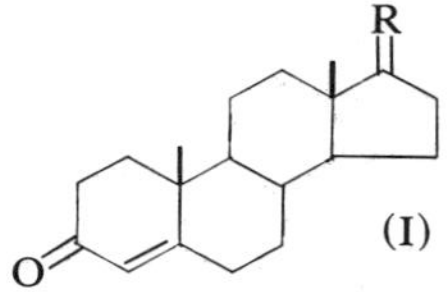 (I)			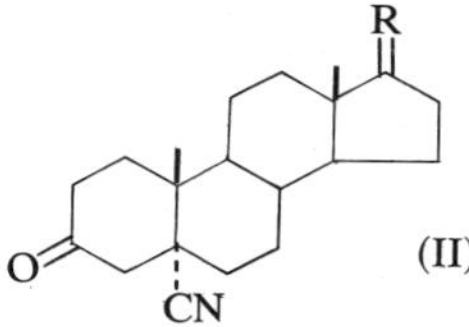 (II), 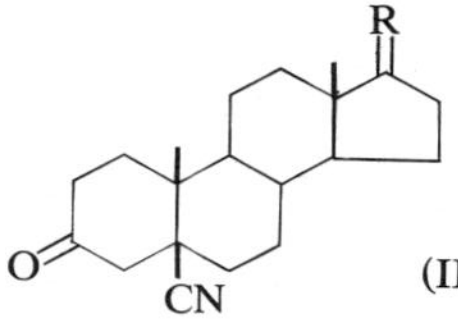 (III)	

TABLE VI. CONJUGATE HYDROCYANATION OF CYCLIC α,β-ETHYLENIC KETONES (*Continued*)

	Substrate	Reagent(s)	Reaction Conditions	Product(s) and Yield(s) (%)	Refs.
C_{19} (*Contd.*)	I, R = H_2	$Ca(CN)_2$	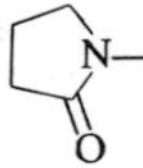 [structure] N—CH_3, 20°	R = H_2, II (43), III (43)	58
	I, R 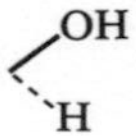 = [structure] OH, H	$Ca(CN)_2$	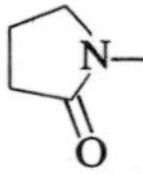 [structure] N—CH_3, 20°	R 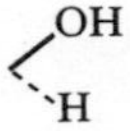 = [structure] OH, H, II (40), III (45)	58
		$KCN-NH_4Cl$	Aq DMF, reflux	II (29), III (24)	141
		$HCN-(C_2H_5)_2AlCl$	THF, room temp	II (40), III (39)	43
		$HCN-C_2H_5AlCl_2$	” , ” ”	II (42), III (34)	43
	I, R 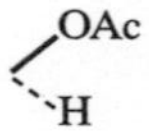 = [structure] OAc, H	KCN	Aq C_2H_5OH, reflux	R 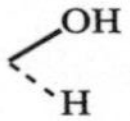 = [structure] OH, H, II (27), III (2)	140
		$HCN-Al(C_2H_5)_3$	THF, room temp	R 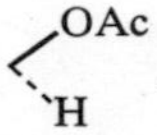 = [structure] OAc, H, II (47), III (37)	43
	I, R = O	$HCN-Al(C_2H_5)_3$	” , ” ”	R = O, II (43), III (36)	43
	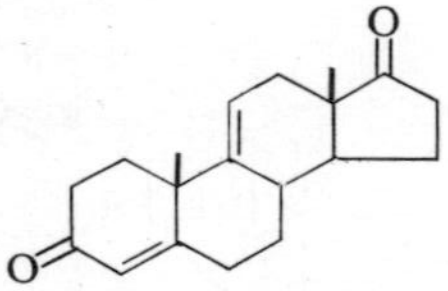 [structure]			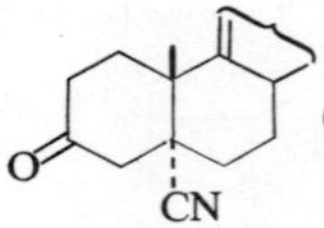 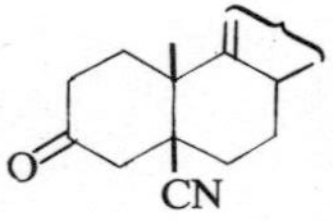 [structure] CN (I), [structure] CN (II)	
		$KCN-NH_4Cl$	Aq DMF, 100°	I (53), II (15)	189
		$HCN-Al(C_2H_5)_3$	THF, room temp	I (76), II (8)	189, 84

Substrate	Reagent	Conditions	Product(s) (Yield, %)	Ref.
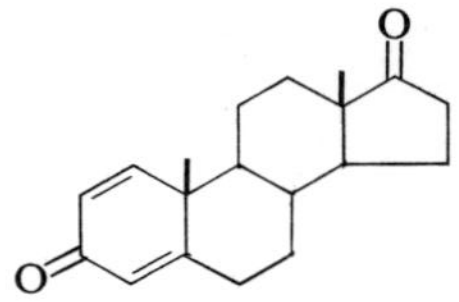			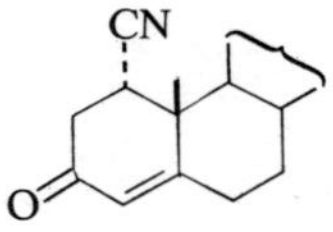 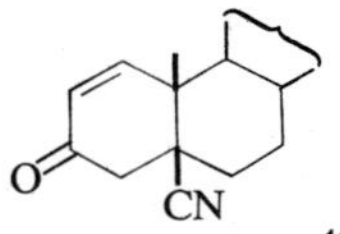 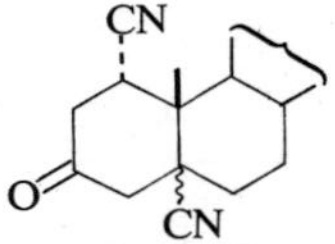 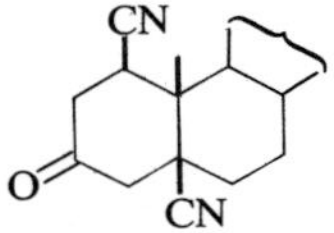 CN, (I), CN (II), CN CN, III (mixture), CN CN (IV)	43
	HCN-$Al(C_2H_5)_3$	THF, room temp	I+II (Main), III+IV (Minor)	
	$(C_2H_5)_2AlCN$	THF-toluene, room temp, 15 min	I (52), II (26)	
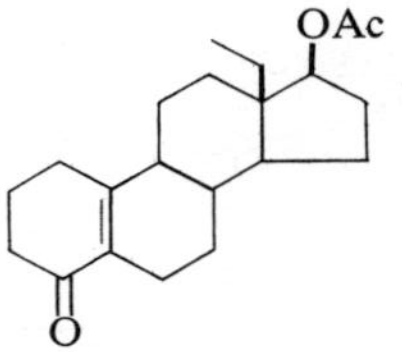 OAc	HCN-$(C_2H_5)_2AlBr$	THF-heptane, room temp	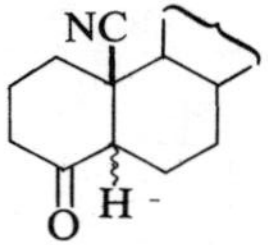 NC, H (75)	197

TABLE VI. CONJUGATE HYDROCYANATION OF CYCLIC α,β-ETHYLENIC KETONES (*Continued*)

Substrate	Reagent(s)	Reaction Conditions	Product(s) and Yield(s) (%)	Refs.
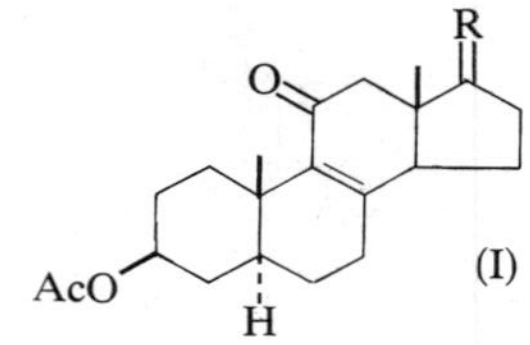 C_{19} (*Contd.*) AcO (I)			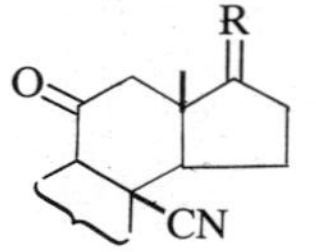 (II)	
I, R = O	$HCN\text{-}Al(C_2H_5)_3$	THF, room temp	II, R = O (65)	113, 43
I, R 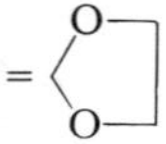 = (ethylenedioxy)	$(C_2H_5)_2AlCN$	C_6H_6, 0°, 10 min[d]	II, R 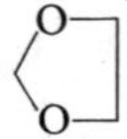 = (ethylenedioxy) (84)	113, 42, 43
(bis-ethylenedioxy dienone)			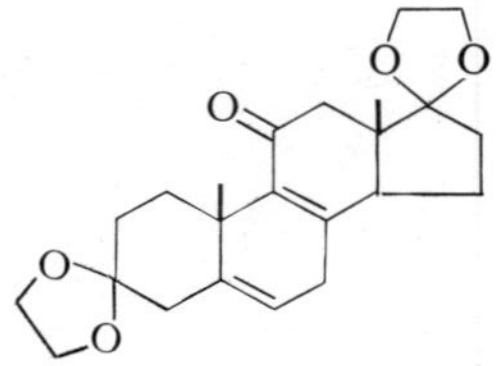 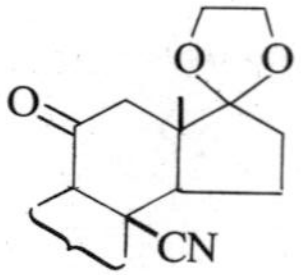 CN (I)	
	$KCN\text{-}NH_4Cl$	Aq DMF, 100°	No reaction	25
	$HCN\text{-}Al(C_2H_5)_3$	THF, room temp	I (74)	25, 42, 84
	$(C_2H_5)_2AlCN$	C_6H_6, 0°, 10 min[d]	I (92)	25, 42

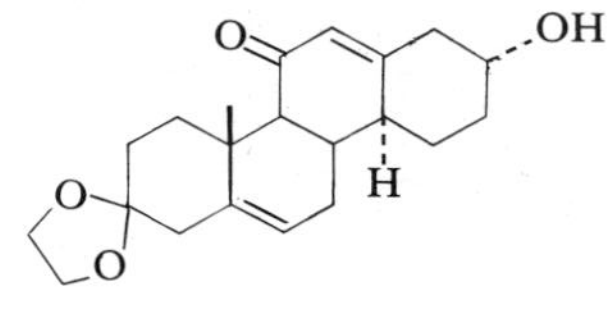

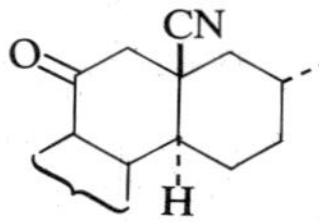

Substrate	Reagent	Conditions	Product(s) (Yield, %)	Ref.
OH			CN OH (I)	
	KCN-NH_4Cl	Aq DMF, 100°	I (90)	80
	HCN-$Al(C_2H_5)_3$	THF, room temp	I (73)	80, 84
R (I)			CN (II),	
			CN (III)	
			NH OH (IV),	

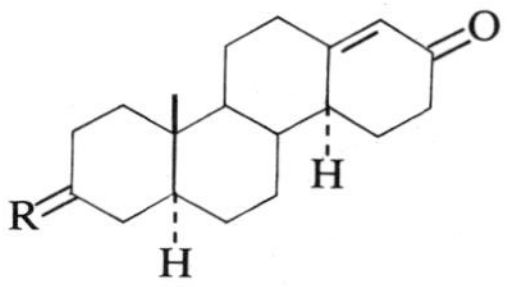

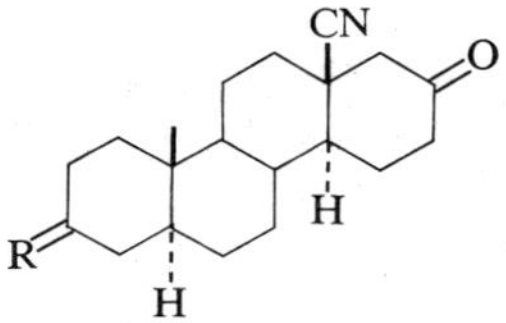

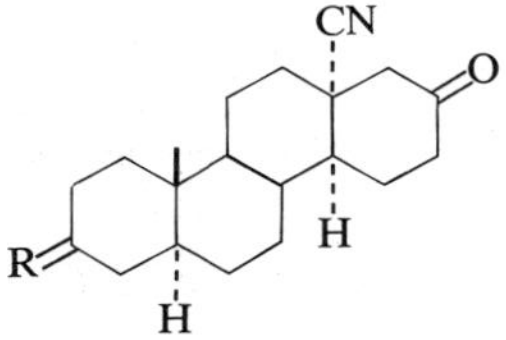

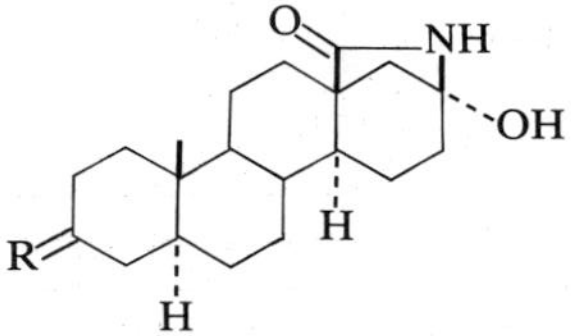

TABLE VI. Conjugate Hydrocyanation of Cyclic α,β-Ethylenic Ketones (*Continued*)

Substrate	Reagent(s)	Reaction Conditions	Product(s) and Yield(s) (%)	Refs.
C_{19} (*Contd.*) 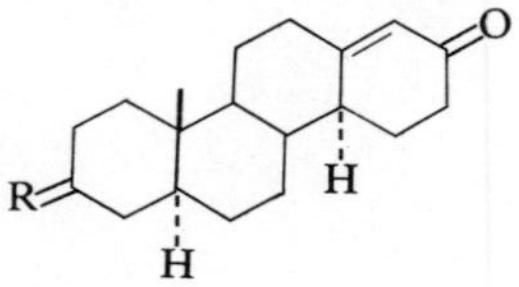 (I) [structure, R=]			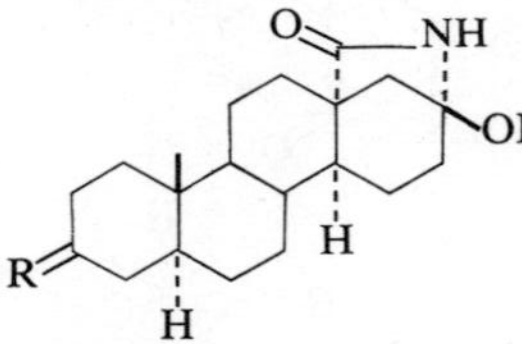 (V) [structure, R=, OH]	
I, R = OH, H 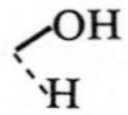	HCN-[N(C_2H_5)$_3$]	Dioxane, room temp	No reaction	78
	"	DMF, 80–90°	R 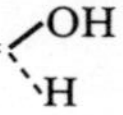 = OH, H,[e] I (11), II (15), III (1)	78
	KCN	Aq CH_3OH, reflux	IV (27), V (19), dimer (7)	103
	KCN-AcOH	Aq C_2H_5OH, 80–90°	II (16), III (7)	78
	KCN-NH_4Cl	Aq CH_3OH, reflux	II (41), III (25)	78
	"	Aq DMF, 100°	II (35), III (11)	78
I, R 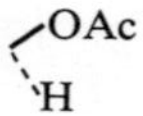 = OAc, H	KCN	Aq CH_3OH, reflux	R 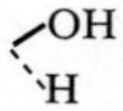 = OH, H, II (29), III (1), IV (trace), V (5), dimer (4)	103
	HCN-Al(C_2H_5)$_3$	THF, room temp	R 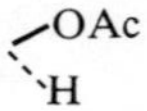 = OAc, H, II (68), III (13)	43, 84
I, R = O	KCN	Aq CH_3OH, reflux	R = O, II (13), IV (9), V (14)	43
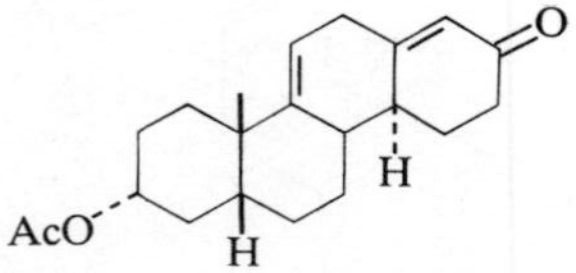 [structure, AcO, H, H]	HCN-Al(C_2H_5)$_3$	THF, room temp	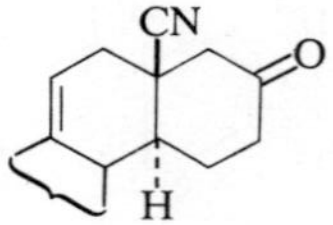 [structure, CN, O, H] (> 55)	152, 84

C_{20}	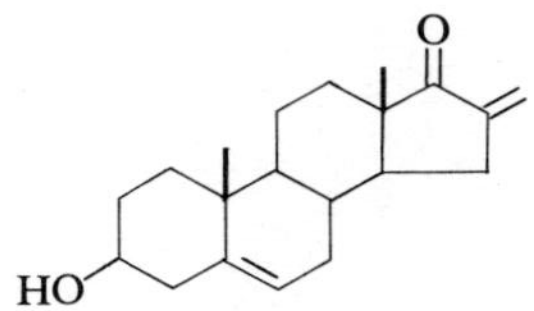	KCN	Aq DMF, room temp	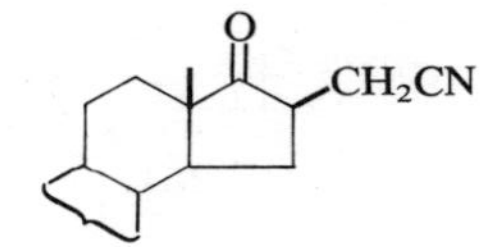 (18),	243, 295
				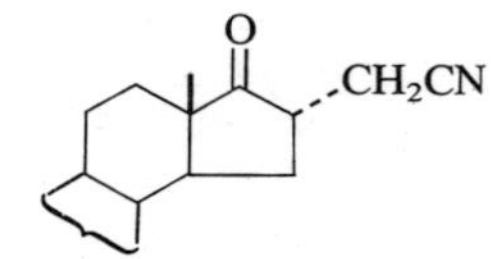 (38)[f]	
	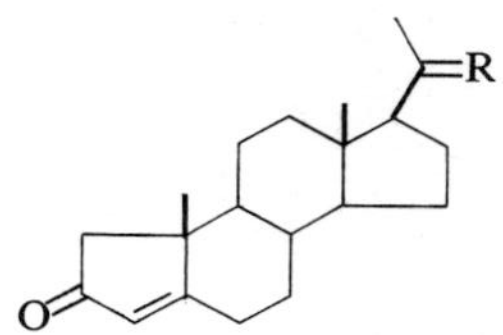 (I)	$KCN\text{-}NH_4Cl$	Aq CH_3OH, reflux	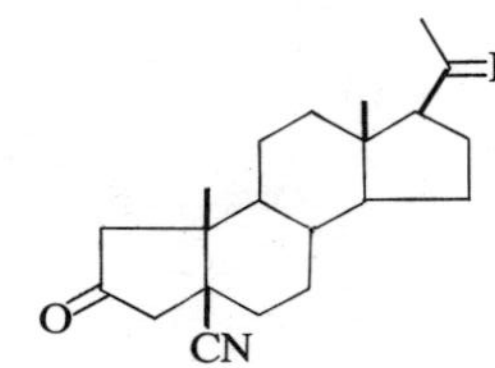 (II)	
	I, R = O			II, R = O (—)	193
	I, R = $\langle^{OH}_{H}$			II, R = $\langle^{OH}_{H}$ (28)	194

TABLE VI. CONJUGATE HYDROCYANATION OF CYCLIC α,β-ETHYLENIC KETONES (*Continued*)

	Substrate	Reagent(s)	Reaction Conditions	Product(s) and Yield(s) (%)	Refs.
C_{20} (*Contd.*) 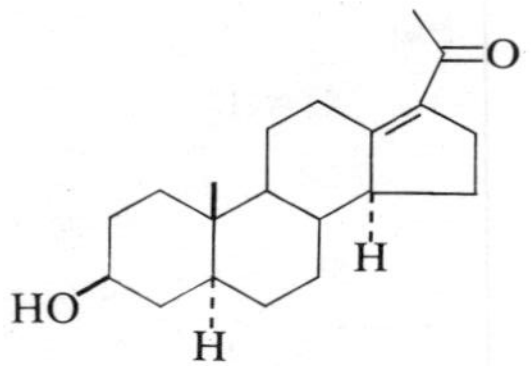	HO, H, H			NC, H (I), NC, H (II)	107
		$HCN\text{-}Al(C_2H_5)_3$	THF, room temp	I (80)	151
		$(C_2H_5)_2AlCN$	C_6H_6—CH_2Cl_2, room temp, 20 min	I (77), II (3)	43
	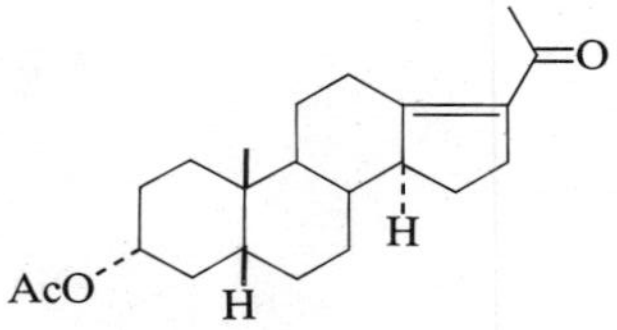 AcO, H, H (I)			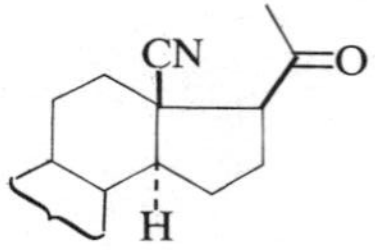 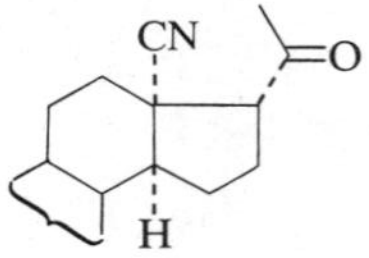 CN, H (II), CN, H (III)	107

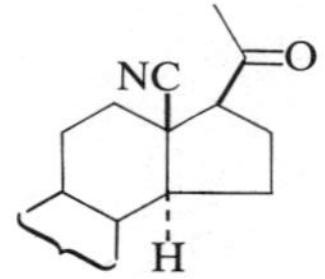

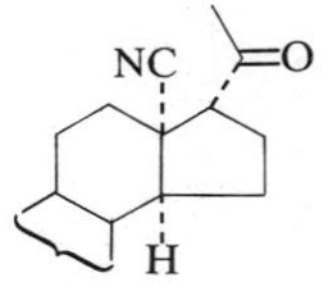

Substrate	Reagent	Conditions	Product(s) and Yield(s) (%)	Refs.
	KCN-NH_4Cl	Aq DMF, 100°	II (25), III (23)	
	HCN-$Al(C_2H_5)_3$	THF-heptane, 25°	II (23), I+III (—)	
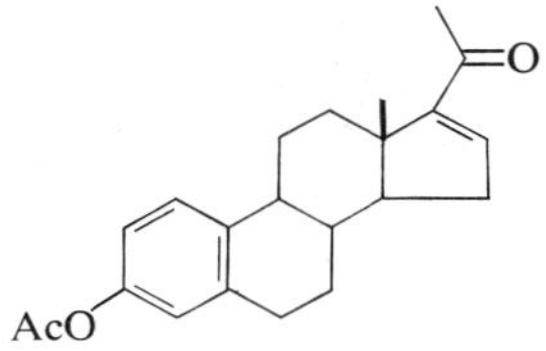	NaCN	CH_3OH, reflux	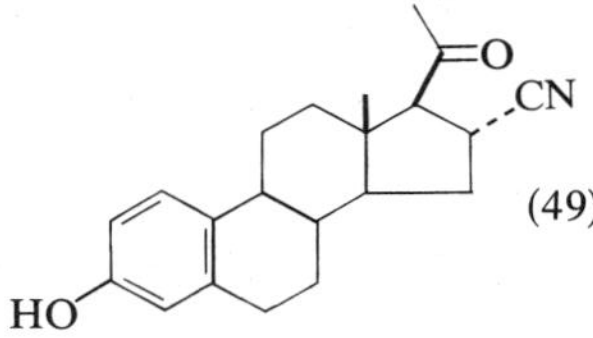 (49)	91
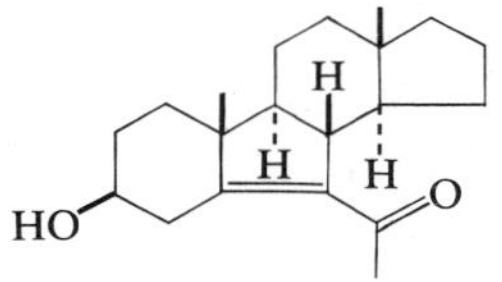	$(C_2H_5)_2AlCN$	C_6H_6-toluene, room temp, 30 min[d]	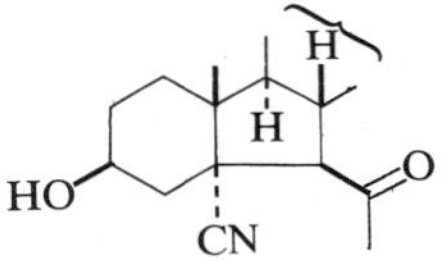 (93)	43
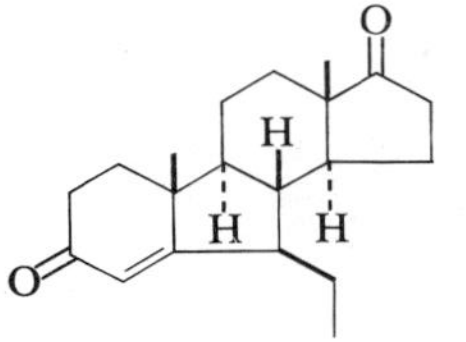	$(C_2H_5)_2AlCN$	THF-toluene, 70°, 1 hr	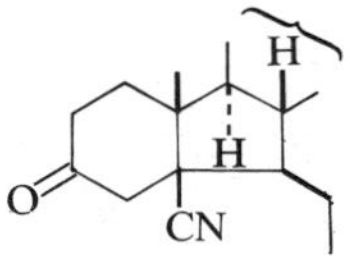 (13)	43
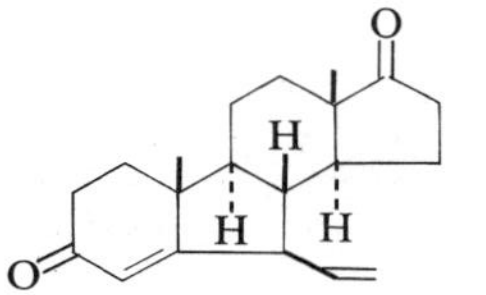			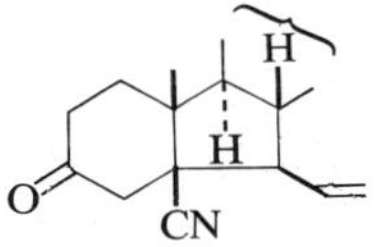 (I)	195

TABLE VI. Conjugate Hydrocyanation of Cyclic α,β-Ethylenic Ketones (*Continued*)

Substrate		Reagent(s)	Reaction Conditions	Product(s) and Yield(s) (%)	Refs.
C_{20} (*Contd.*)		$HCN\text{-}Al(C_2H_5)_3$	THF, room temp	I (16)	
		$(C_2H_5)_2AlCN$	C_6H_6—CH_2Cl_2, room temp, 3 hr	I (61)	
	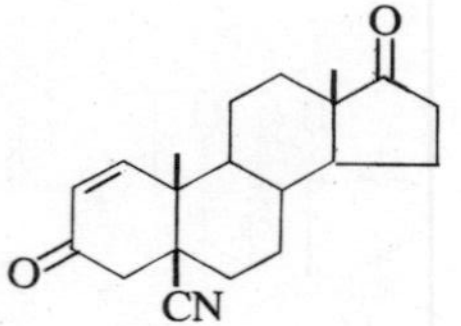	$(C_2H_5)_2AlCN$	THF-toluene, room temp, 5 hr	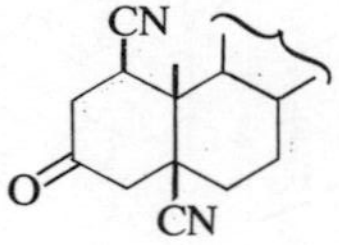 (78)	43
	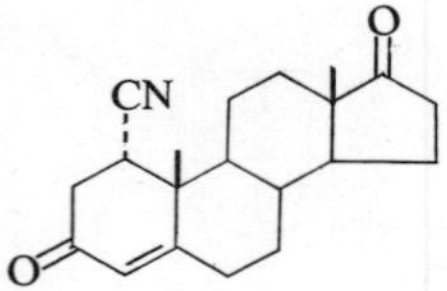	$(C_2H_5)_2AlCN$	THF-toluene, room temp, 5 hr	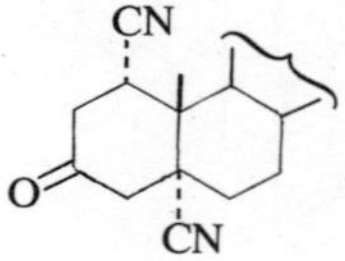 (12), 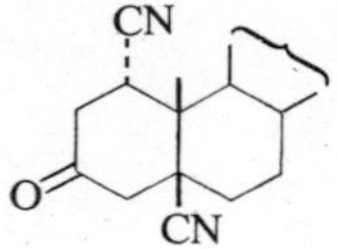 (51)	43
C_{21}	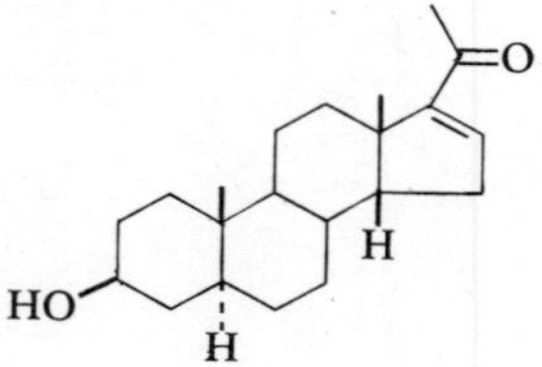	KCN	Aq C_2H_5OH, reflux	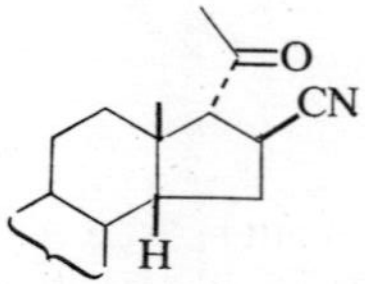 (70)	296

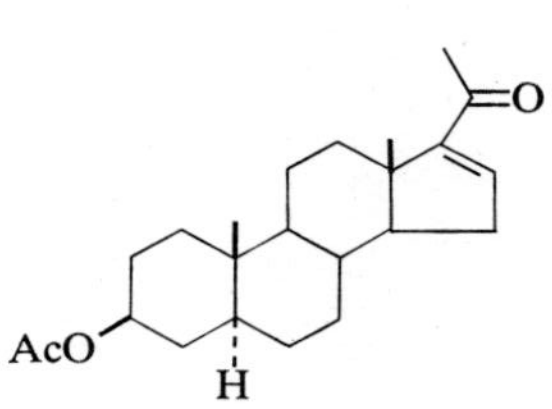

AcO H | KCN | Aq CH_3OH, reflux | HO H (Quant) | 62

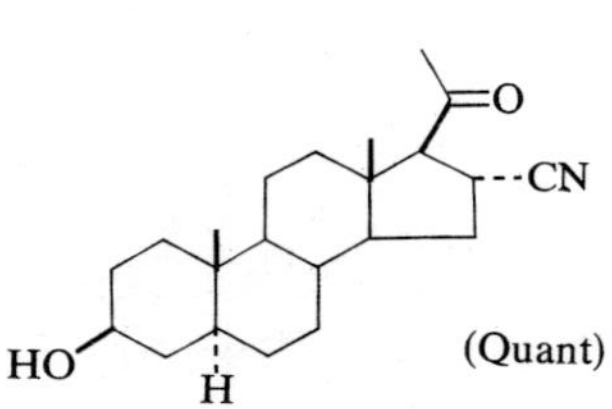

RO (I) R′ | | | RO (II) R′ CN, RO (III) R′ CN, HO (IV) CO_2H

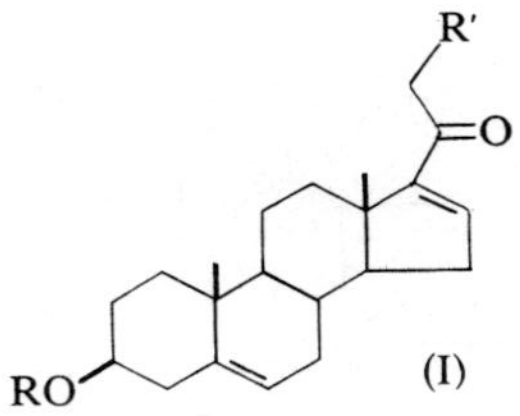

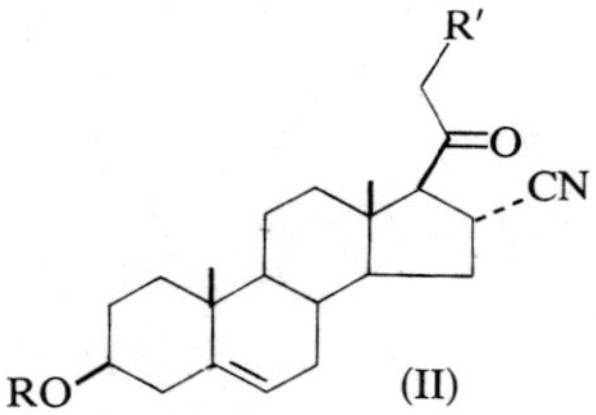

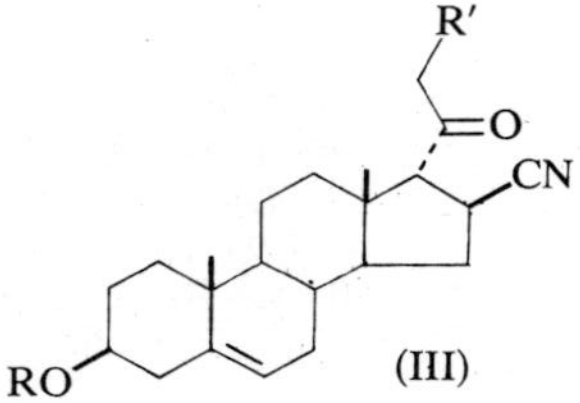

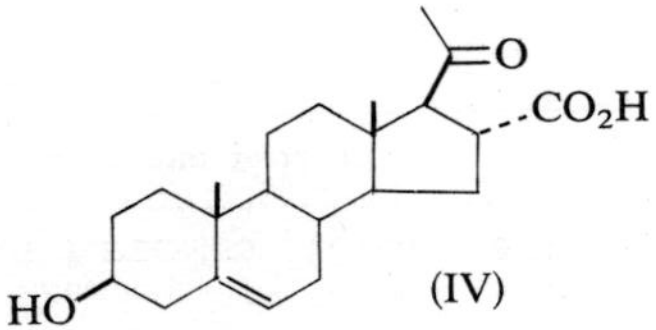

TABLE VI. CONJUGATE HYDROCYANATION OF CYCLIC α,β-ETHYLENIC KETONES (*Continued*)

	Substrate	Reagent(s)	Reaction Conditions	Product(s) and Yield s) (%)	Refs.
C_{21} (*Contd.*)	I, R = R′ = H	KCN	Aq C_2H_5OH, reflux, 2–3 hr	R = R′ = H, II (—)	297
		KCN	Aq C_2H_5OH, reflux, 12 hr	IV (58)	297
		KCN-$AcOC_2H_5$	Aq CH_3OH, reflux	R = R′ = H, II (78), III (1.4)	298
	I, R = Ac, R′ = H	NaCN	CH_3OH, reflux	R = R′ = H, II (67)	91
		KCN-$AcOC_2H_5$	Aq CH_3OH, ”	R = R′ = H, II (68)	155
		HCN-$Al(C_2H_5)_3$	THF, room temp	R = Ac, R′ = H, II (69)	43, 84
	I, R = Ac, R′ = OAc	NaCN	C_2H_5OH, reflux	R = H, R′ = OH, II (12)	91
		HCN-$Al(C_2H_5)_3$	THF, room temp	R = Ac, R′ = OAc, II (72)	43

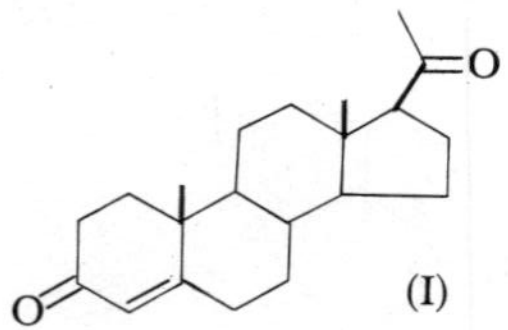

(I)

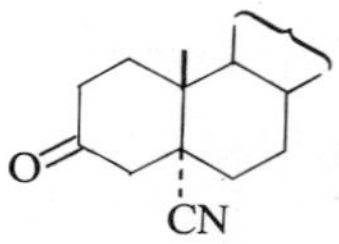

(II), 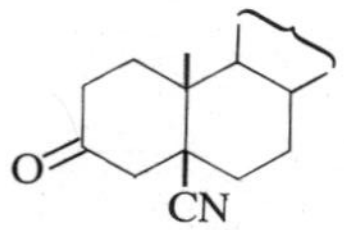 (III),

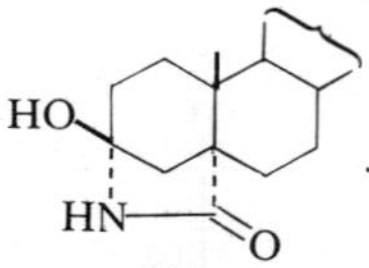

(IV), 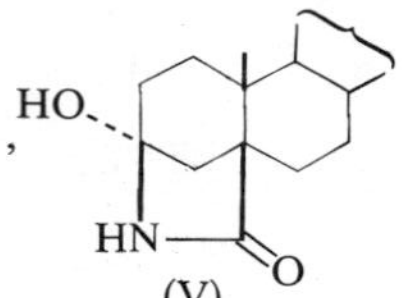 (V)

Substrate	Reagent	Conditions	Product(s) and Yield(s) (%)	Refs.
	KCN	Aq C_2H_5OH, reflux, 2 hr	I (26), II (18), IV (9), V (16)	140
	,,	Aq C_2H_5OH, reflux, 5 hr	IV (40), V (16)	140
	$Ca(CN)_2$	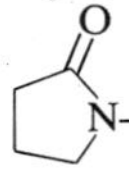 N-methyl-2-pyrrolidone ($N—CH_3$), 20°	II (30), III (39)	58
	$HCN\text{-}Al(C_2H_5)_3$	THF, room temp	II (47), III (23)	43
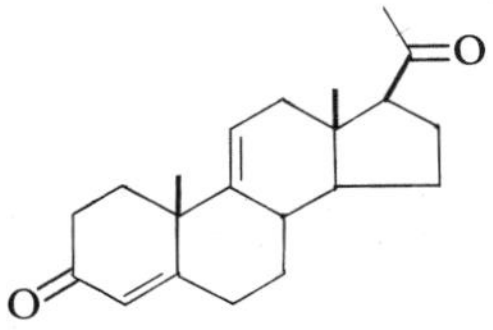	$HCN\text{-}Al(C_2H_5)_3$	THF, room temp	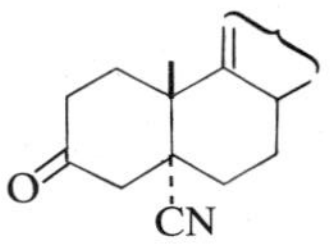 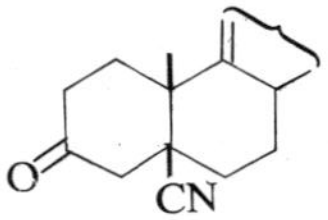 CN (53), CN (2)	43
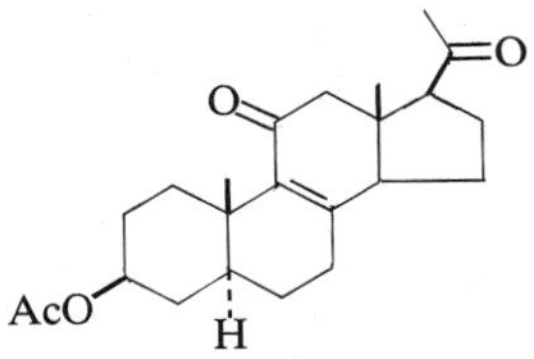 AcO, H	$HCN\text{-}Al(C_2H_5)_3$	THF, room temp	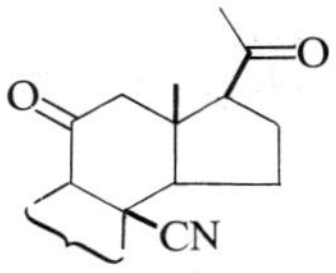 CN (78)	84, 42

R (I) R CN (II)

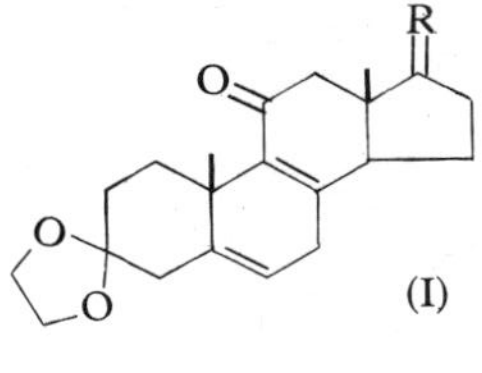

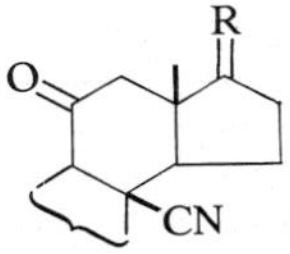

TABLE VI. Conjugate Hydrocyanation of Cyclic α,β-Ethylenic Ketones (*Continued*)

	Substrate	Reagent(s)	Reaction Conditions	Product(s) and Yield(s) (%)	Refs.
C_{21} (*Contd.*)	I, R 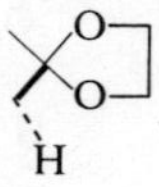 = [structure]	$HCN\text{-}Al(C_2H_5)_3$	THF, room temp	II, R = [structure] (52),	113, 43
				II, R = [structure: OH, CN, H] (8)	
		$(C_2H_5)_2AlCN$	C_6H_6-toluene, 0°; 30 min[d]	II, R 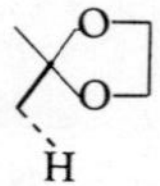 = [structure] (83)	113, 42, 43
	I, R 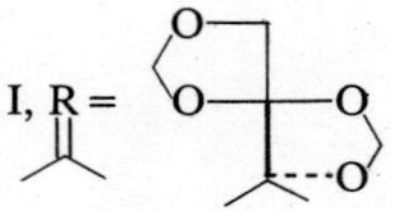 = [structure]	$HCN\text{-}Al(C_2H_5)_3$	THF, room temp	II, R = [structure] (61)	84

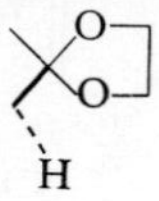

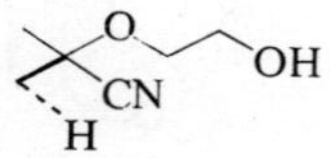

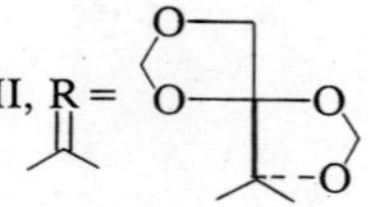

Substrate	Reagent	Conditions	Product(s) and Yield(s) (%)	Refs.
	$(C_2H_5)_2AlCN$	C_6H_6-toluene, 0°, 1 hr[d]	II, 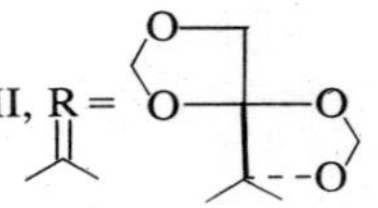 R = (82)	113, 42
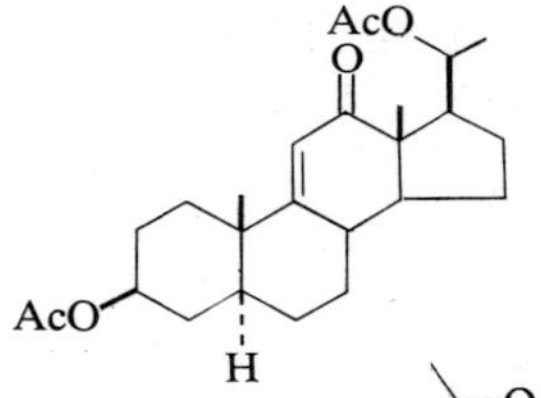 AcO, O, AcO, H	$(C_2H_5)_2AlCN$	C_6H_6-toluene, room temp, 3 hr	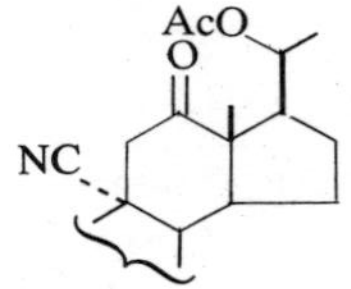 NC, O, AcO (77)	43
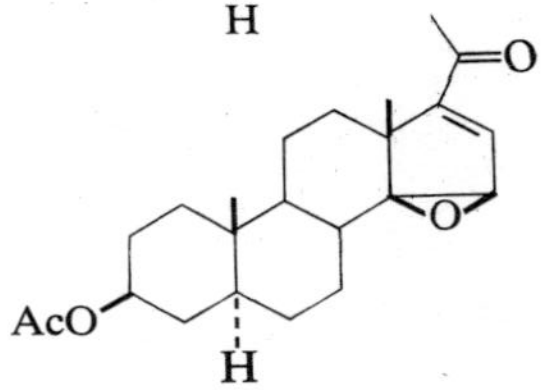 AcO, H, O			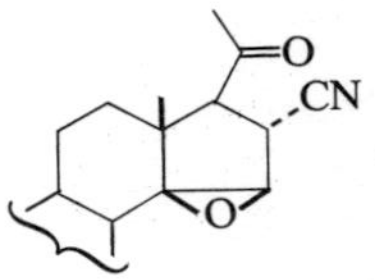 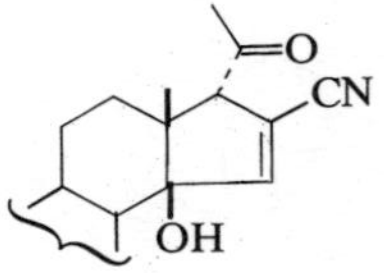 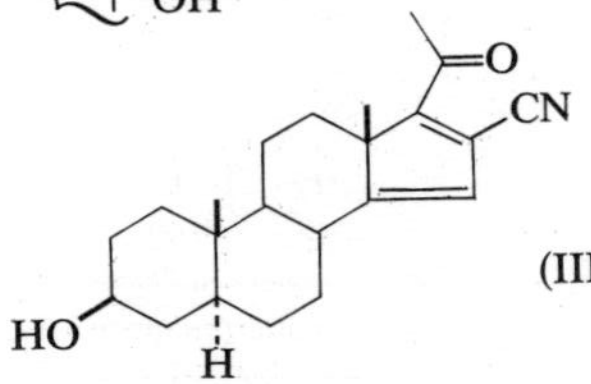 (I), (II), OH, HO, H (III)	299

TABLE VI. CONJUGATE HYDROCYANATION OF CYCLIC α,β-ETHYLENIC KETONES (*Continued*)

	Substrate	Reagent(s)	Reaction Conditions	Product(s) and Yield(s) (%)	Refs.
		NaCN	Aq C_2H_5OH, reflux	III (80)	
		NaCN-NH_4Cl	", room temp	I (79), II (14)	
C_{23}	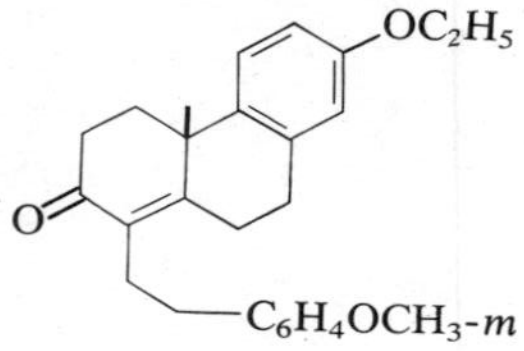 OC_2H_5; O; $C_6H_4OCH_3$-*m*	HCN-$Al(C_2H_5)_2$	THF, —	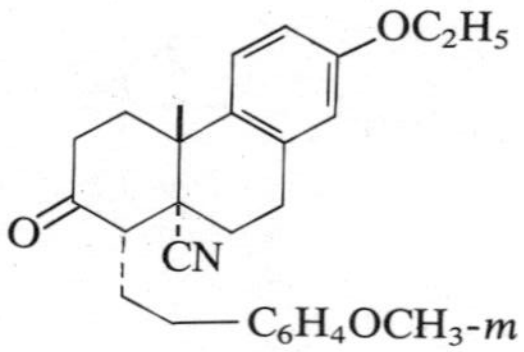 OC_2H_5; O; CN; $C_6H_4OCH_3$-*m* (85)	139
		$(C_2H_5)_2AlCN$	C_6H_6, —	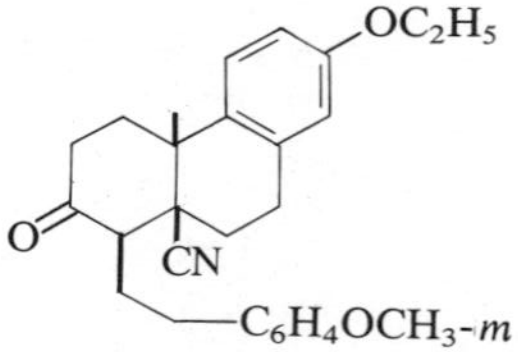 OC_2H_5; O; CN; $C_6H_4OCH_3$-*m* (97)	139
C_{24}	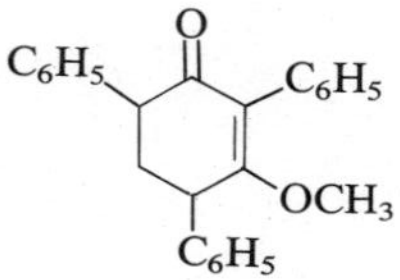 O; C_6H_5; C_6H_5; OCH_3; C_6H_5	$(CH_3)_2C(OH)CN$- [methanolic KOH or $N(C_2H_5)_3$]	No solvent, room temp	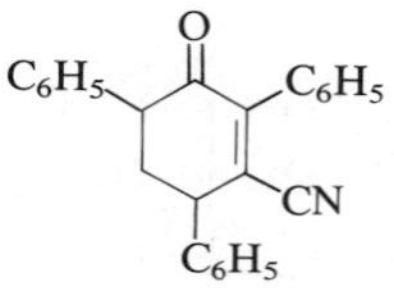 O; C_6H_5; C_6H_5; CN; C_6H_5 (65)	123

	Substrate	Reagent	Conditions	Product(s) (Yield, %)	Ref.
C_{27}	[structure] C_8H_{17}	$(CH_3)_2C(OH)CN$-(Na_2CO_3)	Aq CH_3OH-THF, reflux	I (70) [structure (I): CN, H]	156
		KCN-NH_4Cl	Aq DMF, 100°	I (65)	148
		HCN-$Al(C_2H_5)_3$	THF, room temp	I (90)	43
	[structure] C_8H_{17} (I)			[structures] CN (II), CN (III), HO HN O (IV), HO HN O (V),	

TABLE VI. CONJUGATE HYDROCYANATION OF CYCLIC α,β-ETHYLENIC KETONES (*Continued*)

Substrate	Reagent(s)	Reaction Conditions	Product(s) and Yield(s) (%)	Refs.
C_{27} (*Contd.*) 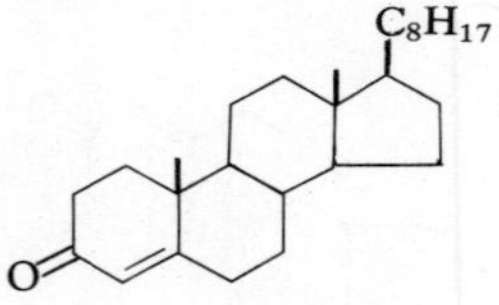 C_8H_{17} (I) (*Contd.*)			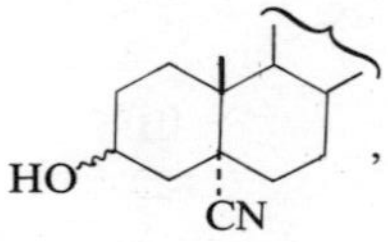 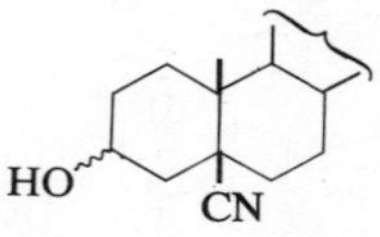 HO, CN (VI), HO, CN (VII)	
	$(CH_3)_2C(OH)CN$-KCN	C_6H_6 or CH_3CN, reflux	No reaction	51b
	$(CH_3)_2C(OH)CN$-KCN-18-crown-6	C_6H_6, room temp	II (7), III (70)	51b
	$(CH_3)_2C(OH)CN$-KCN-18-crown-6	” , reflux	II (15), III (60)	51b
	$(CH_3)_2C(OH)CN$-KCN-18-crown-6	CH_3CN, reflux	II (15), III (61)	51b
	KCN	Aq CH_3OH, reflux	I (18), II (21), IV (2), V (26), dimer (4)	15
		Aq C_2H_5OH, reflux	I (2), II (Trace), IV (24), V (46), IV+V (7)	76
	KCN-18-crown-6	C_6H_6, reflux	No reaction	51b
	$Ca(CN)_2$	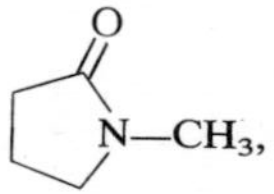 N-methylpyrrolidone (O, N—CH_3), 20°	II (40), III (39)	58

	$KCN-NH_4Cl$	Aq dioxane, 100°	No reaction	79
		Aq DMF, 100°	II (33), III (51)	25, 15
	$KCN-[(n-C_4H_9)_4NBr]$	$CH_2Cl_2-H_2O$, room temp	No reaction	59
	$KCN-(n-C_4H_9)_4NBr$	$C_6H_6-H_2O$, 80°	I + II (Predominant) + III	59
	$MgCNI$[g]	C_6H_6, reflux	I (42), II (9)	87
	$LiAl(CN)_4$	$C_6H_6—C_4H_9OH\text{-}t$, reflux	I (19), II (27), III (30)	87
	$HCN-Al(OC_3H_7\text{-}i)_3$	C_6H_6, reflux	VI (31), VII (21)	87
	$HCN-Al(OC_4H_9\text{-}t)_3$	” , ”	II (27), III (24)	87
	$HCN-Al(C_2H_5)_3$	THF, room temp	II (49), III (41)	25, 84
	$(C_2H_5)_2AlCN$	THF-$(i\text{-}C_3H_7)_2O$, 15°, 4 hr	II (45), III (42)	25
	$(C_2H_5)_2AlCN$	C_6H_6, room temp, 10 min	II (40), III (42)	25
	”	C_6H_6, 25°, 2 min	II:III (44:56, total quant)	25
	”	” , ” 7 hr	II:III (1:9, total quant)	25
(I)			(II), (III)	43
I, 6α-OAc	$(C_2H_5)_2AlCN$	C_6H_6, room temp, 26 min	6α-OAc, II (25), III (31)	
I, 6β-OAc	$(C_2H_5)_2AlCN$	C_6H_6-toluene room temp, 6 hr	6β-OAc, II (15), III (24)	

TABLE VI. CONJUGATE HYDROCYANATION OF CYCLIC α,β-ETHYLENIC KETONES (*Continued*)

	Substrate	Reagent(s)	Reaction Conditions	Product(s) and Yield(s) (%)	Refs.
C_{27} (*Contd.*) 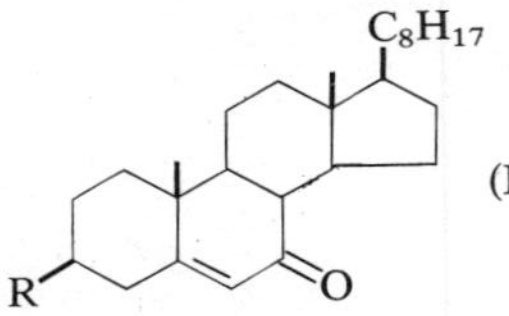	C_8H_{17} (I)			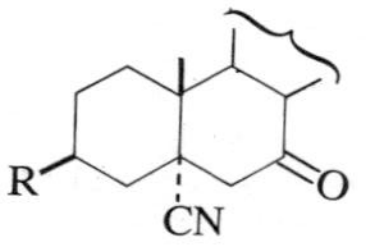 (II)	
	I, R = H	KCN-NH_4Cl	Aq DMF, 100°	II, R = H (51)	90
		HCN-$Al(C_2H_5)_3$	THF, room temp	II, R = H (93)	43
	I, R = OH	KCN-NH_4Cl	Aq DMF, 100°	R = OH, I (16), II (43)	43
	I, R = OAc	HCN-$Al(C_2H_5)_3$-(H_2O)	THF, 25°	II, R = OAc (92–93)	89, 43, 84
	AcO, H	HCN-$Al(C_2H_5)_3$	THF, room temp	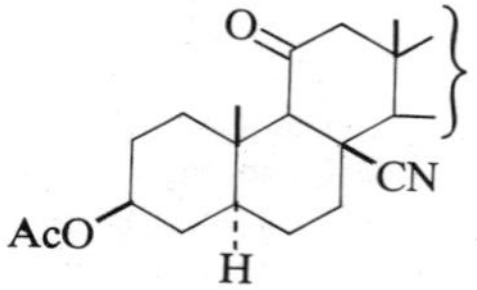 AcO, CN, H (54)	84, 42
	HO, H	HCN-$Al(C_2H_5)_3$	THF, —	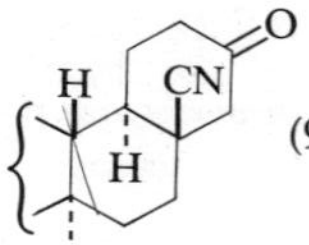 CN, H (90)	222

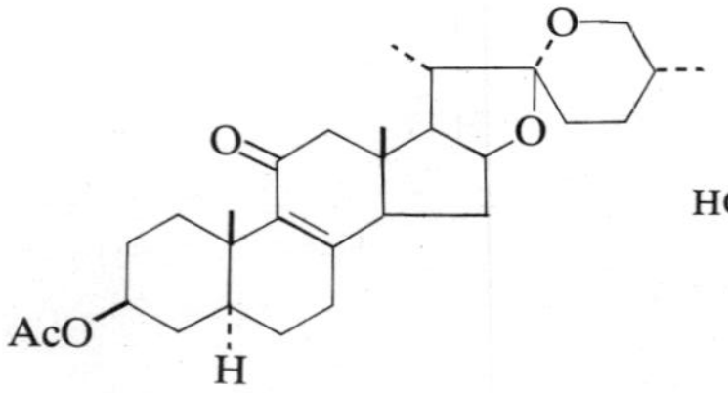

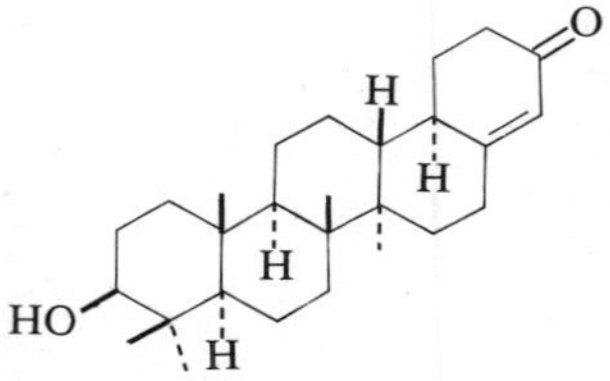

C_{28}	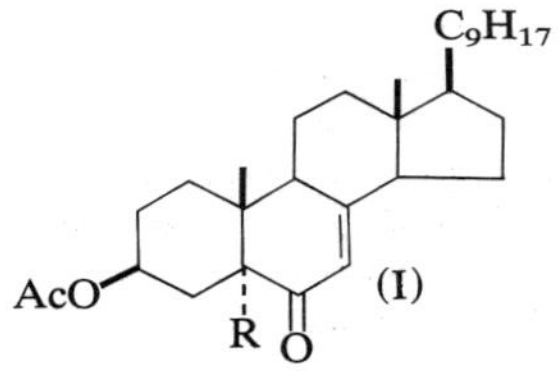 C_9H_{17}, AcO, R, O (I)	$HCN\text{-}Al(C_2H_5)_3$	THF, room temp	No reaction (1,2 addition)	43, 42
	I, R = H, OH, or OAc				
	C_9H_{17}, O, AcO, H	$KCN\text{-}NH_4Cl$	Aq DMF, 100°	No reaction	22
		$HCN\text{-}Al(C_2H_5)_3$	THF, room temp	C_9H_{17}, O, CN (I) I (70)	84, 22, 42
C_{30}	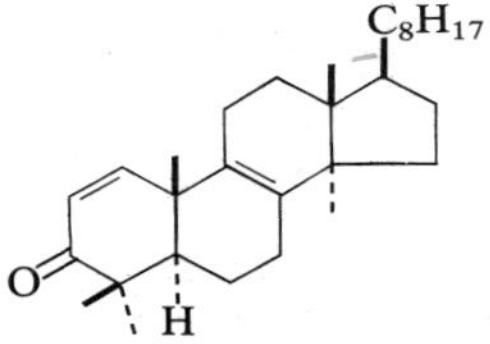 C_8H_{17}, O, H	$NaCN\text{-}NH_4Cl$	Aq DMF, 100°	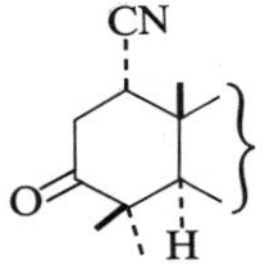 CN, O, H (73)	77

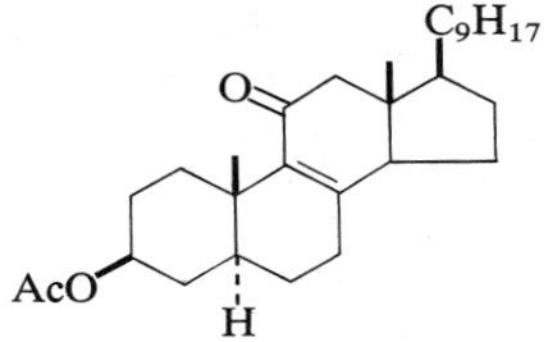

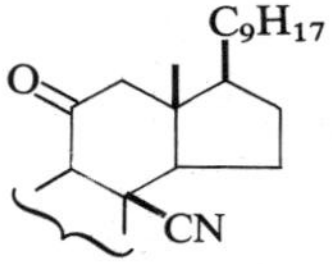

TABLE VI. CONJUGATE HYDROCYANATION OF CYCLIC α,β-ETHYLENIC KETONES (*Continued*)

Substrate	Reagent(s)	Reaction Conditions	Product(s) and Yield(s) (%)	Refs.
C_{30} (*Contd.*) 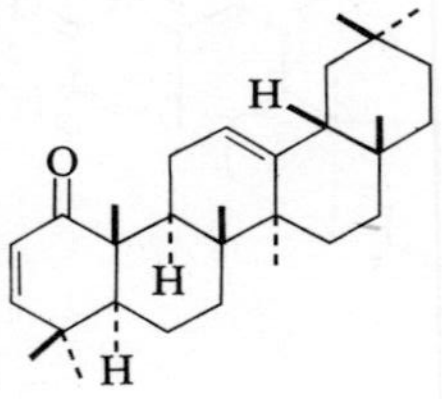	NaCN-NH_4Cl	Aq DMF, 100°,	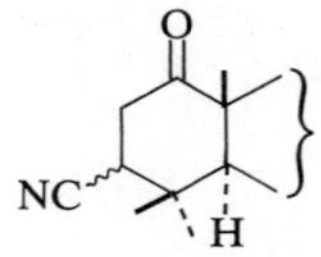 (Major isomer, 35), (Minor isomer, —)	77
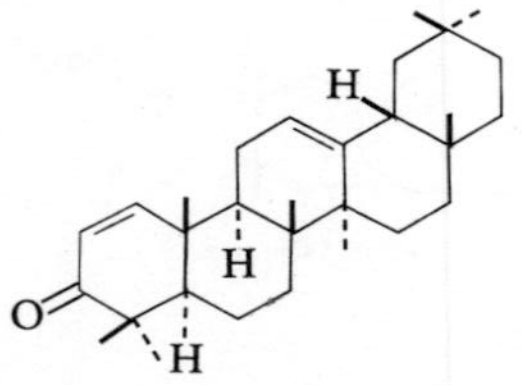	NaCN-NH_4Cl	Aq DMF, 100°	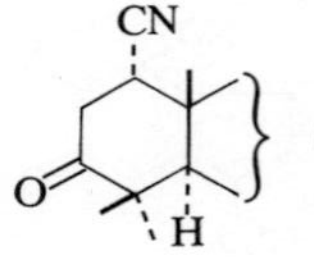 (66)	77

[a] The stereochemistry, though not determined, is clearly *cis* (see discussion on stereochemistry p.319)

[b] The substrate is formed from the corresponding bicyclic diketone by aldol condensation under the reaction conditions.

[c] The products were separated after reacetylation.

[d] Hydrocyanation was repeated on the recovered enone.

[e] The products were separated after ketalization and/or acetylation.

[f] This compound was reported to be obtained in 83% yield, but was assigned the 16β-cyanomethyl structure.[295]

[g] The reagent was prepared *in situ* by reaction of CH_3MgI and HCN.

TABLE VII. Conjugate Hydrocyanation of Conjugated Polyenic Carbonyl Derivatives

	Substrate	Reagent(s)	Reaction Conditions	Product(s) and Yield(s) (%)	Refs.
C_5	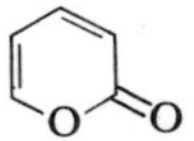 O O	NaCN	DMF, exothermic (25 → 52°)	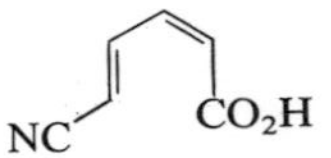 NC CO_2H (53)	154b
	CH_2=CHCH=CHCN (I)	HCN-(KCN)	No solvent, 65–70°	I (—), $NCCH_2CH$=$CHCH_2CN$ (Low)	2
		,,	CH_3CN, 95–100°	NCCH=$CH(CH_2)_2CN$ (12)	
C_6	CH_3CH=CHCH=CHCN	HCN-(KCN)	No solvent, 35–45°	No reaction	2
C_7	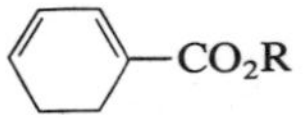 CO_2R (I)			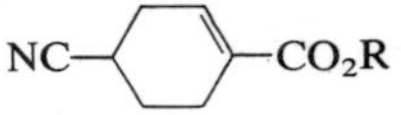 NC– CO_2R (II)	244
	I, R = CH_3	HCN	Pyr, 100°	II, R = CH_3 (18)	
		HCN-(KCN)	CH_3CN, 100°	II, R = CH_3 (50)	
		HCN-(NaOH)	,, , 150°	II, R = CH_3 (49)	
	I, R = C_2H_5	HCN-(K_2CO_3)	DMF, 120°	II, R = C_2H_5 (51)	
		HCN-[$NC_2H_5)_3$]	,, , 140°	II, R = C_2H_5 (56)	
	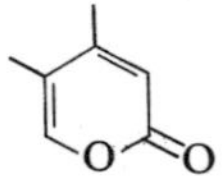 O O	NaCN	DMF, 150°	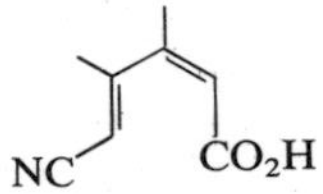 NC CO_2H (54)	154b
	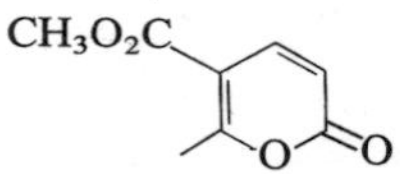 CH_3O_2C O O	NaCN	Aq acetone, room temp	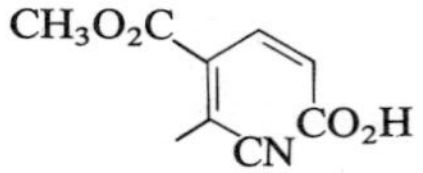 CH_3O_2C CN CO_2H (I, 74)	154b
		,,	DMF, room temp	I (69)	154b

TABLE VII. CONJUGATE HYDROCYANATION OF CONJUGATED POLYENIC CARBONYL DERIVATIVES (Continued)

	Substrate	Reagent(s)	Reaction Conditions	Product(s) and Yield(s) (%)	Refs.
C_8	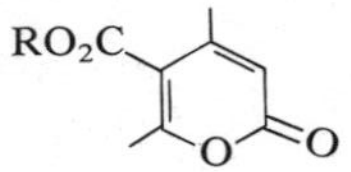 RO_2C, O, O (I)			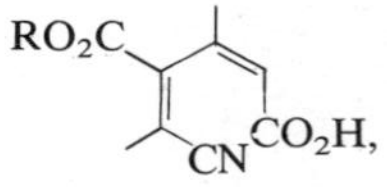 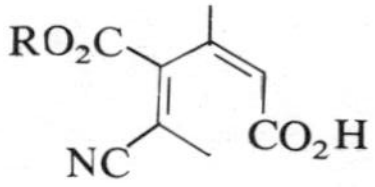 RO_2C, CN, CO_2H, (II); RO_2C, NC, CO_2H (III)	154b
	I, $R = CH_3$	NaCN	Aq acetone, exothermic (5 → 33°)	II, $R = CH_3$ (62)	
	I, $R = C_2H_5$	NaCN	Aq acetone, exothermic (5 → 33°)	II, $R = C_2H_5$ (90)	
		"	DMF, exothermic (20 → 42°)	III, $R = C_2H_5$ (70)	
C_{10}	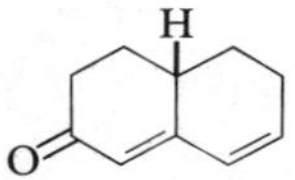 H, O	HCN-Al$(C_2H_5)_3$	$(C_2H_5)_2O$, room temp	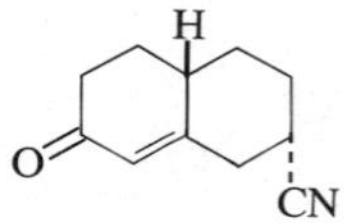 H, O, CN (I, 48)[a]	157
		$(C_2H_5)_2AlCN$	—	I (High)	158
	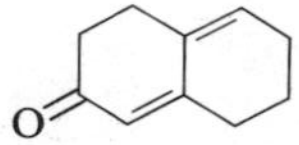 O (I)			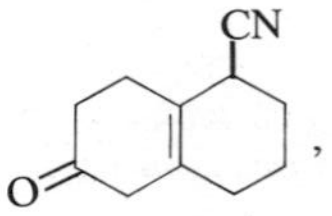 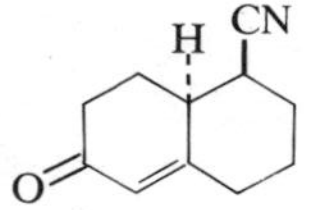 CN, O (II)[b], H, CN, O (III),	94

				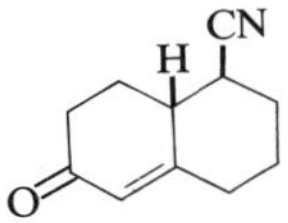 (IV)	
		$KCN\text{-}NH_4Cl$	Aq DMF, 100°	II (Poor)	
		$HCN\text{-}Al(C_2H_5)_3$	THF, room temp, acid workup	II (Quant)	
		”	THF, room temp, basic workup	II (6), III (20), IV (18)	
		$HCN\text{-}(C_2H_5)_2AlCl$	THF, room temp, basic workup	I:II:(III+IV) = 3:6:1 (total —)	
		$(C_2H_5)_2AlCN$	THF, —	No reaction	
		”	C_6H_6, —	II (5)	
C_{12}	$CH_3COCH{=}CHCH{=}CHC_6H_5$	$(CH_3)_2C(OH)CN$-(Na_2CO_3)	Aq CH_3OH, reflux	No reaction	50
	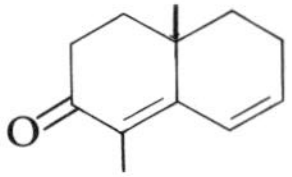 (I)	KCN	Aq C_2H_5OH, reflux	I (4), 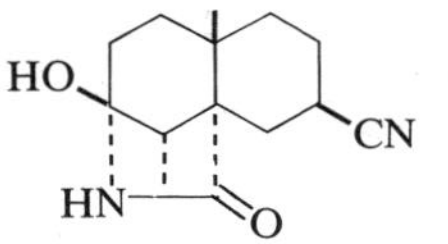 (42), 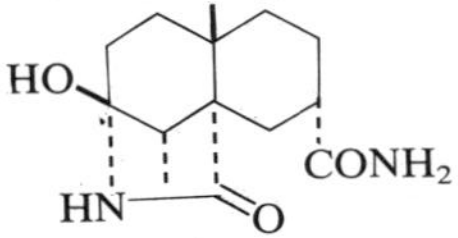 (6)	117
		$KCN\text{-}NH_4Cl$	Aq C_2H_5OH, reflux	I (51), 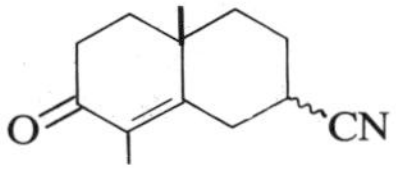 (0.5),	117

TABLE VII. Conjugate Hydrocyanation of Conjugated Polyenic Carbonyl Derivatives (*Continued*)

	Substrate	Reagent(s)	Reaction Conditions	Product(s) and Yield(s) (%)	Refs.
C_{12} (*Contd.*)	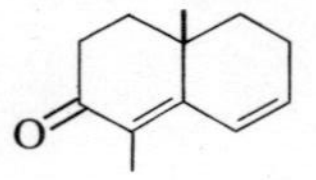 (I)			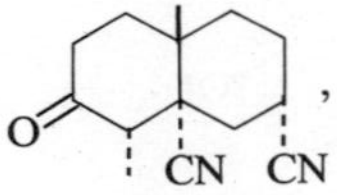 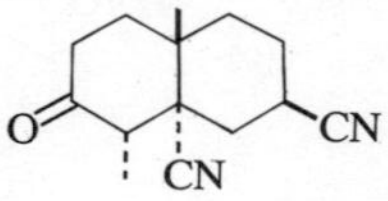 (4), (1.5), 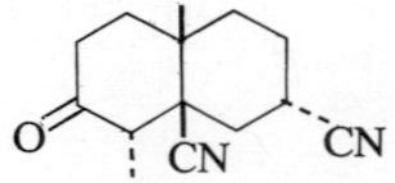 (9)	
	$C_6H_5CH{=}CHCH{=}CH(CO_2CH_3)_2$	KCN	Dry CH_3OH, room temp	$C_{29}H_{29}O_8N$ (dimer, 48)	11
		1. KCN (1 eq) 2. 5% KOH	Aq C_2H_5 OH, reflux Hydrolysis	$C_6H_5CH{=}CHCH(CO_2H)CH_2CO_2H$ (13)	41
		1. KCN (3 eq) 2. 25% KOH	Aq C_2H_2OH, heat Hydrolysis	$C_6H_5CH_2CH(CO_2H){-}CH(CO_2H)CH_2CO_2H$ (—)	10, 41
	$C_6H_5CH{=}CHCH{=}C(CN)CO_2H$	KCN-AcOH	—	Unsuccessful	118
C_{15}	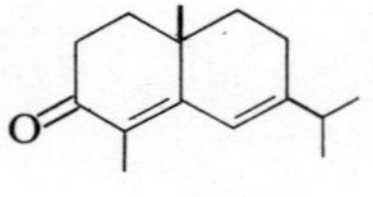 (I)	KCN-NH_4Cl	Aq C_2H_5OH, reflux	I (38), 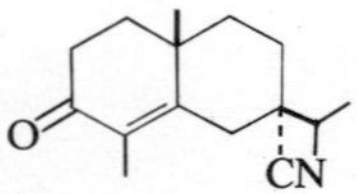 II (36) I+II (11)	134

	Substrate	Reagent	Conditions	Product(s) and Yield(s) (%)	Refs.
C_{18}	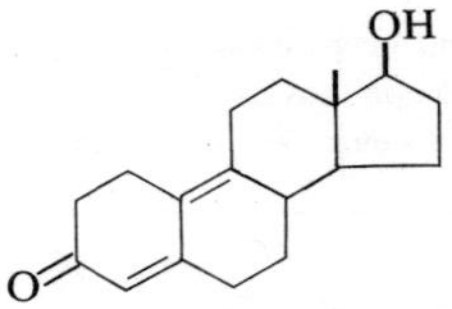	$(C_2H_5)_2AlCN$	THF, —	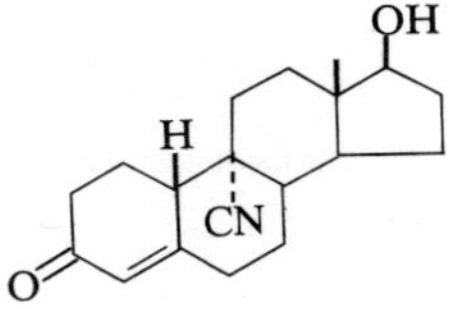 (53)	198
C_{19}	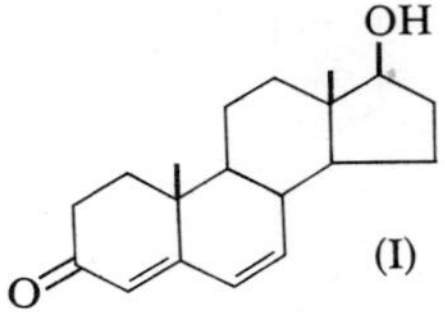 (I)			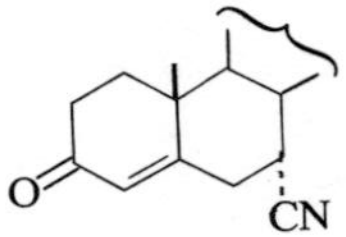 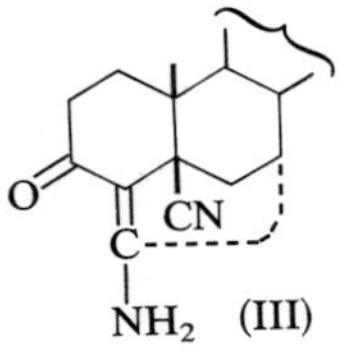 (II), (III)	
		KCN-$AcOC_2H_5$	Aq CH_3OH, reflux	II (6), III (36)	96, 300
		KCN-NH_4Cl	Aq DMF, 100°	II (55), III (10)	43
		HCN-$Al(C_2H_5)_3$	THF, 0°	I (32), II (32)	95
		$(C_2H_5)_2AlCN$	THF-toluene, room temp, 45 min[c]	II (92)	43
C_{20}	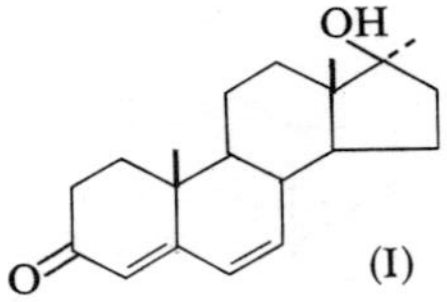 (I)	KCN-$AcOC_2H_5$	Aq CH_3OH, reflux	I (33), 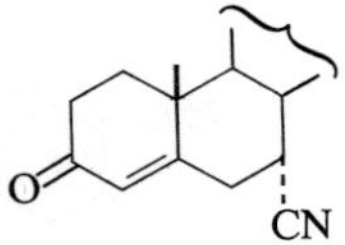 (12), 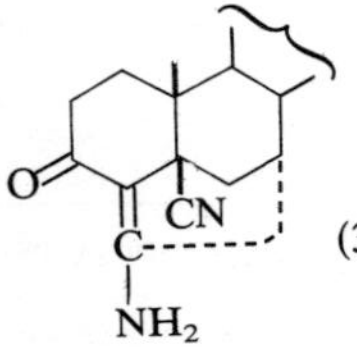 (36)	96, 300

TABLE VII. CONJUGATE HYDROCYANATION OF CONJUGATED POLYENIC CARBONYL DERIVATIVES (*Continued*)

	Substrate	Reagent(s)	Reaction Conditions	Product(s) and Yield(s) (%)	Refs.
C_{21}	$CO(CH{=}CHCH{=}CHC_6H_5)_2$	$(CH_3)_2C(OH)CN$-(Na_2CO_3)	Aq CH_3OH, reflux	No reaction	50
C_{22}	[structure]	KCN-$AcOC_2H_5$	Aq CH_3OH, reflux	[structure: CN, C–NH_2] (42)	236
C_{23}	[structure: C_6H_5, C_6H_5, C_6H_5]	NaCN	DMF, 150°	[structure: C_6H_5, C_6H_5, NC, CO_2H, C_6H_5] (80)	154b
C_{24}	[structure: CH_3O, H]	HCN-$Al(C_2H_5)_3$	THF, room temp	No reaction	198
		$(C_2H_5)_2AlCN$	THF or C_6H_6, —	No reaction	198

[a] The α configuration of the cyano group, though not determined, is based on analogy (see discussion on stereochemistry, p. 311).
[b] For this revised structure see footnote on p. 291.
[c] Hydrocyanation was repeated on the recovered dienone.

Note: References 246–305 are on pp. 475–476.

TABLE VIII. Conjugate Hydrocyanation of α,β-Acetylenic Carbonyl Derivatives

	Substrate	Reagent(s)	Reaction Conditions	Product(s) and Yield(s) (%)	Refs.
C_3	$HC{\equiv}CCO_2CH_3$ (I)	HCN-(KCN)	No solvent, 45°	I (70), $NCCH{=}CHCO_2CH_3$ (24)	2
C_9	$C_6H_5C{\equiv}CCO_2R$ (I)				
	I, $R = CH_3$	KCN	Dry CH_3OH, reflux	$C_6H_5CH(CN)CH(CO_2CH_3)CN$ (II, 19), $C_6H_5C(OCH_3){=}CHCO_2CH_3$ + $C_6H_5COCH_2CO_2CH_3$ (7:3 mixture, total 68)	11
		”	Aq CH_3OH, room temp	I, R = H (48), II (37)	11
	I, $R = C_2H_5$	KCN	C_2H_5OH, reflux	$C_6H_5CH(CN)CH_2CN$ (III, —)	120
		”	H_2O, ”	III (20)	121

TABLE IX. CONJUGATE HYDROCYANATION OF MISCELLANEOUS COMPOUNDS

	Substrate	Reagent(s)	Reaction Conditions	Product(s) and Yield(s) (%)	Refs.
			A. α,β-Ethylenic Aldehydes		
C_3	$CH_2{=}CHCHO$	HCN-(C_2H_5ONa)	Dry C_2H_5OH, reflux	$NCCH_2CH_2CHO$ (—)	199
C_4	$CH_2{=}C(CH_3)CHO$	HCN	DMF, 80°	$NCCH_2CH(CH_3)CH(OH)CN$ (I, 93)	47
		HCN-N($C_2H_5)_3$	CH_3OH, 70°	I (97)	47
	$CH_3CH{=}C(R)CHO$ R = H, Cl	HCN-(aq NaOH)	No solvent, exothermic	$CH_3CH{=}C(R)CH(OH)CN$ (—) R = H, Cl (No 1,4-addition)	301
	$CCl_3CH{=}CHCHO$	KCN	C_2H_5OH, exothermic	$ClCH_2CH{=}CHCO_2C_2H_5$ (—) (No 1,4 addition?)	302
	$CH_2{=}CHCH(OH)CN$	HCN	CH_3OH, 100–105°	$NCCH_2CH_2CH(OH)CN$ (I, 94)	47
		HCN-(Amberlite IRA-400)	CH_3OH, reflux	I (94)	
C_5	$CH_3CH{=}CHCH(OH)CN$	HCN-[N($C_2H_5)_3$]	CH_3OH, 60°	$CH_3CH(CN)CH_2CH(OH)CN$ (90)	47
C_6	$C_2H_5CH{=}C(CH_3)CHO$	HCN	No solvent, 45°	$C_2H_5CH{=}C(CH_3)CH(OH)CN$ (—) (No 1,4 addition)	303
C_8	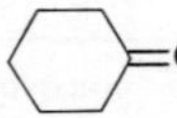 Cyclohexylidene=CHCHO	HCN-($C_2H_5)_2$AlCl	THF, room temp	1,2 Adduct (—), no 1,4 addition	92
		($C_2H_5)_2$AlCN	Toluene, room temp, 2.5 hr	1,2 Adduct (—), no 1,4 addition	
C_{19}	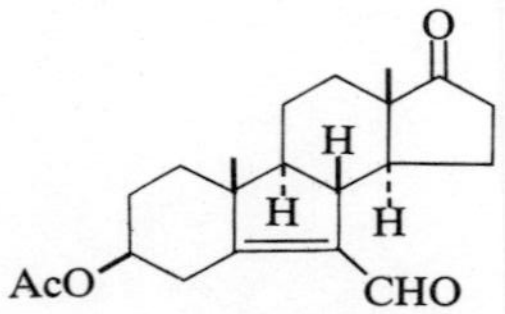 [steroid structure: AcO, 17-keto, CHO, H]	($C_2H_5)_2$AlCN	C_6H_6-toluene, room temp, 2 days, or 50°, 1.5 hr	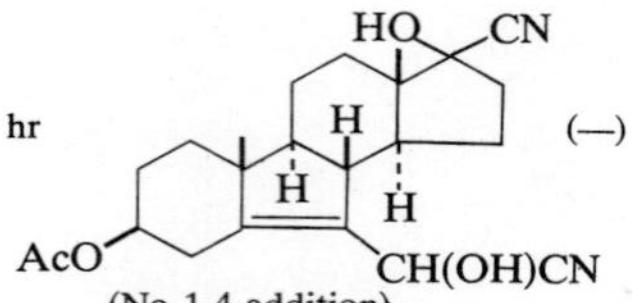 [steroid structure: AcO, HO, CN, CH(OH)CN, H] (—) (No 1,4 addition)	92

C_{20}	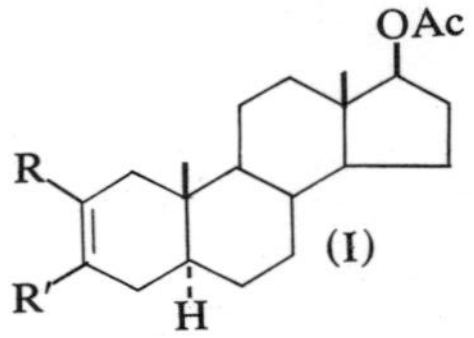 (I) [structure: R, R′, OAc, H]				92
	I, R = CHO, R′ = H	$(C_2H_5)_2AlCN$	C_6H_6, room temp, 17 hr	1,2 Adduct (—), no1,4 addition	
	I, R = H, R′ = CHO	$(C_2H_5)_2AlCN$	C_6H_6-CH_2Cl_2, room temp, 8.5 hr	1,2 Adduct (—), no 1,4 addition	
	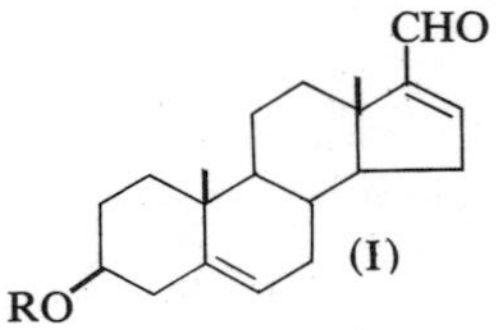 (I) [structure: CHO, RO]			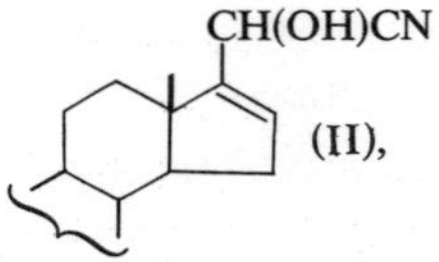 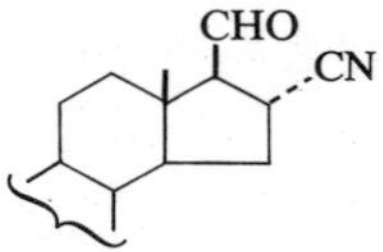 CH(OH)CN (II),	92
				CHO, CN (III)	
	I, R = H	$(C_2H_5)_2AlCN$	Toluene, 0°, 30 min	II (R = H, 84), no 1,4 addition	
	I, R = Ac	$(C_2H_5)_2AlCN$	C_6H_6-toluene, room temp, 5 hr	III (R = Ac, 58)	

B. *α,β-Ethylenic Imino Derivatives*

C_4	$CH_3CH{=}CHCH{=}NC_4H_9$-*t*	1. HCN-$Al(C_2H_5)_3$ 2. Hydrolysis	THF, 0°	$CH_3CH(CN)CH_2CHO$ (15)	92

TABLE IX. Conjugate Hydrocyanation of Miscellaneous Compounds (*Continued*)

B. *α,β-Ethylenic Imino Derivatives* (*Continued*)

	Substrate	Reagent(s)	Reaction Conditions	Product(s) and Yield(s) (%)	Refs.
C_8	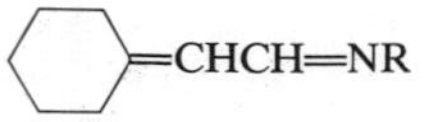 [cyclohexylidene]=CHCH=NR (I)			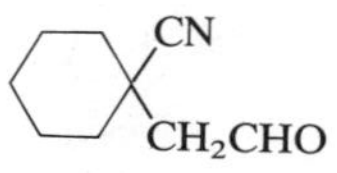 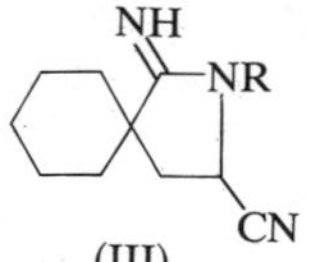 1-cyano-1-(CH_2CHO)cyclohexane (II), [spiro imino-pyrrolidine: NH, NR, CN] (III)	92
	I, $R = t\text{-}C_4H_9$	1. $HCN\text{-}Al(C_2H_5)_3$ 2. Hydrolysis	THF, room temp	II (72), III ($R = t\text{-}C_4H_9$, 11)	
	I, $R = C_6H_{11}$	1. $HCN\text{-}Al(C_2H_5)_3$ 2. Hydrolysis	" , −10°	II (39), III ($R = C_6H_{11}$, 24)	
C_9	$C_6H_5CH{=}CHCH{=}NR$ (I)			$C_6H_5CH_2CH{=}C(OCH_3)NHR$ (II), $C_6H_5(CH_2)_2CONHR$ (III)	
	I, $R = CH_3$	KCN	Dry CH_3OH, reflux	$R = CH_3$, II (50), III (8)	200
		"	90% CH_3OH, "	$R = CH_3$, III (76)	200
	I, $R = C_6H_5$	KCN	Dry CH_3OH, reflux	II ($R = C_6H_5$, 92)	200
		"	95% CH_3OH, "	$R = C_6H_5$, II (60), III (17)	200
		"	90% CH_3OH, "	$R = C_6H_5$, II (25), III (50)	200
	I, $R = C_4H_9\text{-}t$	1. $HCN\text{-}(C_2H_5)_2AlCl$ 2. Hydrolysis	THF, room temp	$C_6H_5CH(CN)CH_2CHO$ (54)	92
	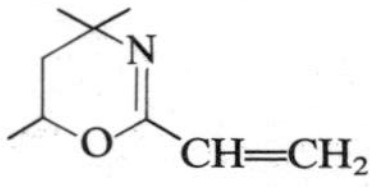 [4,4,6-trimethyl-5,6-dihydro-1,3-oxazine]-2-$CH{=}CH_2$	$HCN\text{-}C_6H_4(NO_2)_2\text{-}m$	AcOH, reflux	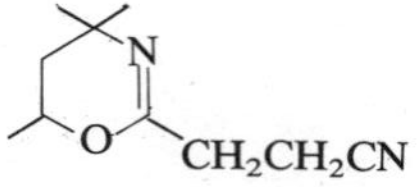 [4,4,6-trimethyl-5,6-dihydro-1,3-oxazine]-2-CH_2CH_2CN (45)	201, 304

Formula	Reactant	Reagent	Conditions	Product(s) and Yield(s) (%)	Refs.
	R-indoleninium, OH^-, $CH_2C_6H_5$ (I)[a]	KCN	Aq DMF, 50°	R-indole, CN, $CH_2C_6H_5$ (II), R-indole, CN, $CH_2C_6H_5$ (III)	203
	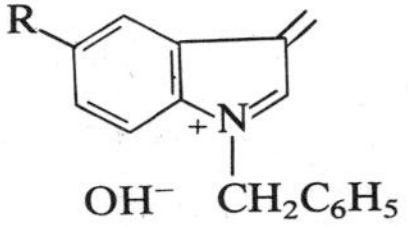 I, R = H I, R = Cl I, R = OCH_3			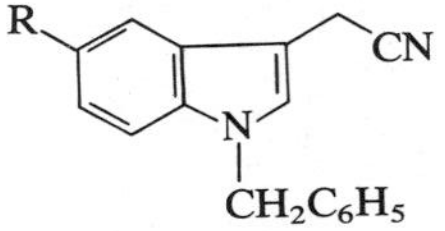 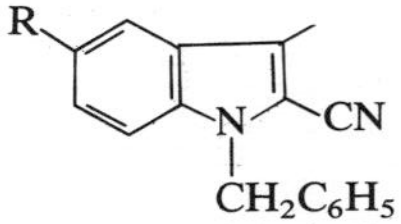 R = H, II (28), III (9) R = Cl, II (32), III (7) R = OCH_3, II (29), III (9)	
C_{10}	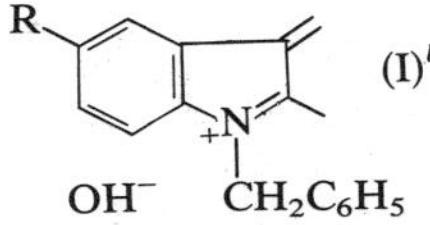 R-indoleninium, OH^-, $CH_2C_6H_5$ (I)[b] I, R = H I, R = Cl I, R = OCH_3	KCN	Aq DMF, 50°	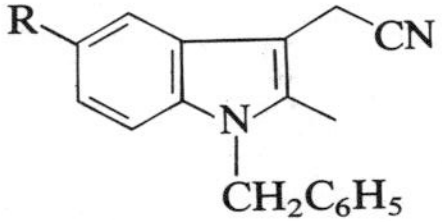 R-indole, CN, $CH_2C_6H_5$ (II) II, R = H (62) II, R = Cl (54) II, R = OCH_3 (67)	203
	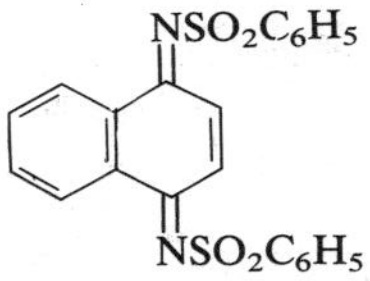 $NSO_2C_6H_5$, $NSO_2C_6H_5$	HCN-[$N(C_2H_5)_3$]	C_6H_6, 0°	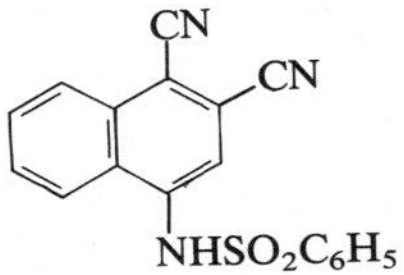 CN, CN, $NHSO_2C_6H_5$ (85)	202

TABLE IX. CONJUGATE HYDROCYANATION OF MISCELLANEOUS COMPOUNDS (*Continued*)

Substrate	Reagent(s)	Reaction Conditions	Product(s) and Yield(s) (%)	Refs.
B. α,β-Ethylenic Imino Derivatives (*continued*)				
C_{18} N ... OH 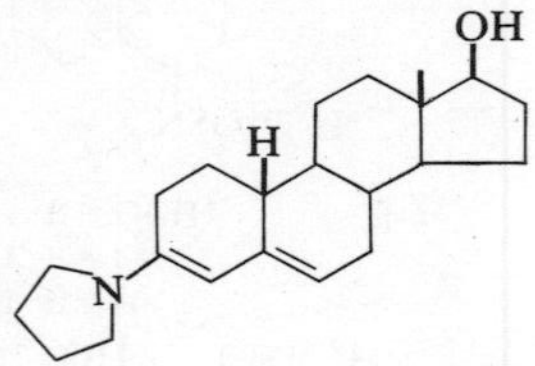 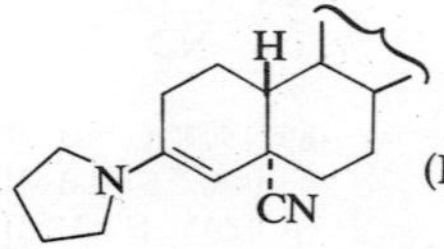 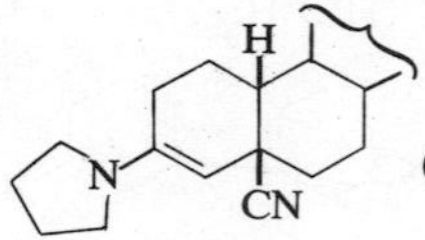			N ... CN (I),[c] N ... CN (II)[c]	43
	$HCN\text{-}Al(C_2H_5)_3$	THF, room temp	I (62), II (15)	
	$(C_2H_5)_2AlCN$	C_6H_6, ", ", 1 hr	I (57), II (Minor)	
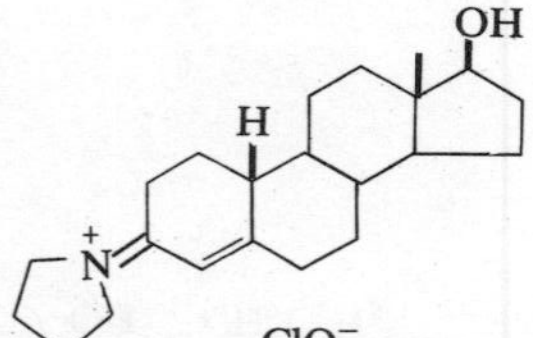 $\overset{+}{N}$ ... OH ClO_4^-	$HCN\text{-}Al(C_2H_5)_3$	THF, room temp	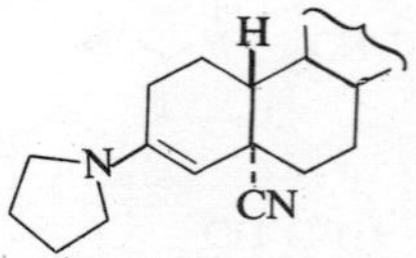 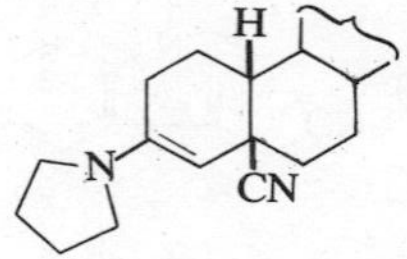 N ... CN (43),[c] N ... CN (19)[c]	43

	Substrate	Reagent	Conditions	Product (Yield, %)	Ref.
C_{19}	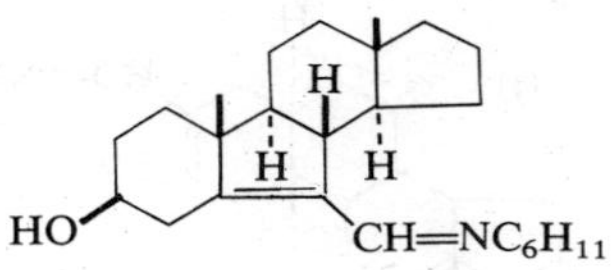 HO, H, H, H, $CH{=}NC_6H_{11}$	$HCN\text{-}Al(C_2H_5)_3$	THF, room temp	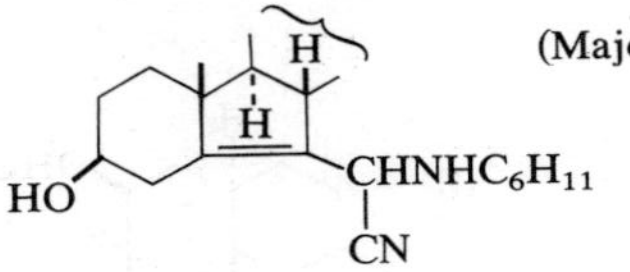 HO, H, H, $CHNHC_6H_{11}$, CN (Major)	92
		1. $HCN\text{-}C_2H_5AlCl_2$ 2. Hydrolysis	", " "	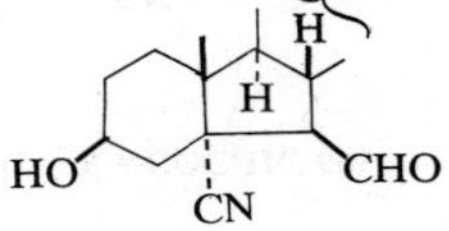 HO, H, H, CN, CHO (55)	92
C_{20}	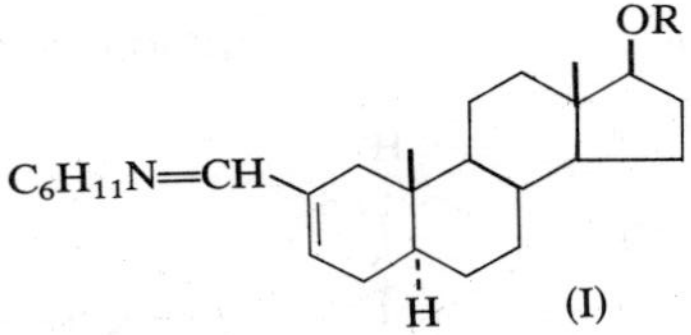 $C_6H_{11}N{=}CH$, OR, H (I)			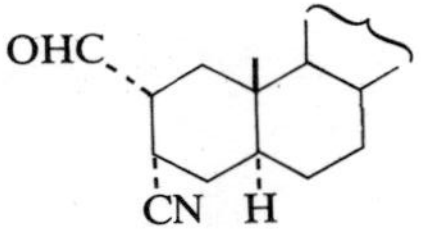 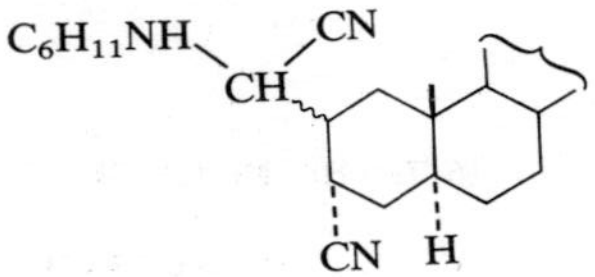 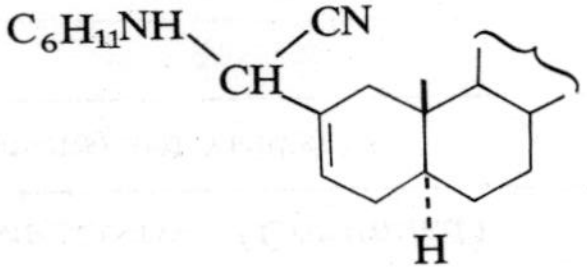 OHC, CN, H (II),	92
				$C_6H_{11}NH$, CH, CN, CN, H (III),	
				$C_6H_{11}NH$, CH, CN, H (IV)	

TABLE IX. Conjugate Hydrocyanation of Miscellaneous Compounds (*Continued*)

Substrate	Reagent(s)	Reaction Conditions	Product(s) and Yield(s) (%)	Refs.
	B. *α,β-Ethylenic Imino Derivatives* (*Continued*)			
C_{20} (*Contd.*) I, R = H	$HCN\text{-}Al(C_2H_5)_3$	THF, room temp	III (R = H, 96)	
	1. $(C_2H_5)_2AlCN$ 2. Hydrolysis	$THF\text{-}C_6H_6$, room temp, 35 min	II (R = H, 67), III (R = H, 9)	
$C_6H_{11}N{=}CH$ … (I)				
I, R = COC_6H_5	$(C_2H_5)_2AlCN$	$THF\text{-}C_6H_6$, room temp, 23 hr	IV (R = COC_6H_5, Major)	
	1. $(C_2H_5)_2AlCN\text{-}(C_3H_7OH\text{-}i)$ 2. Hydrolysis	$THF\text{-}C_6H_6$, 25°, 30 min	II (R = COC_6H_5, 69)	
RN=CH … $OCOC_6H_5$ (I)			OHC … CN … H (II),	92

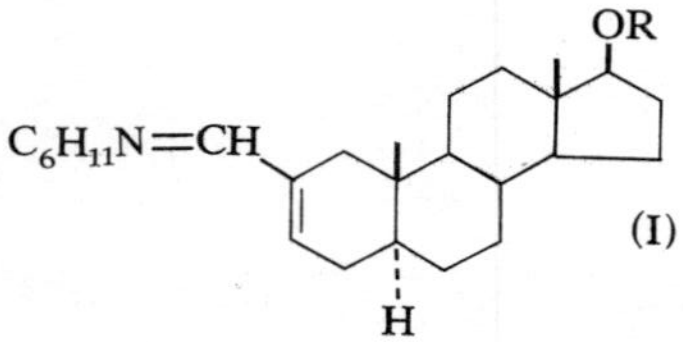

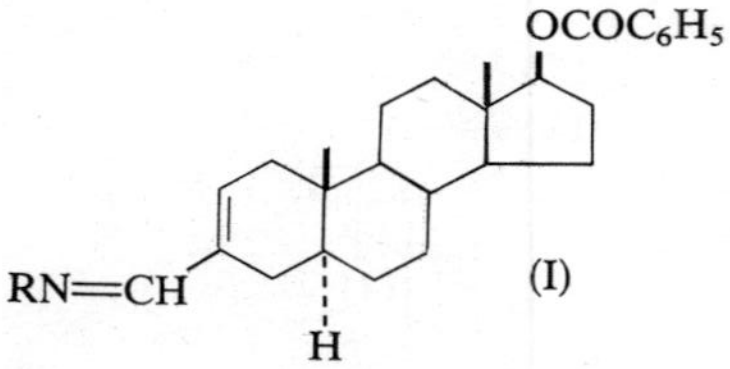

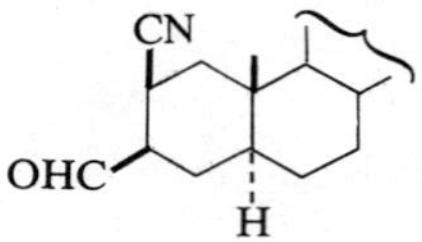

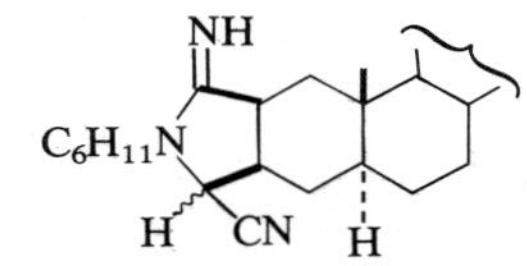

Reactant	Reagents	Conditions	Product(s) (Yield, %)	Ref.
			(III) [structure]	
I, R = t-C_4H_9	1. HCN-Al$(C_2H_5)_3$-(H_2O) 2. Hydrolysis	THF, room temp	II (64)	
I, R = C_6H_{11}	1. HCN-Al$(C_2H_5)_3$ 2. Hydrolysis	THF, 25°	II (11), III (2 isomers, total 19)	
AcO– ... CH=NR (I) [structure]			CHO, CN (II), [structure]; CN, NC_6H_{11}, NH (III) [structure]	92
I, R = C_4H_9-t	1. HCN-Al$(C_2H_5)_3$-(H_2O) 2. Hydrolysis	THF, room temp	II (55)	
I, R = C_6H_{11}	1. HCN-Al$(C_2H_5)_3$ 2. Hydrolysis	THF, 25°	II (25), III (6)	
	1. $(C_2H_5)_2$AlCN-C_3H_7OH-i) 2. Hydrolysis	THF-toluene, 21°, 1 hr	II (47)	

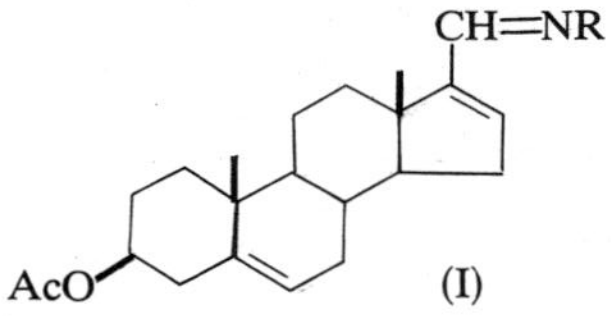

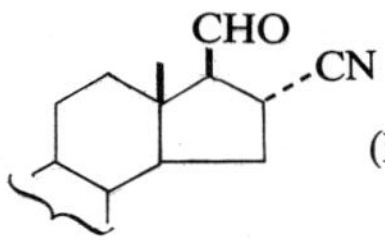

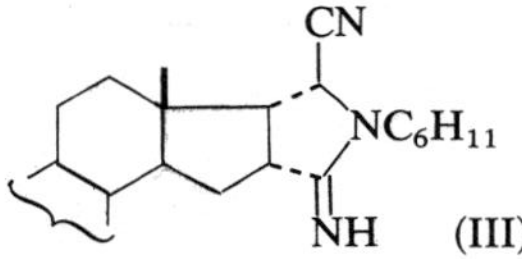

TABLE IX. CONJUGATE HYDROCYANATION OF MISCELLANEOUS COMPOUNDS (*Continued*)

	Substrate	Reagent(s)	Reaction Conditions	Product(s) and Yield(s) (%)	Refs.
C_{27}	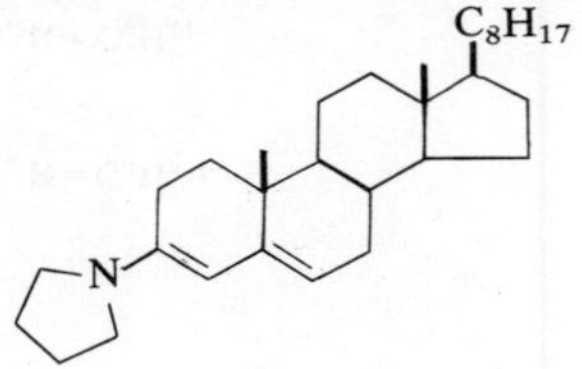 C_8H_{17}	KCN-AcOH	Aq C_2H_5OH, room temp	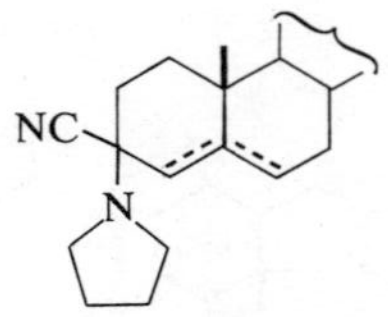 NC (—)	305

[a] The starting materials were prepared *in situ* by oxidation of *p*-$RC_6H_4N(CH_2C_6H_5)CH_2C{\equiv}CH$ with *m*-chloroperoxybenzoic acid followed by rapid rearrangement and cyclization.

[b] The starting materials were prepared *in situ* by oxidation of *p*-$RC_6H_4N(CH_2C_6H_5)CH_2C{\equiv}CCH_3$ with *m*-chloroperoxybenzoic acid followed by rapid rearrangement and cyclization.

[c] The products are separated as cyano ketones after hydrolysis.

Note: References 246–305 are on pp. 475-476.

REFERENCES TO TABLES I–IX

[246] H. Dreyfus, U.S. Pat. 2,481,580 (1949) [*C.A.*, **44,** 653h (1950)].
[247] Ciba Ltd., Swiss Pat. 275,798 (1951) [*C.A.*, **47,** 1732d (1953)].
[248] H.-F. Piepenbrink, *Ann. Chem.*, **572,** 6 (1951).
[249] W. Baker and A. Lapworth, *J. Chem. Soc.*, **127,** 560 (1925).
[250] A. Lapworth, *Proc. Chem. Soc. London,* **19,** 189 (1903).
[251] I. G. Farben AG., Ger. Pat. 427,416 (1926).
[252] Huan, *Bull. Soc. Chim. Fr.*, [5] **5,** 1341 (1938).
[253] E. Hope and W. Sheldon, *J. Chem. Soc.*, **121,** 2223 (1922).
[254] S. S. Sandelin, *Ber.*, **31,** 1119 (1898).
[255] A. A. Akhrem, A. M. Moiseenkov, and F. A. Lakhvich, *Bull. Acad. Sci., USSR (Engl. Transl.)*, **21,** 866 (1972).
[256] S. Wideqvist, *Arkiv Kemi, Mineral. Geol.* [19], **14B,** 1 (1940) [*C.A.*, **35,** 3993[7] (1941)].
[257] J. Staněk and J. Urban, *Collect. Czech. Chem. Commun.*, **15,** 371 (1950).
[258] A. Lapworth and J. A. McRae, *J. Chem. Soc.*, **121,** 2741 (1922).
[259] R. Carrié, *Bull. Soc. Sci. Bretagne,* **37,** 29 (1962) [*C.A.*, **58,** 6665f (1963)].
[260] J. A. McRae and C. Y. Hopkins, *Can. J. Res.*, **7,** 248 (1932) [*C.A.*, **27,** 278 (1933)].
[261] B. B. Corson and R. W. Stoughton, *J. Amer. Chem. Soc.*, **50,** 2825 (1928).
[262] J. P. Almange and R. Carrié, *C.R. Acad. Sci.*, **257,** 1781 (1963).
[263] H. Fischer, P. Hartmann, and H.-J. Riedl, *Ann. Chem.*, **494,** 246 (1932).
[264] J. P. Nallet, R. Barret, C. Arnaud, and J. Huet, *Tetrahedron Lett.*, **1975,** 1843.
[265] W. Schneider, F. Schumann, and G. Modjesch, *Pharmazie,* **25,** 724 (1970) [*C.A.*, **75,** 35676a (1971)].
[266] S. Ruhemann, *J. Chem. Soc.*, **85,** 1451 (1904).
[267] J. A. McRae and A. L. Kuehner, *J. Amer. Chem. Soc.*, **52,** 3377 (1930).
[268] E. Campaigne and W. L. Roelofs, *J. Org. Chem.*, **30,** 396 (1965).
[269] R. K. Ray and B. K. Bhattacharyya, *J. Indian Chem. Soc.*, **23,** 469 (1946).
[270] H. L. Moal, *C.R. Acad. Sci.*, **232,** 736 (1951).
[271] J. Thiele and J. Meisenheimer, *Ber.*, **33,** 675 (1900).
[272] C. F. H. Allen and C. W. Wilson, *J. Amer. Chem. Soc.*, **63,** 1756 (1941).
[273] J. Thiele and F. Günther, *Ann. Chem.*, **349,** 45 (1906).
[274] M. M. Maire, *Bull. Soc. Chim. Fr.*, [4] **3,** 280 (1908).
[275] H. R. Henze, T. R. Thompson, and R. T. Speer, *J. Org. Chem.*, **8,** 17 (1943).
[276] R. C. Atkins and B. M. Trost, *J. Org. Chem.*, **37,** 3133 (1972).
[277] I. N. Nazarov and M. V. Kuvarzina, *Izv. Akad. Nauk SSSR, Ser. Khim.*, **1949,** 299 [*C.A.*, **43,** 6625a (1949)].
[278] H. Rupe and F. Schneider, *Ber.*, **28,** 957 (1895).
[279] B. S. Mehta, K. V. Bokil, and K. S. Nargund, *J. Univ. Bombay,* **9,** 158 (1940) [*C.A.*, **35,** 6945[9] (1941)].
[280] R. Anschütz and W. F. Montfort, *Ann. Chem.*, **284,** 1 (1895).
[281] B. S. Mehta, K. V. Bokil, and K. S. Nargund, *J. Univ. Bombay,* **10,** 137 (1942) [*C.A.*, **37,** 622[4] (1943)].
[282] P. Crabbé, H. Carpio, A. Cervantes, J. Iriarte, and L. Tökes, *Chem. Commun.*, **1968,** 79.
[283] D. K. Banerjee, J. Dutta, and G. Bagavant, *Proc. Indian Akad. Sci.*, **46A,** 80 (1957) [*C.A.*, **52,** 3701b (1958)].
[284] W. C. Agosta and W. W. Lowrance, Jr., *J. Org. Chem.*, **35,** 3851 (1970).
[285] G. A. Berchtold and G. F. Uhlig, *J. Org. Chem.*, **28,** 1459 (1963).
[286] R. Fusco, G. Bianchetti, and M. Dubini, *Rend. Ist. Lombardo Sci. Pt. I,* **91,** 170 (1957) [*C.A.*, **52,** 11762i (1958)].
[287] A. C. O. Hann and A. Lapworth, *Proc. Chem. Soc. London,* **20,** 54 (1904).
[288] A. Lapworth, *J. Chem. Soc.*, **89,** 945 (1906).
[289] A. Borchardt and J. Krupowicz, *Roczniki Chem.*, **40,** 139 (1966) [*C.A.*, **65,** 2302a (1966)].
[290] M. Lasperas, A. Casadevall, and E. Casadevall, *Bull. Soc. Chim. Fr.*, **1970,** 2580.
[291] M. Tichý, A. Orahovatz, and J. Sicher, *Collect. Czech. Chem. Commun.*, **35,** 459 (1970).
[292] W. G. Dauben, R. G. Williams, and R. D. McKelvey, *J. Amer. Chem. Soc.*, **95,** 3932 (1973).

[293] K. E. Harding, R. C. Ligon, T. Wu, and L. Rodé, *J. Amer. Chem. Soc.*, **94,** 6245 (1972).
[294] E. W. Cantrall, R. Littell, and S. Bernstein, *J. Org. Chem.*, **29,** 214 (1964).
[295] K. Brückner, K. Irmscher, F. von Werder, K. H. Bork, and H. Metz, *Chem. Ber.*, **94,** 2897 (1961).
[296] T. Nambara, J. Goto, Y. Fujimura, and Y. Kimura, *Chem. Pharm. Bull.* (Tokyo), **19,** 1137 (1971).
[297] B. Ellis, V. Petrow, and D. Wedlake, *J. Chem. Soc.*, **1958,** 3748.
[298] J. Romo, L. Rodriguez-Hahn, P. Joseph-Nathan, M. Martinez, and P. Crabbé, *Bull. Soc. Chim. Fr.*, **1964,** 1276.
[299] T. Nambara, S. Goya, J. Goto, and K. Shimada, *Chem. Pharm. Bull.* (Tokyo), **16,** 2228 (1968).
[300] R. G. Christiansen and W. S. Johnson, U.S. Pat. 3,200,113 (1965) [*C.A.*, **63,** 13359h (1965)].
[301] C. Moureu, M. Murat, and L. Tampier, *Bull. Soc. Chim. Fr.*, **1921,** 29.
[302] O. Wallach and A. Boehringer, *Ber.*, **6,** 1539 (1873).
[303] G. Johanny, *Ber.*, **23,** 655c (1890).
[304] J. W. Lynn, *J. Org. Chem.*, **24,** 711 (1959).
[305] J. L. Johnson, M. E. Herr, J. C. Babcock, A. N. Fonken, J. E. Stafford, and F. W. Heyl, *J. Amer. Chem. Soc.*, **78,** 431 (1956).

AUTHOR INDEX, VOLUMES 1–25

CHAPTER AND TOPIC INDEX, VOLUMES 1–25

Many chapters contain brief discussions of reactions and comparisons of alternative synthetic methods which are related to the reaction that is the subject of the chapter. These related reactions and alternative methods are not usually listed in this index. In this index the volume number is in BOLDFACE, the chapter number in ordinary type.

SUBJECT INDEX, VOLUME 25

Since the table of contents provides a quite complete index, only those items not readily found from the contents page are listed here. Numbers in BOLDFACE refer to experimental procedures.